Ultrafast Processes in Spectroscopy 1991

Ultrafast Processes in Spectroscopy 1991

Proceedings of the Seventh International Symposium on
Ultrafast Processes in Spectroscopy held in Bayreuth,
Germany, 7–10 October 1991

Emil Warburg — Symposium
EPS — Meeting

Edited by A Laubereau and A Seilmeier

Institute of Physics Conference Series Number 126
Institute of Physics, Bristol and Philadelphia

CODEN IPHSAC 126 1–650 (1992)

British Library Cataloguing in Publication Data are available

ISBN 0-7503-0198-8

Library of Congress Cataloging-in-Publication Data are available

Published under The Institute of Physics imprint by IOP Publishing Ltd
Techno House, Redcliffe Way, Bristol BS1 6NX, England
US Editorial Office: IOP Publishing Inc., The Public Ledger Buildings, Suite 1035, Independence Square, Philadelphia, PA 19106, USA

Printed in Great Britain by Galliard (Printers) Ltd, Great Yarmouth, Norfolk

Dedication

We dedicate this volume to our most distinguished colleague who met an untimely death in 1991:

Sergei Alexandrovitch Akhmanov

He will be sadly missed by all in the field of spectroscopy with ultrashort laser pulses and our deepest sympathy goes to his family.

Preface

Continuing a series of conferences that were organized in Eastern Europe in recent years, the VIIth International Symposium on Ultrafast Processes in Spectroscopy was held for the first time in Western Europe and took place on October 7–10, 1991 in Bayreuth, Germany. Scientists from widely varying disciplines, in physics, chemistry and biology, came together to share their common interest in picosecond and femtosecond phenomena. The symposium attracted approximately 240 participants from 14 European countries and from overseas, 85 of them from Eastern and South-Eastern Europe. These numbers vividly demonstrate the decline of political barriers between East and West and the current need for a profound exchange of ideas among the whole scientific community. The high quality of the research presented in approximately 190 papers, the enthusiasm of the participants and the attractive surroundings combined to provide an enjoyable atmosphere at the symposium.

The present volume reflects the state-of-the-art in the field. Considerable progress is noted in generating powerful ultrashort laser pulses in extended spectroscopic areas and their applications in various fields of science and technology. Numerous scientists from many research groups helped to make the symposium successful. We thank all contributors to the scientific program and the members of the International Program Committee for their helpful advice and the selection of papers. Special thanks are due to Mrs Lenich for making the technical arrangements and to the technical staff of the University of Bayreuth for continuing help.

The meeting has benefited from several financial sources. The generous support of the Emil-Warburg-Stiftung and of the Deutsche Forschungsgemeinschaft were particularly helpful. Financial aids by the Deutscher Akademischer Austauschdienst, W E Heraeus-Stiftung and several important laser companies are also gratefully acknowledged.

A Laubereau
A Seilmeier

Contents

b) Ti:Sapphire Lasers and Parametric Oscillators

c) Dye Lasers

III. Coherent Spectroscopy and Nonlinear Optics

IV. Applications to Solid State Physics and Surface Dynamics

xiv   *Contents*

V. Applications to Semiconductors

a) Carrier Dynamics in Bulk Semiconductors

c) Exciton Dynamics

d) Opto-Electronics

VI. Applications to Molecular Systems

a) Ultrafast Photo-Excitation of Molecules

Inst. Phys. Conf. Ser. No 126: Section I
Paper presented at Int. Symp. on Ultrafast Processes in Spectroscopy, Bayreuth, 1991

1

Femtosecond laser sources for near-infrared spectroscopy

W Sibbett, J F Allen Physics Research Laboratories, Department of Physics & Astronomy, University of St Andrews, St Andrews, Fife KY16 9SS, Scotland.

ABSTRACT: Modelocking schemes by which frequency-tunable, femtosecond pulses can be produced by colour-centre and titanium-sapphire lasers are described. A qualitative, physical explanation is given for the coupled-cavity modelocking and self-modelocking methods that are employed and some relevant experimental data are included.

1. INTRODUCTION

In view of the desirability of spectral versatility in time-domain spectroscopy a major topic of recent research activity has been the development of ultrashort pulse lasers having uncompromised tuning ranges. By exploiting optical nonlinearies that are wavelength non-specific, in contrast to saturable absorbers for example, it has become feasible to generate frequency-tunable femtosecond pulses from a range of laser systems. Significantly, the development of the soliton laser by Mollenauer and Stolen (1984) highlighted the applicability of coupled nonlinear cavities where advantage could be taken of the intensity-induced self-phase modulation effects in anomalously dispersive monomode optical fibres. Follow-up theoretical and experimental research undertaken by Blow and Wood (1988) and Kean et al (1988) confirmed that the more general technique of coupled-cavity modelocking could be applied using a choice of nonlinear elements where the generation of bright optical solitons was shown to be unnecessary.

An alternative modelocking scheme involving a single cavity configuration where the key optical nonlinearies are exhibited by the gain medium or other intracavity components can also be implemented. Within the context of this paper the emphasis of this self-modelocking technique, first demonstrated by Spence et al (1990), will be concentrated on the titanium-sapphire laser and, where appropriate, parallels will be drawn with the coupled-cavity counterparts. Several methodologies whereby pulse evolution can be initiated and/or assisted will be reviewed together with the concept of the pulse shaping and shortening processes that lead ultimately to durations in the femtosecond domain.

The reliable generation of broadly tunable femtosecond laser pulses has direct relevance to a range of time-resolved spectroscopic studies. It is especially noteworthy that pulses from a self-modelocked $Ti:Al_2O_3$ laser oscillator have peak powers ~ 100 kW and so nonlinear optical techniques can be readily applied for frequency up-conversion or down-conversion purposes.

In this paper a qualitative description is advanced to lay the foundation upon which both the basic principles of coupled-cavity modelocking and self-modelocking can be justified and elaborated. At the outset it can be pointed out that all modelocking procedures can be based on one or more of the complement of modulation processes illustrated in Figure 1.

By involving the term, 'self-amplitude modulation', familiar concepts such as passive modelocking can be included. Indeed both saturable absorption and saturable amplification effects can be readily encompassed within this description but the general intention is to consider the amplitude modulation that is produced by the intracavity radiation itself.

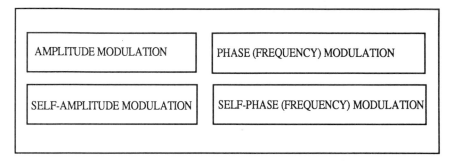

Fig .1 Modulation processes involved in modelocking techniques.

2. COUPLED-CAVITY MODELOCKING

The coupled nonlinear cavity approach is exemplified by the schematic of Figure 2. (Although a nonlinear Fabry-Perot is assumed here, similar physical insights may be applied to alternative configurations such as those involving Michelson and Sagnac interferometers.) The gain medium may be excited at a constant level or by a weakly modulated signal or synchronously by a modelocked pump laser. It follows, therefore, that the radiation that circulates within the main (or master) cavity may take the form of stochastic noise, noise bursts, or the pulses that evolve from synchronous pumping. An intensity component of this radiation propagates within the nonlinear coupled cavity and on its return it thereby interacts with the radiation in the main cavity. An appreciation of this interaction can give important insights into the initial processes that lead to pulse evolution and the subsequent shortening/shaping by which pulse durations into the sub-100 fs region can be obtained.

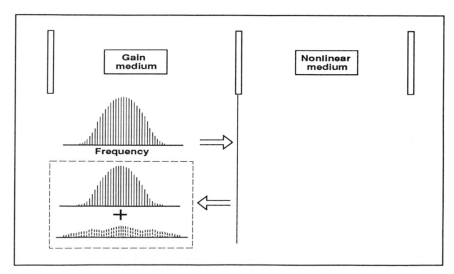

Fig 2. Generalised scheme for coupled-cavity modelocking.

It is instructive to consider the cavity radiations using a frequency-domain description. At optical power levels where intensity-induced nonlinear phase shifts are produced (particularly in the control cavity) the Fabry-Perot resonators have dynamically-varying resonances (or modes). As a consequence of this dynamic process a subset of the many oscillating cavity modes can become sufficiently phase-coupled that *primitive* pulses can evolve with durations in the picosecond regime. (This mechanism has distinct similarities with the modelocking that has been observed by French and co-workers (1990) for linear control-cavity configurations in which the position of one resonator mirror is oscillated). For pulse-shortening into the femtosecond regime it is necessary to have access to significant spectral broadening. This can be conveniently ensured when a nonlinear element such as monomode optical fibre or semiconductor amplifier has been incorporated into the control cavity (Miller and Sibbett, 1988 and Kean et al 1989).

The spectrally extended character of the pulses returning from the control cavity can have a major influence upon the radiation that circulates within the main cavity. For example, a self-phase-modulated control pulse that experiences substantial group velocity dispersion can mean that some of its spectral components actually precede the propagation of the main-cavity pulse through the gain medium. The consequent competitive demand for amplification can therefore precondition the gain-status of the laser medium that is subsequently experienced by the main-cavity pulse. This can imply that preferential gain may enable a pulse to grow with a central wavelength that is significantly shifted (towards the blue or red) from the original noise burst or pulse in the main cavity (eg. see Fig 11(b) of Zhu et al 1991). Another key aspect of pulse refinement relates to the competition for gain that arises in the spectral wing regions of the circulating pulses. To highlight this point, particular reference should be made to the relatively weak but spectrally extended nature of the control-cavity signal that returns to the gain medium in the main cavity (see Fig 2). For relatively weak control-cavity radiation it is not expected that this would greatly affect the intense central region of the pulse spectrum. By contrast, however, this re-entrant control signal amplitude will be sufficient to influence and even dominate the demand for gain at the wings of the circulating pulse spectrum. This constitutes a self-amplitude modulation and so if the configuration affords appropriate frequency and phase matching between the main and control cavities then an enhanced coupling of the longitudinal modes will result. It is for this reason that the interaction that occurs within the gain medium is highlighted in Figure 2. It follows then that the associated pulse shortening will lead to increased peak intensities and this process will ensure that a significantly enlarged laser bandwidth will be engaged in the steady-state condition. This thus explains, in rather simplified terms, how the pulse durations shorten from the picosecond to the femtosecond regime.

The framework within which this explanation has been developed affords consistency with the available experimental data relating to pulse evolution kinetics. For instance, in the case of a coupled-cavity modelocked KCl:Tl colour-centre laser the expected parameter dependencies on intensity, phase and bandwidth have been unambiguously confirmed by the experimental data reported by Zhu et al (1991). It is perhaps worth pointing out that factors such as group velocity dispersion (GVD) mean that the shortest pulses are not necessarily obtained when the lasing bandwidth is maximised. A proper control of the phase coupling of the longitudinal modes for the production of transform-limited pulses usually requires that some intracavity bandwidth restriction is retained (Zhu and Sibbett 1990).

Illustrative examples of interferometric autocorrelations for sub-100 fsec transform-limited pulses produced by coupled-cavity modelocked KCl:Tl colour-centre (Grant and Sibbett 1991) and titanium-sapphire (Spence and Sibbett 1991) lasers are reproduced in Figure 3. With the KCl:Tl colour-centre laser the GVD is relatively small whereas for the Ti:Al$_2$O$_3$ laser counterpart substantial amounts of GVD are accumulated. For this reason the modelocking performance of the latter system can be much improved by incorporating some intracavity GVD-compensation. Given the higher dispersion in the control-cavity fibre around 800 nm and the longer gain crystal for the Ti:Al$_2$O$_3$ laser a properly optimised system requires GVD-compensating prism pairs to be included in both the main and control cavities.

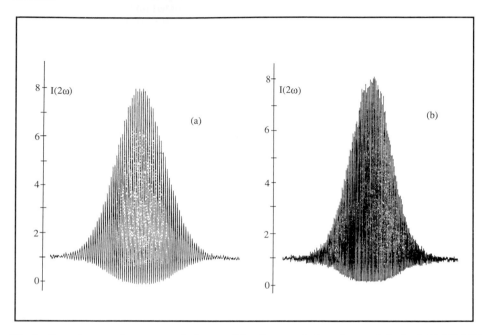

Fig 3. Interferometric autocorrelation profiles for coupled-cavity modelocked (a) KCl:Tl colour-centre laser ($\Delta\tau_p$ = 82 fsec) and (b) Ti:Al$_2$O$_3$ laser ($\Delta\tau_p$ = 95 fsec).

3. SELF MODELOCKING

Although the coupled-cavity modelocking technique has been shown to be very effective and useful the requirement to have interferometrically-matched cavities imposes particular demands with regard to system practicalities. A single-cavity alternative that provides access to exploitable Kerr-type optical nonlinearities is therefore to be preferred. In their original demonstration of this simplified approach for the titanium-doped sapphire laser Spence, Kean and Sibbett (1991) described the process as *self-modelocking* because it was governed principally by intensity-induced nonlinearities.

The initial experimental arrangement comprised a Spectra-Physics Model 3900 Ti:Al$_2$O$_3$ laser that was modified to have an extended cavity length ~ 2 m (see Fig 4). With this resonator configuration it was observed that a self-modelocking operation could be initiated when the beam profile showed evidence for the simultaneous presence of two transverse modes. Because of the Brewster-angled geometry of the Ti:Al$_2$O$_3$ rod, a TEMon mode (eg TEMo5) frequently accompanied the dominant TEMoo mode. For the near-concentric resonator used the frequencies of the longitudinal-modes associated with distinct transverse modes tend to degeneracy.   Significantly, however, the transverse modes are not spatially-degenerate and so there exists a competition for gain in the Ti:Al$_2$O$_3$ laser medium. This can thereby give rise to an effective self-amplitude modulation of the axial modes at the cavity frequency and sufficient phase coupling can thus be initiated to give rise to the evolution of a primitive pulse. Because this pulse has a peak power which is greatly in excess of that of the stochastic noise it can induce optical Kerr type nonlinear effects as it propagates in the laser rod. The nonlinear effects that are particularly relevant are the self-phase modulation and self-focussing that arise in the gain medium. As previously mentioned in the context of coupled-cavity modelocking (see Section 2) the self-phase modulation leads to a spectral broadening within which a substantially enlarged comb of longitudinal modes can be phase-locked. In consequence therefore the duration of the circulating pulse can decrease accordingly and the enhanced peak intensity further promotes pulse-shortening. Provided appropriate intracavity compensation of group-velocity dispersion is included then the pulse duration can decrease into the sub-100 fsec domain (Spence et al 1990). Although the dual-transverse mode operation enables a pulse to evolve initially it is worth emphasising that as the modelocked laser output develops further towards the steady state the beam profile for the femtosecond pulses tends to be refined towards a predominantly TEMoo characteristic.

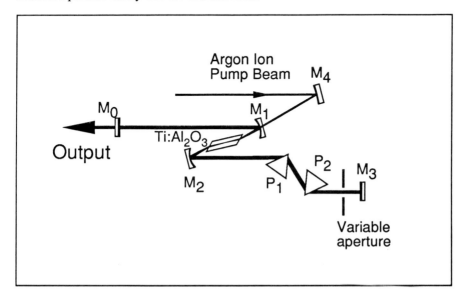

Fig 4. Cavity configuration for a self-modelocked Ti-sapphire laser.

The existence of an accompanying self-focussing effect has a particular relevance to the resonator beam optics. By including a hard-aperture, or relying upon the equivalent "soft-aperture" that is associated with the pump beam in the $Ti:Al_2O_3$ rod, then intensity-induced beam reshaping can improve the steady-state stability of the self-modelocked output. This can be more deliberately exploited by designing a resonator to take advantage of the fact that the beam waist can be smaller in the modelocked operation than in the unmodelocked CW operation (Spinelli et al 1991 and Piché 1991). Such an intensity-induced lensing is of course initially time-varying as the pulse power changes but some constant focal length arises under steady-state conditions. In this respect the losses are reduced when self-focussing is present and thus this can be regarded as a *Kerr-lens-assisted* self-modelocking process. [The Author would contend that the designation of Kerr-Lens Modelocking as introduced by Spinelli and co-workers (1991) is misleading because the modelocking process must be established in order to produce a pulse having sufficient intracavity power to induce significant self-focussing in $Ti:Al_2O_3$]. The self-focussing effect has also been utilized in a microdot-mirror resonator scheme (Gabetta et al 1991). In this demonstration the lensing was induced in a passive medium that was distinct from the gain medium and this approach clearly has applicability to other lasers. The rather more general scheme of using a high-nonlinearity passive medium within a dedicated focussing section of the resonator should also be applicable to a wide range of laser types.

Although the first demonstration of self-modelocking modelocking involved pulse evolution through the interaction of two transverse modes there are several alternative approaches to initiating the phase-locking of longitudinal modes. For example, the inclusion of a weak intracavity saturable absorber can provide sufficient amplitude (loss) modulation at the cavity frequency to give rise to pulse build-up. This scheme which was originally assumed to be conventional passive modelocking (Sarukura et al 1989) is really a variant of self-modelocking where a very weak initiating intensity modulation (a few percent) is sufficient. The main pulse-shortening/shaping process is actually due to the nonlinear effects mentioned above for the self-modelocking technique. Interestingly, however, the modest requirement on saturable absorption enables a low concentration cocktail of dyes to enable a usefully extended tuning range to be ensured. This is in contrast to the more usual modelocking conditions (eg the colliding-pulse modelocked dye laser) which impose an essentially non-frequency-tunable operation. Uncompromised frequency tunability in femtosecond Ti:Al$_2$O$_3$ lasers has also been achieved by other options for initiating the modelocking. These include active cavity-length modulation using intracavity Brewster-angled plates (Spinelli et al 1991), synchronous intensity (gain) modulation using a modelocked pump laser (Spielmann et al, 1991) and regenerative acousto-optic (loss) modulation (Spence et al 1991).

A further refinement to the self-modelocked Ti:Al$_2$O$_3$ laser has also been introduced whereby the frequency of the cavity is referenced to the output of an electronic crystal oscillator (Spence et al 1991). By this means the longer-term pulse timing jitter (or phase noise) has been reduced to 170 fsec (in the 5-50 kHz frequency range) while sub-100 fsec pulse durations are retained (see Fig 5). In lower frequency ranges the jitter is greater (eg 480 fsec in 50-500 Hz, 380 fsec in 500 Hz - 5 kHz) because of noise in the pump laser/power supply units. It therefore follows that to minimize the timing jitter in such femtosecond lasers it is required that the output beams from the pump lasers should be free from any intensity variability or noise.

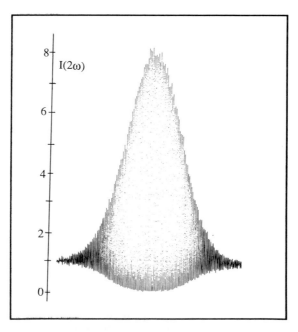

Fig 5. Interferometric autocorrelation profile for pulses from a regeneratively-initiated self-modelocked Ti:Al$_2$O$_3$ laser. ($\Delta\tau_p = 70$ fsec).

(In practical terms this requirement probably implies that frequency-doubled, diode-pumped Nd:YAG (or Nd:YLF) laser will be the best suited sources for the optical excitation of low-phase-noise, self-modelocked Ti:Al$_2$O$_3$ lasers.) Nevertheless in their present state of development the self-modelocked Ti:Al$_2$O$_3$ laser is a reliable source of broadly tunable femtosecond pulses (eg 750-950 nm - Spence, Kean and Sibbett 1991) having relatively high peak powers ~ 100kW without any additional amplifier stages being involved.

With regard to future work in this topic area it is already clear that these techniques for modelocking will be applicable to the new solid-state gain media which are being developed. In particular, the diode-pumpable Cr:LiCAF; Cr:LiSAF and Cr:LiSCAF gain media offer very interesting possibilities as sources for time-domain spectroscopy in the near-infrared spectral region. It is indeed very likely, therefore, that self-modelocking will become a widely applied technique for femtosecond lasers having modest intracavity power levels and the expectation is that pulse durations into the sub-20 fsec will be achievable using suitable broadband media. Such ultrashort-pulse sources will no doubt find many applications in time-domain spectroscopic studies in the future.

4. ACKNOWLEDGEMENTS

Major contributions to the work described in this paper have been made by my colleagues David Spence and Xianong Zhu. Overall funding support from the UK Science and Engineering Research Council is also gratefully acknowledged.

5. REFERENCES

Blow K J and Wood D 1988, J Opt. Soc. Am. B **5** 13.

French P M W, Kelly S M J and Taylor J R 1990, Opt. Lett **15** 378.

Gabetta G, Huang D, Jacobson J, Ramaswamy M, Haus H A, Ippen E P and Fujimoto J G 1991, Technical Digest of Conference on Lasers and Electro-Optics - *CLEO '91* (Optical Society of America 12-17 May 1991, Baltimore M D) paper CPDP8.

Grant R S and Sibbett W 1991, Opt. Commun. **86** 177.

Kean P N, Zhu X, Crust D W, Grant R S, Langford N and Sibbett W 1988, Tech. Dig. *CLEO '88*(25-29 April 1988, Anaheim CA) paper PD7.

Kean P N, Zhu X, Crust D W, Grant R S, Langford N and Sibbett W 1989, Opt. Lett. **14** 39.

Miller A and Sibbett W 1988, J. Mod. Opt **35** 1871.

Mollenauer L F and Stolen R H 1984, Opt. Lett. **9** 13.

Piché M 1991, Opt. Commun. **86** 156.

Spence D E, Kean P N and Sibbett W 1990, Tech. Dig. *CLEO '90* (21-25 May 1990, Anaheim CA) paper CPDP10.

Spence D E, Kean P N and Sibbett W 1991, Opt. Lett. **16** 42.

Spence D E and Sibbett W 1991, J. Opt. Soc. Am. B **8** 2053.

Spence D E, Evans J M, Sleat W E and Sibbett W 1991, Opt. Lett. **16** 1762.

Spielmann C L, Krausz F, Brabec T, Wintner E and Schmidt A J 1991, Opt. Lett. **16** 1180.

Spinelli L, Couillaud B, Goldblatt N and Negus D K 1991, Tech. Dig. *CLEO '91* paper CPDP7.

Inst. Phys. Conf. Ser. No 126: Section I
Paper presented at Int. Symp. on Ultrafast Processes in Spectroscopy, Bayreuth, 1991

9

Femtosecond pulse generation from solid-state lasers

Ch. Spielmann, F. Krausz, T. Brabec, E. Wintner, and A. J. Schmidt

Technische Universität Wien Abteilung für Quantenelektronik und Lasertechnik
Gusshausstr. 27, A-1040 Wien, Austria

Abstract: The performance of broad-band cw solid state lasers (Nd:glass, Ti:sapphire) passively mode locked by various techniques including additive pulse modelocking and Kerr-lens modelocking is analyzed.

1. Introduction

During the last decade the passively mode locked dye laser [1]has been the only widely used source of femtosecond optical pulses. Recent advances in ultrashort pulse generation with solid state lasers opened up a new era in femtosecond pulse technology. We demonstrate different cavity configurations providing efficient passive mode locking of broad-band solid state lasers such as Nd:glass [2,3,4] and Ti:sapphire [5,6,7,8]. In our experiments these lasers are pumped by a krypton and a frequency-doubled, lamp-pumped Nd:YLF laser, respectively. Ultrafast all-optical modulators have been realized using the concepts of additive pulse modelocking (APM) and Kerr-lens modelocking (KLM). The best performance in a low-gain APM system is achieved with a dispersively balanced Michelson interferometer[4]. In this APM laser the Kerr-nonlinearity of optical fibers is utilized for mode locking. In high-power lasers a bulk Kerr-medium in conjunction with an aperture [9] or cavity misalignment (KLM) [8,9] provides a promising alternative for femtosecond pulse production. The powerfulness of these techniques is demonstrated by the generation of stable sub-100-fsec pulses in Nd:glass and Ti:sapphire lasers. Both systems have the potential of being pumped by diode lasers and becoming the future workhorses for femtosecond spectroscopy.

2. Femtosecond Passive Mode Locking by a Nonlinear Interferometer

In this section we report on dispersively compensated additive–pulse mode locking using a Michelson–cavity configuration [4]. Dispersion compensation is shown to improve steady–state laser performance, facilitate self–starting, and reduce relaxation oscillation

instabilities. Using Nd:phosphate glass as the active material we demonstrate the generation of stable femtosecond pulses at considerably reduced self-starting threshold in
comparison with the conventional APM technique.

Consider a laser cavity terminated into a nonlinear Michelson interferometer as illustrated in Fig. 1. This Michelson cavity configuration is superior to the conventional
coupled-cavity arrangement owing to its greater compactness due to a smaller number of cavity components and to a reduction of the physical extension of the system.
Furthermore, the single-beam output leads to the production of more intensive output
pulses. The beam splitter BS transmits a small part of the intracavity power into the
weakly-coupled nonlinear arm M1-BS containing a short length of a single mode optical
fiber. The greater part of the power is fed into the dispersive linear arm M2-BS having
a net negative GVD.

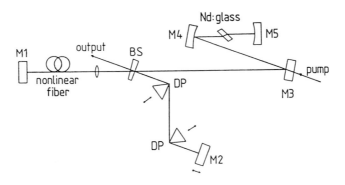

FIG. 1 Schematic diagram of the passively mode-locked laser (M 1-5), high-
reflectivity mirrors at the laser-wavelength; (BS) 16% transmitting beam splitter; (DPs)
dispersive prisms.

At sufficiently high intracavity powers the laser immediately becomes mode locked
when the relative phase is properly adjusted. Without dispersion compensation ($D \approx$
0) the laser produces strongly chirped pulses ($\approx$ 1 ps) as predicted by numerical
investigations.[10] When dispersion in the linear interferometer arm is optimized, nearly
bandwidth-limited femtosecond pulses are generated. Compensation can be achieved by
a single pair of prisms separated by a distance approximately equal to the fiber length
allowing a compact optical setup.

Dispersion compensation also reduces the self-starting threshold significantly. Fig.
2 is a summary of the most important experimental results obtained with different
fiber lengths. The threshold intrafiber average power P_{th} above which mode locking
is self-initiating is depicted by open and full circles obtained with and without GVD

compensation, respectively. P_{th} is reduced by more than a factor of three with the introduction of negative GVD pointing to its significance even in the initial pulse evolution. The fiber length is an important scaling parameter for both threshold power and steady-state pulse duration represented by triangles in Fig. 2.

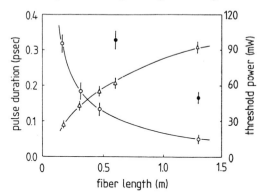

FIG. 2 Steady-state pulse durations (triangles), and intrafiber self-starting thresholds with (open circles) and without (full circles) dispersion compensation as a function of the fiber length.

Long fiber lengths allow low power operation at the expense of an increase in pulse width, whereas short fiber lengths support shorter pulses at the cost of higher thresholds for passive mode locking. In our experiments the shortest pulses have been generated at $l_f = 18$ cm with $D = -3.8 \times 10^{-3}$ ps^2. Careful optimization of system parameters might bring the phase settings for minimum pulse width and pedestal-free pulse generation closer together. Nevertheless, the 88 fs pulses represent the shortest optical pulses produced in a passively mode-locked Nd:glass laser to date. Note that the short fiber length improves the stability of the laser as well. Owing to the reduced extension of the interferometer, stable long-term operation is easily achieved with a servo-loop providing active stabilization.

3.Femtosecond pulse generation from a synchronously pumped Ti:sapphire laser

Owing to its exceptionally broad fluorescence line extending over 400 nm and to other favorable characteristics, Ti:sapphire is an attractive laser material for tunable ultrashort pulse generation and amplification. Passive [5-9] mode locking techniques have resulted in the production of femtosecond pulses as short as 60 $fsec$.[6]

The experimental setup of the laser system[11] described in this is illustrated in Fig. 3. The lamp-pumped cw Nd:YLF laser (Quantronix 4216D) is mode locked by acousto-optic loss modulation. This radiation is frequency-doubled by temperature-phase-matched second harmonic generation in a lithium triborate (LiB_3O_5, LBO) crystal. The Ti:sapphire laser is a five-mirror astigmatically compensated arrangement as shown in Fig. 3. It is based on a previous design producing pulses of 70-80 $fsec$.[8] The Ti:sapphire rod, which also acts as the Kerr-medium, is as short as 8 mm to reduce the positive group delay dispersion in the nonlinear medium, and absorbs $\approx 75\%$ of the pump radiation. So far, all femtosecond solid state systems [3,4,6−9] have employed a pair of prisms made of some heavy flint glass, e.g. SF10 or related materials. These dispersive delay lines offer the possibility of large negative group delay dispersion with relatively small prism separations but exhibit large third-order dispersion. Replacing the heavy glass prisms with F2 prisms (Schott, n=1.61) and minimizing the glass path way in the prisms reduced D_3(third-order dispersion) by more than a factor of 3 compared to the previous system.[8] The reduction of the unfavorable dispersive effects leads to a more or less ideal solitary system with the shortest pulses generated near the dispersive end of the cavity[12,13]. The spatial chirp on the pulse coupled out by OC is eliminated by an identical prism delay line outside the cavity as shown in Fig. 3. In addition, the external delay line offers the possibility of compensating for some residual linear chirp carried by the pulse, and of precompensation for the dispersion of optical components prior to the target when it comes to experiments. In our experiments, wavelength tuning has been accomplished by translating a variable-aperture slit perpendicular to the cavity axis between $P2$ and OC.

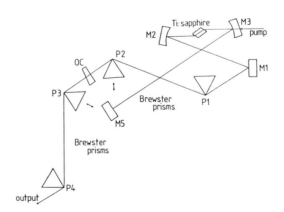

Fig. 3. Layout of the Ti:sapphire laser. OC, 2% transmitting output coupler; M2, M3, curved broadband mirrors with radii of 10 cm; M1, M5 flat broadband mirrors; P1-P4, Brewster-angled prisms made of flintglass (Schott, F2).

Synchronous mode locking is easily obtained by matching the cavity length to that of the pump laser. Using a fast photodiode a stable pulse train well synchronized to the pump laser is observed. As the absorbed pump power is increased above 2.5 W, the noisy autocorrelation trace becomes clean and indicates a dramatic reduction of pulse duration. Once the second threshold is reached, synchronization between the sapphire laser and the pump laser ceases and stable mode locking tolerates considerably more detuning, typically several tens of micrometers. The desynchronization indicates that passive mechanisms take command over mode locking. At an absorbed pump power of 3 W the average output power is 75 mW and the optimum net round-trip group delay dispersion, for which the shortest pulses are produced, has been found to be $D = \partial^2 \phi / \partial \omega^2 = -5 \times 10^2 \ fsec^2$. Under these conditions pulses as short as 33 $fsec$ have been generated. It is important to notice that the laser requires a slight misalignment of the end mirrors to produce femtosecond pulses. However, the misalignment implies only a small (<10%) reduction of average output power without affecting TEM_{00} oscillation. In the synchronously-pumped Ti:sapphire laser, however, femtosecond pulse generation is completely self-starting even at comparatively low (>2.5 W) pump powers and interruption-free stable long-term operation is achieved without any kind of environmental isolation.

The gain modulation in the synchronously pumped Ti:sapphire laser is less than 1%. As a consequence, the presence of efficient passive pulse shaping mechanisms is absolutely necessary for femtosecond pulse formation. Self-phase modulation by the Kerr-nonlinearity of the gain medium in conjunction with negative dispersion is expected to play a dominant role in ultrashort pulse evolution in the presented system. It can be shown generally that pulse propagation through a system with distributed dispersion and distributed self-phase modulation is not capable of compressing a picosecond pulse into a steady-state femtosecond pulse.[13] Therefore some addtitional passive amplitude modulation is necessary for stable operation. Passive amplitude modulation originates from an intensity-dependent gain, which is the consequence of an intensity-dependent overlap between the pump and cavity beams due to self focusing in the gain medium. Cavity misalignment leads to a mismatch between the pump and cavity mode, and may increase it under certain conditions. Our calculations predict also a threshold peak power for femtosecond pulse production in full agreement with the experimental observation.

In conclusion, we have presented guidelines for the design of femtosecond solid state systems, using well-defined, easily obtainable parameters. These design considerations led to the construction of lasers capable of producing optical pulses as short as 33 $fsec$. Owing to the simplicity and solid state nature of the system, the described performance is easily reproduced from day to day, a feature making this novel class of femtosecond sources especially attractive for practical applications.

This work was supported by the Jubiläumsfonds der Österreichischen Nationalbank grant 3828.

References

1. R. L.Fork, B. I. Green, C. V. Shank, Appl. Phys. Lett. **38**, 671 (1981)

2. F. Krausz, Ch. Spielmann, T. Brabec, E. Wintner, and A. J. Schmidt, Opt. Lett. **15**, 1082 (1990)

3. M. Hofer, M. H. Ober, F. Haberl, and M. E. Fermann, "Characterisation of ultrashort pulse formation in passively modelocked fiber lasers," IEEE J. Quantum Electron. (to be published)

4. Ch. Spielmann, F. Krausz, T. Brabec, E. Wintner, and A. J. Schmidt, Appl. Phys. Lett. **58**, 2470 (1991)

5. J. Goodberlet, J. Wang, A. J. Fujimoto, and P. A. Schulz, Opt. Lett. **15**, 1125 (1990)

6. D. E. Spence, P. N. Kean, and W. Sibbett, Opt. Lett. **16**, 42 (1991)

7. N. Sarukura, Y. Ishida, and H. Nakano, Opt. Lett. **16**, 153 (1991)

8. Ch. Spielmann, F. Krausz, T. Brabec, E. Wintner, and A. J. Schmidt, Opt. Lett. **16**, 1180 (1991)

9. L. Spinelli, B. Conillaud, N. Goldblatt, and D. K. Negus, in *Digest of Conference on Lasers and Electro-Optics* (OSA, Washington, D. C.,1991) paper CPDP7

10. T. Brabec, F. Krausz, M. Budil, Ch. Spielmann, and E. Wintner, J. Opt. Soc. Am. B **9**, 1818 (1991)

11. Ch. Spielmann, F. Krausz, T. Brabec, E. Wintner, and A. J. Schmidt, "Generation of 33 femtosecond optical pulses from a solid state laser," submitted to Opt. Lett.

12. T. Brabec, Ch. Spielmann, F. Krausz, "Limitations to pulse duration in solitary lasers," submitted to Opt. Lett.

13. T. Brabec, Ch. Spielmann, F. Krausz, "Mode locking in solitary lasers," Opt. Lett. **16**, December 15 (1991)

Inst. Phys. Conf. Ser. No 126: Section I
Paper presented at Int. Symp. on Ultrafast Processes in Spectroscopy, Bayreuth, 1991

Coupled cavity mode-locking of a neodymium-doped fiber laser

C Unger, G Sargsjan, U Stamm[*] and M Müller

Friedrich-Schiller-University, Faculty of Physics and Astronomy, Institute of Optics and Quantum Elektronics, Max-Wien-Platz 1, Jena, O-6900, Germany

[*] present address: Lambda Physics GmbH, Hans-Böckler-Str. 12, Göttingen, 3400, Germany

ABSTRACT: The mode-locking of a neodymium-doped fiber laser with a linear external cavity incorporating a vibrating mirror was investigated in dependence on the mismatch between main and coupled resonator. The occurence of satellites in the autocorrelation traces allows us to determine the exact matching of the lenghts of both cavities. Under optimum conditions the fiber laser generates pulses with durations shorter than 32 ps.

1. INTRODUCTION

Recently, there has been much experimental (Mark *et al* 1989) and theoretical (Ippen *et al* 1989) research on cavities externally coupled to a main laser. The exploitation of extracavity (Krausz *et al* 1990) or intracavity (Spence *et al* 1991) off-resonant nonlinear media has been demonstrated to lead to mode-locking of various lasers.

In this paper we examine a technique for ps-pulse generation in a neodymium-doped fiber laser mode-locked by feeding back the light from a linear external cavity including only a vibrating mirror. French *et al* (1989) first reported this kind mode-locking of a titanium-doped sapphire laser, which was later applied succesfully for a fiber laser (Wigley *et al* 1990). However, the pulses from the neodymium-doped fiber have still a duration of about 200 ps, whereas the gain bandwidth should support pulse duration of loss than 150 fs.

Our investigation of the dependence of the duration of the laser pulses on the absolute determinable mismatch between the lenths of laser resonator and external cavity shows that the mode-locking is impossible if both cavities are exactly matched. Careful adjustment allows to generate pulses of less than 32 ps by the present mode-locking technique the basis of which is discussed qualitativly.

2. EXPERIMENTAL

The laser configuration used for the present research is illustrated in Fig. 1. The gain medium is a 74 cm long single-mode silica fiber doped to around 400 ppmw with Nd^{3+}. With a laser threshold of 19 mW absorbed power of the cw-argon-ion pump laser a slope efficiency of 13% has been obtained.

The external cavity is formed by a thin glass plate M_3 mounted on a loudspeaker system and mirror M_2. Its length is approximately matched to that of the fiber laser.

3. RESULTS AND DISCUSSION

The cw- and pulse laser spectrum is measured by an optical multichannel analyzer system. The observed structure of the oscillating spectrum (full width at half maximum 12 nm) is assumed to be caused by crystalline clusters in the fiber core (Sargsjan *et al* 1991).

Best results of mode-locking with respect to stability are achieved with the fastest translation speed of the end mirror M_3 (0.33 m/s) which was possible for the present arrangement. The average output power of the fiber laser was the same as under cw-operation

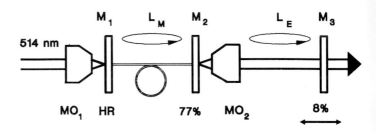

Fig.1. Schematic diagram of the fiber laser cavity
cavity mismatch $\delta L := L_M - L_E$

(36 mW after passing mirror M_3).
With a sinusoidal motion of the loudspeaker (0.33 m/s maximum speed) the laser emits stable trains of mode-locked pulses with a repetition rate of 139 MHz.
The external cavity can be considered as one output mirror of the main laser. The maximum variation of the reflectivity of the given Fabry-Perot-Interferometer is 24%. The corresponding modulation of the envelope of the pulse trains could be observed experimentally at the oscilloscope trace.

While mirror M_3 oscillated the length of the external cavity was varied by moving the whole oscillating system. In Fig. 2 we have depicted three typical autocorrelation traces obtained for different lengths of the external cavity. With increasing mismatch between the laser and external cavity lengths the autocorrelation broadens and satellites appear in the trace.

The variation of the separation between the peak and first satellite of autocorrelation with cavity mismatch is equal to the change of the difference of the round trip times of both cavities. Because of the different optical lengths of main and external resonator the pulses in the external cavity are shifted with respect to the local time of the pulses in the main cavity by an amount of $\delta L/c$, where c is the velocity of

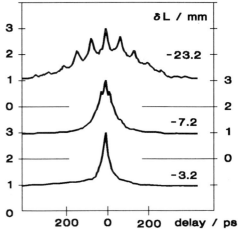

Fig. 2. Autocorrelation traces at different lenghts of the external cavity

light. An extrapolation of both sets of points for negative and positive mismatch, respectively, into the range of mismatch where no satellites are observed allows to determine the point of exact matching of the lengths of both cavities. It is noteworthy that the mode-locking vanishes if the mismatch between laser and external cavity is less than 3.2 mm. This is similar to the observation made by French *et al* (1990) in a Ti:sapphire laser. The full width at half maximum (FWHM) of the autocorrelation is depicted as a function of the cavity mismatch in the lower part of Fig. 3. Assuming a Gaussian shaped pulse we would estimate the shortest pulses to have a duration of less than 32 ps. In the upper part of Fig. 3 we have plotted the contrast ratio of the autocorrelation. If the cavities are closely matched this ratio decreases to values of about or less than 2:1. In the corresponding range of cavity mismatch we regard the mode-locking as vanished.

It is likely that this contrast ratio results from the dynamic evolution of pulses of limited duration (for which a contrast ratio of 3:1 is expected) which break into bandwidth-limited noise (with an contrast ratio 1.5:1).

The region of good mode-locking performance extends over a few mm of cavity mismatch of both sides of the "mode-locking gap". This is a much wider range of cavity mismatch than reported for lasers with APM (Mark *et al* 1989).

For the present type of mode-locking an additional simplification is obtained compared to APM at the expense of pulse duration since no use of an active stabilization of the cavity lengths is necessary.

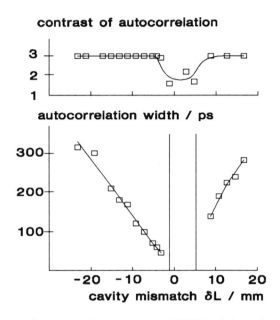

Fig. 3. Contrast ratio (upper) and FWHM of the ACF (lower) in dependence on the cavity mismatch

To discuss qualitatively the nature of mode-locking by the external cavity with vibrating mirror we think mainly about two possible effects which probably occur in combination: (i) The moving mirror M_3 of the external cavity shifts the frequency of the laser pulse during reflection by an amount given in essence by the Doppler effect (Sargsjan *et al* 1991):

$$\Delta\omega = \left[\frac{\dfrac{\tan \phi}{\tan(\omega\, T_E(t))} + \tan^2\phi\; \dfrac{1+r_1^2}{1-r_1^2}}{1+\tan^2\phi} \right] \frac{-2\,\omega\, v_M}{c} \quad \text{with} \quad \tan\phi = -\frac{r_2(1-r_1^2)\,\sin(\omega\, T_E(t))}{r_1(1+r_2^2) - r_2(1+r_1^2)\cos(\omega\, T_E(t))}$$

r_1, r_2 amplitude reflectivities of the mirrors in the external cavity
$T_E(t)$ explicitly time-dependent round-trip time of the external cavity
      $T_E(t) = T_M - h - (2\, v_M/c)\, t$
h timing mismatch between both cavities ($h = T_M - T_E = \delta L/c$)
v_M mirror velocity, ω carrier frequency of the incident pulse.

The term in brackets can reach values in the order of 10 to 100 for typical arrangements and therefore the frequency shift is enhanced. The frequency-shifted and back fed light will be amplified by the broad gain profile in the laser resonator. This process repeats and for the present translation speed after several tens of cavity round-trips this shift is equal to the frequency separation between two adjacent longitudinal modes. Thus one mode after the other may be locked by the vibrating mirror and best mode-locking results should be achieved with the highest translation speed of the mirror M_3.

(ii) Once several modes are locked, a higher peak intensity in the main cavity is obtained. In this case the self-phase modulation due to the Kerr-nonlinearity in the fiber causes a change of the wavelength in the main cavity which may lead to an increase of the reflectivity of the external cavity around the pulse center. This leads to an additional pulse shortening similar to the additive pulse mode-locking (APM) mechanism.

A comparision of the experimental results with values of a simple analytical and numerical model (Müller *et al* 1991) shows that the mode-locking technique can be explained satisfatory by both above mentioned effects.

Obviously, the mode-locking force by Doppler translation is rather weak. However, it plays the role of the starting mechanism of mode-coupling which is nessecary for all passive mode-locking techniques (Sibbett 1991). The pulse duration remains rather long in comparision to APM since the phase shift at the out coupling mirror of the laser cavity varies due to the mirror vibration. If it would be possible to stop the motion instantaneously APM should be observed.

4. CONCLUSIONS

The mode-locking of a neodymium-doped fiber laser with linear external cavity and vibrating mirror was investigated systematically. To our knowledge the pulses with a duration of less than 32 ps are the shortest pulses generated in a fiber laser in this way. It is possible to determine the point of the exact matching of the lengths of both cavities. This mode-locking technique could be qualitatively explained by the action of the Doppler effect in the coupled cavity and by self-phase modulation in the neodymium-doped fiber.

Shorter lengths of the coupled cavity or of the fiber in the main laser could increase the repetition rate by an order of magnitude at moderate high average powers.

The application of the vibrating mirror mode-locking to diode-pumped solid state lasers would yield a source of ultrashort pulses without the need for intracavity modulators and associated rf-signal generation and amplification.

REFERENCES

French P M W, Kelly S M J and Taylor J R 1989 *Opt. Lett.* **15** pp 378-80
Ippen E P, Haus H A and Liu L Y 1989 *J. Opt. Soc. Am. B* **6** pp 1736-45
Krausz F, Spielmann C, Brabec T, Winter F and Schmidt A J 1990 *Opt. Lett.* **15** pp 1082-4
Mark J, Liu L Y, Hall K L, Haus H A and Ippen E P 1989 *Opt. Lett.* **14** pp 48-50
Müller M, Stamm U and Sargsjan G 1991 *Proc. Ultrafast Processes in Spectroscopy*
   (Bristol: IOP Publishing Ltd)
Sargsjan G, Stamm U, Unger C and Müller M 1991 *Proc. Rare Earth-Doped Fibers,*
   Sources, & Amplifiers (SPIE, Boston)
Sibbett W 1991 *Proc. Ultrafast Processes in Spectroscopy* (Bristol: IOP Publishing Ltd)
Spence D E, Kean P N and Sibbett W 1991 *Opt. Lett.* **16** pp 42-4
Wigley P G J, French P M W and Taylor J R 1990 *Electron. Lett.* **16** pp 1238-40

Inst. Phys. Conf. Ser. No 126: Section I
Paper presented at Int. Symp. on Ultrafast Processes in Spectroscopy, Bayreuth, 1991

Numerical analysis of pulse formation in solid-state lasers mode-locked with a linear external cavity

M. Müller, U. Stamm

Friedrich-Schiller-University, Faculty of Physics and Astronomy
Institute of Optics and Quantum Electronics, Max-Wien-Platz 1
O-6900 Jena/FRG

ABSTRACT: Based on a numerical simulation the flying mirror mode-locking mechanism is found to be that of an active mode-locking due to the explicitly time-dependent phase shift by the external cavity involving a slowly moving mirror and a non-zero temporal mismatch to the master cavity. Additional pulse shortening and stabilization is performed by the additive pulse mode-locking mechanism arising from self-phase modulation due to the Kerr-effect in the host crystal of the laser amplifier.

1. INTRODUCTION

Presently a lot of research interest is focused onto the generation of ultrashort light pulses from tunable solid-state lasers for applications in laser spectroscopy. Novel vibronic solid-state laser materials with broad amplification bandwidth offer the potential to generate laser pulses down to a few femtoseconds in pulse width.

In the near infrared spectral region titanium-doped sapphire is up to now one of the most promising solid-state laser materials with a gain profile extended over about 400 nm and a gain maximum at 780 nm.

French et al. demonstrated for the first time mode-locking of a cw titanium-doped sapphire laser using a linear external cavity including only a slowly oscillating mirror (flying mirror mode-locking (FMM)) French et al. (1990a). Cw pumped neodymium-doped fiber lasers have been mode-locked by the FMM technique and it is expected that the same technique will work with other lasers too.

The novel FMM technique is simple to implement, reliable and robust and moreover cw mode-locking by a linear external cavity has the potential for generating pulse trains with a high repetition-rate being governed only by the lengths of the laser and external cavities French et al. (1990b). Therefore the FMM technique may lead to the development of an extremely compact high-repetition-rate picosecond light source.

2. ROUND TRIP MODEL

The electrical field circulating in the master and external cavity is described in the slowly varying envelope approximation (SVEA). Since the broad amplification bandwidth of the considered laser amplifier we use the rate equation approximation (REA) for the homogeneously broadened gain medium consisting of noninteracting effective 2-level-systems with the doping density N embedded in a nonlinear index host medium.

The moving mirror external cavity builds up a Fabry-Perot interferometer acting as an effective outcoupling mirror. The spectral location of the reflectivity maxima of this out-coupling mirror is moving in time since the translation of the external mirror. Therefore and because of the extremely broad gain profile a time-varying actual carrier-frequency of the complete laser field E(z,t) is expected and experimentally established French et al. (1990b).

We involve this slow time variation of the actual laser frequency in a time-dependent linear phase modulation of the complex SVEA amplitude. Under realistic experimental conditions this phase modulation never violates the SVEA. Therefore we carry out our treatment of a FMM laser with a constant carrier-frequency ω_L set to the center of the homogeneous broadened laser line.

Our cavity configuration consists of two coupled uni-directional ring cavities as depicted in Fig.1. The quantities in the master cavity we index with M, the quantities in the external cavity are indexed with E.

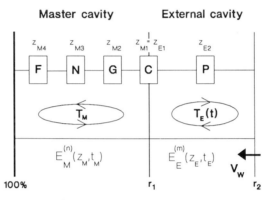

Fig.1 Coupled cavity arrangement

That we consider an electromagnetic field travelling only into one direction should be justified if the length of the amplifying medium is great in comparison to the duration of the pulses which is in a good agreement to the experimental set-up of a fiber laser but somewhat problematic in the case of an usual solid-state laser.

The master cavity of round trip time T_M is assumed to contain corresponding to Fig.1 the following elements: coupling mirror C with field amplitude reflection r_1 at z_{M1}, gain medium G with third order non-resonant nonlinearity due to the host crystal at z_{M2}, noise injection N due to spontaneous emission at z_{M3} and bandwidth-limiting filter F at z_{M4}.

The description of the round trip operator of the master cavity follows the standard way including the dispersion of a nonresonant third order polarization up to the first order for details see Mülller,Stamm (1991a).

Using the REA for describing the homogeneously broadened laser amplifier the electrical field amplitude in round trip n, $z_{\epsilon}[z_{M1},z_{M2}]$ travelling through the laser crystal of length $L=z_{M2}-z_{M1}$ is in the SVEA governed by a shortened amplified wave equation. Assuming the group velocity dispersion in the host crystal to be negligible ($L_D \gg L$) but involving a non-resonant nonlinearity due to nonlinear refractive index of the host crystal the wave equation is easily to be solved. The field after passage through an amplifier like this is given by the amplified and selfphase-modulated initial field. The dynamic of the gain of the laser field as the over the crystal length integrated population of the upper laser level follows from a shortened 4-level rate equation system assuming fast nonradiative decay of the lower laser level and the upper pump level. The influence of spontaneous emission is most simply considered by adding a noise term E_{sp} with the probability density of thermal radiation at each time to the solution of the wave equation. Because of the great gain-bandwidth of titanium-doped sapphire lasers we neglect gain dispersion and use the most simple case of a passive optical filter (with response-time $\tau_f \gg \tau_{21}$) as the bandwidth-limiting element in the master cavity. The filter center is set to the laser line center.

The external cavity including a slowly moving mirror causes a time-dependent phase shift resulting from the time-dependent difference of the round trip times. The time-dependent round trip time of the external cavity is given by

$$T_E(t) = T_M - \delta - \Theta t \tag{1}$$

assuming a linear translating external mirror of velocity v_w giving $\theta = 2v_w/c$.
The round trip numbers n,m and local times η_M, η_E indexes master and external cavity fields at the same external time t, therefore

$$t = nT_M + \eta_M = mT_E(t) + \eta_E . \tag{2}$$

Neglecting constant phase terms and higher order terms in the small quantity θ we get a frequency shift of the field circulating in the external cavity due to reflection at the moving mirror corresponding to

$$\mathscr{E}_E^{(m)}(z_{E2}, \eta_E) = r_2 e^{i[\omega_L \theta (m(T_M - \delta) + \eta_E]} \mathscr{E}_E^{(m)}(z_{E1}, \eta_E) . \tag{3}$$

At the coupling mirror r_1 the master and external cavity field interferes at each time t according to Ippen et al..

$$\mathscr{E}_M^{(n+1)}(z_{M1}, \eta_M) = r_1 \mathscr{E}_M^{(n)}(z_{M4}, \eta_M) + (1 - r_1^2)^{\frac{1}{2}} \mathscr{E}_E^{(m)}(z_{E2}, \eta_E) \tag{4}$$

$$\mathscr{E}_E^{(m+1)}(z_{E1}, \eta_E) = (1 - r_1^2)^{\frac{1}{2}} \mathscr{E}_M^{(n)}(z_{M4}, \eta_M) - r_1 \mathscr{E}_E^{(m)}(z_{E2}, \eta_E)$$

In the FMM situation the Doppler frequency shift $\omega_L \theta$ is small in comparison to the mode spacing frequency, whereas active mode-locking due to phase modulation usually requires matching of the modulator period and the round trip time up to the order of the inverse bandwidth of the master cavity. The same physical reason requires for FMM matching of the Doppler period and multiples, say m-times, of the master cavity round trip time up to the order of m-times of the invers effective bandwitdh of the coupled cavity configuration. This results in an upper limit for the mirror velocity v_w Müller,Stamm (1991b)

$$\frac{\omega_L \theta (v_W) T_M}{2\pi} \leq \frac{\tau_{eff}}{T_M} \tag{5}$$

yielding a Doppler frequency shift $\omega_L \theta$ which is for typical experimental conditions three orders of magnitude lower than the mode spacing. Mode-locking due to such slow phase modulation provides pulse durations in the order of the round trip time, e.g. nanoseconds because the modulator strength is three orders of magnitude smaller then that usually given by phase-modulators working at the mode spacing frequency. Above the limit (5) FMM is only possible at some discret velocities providing exact matching of the Fabry-Perot period and multiples of the master cavity round trip time.
The cavity mismatch δ plays now the keyrole as the control parameter in the FMM technique. By controlling the effective dispersion in the master cavity and the pulse shortening due to the self-phase modulation proper adjustment of the cavity mismatch δ allows to generate

pulses down to several tens of picoseconds in pulse duration. The pulse shortening arising from the self-phase modulation is performed similar to the additive pulse mode-locking mechanism Müller,Stamm (1991b).

An early work of Kravzov et al. reports mode-locking of a single cavity cw Nd-YAG laser by slow translation of one cavity mirror which is called kinematic mode-locking. They found an upper mirror velocity limit which aggrees with our FMM condition (5), but the missing delayed feedback with a δ greater than the invers master cavity bandwidth restricts this technique to the nanosecond time scale for the pulse width.

3. RESULTS OF NUMERICAL SIMULATION

The parameters used in the numerical simulation are close to that of the experimental set-up of a usual titanium-doped sapphire laser : $P=2.5*10^5 s^{-1}$, $T_2=3.2\mu s$, $n_0=1.76$, $A=2.3$, $L=2cm$, $n_2=1.6*10^{-22}m^2V^{-2}$, $r_1^2=0.96$, $\tau_F=1ps$, $T_M=8ns$, $< | E_{sp} | > =0.25Vm^{-1}$ where P stands for the pump-rate and A denotes the absorption of pump radiation.

For the external cavity we use fixed parameters: reflectivity of the moving mirror $r_2^2=0.5$ and mirror velocity $v_w=0.05ms^{-1}$ which are typical to that used by French et al. (1990a) whereas the temporal mismatch of the cavities will be varied.

As French et al. (1990a) found for the titanium-doped sapphire laser as well as Sargsjan et al. for the neodymium-doped glass fiber laser both mode-locked with the flying mirror technique, there exists a characteristic dependence of the pulse width on the temporal mismatch δ between master and external cavity with the significant fact, that for exactly matched cavities $\delta=0$ mode-locking fails.

Using this characteristic dependence of the pulse width on the temporal mismatch δ between master and external cavity as a validity check for our model we compare the numerical results with the experimental data. To this aim we have calculated the FWHM of the laser pulses averaging the intensity distribution over about 7000 round trips. The comparison between the calculated dependence of the pulse width on the cavity mismatch and the experimentally obtained is depicted in Fig.2. The agreement between both curves is suf-

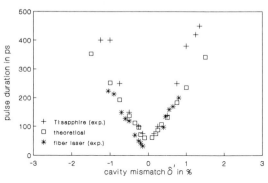

Fig.2 Pulse duration over cavity mismatch δ

ficient and we can conclude the effects being essential for the pulse formation in the FMM process are included in our model. Note, that there is no fit-parameter in the model.

4. REFERENCES

French et al. (1990a)   Opt.Lett.**15**,378-380 (1990)
French et al. (1990b)   to be published in Opt.Lett.
Müller,Stamm (1991a)   submitted to Appl.Phys.
Müller,Stamm (1991b)   to be published
Kravzov et al. (1977)   Sov.Tech.Phys.Lett.**3**,126-130 (1977)
Ippen et al. (1989)   J.Opt.Soc.Am.B **6**,1736-1745 (1989)
Sargsjan et al. (1991)   to be published in Opt.Commun.

Inst. Phys. Conf. Ser. No 126: Section I
Paper presented at Int. Symp. on Ultrafast Processes in Spectroscopy, Bayreuth, 1991

23

Feedback-controlled mode-locking operation of a Nd:YLF laser at 1.047 μm

K Wolfrum and P Heinz

Physikalisches Institut, Universität Bayreuth, D-8580 Bayreuth, Germany

The superior performance of a Nd:YLF laser is demonstrated by feedback-controlled mode-locking (FCM). The larger stimulated emission cross section of the π-polarized (λ=1.047 μm) transition compared to the cross section of the σ-polarized (λ=1.053 μm) transition leads to a lower pump energy threshold. Inserting an anti-resonantly tuned étalon in the cavity yields stable operation with short bandwidth-limited pulses of duration less than 3 ps at a repetition rate of 70 Hz and single pulse energy of 4μJ.

1. INTRODUCTION

A comparison of the σ-transition of Nd:YLF ($\lambda = 1053$ μm) with the π-transition ($\lambda = 1047$ μm) shows two important differences:
i) The stimulated emission cross section of the π-transition is larger by a factor of 1.5, leading to higher amplification at a given pump power or correspondingly to a lower pumping threshold (Bado et al. 1987).
ii) The larger bandwidth of the stimulated emission of the π-transition (Harmer et al. 1967) offers a potential for shorter pulses.
By implementation of a fast electro-optic feedback loop into the laser cavity, the amplitude of the laser pulse can be stabilized for more than one hundred round trips at a nearly constant intensity level with optimum bleaching of the absorber dye and optimum shortening of the laser pulses can be achieved (Heinz and Laubereau 1990)
Further reduction of pulse duration can be achieved by inserting a suitable glass étalon in the cavity, which is tuned to anti resonance. Angle tuning of the étalon is performed by monitoring the autocorrelation function in quasi real-time as is standard for cw mode-locked lasers.

2. EXPERIMENTAL

The experimental setup is depicted schematically in Figure 1. The solid lines indicate optical beam axes; dash-dotted lines and arrows show the paths of electrical signals.
The cavity is formed by two highly reflecting mirrors and an AR coated lens L1. The nonlinear absorber cell NA (absorber dye Kodak 9860) is placed near the focus between the lens and mirror M2. To provide the electrical input signal of the feedback control loop, approximately one percent of the circulating pulse is coupled out by a wedged glass plate GP (single sided AR coated) and monitored by the photodiode PD 2.

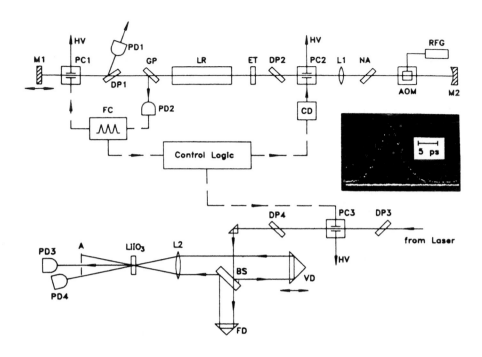

Figure 1.
Schematic of the pulsed Nd:YLF laser with feedback-controlled
active-passive mode-locking and with cavity dump, external pulse selector
and autocorrelation setup:
cavity mirrors M; laser rod LR; étalon ET; acousto-optic modulator AOM;
radio-frequency generator RFG; nonlinear absorber cell NA; Pockels cells
PC; dielectric polarizers DP; photodiodes PD; feedback controller FC;
cavity dumper CD; beam splitter BS; variable delay VD; fixed delay FD;
autocorrelation crystal LiIO3;

Inset: quasi realtime autocorrelation trace monitored by a digital
sampling oscilloscope.

Electro-optic losses are introduced by the help of the double-Pockels
cell PC1 in conjunction with the dielectric polarizer DP1. The latter
also serves as a variable output element for pulse train applications.
The two electrodes of PC1 are biased with a voltage difference of
approximately 1200 V (around the $\lambda/8$ voltage) so that strongly elliptical
polarization of the pulse is produced after a double pass through PC1. As
a consequence several ten percent of the pulse energy is decoupled by
polarizer DP1.
In addition to these constant losses, the feedback controller FC switches
a dynamic voltage (dependent on the detected signal of photodiode PD2)
that is superimposed on the bias voltage of PC1 generating fast
transmission changes of a factor up to 1.8. In this way the feedback
control loop stabilizes the circulating intracavity pulse energy to an
adjustable level of several µJ (Heinz and Laubereau 1990).

The étalon Et (uncoated LaSFN 18, thickness 0.251 mm) is mounted on a standard mirror holder. The repeatability of the tuning angle is estimated to be about ∓ 0.5 mrad.
Pulse durations are measured with a conventional autocorrelation setup using non-collinear second harmonic generation in a LiIO₃ crystal. To observe the autocorrelation in quasi real-time the optical delay is scanned continuously over the entire range (∓ 6 ps) and the detected second harmonic signal is displayed by a digital sampling oscilloscope without normalization nor averaging (Wolfrum and Heinz 1991). In this way two autocorrelation scans per second can be obtained. The spectral intensity distribution is recorded with a 1 m grating spectrometer and a multichannel detector.

3. EXPERIMENTAL RESULTS

Spectral anti-resonance of the étalon is adjusted by angle tuning and can be readily detected either measuring the pump energy at threshold, or monitoring the autocorrelation. To measure the pulse duration quantitatively, an external pulse selector consisting of dielectric polarizers DP3, DP4, Pockels cell PC3 is used (also shown in Figure 1). In this way, pulses at any position of the pulse train can be analysed by the autocorrelation measurement. For the measurements presented here, the last 50 single pulses of the pulse train are selected for autocorrelation; averaging is carried out over 10 consecutive shots. The reported pulse durations t_P (FWHM) refer to a Gaussian pulse shape.

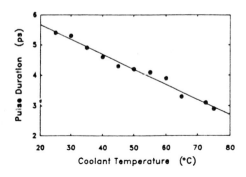

Figure 2
Measured pulse duration of the Nd:YLF laser versus coolant temperature.

For optimum pulse shortening, i. e. maximum broadening of the net gain curve of the laser, the pulse duration is measured as a function of coolant temperature of the laser rod in the range between 25 °C to 75 °C. An example for the autocorrelation signal is shown in Figure 3. The calculated Gaussian curve yields a pulse duration of (2.9 ∓ 0.1) ps while a bandwidth product $\Delta\nu \times \Delta t \cong 0.5$ is found. This result refers to a coolant temperature of 75 °C. The stability of the pulse train, the single pulse energy, and the total energy of the pulse train remain constant for the entire coolant temperature range. We obtain a total output energy of 0.6 mJ for a train of approximately 150 pulses, corresponding to a single pulse energy of 4 μJ at a repetition rate of 70 Hz, with a standard variation of single pulse energy of less than 3 %.

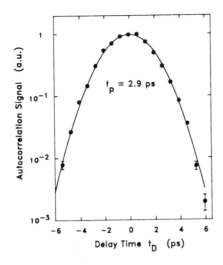

Figure 3
Autocorrelation signal of the FCM laser with anti-resonant étalon and coolant temperature 75 °C; calculated Gaussian curve, experimental points.

The variation of pulse duration and threshold energy with étalon tuning angle is depicted in Figure 4. Maximum broadening, i. e. minimum curvature of the net gain curve, is accomplished when the étalon is tuned to antiresonance. This results in minimum pulse duration and simultaneously maximum pump energy at threshold. The angular distance between the minima of pulse duration agrees with the theoretically calculated value, based on the étalon data.

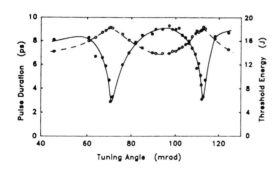

Figure 4
Measured pulse duration (full circles, left-hand ordinate scale) and electric input energy at threshold (open circles, right-hand ordinate scale) versus angle of incidence of the intracavity étalon. Antiresonance with minimum pulse duration corresponds to the maximum threshold energy. The lines are drawn as a guide to the eye.

4. REFERENCES

Bado B, Bouvier M, Coe S J, 1987, Opt. Lett. **12**, 319
Harmer A L, Linz A, Gabbe D R, 1969, J. Phys. Chem. Solids **30**, 1483
Heinz P, Laubereau A, 1990, J. Opt. Soc. Am. B **7**, 182
Wolfrum K, Heinz P, 1991, Opt. Comm. **84**, 290

Inst. Phys. Conf. Ser. No 126: Section I
Paper presented at Int. Symp. on Ultrafast Processes in Spectroscopy, Bayreuth, 1991

Antiresonant-ring mirror passive mode-locking of solid state lasers

A Agnesi, G Gabetta, and G C Reali

Universita' di Pavia, Dipartimento di Elettronica
Via Abbiategrasso 209, 27100 Pavia, Italy

ABSTRACT. Transverse-mode-filtering in the antiresonant-ring mirror passive mode-locking of solid state lasers explain why the pulses are shorter in this case than using a linear cavity with standard contacted dye-cell mirror.

Passively mode-locked solid-state lasers with antiresonant-ring mirror (ARRM) including a dye-cell were proposed by Siegman [1], and were first demonstrated by Vanherzeele et al.[2] effective in generating pulses shorter than with a linear cavity and conventional contacted dye-cell mirror (DCM).

We also demonstrated (Fig. 1) that the pulses from a passively mode-locked flashlamp pumped solid-state laser can be considerably shorter (by a factor of ≈2) using an antiresonant-ring mirror than using a linear cavity with a standard contacted dye-cell mirror [3].

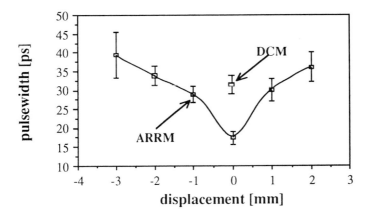

Fig. 1. Experimental result: pulse-width vs. displacement of the dye cell from the center of the interferometer ring. The pulse-width of the linear cavity is also shown.

The equivalent reflectivity of the interferometer, without the dye, was found to be better than 96%. As a matter of fact, a small umbalancement of the ARRM was present and we always detected a small radiation leakage at the exit arm of the interferometer which couples the radiation out of the resonator. Under these conditions, the ARRM smoothly selects the lowest order transverse mode of the cavity, which originates from an interference coupling having the interesting property that the degree of loss increases for higher order transverse modes even though no finite-aperture diffraction effects of the usual sort are involved [4,5]. We did require instead an intracavity pin-hole of the appropriate diameter for single mode operation in the DCM configuration.

Since the first report of similar results by other groups, it appears that no theoretical explanation has survived. To explain the pulse shortening results we propose to consider the effect of a possible transverse multi-mode oscillation of the cavity.

In simulating mode-locking [6], single transverse mode (plane wave) operation is generally assumed, no matter what is the real field oscillating in the resonator, and this leads to approximately equal pulse durations for the two configurations, despite of the increased effective saturation dye intensity in the ARRM scheme. If we think of the possibility that more than the single transverse TEM_{00}-mode be present during the build up of the mode-locked pulse, we should consider the frequency difference between the mnq- and the 00q-mode [7]:

$$\Delta v_{mn} = \frac{(m+n)}{u} \frac{cos^{-1}\sqrt{g_1 g_2}}{\pi} \qquad (1)$$

In (1) g_1 and g_2 are the usual resonator parameters, $u=2L/c$ is the round-trip time, and L is the resonator length. For one such component present in addition to the 00-mode, the field distribution can be written as

$$U(r,\theta,t) = U_{00}(r,\theta) + U_{mn}(r,\theta) \cos(2\pi\Delta v_{mn}t) \qquad (2)$$

It is easy to verify that even the presence of a single extra transverse mode actually makes the spatial field distribution vary cyclically over few round-trips (fig. 2), lowering the compression capabilities of the dye with respect to the the case of the single 00-mode. The beats enhance the peak intensity of the field distribution anticipating the Q-switching action, and cause different local field fluctuations that, due to the strongly nonlinear interaction with the dye, change the temporal evolution of the field slices across the transverse pulse profile and integrate in time to a longer duration.

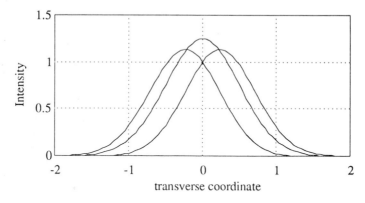

Fig. 2. Transverse mode beating TEM_{00} (80%) + TEM_{01} (20%) during one beat period.

To clarify these points, we performed a model calculation with a mixture of the first two azimuthally symmetric modes. The total field is written as:

$$U(\rho,t) = [U_{00}(\rho) + U_{10}(\rho)\cos(\omega t)] \, \alpha(t) \qquad (3)$$

where $U_{00}(\rho) = \varepsilon_1 exp(-\rho^2)$, $U_{10}(\rho) = \varepsilon_2 \rho exp(-\rho^2)$, $\rho = r/w_0$ is an adimensional transverse coordinate, ε_1, ε_2 and $\alpha(t)$ are the fractional contents of the two modes and a time-dependent normalization factor, respectively, to be determined so that the total energy **E** (=1) sums up to

E_1, contributed by U_{00}, plus E_2, contributed by U_{10}, and ω is the beat frequency. We found useful to quantify the transverse mode mixture by means of the parameter $\eta = E_2/E$ representing the fraction of the TEM_{10} mode content in the circulating pulse shape. The transverse beat oscillation was assumed to have a transverse periodicity of approximately N=15 round-trips. The modulation factor of the laser field, as a function of the transverse coordinate ρ and time, is given by:

$$Q(\rho,t) = \left\{ 1 + \left[\frac{2\eta}{\sqrt{\pi}(1-\eta)} \right]^{1/2} \rho \cos\left(\frac{2\pi t}{Nu}\right) \right\} \left[1 - \eta \sin^2\left(\frac{2\pi t}{Nu}\right) \right]^{-1/2} \quad (4)$$

In a proper simulation of the mode-locking dynamics, which is under development, it is important to take diffraction into account, since it is what constantly correlates the field at different coordinates ρ's, which would otherwise evolve in an independent way resulting in time-shifted mode-locked pulse slices. In a first approach, we performed a simpler one dimensional simulation at the fixed coordinate $\rho = 0$, which should give a lower bound for the time duration since from (4) it is seen that for $\rho > 0$ the modulation increases as more energy is distributed over larger areas (more TEM_{10}).

In our simulations we used the physical model of passively mode-locked solid state lasers described in [8], adopting the numerical scheme of the characteristics to take the coherent interaction of the colliding pulses in the dye-cell into account and solving a rate equation system to describe the amplification of the circulating pulse. Using the statistical model of [8] to get the character of the field rising from the spontaneuos emission noise and the beginning linear amplification stage, it is possible to start the simulation at the beginning of the nonlinear amplification stage. After each round-trip, the laser field was multiplied by the factor $Q(\rho=0,t)$ at its corresponding time to take the effect of on-axis mode beating into account for several η values. We report in fig. 3 the results of the computation, in which the parameters were tuned to get, for $\eta = 0$, a consistent agreement with the time duration and energy content of the pulses in the conditions of optimum mode-locking for the cavity with ARRM. It is seen that a progressive contamination by TEM_{10} monotonically lengthens the pulse, reaching the duration of 29 ps when $\eta = 0.5$, i.e. half of the energy is contributed by TEM_{10}.

In conclusion the following picture emerges:

The slightly unbalanced ARRM smoothly selects the lowest order transverse mode of the cavity with no finite-aperture diffraction effects of the usual sort involved. These finite-aperture effects are precisely the origin of higher order transverse mode contamination in a standard linear cavity, and, according to the above reasonings, explain why longer pulses are obtained in the DCM arrangement. The multi-transverse mode operation is thus seen not only to affect the optical quality of the laser beam but also to act more deeply in determining the evolution of the passive mode-locking regime. It also seems that the dye-cell placement within the ARRM only slightly affects its performances, and it can more simply be incorporated into the other mirror, with a less critical mode-locking operation for the cavity with the ARRM.

Acknowledgments

We appreciate a very useful discussion with Prof. A.E.Siegman, who also brought to our attention the references [4] and [5].

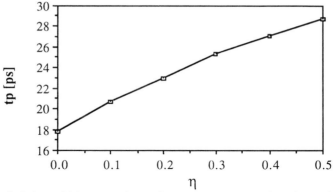

Fig. 3. Pulse-widths vs. E_2/E obtained from mode-locking simulations.

References

[1]   A.E.Siegman, IEEE J. Quant. Electr., **QE-9**, 247 (1973)
[2]   H.Vanherzeele, J.L.Van Eck, A.E.Siegman, Appl. Optics, **20** (1981), 3483 (1981)
[3]   G.P.Banfi, G.Gabetta, P.G.Gobbi, G.C.Reali, Sov. J. Quant. Electr., **20**, 1203
       (1990)
[4]   S.-C. Sheng and A.E.Siegman, J. Opt. Soc. Am., **66**, 1032 (1976)
[5]   M. McGeoch, Opto-Electronics, **1**, 85 (1970)
[6]   J.A.Fleck, Phys. Rev. **B1**, 84 (1970)
[7]   A.E.Siegman, "Lasers", University Science Books, Mill Valley, CA (1986)
[8]   J.Herrmann and B.Wilhelmi, "Lasers for Ultrashort Light Pulses", North Holland
(1987)

Inst. Phys. Conf. Ser. No 126: Section I
Paper presented at Int. Symp. on Ultrafast Processes in Spectroscopy, Bayreuth, 1991

The mode-locking technique using intracavity frequency doubling

Stankov K A

Faculty of Physics, University of Sofia
BG-1126 Sofia, Bulgaria

ABSTRACT: The recent advances in the mode locking based on the frequency-doubling nonlinear mirror are reviewed. The advantages of this new technique as all-solid-state, operating in wide spectral regions and providing synchronous second harmonic output are indicated. The limitations arising from the group-velocity mismatch in the intracavity frequency doubling crystal are also indicated. Experiments with pulsed lasers and theories describing the operation of pulsed and CW lasers are reviewed and a design concept for high-performance, all-solid-state modelocked pulsed laser is presented.

1. INTRODUCTION

It has been known since the first experiments with intracavity frequency doubling that the latter is associated with losses which grow as the light intensity increases. The second harmonic radiation which leaves the laser cavity represents intensity-dependent loss, resulting in smoothening of the pulse form of the laser generation or pulse lengthening in mode-locked laser (Smith 1970, Hitz et al 1971, Siegman et al 1980). Only recently it has been realized that just the opposite behaviour can be achieved, by recognizing the importance of the phase difference $\delta\phi=\phi_2-2\phi_1$ between the phases of the fundamental, ϕ_1 and the second harmonic, ϕ_2, in the laser cavity (Stankov 1988). Depending on the phase relationship, the interaction of the light waves may be in opposition (Stankov 1991a), and thus the intensity-dependent losses may either increase, or decrease with the increase of the light intensity. In the latter case, an intensity dependent positive feedback is developed, providing mode-locking in a laser with intracavity frequency doubler. This type of feedback has been realized in the form of a nonlinear output mirror, composed of a frequency doubling crystal and a dichroic mirror, totally reflecting at the second harmonic, known as the frequency doubling (second harmonic) nonlinear mirror. We briefly present the frequency doubling nonlinear mirror concept and review recent experiments and theories, related to this modelocking technique.

2. THE FREQUENCY-DOUBLING NONLINEAR MIRROR

This nonlinear optical device was introduced in 1988 (Stankov 1988a) as a combination of a second harmonic generator SHG and a dichroic mirror DM, as depicted in Figure 1. Means for controlling the phase difference $\delta\phi$ are provided in the form of a transparent phase plate PP which can be tilted in order to vary the thickness. A simpler way is to use the dispersion of the air refraction index, as has been done in most experiments, by changing the distance between the nonlinear crystal and the dichroic mirror. The nonlinear reflection coefficient R_{NL} depends on the reflectivities of the dichroic mirror at the fundamental and the second harmonic wavelength, R1 and R2 respectively, and the second harmonic conversion efficiency η, (Stankov 1988, Stankov 1991a):

$$R_{NL} = B\{1-\tanh^2[\sqrt{B}.\text{arctanh}\sqrt{\eta} \pm \text{arctanh}(\sqrt{\eta R2/B})]\} \qquad (1)$$

with $B = \eta R2 + (1- \eta)R1$. In these expressions, the sign (-) refers to a phase difference $\delta\phi=\pi/2\pm(2m+1)\pi$, (m=0,1,2,3,...) and increasing reflectivity, while the sign (+) is valid for a phase difference $\delta\phi=\pi/2\pm2m\pi$ and decreasing reflectivity.

The nonlinear reflection which grows with the light intensity is the base for modelocking of a laser, since it provides positive intensity-dependent feedback (Stankov 1988). The highest nonlinearity is achieved for R2=1. However, lower reflectivity at the second harmonic wavelength may result in reflection which maximizes at certain power level, with potentially interesting application as a power limiter (Stankov 1991b). The nonlinearity in reflection increases as the reflectivity R1 decreases.

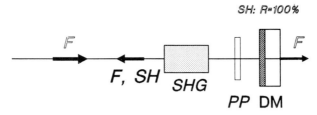

Figure 1. The frequency-doubling nonlinear mirror.

At a first glance it seems that the frequency-doubling nonlinear mirror is the ideal modelocking device, since it has a number of excellent features. First of all, it is all-solid-state, durable, provides considerable amount of usable second-harmonic radiation. In contrast to saturable absorbers, it can be used in extremely wide spectral ranges and the time response seems to be instantaneous. However, a more detailed analysis leads to the conclusion, that the operation is limited by the group-velocity mismatch (GVM) in the process of the second harmonic generation, or equivalently, by the limited phase-matchable bandwidth. If the latter is smaller than the laser bandwidth, the nonlinear mirror will act as a spectral filter, determining a limit for the minimum achievable pulse duration. A recent nonstationary analysis (Stankov et al 1991c) has identified this limit. In terms of group-velocities at the fundamental and the second harmonic, u_1, u_2, and crystal length L_{cr}, it is given approximately by $\tau_p=3L_{cr}(1/u_1 - 1/u_2)$. In the near infrared for some crystals there is also group-velocity matching. This would allow generation of several tens of femtoseconds long pulses, provided group-velocity compensated cavity is used. In the visible, however, the group-velocity mismatch may induce severe limitations on the pulse duration, if long and strongly dispersing crystals are used (Hamal 1991).

2. THEORETICAL MODELS

Shortly after the demonstration of modelocking of Nd:YAG laser with an intracavity frequency doubling crystal (Stankov et al 1988), a theory for the operation of a continuous-wave laser incorporating the second-harmonic nonlinear mirror was published by Barr (1989). The calculated pulse shape was slightly different from that of the originally assumed gaussian one. The pulse length reduction in an actively modelocked laser due to the presence of the nonlinear mirror has been estimated from 72.5 ps down to 21.3 ps, assuming R1=0.8, R2=1 and peak conversion efficiency of 10%. Later an improved model was presented by the same author (Barr 1991). The new model has accounted for the saturation effects in the active medium, the dynamic change of the reflectivity due to the increased conversion efficiency in the process of the pulse shortening and the Q-switching action. The influence of the fluorescence lifetime was analysed and the observed threshold second harmonic efficiency was shown to be in agreement with the observed laser behaviour. A more elaborated CW model was presented by Petrov et al (1990a) and refined later (Petrov et al 1990b). The model has been applied to find the optimum conditions for additional pulse shortening in terms of nonlinearity of the mirror and/or the focussing parameter.

The fluctuation approach was adopted by Huo et al (1990), who were able to analyse a number of important laser design parameters, responsible for attaining optimized modelocking performance. The second-harmonic enhancement ratio was used as a measure of the quality of the modelocking.

An analysis of the operation of the second-harmonic nonlinear mirror in the frequency domain was first presented by Stankov et al (1991b). The process of modelocking was identified as generation of second harmonic and frequency mixing in the first pass of the laser radiation through the nonlinear crystal and difference-frequency generation in the second pass. These two stages of the interaction of the longitudinal modes in the nonlinear crystal result in generation of side-band frequencies, equally spaced at c/2L. At the same time, the phase differences between each two adjacent modes converge to a certain value, what is actually the essence of the modelocking. For the first time, it has been possible to follow the evolution of the amplitudes and the phases of the individual modes in the process of modelocking. The model accounts for the lineshape of the active medium and the saturation effects.

At present, it seems that the theories developed for a laser modelocked with the second-harmonic nonlinear mirror explain satisfactorily most of the experimentally observed features. However, such effects as transverse laser field distribution, diffraction effects, cavity dispersion (responsible for nonequidistant mode spacing) have not been yet considered. Also, the models do not take into account the influence of the group-velocity mismatch in the frequency-doubling crystal.

3. EXPERIMENTAL RESULTS: DESIGN CONCEPT FOR HIGH PERFORMANCE MODELOCKED LASER

Shortly after publishing the proposal for the frequency doubling nonlinear mirror, an experiment for modelocking of a pulsed Nd:YAG laser was successfully realized (Stankov et al 1988). An intracavity telescope was employed, providing high conversion efficiency into second harmonic in a 6-mm long KTP (oe-e type) crystal. The generated pulses were approx. 100 ps long, exhibiting pronounced substructure. Refinements of the experiment and using ß-BaB$_2$O$_4$ crystal have produced 25 ps pulses (Stankov 1991c). The shortest pulse duration of 15 ps achieved to date with this technique is from Nd:YAlO$_3$ laser operating at 1.34 μm (Stankov et al 1991a). In this experiment, two laser transitions of the same laser were modelocked, using a single LiIO$_3$ crystal. At the other transition (1.08 μm) the pulse duration was 40 ps. The experiment is remarkable by the fact that for the first time quite differing laser transitions were passively modelocked using the same modelocking device. Useful information regarding the dynamics of the modelocking process in such laser has been obtained. A strong evidence was found that the pulse duration was limited by the premature pulse train termination.

In a very recent experiment, the same frequency doubling 30°-cut LiIO$_3$ crystal was used to modelock a Ti:Sapphire laser (Hamal 1991). Tuneable ultrashort pulses were generated in the range 792-811 nm, by synchronously tuning the dispersive element (2-mm thick crystal-quartz plate) and tilting the frequency doubling crystal. The pulse duration was quite long - 100 to 120 ps, due to the large dispersion and the long crystal.

The nonlinear mirror arrangement has been used also in a coupled-cavity configuration (Barr et al 1990) producing pulses 30 ps long.

Here we present a design concept for high performance all-solid-state pulsed modelocked laser which is under development, Figure 2. The concept is based on the experience gathered to date from the experimental observations and the theoretical analyses. First of all, such laser will include an active modelocker AOM for reproducible and reliable operation. At the same time, the active modelocking will increase considerably the frequency-doubling efficiency, allowing use of short nonlinear crystals. Thus, the group-velocity mismatch will be substantially reduced, allowing generation of shorter pulses. Further reduction of the GVM is possible by using a compensating birefringent plate CP which

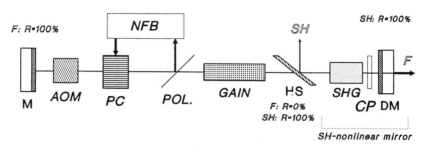

Figure 2. Concept for a high performance modelocked laser.

induces an opposite time delay with respect to that accumulated in the nonlinear crystal (Stankov et al 1991d). In lasers with long relaxation time active medium, a negative feedback NFB will provide long pulse train evolution allowing for ultimate pulse shortening. Finally, a harmonic separator HS will extract a pulse train at the second harmonic wavelength. Such laser should be capable of producing high power subpicosecond pulses at the fundamental and the second harmonic from lasers with broadband gain profile.

4. CONCLUSION

The modelocking technique based on intracavity frequency doubling has entered the mature stage of the theory and experiment. The attractive features of this technique can be used to develop a reliable and versatile source of high power ultrashort laser pulses.

References:

Smith R G 1970, *IEEE J. Quant. Electron.*, **QE-6** 215
Hitz C B and Osternik L M 1971, *Appl. Phys. Lett.* **18** 378.
Siegman A E and Heritier J -M 1980 *IEEE J. Quant. Electron.*, **QE-16** 324.
Stankov K A 1988, *Appl. Phys.* **B45** 195.
Stankov and Jethwa J 1988, *Opt. Commun.*, **66** 41.
Stankov K A 1991, *Appl. Phys.* **B52** 158
Stankov K A, Kubecek V and Hamal K 1991a, *IEEE J. Quant. Electron.* **QE-27** 2135,
Hamal K, Stankov K A, Jelinkova H, Prochazka I and Koselja M 1991, *Proceedings of UPS'91, Bayreuth, Germany*, paper Tu 3-3
Petrov V P and Stankov K A 1990, *Appl. Phys.* **B50** 409.
Petrov V P, Rudolf W and Stoev V D 1990 *Revue Phys. Appl.* **25** 1239
Chong-ru Huo and Zhen-he Zhu 1990, *Opt. Commun.*, **79** 328
Stankov K A, Tzolov V P and Mirkov M G 1991b, *Opt. Lett.* **16**, 639
Barr J R M 1989, *Opt. Commun.* **70** 229
Barr J M R 1990 *Opt. Commun.* **81** 215
Stankov K A, Tzolov V P and Mirkov M G 1991c, *Opt. Lett.* **16** 1119.
Stankov K A 1991b, *unpublished results*
Stankov K A 1991c, *Appl. Phys. Lett.* **58** 2203
Barr J R M and Hughes D W 1990 *J. Mod. Opt.* **37** 447
Stankov K A, Tzolov V P and Mirkov M G 1991d, *submitted for publication*

Inst. Phys. Conf. Ser. No 126: Section I
Paper presented at Int. Symp. on Ultrafast Processes in Spectroscopy, Bayreuth, 1991

Mode-locking of a Nd:YAG laser by a frequency-doubling crystal and saturable absorber

Zhen–guo Lü , Qi WU, Qing–xing LI and Zhen–xin YU

Institute for Laser and Spectroscopy, Zhongshan University, Guangzhou 510275, P.R.China

ABSTRACT: The modelocking of a pulsed Nd:YAG laser using the combined pulse shortening mechanism of a nonlinear mirror based on a frequency–doubling crystal and saturable absorber is reported. The average pulse duration is 35ps, and the pulses are well–shaped. Theoretical analysis concerning the mechanism is also presented.

1. INTRODUCTION

The nonlinear mirror formed by a frequency–doubling crystal and a dichroic mirror with proper separation has been used as a passive modelocking element in several kinds of laser caveties by Stankov (1989). Barr (1989a), Carruthers (1990) respectively. Compared with conventional modelocking technique, this novel technique has some advantages such as simplicity, durability and fast response. As observed by Stankov(1988) and Lü (1991a) when the modelocking probability approach 100%, satellites of considerable amplitude and double even triple pulse trains appeared. This phenomenon adversely limits the application of the modelocking technique.

We put a saturable absorber cell adjacent to the total reflector on the basis of nonlinear mirror modelocking configuration. This modification is aimed at generation of structure free and sigle train pulses. Indeed, in the case of nonlinear mirror modelocking, small perturbations may grow up becasue of the high gain and low loss in the cavity. In section 2, we will study the differential equation for the pulse envelop and give the expression for the pulse width. Section 3 is our experimental results which support our predictions.

2. THEORETICAL ANALYSIS

Consider the situation shown in Figure 1, let's first remove the frequency–doubling crystal. According to the modelocking theory with fast saturable absorber of H.A.Haus (1975), the envelop of the pulse $V(t)$ satisfies the foundamental equation

$$[1 + \frac{Q}{Q_A(T)} - g(1 + \frac{1}{\omega_L^2}\frac{d^2}{dt^2}) + \frac{g+\delta}{\omega_L}\frac{d}{dt}]V(t) = 0 \qquad (1)$$

where $\dfrac{Q}{Q_A(t)} = \dfrac{2L}{(\omega_0/2Q)T_R}$, $g = \dfrac{2G(\omega_0)}{(\omega_0/2Q)T_R}$, $\delta = \dfrac{\omega_L\delta_T}{(\omega_0/2Q)T_R}$, $\omega_0T_R/(2Q)$ represents the

loss of the cavity; $G(\omega_0)$ is the saturated gain at the center of the gain curve; ω_L is the linewidth of the gain medium, δT is the additional advance or retardation resulted from the change of the pulse shape; and L(t) is a power dependent absorption coefficient.

Then, we put again the frequency–doubling crystal in its place so that a nonlinear mirror is formed. The foundamental equation for this situation can be easily obtained if we note that the behaviour the nonlinear mirror is somewhat similar to that of a saturable absorber. Therefore eq.1 applies to our problem with some modifications on L(t), which should include the contribution of the nonlinear mirror.

Fig. 1. Schematic of laser modelocking with combination of nonlinear mirror and saturable absorber. NLC: nonlinear frequency–doubling crystal; S.A.: saturable absorber.

For the saturable absorber with fast relaxation time, the absorption coefficient is

$$L_{SA}(t) = \sigma_A \theta_A A_A n_e \left(1 - \frac{|V(t)|^2}{P_A} \right) \qquad (2)$$

where σ_A is the optical cross section of the absorbing partical; θ_A is the length of the absorber; A_A is the cross section of the optical beam within the absorber; n_e is the equilibrium population difference; P_A is the saturation power for the absorber.

According to the work of Barr (1989b), the reflectivity of the nonlinear mirror can be approximately expressed as

$$\gamma_{NL} = \sqrt{R_\omega} \, exp \left\{ \frac{1}{2} \sigma \eta_0 \frac{|V(t)|^2}{V_0^2} \right\} \qquad (3)$$

where R_ω represents the reflectivity of the dichroic mirror at foundamental wavelength; σ is a positive parameter relative to R_ω ; η_0 is the second harmonic conversion efficiency at the peak of the pulse; and V_0 is the amplitude of the pulse. It follows from eq.3 that the nonlinear mirror can be viewed equivalently as the combination of a saturable absorber with absorption coefficient

$$L_{NLM}(t) = - \frac{1}{2} \sigma \eta_0 \frac{|V(t)|^2}{V_0^2} \qquad (4)$$

and a mirror with reflectivity R_ω, which has been taken into account in the cavity loss $\omega_0 T_R / (2Q)$. Therefore, the total nonlinear absorption coefficient is

$$L(t) = \sigma_A \theta_A A_A n_e \left(1 - \frac{|V(t)|^2}{P_{AN}} \right) \qquad (5)$$

where $1 / P_{AN} = 1 / P_A + \sigma \eta_0 / (2 V_0^2 \sigma_A \theta_A A_A n_e)$

Now we look for the solution of eq.1. The pulse must pick their repetition period so that $g + \delta = 0$. The remaining equation is

$$\left[1 + q - g(1 + \frac{1}{\omega_L^2} \frac{d^2}{dt^2}) \right] V(t) - q \frac{|V(t)|^2}{P_{AN}} = 0 \qquad (6)$$

where $q = 2 \, \sigma_A A_A n_e / [(\omega_0 / 2Q) T_R]$ is the small–signal inverse Q of the saturable absorber normalized to the cavity Q. The solution is soliton like and has the time dependence

$$V(t) = V_0 \, sech(t / \tau_p) \qquad (7)$$

where we have approximately

$$\tau_p = 4 \frac{P_{AN}}{P_L} \frac{(1+q)^2}{q(1+q-g_0)} \cdot \frac{1}{\omega_L T_p} \tag{8}$$

where P_L is the saturation power for the gain medium, and T_p is the pulse repetition time. According to eq.8, the combined action of nonlinear mirror and saturable absorber relsults in shorter pulsewidth τ_p than the case with mere saturable absorber because $P_{AN} < P_A$. In general, shorter pulsewidth is obtained by means of increasaing the concentration of the saturable dye, but this raises the lasing threshold and may cause damage in the laser medium. The application of the saturable absorber in Figure 1 is two—fold. In the first place, perturbations are prevented from growing up, hence structure—free modelocked pulse are generated. Next, the pulse in the cavity becomes stable more quickly than the case without the saturable absorber, this is of great significance for pulsed lasers.

3. EXPERIMENTAL RESULTS AND DISCUSSINS

The experimental setup is show in Figure 2. Applying the work of Lü (1991b), we use a 50mm long

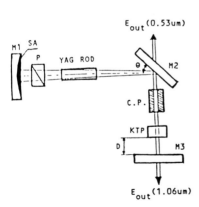

Fig. 2. Experimental set up. P–polarizer; M_1–totalreflector with curvature of 2m; M_2– harmonic splitter; M_3–dichroic output mirror (total reflector for the second harmonic wave and 16% for the laser wavelength).

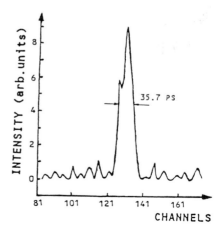

Fig. 3. Streak—camera record of an individual ultrashort laser pulse at 0.53μm revealing pulse duration of 35.7 ps (FWHM). the sweep rate is 5.1 psec / channel.

capillary C.P with 1.90mm internal diameter to provide single transversal—mode operation as well as high intensity beam for efficient second—harmonic generation in the KTP crystal, which satisfies type II (oe–e) phase matching condition. The laser is operated at a 1Hz repetition rate. The saturable absorber is pentamethylidyne dye solution with 75% transmissivity, the solvent is CS_2.

The distance between the KTP crystal and the dichroic mirror is set to be 51mm, which is the optimum position for modelocking. The KTP crystal is adjusted to satisfy phase matching condition. When modelocking does not occur, the output energy are $E_{out}(1.06\mu m) = 1.4mJ$, $E_{out}(0.53\mu m) = 30\mu J$. When the KTP crystal is tuned to the optimum phasematching angle, the output enegy at both foundamental wavelength and second harmonic wavelength increase by large scale: $E_{out}(1.06\mu m) = 5.3mJ$, $E_{out}(0.53\mu m)$ is in the range of 0.8mJ to 2.1mJ. A 100% modulated pulse—train is observed, and its envelope is approximately 200ns, the period of the modelocked pulse—train is measured to be 10.7ns, which is in accordance with the cavity length of 160cm. The pulsewidth measured with a Streak camera (Hamamatsu Temporal Disperser C 1587) are between 28ps and 50ps (FWHM). Figure 3 shows the streak—camera record of an individual modelocked pulse. Multitrain or "satellites" have not been observed.

The modelocking probablility is not high. This is partly due to the instability caused by the solvent according to Xie et al (1980). More stable modelocking operation calls for better dye solvent. Because the saturable absorber in the cavity raises the lasing threshold, it followes from the work of Barr (1989b) that pulse compression is less effective than the case with lower threshold. However, the idea of using thin dye solution to constrain the satellites and multitrain also applies to other modelocking configurations such as coupled cavity modelocking scheme. Further studies are still going on.

In conclusion, we have generated structure free ultrashort pulses in a new way. The pulsewidth is shorter than that obtained by using saturable absorber but slightly longer than that by using the nonlinear mirror.

REFERENCES:

Barr J R M 1989a Opt. Commun. **70** pp 229—4

Barr J R M and Hughes D W 1989b Appl. Phys. B **49** pp 323—4

Carruthers T F and Duling I N 1990 Opt. Lett. **15** pp 306—3

Haus H A 1975 J. Appl. Phys. **46** pp 3049—6

Lü Z G , Li Q X and Yu Z X 1991a "Study of a pulsed Nd: YAG Laser modelocked by a KTP crystal ", to be published in 《Acta Optica Sinica》 of China.

Lü Z G, Fu C H, Zhou J Y, Li Q X and Yu Z X 1991b 《Acta Optica Sinica》 of China **15** pp 232—2

Stankov K and Jethwa J 1988 Opt. Commun. **66** pp 41—4

Stankov K 1989 Opt, Lett. **14** pp 359—3

Xie Z M and Chen S H 1980 《J. Chinese Lasers》 **7** pp 57—1

Inst. Phys. Conf. Ser. No 126: Section I
Paper presented at Int. Symp. on Ultrafast Processes in Spectroscopy, Bayreuth, 1991

39

Chirped pulse formation in mode-locked solid state lasers with intra-cavity harmonic generation

S F Bogdanov, G I Onishchukov, P G Konvisar, S D Ryabko, S R Rustamov, A A Fomichov

Moscow Institute of Physics and Technology, Laser Center, Moscow

ABSTRACT: The system of differential rate equations describing the generation of chirped Gaussian ultrashrot pulses in lasers with homogeneously broadened active media with active mode-locking and intracavity second harmonic generation has been obtained. It is shown that the wave detuning in nonlinear crystal causes chirp which has been observed in our experiments and shift of central freqency. Effect of self compression when the pulse duaration is less than in laser without nonlinear crystal can, at least principally, be observed under certain conditions.

1. INTRODUCTION

One of the ways to increase the efficiency of nonlinear frequency convertion of laser radiation is the intracavity second harmonic generation. Though presence of nonlinear element in the cavity of mode-locked laser essentially changes parameters of generated ultrashort pulses. The results of our theoretical and experimental investigation show that in this case it is possible to obtain pulses with tunable (value and sign) linear chirp. This feature is very important in study of coherent processes such as foton echo and further frequency convertion, in particular, in parametrical oscillators.

2. THEORETICAL CONSIDERATION

Theoretical treatment of the problem is based on the time-spatial approach suggesed by Kuizenga and Siegman (1970) and Folk (1975). Assuming that Gaussian laser puls can be described by

$$A_1(t,r) = A_{10}\exp[-(\gamma - i\varepsilon)t^2/2 - r^2/w_{01}^2]$$

we developed the system of differential rate equations describing the generation of chirped Gaussian ultrashort pulses in laser with a homogeneously broadened active media with active mode-locking and intracavity second harmonic generation (Bogdanov and Konvisar 1990):

$$\frac{dK}{d\tau} = B_2 - \frac{g\mu^2}{2(1+\eta^2)^3} [(K^2-E^2)(1-3\eta^2)-2EK\eta(3-\eta^2)] - \pi^{1/2}\Gamma IK^{3/2}\Sigma;$$

$$\frac{d\phi}{d\tau} = \frac{g\mu}{4(1+\eta^2)^2} (1-\eta^2) - \frac{g\mu E\eta}{2K(1+\eta^2)^2} - \frac{B_2}{2K} + \pi\frac{\Delta f}{f_m};$$

$$\frac{dI}{d\tau} = I(g'-\alpha -2B_0 - \frac{B_2}{2K} - 2\pi^{1/2}\Gamma IK^{1/2}\Sigma);$$

$$\frac{dE}{d\tau} = \frac{g\mu^2}{(1+\eta^2)^3} [KE(1-3\eta^2)+\frac{1}{2}(K^2-E^2)\eta(3-\eta^2)] - 2(2\pi)^{1/2}\Gamma IK^{3/2}\Lambda;$$

$$\frac{d\eta}{d\tau} = -\frac{\eta g\mu^2}{2(1+\eta^2)^2} \frac{K^2+E^2}{K} - \mu\frac{EB_1}{2K};$$

$$\frac{dg}{d\tau} = a(g_0-g-Ig'),$$

where

$$g'= \frac{g}{1+\eta^2} \left[1- \frac{\mu^2(K^2+E^2)(1-3\eta^2)}{4K(1+\eta^2)^2} \right];$$

$$\mu = \frac{2\omega_0}{\Delta\omega} ; \quad K = \frac{2\gamma}{\omega_0^2} = \frac{2\ln2Q^2}{\pi^2} ; \quad \eta = \frac{2\Delta\omega_s}{\Delta\omega} ; \quad E = \frac{2\varepsilon}{\omega_0^2} ;$$

$B_0 = \delta sin^2\phi$; $B_1= \delta sin2\phi$; $B_2= \delta cos2\phi$; $Q = T_0/\tau_p$; $T_0=1/f_0=2L/c$;

$\Sigma=sin^2\Delta/\Delta^2$; $\Lambda=(1-sin\Delta \ cos\Delta/\Delta)/\Delta$; $\Delta=\Delta kl/2$-wave detuning;

Γ- nonlinear coupling coefficient; L-resonator length;

l- nonlinear crystal length; τ_p-pulse length;

α - effective double-pass losses;

g_0,g -unsaturated and saturated double-pass gain coefficient;

$a = T_0/T_1$; T_1-longitudial relaxation time;

$\Delta f = f_m-f_0$-frequency detuning; δ-modulation index;

ϕ-pulse transmission phase through modulator;

$f_m/2$ - modulator drive frequency;

I- intracavity fundamental wave intensity;

$\Delta\omega_s$ - shift of the central freqency;

$\Delta\omega$ - homogeneous linewidth of the laser transition;

Some results obtained by numerical solution of this system are shown on the figure 1. The dependences of frequency deviation $D =\pi E f_m^2$ (a), central frequency shift $\Delta\omega_s$ (b) and ultrashort pulse duration (c) as functions of the wave detuning in nonlinear crystal for Nd:YAG laser are shown.

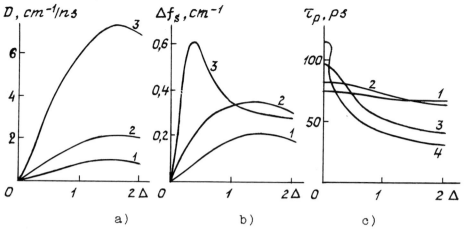

Fig.1. Typical dependences of frequency deviation (a), frequency shift (b), and pulse duration for 1.- $\Gamma=0,5\ 10^{-6}$; 2.- $\Gamma=10^{-6}$; 3.- $\Gamma=0,3\ 10^{-5}$; 4.- $\Gamma=0,5\ 10^{-5}$.

The wave detuning in nonlinear crystal causes chirp of the pulses as well as frequency shift which increase with growth of the nonlinear coupling coefficient. Curves shape is defined by the shape of the function $\Lambda(\Delta)$ for small values of Γ. The change of wave-detuning sign results in change of both chirp sign and frequency shift sign.

The appearence of chirp can cause an effect of self compression when the pulse duration is less than one in laser without nonlinear cristal can be observed. The more value of the nonlinear coefficient is, the more self compression can be achieved. However, solutions of the equations become unsingle-valued for large values of nonlinear coefficient, as shown on figure 3b for $\Gamma=0,5\ 10^{-5}$. In this case both chirp and frequency shift increase ($E{\approx}K$, $2\pi\Delta\nu_0{\approx}\Delta\omega$) that can result in essential change of the pulse shape, and that Gaussian pulse shape approach can be uncorrect.

Thus, the wave detuning in nonlinear crystal causes chirp, shift of central frequency and effect of self compression when the pulse duration is less than one in laser without nonlinear crystal can be observed under certan conditions.

3. EXPERIMENTAL RESULTS

Experiments have been made with CW-pumped Q-switched and active mode-locked Nd:YAG laser "ARGO" (Onishchukov et al 1988). Such crystals as LiJO, KTP, $Ba_2NaNb_5O_{15}$ were used for intracavity second harmonic generation.

Mode diameter of the beam was about 100 μ in LiIO₃, and 500 μ in KTP. Second harmonic radiation energy of pulse train was 50–200 μJ, duration – 0,15–0,5 μs, and ultrashort pulse duration was 200–500 ps, peak power – 10 kW.

All experiments were made with the laser working in preliminary free generation regime (Onishchukov et al 1986). The chirp measurements of ultrashort pulses were carried out by dynamic interference method (Treacy E B 1971) with the aid of electronic-optical chamber "Agat SF-3".

 Typical dependence of chirp (1) and power of second harmonic radiation (2) for 20 mm LiJO (a) and 5 mm KTP (b) are shown on figure 2.

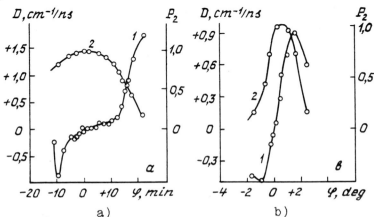

a)   b)
    Fig.2. Typical dependences of chirp (1) and normalized output power of second harmonic radiation (2) for LiIO₃ (a) and KTP (b).

Chirp dependence on wave detuning is in a good qualitative agreement with theoretical calculations, although experiments have been made for Q-switch and mode-locked laser and equations have been obtained for mode-locked laser only.

High linearity of chirp made it possible to compress pulses by factor of 8–10 with the use of grating dispertion optical delay. Relatively small value of chirp but large value of the bandwidth were the reason of high accuracy of dispertion delay adjustment with the large angle of beam incidence on the grating.

Bogdanov S F, Konvisar P G 1990 Bull.Acad. Science. USSR **12**
Folk J 1975 IEEE J. Quantum Electron. **1**₂₁
Kuizenga D J, Siegman A E 1970 IEEE J.Quantum Electron. **11**₆₉₄
Onishchukov G I,Ryabko S D, Fomichov A A, Lavrovskaya O I 1988 Sov. J. Quantum Electron. **5**
Onishchukov G I, Stelmah M F, Fomichov A A 1986 Bull.Acad. Science USSR **6**
Treacy E B 1971 J. Appl. Phys. **42** ₃₈₄₈

Inst. Phys. Conf. Ser. No 126: Section I

Paper presented at Int. Symp. on Ultrafast Processes in Spectroscopy, Bayreuth, 1991

43

A method for compensating the group-velocity mismatch in the frequency-doubling modelocker

K A Stankov, V P Tzolov and M G Mirkov

Faculty of Physics, University of Sofia
BG-1126 Sofia, Bulgaria

ABSTRACT: A birefringent plate can be used to introduce a time delay between the fundamental and the second-harmonic pulses which is opposite in sign to that accumulated in the frequency-doubling crystal. Thus, the deteriorating effect of the group-velocity mismatch on the frequency-doubling modelocker time response can be considerably reduced. In the new arrangement the modelocker may provide generation of sub-picosecond laser pulses.

1. INTRODUCTION

The intracavity frequency doubling was used recently to achieve passive mode-locking (Stankov 1989, Stankov 1991, Stankov et al 1991a). Pulses 25-ps long from Nd:YAG laser (Stankov 1991) and 15 ps at 1.34 μm from a Nd:YAP laser (Stankov et al 1991a) were generated using as an output coupler the combination of a frequency-doubling crystal and a dichroic mirror, known as the second-harmonic (frequency-doubling) nonlinear mirror. The pulse duration was not limited by the mode-locking device, but by the premature pulse train termination (Stankov et al 1991a). A fluctuation model (Chong-xiu Hou et al 1990), and a frequency-domain analysis (Stankov et al 1991b) describing the mode-locking process in a laser mode-locked with the second-harmonic nonlinear mirror were recently developed.

The group-velocity mismatch in the process of generation of the second harmonic in the nonlinear crystal was pointed out as the only principal limitation for the time response of the new mode-locking device. A recent nonstationary analysis accounting for the group-velocity mismatch has evaluated the pulse shortening limit of the frequency-doubling nonlinear mirror (Stankov et al 1991c). For low conversion efficiency the limit is given approximately by $3\tau_{cr}$, with τ_{cr} being the nonstationary crystal parameter, defined as $\tau_{cr}=L_{cr}(1/u_1-1/u_2)$ (Akhmanov et al 1968). In this expression u_1 and u_2 denote the corresponding group velocities of the fundamental and the second harmonic and L_{cr} is the crystal length. In the near infrared region for highly efficient frequency doublers such as BBO, LBO, having length of several millimetres, the limit is around 1 ps, but in the visible it may be considerably longer, due to the high dispersion.

Here we propose a method for reducing the deteriorating effect of the group-velocity mismatch on the frequency-doubling nonlinear mirror performance. The method utilizes an additional birefringent plate in which the group velocities for the fundamental and the second harmonic are inverted with respect to those in the frequency-doubling crystal. Thus a time delay, opposite in sign to that accumulated in the frequency-doubling crystal, is created. The analysis is concentrated on type I phase-matched second harmonic generation (*oo-e* interaction), since use of type II crystal may limit the laser bandwidth (and hence generation of shorter pulses) because of formation of a birefringent filter inside the laser cavity.

II. THE COMPENSATED ARRANGEMENT

The new nonlinear mirror arrangement is illustrated in Fig.1, which depicts the standard configuration as composed of a frequency-doubling crystal FDC and a dichroic mirror DM. In addition, a birefringent compensating plate CP is inserted between the nonlinear crystal and the dichroic mirror.

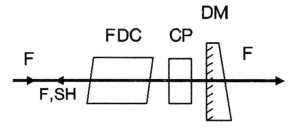

Fig.1. Configuration for reducing the influence of the group-velocity mismatch in the second-harmonic nonlinear mirror by using a compensating birefringent plate.

In the simplest case the plate is cut parallel to the optical axis. The ordinary and extraordinary light pulses propagate with u_o and u_e group velocities, respectively. If we consider *oo-e* type frequency doubling in the nonlinear crystal FDC, the second harmonic and the fundamental radiation will exit the nonlinear crystal polarized in mutually perpendicular planes. For suitable orientation of the compensating plate the fundamental and the second harmonic propagate as *o* and *e* (or *e* and *o*) rays correspondingly. The time delay τ_d between the fundamental and the second harmonic pulses will be determined by the thickness of the plate L_{cp} and the relevant group-velocities: $\tau_d = L_{cp}(1/u_o - 1/u_e)$. The sign of the time delay is determined by the orientation of the polarization plane of the light waves (fundamental and second harmonic) with respect to the axes of the compensating plate, while its value is given by the group velocities in the plate and its thickness. For negative birefringent crystals (see below) the sign of time delay can be easily changed simply by 90° rotation of the plate, since in this case the second harmonic and fundamental pulse exchange group velocities in the plate. By tilting the compensating plate, the time delay can be varied within a limited range.

This configuration of the nonlinear mirror provides partial reduction of the group-velocity mismatch in the double pass through the nonlinear crystal. Intuitively, the optimum will be reached if the time delay τ_d in the transit through the compensating plate is opposite in sign and equal in magnitude to the group-velocity mismatch in the frequency-doubling nonlinear crystal τ_{cr}, i.e. $\tau_d = -\tau_{cr}$. We show, that in some cases this is not the optimum relation, in sense that the pulse shortening of the nonlinear mirror may be improved for a time delay differing from τ_{cr}.

III. ANALYSIS OF THE COMPENSATED NONLINEAR MIRROR ARRANGEMENT.

An analysis of the pulse shortening for a single reflection by the nonlinear mirror configuration shown in Fig.1 was done using nonstationary model for the interaction between the fundamental and the second harmonic pulses in the nonlinear crystal (Akhmanov et al 1968), in a manner similar to that given by Stankov et al (1991c). Assuming gaussian pulse shape, the fundamental and the second harmonic pulse forms were calculated after a pass of the input pulse through the nonlinear crystal in direction toward the dichroic mirror, after reflection by the dichroic mirror, and for a second pass through the nonlinear crystal. The dichroic mirror was assumed to be a total reflector at the second harmonic wavelength (R2=1), and having reflectivity at the fundamental R1=0.25 (Stankov 1989, Stankov 1991).

We define a parameter $c= \tau_d/\tau_{cr}$, expressing the value of compensation due to the presence of the compensating plate. For $c=0$, there is no compensation ($\tau_d=0$) and the nonlinear mirror configuration is identical to the standard one. For $c=1$, there is an exact compensation, i.e. the time delay between the fundamental and the second harmonic pulses resulting from the compensating plate is equal in magnitude and opposite in sign to that induced by the frequency-doubling crystal.

The system of differential equations describing the interactions of the light waves in the nonlinear mirror was solved numerically with several variable parameters: the ratio of the doubled crystal time parameter to the pulse duration $2\tau_{cr}/\tau_p$, the compensation $c=\tau_d/\tau_{cr}$ (since the interaction between the light waves takes place in the double transit through the crystal), and the conversion efficiency into second harmonic η of the peak of the pulse in a single pass through the crystal. The latter is assumed for a stationary interaction to characterize conveniently the frequency doubling process.

IV. RESULTS AND DISCUSSIONS

We illustrate some of the computed results in the following Figures. We have first investigated the relative pulse shortening, defined as $(\tau_p-\tau)/\tau_p$, as a function of the ratio $2\tau_{cr}/\tau_p$, for several values of the compensation c. Here τ_p denotes the initial pulse length and τ is the final one, both measured as the full width at half maximum (FWHM). Fig.2 refers to 40% of the conversion efficiency into second harmonic. The standard frequency-doubling mode-locker configuration is presented with the curve with $c=0$. The pulse shortening for the values of the argument approaching 0 refers to the quasi-stationary case, i.e. $\tau_p>>2\tau_{cr}$.

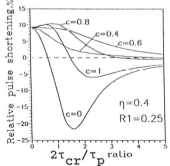

Fig.2. Relative pulse shortening (FWHM) for gaussian pulse

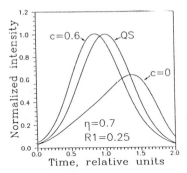

Fig.3. Shapes of fundamental pulse after reflection

The first important and somewhat unexpected result is that the exact compensation is not the most favourable one. Undercompensation can be clearly advantageous. In a wide range of conversion efficiencies (from 0.1 to 0.9) and $2\tau_{cr}/\tau_p$ ratios, the optimum compensation is approximately 0.6. The pulse shortening can be still effective even for pulse durations more than five times shorter than the doubled nonstationary crystal parameter, $2\tau_{cr}$. For the practice this means that the modification of the nonlinear mirror can provide pulse shortening well below 1 picosecond. To give an example, let us consider 5-mm long BBO (ß-BaB$_2$O$_4$) frequency-doubling crystal which has around 1.06 μm wavelength (Nd:doped lasers) group-velocity mismatch of 85 fs/mm. In a double transit, τ_{cr}=-850 fs time delay between the fundamental and the second harmonic will be accumulated. The pulse shortening will be satisfactorily good for pulses more than five times shorter than the doubled nonstationary parameter, i.e. as short as $\tau_p = 2\tau_{cr}/5 = 170$ fs. Use of LBO crystal may be advantageous, since it has approx. 1.5 times lower group-velocity mismatch.

We have also investigated the pulse forms of the reflected pulses. Since the features are more pronounced for higher second-harmonic conversion, an illustration is given in Fig.3, assuming 70% SH efficiency and for pulse duration $\tau_p = 2\tau_{cr}$. The three curves represent the quasi-stationary case (curve QS), the non-compensated arrangement ($c=0$) and the optimum   compensation ($c=0.6$). Note that the resulting pulse for the compensated arrangement is even shorter as compared to the quasi-stationary case. A noticeable shifting of the pulse maximum position on the relative time scale occurs when the interaction is nonstationary.

For pulsed lasers, high conversion efficiency can be achieved with shorter crystals and higher pulse shortening than in the above example can be thus attained. Especially advantageous may be the combination with active mode-locking (Buchvarov et al 1991) The high conversion efficiency, attainable in this case, may be achieved with very thin crystals.   Thus, the limit of the ultrashort pulse duration may be reduced below 100 femtoseconds.   Utilization of dispersion-compensated cavities will be necessary in this case.

We point out also other possibilities for compensation of the group-velocity mismatch effects, as for example, angularly dispersed frequency-doubling arrangement (Martinez 1989, Szabo et al 1990) which allows efficient frequency doubling of up-to 10-femtosecond pulses in relatively long nonlinear crystals.

V. CONCLUSION

In conclusion, we have described an arrangement for partial compensation of the group-velocity mismatch in the frequency-doubling nonlinear mirror. Thus, the pulse-shortening capabilities are improved, which would allow generation of ultrashort laser pulses from mode-locked lasers at sub-picosecond time scale.

The described arrangement may also find application in efficient double-pass frequency doubling of femtosecond pulses in relatively long crystals. For the purpose, the dichroic mirror should be replaced by a total reflector at both, fundamental and second harmonic wavelengths. The advantage will be not only partial compensation of the group-velocity mismatch, but also of the walk-off effect.

REFERENCES:

Stankov K A 1989 *Opt. Lett.* **14**, 359
Stankov K A 1991 *Appl. Phys. Lett.* **58**, 2203
Stankov K A, Kubecek V and Hamal K 1991a *IEEE J. Quant. Electron.* **QE-27** 2135
Chong-ru Hou and Zhen-he Zhu 1990 *Opt. Commun.* **79** 328
Stankov K A, Tzolov V P and Mirkov M G 1991b *Opt. Lett.* **16** 639
Stankov K A, Tzolov V P and Mirkov M G 1991c *Opt. Lett.* **16** 1119
Akhmanov S A, Chirkin A S, Drabovich K N, Kovrigin A I, Khokhlov R V, and
    Sukhorukov A P 1968 *IEEE J. Quant. Electron.* **QE-4**, 598
Buchvarov I Ch, Stankov K A and Saltiel S M 1991 *Opt. Commun.*, **83**, 241
Martinez O E 1989 *IEEE J. Quant. Electron.*. **25** 2464
Szabo G and Z. Bor Z 1990 *Appl. Phys.* **B 50**, 51

Inst. Phys. Conf. Ser. No 126: Section I
Paper presented at Int. Symp. on Ultrafast Processes in Spectroscopy, Bayreuth, 1991

Study of ultrashort pulse generation in a pulsed Nd:YAG laser with an auxiliary nonlinear cavity

Zhen—guo Lü , Qi WU, Qing—xing LI and Zhen—xin YU

Institute for Laser and Spectroscopy, Zhongshan University, Guangzhou 510275, P.R.China

ABSTRACT: A systematical study concerning the mode—locking of a pulsed Nd:YAG laser based on a nonlinear auxiliary cavity containing a nonlinear frequency—doubling crystal KTP is reported in this paper. Our studying results show that the experimental results conincide with the theoretical calculations dramatically. Finally, we have also analyzed and discussed the dependence of laser output characteristic on some parameters.

1. INTRODUCTION

The mode—locked Nd:YAG laser is an important sourse of high—power infrared pulses, it is often used for efficient frequency doubling to the green and for pumping other short—pulse systems. For this reasons there is great interest in improving the mode—locking technique for the Nd:YAG laser. Recently, a unifying, time—domain description for ultrashort pulse generation in lasers with auxiliary nonlinear cavities was presented by Ippen (1989) , called additive pulse mode—locking (AMP). It appear to explain experimental results obtained in both soliton lasers of Mollenaur (1984) and those coupled to fibers with normal, pulse spreading dispersion of Kean (1989). Additive mode—locking utilizes self—phase modulation in an auxiliary cavity to produce shortening in the main laser by coherent interference at the coupler mirror. The APM shortening mechanism is affected by, but does not rely on, dispersion in the auxiliary cavity. In this paper we expand this new technique to the modelocking of a pulsed Nd:YAG laser using a different form of nonlinearity, namely that of second harmonic generation in an auxiliary nonlinear cavity.

Stankov (1989) and Lü (1991) report that using intracavity second harmonic generation (SHG) can modelock a Nd:YAG laser. But the inclusion of an SHG crystal within the laser cavity introduces loss and may adversely affect the modelocking process by acting as an etalon and limiting the laser bandwidth. To avoid these problems SHG modelocking in a coupled cavity was examined. In this paper, we have reported systematically study concerning the modelocking of a pulsed Nd:YAG laser based on a nonlinear auxiliary cavity containing a SHG crystal KTP. Our studying results show that the experimental results agree with the theoretical results dramatically. Finally, we have also discussed the dependence of laser output characteristics on some experimental parameters.

2. THEORY

Because this paper has limited space, here we have cancelled process of detailed theoretical analysis and discussion. If you are interested in our theoretical analysis, please refer to the work of WU (1991).

3. EXPERIMENTAL TECHNIQUE

We have used a pulsed Nd:YAG laser with AR coated rod (80mm long, 7mm diameter), pumped by one linear flash lamp (see Figure 1). C.P. is a 50mm long, 1.90mm in internal diameter capillary that is employed as the transverse mode control element. According to the work of LÜ (1991b) we know that this capillary can provide very good single transversal mode operation as well as a high intersity beam for efficient second harmonic generation in the nonlinear crystal KTP. P is a polarizer that was placed between

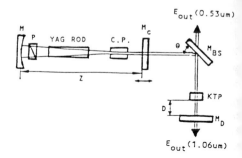

Fig. 1. Experimental set up.

the laser rod and the cavity end mirror M to provide a definite polarizationof the laser radiation and easy orientation of the frequence. For SHG generation we used a 6mm long KTP crystal, cut for oe–e type interaction. The both sides of the crystal KTP are antireflection coated for the fundamental wave and the second harmonic wave. The main cavity length and the auxiliary cavity length are both about 1160mm. The end mirror M of the main cavity is a 2000mm radius of curvature total reflector at $1.064\mu m$. The reflectivity of the plane couple mirror M is 50% at $1.064\mu m$. The plane mirror M is a beam splitter (total reflector at $1.064\mu m$, fully transmitting at $0.532\mu m$ when angle θ is 45°). The plane mirror M is a dichroic mirror (total reflector at $0.532\mu m$, 10% reflection at $1.064\mu m$).

In our experiment, the output characteristics of this modelocking laser were measured with an energy meter and a combination of a PIN diode (its risetime <500PS) and Tektronic 485 oscilloscope. The individual–pulse duration of the modelocked pulse was measured by using a Hamamatsu Temporal Disperser C1587 Model Streak Camera.

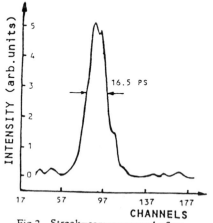

Fig.2. Streak–camera record of an individual ultrashort laser pulse at $0.53\mu m$ revealing pulse duration of 16.5ps (FWHM). the sweep rate is 0.92 psec / channel.

4. RESULTS AND DISCUSSION

The pulse Nd:YAG laser was operated at 1Hz repetition rate. When the frequency–doubling crystal KTP in the auxiliary cavity was carefully tuned to the optimum phase–matching angle and modelocking was not observed, the average output energy at $1.06\mu m$ andat $0.53\mu m$ were about 16.7mJ and $84\mu J$ respectively. At the same time the laser generated a single, bell–shaped pulse of 450ns duration

FWHM). Careful adjustment of the crystal–mirror seperation and when the distance D between the crystal KTP and the dichroic mirror M was in the range 42mm–60mm, a 100% modulated train of pulses are observed. The amplitudes of the laser pulses both at 1.06μm and at 0.53μm increased very large and the duration of the pulse–train envelope at 1.06μm was limited to approximately 250ns. The measured pulse train period at 7.7ns was in accordancewith the cavity length of 1160mm. Under the optimal modelocking operation condition, the pulse width at 0.53μm and at 1.06μm were measured by using a Hamamatsu Temporal Disperser C1587 Model Streak Camera to be in the range 12.6ps–19.8ps and 17.8ps–28.0ps respectively. At the same time the average output energy at 0.53μm and at 1.06μm were 0.98mJ and 14.8mJ respectively. Figure 2 was a streak camera record of an individual ultrashort laser pulse at 0.53μm revealing a pulse duration of 16.5ps (FWHM).

In our experiment, the laser gain bandwidth corresponded to Nd:YAG, was 120GHz; $R_C = 50\%$; $R_D = 10\%$; The conversion efficiency η from the foundamental wave to the second harmonic wave on the first pass through the frequency–doubling crystal KTP was in the rang 4.2%–13.4%. According to the work of Lü(1991c), we can calculate that was 1.12 and g values, corresponding to were 6% and 19%, were 0.97 and 0.83 respectively. Using the equation (11) of paper of WU (1991) we cancalculate that this modelocking pulse duration at 1.06μm was in the range 14.9ps–26.5ps. This theoretical calculation results agree with our experimental results dramatically.

Here we have analyzed and discussed the effects of varying some experimental parameters on the output properties of this modelocking laser. Figure 3 was modelocking probability as functions of operation voltage for the pulsed Nd:YAG laser (a) and the distance D between the nonlinear crystal KTP and the dichroic mirrorM (b), rspectively.

Another, cavity length detuning between the main cavity and the auxiliary cavity can significantly affect the output properties at this modelocking laser. Our experimental results show that the phenomenon of modelocking can be observed only when the optical length detuning between the two cavities is less than 0.180 mm.

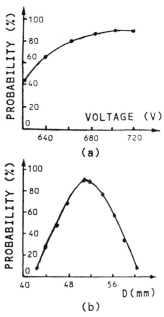

Fig. 3. Modelocking probability as functions of operating voltage for the pulsed Nd:YAG laser (a) and the distance between the nonlinear crystal KTP and the dichroic mirror M_D (b) respectively.

5. CONCLUSION

In this paper we have reported a systematical study concerning the modelocking of a pulsed Nd:YAG laser based on a nonlinear auxiliary cavity containing a frequency–doubling crystal KTP. First, an auxiliary coupled–cavity modelocking scheme has been demonstrated using second–harmonic generation in KTP as the only pulse shortening process. Moreover, the effect of various parameters on the characteristics of this additive pulse mode–locking technique was investigated. Our experimental results show that for certain applications coupled–cavity second−harmonic generation modelocking may prove a useful and reliable method of modelocking. The observation that the modelocking occurs with a fixed phase relation between the two cavities is in agreement with our theoretical analysis. The ultrashort pulse length limitation of this type of modelocking will be limited by group velocity dispersion in KTP. In this paper, we have also analyzed the dependence of laser output characteristic on some parameters. Finally, a stable mode−locked pulse duration at 0.5μm was measures by using Streak camera to be in the range 12.6ps−19.8ps and the laser output average energy at 0.53μm was 0.98mJ.

REFERENCES

Ippen E P, Haus H A, and Liu L Y 1989 J. Opt. Soc. Am. B **6** pp 1736−10

Kean P N, Zhu X, Crust D W, Grant R S, Langford N and Sibbert W 1989 Opt. Lett. **14** pp 359−3

Lü Z G, Li Q X and Yu Z X 1991a "Study of a pulsed Nd:YAG laser modelocked by a KTP crystal", to be published in 《Acta Optica Sinica》 of China

Lü Z G, Fu C H, Zhou J Y, Li Q X and Yu Z X 1991b 《Acta Optica Sinica》 of China **15** pp 232−3

Lü Z G, Li Q X and Yu Z X 1991c "Analysis of dynamic process of modelocked laser based on a novel nonlinear mirror" to be published in 《Acta Optica Sinica》 of China

Mollenaur L F and Stolen R H 1984 Opt. Lett. **9** pp 13−3

Stankov K A 1989 Opt. Lett. **14** pp 359−2

Wu Q, L Z G, Li Q X and Yu Z X 1991 "Theoretical analysis of modelocking process in an auxiliary laser" to be published in 《Applied Lasers》 of China.

Inst. Phys. Conf. Ser. No 126: Section I
Paper presented at Int. Symp. on Ultrafast Processes in Spectroscopy, Bayreuth, 1991

Femtosecond passively mode-locked Ti:sapphire lasers

Patrick Georges, Thierry Lépine, Gérard Roger and Alain Brun

Institut d'Optique Théorique et Appliquée
CNRS U.A. 14
BP 147 91403 ORSAY-FRANCE

Abstract: We present some results about the generation of pulses in Ti:Sapphire. Active and passive mode-locking techniques have been studied yielding the production of picosecond and femtosecond pulses when an intracavity group velocity dispersion compensation (prisms) is used. Pulses as short as 75 fs have been obtained and we show that the temporal profile is close to a gaussian.

Since its first lasing demonstration by Moulton et al. in the beginning of the 80's [1], the Titanium Sapphire ($Ti^{3+}:Al_2O_3$) has proved to be a very attractive near infrared laser materials. This crystal combines a large fluorescence bandwidth (from 650 nm to 1100 nm) which is more important than the infrared dyes and a high saturation fluence (1 J/cm^2). Moreover its absorbtion bandwidth in the blue-green region (400-600 nm) allows it to be pumped by cw Argon ion or pulsed Nd-Yag (SHG) lasers. Its large fluorescence bandwidth has been exploited to produce ultrashort pulses by using several mode-locking techniques (active or passive).

We report here some results we have obtained first with an actively mode-locked Ti:Sapphire laser that used an acousto-optic mode-locker to produce the pulses. With an intracavity group velocity dispersion (GVD) compensation this laser produces subpicosecond pulses. We than have studied a passively mode-locked Ti:Sapphire laser by using a saturable absorber and that produces sub-100fs pulses when an intracavity GVD compensation is used.

The easier solution to generate pulses with a laser is the active-mode-locking. So we have started our study on pulsed Ti:Sapphire by developing an actively mode-locked laser [2,3]. The cavity is very classical : it is a four mirror linear astigmaticaly compensated cavity (Figure 1).The concave mirrors M_2 and M_3 have a 100 mm radius of curvature and are used off axis in order to compensated the astigmatism introduced by the brewster angle cut crystal. The pump beam is focused on the crystal with a 100 mm focal length lens. With 6 watts of pump power we obtained 250 mW average power at 125 MHz.

Pulse duration is measured by using a standard noncollinear second-harmonic-generation autocorrelator with a KDP as the doubling crystal. Spectra are recorded using a spectrometer coupled with a photodiode array. The RF power injected in the mode-locker was around 2 Watts.

Due to the large fluorescence bandwidth of the Ti:Sapphire we expected to produce very short pulses. But in fact it was impossible to produce pulses shorter than 30 ps when one want to have very stable pulses. By changing sligthly the cavity length of the laser in order to mismatch the frequency given the cavity round trip from the frequency of the mode-locker, it was possible to obtain pulses around 10 ps. But in that case the laser operated in a Q-switched regime and the stability andthe repetability of the performances were poor. By looking at the spectrum of the 30 ps pulses we observed that the pulses were not limited by the Fourier transform. In fact the time bandwidth product was 10 times the theory. That means that the pulses exhibited a strong chirp.

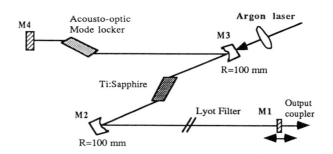

Fig. 1 : Schematic of the actively mode-locked Ti:Sapphire laser.

In order to avoid such a chirp and to produce shorter pulses, we have inserted in the cavity a system of two high index prisms (n=1.76 at 800 nm) that introduces a negative GVD. This negative dispersion will compensate the positive dispersion of the Ti:Sapphire rod and the acousto-optic crystal. The figure 2 shows the dispersion compensated cavity with the prisms.We also removed the Lyot filter and used a slit after the second prisms to control the laser wavelength.

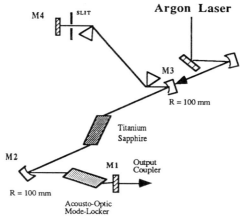

Fig. 2 : Schematic of the dispersion compensated actively mode-locked Ti:Sapphire laser.

With this cavity we produced pulses as short as 300 fs at around 790 nm with around the same energy level. In that case the pulses are limited by the Fourier transform and the time-bandwidth $\Delta\tau.\Delta\nu = 0.43$ near the theory (0.44) for gaussien pulses. The figure 3 shows the autocorrelation and the spectrum of the pulses we produced. But due to the length of the mode-locker crystal (65 mm) we have to operate with the prisms separated by 57 cm. We think that is a reason why we did not produce shortest pulses because the high order dispersion of the prisms sequence was to important. We also observed that we could obtain these pulses even with a very low RF power in the mode-locker. Moreover when the laser was producing the pulses it was possible to turn off the RF power in the mode-locker without stopping the pulses emission. But when something disturbed the laser (a dust for exemple) the laser stopped and did not restared itself. To start the laser one needed to use the mode-locker. That means that the mode-locker is only for starting the laser and these results have to be correlated to the work of people from the University of Saint Andrews who first observed mode-locking in a Ti-Sapphire without any mode-locker in the cavity [4]. Since that people from Coherent have analysed this behavior and have explain it as Kerr Lens Mode-Locking[5]. In summary they have shown that the mode-locking is due to a change of the cavity spatial mode (due to a lens which appears in the crystal by non linear effect) and the laser only needs a starter which initiates the production of the pulses. The starter can be a mode-locker like in our case but it can also be a moving mirror [6] and also a saturable absorber.

The active mode-locking technique needs a fine adjustement of the cavity length in order to match the frequency injected in the acousto-optic crystal. As we have found that the mode-locker is only used to initiate the pulses, we have try to produce pulses by using a saturable absorber. The figure 4 shows the modified cavity. We removed the mode-locker and added two 50 mm radius of curvature mirrors and a dye jet.

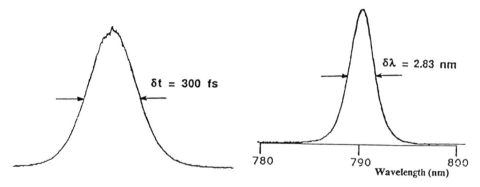

Fig. 3 : Autocorrelation trace and spectrum of the pulses produced by the dispersion compensated actively mode-locked Ti:Sapphire laser.

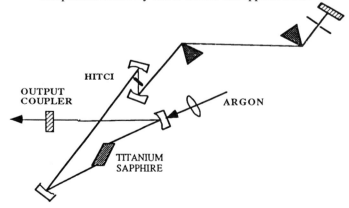

Fig. 4 : Schematic of the passively mode-locked Ti:Sapphire laser.

The saturable absorber dye is HITCI with a concentration around $5 \; 10^{-5}$ Mol/l in ethylene glycol corresponding to only a few trail of saturable absorber. The distance between the prisms is reduced to 25 cm because we only have the Ti:Sapphire crystal which introduces positive GVD. In order to control the pulse spectrum we used, like previously, a slit after the second prism. The length of the cavity is 1.5 m corresponding to a 100 MHz repetition rate. With a pump power of 7.5 W on the crystal, we obtained an average power of 250 mW around 807 nm and the threshold was near 5 W. Femtosecond pulses have been obtained for average output power between 100 mW and 250 mW. Below 100 mW, the laser produced large picosecond pulses and above 250 mW, a satellite pulse was observed at around one picosecond from principal pulse. The figure 5a shows the autocorrelation trace of the pulses we have obtained and the fit of the autocorrelation with a gaussian pulse shape. The figure 5b shows the corresponding spectrum of the pulses with also the best fit with a gaussian shape. The spectrum width is 12.5 nm centered at 807 nm. Considering that the best fit of the autocorrelation trace is for a gaussian pulse shape, that the spectrum also fits very well with a gaussian and that the product time-bandwidth $\Delta\tau.\Delta\upsilon = 0.435$, we can deduce that the temporal shape of the pulses produced by our laser is close to a gaussian. The pulse train is very stable (the fluctuations are less than 1 % when the pulse train is recorded with a fast photodiode) and the long term stability is excellent. For example the mode-locked operation has been obtained over more than one week without adjusting (even delicately) any cavity or pump mirror control. We only have to wait the warm-up time of our Argon ion laser, and after this time the laser produced very stable and short pulses.

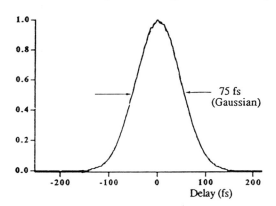

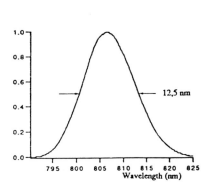

Fig. 5a: Autocorrelation trace of the 75 fs pulses (assuming a gaussian pulse pulse shape) obtained.

Fig. 5b: Corresponding spectrum and its gaussian fit (dashed line) of the 75 fs pulses.

This work was partially supported by the Ministère de la Recherche et de la Technologie.

References:

1.   P.F. Moulton, Solid State Research Quartly Technical Summary Report, MIT Lincoln Laboratory, (May 1 - July 31, 1982) pp. 15-21
2.   J.D. Kafka, A.J. Alfrey, and T. Baer, Ultrafast Phenomena VI (Springer Verlag, Berlin 1988), p. 64.
3.   J. Goodberlet, J. Wang, and J.G. Fujimoto, Opt. Lett. **14**, 1125 (1989).
4.   D.E. Spence, P.N. Kean, and W. Sibbett, Opt. Lett. 16, 42 (1991)
5.   D.K. Negus, L. Spinelli, N. Goldblatt and G. Feugnet, Proccedings of the Advanced Solid State Laser, Hilton Head Island (March 1991).
6.   Mira, technical bulletin, Coherent Inc, Palo Alto.

Inst. Phys. Conf. Ser. No 126: Section I
Paper presented at Int. Symp. on Ultrafast Processes in Spectroscopy, Bayreuth, 1991

Cavity considerations and intracavity second harmonic generation in Kerr-lens mode-locked titanium-doped sapphire lasers

D. Georgiev, U. Günzel, J. Herrmann*, V. Petrov, U. Stamm**, K.-P. Stolberg

Friedrich Schiller University, Faculty of Physics and Astronomy, Institute of Optics and Quantum Electronics, Max-Wien-Platz 1, O-6900 Jena, Germany

*Institute of Optics and Spectroscopy, Rudower Chaussee 6, O-1199 Berlin, Germany

**present address: Lambda Physik GmbH, Hans-Böckler-Str.12, 3400 Göttingen, Germany

ABSTRACT: Several cavity configurations have been used to generate mode-locked picosecond and femtosecond pulses in titanium-doped sapphire lasers. The inclusion of a frequency doubling crystal in the cavity was found to stabilize the Kerr-lens mode-locking operation. Simple analytical ABCD-matrix calculations allow to give criteria for optimum nonlinear absorption in Kerr-lens mode-locked four mirror cavities.

1. INTRODUCTION

Recently, a novel mode locking technique in the femtosecond region has been demonstrated for cw pumped titanium-doped sapphire lasers employing the self-focusing effect within the laser crystal due to the nonlinear optical change of the refactive index of the material (Spence *et al* 1991). The mode locking can be achieved by appropriate cavity alignment. Introducing an additional intracavity aperture at a location where the laser beam diameter decreases for increasing intensity allows to optimize the mode-locked operation (Spinelli *et al* 1991). In such so-called Kerr-lens mode-locked lasers the aperture in combination with the Kerr-lensing acts similar as a fast saturable absorber. The cross section of the absorber may be easily adjusted by changing the size of the aperture. Using a prism pair to compensate for the group velocity dispersion transform limited pulses with durations of less than 80 fs at output powers of 0.8 W have been reported.

In the present paper we give experimental results about Kerr-lens mode-locking achieved in different cavity configurations without and with extracavity and intracavity chirp compensation. The incorporation of a nonlinear second harmonic crystal does not destroy but stabilizes the the mode-locking and allows to generate femtosecond pulses both at the fundamental and the second harmonic frequency. Calculations are carried out to investigate the optimum conditions for Kerr-lens mode locking in solid-state lasers by a simple time-independent Gaussian beam analysis.

2. SELF MODE LOCKING OF TITANIUM-DOPED SAPPHIRE LASERS

In the first experiments reported here we used a 3-mirror folded astigmatically compensated cavity the schematic of which is shown in Fig. 1 (design 1). The pump laser beam (10 W of all line operated SP 171 argon ion laser) is focused through mirror M2 into the 20 mm long Brewster angle cut 0.3 % titanium-doped sapphire crystal (Union Carbide) which absorbs nearly 80% of the pump radiation. The mirrors are highly reflecting between 700 and 900 nm. Mirror M1 has 20 cm radius of curvature, M2 10 cm and the outcoupling mirror M3 with 92% reflectivity is flat. The cavity length for all cavity configurations to be discussed in the following was between 1.5 and 1.9 m.

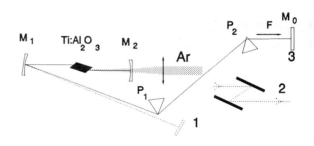

Fig .1 Caviy configuration of three mirror Kerr-lens mode-locked Ti-sapphire laser. 1 without GVD compensation, 2 - with external compression, 3 - with intracavity GVD compensation.

If no tuning element is inserted into the cavity the laser emits cw radiation at about 780 nm with 1.5 to 2.0 W average power and can be wavelength tuned by translating the laser crystal between mirrors M1 and M2. Self mode-locking of the laser is accomplished by precise cavity alignment accompanied with a considerable power reduction to about 1 W (Spence et al. 1991), but it could be started (or stopped) by any sudden physical shock. Although no special slit is required to sustain the pulse generation we suppose that other cavity elements (e.g. the pumped volume in the laser crystal) play the role of soft apertures. The pulse characteristics are determined by standard background-free autocorrelation and spectral measurements using a Fabry-Perot interferometer and a grating spectrometer with optical multichannel detector. Typical autocorrelation traces were measured to have a full width at half maximum (FWHM) between 5 and 7 ps. The pulse duration-bandwidth product is about 4 indicating a strong chirp which should allow to compress the pulses extracavity with an arrangement having suitable group velocity dispersion (GVD).

To prove this a grating compressor with a 651 lines/mm grating (design 2) was used after the laser output and a short study was made of the dependence on the pulse duration after the grating compressor on the distance between the gratings. Minimum pulse durations of about 600 fs assuming a Gaussian shape were measured for a group velocity dispersion of the arrangement of about -0.66 ps^2.

In a third experiment two SF10 prisms with a distance of 76 cm between them have been inserted into the laser cavity as intracavity chirp compensation (configuration 3 in Fig.1). At 1.0 to 1.2 W average power pulses with an autocorrelation width of 180 fs are generated by the laser. The laser operates nearly in TEM_{00} mode and the laser operation is more stable than in the cavity without prism pair. All the given results are reproduced by four mirror cavity arrangements depicted in Fig. 2 (configuration 1).

To use the high intracavity power of the femtosecond pulses for frequency conversion we have extended the four mirror dispersion compensated cavity from Fig. 2 by replacing one of the end mirrors by a pair of focusing mirrors M3 and M4. Mirrors M1, M2, M3 (10 cm radius of curvature) and M4 (5 cm of curvature) have a reflectivity of >99% between 700 and 900 nm (Layertec). M3 and M4 are specially designed for high transmission (90%) in the region of the second harmonic around 400 nm. The output coupler M0 reflects 95%. Again after careful adjustment the laser operates in the self mode-locked regime. Since the self mode locking is obtained mainly due to the self-focusing effect in the laser crystal the generation of the second harmonic could be expected to destroy the mode locking because of the nonlinear losses at the fundamental which increase with intensity. However, if we insert a 5 mm $LiJO_3$ crystal phase matched to the fundamental we observe rather unexpected a substantial stabilization of the mode locking regime. The necessary resonator alignment reduces the fundamental power to about 250 mW. Mode locking could be achieved by adjusting only the folding section of Fig. 2 and the SHG crystal. Abrupt increase in the second harmonic power is observed when the laser enters the mode-locked regime. The average second harmonic power was measured to 10 mW corresponding to a conversion efficiency of 0.2%. A typical fringe resolved autocorrelation trace is depicted in Fig. 3. The FWHM of 190 fs corresponds to 125 fs pulse length if a

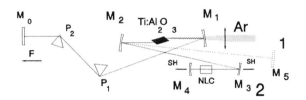

Fig. 2 Improved mode locking Ti-sapphire laser by intracavity SHG. 1 - the initial cavity configuration. 2 - the additional focussing section for the crystal.

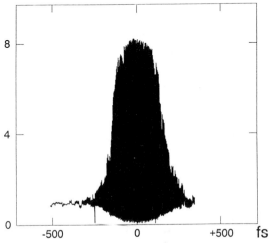

Fig. 3 Fringe resolved autocorrelation trace indicating weak chirping of the pulse and high laser stability.

sech²-intensity profile is assumed. The pulse spectrum is nearly symmetric with a spectral width of 7.5 nm. This results in a pulse duration-bandwidth product of 0.43. Since on the other hand the fringe-resolved correlation trace does not indicate a strong chirp a more square-like pulse shape is presumable. The second harmonics spectral width of 3.4 nm would support about 70 fs bandwidth limited pulses but it is probable that some phase modulation occurs due to the rather thick SHG crystal. Further experiments are currently in preparation to measure the pulse width at the second harmonic frequency directly and to understand the stabilizing action of the frequency conversion.

3. CAVITY CONSIDERATIONS

To give a simple approach to the cavity optimization of a Kerr-lens mode-locked laser let us consider the schematic of a cavity as depicted in Fig.4 which is equivalent to the most of the experimentally used titanium-doped sapphire laser cavities for Kerr lens mode-locking. The equivalent cavity scheme consists of four mirrors. The Brewster-angle cut laser crystal is placed between two curved mirrors M_2 and M_3 of focal width f. The beam waists at the end mirrors M_1 and M_4 and inside the laser crystal are w_1, w_2, and w_3, respectively. Inside the laser crystal the increase in temperature ϑ by the pump radiation will give rise to a thermal lens and the laser beam causes an intensity

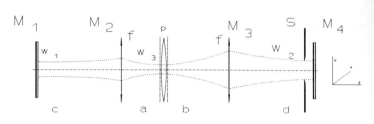

Fig. 4 Equivalent cavity configuration for a four mirror cavity. p- Kerr-lens, S-aperture.

dependent lens due to the Kerr effect. Just in front of mirror M_4 an aperture is located. We make a simple Gaussian beam analysis including the temperature and intensity dependence of the refractive index in the laser crystal as

$$n(x,y) = n_0 + \frac{\partial n}{\partial \vartheta} \, \Delta \vartheta(x,y) + n_2 \, I_L(x,y)$$

Straight forward one can derive analytical expressions which allow to minimize the equivalent saturation intensity of the Kerr absorber (Georgiev et al 1991)

$$I_{SAT} = -\frac{q_l}{q_{nl}}\left(\frac{1}{w_2}\frac{\partial w_2}{\partial I}\right)^{-1} ,$$

where q_l and q_{nl} are the linear and nonlinear loss coefficients introduced by the intracavity aperture which depend only on its geometrical shape and $\partial w_2/\partial I$ is the derivative of the waist at mirror M4 with respect to the pulse intensity I.

In Fig. 5 is depicted the derivative $\partial w_2/\partial I$ in dependence on the distance between the crystals beam entrance face and the focusing mirrors M2. Obviously the nonlinear absorption coefficient very sensitively changes in the considered range. Optimum Kerr-lens mode locking can be achieved only by proper choose of the distance crystal-mirror. The region where $\partial w_2/\partial I$ becomes positive is that where defocusing occurs. This range can not be used for mode locking in titanium-doped sapphire.

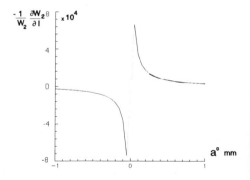

Fig. 5 Intensity dependent change of the beam diameter versus crystal position.

REFERENCES

Spence D E, Kean P N and Sibbett W 1991 *Opt.Lett* **16** pp 42-44
Spinelli L, Couillaud B, Goldblatt N and Negus D, *CLEO'91* CPDP7-1/581
Georgiev D, Herrmann J and Stamm U, *Opt.Commun.* submitted

Inst. Phys. Conf. Ser. No 126: Section I

59

Paper presented at Int. Symp. on Ultrafast Processes in Spectroscopy, Bayreuth, 1991

Mode-locking of a flashlamp pumped Ti:sapphire laser using the frequency doubling nonlinear mirror

K Hamal, K A Stankov*, H Jelinkova I Prochazka and M Koselja[+]

Czech Technical University, Brehova 7, 115 19 Prague 1, Czechoslovakia
*University of Sofia, BG-1126 Sofia, Bulgaria
[+]Monokrystaly Turnov, 511 19 Turnov, Czechoslovakia

ABSTRACT: A flashlamp pumped $Ti:Al_2O_3$ laser has been mode locked using a $LiIO_3$ intracavity frequency doubler in a nonlinear mirror arrangement. The long pulse duration of 100 ps was due to the very large group velocity mismatch in this particular frequency doubling crystal. The principle of operation of the nonlinear mirror allows ultrashort pulse generation in the wide tuning range of this laser.

I. INTRODUCTION

The principle of operation of the frequency doubling nonlinear mirror (Stankov 1988, Stankov et al 1988) allows mode locking in extremely wide spectral ranges, limited only by the available nonlinear crystals. It was recently demonstrated (Stankov et al 1991a) that a single $LiIO_3$ frequency doubling crystal can mode lock quite differing with respect to the wavelength transitions (1.08 μm and 1.34 μm of $Nd:YAlO_3$ laser and 1.66 μm of $Er:YAlO_3$ laser), thus demonstrating the advantage of the frequency doubling mode locker with respect to the saturable absorbers.

Here we report on application of the frequency doubling nonlinear mirror to mode lock *a flashlamp pumped* $Ti:Al_2O_3$ laser. To our knowledge, there is only one report on mode locking a flashlamp pumped Ti:Sapphire laser (Demchuk et al 1990), in which two different dyes were used to mode lock in the ranges 750 - 765 nm and 780 - 790 nm. We remind, that the Kerr-lense mode locking has been exclusively applied to continuous-wave Ti:Sapphire lasers (Coherent 1991, Spectra-Physics 1991). This technique may encounter serious difficulties in flashlamp-pumped lasers due to lower nonlinearity, which would not allow complete locking for the short time of the pulse train evolution or may lead to damage of the laser rod due to self focusing in pulsed lasers.

II. LASER CONFIGURATION

The laser is shown schematically in Figure 1. The laser cavity was composed of a flat total reflector (720-830 nm) and a dichroic output coupler. For the latter, we had a choice of three types of flat mirrors, each having high reflectivity at the second harmonic in the range 360 - 440 nm and reflectivity at the fundamental (750 - 840 nm) 25, 50 and 70% correspondingly. As a tuning element we have employed a Brewster angle flint prism or alternatively, birefringent quartz plates of 0.5, 1 and 2 mm thickness. A circular aperture provided operation in a single transversal mode. The active medium in the first experiments was a 150 mm long, 7 mm diameter $Ti:Al_2O_3$ rod, pumped by 400 μs pulses of 300 J energy by a single flashlamp in an elliptical pump cavity. For better efficiency and higher intensity, it was later replaced by a laser rod 5x85 mm, pumped by 120 J (effective) by two flashlamps with a pump pulse of much shorter pulse duration (15 μs FWHM). The mode locking was provided by insertion between the laser rod and the dichroic output coupler of a 2-cm long, 30°-cut LiIO$_3$ crystal. The latter has been designed for frequency doubling of 1.06 μm neodimium lasers. Its aperture of 10x10 mm has allowed angle phase matching in the whole oscillation band of the Ti:Sapphire laser. The same frequency doubling crystal was used to mode lock various laser in the infrared (Stankov et al 1991a). The tilt of the crystal has eliminated etalon formation by the parallel crystal faces.

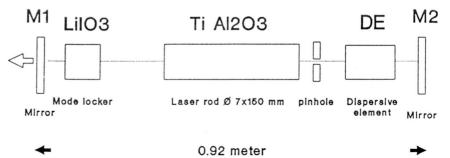

Figure 1. Ti:Sapphire flashlamp-pumped laser mode locked by a second-harmonic nonlinear mirror. M1 - output dichroic mirror, totally reflecting the second harmonic; M2 - total reflector at the fundamental.

The time parameters of the laser generation were measured by fast oscilloscopes and a streak camera. The energy output was controlled by PJR-M type Scientech energy meter with ED-100 probe and the laser wavelength was monitored by a simple grating monochromator. The laser was operated at 0.5 Hz repetition rate.

III. LASER PERFORMANCE AND DISCUSSION

In free running generation the laser has delivered 6 to 10 mJ energy in a single transversal mode. The mode locking was activated by adjusting the proper phase-matching angle and by translating the crystal with respect to the dichroic mirror in order to attain the proper phase difference (Stankov et al 1988) which facilitates increasing intensity-dependent reflection. We have optimized the various laser components in order to achieve stable mode locking. First of all, the narrowing of the laser spectrum was found to be of great importance for achieving mode locking. With a single 60° flint prism the mode locking was incomplete, as could be judged from the resulting modulation depth which was about 80%, and from the frequent appearance of satellites of considerable amplitude. By replacing the prism by a single Brewster angle crystal quartz plate forming a birefringent filter we were able to achieve better modulation depth, which was increasing by increasing the quartz plate thickness in the order 0.5, 1 and 2 mm. The mode locking was notably improving as thicker plate was used.

We have also experimented with dichroic mirrors of various reflectivities as described above. The optimum one was found to be 50%. Lower reflectivity (25%) has resulted in very high threshold, higher one (70%) - in reduced modulation depth.

Typical pulse trains had durations of about 8 to 10 μs. When the nonlinearity of the mode locker was insufficient as in the case of 70% output mirror reflectivity or thin birefringent plates, we could observe relatively long build-up time, as shown in Figure 2a. As was shown by Stankov (1988), the nonlinearity in reflection is lower for higher reflectivity of the dichroic mirror at the fundamental wavelength and for lower conversion efficiency. In case of thin quartz plates, the bandwidth of the laser generation considerably exceeds the phase matching bandwidth of the frequency doubling crystal thus reducing the conversion efficiency. We have roughly estimated the bandwidth (FWHM) of the laser operated with 2-mm thick quartz plate from the dispersion of the phase matching angle which was -1.14

mrad/nm at 800 nm. For the purpose, the crystal was angle-tuned and from the points for 50% decrease of the second harmonic intensity, the spectrum width was determined taking into account the quadratic dependence of the frequency doubling efficiency on the light intensity. The accuracy of this nonlinear spectrometer was determined by the frequency doubling band width of the nonlinear crystal (0.125 nm), which was considerably smaller than the measured laser band width of 3 nm.

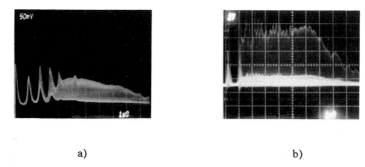

a)   b)

Figure 2. Pulse train forms: a) in case of insufficient nonlinearity; b) 100% modulation with R1=50% output coupler and 2 mm quartz plate.

The individual pulse duration was measured using a streak camera, by switching out a part of the pulse train (3-4 pulses) externally by means of a Pockels cell. The pulse duration was typically 100 - 120 ps.

Tuning of the generated ultrashort light pulses was achieved by rotating the quartz plate and tilting synchronously the frequency doubling crystal. In this way, we could mode lock the laser at an arbitrary wavelength within the range 792-811 nm. The tuning range was limited by the free spectral range of the 2-mm thick birefringent filter.

As was mentioned above, the phase-matchable band width was smaller than the laser band width, the latter being determined by the thickness of the quartz plate. This has allowed us to obtain tuneable ultrashort pulses *within* the laser bandwidth of the free running operation, for a fixed position of the birefringent filter, simply by tilting the nonlinear crystal. The tuning range was limited in this case to 1.5 nm.

The laser has produced simultaneously with the fundamental radiation a pulse train at the second harmonic wavelength with total energy of 20 μJ, which was emitted through the total cavity reflector, mirror M2. Accounting for reflection losses due to various cavity components and absorption in the laser rod, we have calculated 2 mJ total energy of usable blue radiation, which can be extracted by means of harmonic separator or polarizer.

The relative long pulse duration obtained in this experiment was not determined by the limited time for pulse train evolution, as was observed in lasers with high gain (Stankov et al 1991a). The limiting factor for this arrangement was obviously the narrow frequency doubling band width of the $LiIO_3$ crystal (approx. 0.063 nm for double pass), or equivalently, the large group-velocity mismatch (GVM) in this wavelength region. A nonstationary analysis of the second-harmonic mode locker performance (Stankov et al 1991b) predicts that the pulse shortening vanishes for pulses shorter than approximately $3L_{cr}(1/u_1 - 1/u_2)$, where L_{cr} is the crystal length, u_1, u_2 are the group-velocities at the fundamental and second harmonic respectively. At 800 nm for this particular crystal of 23 mm effective length (when tilted for phase matching) this limit is approx. 36 ps (group-velocity mismatch –518 fs/mm). Having in mind that the pulse shortening capability decreases rapidly when the pulse duration approaches this limit, it is easy to explain the observed long pulse duration. Note that a simple estimation of the expected pulse duration based on the phase matchable bandwidth gives a pulse duration of about 15 ps, which is considerably shorter than the result based on the nonstationary analysis.

Using less dispersive crystals such as $\beta-BaB_2O_4$ (BBO) or LiB_3O_5 (LBO) would allow generation of considerably shorter pulses.

We point out, that even with this long pulse duration, the available peak power is as high as that obtainable from CW femtosecond Ti:Sapphire lasers. Therefore, a further decrease of the pulse duration of the flashlamp–pumped mode locked Ti:Sapphire laser would make it competitive for application in various nonlinear optical experiments.

IV. CONCLUSION

We have presented the result of the first experiment for mode locking a flashlamp pumped $Ti:Al_2O_3$ laser using the frequency doubling nonlinear mirror. The advantages of this approach are demonstrated by producing tuneable ultrashort light pulses simultaneously at the fundamental and the second harmonic wavelength by an all–solid–state technique. Further improvements will reduce the pulse duration by employing less dispersive frequency doubling crystal and GVM compensating arrangement.

REFERENCES:

Stankov K A 1988 *Appl. Phys.* **B 45** 191
Stankov K A and Jethwa 1988 *Opt. Commun.* **66** 41
Stankov K A, Kubecek V and Hamal K 1991a *IEEE J. Quant. Electron.* QE–27 2135
Demchuk M I, Demidovich A A, Zhavoronkov N I, Mikhailov V P, Shkadarevich A P
    Ishchenko A A and Sopin A I 1990 *Sov. J.Quant. Electron* **20** 93
Stankov K A, Tzolov V P and Mirkov M G 1991b *Opt. Lett.* **16** 1119
Coherent 1991 Mira Model 900–F laser data sheet
Spectra–Physisc 1991 Tzunami laser data sheet

Inst. Phys. Conf. Ser. No 126: Section I
Paper presented at Int. Symp. on Ultrafast Processes in Spectroscopy, Bayreuth, 1991

Generation of ultrashort pulses from the visible to the infrared

R. Laenen, H. Graener and A. Laubereau
University of Bayreuth, Physics Department
D-8580 Bayreuth, Germany
Telephone: (49-921)-553182
FAX-Number: (49-921)-553172

ABSTRACT: We report on the generation of fs to sub-ps pulses by optical parametric oscillation with broadband tunability from the visible to the infrared. Pumping with the frequency doubled pulse train of a Nd:glass laser, we achieve continuously tunable parametric pulses from 0.7 to 1.8 μm with an almost constant pulse duration of 200 $\pm$ 50 fs. A special feature is observed near degeneracy where we obtain pulses as short as 65 fs by the help of chirp reversal and self compression. Further down-conversion of the OPO pulses with the fundamental of the glass laser yields tunable pulses in the infrared from 3.3 to 10 μm.

1. INTRODUCTION

Parametric 3-photon interaction offers the possibility to generate light pulses in extended tuning ranges from the UV to the IR and with large single pulse energies. The simple frequency adjustment by rotating the nonlinear crystal is an important advantage of such a system. Working with an optical parametric oscillator (OPO) and synchronous pumping allows to generate shorter pulses. In this way a versatile tool for time resolved spectroscopy can be accomplished with good long term stability.

2. EXPERIMENTAL SETUP

As a pump source we use microsecond pulse trains of 0.8 ps generated by a frequency-doubled FCM-Nd:glass laser with repetition rate of 7 Hz. The optical parametric oscillator consists of two concave metal mirrors, internal lens, dichroic output coupling mirror and a BBO-crystal in the focus of the resonator.[1] The non-collinear pump geometry of the OPO ensures that only one of the two generated parametric pulses is coupled back in the resonator; i.e. the OPO is singly resonant. The pump pulse is retro-reflected by a separate mirror with an adjustable time delay for a second pass through the nonlinear crystal. With this special scheme it is possible to compensate for the group velocity mismatch between pump and parametric pulse.

3. TUNABLE FEMTOSECOND PULSES

By rotating the crystal and readjusting the mirror positions to maintain a constant cavity transit time the OPO output is continuously tunable from 0.7 μm to 1.8 μm. An approximately constant pulse duration of 160 to 260 fs is measured in this large spectral range (except close to degeneracy). The experimental results on the pulse duration are shown in Fig. 1. The spectral width of the OPO output is measured to be 100 to 200 cm^{-1}, resulting in a bandwidth – pulse duration product of approximately 0.6. Single pulse energy is 40 nJ.[1]

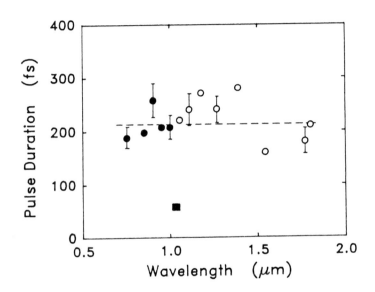

Fig. 1: The measured OPO pulse duration as function of the wavelength from autocorrelation data (sech^2-pulse shape)

4. CHIRP REVERSAL AND SELF-COMPRESSION

Close to degeneracy a striking self-compression effect occurs with shorter OPO pulses of 65 ± 7 fs at 1.076 μm.[2] It is proposed that the phenomenon is due to a transfer of the frequency chirp of the pump pulse to the parametric emission with simultaneous chirp-reversal:[3] i.e. enlarged down-chirp of the idler component. As a result the pulse can be self-compressed in the OPO via the normal dispersion of the cavity elements. The resulting pulse duration by optimum compression is given by:

$$t_i(\lambda_i) = t_i[(1-4\phi_i(\lambda_i)\nu_o)^2 + (8\ln2\ \nu_o/t_i^2)^2]^{1/2} \tag{1}$$

$\phi_i(\lambda_i)$ denotes the idler chirp (depending on the crystal dispersion at λ_i and the pump chirp).[3] t_i is the initial idler duration (250 fs) and ν_o is a measure of the amount of normal dispersion inside the OPO required for optimum compression. The experimental idler pulse duration as measured by the autocorrelation technique and the calculated behaviour from Eq. 1 are shown in Fig. 2. Our experimental data are in semi-quantitative accordance with this theoretical estimate.

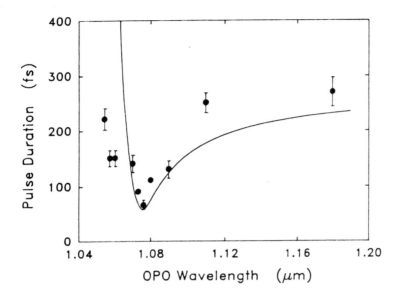

Fig.2: Measured pulse duration of the parametric pulses near degeneracy and the calculated dependence from a simple theoretical model (solid line).

5. TUNABLE PULSES IN THE IR BY DOWN CONVERSION

The pump pulse for difference ·frequency generation is selected electro-optically from the amplified pulse train at 1.054 μm and subsequently amplified to a pulse energy of about 200 μJ. This pulse is divided into two parts: 20% of the energy is used to amplify parametrically a late pulse of the OPO pulse train in an $AgGaS_2$ crystal. The remaining 80% of the laser energy generates together with the amplified parametric pulse tunable radiation in the infrared from 3.3 to 10 μm by down conversion in a second $AgGaS_2$ crystal. At a wavelength of 1.54 μm we measure the overall ampli-fication of the OPO pulse to be in the range of 100 to 300 which corres-ponds to a maximum energy conversion of 12%. The pulse duration varies between 0.4 to 0.7 ps as function of the time delay between the pump and the parametric pulse, while the spectral width $\Delta\nu$ is determined to 26 ± 6 cm^{-1}. For the corresponding idler component at 3.34 μm the data are:

energy conversion: 6%
pulse duration $t_p \geq$ 0.6 ps (determined by crosscorrelation signal – idler)
frequency width $\Delta\nu$ = 17 ± 4 cm^{-1}
Results on the crosscorrelation between signal and idler at the mentioned wavelengths are depicted in Fig. 3. From the full width of the correlation curve and the duration of the signal pulse we determine an minimum idler pulse duration of 0.64 ± 0.08 ps.

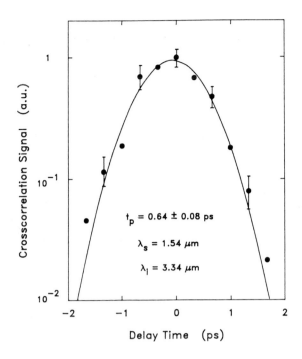

Fig. 3: Experimental cross-correlation curve between signal and idler pulses at 1.54 μm and 3.34 μm leading to an idler pulse duration of 0.64 ± 0.08 ps.

6. CONCLUSION

We have demonstrated continuously tunable pulses from 0.7 to 1.8 μm by synchronous pumping of a singly resonant OPO applying a novel double pass pumping scheme for partial compensation of group velocity dispersion. An almost constant pulse duration of 200 fs and a maximum pulse energy of 4 10^{-8} J are achieved. Close to frequency degeneracy of signal and idler we get the shortest pulses from an OPO with duration of 65 fs. Down conversion between a pump laser pulse and the OPO output offers tunable subpicosecond pulses in the mid-infrared from 3.3 to 10 μm which are well suited for spectroscopic applications.

1   R. Laenen, H. Graener and A. Laubereau, Optics Lett. 15, 971 (1990)
2   R. Laenen, H. Graener and A. Laubereau, J. Opt. Soc. Am. B8 (5), 1085 (1991)
3   V. Vasilyauskas, A. Piskarskas, V. Sirutkaitis, A. Stabinis and Y. Yankauskas, Bull. Acad. Ser. UdSSR 50(6), 32 (1986)

Inst. Phys. Conf. Ser. No 126: Section I
Paper presented at Int. Symp. on Ultrafast Processes in Spectroscopy, Bayreuth, 1991

Parametric generation of femtosecond pulses by LBO crystal in the near IR

S A Akhmanov, I M Bayanov, V M Gordienko, V A Dyakov, S A Magnitskii, V I Pryalkin, and A P Tarasevitch

Nonlinear Optics Laboratory, Moscow State University, Moscow, 119899, USSR

The parametric down conversion of high intensity femtosecond laser pulses is discussed. An injection-locked parametric oscillator, based on a new nonlinear crystal LBO, is presented. With 1 mJ, 400 fs pump pulses we observe 25% quantum efficiency. With the injection at $\lambda = 1.08\,\mu m$ the oscillator is tuned throughout the range 1.2-1.5 μm by changing the pump frequency from 0.57 to 0.63 μm. The surface breakdown threshold of LBO was measured to be $3.5 \cdot 10^{13} W/cm^2$ for 400 fs pulses.

1. INTRODUCTION

The ultrahigh brightness laser systems based on the amplification of femtosecond pulses in broadband amplifiers have been developed in the UV - λ=0.308, 0.248 μm (excimer), at λ=0.6-0.8 μm (dye, Ti:Al$_2$O$_3$), and at $\lambda \approx 1\,\mu m$ (glass amplifiers). At the same time it is highly desirable to develop frequency-tunable sources of superintense field in other spectral ranges.

To our point of view optical parametric oscillators (OPOs) are one of the most promising tools of solving this problem. We report the development of an injection-locked OPO, based on a new wide-aperture nonlinear crystal LBO, pumped by a high-power femtosecond pulses from a dye system. The choice of LBO is determined by its unique characteristics. Measurements, carried out by Chen et al (1989), Lin et al (1990), Kato (1990), Dyakov et al (1991) have shown the extremely broad spectral range of transparency of this crystal (0.16 - 2.6 μm), relatively strong nonlinearity (d$_{32} \approx$ 3d$_{36}$ KDP), and very high optical breakdown threshold (3 times higher than that of KDP).

Ebrahimsadeh et al (1990) demonstrated a first, to our knowledge, OPO on LBO crystal, pumped by UV nanosecond pulses of an excimer laser, tunable from the UV to the IR.

At the same time femtosecond pulse generation in OPO was demonstrated by Laenen et al (1990a, 1990b), Gagel et al (1990), Fickenscher et al (1990), Danielius et al (1989).

In this paper we discuss femtosecond pulse generation at very high pump intensities (up to breakdown threshold). Our first results with LBO pumping at λ = 308 nm and pulse duration of 300 fs were reported at CLEO 91 (Magnitskii et al (1991)). Here we present our results with pumping at $\lambda = 0.57 \div 0.63\,\mu m$.

2. BASIC PARAMETERS

Let us consider a single crystal travelling wave parametric amplifier. It is the group velocity dispersion that limits the efficiency of parametric conversion of short laser pulses (Akhmanov et al (1991a)). For parametric conversion of high-power pulses at least 4 basic parameters should be taken into account: an avalanche breakdown threshold - I_{br}, nonlinear length - L_{nl}, group velocity mismatch length - L_g, and B - integral. Here

$$L_{nl} \cong (k\chi^{(2)}E_p)^{-1}, \qquad L_g = \frac{\tau}{U_p^{-1} - U_{s,i}^{-1}}, \qquad B = \frac{2\pi}{\lambda} \int_0^L n_2 I_p \, dz$$

where k is a wave vector, $\chi^{(2)}$ is a nonlinear susceptibility, E_p, I_p - pump field and intensity, τ - pulse duration, U_p, U_s, U_i - group velocities of the pump, signal, and idler pulses respectively, L - crystal length.

One can see that with pump intensity growing L_{nl} diminishes. The lower limit for L_{nl} is determined by I_{br}. As soon as for an avalanche breakdown $I_{br} \sim (\tau^{-1})$,
$L_{nl,min} \sim (I_{br})^{-1/2} \sim (\tau)^{1/2}$
Thus one can obtain an extremely short L_{nl} for high power femtosecond pulses. This seems to be the way to make L_{nl} so small (smaller than L_g) that one may ignore the group velocity mismatch. But as L_g diminishes faster then L_{nl} with shortening the pulse ($L_g \sim \tau$), for pulses short enough we shall have $L_g < L_{nl}$, and the efficiency of parametric conversion will be small.

If we take into account self-focusing and fix a certain value of B, we also get a limit for pump intensity:

$$I_{max} \cong \frac{\lambda}{2\pi} \frac{B}{n2} (L)^{-1}, \text{ and for } L \cong L_g, \quad I_{max} \sim (\tau)^{-1}.$$

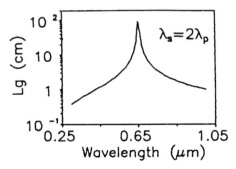

Fig. 1
LBO group velocity mismatch length.

So the problem is to determine the pulse duration range of effective parametric conversion. And to do this one have to know the breakdown threshold, group velocity dispersion, effective nonlinearity and self-focusing threshold.

3. NUMERICAL CALCULATIONS

We carried out computer calculations of tuning curves, group velocity dispersion and conversion efficiency of LBO crystal. This results were obtained starting from the Sellmeier equations and nonlinear coefficients from the papers of Chen et al(1989),Lin et al (1990), Kato (1990), Dyakov et al (1991). The group velocity mismatch curve is shown in Fig.1. It is important to note, that this curve indicates the possibility of group synchronism at $\lambda \approx 0.6 - 0.7 \mu m$.

4. EXPERIMENTAL

Our experimental setup produces ~ 400 fs, 1 mJ pulses, tunable over the region $0.57 - 0.63 \mu m$. It is based on a hybrid mode-locked dye laser pumped by the second harmonic of a pulsed active-passively mode-locked Nd:YAP picosecond laser with feedback control (described in detail by Bayanov et al (1990) and Akhmanov et al (1991b)).The latter generates 5 μs trains of 40 ps pulses (~ 500 pulses in each train) at 2 Hz repetition rate. The dye laser output pulse train with the energy ~ 1 nJ in each pulse is directed into a three stage dye amplifier, pumped by a single picosecond pulse selected from the pulse train of the Nd:YAP laser (amplified and frequency-doubled). The total pump energy is 30 mJ.

The dye laser was tuned from 0.57 to 0.63 μm with the help of Lio plate. The dye solutions in the amplifier stages were changed to cover all this frequency range. An output pulse from the dye amplifier was focused by a 1 m lens into the LBO crystal.

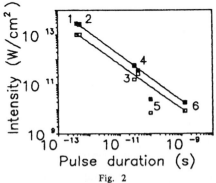

Fig. 2
Surface damage thresholds of LBO (filled squares) and KDP (opened squares) crystals. $1 - \lambda = 308$ nm, $\tau = 300$ fs; $2 - \lambda = 616$ nm, $\tau = 400$ fs; $3 - \lambda = 539$ nm, $\tau = 30$ ps; $4 - \lambda = 1079$ nm, $\tau = 40$ ps; $5 - \lambda = 1064$ nm, $\tau = 0.1$ ns, (Chen et al (1989)); $6 - \lambda = 1064$, $\tau = 1.3$ ns,

The damage threshold was determined simultaneously for LBO and KDP crystals. The measurements were held at different wavelengths and pulse durations. The breakdown onset was determined in a long series of shots. The results are presented in Fig. 2. Points 1, 2, 3, 4 were obtained with the dye system second harmonic and fundamental, Nd:YAP second harmonic and fundamental

respectively. Points 5 and 6 were taken from the papers of Chen et al (1989), Lin at al (1990). All points except 5 lie very well on $1/\tau$ line, revealing $I_{br,LBO}$ about 3 times higher then $I_{br,KDP}$. Note that points 5 also give $I_{br, LBO} / I_{br, KDP} \approx 3$.
Starting with the damage and self-focusing thresholds, L_g, and LBO effective nonlinearity we can

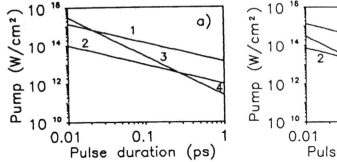
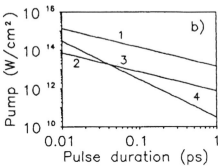

Fig. 3
Pulse duration range of efficient parametric amplification, a) $\lambda = 308$ nm, b) $\lambda = 616$ nm. 1 - surface damage threshold intensity; 2 - self-focusing intensity; 3 - 20% efficiency pump intensity; 4 - efficient parametric amplification area.

define the time duration range of efficient parametric conversion.
The self-focusing threshold was measured at $\lambda = 308$ nm and 616 nm. Fig. 3a and 3b were plotted for $\lambda = 308$ nm and 616 nm respectively $(\lambda_s = 2\lambda_p, L = L_g)$. Here 1 is surface damage threshold intensity, 2 - self-focusing intensity, 3 - 20% efficiency pump intensity. From this figures we can define the range of efficient parametric amplification - 4. One can see that in spite of the fact that 308 nm pumping allows to cover much wider frequency range, for $\lambda_p = 616$ nm we obtain a range of efficient conversion stretching up to much shorter pulses.
The results of our experiments on parametric conversion with $\lambda_p = 616$ nm are depicted in Fig. 4, and 5. We used a 0.9 cm long $\theta = 85°$, $\varphi = 9°$ cut LBO crystal. We also injected an energetic wave at $\lambda = 1.08$ μm (40 ps) in the crystal (signal wave) to achieve a strongly nonlinear regime of parametric conversion (>10%).
For 1 TW/cm^2 pumping we achieve 10% efficiency of conversion into the idler pulse. It corresponds to 25% total quantum efficiency. The autocorrelation functions measured for the signal pulse show the pulse duration of 400 fs. It should be noted than for pumping intensities above 1 TW/cm^2 one can observe wings on the correlation function, indicating the distortion of the pulse. The idler wave was tuned throughout the range 1.2 + 1.5 μm with approximately the same convertion efficiency by changing the pump frequency from 0.57 to 0.63 μm.

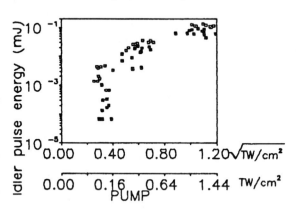

Fig. 4
Idler pulse energy versus pump power at different injection levels: opened squares - 3 GW/cm^2, filled sqares - 0.3 GW/cm^2.

5. CONCLUSION

We have demonstrated femtosecond parametric converter on LBO crystal tunable from 1.2 to 1.5 μm with 25% quantum efficiency. We have measured the surface damage threshold of this crystal to be 3 times higher than that of KDP at different pulse durations. We have calulated group velocity dispersion curve and efficiency of parametric conversion in LBO crystal.

It should be stressed, that this crystal may be very useful for parametric conversion of high power femtosecond pulses. The use of dye amplifiers, of course, limits the pump energy at the level of several millijouls. But relatively large L_g at $\lambda \approx 0.8$ μm indicates that this is a proper crystal for the conversion of much more energetic Ti:Al$_2$O$_3$ radiation. According to our results for LBO crystal with the cross-section of 1 cm^2 one can expect the parametric generation of femtosecond pulses with the power up to 1 TW.

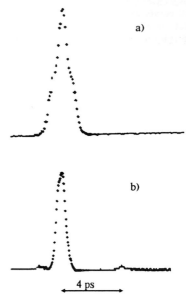

Fig. 5
Autocorrelation function of the signal pulse, a) $I_p = 1.1$ GW/cm^2, b) $I_p = 0.7$ GW/cm^2.

References

Akhmanov S A, Vysloukh V A and Chirkin A S 1991a *Optics of femtosecond pulses* (New York: AIP)

Akhmanov S A, Bayanov I M, Gordienko V M et al 1991b *Sov. J. Quantum Electron.* **21** p 248

Bayanov I M, Biglov Z A, Gordienko V M et al 1990 *Izvestia Acad. Nauk SSSR Ser. Fiz.* **54** p 161

Chen C, Wu Y, Jiang A et al 1989 *J. Opt. Soc. Am.* **B6** p 616

Danielius R, Grigonis R, Piskarskas et al 1989 *Proc. 6th UPS Int. Symp.* (Berlin: Springer Verlag) p 40

Dyakov V A, Dzhafarov M Kh, Lukashev et al 1991 *Sov. J. Quantum Electron.* **21** p 339

Ebrahimsadeh M, Robertson G and Dunn M 1990 *CLEO Techn. Digest* (Washington: OSA) postdeadline paper CPDP26

Fickendhcer M, Purucker H and Laubereau A 1990 *Appl. Phys.* **B51** p 207

Gagel R, Angel G and Laubereau A 1990 *Opt. Commun.* **76** p 239

Kato K 1990 *IEEE J. of QE* **QE26** p 1173

Laenen R, Graener H and Laubereau A 1990a *Opt. Letters* **15** p 971

Laenen R, Graener H and Laubereau A 1990b *Opt. Commun* **77** p 226

Lin S, Sun Z, Wu B et al 1990 *J. Appl. Phys.* **67** p 634

Magnitskii S A, Akhmanov S A, Bayanov I M et al 1991 *CLEO Tech. Digest* (Washington: OSA) paper CTuB4

Inst. Phys. Conf. Ser. No 126: Section I
Paper presented at Int. Symp. on Ultrafast Processes in Spectroscopy, Bayreuth, 1991

71

Picosecond BBO optical parametric oscillator

V Kubecek[1], Y Takagi[1], K Yoshihara[1], and G C Reali[2]

[1] Institute for Molecular Science
Myodaiji, Okazaki 444, Japan

[2] Universita' di Pavia, Dipartimento di Elettronica
Via Abbiategrasso 209, 27100 Pavia, Italy

ABSTRACT. We report here our recent results of the operation of a singly resonant Optical Parametric Oscillator based on the use of a BBO crystal and synchronously pumped by the third harmonic of a passive negative feedback mode-locked Nd:YAG laser.

Optical parametric generation in a nonlinear crystal provides a most attractive method to produce widely tuneable coherent radiation. In the past, insufficient nonlinearities together with the low damage thresholds of the available nonlinear crystals have prevented the common use of this scheme. The recent advances in material production and the great improvements in stability of the pumping laser sources have renewed the interest in optical parametric oscillators (OPO).

The work in pulsed systems has been mainly concentrated on the use of BBO crystals, which often have been pumped by an ultraviolet beam since in this case the output can cover a larger fraction of the visible range in addition to the near infrared. Both OPO's generating nanosecond [1], picosecond [2], and femtosecond [3] pulses, and optical parametric amplifiers (OPA's) generating picosecond pulses [4] have been demonstrated using BBO.

We report here the operation of a picosecond BBO OPO [6] synchronously pumped by the third harmonic of a passive negative feedback mode-locked (PNFM) Nd:YAG laser [5]. Such a laser pump is very attractive for synchronous pumping of laser-active media and nonlinear crystals for the generation of ultrashort pulses, since it generates long trains of short, energetic and stable pulses in several crystalline Nd:doped materials. For example, from a PNFM Nd:YAG laser trains of 50 to over 100 pulses, depending upon the operating conditions, with energy up to 10 μJ per pulse and pulse duration less than 10 ps are directly generated from the oscillator.

The experimental arrangement of the system is shown schematically in Fig. 1.

The pulse train from PNFM Nd:YAG laser, with duration of 600 ns and containing approximately 80 pulses with total energy of 1 mJ, was amplified by three Nd:YAG single-pass amplifiers to an energy of 20 mJ, and the third harmonic was generated using two KDP Type II crystals with conversion efficiency of approximately 10 %. The 355 nm beam, having total energy in excess of 2 mJ and single pulse width of 9 ps, was focussed, using a 1 m focal length spherical lens, into a 5 x 7 x 7 mm^3 BBO crystal (supplied by Fujian Castech Crystals, Inc.) cut for Type I phase-matching at $\theta = 31°$. The pump beam at the BBO crystal was detected by a CCD TV camera, and from its bell-shaped profile the 10% intensity level diameter was measured to be ≈ 300 μm. When the 355 nm pump beam was focussed into the BBO crystal, a bright parametric fluorescence (signal wave) was observed, which did not propagate in the same direction of the pump beam, but at a certain angle. This angle was measured to be 95 mrad. The non-collinearity of the pump and parametric fluorescence was a reason for the non-collinear pumping geometry, used with an angle of 95 mrad between the pump beam and the resonator axis. The resonator of the OPO consisted of two identical dielectric mirrors M3 and M4, each having a radius of curvature of 1 m and high reflectivity (>99%) from 405 to 690 nm. The mirror M4 was mounted on a precision translation stage to allow the fine tuning of the cavity length. The beam waist of this confocal resonator had a diameter of 400 μm which approximately matched the pump beam diameter at the focus of spherical lens L .

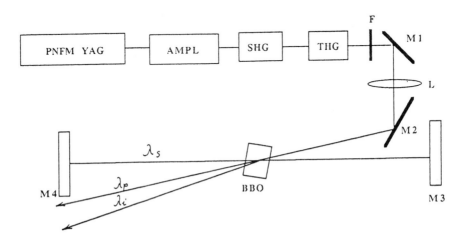

Fig. 1 Schematic of the singly resonant OPO in noncolinear geometry.

The pumping threshold for parametric generation was 150 μJ, corresponding to a peak pumping intensity of ≈0.3 GW/cm^2. At a pump energy 3 times above this threshold, 45 μJ of energy was obtained at 900 nm, corresponding to a conversion efficiency into the infra-red idler wave of 10 %. was obtained. For this case, Fig. 2a shows the undepleted pump train detected after one pass through the BBO crystal when the parametric oscillation was blocked and there was no generation. Fig. 2b shows the depleted pump at the same position in the case of parametric oscillation, while the generated signal wave train is shown in Fig. 2c. These oscillograms show an optical parametric oscillation build up time of 150 ns, measured at the time of half the peak intensity. This build up time is longer at lower pump levels. From this figure we also know that single pulse internal conversion efficiency gets as high as 70 %, while the overall internal efficiency, integrated over the whole train, is 26%.

The pulse width of the pump and signal pulse from OPO was measured using a streak camera Hamamatsu, Model HTV C1370-01, with a resolution of 1.8 ps. When the pulse width of the pumping pulse was 9 ps, the pulse width, measured at the signal wavelength of 600 nm, was found to be 3.1 ps.

The tuning of the OPO was achieved by tilting the BBO crystal with respect to the axis of the resonator. The shortest measured signal wavelength was 407 nm. This wavelength limitation was due to the geometrical dimensions of our crystal. Near degeneracy we were able to generate a signal wave up to 690 nm, where the limit was set by the reflectivity of our mirrors. The calculated idler wave tunability corresponding to measured signal wavelengths is from 731 nm to 1893 nm. The spectral width of the signal wave for pumping twice above the threshold , as measured using a monochromator and CCD TV camera, is shown for different wavelengths in Fig. 3. We can see that near the degeneracy the spectral width is 1.4 nm, but further from this point the spectral width decreases to 0.5 nm, reaching a time-bandwidth product of ≈ 4 at 499 nm. This is a common behavior of a type I interaction.

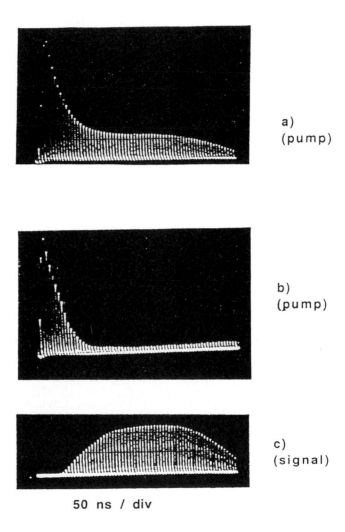

a)
(pump)

b)
(pump)

c)
(signal)

50 ns / div

Fig. 2. Oscilloscope traces of the undepleted pump (a), depleted pump (b), and signal wave parametric pulse (c) trains. Pumping level was 3 times above the oscillation threshold.

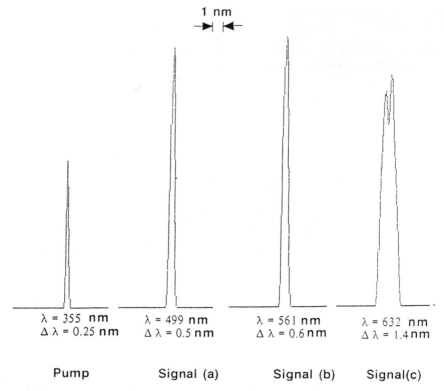

1 nm

λ = 355 nm   λ = 499 nm   λ = 561 nm   λ = 632 nm
Δλ = 0.25 nm   Δλ = 0.5 nm   Δλ = 0.6 nm   Δλ = 1.4 nm

Pump   Signal (a)   Signal (b)   Signal(c)

Fig. 3. Spectral widths of the OPO signal wave for different wavelengths at twice above threshold pumping.

References

[1] Y. X. Fan, R. C. Eckardt, R. L. Byer, J. Nolting, and R. Wallenstein, Appl. Phys. Lett., 53, 2014, 1988
[2] S. Burdulis, R. Grigonis, A. Piskarskas, G. Sinkevicius, V. Sirutkaitis, A. Fix, J. Nolting, and R. Wallenstein, Opt. Comm., 74, 398, 1990
[3] R. Laenen, H. Graener, and A. Lauberau, Opt. Lett., 15, 971, 1990
[4] J. Y. Huang, J. Y. Zhang, Y. R. Shen, C. Chen, and B. Wu, Appl. Phys. Lett., 57, 1961, 1990
[5] A. Del Corno, G. Gabetta, G. C. Reali, V. Kubecek, and J. Marek, Opt. Lett., 15, 734, 1990
[6] V. Kubecek, Y. Takagi, K. Yoshihara, and G. C. Reali, CLEO Technical Digest, paper CThR10, 8, 434, 1991

Performance of a barium borate parametric oscillator in the picosecond regime with different cavity configurations

G.P.Banfi, M.Ghigliazza and P.Di Trapani

Dipartimento di Elettronica, Universitá di Pavia, 27100 Pavia, Italy

ABSRACT: We report on the generation of tunable picosecond pulses through a parametric oscillator employing a beta barium borate crystal sincronously pumped by the SH of an active-passive mode-locked Nd-Yag laser. Linewidth, efficiency and beam quality obtained with different arrangements are presented and compared.

The mode-locking regime appears particulary suitable to optical parametric oscillators and amplifiers since the signal can achieve an high gain per pass due to the high pump intensity that can be supported by the crystals with short pulses. This is particulary true for the new crystals such as BBO, which offers high nonlinear coefficients, transparency and phase matching on a wide frequency range, and at the same time a very high damage treshold[1]. Successfull parametric conversion with BBO has been reported in picosecond and even in femtosecond regime, both with oscillators (OPO)[2-4] and travelling wave configurations [5-6], and new recent results are presented also in a few papers at this conference. In the picosecond regime we here consider, the high gain/pass allows the BBO-OPO to reach the threshold sincronously pumping the crystal for just a few round-trips and hence to be driven by the harmonic of a conventional pulsed modelocked Nd-Yag laser, which gives a train of few pulses, each of 30 ps duration[3]. Simplicity, easy of alineament and syncronization, energy of a fraction of a mj in the pulses tunable down to 3 μm are some attractive features of this approach. We here report the main results of a rather sistematic investigation of this scheme, which include bandwidth, threshold energy, conversion efficiency.and also some data on spatial quality of the converted beam. This last one is a crucial aspect which has not jet been much investigated, but which is relevant in many cases, in particular for the characterization of the nonlinear optical properties of materials, an application that partly motivated the present work.

The pump source is a stable Nd-Yag oscillator, active-passive mode-locked, followed by an amplifier. After doubling in KDP, the SH has a train envelope of 7-8 pulses at FWHM, each pulse of about 25 ps duration. The BBO crystal, 10 mm long, was cut at 23° , for type I SH p-m at 1.06 μm. The 1.5 m long OPO cavity was designed to trasmit the IR idler, and the mirrors were manufactured with high reflectivity between 650 and 1020 nm , the range of the resonating signal wave. The front output mirror was either a plane or an R=2 m concave reflector; back mirror was either a plane reflector (fig.1a) or a 600 l/mm grating operating in 2nd or 3rd order (fig.1b). In all four arrangments phase-matching was collinear. On entering the BBO crystal the pump train has a spatial profile that could be reasonably fitted by a Gaussian beam with a waist of 0.8 mm (radius at $1/e^2$ intensity). When unspecified, the data reported are taken with a pump energy of 7-8 mJ, always on entering the crystal, to which corresponds an intensity of about 3-4 GW/cm^2.

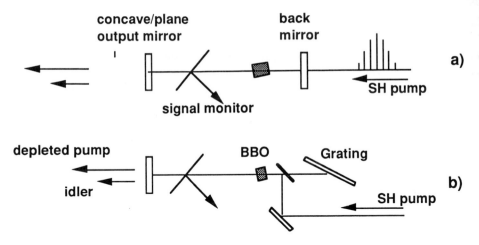

Fig1- experimental layout

Diagnostic included fast Ge and Si diodes, connected to a Tektronik Transient Digitezer 7912AD for simultaneous recording of the the impinging and depleted pump train/or signal (idler) train. Example of an on line recording of the evolution of conversion during the pump train is given in Fig 2. For the beam shape, we made use of a computer interfaced CCD camera.

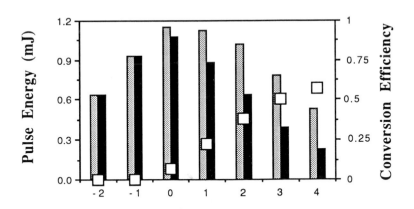

Fig. 2. Impinging pump pulses (gray), depleted pump pulsed (black) and conversion efficiency at λ=740 nm. Plane-plane cavity. Pump energy: 6 mJ

Round trip losses increase progressively going through the following sequence of configurations for the cavity reflectors: a) concave and plane mirrors, b) plane -plane, c) concave mirror and grating, d) plane mirror and grating. Than, as expected, the plane-concave arrangement has the higher average conversion, the lower threshold energy, and it shows a parametric oscillation already fully developed before the peak of the pump train. Decreasing pump energy , introducing losses and tuning the wavelengths away from degeneracy delays the appearance of the parametric generated pulses respect to the pump train. Going then from a) to d) the threshold energy increases from 1.2 to 3 mJ (at λ_s=0.9 μm), while the average conversion efficiency (on both signal+ idler) decreases from 45% to 12%.

The high gain at which the OPO operates is such that some oscillation could still be noticed introducing in the arrangement a) a loss per round trip as high as 10^4. Observations on thresholds and onset of oscillations are in broad agreement with simple extimations of the gain in a single pass. Typical values of the maximum pulse conversion range, accordingly to energy and λ_s , from 60 to 75% in configurations a) and b), and are about 30-40% in c) and d). We noticed that the peak pulse conversion is not much correlated to cavity losses, and we are tempted to speculate that it is more related to the bandwidth of the the converted pulses, increasing with a broad spectra.

A comparison of the bandwidth $\Delta\lambda_s$ observed in the arrangements with no grating is shown in the table below. We have also shown the results we obtained in a travelling wave configuration, with a single SH pulse pumping 2 crystals in series.

configuration	travelling wave superfluorescence.	plane-concave cavity	plane-plane cavity
$\Delta\lambda$ (nm)	8.5	4.8	3.5

The data are given for λ_s=740 nm. Approaching degeneracy , as expected in type I p-m , we observe much larger widths (few tens of nm for $\lambda_s = 1$ μm), but the bandwidths in the different arrangements are about in the same ratio as in the table above. From the same table one can notice that replacing the curved mirror with the flat one sligthly reduces the spectral bandwidth. The effect should be due to the rejection of the more spread out signal components which would phase-match for wavelengths different from the axial ones.

Much narrower bandwidth are obtained inserting the grating, as summarized here below where we report the linewidths of the idler wave when tuned at λ_i=1160 nm. In any case, with a grating, the width $\Delta\lambda_s$ of the resonating signal wave is independent from the chosen wavelength , so that $\Delta\lambda_i$ at the different wavelengths can be easily predicted.

configuration	concave output mirror	concave output mirror +8x beam expander	plane-plane cavity
$\Delta\lambda_i$ (nm)	0.83	0.35	0.38

We note the higher selectivity of the plane-plane cavity. To obtain comparable bandwidths with the concave-plane cavity required the insertion in the resonator of a beam expander in front of the grating. The frequency widths with the plane-plane cavity are $\Delta\nu$=5, 9, 8.5 $\times 10^{10}$ Hz, respectively for the pump at 1.06 , the idler at 1.16 and the signal at 0.98 μm. For a comparison, we remind that for a transform limited pulse of 20 ps one has $\Delta\nu = 2.2 \times 10^{10}$ Hz.

The beam quality appears to depend mainly from the conversion rate and to be rather independent from cavity configuration. Due to sensitivity limitation at longer wavelengths of the CCD camera, the observation was limited to the signal wave; only around degeneracy also the idler wave could be directly observed.

Even if we always observed smooth far field patterns, we estimated the beam divergence to be no less than 3 x the diffraction limit when pulse conversion was at 70 % level, as it could be achieved in the configurations a) and b)) Much better spatial quality of the beam was noticed at low conversions. Beside the observation of the far field patterns, this fact was confirmed by the M^2 test [7] performed on the configuration d). In fact the test showed that the M^2 (practically a measure of how many times the beam exceeds the diffraction limit) increases from 1.7 to about 2.5 as the maximum pulse conversion moves from 15-20% to 40 %. The main difficulty in keeping a low conversion arises from the instability which occurs when the pump energy is not much above threshold.To operate in this regime a more stable pump laser would be most welcome.

When the aim is to provide a tunable picosecond pulse for spectroscopic use, the choise between an OPO and a travelling wave configuration is often embarrassing. In most cases the advantage of a better control of the linewidth and of the spatial profile, that can be obtained with an oscillator, is offset by the difficulty of extracting a single pulse from the parametrically converted train. The usual approach to single pulse extraction is through a Pockel cell driven by an high voltage syncronized pulse, produced either by a spark gap or by an avalanche transistors chain triggered by a photodiode. With a tunable wavelength, a similar set -up not only would require continuous adjustements, but it would become rather complicated in IR, where its feasibility is even questionable.

We believe that a simple way to obtain a single pulse starting from the parametric oscillator is to have it followed by a low gain parametric amplifier stage pumped by a single SH pulse, with single pulse extraction being done at 1.06 μm. When seeded with the train of the idler (signal) wave from the oscillator, the parametric amplifier will provide a single pulse at the signal (idler) wavelength. Preliminary measurements we performed on this scheme appear encouraging.

We thank prof. A.Piskarskas for discussions and suggestions.

We acknowledge that this work has been supported by the Italian Research Council (CNR) in the frame of the Progetto Finalizzato Telecomunicazioni.

REFERENCES

1- D.Eimerl, L.Davis, S.Velsko, E.K.Graham, and A.Zalkin, J.Appl. Phys 62 1968 (1987)

2- L.J. Bromley, A. Guy and D.C. Hanna , Opt. Comm **67,** 316 (1988)

3- S.Burdulis, R.Grigonis, A.Piskarkas, V.Sinkevicius, V.Sirutkaitis, A.Fix, J. Nolting and R.Wallenstein, Opt. Comm. **74**, 398 (1990)

4- R.Laenen, H. Graener, and A. Lauberau, Opt.Lett. **15**,971(1990)

5- J.Y.Huang, J.Y.Zhang, Y.R.Shen, C.Chen and B.Wu, Appl.Phys.Lett.**57**, 1961(1990);

6-R. Danielius, A. Piskarskas, D. Podenas,P. Di Trapani, A. Varanavicius and G.P.Banfi to be pubblished in Opt. Comm.

7- A.E. Sigman, "New development in laser resonators", SPIE Proc.Vol 1224 , 2 (1990)

Inst. Phys. Conf. Ser. No 126: Section I

Paper presented at Int. Symp. on Ultrafast Processes in Spectroscopy, Bayreuth, 1991

Multiphoton ionization of atoms with parametrically amplified femtosecond infrared pulses

W. Joosen,[a] **P. Agostini,**[a] **G. Petite,**[a]
J.P. Chambaret,[b] **and A. Antonetti**[b]

[a] Service de Recherches sur les Surfaces et l'Irradiation de la Matière, Centre d'Etudes de Saclay, F-91191 Gif-sur-Yvette, France.

[b] Laboratoire d'Optique Appliquée, Ecole Polytechnique, Ecole Nationale Supérieure de Techniques Avancées, F-91120 Palaiseau, France.

Abstract We developed a femtosecond type I parametric amplifier consisting of two β-BBO crystals pumped with an amplified 615-nm colliding pulse mode-locked dye laser. Tunability throughout the near infrared and focusable intensities close to 10^{13} W/cm^2 are demonstrated, permitting six-photon ionization of Mg at 830 nm.

1. INTRODUCTION

Multiphoton ionization (MPI) of atoms has usually been shown with mJ pulses of the fundamental and the (second, third, and fourth) harmonics of a Q-switched or mode-locked Nd:YAG laser. More recently, excitation with relatively low energetic femtosecond (fs) dye-laser pulses was performed, revealing previously unresolved fine structure in the photoelectron spectra (Agostini 1989, Freeman 1987). An optical *parametric* excitation source permits much larger and continuous tunability, and would be useful to investigate (i) the effect of n, n+1, n+2,... -photon resonances on the ionization probability and (ii) the frequency dependence of branching ratios of distinct MPI processes.

This paper describes a fs optical parametric amplifier (OPA) (Section 2) tunable throughout the near infrared (ir). In Section 3 the output diagnostics are summarized, emphasizing the focusable intensity, which is the relevant parameter for MPI. In Section 4 MPI of magnesium is shown and further possible investigations are indicated.

2. BBO OPTICAL PARAMETRIC AMPLIFIER

The OPA consists of two β-barium borate (BBO) crystals (length 0.5 cm (BBO I) and 0.7 cm (BBO II), aperture 0.25 cm^2) cut for Type I collinear phasematching (Fig.1). The prominent characteristics of BBO for this device are the large effective nonlinearity (1.6 pm/V), the high damage threshold for fs pulses (> 1 TW/cm^2), and

the small group velocity (GV) mismatch between the pump, signal and idler fields
(< 20 fs/mm) (Joosen 1991a).

The OPA is pumped by ~ 0.6 mJ pulses of an amplified colliding pulse mode-
locked (CPM) dye laser. Thirty percent is used for BBO I, which serves as a preamplifier for BBO II. A typical preamplified ir pulse is shorter than 200 fs with an energy
between 0.1 μJ and 1.0 μJ in a diffraction-limited beam profile (Joosen 1991b). Its
bandwidth depends on the pump intensity I_P and remains smaller than 40 nm (at
830 nm) for $I_P < 10$ GW/cm^2.

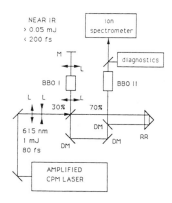

Fig.1 Scheme of the parametric preamplifier (BBO I) and amplifier (BBO II)
pumped with amplified CPM laser pulses; M : mirror, L : lenses (double-sided
arrows), DM : dichroic mirror, RR : retroreflector.

The near ir pulses are injected simultaneously with the remaining part of the
pump (~ 0.4 mJ) in BBO II. The amplified parametric beam is spectrally filtered
from the depleted pump and focused into an ion spectrometer, crossing an effusive
Mg atom beam at an angle of 90 degrees.

3. OUTPUT DIAGNOSTICS

Fig.2(a) shows the parametric amplification of the signal in BBO II as a function
of the pump delay. The ir pulse from BBO I is amplified by two orders of magnitude
and enhances the conversion efficiency in BBO II typically by a factor of three. For
signal optimization at 830 nm we obtain pulses of 108 μJ and an overall energy
conversion efficiency of 25%, which is comparable to those of picosecond (ps) BBO
OPA's (Huang 1990).

The large full width at half maximum (FWHM) of the amplification curve (0.65
ps without deconvolution)) is caused by dispersive broadening of the pump and its
asymmetry reflects the positive GV dispersion between 615 nm and 830 nm. For
optimum parametric amplification, i.e., with the largest effective interaction length,
the pump predelay is estimated to be 150 fs. Parametric gain narrowing (Laenen
1990) yields a short pulse duration of 150 fs as derived from the hyperbolic secant
squared fit to the autocorrelation trace (Fig.2(b)). The time-bandwidth product is
2.23 and can be reduced to the transform limit at the expense of a proportional
signal-energy reduction.

The amplification stage BBO II is pumped at intensities between 10 and 15 GW/cm^2 in order to saturate the gain. This turns out to be somewhat detrimental for the beam quality. Gaussian fits of the intensity profiles yield a 3 times diffraction-limited OPA beam, i.e., 2.3 times worse than the pump. The pulse and beam diagnostics predict a focusable intensity of 3.7 10^{12} W/cm^2 for a 16-μJ, *transform-limited* 150-fs pulse with an f/10 optic. Such an intensity is relevant for MPI in the ir of alkalines and earth alkalines and for photon detachment of negatively charged ions (Davidson 1991).

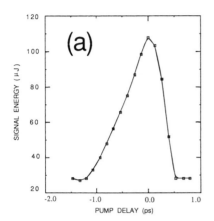 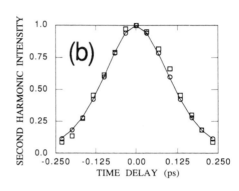

Fig.2 (a) Output 830-nm signal energy of the OPA as a function of the pump delay in BBO II (taken zero for optimum amplification) and (b) second order autocorrelation trace of the amplified pulse (squares) fitted to a secant hyperbolic squared distribution (circles).

4. SINGLY IONIZED MAGNESIUM

The prediction of the focusable intensity from the above diagnostics is confirmed by the generation with the OPA beam of Mg ions, which show up in the time-of-flight (TOF) and mass spectra: First the TOF of Mg$^+$ and Mg^{++} ions are obtained from MPI with the depleted 615-nm pump, which is almost collinear with the ir signal beam. The latter is then separated with a broadband, 830-nm interference filter (FWHM 34 nm). In the resulting TOF spectrum only the Mg$^+$ peak occurs, which is shown to disappear upon detuning BBO II. The corresponding mass spectrum is given in Fig.3.

Further extensions and improvements of this prototype BBO OPA may include increasing the pump energy (up to 2 mJ), the use of larger-aperture nonlinear crystals and optimum balancing of the parametric gain against the self-phase modulation. When the focusable intensity can be increased by a factor of ten, one could study 5- up to 8-photon resonantly enhanced MPI of Mg (Hou 1990) with one single excitation source. The branching ratio of direct and sequential double ionization of Xe as a function of frequency is another intriguing item to investigate (L'Huillier 1983).

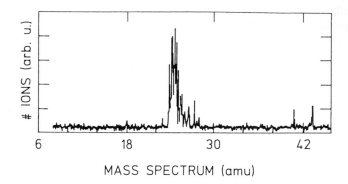

MASS SPECTRUM (amu)

Fig.3 Magnesium ion mass spectrum from MPI with a parametrically amplified beam at 830 nm. Six-photon absorption is required to overcome the 7.64-eV ionization potential.

ACKNOWLEDGEMENT

The authors acknowledge the financial support of the European Community (E.E.C. Contract SC1000103), to which they are greatly indebted.

REFERENCES

Agostini P, Antonetti A, Breger P, Crance M, Migus A, Muller H G, and Petite G 1989 *J. Phys.B* **22** 1971

Davidson M D, Muller H G, and van Linden van den Heuvell H B 1991 *Phys. Rev. Lett.* **67** 1712

Freeman R R, Bucksbaum P H, Milchberg H, Darack S, Schumacher D, and Geusic M E 1987 *Phys. Rev. Lett.* **59** 1092

Hou M, Breger P, Petite G, and Agostini P 1990 *J. Phys. B* **23** 583

Huang J Y, Zhang J Y, Shen Y R, Chen C, and Wu B 1990 *Appl. Phys. Lett.* **57** 1961

Joosen W, Agostini P, Petite G, Chambaret J P, and Antonetti A 1991a *Post-deadline Paper* in *Conference on Lasers and Electro-Optics*, (Optical Society of America, Washington D.C., 1991), pp. 618-619

Joosen W, Bakker H J, Noordam L D, Muller H G, and van Linden van den Heuvell H B 1991b *J. Opt. Soc. Am.B* **8**, xxxx

Laenen R, Graener H, and Laubereau A 1990 *Opt. Lett.* **15** 971

L'Huillier A, Lompré L A, Mainfray G, and Manus C 1983 *Phys. Rev.A* **27**, 2503

Coherent photon seeding: A scheme for ultra stable ultrashort pulse generation

H. P. Weber, W. Hodel, J. Q. Bi, P. Beaud, and D. S. Peter

Institute of Applied Physics, University of Bern, Sidlerstr. 5, CH-3012 Bern, Switzerland

ABSTRACT: We recently demonstrated a simple self stabilization technique called coherent photon seeding in a synchronous pumped laser. The application of this method in a nonlinear mirror hybridly mode locked laser results in a further pulse shortening and mode locking improvement. The laser output pulses are insensitive to cavity length variations of up to 40 μm and the laser works without any readjusting over many hours.

1. INTRODUCTION

Synchronous pumping is a powerful version of active mode-locking for the generation of ultrashort laser pulses. However, it is well known that the stability of the resulting pulse trains is quite poor: the pulses usually suffer from large fluctuations in width, height and therefore pulse energy (*Catherall et al 1986, Kluge et al 1984*). Recently we have reported on a new experimental technique (*Beaud et al 1990*) to stabilize synchronously pumped mode-locked (SPML) lasers which we called "Coherent Photon Seeding" (CPS). We have also analyzed the CPS technique in detail by using numerical simulations (*Bi et al 1991*). The numerical results were in full agreement with the experiments and confirmed our first intuitive interpretation of how the CPS technique works. The technique has also been applied to different lasers (*Hooker et al 1991, McCarthy et al 1991*) and numerically studied by another author (*New 1990*). The basic idea of CPS can be understood when considering the dynamics of synchronously pumped lasers. Due to the gain dynamics the best performance of SPML lasers is obtained for a slight mismatch between the pump laser and the main laser cavity. Because the leading edge of the laser pulse is preferentially amplified in the gain medium, the pulse is slightly accelerated during each pass through the amplifier. In order to maintain the pulse repetition rate the laser cavity must thus be slightly longer than the pump laser cavity. As a consequence the laser pulse envelope in the main cavity moves a bit faster than the background radiation which originates from spontaneous emission. Noise components initially located ahead of the pulse then migrate into the pulse and are progressively amplified which leads to large variations of the pulse profiles. The CPS technique uses an extremely weak (10^{-10}-10^{-6}) coherent signal which is fed back into the main laser cavity slightly advanced in time and overpowers the influence of spontaneous emission. In this way the laser fluctuations are suppressed to a large extent and the laser stability is greatly improved. We will present experimental evidence for this stabilization technique for the case where (i) the seeding signal is derived from an external coupled cavity and (ii) from a slightly modified out coupling mirror (compact design of CPS).

For purely synchronous pumping the rise time of the laser pulses depends on the duration of the pump pulses, whereas the fall time is determined by gain saturation. In order to obtain even shorter pulses active mode-locking is often combined with passive mode-locking using saturable absorbers (hybrid mode-locking). The saturable absorption leads to a steepening of the leading edge of the laser pulse and results in a further pulse shortening. The most commonly used saturable absorbers are organic dyes. However, the applicability of this technique is limited to wavelengths where suitable organic dyes are available. The advanced technique of semiconductor band gap tailoring offers the possibility to use these materials in combination with many different gain media to cover a large wavelength range. We will show that a bulk semiconductor material can be used successfully as a saturable absorber which considerably simplifies the laser set-up. Finally, we demonstrate the applicability of the CPS technique to a hybridly mode-locked laser. In this case CPS not only leads to a mode-locking improvement, but the laser pulses become insensitive to unusually large cavity length variations (up to 40 μm). As a result the laser works stable over many hours without any readjustments.

2. COHERENT PHOTON SEEDING FOR A SYNCHRONOUSLY PUMPED LASER

The CPS technique was applied to a synchronously pumped infrared dye laser (IR dye No 26). The dye laser is pumped by the compressed (fiber/grating-pair compressor) 4 psec pulses from a commercial Nd:YAG laser. It delivers sub-psec pulses with kW peak power in the wavelength range from 1250 nm to 1350 nm. Figure 1 shows the two configurations used to realize CPS.

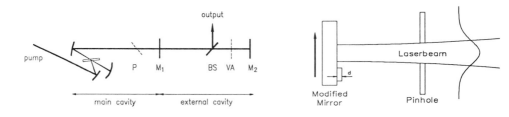

Figure 1. The two basic configurations used for the CPS technique: an external coupled cavity (left) and a modified out coupling mirror (right).

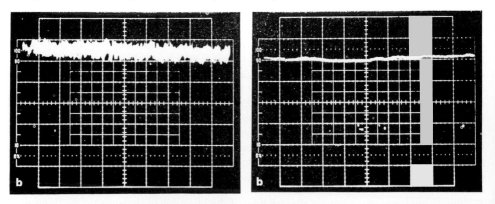

Figure 2. The laser output power without feedback (left) and with a feedback of 10⁻¹⁰ (right).

In the first experiments the weak seeding signal was coupled back into the main laser from an external coupled cavity containing a variable attenuator to change the feedback intensity level. Figure 2 compares the output power on a slow time scale (100 μsec/division) of the laser without feedback (left) and with a feedback of 10^{-10} (right).

The laser energy fluctuations decrease from 5.7% rms to 1% rms by applying the CPS technique. Figure 3 shows the energy fluctuations and the resulting pulse durations as a function of the feedback level.

The Figure clearly demonstrates how powerful the CPS technique is: for feedback levels between 10^{-9} and 10^{-6} the energy fluctuations are reduced by at least one order of magnitude while the pulse duration changes only slightly.

A disadvantage of using an external cavity is that slow fluctuations of the relative length of the two cavities (e.g. due to mechanical vibrations and/or thermal drifts) induces an undesired frequency hopping of the laser output. In order to eliminate this frequency hopping and to improve the long term stability of the laser an inte-

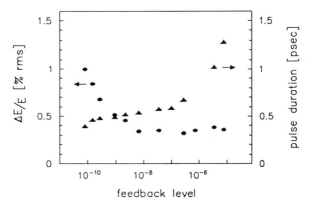

Figure 3. Pulse energy fluctuations and pulse duration as a function of the feedback level.

grated version of the CPS technique was also developed (*Peter et al 1991*). In this compact design the seeding pulse is derived from the reflection of a glass plate of appropriate thickness d (Figure 1, right) which is glued onto the peripheral part of the laser out coupling mirror. The feedback level can be varied by simply moving the mirror perpendicular to the laser beam so that the glass plate comes closer to the beam centre. Since only a very small reflection is required the spatial beam quality is not degraded in any way. This compact CPS version leads to the same results as the coupled cavity technique but in addition the long term stability is now greatly improved. It should also be mentioned that the laser pulses can very well be fitted by a sech² pulse form and that the time-bandwidth product is typically 0.34 (for an ideal sech² pulse it has a value of 0.32). This indicates that the CPS technique leads to the generation of stable trains of close to bandwidth limited pulses which is an important property for many applications. The wavelength tunability is maintained over the entire tuning range of the dye.

3. HYBRID MODE-LOCKING USING A NONLINEAR MIRROR

The hybrid mode-locking experiments were performed with a synchronously pumped Styryl 9 dye laser. Using the compressed (4 psec) and subsequently frequency doubled Nd:YAG laser pulses as the pump, the dye laser delivers pulses with a duration of approximately 900 fsec in the wavelength range from 800 nm to 850 nm. The time bandwidth product is typically 0.64 which indicates that the pulses are not transform limited. In order to reduce the pulse duration we have designed a nonlinear mirror which consists of a 200 nm thin layer of AlGaAs (5% Al content) grown on a linear Bragg mirror. The AlGaAs layer acts as a saturable absorber (the absorption recovery time is in the order of a few nsec which is shorter than the cavity round trip time) which gives rise to an intensity dependent reflection of the mirror. The mirror reflection as a function of wavelength is shown in Figure 4 for two intensity levels of interest.

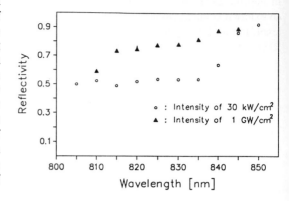

Figure 4. Reflectivity of the nonlinear mirror as a function of wavelength for two intensity levels where the higher intensity corresponds to the laser intracavity intensity.

Figure 4 demonstrates that the intensity induced change in reflectivity is as high as 30% for certain wavelengths. If the NLM is incorporated as the main reflector in the dye laser resonator, the pulse duration is reduced considerably. Figure 5 shows the output power and the pulse durations as a function of wavelength. Pulse shortening is achieved in the wavelength range from 832 to 850 nm where the nonlinearity of the mirror is high. Optimum operation is achieved near the band edge of the AlGaAs material: at 836 nm the pulse duration is 220 fsec and the corresponding time bandwidth product is 0.44. This indicates that NLM not only induces a pulse shortening but also leads to a mode-locking improvement.

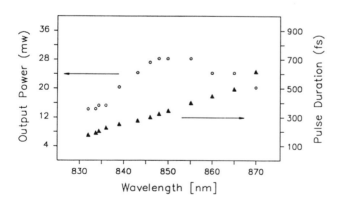

Figure 5. The pulse duration and the average output power of the hybridly mode-locked Styryl 9 dye laser as a function of wavelength (pump power is 200 mW).

4. CPS APPLIED TO A HYBRIDLY MODE-LOCKED LASER

Applying the Coherent Photon Seeding technique to the hybridly mode-locked Styryl 9 dye laser described in the previous paragraph we obtained interesting new results. For comparison the typical autocorrelation traces with and without seeding are shown in Figure 6. Without CPS the laser generates acceptable mode-locked pulse trains only for two distinct cavity length settings which are 30 μm apart. Furthermore, optimum operation of the laser is found for only one (+30 μm) of these values. Assuming a sech² pulse shape the pulse duration is 184 fsec and the time bandwidth product is 0.4. However, these pulses are very sensitive to a cavity length detuning in the order of 1 μm and stable operation requires adjustment typically every several minutes.

The laser performance for different cavity length detunings changes dramatically when a weak feedback is applied. The most striking feature is that clean mode-locking is achieved in a region of resonator length (from -10 μm to + 30 μm) where before no or only unsatis-factory results were obtained. Further the autocorrelation profiles experience no deterioration and their shape and width are almost identical within this region. This insensitivity to cavity length varia-tions of up to 40 μm allows an extremely stable laser operation over many hours without any readjustment of the laser itself (although the pump laser has to be readjusted). Another important aspect is that the application of the feedback results in a further pulse shortening and mode locking im-

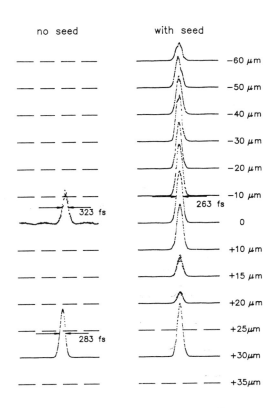

Figure 6. The autocorrelation traces without (left) and with CPS (right) of the hybridly mode-locked Styryl 9 dye laser pulses for different cavity length settings. The feedback level is 1.2X10³ and the seed pulse is delayed by 10 ps with respect to the main pulse.

provement. The autocorrelation trace at position 0 shown in Fig.6 (right) is 263 fs wide and its corresponding spectrum width is 4.83 nm. Assuming a sech² pulse shape the pulse duration is 170 fs and the time bandwidth product is 0.36 which is very close to the ideal value of 0.32.

Similar to the case of CPS in pure SPML lasers, the above mentioned effects are observed only for certain intensity levels of the feedback. The feedback level must lie within the range from 2X10⁻³ to 5X10⁻⁴. The insensitivity to cavity length variations and the pulse shortening described above are observed when the feedback pulse is reinjected into the main cavity 300 fs up to 30 ps behind the master pulse. At first sight these results seem to contradict our interpretation of the CPS technique given in the first paragraph of the paper. There we pointed out that the feedback pulse must be reinjected slightly in advance of the master pulse. We believe that the loss caused by the nonlinear mirror considerably modifies the relative moving of the pulse profile with respect to the noise background. Thus, in contrast to SPML lasers, the moving of the pulse profile with respect to the noise background is backward in the hybridly mode locked laser at its optimum operation. The feedback pulses influence the laser in such a way that they are amplified and eventually evolute into the master pulse from the rear. The phenomena might be general for CPS application in hybridly mode locked lasers.

5. CONCLUSIONS

We have shown that the CPS technique can be successfully used to stabilize synchronously pumped lasers both without and with saturable absorber in the cavity. Particularly, for hybridly mode-locked system the technique leads to an arrangement that is unexpectedly uncritical in adjustment and remains in an extremely stable condition for many hours without any readjusting. In addition, we demonstrated an integrated bulk semiconductor nonlinear mirror which can be used as a saturable absorber inside the laser cavity.

References

Catherall J M and New G H C 1986 *IEEE J. Quantum Electron.* **QE-22** 1593
Kluge J, Wiechert D, and von der Linde D 1984 *Opt. Commun.* **51** 271
Beaud P, Bi J Q, Schütz J, Hodel W, and Weber H P 1990 *Ultrafast Phenomena VII - Springer Series in Chemical Physics* **53** (Berlin: Springer Verlag) pp 23-25, *Opt. Commun.* **80** 31
Bi J Q, Hodel W, and Weber H P 1991 *Opt. Commun.* **81** 408
Hooker C J, Lister J M D, and I. N. Ross I N 1990 *Opt. Commun.* **80** 375
N. McCarthy and D. Gay, *Opt. Lett.* **16** 1004
New G H C 1990 *Opt. Lett.* **15** 1306
Peter D S, Beaud P, Hodel W, and Weber H P 1991 *Opt. Lett.* **16** 407

Femtosecond pulse generation in a cw pumped passive mode-locked linear rhodamine 6G—DODCI dye laser

A Penzkofer and W Bäumler

Naturwissenschaftliche Fakultät II – Physik, Universität Regensburg,
D-W-8400 Regensburg, Fed. Rep. Germany.

ABSTRACT: The femtosecond pulse generation in a cw pumped linear passive mode-locked rhodamine 6G - DODCI dye laser is studied. The laser is operated without and with a prism pair. The DODCI concentration is varied and the absorber jet is detuned from the CPM postion. The influence of the prism pair positioning is investigated. A fast partial absorption recovery of DODCI is necessary for sufficient background suppression. With the prism-pair balanced oscillator soliton-like stable pulses of 50 fs duration were generated independent of detuning the absorber jet out of the CPM position.

1. INTRODUCTION

The femtosecond pulse generation in cw laser pumped colliding pulse mode-locked (CPM) ring dye lasers is well established (Shank 1988). Antiresonant ring linear CPM dye lasers were operated with similar performance data as ring CPM dye lasers (Diels 1990). Here a cw argon ion laser pumped linear rhodamine 6G - DODCI femtosecond dye laser is investigated experimentally and theoretically. The schematic experimental setup is shown in Fig.1. Colliding pulse mode-locking is achieved by placing the DODCI loss jet in the center of the resonator. The laser is detuned from the CPM - position by moving mirror M1.

2. THEORETICAL ANALYSIS

The pulse development and background suppression in multiple transits through the saturable absorber is illustrated in Fig.2. The pulse shaping action of the saturable absorber and of the gain medium are considered. The dotted line shows the input pulse shape. The saturable absorption dynamics of DODCI is discussed by Penzkofer and Bäumler (1991). At the laser wavelength of $\lambda_L = 620nm$ thermally elevated N-isomers and P-isomers of DODCI contribute to the absorption. The $S_1 - S_0$ relaxation times are $\tau_N = 1.3ns$ and $\tau_P = 1.4ns$ (Bäumler and Penzkofer 1990). A fast partial absorption recovery (time constant τ_{rec}) is present in both isomers, because in the P-isomers there occurs a fast relaxation ($\tau_{rec} = 0.95ps$, Angel et al 1989) out of the populated Franck Condon state, and in the N-isomers there occurs a fast refilling of the emptied S_0-state by spectral cross-relaxation. In Fig.2 the solid curve is calculated for $\tau_{rec} = 1ps$ while the dashed curve is calculated for $\tau_{rec} = 10ns$ (100 round-trips, small-signal transmission of

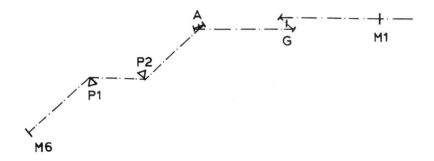

Fig. 1. Schematic laser arrangement.

absorber $T_0 = 0.95$). The fast partial absorption recovery of the slow saturable absorber is necessary for sufficient background suppression and the formation of femtosecond pulse trains.

The steady-state pulse duration is determined by equating the pulse shortening and pulse broadening within a single round-trip. Pulse shortening is caused by saturable absorption of the mode-locking dye. The gain depletion of the lasing dye acts slightly pulse broadening. Without a prism pair in the resonator the pulse broadening is caused mainly by the positive group velocity dispersion (GVD) of the positive self-phase modulated (SPM) pulses in the gain and loss jet. The solvent contributions (ethylene

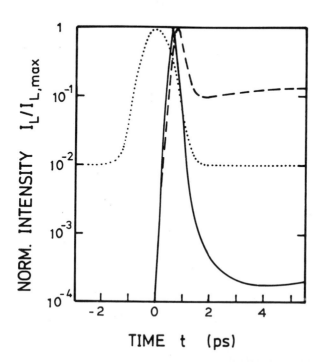

Fig. 2. Transient pulse shortening and background suppression.

glycol are dominant for the GVD (dispersion of refractive index) and the SPM (optical Kerr effect, $n_2 > 0$). With a prism pair in the resonator the laser performance depends on the prism separation (30 cm in our case) and on the apex heights h_1 (0.68 mm in our case) and h_2 of the laser beam passing through the prisms. The behaviour is illustrated in Fig.3. The spectral transit time lags per round-trip for the prism pair, $\partial t_{pr}/\partial \tilde{\nu}_L$, and of the total resonator, $\partial t_{tr}/\partial \tilde{\nu}_L$, including the GVD of the gain and loss jet (thicknesses $d_A = 35\mu m$ and $d_G = 250\mu m$) are shown by the dashed and solid curves in Fig.3a, respectively. The laser pulse duration is given by

$$\Delta t_L \approx |\ \Delta t_{ini} + \delta t_{tr} - \delta t_{AG}\ |$$

where Δt_{ini} is the inital pulse duration, $\delta t_{tr} \approx (\partial t_{tr}/\partial \tilde{\nu}_L)\Delta\tilde{\nu}_L[ln(\rho_1^{-1})]^{-1}$ is the total transit time lag (ρ_1 is output mirror reflectivity), and δt_{AG} is the combined temporal pulse shortening of the absorber and the gain medium. The spectral width of the laser is given by $\Delta\tilde{\nu}_L \approx [(\Delta\tilde{\nu}_L^{bwl})^2 + (\delta\tilde{\nu}_c)^2]^{1/2}$ where $\delta\tilde{\nu}_c$ is the frequency chirp caused by the SPM ($\delta\tilde{\nu}_c \propto \Delta t_L^{-1}$).

The steady-state condition requires $\Delta t_L = \Delta t_{ini}$. Three regions have to be distinguished: (i) In the soliton-like pulse formation region I (Martinez et al 1985) the negative GVD interacts with the positve SPM and it is $\delta t_{tr} = -2\Delta t_L + \delta t_{AG}$. Stable femtosecond pulse trains are generated. At the prism-pair balanced position h_{2m} it is $\Delta t_{ini} = \Delta t_{min} \approx 0.5/\Delta\nu_{AMP}$, where $\Delta\nu_{AMP}$ is the spectral width of the amplification profile. The shortest stable pulses are generated.
(ii) In region II the pulse compressive negative GVD tries to generate pulses of $\Delta t_L < \Delta t_{min}$ and $\Delta\tilde{\nu}_L > \Delta\tilde{\nu}_{AMP}$. The laser falls below threshold and restarts again (self-quenching laser operation). At $h_2 = h_{2c}$ it is $\delta t_{tr} = 0$.
(iii) In region III the pulse broadening by the positive group velocity dispersion is counteracted by the saturable absorber pulse shortening, i.e. $\delta t_{tr} = \delta t_{AG}$. The laser stability is moderate.
In the transient femtosecond pulse formation start-up process the initial pulse duration is approximately given by $\Delta t_{ini} = \Delta t_{min}$ (duration of statistical fluctuating spontaneous emission, Glauber 1972). It is modified to the steady-state value by the laser build-up process.

3. EXPERIMENTAL STUDIES

Without the prism pair in the resonator pulse durations of 140 fs were obtained for a DODCI small-signal transmission of $T_0=0.97$ at the lasing wavelength of $\lambda_L = 620nm$. Detuning the loss jet from the center position resulted in the formation of trailing pulse tails of duration up to 900 fs.

The dependence of the femtosecond laser pulse duration and spectral width on the prism pair positioning is illustrated in Fig.3b and 3c for $T_0=0.97$. At the prism-pair balanced position h_{2m} stable pulses of 50 fs duration were generated ($\lambda_L = 620nm$) independent of the absorber jet detuning from the resonator center over distances of many centimeters. Decreasing the DODCI concentration from $3 \times 10^{-4}mol/dm^3$ to $3 \times 10^{-5}mol/dm^3$, the pulse duration increased from 50 fs to 110 fs and the peak laser wavelength shifted from 620 nm to 610 nm.

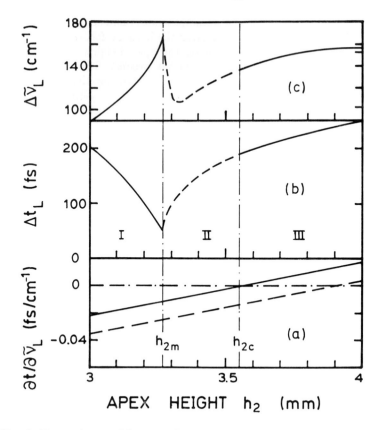

Fig. 3. Dependence of laser performance on prism pair positioning.

The advantages of the described linear femtosecond dye laser compared to the ring CPM dye laser are its easy alignment and the reduced number of optical components.

REFERENCES

Angel A, Gagel R and Laubereau A 1989 *Chem. Phys.* **131** 129

Bäumler W and Penzkofer A 1990 *Chem. Phys.* **142** 431

Diels J C 1990 *Dye Laser Principles with Applications* eds F J Duarte and L W Hillman (Boston: Academic Press) pp 41-132

Glauber R J 1972 *Laser Handbook, Vol. 1* eds F T Arecchi and E O Schulz-DuBois (Amsterdam: North Holland) Ch A1

Martinez O E, Fork R L and Gordon J P 1985 *J. Opt. Soc. Am.* **B2** 753

Penzkofer A and Bäumler W 1991 *Opt. Quant. Electron.* **23** 439

Shank C V 1988 *Ultrashort Laser Pulses and Applications* ed W Kaiser (Berlin: Springer-Verlag) pp 5-34

Prismatic pulse compressor for synchronously pumped mode-locked lasers

K Osvay, Z Bor, A Kovács, G Szabó, B Rácz, H A Hazim and O E Martinez[1]

Department of Optics and Quantum Electronics, József A. University, Dóm tér 9, H-6720 Szeged, Hungary
[1]Departamento de Fisica, Comision Nacional de Energia Atomica Av. del Libertador 8250, 1429 Buenos Aires, Argentina

ABSTRACT: In a synchronously pumped mode locked laser the group velocity dispersion due to a prismatic pulse compressor can be continuously altered while the cavity round trip time is being kept constant. This effect is achieved by moving one of the prisms along a special direction whose existence is proved both experimentally and theoretically.

1. INTRODUCTION

The requirements which should be fulfilled for the operation of synchronously pumped mode locked dye lasers (SPML) are the precise adjustment of cavity length and the introduction of a given amount of negative group velocity dispersion (NGVD) into the laser cavity. NGVD obtained by a pair of Brewster prisms (see inset of figure 1) is usually adjusted by translating one of the prisms (Martinez *et al* 1984, Taylor *et al* 1989, Angel *et al* 1989) along the axis normal to its base (direction X). Since both the transit time and NGVD are changed by this method simultaneously, the optimization of SPML is only successively possible.

In this paper the existence of a direction is demonstrated, along which translating one of the prisms the NGVD can be adjusted without changing the transit time.

2. EXPERIMENTS

The effect of prism translation along two perpendicular directions X, Y on the NGVD and on the transit time was investigated (see figure 1) using the time of flight interferometer (TOFI) detailed by Bor *et al* (1989). The pulse compressor is composed of a pair of 60° fused silica prisms oriented at minimal deviation angle. The TOFI was illuminated by a broadband ($\delta\lambda \simeq 10$nm) dye laser pumped by a N_2-laser. In our experiments the relative changes of the transit time were measured.

The measurements, whose results can be seen in figure 2a and b, carried out at the wavelength of 620nm have revealed that the change of the transit time ΔT caused by shifts into X and Y directions can be described by

$$\Delta T = a \cdot X + b \cdot Y \qquad (1)$$

where $a = 2010$fs/mm and $b = 39.14$fs/mm.

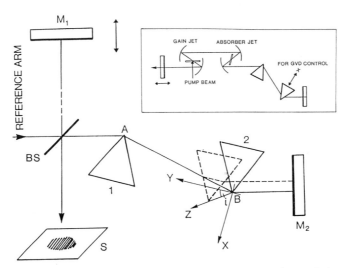

Figure 1. TOFI used for measuring the transit time of the pulse through the compressor (BS: beamsplitter, M1, M2: mirrors, S:screen, $\alpha=60°$, quartz prisms; AB=50.7cm). The inset shows a SPML incorporating a prismatic compressor.

It means, there is a certain direction Z (called isochronic, defined by $tg(i)=X/Y=-b/a$, see figure 1), along which the transit time remains constant. Note that the isochronic direction is nearly parallel to the base of the prism, that is, the translating direction X which is commonly used by e.g. Martinez *et al* 1984, Dawson *et al* 1986 and Taylor *et al* 1989 is a very

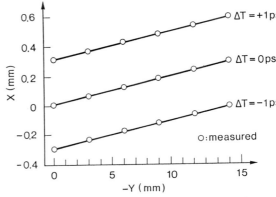

Figure 2a. The lines of the equal transit times vs the position of the prism 2.

unfavourable one because it gives rise to a large change in the transit time of a pulse. Figure 3 shows the wavelength dependence of the transit time which was measured by tuning the dye laser. It can be seen in this figure, while the prism 2 is shifted along the isochronic direction (curves 1,2,3), at $\lambda_0=620$nm the propagation time is not modified but the NGVD, which is given by the slopes of the curves, is changing. Moving

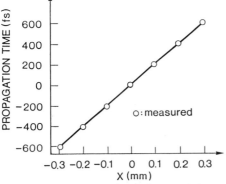

Figure 2b. The dependence of the propagation time on the shift of the prism 2 along the X axis.

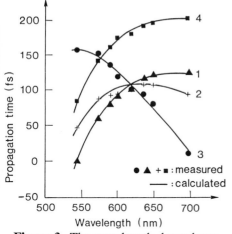

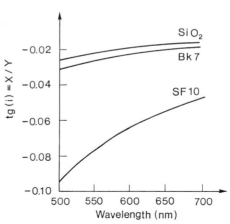

Figure 3. The wavelength dependence of the transit time in 4 different positions of the second prism. (1-3: ΔZ=0mm, 4.201mm, 15.003mm, resp.; 4: ΔX=0.04mm)

Figure 4. Isochronic direction vs wavelength for compressors consisting of Brewster prisms made of fused silica, BK7 and SF10.

the prism 2 along the X direction (curve 4) the transit time is different from that of curve 1, but the values of NGVD are nearly the same.

3. CALCULATIONS

The corresponding computations were carried out using geometrical optical ray tracing. As it can be seen in figure 3, the agreement between the measured and calculated data is excellent.

Detailed calculation showed, that for a pair of Brewster prisms, which is of special interest in practice, the isochronic direction can be given by

$$tg(i) = X/Y = \frac{(n^2 - 1) \cdot dn/d\lambda}{n^2 - 1 - 2 \cdot n \cdot \lambda \cdot dn/d\lambda} \qquad (2)$$

where $dn/d\lambda$ is the dispersion of the material of the prism.

The derivation of the equation (2) is rather lengthy, it requires the calculation of the derivatives of the propagation time over the parameters X and Y. The validity of equation (2) was also checked by numerical ray tracing as described above. Figure 4 shows $tg(i)$ as a function of the wavelength for pairs of Brewster prisms made of fused silica, BK7 and SF10. For practical purposes $tg(i)$ can either be determined by using eq.2 or it can directly be measured.

4. CONCLUSIONS

We have pointed out that constant transit time and controllable group velocity dispersion can be achieved in synchronously pumped mode locked lasers containing intracavity prismatic pulse compressor, by moving one of the prisms along a special direction (isochronic direction). For practical cases this direction is nearly parallel to the base of the prism. The isochronic direction can be simply measured or calculated by using equation 2. The existence of the isochronic direction can greatly facilitate the adjustment of synchronously pumped mode locked lasers.

ACKNOWLEDGEMENTS

This publication based on work sponsored by the Hungarian-U.S.Science and Technology Joint Fund in cooperation with National Science Foundation and Hungarian National Academy of Sciences under Project 061/90. This report has been also supported by the OTKA Foundation of the Hungarian National Academy of Sciences under Grant no. 3055.

Angel G, Gagel R, Laubereau A 1989 *Opt.Lett.* **14** 1005
Bor Zs, Gogolak Z, Szabó G 1989 *Opt.Lett.* **14** 86
Dawson M D, Bogess T F, Garvey D 1986 *Opt. Comm.* **60** 79
Martinez O E, Gordon J P, Fork R L 1984 *J.Opt.Soc.Am.* **A1** 1003
Taylor A J, Roberts J P, Gosnell T R, Lester C S 1989 *Opt.Lett.* **14** 444

Inst. Phys. Conf. Ser. No 126: Section I
Paper presented at Int. Symp. on Ultrafast Processes in Spectroscopy, Bayreuth, 1991

A simple technique for generation of frequency shifted femtosecond pulses

D. Grosenick, F. Noack, F. Seifert, B. Wilhelmi [*]

Zentralinstitut für Optik und Spektroskopie, Rudower Chaussee 6,
O-1199 Berlin, Germany
[*]Jenoptik Jena GmbH, O-6900 Jena, Germany

ABSTRACT: Fs pulses at a new wavelength are generated in a short dye cell pumped by amplified fs pulses of a CPM ring laser. The generation process and the influence of various parameters on the evolution of these pulses are investigated by spectral and temporal measurements.

1. INTRODUCTION

Femtosecond excite and probe techniques are widely used for investigations in many fields of science, material research and communication technology. Most of the applications require strong and synchronous fs pulses at two different wavelengths. For this reason there is an increasing interest in simple and efficient methods for wavelength conversion of femtosecond pulses.

Frequency conversion by fs-continuum generation and subsequent reamplification of the desired part of the white light continuum (WLC) is one way to get fs pulses at different wavelengths, which is from the experimental point of view rather complex and expensive (Migus et al 1985, Heist et al 1990).

Another approach uses travelling wave amplification (TWA) of amplified spontaneous emission (ASE) (Hebling and Kuhl, 1989) or WLC (Klebniczki et al 1990). These travelling wave arrangements are normally pumped transversally which leads to an inhomogeneous spatial profile of the amplified pulse. For generation of very short pulses by TWA one has to give special attention to exact velocity matching of the pump pulse and the amplified signal.

We have investigated the evolution of femtosecond light pulses in a 1 mm long dye cell, longitudinally pumped by amplified fs light pulses from a CPM laser which proves to be a further simplification compared to typical TWA arrangements. This simplification is paid by the loss of tunability.

Pumping the dye cell not only excitation of the dye molecules but also continuum generation takes place. A small part of the continuum generated near the entrance is reflected from the output window back into the cell and provides a seed pulse at the dye fluorescence wavelength. After one roundtrip time we get an amplified pulse at the wavelength determined by the dye with about the same time duration as the input pulse. Depending on dye concentration and focussing conditions further pulses are more or less pronounced.

2. EXPERIMENT

Figure 1 shows the experimental scheme. The fs pulses of a CPM ring laser are amplified in a two stage amplifier pumped by a Nitrogen laser NIL 1000 (6 mJ, 6 ns, 337 nm). The output pulses have a duration of about 100 fs and an energy of nearly 5 μJ. These pulses we splitted into two equal parts which were used to pump the dye cell and to measure the duration of the frequency shifted pulses generated in the dye cell by sum frequency generation, respectively.

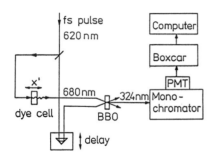

Fig. 1. Experimental setup

The dye cell was a 1 mm thick quartz glass cuvette filled with various solutions of Oxazine 1 or Rhodamine 700 in ethylene glycol, methyl alcohol or ethyl alcohol. The cell windows are about 1 mm thick. The pump pulse was focused by a lens with 16 cm focal length. The beam waist was measured to be 35 μm. To vary the diameter of the pump pulse and the pump intensity the dye cell was mounted on a translation stage.

3. RESULTS AND DISCUSSION

The measurements have shown that the temporal structure of the new radiation generated in the dye cell depends first of all strongly on the focussing conditions. Figure 2 shows the temporal behaviour of the new signal for the dye cell arranged some millimeters before the focus of the pump beam (fig. 2a.), in the focus (fig. 2b.) and behind the focus (fig. 2c.). In this case (low concentration solution, 5×10^{-4} molar of Oxazine in ethylene glycol) an emission wavelength around 680 nm was obtained.

For cell position in the focus only a single pulse at zero time delay appears. Our investigations have shown that this pulse originates from the continuum generated in the solvent and in the dye cell windows.

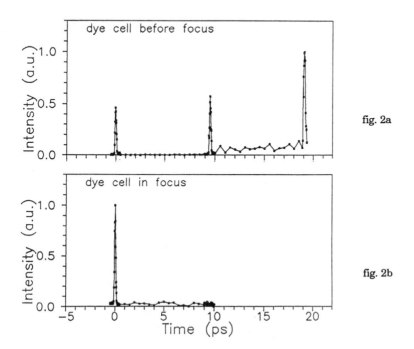

fig. 2a

fig. 2b

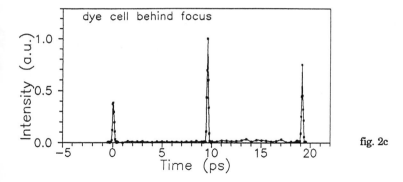

fig. 2c

Fig. 2. Temporal behaviour of the signal

With the dye cell some millimeters outside the focus a group of fs pulses is generated. The time intervals between adjacent pulses correspond to the roundtrip time inside the dye cell. A very small portion of the pulse leaving the dye cell at time zero is retroreflected at the inner dye cell surface (reflectivity about 9×10^{-5} for ethylene glycol) and amplified along the dye cell length, then it is again slightly reflected and amplified and leaves the dye cell after one roundtrip. If the amplification is not essentially reduced by gain depletion this process goes on resulting in further pulses.

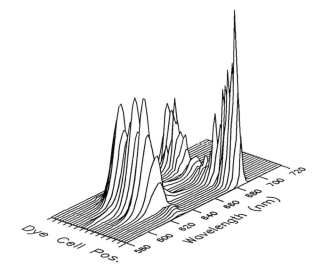

In figure 3 the signal spectrum and the transmitted pump light are depicted as functions of the dye cell position with respect to the focus. Near the focus the pump light absorption is saturated and a considerable part of the pump pulse passes the cell. Also a broadening of the spectrum around 620 nm due to continuum generation is visible. For the dye cell standing before or behind the focus most of the pump pulse energy is absorbed by the dye and there is a strong signal emission around 680 nm (for Oxazine 1). For Rhodamine 700 the signal emission occurs around 700 nm.

Fig. 3. Signal spectrum and transmitted pump light for different dye cell positions

The feedback at the inner dye cell surfaces can be increased using other solvents. With ethyl alcohol we get a reflectivity of 1×10^{-3}. In comparison with ethylene glycol as solvent the pulse occurring after one roundtrip is increased. A still higher reflectivity can be got by methyl alcohol (reflectivity 2×10^{-3}). In this case we observed ASE between the main signal peaks.

The height of the signal pulses also depends on the dye concentration. Increasing the concentration the peak appearing after one roundtrip becomes larger. For a 1.6×10^{-3} molar solution of Oxazine 1 in ethyl alcohol we got a peak ratio 1:7:1 for the first three peaks.

Slightly tilting the dye cell the peak at time zero has the same direction as the pump pulse whereas the next pulses are emitted perpendicular to the dye cell windows. This gives the possibility to select the pulses of interest.

A typical cross correlation function of the signal pulses emitted by the dye cell is given in figure 4. This pulse was generated using Rhodamine 700 in ethylene glycol. A slight asymmetry is remarkable. The FWHM of this cross correlation function corresponds to the pump pulse autocorrelation function.

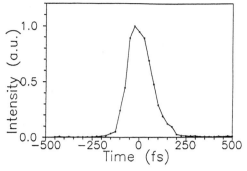

Fig. 4. Cross correlation function of the signal pulse

4. CONCLUSIONS

We have demonstrated a simple method for the generation of wavelength shifted fs pulses. The dye cell output consists of a group of fs pulses with a distance between adjacent pulses corresponding to the dye cell roundtrip time. The peak after one roundtrip can be enlarged by changing feedback and dye concentration. A further improvement should be achievable by optimizing the arrangement (focus parameter, concentration, sample thickness, reflectivity).

A variation of the wavelength is possible by using different dyes. As the process under consideration needs a seed pulse the wavelength range is limited by the continuum generation in the dye cell.

5. REFERENCES

Hebling, J. and Kuhl, J. 1989 Opt. Commun. **73** 375
Heist, P., Rudolph, W., and Wilhelmi, B. 1990 Exp. Tech. Phys. **38** 163
Klebniczki, J., Hebling, J., and Kuhl, J. 1990 Opt. Lett. **15** 1368
Migus, A., Antonetti, A., Etchepare, J., Hulin, D., and Orszag, A. 1985 J. Opt. Soc. Amer. B **2** 584

Inst. Phys. Conf. Ser. No 126: Section I
Paper presented at Int. Symp. on Ultrafast Processes in Spectroscopy, Bayreuth, 1991

Ultrashort pulse generation in a DFB-laser with a saturable absorber

A A Afanas'ev, M V Korol'kov, T V Veremeenko

Institute of Physics, BSSR Academy of Science,220602 Minsk, USSR

ABSTRACT: A quantitative experiment has been performed to investigate the nonstationary regime of operation of a laser based on a mixture of two dyes (active and absorbing) with distributed feedback formed by spatially periodic pump.The region of values of DFB- laser pump power has been determined in which an energy- and duration-stable regime of generation of a single frequency-tuned ultrashort light pulse is attained. It is shown that for typical values of the pump parameters and dye mixture the generation pulse duration is shorter by a few times than the length of the pulse generated by a DFB-laser whose active medium contains no saturating absorber.

1. INTRODUCTION

Distributed   feedback (DFB) dye lasers with spatial-periodic pumping are able to transform the nanosecond pump pulses to generation pulses with duration $\geq$ 10 ps. However, sufficiently active spectral-limited pulses of picosecond duration are generated only in a relatively narrow range of pump energy near the double-spike generation threshold, the energy and time instabilities of such generation mode as had established in the works of Rubinov et al (1987) and Bor et al (1986) being not high.The possibility of getting rid of the subsequent spikes in the generation pulse by using shifted in time additional pulses at the pump as had proposal by Bor et al (1986) or generation frequency (Raksi 1988), as well as the use of a ring cavity as in the work of Afanas'ev et al(1989) with that end in view greatly complicates the laser system and diminishes its efficiency.

In the present paper, the possibility is demonstrated of stabilization of the time and energy characteristics of a DFB- laser, as well as widening of the dynamic range of pump power for single- spike generation with additional shortening of the spike duration due to the introduction the solution of the active medium of the saturable absorber at the frequency of generated radiation.

2. BASIC EQUATIONS

The numerical analysis of the generation kinetics of a DFB- laser based on a mixture of two dyes – active (A) and absorbing (B) - at the generation frequency is based on the equations of the one- dimensional model developed by Apanasevich et al (1987 ,1989)[*] :

$$\left(\pm \frac{\partial}{\partial x} + \mu \cdot \frac{\partial}{\partial t}\right) \cdot E_{\pm} = G \cdot {}_a L \left[\left(y_o^a - \gamma \cdot \varkappa \cdot y_o^b\right) \cdot E_{\pm} + \left(y_{\pm1}^a - \gamma \cdot \varkappa \cdot y_{\pm1}^b\right) \cdot E_{\mp}\right], \quad (1)$$

$$\frac{\partial y_m^a}{\partial \tau} = H_o \cdot \delta_{om} + \eta \cdot H_o \cdot \delta_{1m}/2 - A_o \cdot y_m^a - A_+ \cdot y_{m+1}^a - A_- \cdot y_{m-1}^a, \quad (2)$$

$$\frac{\partial y_m^b}{\partial \tau} = \xi \cdot \delta_{om} - B_o \cdot y_m^b - B_+ \cdot y_{m+1}^b - B_- \cdot y_{m-1}^b, \qquad m=0,\pm1,\dots , \quad (3)$$

where $E_{\pm}$ - generation wave amplitudes normalized to the saturating amplitudes, y_m^a, y_m^b are amplitudes of population grating of working transition of the active and absorbing dyes, $H_o = W_o T_1^a$, $W_o(\tau)$ is the pump rate amplitude, η is the visibility of the pump field, δ_{jm} is the Kroneker symbol, $x = z/L$ and $\tau = t/T_1^a$ are normalized coordinate and time, L is active medium length ; $\mu = L/(V \cdot T_1^a)$; $G_a \cdot y_o^a$ - is the unsaturated gain $(y_m^{a,b})^* = y_{-m}^{a,b}$. In equations (1)-(3) the following notation is used:

$$A_o = 1 + H_o + \left(|E_+|^2 + |E_-|^2\right) ,$$

$$A_+ = A_-^* = \eta \cdot H_o/2 + E_+^* \cdot E_- ,$$

$$B_o = \xi + \gamma \cdot \left(|E_+|^2 + |E_-|^2\right) ,$$

$$B_+ = B_-^* = \gamma \cdot E_+^* \cdot E_- ,$$

$\varkappa = N_b/N_a$ - ratio of concentration of dye molecules of the mixture,

$\gamma = \sigma_{41}^b/\sigma_{32}^a$ -ratio of the sections of the corresponding transitions,

$\xi = T_1^a/T_1^b$ -ratio of the relaxation times of population of dyes molecules.

Equation (1) - (3) have been obtained under the assumption that DFB is due to the first-order Bragg diffraction on the main grating provided that $\omega = \omega_o = \omega_{32}^a = \omega_{41}^b$ and $t_p, t_r, T_1^{a,b} \gg T_2^{a,b}$ where ω is the generation frequency , ω_o is the fundamental Bragg frequency, ω_{32}^a and ω_{41}^b are the frequencies of the working transition of molecules of the dyes (A and B),

---

[*] The influence of the absorbing dye on the DFB- laser generation kinetics was discussed by Hebling (1988) from the point of view of the concentrated model.

t_p and t_r are pump and generation pulse durations, $T_2^{a,b}$ are the transverse relaxation times.

The boundary and initial conditions of equations (1) – (3) are of the form:

$$E_\pm(\mp 0.5) = E_{\pm o}(\tau); \quad y_m^a(\tau=-\infty) = 0; \quad y_m^b(\tau=-\infty) = \begin{cases} 1, m=0 \\ \\ 0, m\neq 0. \end{cases} \quad (4)$$

3. RESULTS OF THE NUMERICAL EXPERIMENT

As the result of the numerical solution of equation (1)-(3), it is shown that the DFB-laser generation kinetics strongly depends on the parameters of the absorbing dye and the properties of pumping. Fig.1 shows the time dependencies of the generation intensity and normalized coefficients of amplification (dye A) and absorption (dye B) for the pump power of Gaussian form $H_o(\tau) = H_o \exp\{-(\tau/\tau_p -2)^2\}$ at $\tau_p=0.4$, $\xi=13.3$, $G_a L=42$, $\gamma=5$, $\mu=3.6 \cdot 10^{-3}$, $H_o = 0.195$, $E_{\pm o}=10^{-3}$ where $\tau_p = t_p/T_1^a$. For comparison Fig.1 gives analogous dependencies of the characteristics of a DFB- laser whose active medium contains no saturating absorber, i.e at $\kappa = 0$. From comparison of the curves of Fig.1 it follows that the saturating absorber changes the DFB- laser generation as follows:

    -the number of generation spikes decreases;

    -their peak power significantly increases and duration decreases;

    -the time interval between the neighboring spikes increases and the dynamic range of pump powers for generation of a single pulse widens.

In the region of parameters under consideration, the saturation of the active medium by the pump pulse is determined by its energy ($t_p < T_1^a$). And, depending on the generation pulse duration, different regimes of saturation of the active and absorbing dyes are possible. So, in the case of $t_r < T_1^{a,b}$ the saturation of dye B

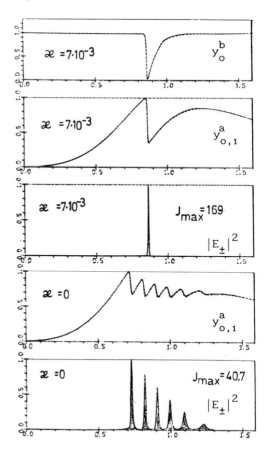

Fig. 1. The dependence of the generation kinetics on the concentration of dye "B".

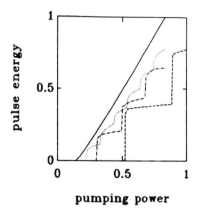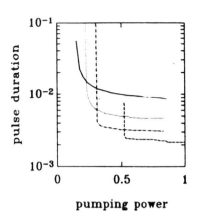

Fig. 2. The variations in the pump energy dependences of the energy generation and first spike duration with growing absorber concentration $\mathcal{x}$ = 0 (——), 0.007 (····), 0.014 (–·–·–), 0.03 (–––––).

absorption and dye A amplification is determined by the generation pulse energy. In the intermediate region $T_1^b < t_r \ll T_1^a$ (which is possible at $\xi \gg$ 1), the saturation of dye B absorption is determined by its instantaneous power.

The relations of the active and absorbing dyes parameters have been established when both are achieved the generation mode stable in energy and duration and the additional two- and three-fold reduction of the spike, generation duration as compared to the absorber – free DFB-laser. The 200 times total reduction of pulse duration was achieved. The Fig.2 illustrates these variations in the pump energy dependencies of the energy generation and first spike duration with growing absorber concentration.The sharp jumps in the dependence of output radiation energy on the pump power are due to the excitation of the subsequent generation spikes.

Thus, the introduction of an absorbing due into the active medium of a DFB- laser leads to additional Q-switching of the distributed Bragg cavity, which significantly changes the generation kinetics. As a result, at the certain parameters of amplifying and absorbing dyes in a wide range of pump power variation the DFB- laser can serve as a stable source.

Afanas'ev A A, Zaporozhchenco V A, Kachinskii A V, Korol'kov M V and Chekhlov O V 1989 Kvant.Elektr. **16** 1827
Apanasevich P A, Afanas'ev A A and Korol'kov M V 1987 IEEE J. Quantum. Electron. **QE23** 533
Apanasevich P A, Afanas'ev A A, Korol'kov M V and Veremeenko T V 1989 Izv. AN SSSR, ser.fiz.**53** 1026
Bor Z, Muller A 1986 IEEE J. Quantum Electron. **QE22** 1524
Hebling J 1988 Appl.Phys. **B47** 267
Raksi F 1988 Appl.Phys. **B47** 91
Rubinov A N, Bushuk B.A.and Berestov A L 1987 .Kvant.Elektr. **14** 906

Passively mode-locked dye laser with spatial dispersion in the gain medium

N. I. Michailov

Faculty of Physics, Sofia University, 1126 Sofia, Bulgaria

ABSTRACT: The advantages of mode-locking with spatial dispersion in the gain medium are demonstrated. The technique is applied to a passively mode-locked dye laser. Under the same experimental conditions pulse reduction by an order of magnitude is achieved as compared to the standard case of no dispersion. Pulses shorter than 100 fs are generated in a simple linear resonator at pump powers far above the threshold. Wavelength tuning is obtained over a range of about 10 nm.

1. INTRODUCTION

The mode-locked lasers are nowadays the most convenient tool to generate short light pulses. The mode-locked systems in widespread use as dye and solid-state lasers have a homogeneously broadened gain medium. The physical mechanisms that determine the limit of achievable pulse duration in the various mode-locked lasers are of different nature. Nevertheless, the longitudinal mode-competition in the case of a homogeneous broadening is the most common mechanism that restricts the oscillating bandwidth and counteracts the pulse shortening caused by the mode-locking element. On the contrary, in the case of an inhomogeneous laser more or less the full oscillating spectral width can be easily mode-locked by only applying sufficient modulation signal.

It was recently shown by Danailov and Christov (1989, 1990) that the longitudinal mode competition in a homogeneous laser can be strongly reduced if the spectral components of the radiation are spatially dispersed in the gain medium. The result is a significant broadening of the oscillating spectrum of a free-running laser as it has been demonstrated for a pulsed dye laser. This effect might be expected to favor the mode-locking mechanism in a homogeneous laser resembling the mode-coupling behavior in the inhomogeneous case. This may allow in particular more effective exploitation of the laser gain bandwidth towards generation of shorter pulses. Promising candidates are the lasers generating pulses of duration well bellow the limit of the inverse gain bandwidth (e.g. synchronously pumped dye lasers and broadband solid-state lasers). Here we report on experimental demonstration of the advantages of mode-locking with spatial dispersion in the gain medium (MLSD) as employed in a passively mode-locked dye laser. It is shown that this technique can be used to improve the mode-locking performances of lasers which anyway generate extremely short pulses approaching the limit of the inverse gain bandwidth such as dye lasers mode-locked by a saturable absorber.

2. CAVITY DESIGN

Spatial dispersion in the gain medium of a c.w. dye laser can be created within the conventional resonator concept (Michailov submitted). One possible configuration is shown in Figure 1. A standard folded four mirror asymmetric cavity is composed of the curved mirrors M2, M3 and two plane mirrors M1 and M4. The prism P1 located at a distance *l* from one of the folding mirrors creates angular dispersion of the intracavity radiation. The cavity can be specifically designed so that the spectral components diverging approximately from the apex of the prism (an aperture is placed as shown in the figure) are collimated by the folding mirrors. At the same time the beam waist position of an intracavity mode (where the gain jet is located) is away from the intersection point C of the spectral components. Thus, the different spectral components are brought in shifted spatial regions in the gain medium and the "spatial chromaticity" of the resonator is preserved after one round trip. The cavity configuration is illustrated in more details in Figure 1b where part of the equivalent unfolded scheme with intracavity lenses is shown.

Referring to Figure 1, a passively mode-locked dye laser configuration can be arranged if the absorber mirror section given in the inset *I* in Figure 1 is inserted in the long leg of the cavity so far considered. According to the design approach suggested by Michailov (1988) the cavity can be properly rearranged so that the beam properties in the gain folded section can remain unaltered. The prism pair P2P3 given in the inset *II* could also be added for correction of the group velocity dispersion (GVD) and to compensate for the transverse displacement of the radiation.

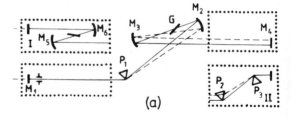

(a)

3. THE MODE-LOCKED LASER

The passively mode-locked dye lasers generate extremely short pulses (below 100 fs) in the well known colliding pulse mode-locking (CPM) ring configuration. For pulses in the fs-time domain the combined action of the intracavity self- phase modulation and GVD has been demonstrated to play an important role in the pulse shortening. The nonlinearities associated with the exploitation of this shaping mechanism ultimately limit the achievable pulse durations (Haus and Silberberg 1986). The primary purpose of this work was to demonstrate how the reduction of the longitudinal mode competition can be used to favor the basic mode-locking mechanism (pulse shaping by saturable gain and loss in the passive case).

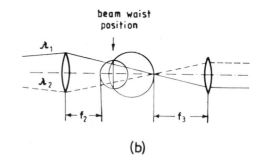

(b)

Fig. 1. (a) Dispersive resonator schemes of a CW dye laser (M1 P1 M2 M3 M4) and a linear cavity passively mode-locked dye laser (inset *I* inserted). (b) Equivalent unfolded scheme of the gain mirror section and stability circles diagram (Peterson 1979). The beam paths of two spectral components are indicated by solid and dashed lines.

Therefore we used the simple linear configuration shown in Figure 1 although with the prism pair P2P3 inserted in the cavity this configuration can be reconstructed to a ring one. For the same reason very low absorber concentrations were used yielding pulse durations in the picosecond range in the standard case of no dispersion.

Mode-locking was first achieved in the conventional laser configuration with no spatial dispersion (the prism P1 removed from the cavity) and with the prism pair P2P3 inserted. Some of the resonator parameters are: mirror radii of curvature R(M2, M5, M6) = 3.4 cm, R(M3)=5.4 cm; length of the cavity legs M1M2=200 cm, M3M4=136 cm; l=50 cm. Rhodamine 6G was used as a gain medium and the new markedly stable styryl dye TCETI-tetrafluoroborate (Michailov et al 1990) as saturable absorber. The output mirror has 3% transmission. For low absorber concentrations ($\sim 10^{-4}$M) pulse durations of 1-2 ps were measured with a corresponding spectral width of 1.2 nm FWHM.

The main question to be answered if the technique of MLSD is to be employed in a mode-locked laser is how large the dispersion should be so as to take advantage of the dispersive scheme. We characterize the spatial dispersion by a parameter δ called here "spectral content of the resonator mode". This is the width of the spectral packet confined in the transverse size of a resonator mode in the dispersive area. It was found (Michailov submitted) that the necessary value of the dispersion can be estimated in practice by correlating the spectral width of the pulses delivered by the given mode-locked laser in the case of no dispersion with the spectral content of the resonator mode. As a good approximation, the value of the parameter δ should not exceed the spectral width of the pulses supported by the standard mode – locking technique. In other words the standard mode-locking mechanism should be able to phase match the spectral content of the resonator mode in the dispersive area. The parameter δ for the resonator configuration shown in Figure 1 is given by (Michailov submitted):

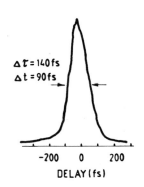

$$\delta=-(\lambda/k\pi d_1)^{1/2}(dn/d\lambda)^{-1}. \qquad (1)$$

The factor k varies from 20 ($l\ll 2d_1$) to 5 ($l\simeq d_1$) where d_1=M1M2 is the length of the longer cavity leg. In (1) λ is measured in nanometers, d_1 in centimeters and $dn/d\lambda$ in μm^{-1}.

A dispersive prism made of Soviet glass TФ12 was used in the experiment yielding a value of the parameter δ=1.5 nm. When the prism was inserted and the cavity realigned according to Figure 1 a dramatic reduction of the pulse duration was achieved . An auto-correlation trace and pulse spectrum are shown in Figure 2.

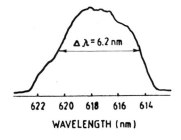

Fig. 2. Autocorrelation trace and corresponding spectrum of pulses generated at a pump power (3 W) twice the treshold.

Assuming sech^2 pulse shape, these correspond to an almost transform limited pulse of 90 fs. It was possible to adjust the intracavity GVD to an optimum value when the prism pair P2P3 was removed from the cavity. The pulse duration changes almost symmetrically around that value of the GVD (as varied by the prism P1) which yields the shortest pulses.

As an important advantage of the spatially dispersive scheme, the laser was able to support stable pulses unaffected in duration at pump powers far above the threshold. Neither pulse broadening nor multiple pulsing was observed when the pump power was raised from the threshold (1.5 W) up to the maximum power of 3 W of our Ar-ion laser. Output powers as in the case of a hybridly mode-locked dye laser might be expected if sufficient pump powers are available. Wavelength tuning to a certain extent was also possible owing to the low absorber concentration. Under the conditions of the experiment the laser was tuned over a range of about 10 nm with minor changes in the pulse duration after correction of the GVD. The tuning was accomplished by "scanning" the pump beam across the dispersive area in the gain jet.

4. CONCLUSION

The advantages of mode-locking with spatial dispersion in the gain medium are demonstrated for the first time. The operating characteristics of the passively mode-locked dye laser presented here together with the simple linear geometry of the resonator make the dispersive scheme attractive and convenient for implementation on the commercial dye lasers. The results are encouraging in the expectations that the technique of MLSD can be employed to improve the mode-locking performances of other lasers with homogeneously broadened gain medium. The cavity design suggested can be used to apply the technique to synchronously pumped dye lasers and mode-locked Ti:sapphire lasers.

REFERENCES

Danailov M B and Christov I P 1989 *Opt.Comm.* **73** 235
Danailov M B and Christov I P 1990 *Appl.Phys.* **B 51** 300
Haus H A and Silberberg Y 1986 *IEEE J.Quantum Electron.* QE-22 325
Michailov N I 1988 *Opt.Quantum Electron.* **20** 175
Michailov N I, Deligeorgiev T G, Christov I P and Tomov I V 1990 *Opt. Quantum Electron.* **22** 293
Michailov N I *J.Opt.Soc.Am.* **B** submitted
Peterson O G 1979 in *Quantum Electronics* v.15 of *Methods of Experimental Physics* ed C L Tang (New York: Academic Press) p.338

Inst. Phys. Conf. Ser. No 126: Section I
Paper presented at Int. Symp. on Ultrafast Processes in Spectroscopy, Bayreuth, 1991

109

Investigation of chirped pulse amplification in a dye amplifier system

O. Kittelmann, G. Korn, J. Ringling, F. Seifert

Zentralinstitut für Optik und Spektroskopie, Rudower Chaussee 6, O-1199 Berlin, Germany

ABSTRACT: The results of amplification of fiber chirped pulses from a Ti:Sapphire oscillator are reported. The amplified spectra show distortions on the blue side. This distortions depends on the signal level and can be explained by gain depletion of the upchirped pulses in the dye amplifier system. FFT-calculation shows a low influence of these effects on the compressed pulse quality.

1. INTRODUCTION

Chirped pulse amplification (CPA) has been very successfully applied in the design of ultra-high brightness solid state laser sources (P. Main, 1988).
In general chirping and stretching before and compression after amplification reduces the nonlinear effects in the amplifier chain and leads to a drastic increase of focusable power. In the paper of R. Nagel (1990) a linear pulse stretching, amplification and recompression for 30fs pulses were used. Stretched pulse amplification in dye amplifiers with fiber chirped pulses was demonstrated in (G. Boyer, 1988). The compressed pulse width was 16fs. In addition to the intensity reduction in the optical components in dye amplifiers we are mainly interested in the specific problems of propagation of chirped pulses in highly depleted gain media. We adapted the CPA-technique for our dye amplifier system to investigate the compressibility of strongly chirped pulses after propagation through highly saturated amplifiers.

2. EXPERIMENTAL SETUP

The experimental setup is shown on fig.1. As pulse oscillator we use a Ti:Sapphire laser ("Tsunami" Spectra Physics) pumped by 9W of Argon-Ion Laser (model 2040E Spectra Physics).
The typical average output power of the Ti:Sapphire laser is about P_{out}=1.2W with a pulse duration of τ_p=1.0ps. The oscillator radiation was coupled in to a 50cm long single mode fiber.
These low energy pulses are amplified in a two stage dye amplifier pumped excimer laser (EMG-150 Lambda Physik). The first stage contains Rh 700 and the second Rh 800 as amplifying media.
The ASE is suppressed by the use of an 1mm colour filter (Schott RG 850) acting as an fast saturable absorber. The second stage is followed by a prism pair to compensate phase modulations, which arise from self phase modulation in the fiber and linear and nonlinear interaction during the process of amplification. The measurements of pulse width we performed with standard background free autocorrelation measurement in a BBO crystal.

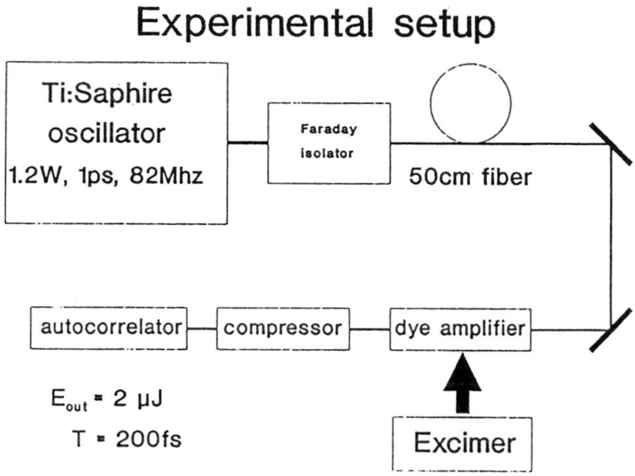

Experimental setup

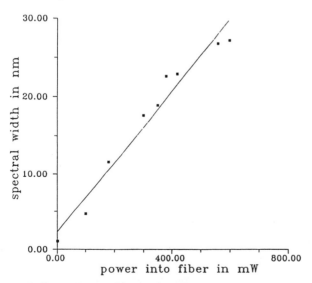

Fig. 1. Experimental setup

3. RESULTS AND DISCUSSION

Figure 2 shows the dependence of the maximum bandwidth on the average power coupled into the fiber. The maximum bandwidth is about 27nm for an 600mW input power. The interferometric autocorrelation traces show a strong chirp for pulses having passed the fiber. Pulses with an average input power of 400mW could be compressed to 60fs by an prism compressor.

The typical average input power of chirped pulses in the dye amplifier chain was 100mW.

Fig. 3 shows the spectra of the nonamplified pulse (dashed line) at the end and of the amplifier chain. The nonsymmetrical shape results from the action of the RG 850 filter. As a solid line the spectrum of the amplified pulse (output energy 2µJ) is depicted. The blue part of the spectrum shows a lower amplification in comparison with the red part, although the used amplifying dye (Rh 800) has its maximum amplification near 736nm (see fig. 4).

Fig. 2. Dependence of spectral width on average power coupled into the fiber

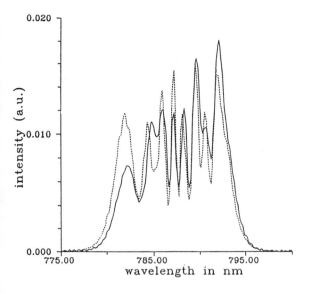

Fig. 3. Spectra of non-amplified (dotted line) and amplified chirped pulse (solid line)

The spectral distortion of the amplified spectra depends on the signal level and can be explained by gain depletion of the up-chirped pulse having passed the fiber. The red part of the radiation enters the amplifier stages first and reduces due to gain depletion the amplification for the following "blue parts" of the pulse spectrum.

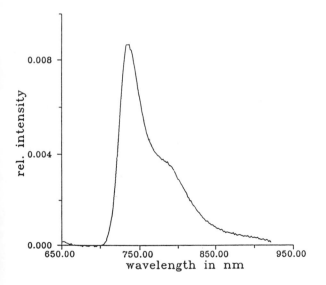

Fig. 4. Fluorescence spectrum of Rh 800

The pulse width of the compressed pulses is 200fs and therefore exceeds the bandwidth limit by an factor of 2.

The energy density in the second amplifier stage is 6mJ/cm^2 taking into a count the beam diameter in the cell. This value exceeds saturation energy by a factor of 3.

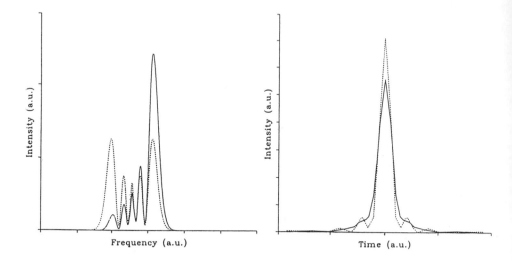

Fig. 5a Calculated spectra of fiber chirped gaussian pulse. Dotted line without distortion ; solid line distorted spectrum.

Fig. 5b Correspondimg to 5a compressed pulse shapes assuming ideal compressor.

The fig. 5a (dotted line) shows the calculated spectrum of a gaussian pulse chirped by self phase modulation in a fiber using FFT. To demonstrate the influence of an asymmetric spectrum on pulse compression and pulse quality we calculated the pulse shape with an ideal compressor. Slight changes are seen at fig. 5b for an asymmetric spectrum. The influence of pulse changes due to nonlinear interaction during amplification can be studied only by solving the nonlinear Schrödinger equation (Heist, 1990).

4. REFERENCES

Boyer, G. et. al. 1988 Applied Physics Letters 53 823
Gagel, R., Angel, G., Laubereau, A. 1990 Verhandlungen der Deutschen Physikalischen Gesellschaft 3 465
Heist, P., Rudolph, W., Wilhelmi, B. 1990 Experimentelle Technik der Physik 38 163
Main, P., Strickland, D., Bado, P., Pessot, M., Mourou, G. 1988 IEEE Journal of Quantum Electronics QE-24 398

Inst. Phys. Conf. Ser. No 126: Section I
Paper presented at Int. Symp. on Ultrafast Processes in Spectroscopy, Bayreuth, 1991

Femtosecond second harmonic generation from Al at the laser intensity level up to 10^{14} W/cm²

S.V.Govorkov, N.I.Koroteev, I.L.Shumay

Physics Department and International Laser Center,
Moscow State University, Moscow 119899 U.S.S.R.

ABSTRACT: We observed specularly reflected Secon Harmonic of 120-fs laser pulses from Al target at the intensity level up to 10^{14} W/cm². Intensity dependence of Second Harmonic exhibits subquadratic rise. This fact was examined theoretically in terms of dense surface plasma dynamics. An important information on plasma parameters can be extracted from theoretical fit of the data.

1. INTRODUCTION

The physics of superintense pico- and femtosecond laser pulse interactions with solids is attracting great current interest. The development of powerful femtosecond laser systems has offered an unique opportunity to study the properties of laser-created plasma of the density comparable with the solid-state density and the temperature of the order of tens and hundreds electron-volts.

Linear optical reflection was carefully studied in a number of works (Milchberg et al 1989a, 1989b, Murnane et al 1989, Kieffer et al 1989a, 1989b, Landen et al 1989, Fedosejevs et al 1990). The gradual decrease of the reflection coefficient from the metal target at intensity level up to 10^{14} W/cm² as well as "saturation" effect were observed. The spectral shift of the reflected radiation is consistent with the plasma expansion velocities $10^6...10^7$ cm/sec which corresponds to the electronic temperature of the order 100 eV. Thus, the plasma scale length is less than the laser wavelength. The characteristic dependence of resonant absorption on the angle of incidence and polarization of the probe pulse were used to determine the density scale length (Landen 1989, Fedosejevs 1990).

Commonly considered theoretical model is based upon the assumption of Drude-like linear optical properties of expanding plasma layer with relevant density and dumping constant. The results obtained by different groups in this way are in reasonable agreement.

Recently, reflected Second Harmonic (SH) was observed (Engers et al 1991) from solid targets at intensity level up to 10^{16} W/cm². Spectral and temporal behavior of SH demonstrates some new interesting features.

In present paper we report on the measurements of specularly reflected SH of 120 fsec amplified laser pulses from a smooth Al target at intensity level up to 10^{14} W/cm². We concentrated

on the intensity dependence of SH and employed a simple model
of free electron gas to explain the features observed experimentally.

2. EXPERIMENT

A standard Colliding-Pulse Mode-Locked laser was used as a
source of the pulses with duration of 80 fsec and wavelength
around 615 nm. Three stage amplifier was pumped by 4 nsec pulse
frequency doubled YAG:Nd laser. After dispersion compensation
the output pulses had a duration 120 fsec, energy up to 0.1 mJ
and repetition rate 5 Hz. Special care was taken to prevent the
effect of Amplified Spontaneous Emission (ASE) on our measurements, namely, to avoid production of the long scale length
plasma before the main femtosecond pulse arrives. By proper filtering the ASE/pulse ratio was reduced down to 10^{-3} level. Additional reducing of the ASE/pulse intensity ratio on the surface
of Al target was achieved due to the difference in spot diamete
of the pulse and ASE. In any case, we have not observed any damage of the sample surface if the CPM output was blocked.

The laser pulses were focused onto the surface of the targe
to a spot diameter of few tens microns at the angle of incidenc
50°. The absolute calibration of intensity was performed by the
procedure based on the known damage threshold of high quality
Si surface. The Al sample was a 1-μm thick film evaporated onto
the glass substrate. It was moved after each laser shot in orde
to expose a fresh area to the next pulse.

We observed strong reflected UV radiation in the direction o
specularly reflected laser beam. We note,that since the reflected laser beam has an angular distribution similar to that of
specularly reflected beam, it means that laser-induced damage
of the surface occurs after the pulse is off. The morphological
changes of the surface can be easily seen afterwards as a
craters. The angular distribution of the UV radiation turned ou
to consist of the two components. First one is well collimated
component with divergence close to that of reflected laser beam
and the second is   nearly isotropic background with intensity
approximately ten times lower.

Then we placed the monochromator entrance slit to the direction of specularly reflected UV radiation and measured its
spectrum. Again, it consisted of strong component with the
frequency equal to the doubled fundamental frequency within the
spectral resolution (1 nm) and broadband background.
Therefore, we come to the conclusion that there exists strong
second harmonic emission accompanied by incoherent radiation
from hot dense plasma on the surface.

With incident fundamental pulse being **p**-polarized, the SH
was predominantly **p**-polarized.

The experimental dependence of the SH intensity on the
intensity of incident fundamental pulse exhibits the subquadratic law. Fig.1a illustrates the effective nonlinear susceptibility (dots) versus laser intensity I_L calculated from SH
intensity data by simply dividing $I(2w)/I_L$. Here we take into
account that the deviation from quadratic law cannot be attributed to the scattering of reflected radiation due to morphological changes of the surface. Simultaneously we obtained the

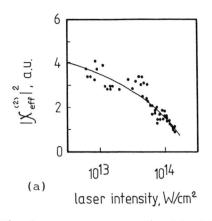

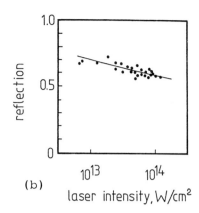

Fig.1. a - square of effective nonlinearity and b - linear
reflectivity versus laser intensity. Dots - experiment, lines -
numerical fit (see the text).

decrease of linear reflection R (Fig.1b) which is similar to
that observed previously (see for example Milchberg et al 1989).

3. COMPUTER MODELING AND DISCUSSION.

Our theoretical analysis of the SHG from Al surface under superintense irradiation is based upon the set of assumptions
concerning the structure of surface layer inferred in earlier
works. We considered the surface of Al target as a layer of
rapidly expanding plasma with the steep density gradient, the
density ranging from the bulk value to zero in the vacuum.
    There are basically several sources of the second-order non-
linearity of the solid-state plasma. First and commonly considered source originates from hydrodynamical behavior of the
free electron gas and Lorenz force (Shen 1984, Sipe et al 1980).
Another possible terms are "collision-induced" nonlinearity
(Kaw, Mittal 1968), interband electron transitions (Govorkov et
al 1991), relativistic effects (Majid et al 1990). We consider
below the simple model of free electron nonrelativistic gas.
In so doing one has to solve equation of motion coupled with
continuity equation and set of Maxwell equations. In the second
approximation one can obtain the expression for the second-
order nonlinear current as follows:

$$\vec{J}(2\omega) = i(ne^3/4m^2\omega^3)\,\vec{\nabla}(\vec{E}\cdot\vec{E}) + i(e^2/m^2\omega^2)(\omega+i\nu)^{-1}(1-\omega_p^2/\omega(\omega+i\nu))^{-1}(\vec{\nabla}\vec{n}\cdot\vec{E})\vec{E} \quad (1)$$

Here $w_p^2 = 4\pi ne^2/m$ is the plasma resonance frequency, w is laser
frequency, E is electric field at fundamental frequency.
    It can be seen from (1) that the SH radiation generated in
the plasma transition layer between vacuum and unperturbed bulk
is sensitive to the details of the space variation of plasma
density and electric field along the surface normal. Since the
thickness of   this layer is comparable with the skin-depth,
these details could not be automatically ignored. Also important is the strong dependence of nonlinear susceptibility on
the plasma collisionality as can be seen from the resonant term

of (1).

In order to carefully take into account these details, we performed the calculations of electric field in the plasma transition layer using Helmholtz equation and the Drude model of the linear susceptibility ϵ , similar to employed in previous works (see Milchberg et al 1989, Kieffer et al 1989). Main two reasons of variation of the effective second-order nonlinearity are changes of plasma collisionality and growing plasma scale length as intensity increases. A possible source of variation of nonlinearity is also change of plasma density due to multiple ionization of Al atoms. Calculations taking into account impact ionization (Milchberg et al 1989) reveal the increase of plasma density as the electronic temperature exceeds 40 eV. Simple estimates of threshold intensity of tunnel ionization yield the value of 10^{14} W/cm^2 (Koroteev, Shumay 1991). Thus parameters being realized in our experiment are close to but do not exceed these values. Therefore, in the first approximation the variation of plasma density may be neglected but it should be taken into account in more detailed analysis.

Following the results of previous works we assumed the electron concentration n profile to be exponential (which is consistent with one-dimensional isothermal expansion):

$$n(z) = n_s \exp(-z/L)$$

(2)

where z is coordinate along the surface normal directed from the bulk to the vacuum, and the plasma scale length L is a product of the pulse duration τ_p and the velosity of expansion V_{exp}. The velocity V_{exp} is given by:

$$V_{exp} = (2/(\gamma-1))(ZkT_e/m_i)^{1/2}$$

(3)

where γ is the heat capacities ratio, m_i is an ion mass, Z is the ion charge and T_e is electronic temperature.

In the case of **p**-polarized incident laser beam the Helmholtz equation for magnetic field B should besolved:

$$d^2B/dz^2 - \frac{1}{\epsilon}(d\epsilon/dz)(dB/dz) + k_0^2(\epsilon - \sin^2\theta)B = 0$$

(4)

where k_0 is the wavevector in the vacuum, θ is angle of incidence. Magnetic field vector is normal to the plane of incidence and coincides with the OX axis. Dielectric constant was calculated using Drude formula:

$$\epsilon = 1 - \omega_p^2(1 - i\nu/\omega)/(\omega^2 + \nu^2)$$

(5)

where ν was taken to be density dependent: $\nu = \nu^* \omega n(z)/n_s$
The electric field can be easily calculated from B(z,y):

$$\bar{E} = (ic/\omega)(1 - \omega_p^2/\omega(\omega + i\nu))^{-1} \text{rot} \bar{B}$$

(6)

The calculations of y- and z-components of electric field in the plasma transition layer reveal that Z-component demonstrates resonant behavior in the "critical" plane due to resonant coupling of normal component of laser electric field to the oscillations of plasma density. The y-component behaves much more smoothly. Similar calculations performed in the case of **s**-polarized incident radiation yielded simply exponential decay of electric field amplitude towards the bulk due to the skin-effect.

The next step is calculation of nonlinear current using (1).

Obviously, due to the fact that first term in (1) is proportional to the electric field gradient as well as due to resonant
behavior of the second term, the nonlinear current is located
mainly in the thin layer around the critical plane and thus
forms a "screen" emitting SH radiation. We also found that the
contribution of z-component of second term of (1) dominates in
the whole range of parameters used in our calculations.

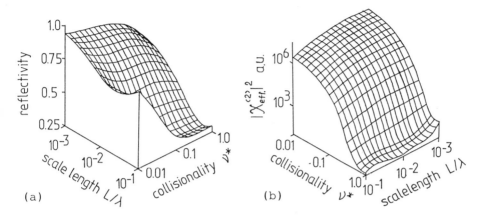

(a)   (b)

Fig.2.Calculated reflectivity (a) and effective nonlinearity
(b) as a functions of collisionality and scale length.

Fig.2 illustrates the calculated dependencies of reflectivity
and nonlinearity on the parameters of interest. We note that
the main features of the behavior of R are similar to that
observed previously (see introduction). Absorption of laser
pulses is monitored by resonant absorption (in P-polarization)
and inverse bremsstrahlung effect. Expansion of the surface
layer causes the shift of the angle of maximum resonant absorption from approximately 80° for abrupt interface to 50° at
L/λ =0.1 (Landen et al 1989), thus leading to the decrease of
reflectivity.
     The effective nonlinearity turnes out to be dependent on
scale length and extremely strongly dependent on collisionality.
To our opinion, it´s due to gradient and resonant nature of the
nonlinear polarization as mentioned above. We also conclude
that the sensitivity of SH reflection to the plasma collisionality can be effectively employed to obtain these parameters
from theoretical fit.
     First, we fitted the experimental data on the linear reflec-
tion (see Fig.1b). Since R weakly depends on ν^{*} in the range
of L/λ estimated roughly from previuos works, we can obtain
fit of $R(I_{L})$ varying the L/λ independently. In so doing we
found values of L/λ corresponding to $I_{L} = 10^{13}$ and 10^{14} W/cm^{2}
equal to $5.6 \cdot 10^{-3}$ and $1.0 \cdot 10^{-2}$. These values infer the electronic temperature T_{e}=8eV and 27eV correspondingly, taking Z=3.
We note that these observations are in reasonable agreement
with the values, obtained by Milchberg(1989a), Fedosejevs(1990)
taking into account the difference in wavelengths and pulse
durations.
     Second, as the L parameter is now fixed, we are able to

fit SH data by varying parameter ν^*. Since the absolute value of SH intensity cannot be directly related to this parameter, only relative change of ν^* can be extracted form the fit. It was found that for three reasonable values $\nu^*(10^{13}\text{W/cm}^2)=0.032$, 0.1 and 0.32 it should change by 12.5%, 6% and 15% correspondingly as intensity increases up to 10^{14}W/cm^2. This increase of ν^* can be related to the formula employed by Kieffer et al (1989b) relevant at moderate temperatures and taking into account electron-phonon coupling: $\nu^* \propto T_i (E_f + kT_e)^{-1.5}$, where Fermi energy $E_f = 12$ eV. Thus, we have $T_i \propto T_e^\delta$, where $\delta = 0.93$; 0.88 and 0.96 correspondingly. These observations are consistent with the assumption of delayed rise of ionic temperature.

4. CONCLUSION

In conclusion, we have observed Second Harmonic Generation from dense hot plasma at the Al surface created by 120-fsec pulse. Simple theoretical model based on the description of surface plasma as free electron collisional gas demonstrates high sensitivity of SH intensity to plasma scale length and especially to the plasma collisionality.

Engers T, Fendel W, Schuber M, Schulz H and von der Linde D 1991 Phys.Rev.**A43** 4564
Fedosejevs R, Ottmann R, Siegel R, Kuhnle G, Szatmari S and Schafer F P 1990 Phys.Rev.Lett. **64** 1250
Govorkov S V, Koroteev N I, Petrov G I, Shumay I L and YakovlevV V 1991 JOSA **B8** 1023
Kaw P K, Mittal R S 1968 J.Appl.Phys. **39** 1975
Kieffer J C et al 1989a IEEE J. of Quant.Electr. **25** 2640
Kieffer J C et al 1989b Phys.Rev.Lett. **62** 760
Koroteev N I and Shumay I L 1991 Physics of High-Power Laser Radiation (Moscow: Nauka) (in Russian)
Landen O L, Stearns D G, Campbell E M 1989 Phys.Rev.Lett. **63** 1475
Majid F et al 1990 Nuovo Cim. **D12** 293
Milchberg H M, Freeman R R, Davey S C and More R M 1989a Phys. Rev.Lett. **61** 2364
Milchberg H M and Freeman R R 1989b JOSA **B6** 1351
Murnane M, Kapteyn A C, Falcone R 1989 Phys.Rev.Lett. **62** 155
Shen Y R 1984 The Principles of Nonlinear Optics (New-York: Wiley) pp 541-54
Sipe J E, So V C Y, Fukui M, Stegeman G I 1980 Phys.Rev. **B21** 4389

Inst. Phys. Conf. Ser. No 126: Section I
Paper presented at Int. Symp. on Ultrafast Processes in Spectroscopy, Bayreuth, 1991

Femtosecond x-ray emission from laser irradiated Al targets

A.Mysyrowicz, J.P. Chambaret, A. Antonetti,

Laboratoire d'Optique Appliquée Ecole Polytechnique-ENSTA,
Batterie de l'Yvette, 91120 Palaiseau (France)

P. Audebert, J.P. Geindre, J.C. Gauthier,

Laboratoire de Physique des Milieux Ionisés, Ecole Polytechnique, 91128 Palaiseau
(France)

ABSTRACT: Solid aluminum targets irradiated with intense ($I_0 \sim 10^{17}$ W/cm^2) femtosecond optical pulses are shown to emit strong Kα lines in the keV spectral region. An analysis of the spectra, based on hydrodynamic simulations, indicates that the X-ray emission has a duration comparable to the incident optical pulse. Peak X-ray emission intensities are in the range 1.-5. 10^{12} W/cm^2.

In this paper, we report results on X-ray radiation from targets. consisting of thin Aℓ films deposited on fused silica, irradiated with intense femtosecond visible radiation. We find evidence of a two-temperature plasma from a detailed analysis of several spectra obtained in the 7.5 - 8.5 wavelength range. Helium-like resonance lines and dielectronic satellites give the spectral signature of a hot dense plasma of about 0.5 keV electron temperature. Equally strong Kα lines from a distribution of low-charge Aℓ ions reveal the existence of a dense, much colder plasma which is excited by an important flux of electrons expelled from the primary hot interaction region. The cold and hot plasmas are in close contact, separated by an abrupt interface. An important conclusion deduced from our analysis is that the Kα X-ray emission is very brief, lasting essentially as long as the incident optical pulse. This is due to efficient cooling of the dense hot plasma which is in close contact with a cold solid.

The femtosecond laser consisted of a CPM dye oscillator and amplifier system operating at λ = 620 nm with a repetition rate of 20 Hz and an energy per pulse of 1.5 mJ. The laser was focussed on flat targets at normal incidence with a 70 mm focal length lens. The focal spot was carefully examined in an equivalent plane magnified by a factor 200. Eighty percent of the laser energy was found to be confined in a focal spot of 3 mm radius. The temporal shape of the laser radiation was examined with a third-order induced index-grating correlation technique over a dynamic range of 10^5. The autocorrelation signal was very well fitted by $I(t) = I_0 \, \text{sech}^2 \, (1.76t/\Delta t)$ with $\Delta t = 80$ fs, resulting in a peak intensity $I_0 \sim 10^{17}$ W/cm^2 in the focal spot.

An important consideration in the physics of femtosecond laser-produced plasmas is the possible existence of a background of nanosecond duration, preceding the main pulse, due to amplified spontaneous emission (ASE) originating in the oscillator and the dye amplifier chain. Prepulse intensities even 10^{-5} below the main pulse intensity can be sufficient to ablate the target and form a low density preplasma which acts as a shield, preventing interaction of the main pulse with the solid. In order to detect the eventual presence of a low density preplasma, we have performed side view time-resolved Schlieren experiments capable of detecting the critical density of an expanding plasma with a spatial resolution of 1mm and a time resolution of 100 fs. The critical density of the plasma was less than 1 μm away from the solid surface at the peak of the main pulse, 'J.P. Geindre, 1991'. This excludes the presence of a preplasma in our experiments.

The targets consisted of thin Aℓ films, of different thicknesses, deposited on flat fused silica substrates. The samples were mounted inside a vacuum chamber on a carefully aligned X-Y translational stage in order to expose a fresh surface to each laser shot. An estimate of the absorbed laser energy was obtained by measuring the ratio of incident to reflected intensity at normal incidence within an aperture of 0.5 f number. Typically 20% of the main pulse energy was found to be transferred to the target. The time-integrated emission from the plasma was analyzed in the 1.4 - 1.6keV X-ray region using a von Hamos spectrograph equipped with a pentaerythritol crystal (PET) with 100 mm curvature radius. The spectrum was recorded on SB 392 photographic films covered with a 20 mm thick foil of Beryllium in order to eliminate visible radiation. Film density could be converted to X-ray emissivity by using the known crystal reflectivity, the filter transmission and the spectrograph geometry.

Figure 1 a) shows the spectrum obtained with an aluminum layer of 300 nm resulting from the accumulation of $1.5.10^4$ shots. From such spectra, the existence of two emitting plamas with different temperatures can be directly inferred. The emission at 0.775 nm and around 0.782 nm can be identified as being due to the 1s2p ^{1}P - 1s^2 ^{1}S transition from helium-like Aℓ$^{11+}$ and from Aℓ$^{10+}$ lithium-like ions, respectively. From the ratio of the lithium-like to helium-like line intensities, an electron temperature of Te > 500 eV can be estimated. The hot plasma electron density can be inferred from a lineshape analysis, assuming broadening due to a Stark effect from averaged local microfield and Doppler effects . The electron density which gives the best fit over the whole line profile is Ne ~ 10^{23} cm^{-3} (see Fig. 1 right). This high density value is also consistent with the absence of the intercombination line 1s2p ^{3}P - 1s2 ^{1}S of Aℓ $^{11+}$ at 0.7807 nm, a fact which can be attributed to collisional quenching of the metastable triplet state, 'J. Dubeau et al., 1982'.

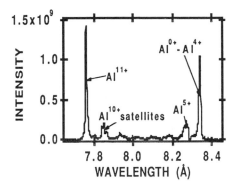 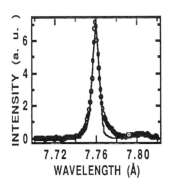

Fig. 1. *(left): Time-integrated emission of an Aℓ film (thickness = 300nm) deposited on fused silica substrate, irradiated with laser pulses at 620 nm, with pulse duration 80 fs and peak intensity I_o~10^{17}W/cm^2. The origin of the emission lines is discussed in the text. Fig. 2 (right)shows a theoretical fit of the emission from He-like Aℓ ions, assuming a plasma density $N_e = 10^{23}$cm^{-3}.*

On the other hand, lines at 0.825 nm and 0.834 nm indicate that a cold plasma with a temperature of a few eV is generated by energetic electrons produced during the laser interaction with the hot plasma. The line at 0.834 nm results from the (2p-1s) Kα transition of aluminum ions in a charge state between neutral (Aℓ$^{0+}$) and fluor-like-Aℓ$^{4+}$, whereas line at 0.825 nm is due to 2p - 1s emission in oxygen-like Aℓ$^{5+}$. From the measured ratio of these lines, taking into account the branching ratio of the fluorescence and Auger decays for all upper levels which play a rule in the Kα line formation, we have obtained the relative populations of the different aluminum ions from which we have calculated the average charge of the plasma. Using a local thermodynamic equilibrium calculation of the charge state at near solid density, we also obtain an electronic temperature T_e 15 - 20 eV. Earlier work has shown that Kα emission is the signature of impact ionization by fast electrons of kinetic energy above the K shell ionization energy which originate in the laser plasma primary interaction region 'J.D. Hares et al., 1979'.

In order to gain a better understanding about the energy deposition in the target, we have measured the intensity of the emission from helium-like aluminum and silicon ions in targets consisting of layers with different thicknesses of aluminum deposited on fused silica substrates. Figure 2 shows the relative intensities of the lines of $Si^{2+}(1s2p - 1s^2)$ and of $A\ell^{11+}$ ($1s3p - 1s^2$) together with the intensity of the Kα emission of silicon at 7.12 Å, as a function of aluminum thickness. From such measurements, it is possible to extract informations about the thermal conduction lengths in our experimental conditions. The depth of the hot plasma is of the order of 100-150 nm, as shown by the decrease of the aluminum emission accompanied by the increase of the corresponding silicon emission. A larger energy penetration depth of 250 nm has been obtained with a KrF laser 'Zigler et.al., 1984', but with a longer (600 fs) laser pulse duration.

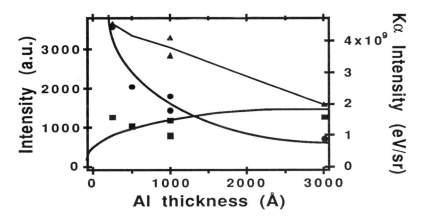

Fig. 2. *Intensity of emission from He like ions of Aℓ (triangles) and Si (dots) and from Kα emission of Si as a function of Aℓ thickness. Samples consisted of Aℓ films of different thickness deposited on a SiO$_2$ substrate.*

The Kα line of silicon from the fused silica substrate was also observed in our multilayer experiment. Its intensity (triangles in Fig.2) shows a decrease of a factor of two when the aluminum thickness increases from 25 to 300 nm. For a 300 nm aluminum target (as in Fig.1) and assuming that the hot plasma thickness is only 100 nm, this implies that the energy of the hot electrons responsible for Kα emission is rather low, i.e. below 20 keV, in view of their high attenuation over the remaining 200 nm of the cold target.

The experiments discussed above were modeled using the FILM simulation code. This model incorporates electron and ion Lagrangian hydrodynamics and time dependent configurations of atomic physics 'J.C. Gauthier et al., 1983'. It successfully describes the physics of laser-matter interaction in the nanosecond time scale. The classical treatment of laser absorption and the local analysis of thermal conduction have been modified to deal with the subpicosecond laser regime as follows. The equations for absorption in a solid were solved at each time step on the basis of the Drude model for conductivity using an accurate value of the electron-ion collision frequency and the resulting deposited energy was adjusted so as to reproduce the measured total laser absorption of 20-30%, 'H.M. Michigan et al., 1989'. A non local treatment of thermal conduction is also essential in subpicosecond simulations. Due to the small hydrodynamical expansion, the scale length of the electron temperature gradient is much smaller than the thermal electron mean free path. In the FILM code, thermal conduction inside the solid is treated within the framework of delocalized heat transport theory, 'J.F. Luciani,et al , 1983'.

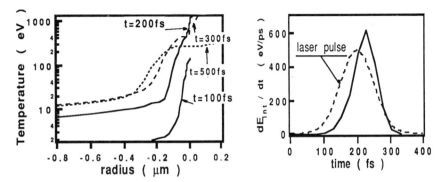

Fig. 3 *(left) Numerical computation by FILM code of the temperature profile within $A\ell$ at different time delays. The unperturbed $A\ell$ /vacuum interface lies at position X=0. The laser impiges from the right. Fig. 3 (right) shows the calculated time evolution of the internal energy inside the target which represents the Kα pulse duration. so shown is the incident laser pulse(dotted line).*

The simulations were done for a solid aluminum target illuminated at normal incidence with a gaussian laser pulse of 100 fs FWHM and a laser intensity of 10^{17} W/cm^2. Results of the computations are shown in Fig.3 where the electron temperature profile inside the target is plotted at various time delays. As can be seen, one can identify two main excited regions. A first region ranging up to 100 nm from the vacuum-metal interface is heated very rapidly up to temperature about 1200eV and then experiences a temperature decrease down to Te ~ 250 eV through plasma expansion towards the vacuum and heat conduction towards the interior of the target. The second region, separated from the hot plasma by an abrupt interface has a much lower temperature of Te ~ 10-15 eV. Fast electrons from the high energy tail of the Maxwellian electron energy distribution are not confined by multiple scattering in the hot plasma region but penetrate deeply into the solid prior to the heat front, simultaneously creating a cold dense plasma and generating $K\alpha$ line emission via impact ionization of deep core levels.

This model simulation reproduces well the experimental results including the dimensions of the hot and cold plasma regions and their respective temperatures. An important conclusion of the calculations concerns the duration of the $K\alpha$ emission. In the low temperature plasma region, the only significant energy source arises from quasiballistic electrons which escape from the hot plasma layer. As noted before, these electrons have a mean free path which greatly exceeds the electron temperature gradient scale length. As a result, the time duration of the source (and thus the $K\alpha$ line duration) can be obtained by taking the time derivative of the internal energy in the cold plasma. The result is shown in Fig.3 The total time duration of the $K\alpha$ a X-ray pulse is not significantly longer that the heating laser pulse.

Knowing the $K\alpha$ pulse duration, it is possible to estimate the $K\alpha$ intensity. The measured yield of $K\alpha$ emission is approximately 2.10^{-8} J per shot. Assuming that the surface of emission of the $K\alpha$ lines is comparable to the focal point size, this corresponds to an intensity of 10^{12} W/cm^2. This is well in the range where nonlinear X-ray spectroscopy should become feasible, provided adequate X-ray optics is available to collect the X-ray emission over a sufficient solid angle.. We also note that the electron-mediated excitation of $K\alpha$ lines in a dense plasma could prove useful as a potential pumping source for an X-ray laser scheme.

REFERENCES

J.P. Geindre, P. Audebert, J.C. Gauthier, R. Benattar, J.P. Chambaret,
    A. Mysyrowicz, A. Antonetti, to be published (Optics Comm.).

J. Dubau and J. Volonte, Ann. Phys. Fr. 7, 455 (1982).

J.D. Hares, J.D. Kilkenny, M.H. Key and J.G. Lunney, Phys. Rev. Letters 42, 1216 (1979).

A. Zigler, P.G. Burkhalter, D.J. Nagel, M.D. Rosen, K. Boyer, G. Gibson, T.S. Luk,
    A. McPherson and C.K. Rhodes, Appl. Phys. Lett. 59, 534 (1991).

J.C. Gauthier, J.P. Geindre, N. Grandjouan and J. Virmont, J. Phys. D 16, 321 (1983).

J.F. Luciani, P. Mora and J. Virmont, Phys. Rev. Letters 51, 1664 (1983).

H.M. Milchberg, R.R. Freemann, J. Opt. Soc. Am. B 6, 1531 (1989).

Inst. Phys. Conf. Ser. No 126: Section I
Paper presented at Int. Symp. on Ultrafast Processes in Spectroscopy, Bayreuth, 1991

First results on the way to a ps multiterawatt glass laser system using fiberless CPA-technique

F.Billhardt, P.Nickles, I.Will

Central Institute of Optics and Spectroscopy,
Rudower Chaussee 6, Berlin O-1199

Nd-glass is a suitable material for generation of ps- and subps-pulses due to its large fluorescence bandwidth. But an amplification of such short pulses to Multiterawattlevel is made difficult by intensity dependent nonlinear effects. Therefore we apply a fiberless CPA-technique.The great potential of this method for the generation of ps pulses with extremely high intensities is shown.

1. INTRODUCTION

The development of a ps-multiterawatt glass laser system in order to applicate energetic ps pulses for laser-matter inter- action investigations at intensities of $I \geqslant 10^{18}$ W/cm^2 or in case of extended focal areas of $I \geqslant 10^{14}$ W/cm^2 is the aim of our work.

It is well known, that glass lasers using chirped pulse ampli- fication technique permit the production of ps-multiterawatt pulses (Maine et al 1988), but the till now used CPA method with a fiber-grating combination as pulse stretcher difficults a perfect compression owing to a nonlinear frequency chirp and a square top pulse envelope. Therefore we propose a fiberless CPA-technique using a broadband ps-start pulse, which shall overcome the disadvantages mentioned above and lead to a per- fect pulse compression.

In detail we report on first results of the development of our system. We describe the first stage of the laser consisting of a passively mode-locked Nd-glass oscillator with negative feedback control (NFC), a fiberless pulse stretcher, one re- generative amplifier followed by a linear one and the final grating compressor.

2. LASER SYSTEM AND EXPERIMENTAL SETUP

Figure 1 shows the principle scheme of a fiberless CPA-laser system. The negative feedback controlled oscillator (Korn et al 1991) has a hemispherical resonator, with a mode-locking dye cell contacted to the curved 100% mirror. Dye 3274 is used as saturable absorber. The reflectivity of the output mirror is 0.5. A 135 mm long Nd-phosphate glass rod (GLS-21) with 8 mm diameter acts as the active element. NFC is performed by a single crystal Pockels cell. An intracavity telescope improves modelocking and the additional NFC loop stabilizes and shortens the ps-pulse. The oscillator produces a pulse train of about 500 ps-pulses each of them having an energy of 2 uJ and a bandwidth of 45 - 65 Å. A single pulse is selected from the pulse train by a pulse picker Pockels cell.

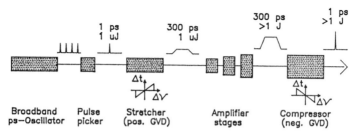

Fig. 1. Principle scheme of a fiberless CPA laser system

In order to avoid self focussing effects by amplifying short pulses we apply a CPA technique. Due to the large bandwidth of generated pulses the ps pulse is stretched in a fiberless grating telescope combination for more than two orders of magnitude. According to Martinez (1987) two gratings (Jobin Yvon 1740 l/mm) are placed within the focal points of the telescope (f = 735mm) resulting in a linear positive group velocity dispersion (chirp).

The stretched pulse is then injected into the regenerative amplifier, consisting of only one crystal polarizer, two Pockels cells and a 135 mm long, 10 mm diameter Nd-phosphate glass rod. The switch-in Pockels cell turns the polarisation of the pulse making roundtrips in the cavity possible. After amplification the second Pockels cell rejects the amplified pulse at a fixed intensity level. For stable action and high fundamental mode volume we use a hemispherical resonator of about 3 m length. A 80% output mirror allows an easy check of the amplifying behavior.

After propagating a hard aperture the pulse is relayed into the linear amplifier by an air spatial filter and then by a vacuum spatial filter to the grating compressor. We use the same gratings as in the stretching configuration but now in a

parallel alignment. A repetition rate of about 0.5 Hz of this
first stage permits an easy alignment for optimum pulse recompression.

3. RESULTS

Figure 2 shows a typical autocorrelation trace and the spectral shape of a pulse selected from the pulse train of the
negative feedback controlled oscillator described above. The
bandwidth and the pulse length changes from 40 Å and 1.5 ps at
the leading part to 80 Å and 2.5 ps at the trailing part of
the pulse train. A modulation of the pulse spectrum at the
trailing part points to self phase modulation, accumulating
during the roundtrips of the pulse in the resonator. The autocorrelation trace is measured by means of a noncollinear SHG
method.

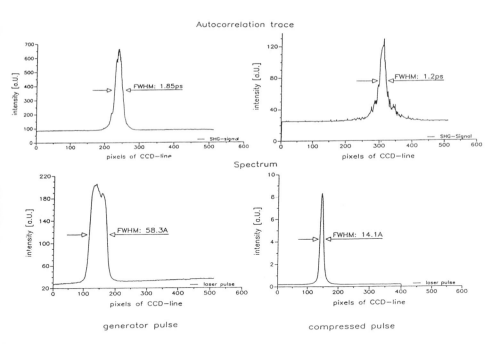

Fig. 2 shows the autocorrelation trace and the spectrum of the
start and the recompressed amplified pulse.

The regenerative amplifier has a high amplification. Whereas
the pulse after the stretching stage has an energy of 0.7 µJ
the output energy of the amplified pulse is about 14 mJ, corresponding to a net amplification of 2×10^4. This high amplification factor leads to a remarkable spectral gain-narrowing,
so that the broad bandwidth of the start pulse is reduced to
about 14 Å. Measurements of the chirp by projecting the pulse

spectrum directly on the slit of a streak camera have shown a
very good linearity of it and confirmed our expectations on
the fiberless CPA-technique. Figure 3 shows the dependence of
the wavelength on time. The chirp is 0.05 Å/ps. Pulse length
was determined to 280 ps measured with a HADLAND Imacon 500
streak camera.

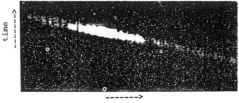

wavelength

Fig. 3 shows the linearity of the chirp

The linear amplifier delivers then an output energy of 120 mJ.
A final double grating compressor completely compensates the
positive chirp and compresses the amplified long pulse to a
short pulse of 1.2 ps duration according to 14 Å bandwidth and
an energy of 80 mJ. Figure 2 b shows the autocorrelation trace
and the corresponding spectrum of the compressed pulse. The
beam divergence is with 3×10^{-4} nearly diffraction limited.

4. SUMMARY

The first results show the great potential of the proposed
fiberless CPA technique using broadband radiation from a glass
laser (or other broadband solid state lasers i.e. Ti-sapphir)
for the generation of pulses with extremely high intensities.
Our next attempts are concerned with the important question of
contrast and shape of the pulse. In the near future we upgrade
the system by further small amplifiers and inject the radiation in the existing large aperture glass amplifier chain
(60 mm diameter) in order to get ps-multiterawatt pulses.

This work was supported in part by the Minister of Science and
Technology of the FRG under project number 211-5291-LAS1991.

REFERENCES

Korn G., Billhardt F., Kalashnikov M., Nickles P., Will I.,
   1991 *Optics Communications in press*
Maine P., Strickland D., Bado P., Pessot M., Mourou G. 1988
   IEEE J. of Quant. Electr., Vol. QE-24, No 2, p 398-403
Martinez O.E. *IEEE J. of Quant. Electr.,*
   Vol. QE-23, No 1, p 59-64

Inst. Phys. Conf. Ser. No 126: Section I
Paper presented at Int. Symp. on Ultrafast Processes in Spectroscopy, Bayreuth, 1991   129

Nonlinear quantum electrodynamics with high power ultrashort laser pulses: possibility of experimental studies

P G Kryukov

P N Lebedev Physics Institute, USSR Academy of Science, 117924 Moscow, USSR

ABSTRACT: Prospects for experimental studies of nonlinear quantum electrodynamics with high power ultrashort laser pulses are presented. The most intensive lasers permit to achieve fields strength of about 10^{11} V/cm. In the rest frame of a 50-GeV electron the field strength is then about 10^{16} V/cm. Fields of this strength will manifest various nonlinear quantum electrodinamics effects. Proposed experiments are based on bringing a high-energy electron beam into collision with a high-intensity laser beam.

1. INTRODUCTION

Quantum electrodynamics (QED) is the most highly advanced part of quantum field theory, distinguished by the simplicity of its basic principles and thoroughness of its mathematical formalism. It is now being verified in the short-range region, where it has reached the limit beyond which nonelectromagnetic interactions must be taken into account. The number of theoretical studies in this direction is rapidly increasing. At the same time it becomes particularly important to investigate it under other extremal conditions - those of strong external fields and nonlinear effects.

The possibility to study the QED processes in a strong laser field becomes in principle feasible in recent years. This is due to the progress in obtaining femtosecond laser beams with the power $>10^{12}$ W and the appearance of electron accelerators with necessary beam parameters.

2. NONLINEAR EFFECTS IN QUANTUM ELECTRODYNAMICS

A strong laser pulse field gives a chance to study the following effects (Ritus 1985, Mc Donald 1985):
1) Nonlinear Compton scattering, i.e. the absorption by an electron of n laser photons of frequency ω in the reaction $\quad n\hbar\omega + e \rightarrow \hbar\omega' + e'$
2) Electroproduction of $e^+ e^-$ pairs $e \rightarrow e'e^+ e^-$.
3) Cherenkov radiation $e \rightarrow e'n\gamma$.
4) γ-quantum $e^+ e^-$ pairs production $\quad \gamma \rightarrow e^+ e^-$, $\gamma \rightarrow \gamma' e^+ e^-$.
5) γ-quantum splitting $\gamma \rightarrow \gamma_1 \gamma_2$
6) Light by light scattering $\gamma_1 + \gamma_2 \rightarrow \gamma_3 + \gamma_4$ and some other effects.

The characteristic value of the electromagnetic field strength in the QED is $\vec{E_0} = m^2 c^3/eh = 1.32$ x 10^{16} V/cm (equivalent 4.41 x 10 gauss). This field performs a work mc^2 over a Compton wavelength of an electron. The most intensive lasers permit to achieve fields strengths of about 10^{11} V/cm. In the rest frame of a 50-GeV electron the field strength is then about 10^{16} V/cm. In the nature such strong fields can be found in the vicinity of pulsars and "black holes". We propose to undertake experiments based on bringing a high energy electron beam onto collision with a high intensity laser beam (an electron-photon collider).

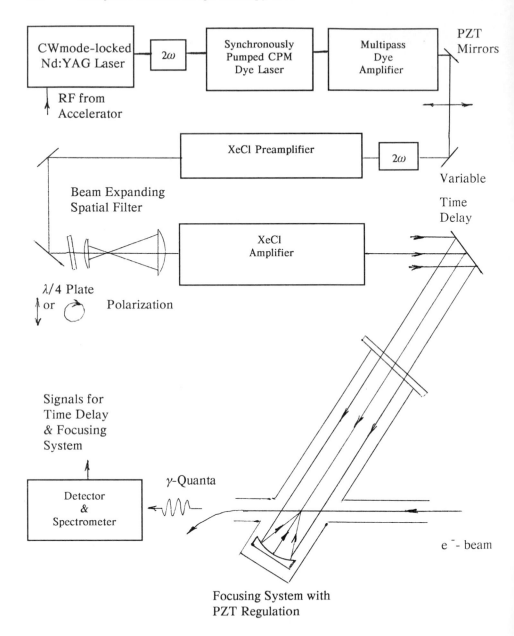

Figure 1. Block diagram of the proposal experiment. A hybrid CPM synchronously pumped dye laser emits 0.5 nJ, 100 fs pulses at 616 nm. After multipass amplification and frequency doubling energy increases up to 1-5 μ J at 308 nm. Ouitput pulse energy about 100 mJ is obtained after the 3x4 cm XeCl excimer amplifier.

The first experimentally investigated effect would be nonlinear Compton scattering. For electron beam energy 50 GeV energy of the backscattered photon is about 37 GeV.

The total probability of the process depends on two invariant parameters:

$$\xi = e\vec{F}/m\omega, \qquad \chi = e\sqrt{(\vec{F}_{\mu\nu}\ \vec{P}\ \nu)^2}/m^3,$$

where $\vec{F}$ is the amplitude of the wave field, ω is the wave frequency, $\vec{F}\mu\nu$ is the wave field strength tensor, and $\vec{P}\nu$ is the electron momentum (used units in which $h = c = 1$). Parameter ξ is characterized the laser intensity. An electron placed in a wave of field-strength $\xi \sim 1$ achieves relativistic velocities during its transverse oscillation and the field performs a work mc^2 over a laser wavelength. Parameter χ is characterized the Lorentz factor of the electron. Optimal condition is $\xi \sim \chi \sim 1$.

Note that a "γ-laser" would be realised in the proposed experiments as a technical byproduct of this effect. Parameters of quality for the backscattered photon beam are:
- high flux (calculated number of $h\omega$'-photons is about 10^3, i.e. flux energy is about 10^3 x 37 GeV = 5.9 μJ/pulse);
- nearly monochromatic energy spectrum;
- angular divergence $mc^2/h\omega$' < 10^{-5}.
The scattered radiation may be interpreted as an undulator radiation. For $\xi \sim 1$ the oscillations of electrons are strong enough and laser wave is an "undulator" or "wiggler" magnet of periodicity λ.

Another nonlinear effect is the production of an electron-positron pair: $nh\omega + h\omega' \rightarrow e^+e^-$
This processes also is governed by two invariant parameters ξ and χ, but it is need threshold energy of a electron beam (about 70 GeV for real laser parameters).

3. THE POSSIBILITY FOR EXPERIMENT

The program of a possible experiment is included two main problems:
1. Production of a laser beam for which the field corresponding to $\xi \sim 1$. It may be a terawatt power laser, based on the amplification of ultrashort pulses in large-aperture excimer amplifiers (Taylor et al 1990, Watanabe et al 1990) or chirped-pulse amplification (Sauteret et al 1990).
2. Synchronization of the laser and the accelerator. We emphasize that the laser-accelerator synchronization is a vital performance of an experiment. A block diagram of the proposed experimental set-up is given in Figure 1.

I thank V I Ritus, A I Nikishov and V I Sergeenko for fruitful discussions.

REFERENCES

Mc Donald K T 1985 *AIP Conf. Proc., N 130* ed C Josh and T Katsouleas (New York), p.23
Ritus V I 1985 *Jornal of Sov. Laser Research* 6, N 5, 3.
Talor A J et al 1990 *Optics Letters* 15 39.
Watanabe M et al 1990 *ibid* 15 845.
Sauteret C, Mainfrau G and Mourou G 1990 *Laser Focus World* !0 85

Inst. Phys. Conf. Ser. No 126: Section II
Paper presented at Int. Symp. on Ultrafast Processes in Spectroscopy, Bayreuth, 1991

Ultra-sensitive detection with ultra-short pulse ring lasers

Jean-Claude Diels, Ming Lai, and Michael Dennis
Department of Physics and Astronomy
The University of New Mexico, Albuquerque, NM 87131

Abstract

Operation of a fs dye laser as an unbiased rotation sensor is demonstrated

1   Introduction

The passively mode-locked dye lasers have played a key role in the development of femtosecond technology. In addition to its record setting properties as ultrashort source, due partly to its large bandwidth, this laser has also remarkable properties at low (i.e. kHz and less) frequencies that have been mostly overlooked. These properties are based on the absence of coupling between the frequencies of the oppositely directed trains of pulses in the ring cavity. The applications are in the area of sensitive detection of anisotropy in index, air flow velocity, magnetic field sensing, linear velocities, etc ... We discuss here the application to rotation sensing which involves the basic laser without addition of any intracavity element.

2   Basic properties of laser gyros

Conservation of angular momentum is the basic principle underlying all rotation sensing. It applies as well to mechanical as optical rotation sensors, but the orders of magnitude are significantly different. For light, the angular momentum of N photons, of angular frequency ω_ℓ, propagating along a circular path of radius R, is equal to the quantum unit $\hbar$ times the number of optical cycles in the ring of light $(\omega_\ell R)$ times the number of photons N. In physical units, this is still a small number. For one Joule of optical energy at a visible wavelength, and a radius $R = 1$ m, the angular momentum is only 10^{-8} J.s, which corresponds to the inertia of a mass of 1 g spinning around a radius of 1 cm at 1 mm/s! Yet, the optical gyro has the advantage over the mechanical one that the ratio of the friction to the inertial couple can be made smaller.

The "ring of light" used in a laser gyro is a standing wave pattern produced by two counterpropagating waves of equal amplitude and frequency. That standing wave pattern is fixed in an absolute frame of reference. An observer in a frame of reference rotating at the angular velocity Ω will see the fringes of spacing $\lambda/2$ pass by at a frequency $\Delta\nu$

$$\Delta\nu = \frac{2R\Omega}{\lambda} = \frac{4A}{P\lambda}\Omega = \mathcal{R}\Omega \tag{1}$$

In the rotating frame of reference, the gyro response $\Delta\nu$ can be measured as a beat note between the two oppositely directed light waves. The ratio of the response $\Delta\nu$ to the rotation rate Ω is called the scale factor $\mathcal{R}$. The expression of $\mathcal{R}$ as function of the ratio of the area A to the perimeter P is quite general, independent of the ring geometry [1].

Just as in a mechanical gyro, there will be "friction" between the absolute rest frame and the laboratory frame, that will tend to "drag" the standing wave pattern into the rotating motion of the laboratory frame. In optics, this friction results from exchange of angular momentum through scattering. The critical momentum is the one needed to impart a rotation Ω to the standing wave of N counterpropagating photons, or $N\frac{\hbar\omega_\ell}{c}\frac{R\Omega}{c}$. An important parameter of a laser gyro is the coefficient $\tilde{r} = r\exp(i\epsilon)$ ($r \ll 1$) [2], which represents the scattering of one of the beams (i.e. $\tilde{r}E_1$) into the other (E_2) during each cavity round-trip time t_c. Because of scattering, there will be a phase shift ψ between the two beams, representing a momentum transfer $Nhd\psi/d(ct)$. Since r is the upper limit of the phase shift /round-trip $d\psi/d(t/t_c)$, we can estimate that the standing wave pattern will be dragged into the rotating frame for rotation rates smaller than Ω_c given by:

$$N\frac{\hbar\omega_\ell}{c}\frac{R\Omega_c}{c} = N\frac{\hbar}{c}\frac{r}{t_c} \tag{2}$$

From Eq. (2), one can deduce a general upper limit for the lock-in rotation rate Ω_c, which is written below in a form valid for all geometries [3]:

$$\Omega_c = \frac{rc\lambda}{2A} \tag{3}$$

3 Passively mode-locked ring lasers as rotation sensors

3.1 Is a fs dye laser a reasonable choice?

Intuitively, the idea of using an ultrashort pulse source for measuring low frequencies does not seem reasonable. In the case of femtosecond lasers, one has to use optical delays with a few micron accuracy to ensure that the output pulses (corresponding to the oppositely traveling waves inside the cavity) overlap in the detector. The signal to be measured is the difference between the two carrier frequencies of the two counterpropagating pulses.

Since this difference is much smaller than the pulse bandwidth, it has to manifest itself at the detection as a slow modulation of the envelope of the pulse train. In the case of a fs laser, observation of a beat note in the range of 100 Hz implies that we could measure a difference in carrier frequencies of the counterpropagating pulses as small as 10^{-11} of the pulse bandwidth!

The dye laser is a homogeneously broadened laser, which would in itself exclude it from consideration as a laser gyro candidate. Mutual saturation in the gain medium makes an homogeneously broadened laser generally unstable for balanced bidirectional operation (the laser tends to become unidirectional). In the case of passively mode-locked dye lasers, the counterpropagating pulses traverse the gain medium at equal time intervals (1/2 the cavity round-trip time). Therefore, the coupling leading to unidirectional operation in rotating homogeneously broadened ring lasers is eliminated.

The only known advantage [4] of an ultrashort pulse laser for gyro operation is a potential reduction in the lock-in threshold defined in Eq. (3). Indeed, the scattering that can contribute to the coupling between the two counterpropagating waves is limited to the two regions where the pulses overlap. In the passively mode-locked laser, the pulses overlap in the absorber jet, and in a volume of air at the opposite point of the ring cavity. A strong coupling between the two waves is needed in the absorber jet, to force the pulses to meet at that location. Moreover, the saturable absorber jet contains undissolved particles, impurities and defects that will contribute to scatter radiation from one beam into the other.

To establish a *lower limit* for the scattering coefficient r introduced by the saturable absorber jet, we performed a simple measurement on a jet of *pure ethylene glycol*. The beam of an argon ion laser is focused onto the jet by a 2.5 cm focal distance lens. The geometry and aperture sizes are carefully adjusted to eliminate backscattering from all surfaces (lens and jet). The minimum backscattering radiation from the jet is $1.5 \ 10^{-6}$ of the incident intensity. This value correspond to a field backscattering coefficient of $r = 1.2 \ 10^{-3}$. According to Eq. (3), such a value for the backscattering coefficient corresponds to a lock-in rotation rate of $5^0/s$ for a ring of $1.4m^2$ area, at a wavelength of 620 nm. The actual lock-in rate should be much larger, in view of the additional contribution from the absorber dye.

3.2 Femtosecond ring laser gyro

Since the rotation measurement is based on the frequency difference between that of the two counterpropagating pulses, it is desirable to have bandwidth limited pulses circulating in the cavity. It has been shown [5] that a crucial parameter in the operation of a fs dye laser is the ratio of the spot sizes in the absorber and amplifier jets. Tight focusing in the absorber (relative to the gain jet) results in peak intensities sufficiently large for the pulse to have the phase modulation dominated by Kerr effect (positive chirp, which needs to be compensated by negative dispersion). After

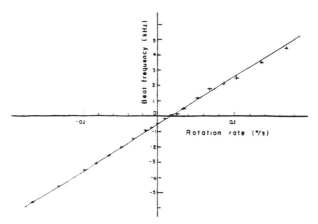

Figure 1: Plot of the beat frequency versus rotation rate of the ring laser

pulse compression, the short pulse has an asymmetrical spectrum indicative of strong residual chirp. In contrast, bandwidth limited pulses with a symmetrical spectrum can be obtained with roughly equal spot cross sections in the gain and absorbing media. In the latter case, the phase modulation is dominated by off-resonance saturation of the absorber, resulting in a downchirp (which has to be compensated by positive group velocity dispersion) [5,6].

A large square cavity as required for the laser gyro implies large angles of incidence on curved optics, hence strong astigmatism. Choosing the absorber jet as origin, the complete round-trip ABCD matrix of the ring was calculated using the algebraic manipulation language MACSYMA. The expressions for the matrix elements are used in a FORTRAN optimization program, which scans all possible values of the folding angles and intermirror distances to determine the stability ranges (distances between pairs of curved mirrors) and the optimum condition for which a round focal spot is obtained in the absorber [5]. The resulting cavity has an area of $1.42 m^2$ and a perimeter of 5.09 m. The scale factor $\mathcal{R}$ is 0.79915 MHz/(radian/s) or 31.40 kHz/(^{0}/s).

The gyro response of this mode-locked dye laser operating at 620 nm with Rh6G and DODCI as saturable absorber is shown in Fig. 1. The data confirm the absence of measurable dead band noted on measurement of Fresnel drag [7]. By careful alignment of the cavity, the natural bias [8] of the laser can be eliminated up to the uncertainty due to the earth rotation rate ($0.035^0/s$). It does however remain a cause of error in the measurement rather than a convenience [9], as long as it can not be compensated in a systematic fashion (i.e. independently of the knowledge of the rotation rate).

The slope of the data corresponds exactly to the predicted scale factor. No departure from linearity can be observed within the accuracy of the measurement. The main causes of error in the data are vibrations of the components of the measuring interferometer (the gyro effect is rather insensitive to small vibration of the laser base or components) and inaccuracies of the drive mechanism at lowest speeds.

3.3 In search of a dead band

The waveform of the beat note is approximately sinusoidal down to the lowest frequencies observable. Even at the lowest frequencies, we do not observe any modulation in the amplitude of any of the beams. These two observation concur to indicate a very low threshold for lock-in.

In order to be able to observe the lock-in characteristics of this laser, we introduced a scattering element (antireflection coated glass window) at the pulse crossing point opposite to the absorber jet. The amount of coupling was adjusted by translating the glass scatterer through the crossing point, resulting in a conventional dead band in the gyro response. At beat frequencies close to the lock-in threshold, the waveform is distorted, and there is strong amplitude modulation of each of the counter-rotating pulse train.

There are three mechanisms that contribute simultaneously to a reduction of the dead band in the fs laser gyro: transit time broadening, Doppler shift of the scattering, and degenerate four wave mixing. The coupling between counterpropagating pulses should be reduced by transit time broadening of the scattering spectrum (the time of flight of the particles through the interaction region is shorter than the period of the beat note). The angle of the absorber jet cannot be controlled with such accuracy as to rule out a small tilt with respect to the normal of the cavity plane. As a result of a horizontal velocity component of the dye flow greater or equal to 0.12 m/s, the backscattering is Doppler shifted by at least 0.12 MHz – an amount larger than the dead band –, and the lock-in should be drastically reduced [10]. The Fresnel drag associated with that longitudinal velocity contributes only 30 Hz to the bias. Intracavity four wave mixing contributes also to a reduction of the dead band [11]. In the particular case of the passively mode-locked dye laser, a large (of the order of several %) two-photon resonant phase conjugated coupling has been demonstrated [12]. The mutual saturation itself contributes to a frequency preserving coupling of one beam into the other. While smaller than the two photon resonant coupling, the single photon resonant wave mixing has nevertheless been shown to be large enough to lead to the generation of a satellite pulse in a passively mode-locked laser [5]. A rigorous theoretical modeling should take into account the coherent interaction between the counterpropagating fields, with the dye represented by a three level system [12], as well as the coupling induced by mutual saturation.

3.4 Fundamental limitations

Let us consider for simplicity a circular fs laser gyro of radius R. Assuming the group velocity is not affected by the rotation, the maximum rotation velocity Ω_{max} ensuring overlap of the pulses of duration τ_p in the absorber jet and at the detector is:

$$\Omega_{max} \approx \frac{c\tau_p}{2Rt_c} \tag{4}$$

A similar limitation can be found in the frequency domain. The modes of the mode-locked laser cover a spectral bandwidth of $1/\tau_p$. At the largest rotation rate Ω_{max}, the difference between beat notes $\Delta\nu_1 = 2R\Omega_{max}/\lambda_1$ and $\Delta\nu_2 = 2R\Omega_{max}/\lambda_2$ measured at the two extreme wavelengths λ_1 and λ_2 of the laser spectrum should not exceed the cavity mode spacing $1/t_c$. This condition leads also to Eq. (4) for the maximum rotation rate. In the case of the fs laser gyro demonstrated here, $\Omega_{max} \approx c^2\tau_p/(4A) = 1600s^{-1}$, for a 100 fs pulse duration. As a rotation sensor, this laser has a dynamic range from $0.005^0/s$ to $91000^0/s$.

4   Acknowledgments

This work was supported by the National Science Foundation, under grant number ECS-88 02530

References

[1] E. O. Schulz-Dubois. *IEEE J. of Quantum Electronics*, QE-2:299–305, 1966.

[2] F. Aronowitz and R. J. Collins. *Journal of Applied Physics*, 41:130–141, 1970.

[3] J. R. Wilkinson. *Ring Lasers*, pages 1–103. Pergamon Press, New York, 1987.

[4] S. I. Wax and M. Chodorow. *IEEE J. of Quantum Electronics*, QE-8:343–361, 1972.

[5] J.-C. Diels. *Dye lasers principles: with applications*, chapter Femtosecond dye lasers, pages 41–132. Academic Press, Boston, 1990.

[6] J.-C. Diels, J. J. Fontaine, I. C. McMichael, and F. Simoni. *Applied Optics*, 24:1270–1282, 1985.

[7] M. L. Dennis, J.-C. Diels, and M. Lai. *Optics Letters*, 16:529–531, 1991.

[8] F. Salin, P. Grangier, P. Georges, G. Le Saux, and A. Brun. *Optics Lett.*, 15:906–908, 1990.

[9] D. Gnass, N. P. Ernsting, and F. P. Schaefer. *Applied Physics B*, B-53:119–120, 1991.

[10] R. A. Patterson, B. L. Jung, and D. A. Smith. In S. F. Jacobs, M. Sargent III, M. O. Scully, J. Simpson, J. Sanders, and J. E. Killpatrick, editors, *SPIE Proceedings*, pages 78–84, 1984.

[11] J.-C. Diels and I. C. McMichael. *Optics Letters*, 6:219–221, 1981.

[12] J.-C. Diels and I. C. McMichael. *Journal of the Opt. Soc. of Am.*, B3:535–543, 1986.

Novel laser schemes in gases and vapors pumped by short pulse high power lasers

A. Tünnermann, K. Mossavi, and B. Wellegehausen

Institut für Quantenoptik, Universität Hannover
3000 Hannover 1, Welfengarten 1, Germany

ABSTRACT: By near resonant two photon excitation of xenon 6p levels with radiation from a femtosecond KrF excimer laser, stimulated emissions on 6p - 6s transitions and broadband continuum emissions in the visible ($\sim$ 600 nm - 850 nm), uv ($\sim$ 200 nm - 400 nm) and vuv (around 148 nm) have been obtained, with peak powers up to 40 MW. The 6s - 6p lines are due to ASE and hyper - Raman processes, while the broadband emissions result from noncollinear four wave mixing schemes. The process has been used to generate tunable fs radiation in the vuv and at specific excimer laser wavelengths. After amplification, mJ pulses at the ArF-, XeCl-, and XeF- excimer laser wavelengths have been obtained.

INTRODUCTION

In the interaction of high peak power laser radiadion with gases and vapors a variety of excitation processes occur, which may be used for the realization of novel laser schemes. Of special importance are multiphoton processes, which lead to the excitation of high lying levels, to the dissociation of molecules and the production of excited fragments (Hanna 1979, Mittleman 1982) or to ionization and plasma production (Chin 1984), allowing, for example, plasma recombination lasers. A first laser of this type could recently be operated by Tünnermann (1991) in atomic cadmium upon two photon ionization of cadmium vapor with fs - KrF excimer laser radiation. Due to the interaction with the pump radiation also nonlinear polarizations are generated, which can be applied for the generation of short wavelength radiation by frequency up conversion. Well known mechanisms are the anti - Stokes Raman and the hyper - Raman process or four wave mixing schemes. Especially two photon resonant four wave sum and difference frequency mixing processes have proven as a powerful tool for the generation of tunable vuv radiation (Hilber 1987, Wunderlich 1990).

1992 IOP Publishing Ltd

In this contribution we will report on hyper Raman oscillation and, to our knowledge, on first four wave mixing experiments in xenon by excitation with fs pulses. A detailed review of these experiments is in preparation.

For the excitation of the xenon, a fs KrF (248.5 nm) excimer laser system has been used, described in detail in the work of Szatmari (1988). The laser system delivers pulses of about 10 mJ in 360 fs. The laser radiation is focussed into the xenon cell (about 20 cm length) and the emissions from the cell have been observed and analyzed, parallel (forward), antiparallel (backward) and perpendicular to the direction of the pump laser.

With the excimer laser radiation, initially a near resonant two photon excitation of xenon 6p - levels is performed, as indicated in the level scheme (Fig. 1). At intensities above about 10^{10} Wcm^{-2}, stimulated emission on some 6p - 6s transitions and intense broadband emissions in the visible, uv and vuv spectral range are observed, which result from different processes, briefly described and discussed in the following.

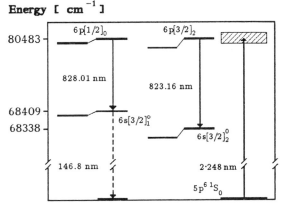

Fig.1: Level scheme of atomic xenon (part) with relevant excitation and emission processes. The 6p $[1/2]_0$ and 6p $[3/2]_2$ levels are shifted into the two-photon resonance by the ponderomotive potential. The hatched area indicates the spectral bandwidth of the pump laser

EXPERIMENTS AND DISCUSSION

A) Infrared line emissions at 823 nm and 828 nm.
These emissions result from amplified spontaneous emission (ASE) on 6p - 6s transitions (Fig. 1). The starting levels, initially off resonant by about 360 cm^{-1} and 1270 cm^{-1}, are shifted into resonance at the used pump intensities by the ponderomotive potential. The ASE emissions are superimposed by resonant hyper - Raman emissions, indicated by the spectral bandwidth, which is determined by the pump laser. Output energies of more than 45 µJ have been measured.

B) Broadband visible (600 nm - 850 nm) and vuv (around 148 nm) emission. These emissions preferentially occur parallel to the pump beam and are optimum around xenon pressures of 80 Torr (Fig. 2). The visible radiation is emitted into a cone of about 4^0, slightly increasing with pressure. This conical emission, a clear correlation between the visible and the vuv emission, and the energy balance $2\omega_p = \omega_{vis} + \omega_{vuv}$ suggest a noncollinear four wave mixing scheme as indicated in Fig. 3. Start photons for the strong visible emission thereby seem to be supplied by ASE and a hyper Raman – type process in molecular xenon, corresponding to the atomic emission scheme. Molecular fluorescence can clearly be observed perpendicular to the pump beam. The pulse duration for the broadband visible emission was measured by an auto-correlator to about 400 fs, while for the vuv emission a pulse width below 10 ps could be estimated so far. Output energies up to 70 µJ and 30 µJ have been detected, respectively.

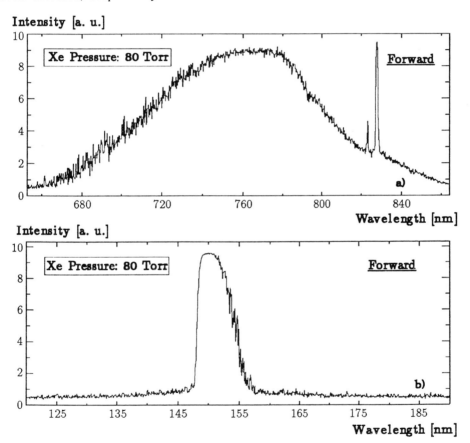

Fig.2: Broadband emission for two-photon excited xenon (parallel to the pump beam direction); a) visible emission with superimposed atomic lines, b) vuv emission.

C) Broadband uv (200 nm – 400 nm) emission.
This emission in the vicinity of the pump radiation (248.5 nm) has its spectral maximum around 280 nm. The optimum output energy of about 60 µJ at 5.5 mJ pump energy was obtained at xenon pressures around 600 Torr. The pulse width was determined to about 400 fs. This uv radiation is only emitted

in forward direction. So far achieved results suggest in this case a nearly degenerate four wave mixing process, starting from parametric fluorescence (PFWM).

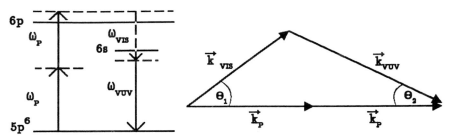

Fig.3: Noncollinear four wave mixing scheme.

D) Generation of tunable fs radiation in the uv and vuv.

The above described four wave mixing process in xenon has been applied to generate tunable fs radiation in the uv and vuv by injecting an additional pump field (ω_i in Fig. 3). As pump radiation for the injection, ns dye laser pulses or fixed frequency pulses from another excimer laser have been used. So far, in the vuv spectral range between 150 nm and 160 nm fs pulses at output energies of several μJ have been generated. Using excimer laser radiation at 351 nm (XeF) we could generate radiation at 193 nm (ArF) and vice versa. By further amplification, fs high peak power pulses at these excimer wavelengths with energies in the mJ range have been obtained.

PERSPECTIVES

To our knowledge, the described experiments represent a first example for the generation of powerful fs pulses and broad continua by four wave mixing schemes. Using other atoms as Kr, Hg or also molecules like CO, perspectives for further powerful emissions also at shorter wavelengths are given.

REFERENCES

Chin S L and Lambropoulos P 1984 *Multiphoton Ionization* (New York: Academic Press)

Hanna D C, Yuratich M A and Cotter D 1979 *Nonlinear Optics of Free Atoms and Molecules* (Berlin: Springer Verlag)

Hilber G, Lago A and Wallenstein R 1987 *J. Opt. Soc. Am. B vol.4, no.11, 1754*

Mittleman M H 1982 *Theory of Laser–Atom Interactions* (New York: Plenum Press)

Szatmari S and Schäfer F P 1988 *Opt. Commun. (68), 196*

Tünnermann A, Henking R and Wellegehausen B 1991 *Appl. Phys. Lett. 58 (10), 1004*

Wunderlich R K, Garrett W R, Hart R C, Moore M A and Payne M G 1990 *Phys. Rev. vol.41, no.11, 6345*

Inst. Phys. Conf. Ser. No 126: Section II
Paper presented at Int. Symp. on Ultrafast Processes in Spectroscopy, Bayreuth, 1991

Ultrafast time-resolved laser scanning microscopy

H Bergner, U Stamm, K Hempel, M Kempe, A Krause, H Wabnitz

Friedrich Schiller University, Faculty of Physics and Astronomy,
Institute of Optics and Quantum Electronics,
Max-Wien-Platz 1, O-6900 Jena, Germany

ABSTRACT: We discuss possible applications and limits of experimental equipments consisting of a combination of mode-locked lasers with laser scanning microscopes to image the "micro-world" with ultrafast photography. To this aim examples for time-resolved investigations with laser scanning microscopy in semiconductor physics and biology are given.

1. INTRODUCTION

Since the development of mode-locked ps and fs lasers ultrafast spectroscopy has been successfully applied to study the dynamics of elementary excitations of various kinds of matter on a time-scale down to 6 fs (Knox et al. 1987). On the other hand the recent development of laser scanning microscopes have led to powerful tools to image various spatially dependent quantities with a spatial resolution of fractions of 1 μm (Wilson and Sheppard 1984). Since in laser scanning microscopy the object is illuminated sequentially point by point by the focused laser beam the combination of a laser scanning microscope with a mode-locked laser could lead to an equipment which allows to image microscopic objects with a high spatial and temporal resolution.

In the present paper we describe two arrangements for time-resolved microscopy in biology and integrated circuit testing. In the first part of the paper we describe three different methods for the dynamic test of high speed integrated circuits (ICs) which base on a modification of the stationary optical beam induced current method (OBIC) (H. Bergner et al. 1989). In the second part we describe applications of time-resolved fluorescence laser scanning microscopy based on the combination of a laser scanning microscope with a time-correlated single photon counting equipment (T.Damm et al. 1990).

2. TIME-RESOLVED OPTICAL BEAM INDUCED CURRENT METHOD

Recent advances in the development and preparation of fast semiconductor devices and integrated circuits (ICs) have provided sources of electrical signals up to the GHz region. For the dynamic failure analysis and chip verification of such devices it is of importance to get knowledge about the propagation of electrical signals through the device with a lateral spatial resolution higher than typical structure widths. Since it obviously becomes difficult to apply standard electrical measuring techniques to advanced ICs it may be of advantage to use time-resolved laser scanning microscopy for the dynamic investigation of these ICs. Fig. 1 shows a schematic diagram of the picosecond OBIC arrangement. The output of a cw mode-locked argon-ion laser with a pulse duration of 100 ps, a repetition rate of 123.2 MHz, and an average power of 500 mW, is split into two parts of different intensity. The more intense excitation pulses are focused onto a fast photodiode or they are launched via separate mirrors or via the scanner mirrors into the LSM. The probe beam travels through an acousto-optic modulator which

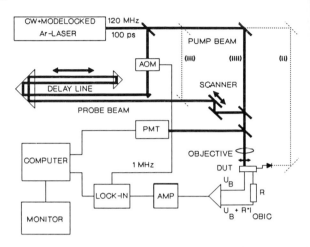

modulates the beam at a frequency of 1MHz. After that it is directed via an optical delay line into the laser scanning microscope where it generates a weak photocurrent the amount of which depends on the internal electrical field at the illuminated area. In dependence on the long distance objective employed in the microscope we have a lateral optical resolution down to 0.6 μm. It is noteworthy that the intensity of the probe beam should be attenuated carefully by neutral density filters to ensure the optically generated charge carriers do not change the operation of the

Fig. 1 Laser scanning microscope with time-resolved OBIC

DUT. That implies the laser induced current is always small in comparison with the operation current of the DUT. To record this small photocurrent on the background of the high operation current a standard lock-in detection at 1 MHz is applied. The output signal of the lock-in amplifier is digitized and stored for every spatial raster-point in the memory of a PC. The scanning frequency (point by point) of the probe beam across the DUT is chosen to be with 6 kHz small compared with the modulation frequency and therefore at each raster position stationary conditions are maintained. Four different experiments have been carried out to measure the picosecond OBIC images of the DUT:

(i) Stationary OBIC-image

The pump beam is blocked and only the modulated probe beam is scanned across the IC. The reflected light as well as the laser induced current is measured and thus the reflection image as well as the conventional OBIC image is obtained in the scanning microscope.

(ii) External generation of electrical pulses

The pump pulses are focused onto a fast photodiode to generate short electrical pulses. After amplification

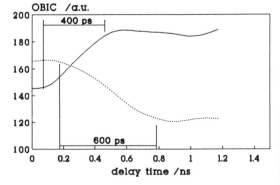

Fig. 2 Time dependent OBIC signal at two transistors

these electrical pulses are sent to the input of the DUT. The modulated probe beam is coupled into the LSM and scanned across the IC and the OBIC image is detected. Varying the delay between the synchronized electrical input pulses and optical probe pulses by the optical delay line the time-evolution of the OBIC is sampled with a temporal resolution limited in essence by the duration of the electrical pulses. Obviously this method has the disadvantage that for very fast semiconductor devices an always even faster switch or photodiode has to be used to generate the necessarily short electrical input pulses for the DUT. Moreover after the propagation of the electrical pulse through the IC the pulse will be more or less stretched dependent on the dispersion of the different elements which have been passed. Therefore the time-resolution after a certain propagation of the electrical pulse through the IC is not better than the switching time of the slowest element passed and the response of later elements can only be determined if these are slower.

(iii) Internal generation of electrical pulses with scanning
Both pump and probe pulses are collinearly launched versus the scanning mirrors into the microscope and the OBIC images are recorded. A variation of the OBIC signal with the optical delay between pump and probe pulses yields a map of all circuit elements which can be switched by intense laser illumination. Moreover, the response of each of these "active" elements to the very short optical pulse can be evaluated from these image series.

(iiii) Internal generation of electrical pulses at a fixed point
The excitation beam with a properly high intensity is focused via separate mirrors onto one of the blocked pn-junctions in the IC. The output of the device is switched with the pulse repetition rate. The propagation of the optically induced charge carriers can be followed by scanning the probe beam across the IC and detecting the OBIC images for various values of the optical delay between pump and probe pulses. This measurement overcomes the disadvantages of method (ii) since short electrical pulses can be generated not only at the input but at various points on the IC. The time-resolution of these experiments is given by the duration of the probe light pulses as well as the response of the material to the excitation pulses.

As an example we give in Fig. 2 the OBIC signal versus optical delay at two transistors of a linear amplifier circuit in ECL technology. From this Figure we can derive the rise time of both elements to 400 ps and to 600 ps, respectively. Moreover, we find a delay in the switching front between both transistors of about 100 ps.

3. TIME-RESOLVED FLUORESCENCE MICROSCOPY

Because of the sensitivity of fluorescence methods in general and time-correlated single photon counting in particular, time-resolved fluorescence microscopy seems to be favoured for the investigation of kinetic processes in biological samples in a microscopic range. In Fig. 3 a scheme of the time-resolved fluorescence microscopy arrangement is depicted. Excitation pulses are provided by a mode-locked argon-ion laser. Fluorescence from the object is collected in backward direction and reaches the image plane after elimination of scattered light by filters. A photomultiplier connected with a time-resolved single photon counting system detects single photon events with a maximum count rate of 200 kHz and a time-resolution (after deconvolution) of 50 ps.

We measure simultaneously at every raster point the average fluorescence intensity and the overall number of photon counts $N_{1,2}$ within two neighbouring time intervals of equal width ΔT properly chosen on the time scale of the fluorescence decay by $\tau = \Delta T / \ln(N_1/N_2)$. Thus we are able to construct in addition to the fluorescence image of the sample the relaxation-time- image ("τ-image"). In Fig. 4 we give an example of the fluorescence intensity and decay time in successive line scans with 1 μm separation of an Ehrlich-Ascites carcinoma cell incubated with HpD. It is visible that distinct variations of s occur within the cell. Concerning the localization of regions of larger and smaller decay time there is no strong correlation between τ and I_F. Additional structures occur in the decay image. Domains of longer decay time exist particularly at the outer cell membrane whereas in the intensively fluorescing nucleus faster decay components are dominating. Referring to this example it can be expected that such measurements could be helpful for the further elucidation of intracellular distribution and photophysical properties of photosensitizers and their compounds as monomers, aggregates, and their reaction products.

4. CONCLUSIONS

We have demonstrated time-resolved methods of laser scanning microscopy for the dynamic testing of integrated circuits and for fluorescence investigations of biological samples. The connection of measureming techniques with high spatial and high temporal resolution opens the way to new informations on the micro-world.

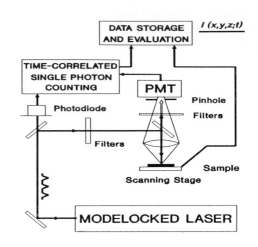

Fig. 3 Scheme of a time-resolved fluorescence microscopy

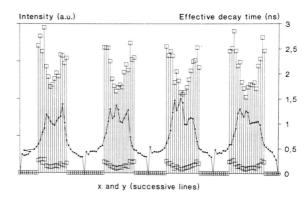

Fig. 4 Histograms of successive line scans of fluorescence intensity (•) and fluorescence lifetime (□)

REFERENCES

Bergner H, Damm T, Stamm U and Stolberg K P 1989 Int. Journ. Optoelectron. **4** 583
Damm T, Kempe M, Stamm U, Stolberg K P and Wabnitz H 1990 Proc. SPIE **1403**
    Laser Application in the Life Science Moscow
Fork R L, Cruz C H, Becker P C and Shank C V 1987 Optics Letters **12** 483
Wilson T and Sheppard C 1984 *Theory and Practice of Scanning Optical*
    Microscopy, Academic Press London

Inst. Phys. Conf. Ser. No 126: Section II
Paper presented at Int. Symp. on Ultrafast Processes in Spectroscopy, Bayreuth, 1991

An all solid state picosecond photon counting system for spectroscopy

I.Procházka[1], K.Hamal[1], B.Sopko[1], J.Říčka[2], M.Hoebel[2]

[1]Czech Technical University, Brehova 7, 115 19 Prague, Czechoslovakia
[2]University of Bern, Institute of Applied Physics, Bern Switzerland

ABSTRACT

We are reporting on the novel design of an all solid state photon counting package and its applications in ultrafast spectroscopy. Developing the new active gating and quenching circuit of the diode we achieved the short and precisely defined detector dead time, the fast gate response and simultaneously the circuit simplicity and compactness.

1. PRINCIPLE OF OPERATION

The application of the solid state detectors having the time resolution of 400 picoseconds and operating at the temperature -60 Centigrade has been reported [1]. The single photon diode operating at the room temperature and having a diameter of 8-40 microns has been reported by S.Cova [2]. In our design we focused on the diode structure design and the chip manufacturing technology, our latest technology achievements permitted to increase the diode active area diameter up to 100 microns while still maintaining an acceptable dark count rate at a room temperature and a fast response, as well. The photon counting package consists of the in house manufactured Single Photon Avalanche Diode (SPAD) and the gating and quenching electronics. The SPAD is a diode structure manufactured using a conventional planar technology on Silicon. The single photon sensitivity is achieved by biasing the diode above the junction break voltage. In this stage, the first absorbed photon is capable of triggering the avalanche multiplication of carriers - the fast risetime current pulse is generated. Its amplitude is limited only by the external circuit connected to the diode. Following the idea of S.Cova [2] we developed the active quenching circuit, which decreases the voltage applied to the diode bellow the break voltage as soon as an avalanche current buildup is detected. The diode voltage is kept bellow the break for the fixed interval adjustable within a few tens of nanoseconds to units of microseconds. Then, the diode is biased above the break, again.

Two principal mode of operation are available :
1. Gated mode : the external electrical gate / pre-trigger signal is applied a short time before the expected arrival of the photon to be detected. In the "gate off" state, the detector sensitivity is decreased about 10^9 times. There are several advantages of this mode : the possibility to study the weak optical signal short after a strong one, namely in fluorescence studies. The detection chips active area diameters of up to 100 microns may be used.
2. Continuous counting mode : the gate signal is permanently on. The detector may be used for the count rate reaching 10^7 counts / second. Due to the lack of the external gating this

mode is very easy to use. However, only the detection chip having active area not bigger than 40 microns in diameter may be used.

2. CONSTRUCTION

The SPAD together with the active quenching and gating circuits are housed in a compact cylindrical housing 40 millimeters in diameter, 120 millimeters long with the detector in the center of the front face. The optional input collecting optics accepting a collimated beam of up to 14 millimeter in diameter and a thermoelectric cooler may be included. On the package output the user gets the uniform pulses : NIM or TTL or ECL or request. The external GATE signal may be applied. The "Gate ON" delay is shorter than 25 nanoseconds. The external biases +/-6 Volts and -30 Volts are needed only.

Detecting a single photon the diode generates an uniform electrical signal on low impedance of the amplitude of Volts. Thanks to this extremely high gain within the diode itself no signal amplification and discrimination is needed in the detector package. Omitting the signal amplification and discrimination, usually needed for standard photon counting systems, the excellent system temporal stability is achieved. The internal delay is changing less than 5 picoseconds within hours. Thanks to low power and low voltages, the power supplies may be simply stabilized and compensated within the wide temperature range. Special voltage stabilizer has been developed to maintain the diode bias voltage within the temperature range of -60 to +100 Centigrade.

3. APPLICATIONS AND PERFORMANCE

The application of this photon counting package in the subcentimeter satellite laser ranging and in the space born laser altimetry has been reported [3,4]. In all the applications the detector simplicity, compact design, wide temperature range and a low voltage and power requirements are attractive features.

The spectroscopy application performance tests have been carried out at the Applied Physics Institute of University of Bern. For this tests the 20 microns diameter diode biased 0.3 to 1.5 Volts above break has been used. The detector has been interfaced to the 50 micron diameter multimode fiber using the

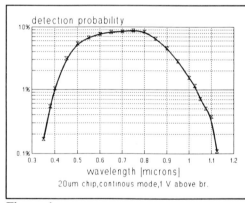

Figure 1 Detection probabiliy versus wavelength

calibrated light source and the calibrated set of filters the detection probability has been found to reach 10 percent at the wavelength 0.8 micron when biasing the diode 1 Volt above the break voltage. Biasing the diode 3-4 Volts above the break voltage, the detection probability may be increased 2-3 times while the dark count rate is increased at the same factor. The detection sensitivity spans over the interval between 0.35 and 1.1 micron, the curve of the sensitivity versus wavelength is plotted on Figure 1.

Using the 300 femtoseconds pulses at the 0.8 micron wavelength, the timing resolution has been found to be better than 100 picoseconds at FWHM, see Figure 2. One has to keep in mind, that the resulting curve is a convolution of the laser pulse length (which is negligible in this experiment), the time interval measuring electronics and the detection package contribution. Deconvoluting the electronics contribution, one gets the timing resolution of the detector package better than 60 picoseconds at FWHM. The detection dead time is adjustable to the values lower than 100 nanoseconds in this set up, the counting linearity is plotted on Figure 3.

Using the picosecond diode laser pulser the dynamical range /the detection delay versus an optical signal strength/of the detector has been tested. For most of the conventional photon counting systems the detection delay is substantially lower for multiphoton detection. In our measurements we did find, that for the signal strength 1 to 30 photons detected, the detector delay changes less than +/-13 picoseconds.

4. CONCLUSION

An all solid state photon counting system described above might find its application in spectroscopy, fluorescence studies, Raman spectroscopy and others, where the optical signal may be focused on a small spot. The detector simplicity, picosecond stability and resolution, dynamical range and count rate are the most attractive features.

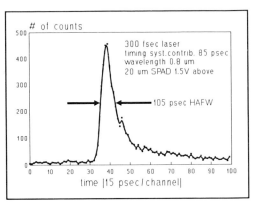

Figure 2 Photon counting time resolution

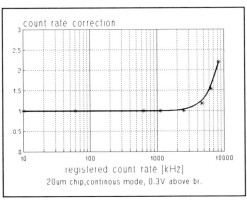

Figure 3 Photon counting linearity

References

1. S.R.Bowman,et al,proc. of the VIth International Workshop on Laser Ranging Instrumentation, Antibes, France, 1986
2. S.Cova et al,IEEE J.Q.Electronics, QE-19,630(1983)
3. I.Prochazka et al, proc. of the Conference on Lasers and Electro OpticsCLEO'90, Anaheim, California, May 1990
4. S.Pershin et al, proc. of the Conference on Lasers and Electro Optics,CLEO'91, Baltimore, MD, May 1991

Inst. Phys. Conf. Ser. No 126: Section II
Paper presented at Int. Symp. on Ultrafast Processes in Spectroscopy, Bayreuth, 1991

Best estimates of exponential decay parameters and the design of single-photon-counting experiments

Malte Köllner

Physikalisch-Chemisches Institut, Im Neuenheimer Feld 253, D-W-6900 Heidelberg

ABSTRACT: On planning experiments for fluorescence-lifetime measurements with minute samples the question of ultimate sensitivity arises. To answer this question, we will find out how many photons one needs to achieve a certain accuracy of the lifetime measurement. Using the maximum-likelihood method and the multinomial distribution, the accuracy of the lifetime estimation can be expressed explicitly as a function of the signal strength. This allows one to determine the required number of photons.

1. INTRODUCTION

Time-correlated single-photon-counting is an important technique for measuring lifetimes of fluorescing species. In the last years, the interest in lifetime measurements on minute samples has been growing. To find the ultimate limit of sensitivity, we will calculate how many photons one needs to achieve a certain accuracy of the lifetime measurement, since the limit of sensitivity is determined by the limited amount of information contained in weak signals. Our general approach will therefore be
  a) to find an optimum estimation procedure.
  b) to determine the error of the estimation as a function of the signal strength.
  c) to calculate the signal strength for the required accuracy.

2. ESTIMATOR AND DESIGN

Many methods of data analysis for time-correlated single-photon-counting experiments have been proposed. In general, they have specific advantages like simplicity, exactness or short calculation times on a computer. In this work we have to focus our attention on estimation procedures that give the highest accuracy, in other words, those procedures that need only very little data to obtain fairly exact results. Only this approach will allow us to predict the very limit of sensitivity and to show that an enhanced sensitivity in a number of experiments can be achieved.

In time-correlated single-photon-counting experiments, we are confronted with the problem of estimating a parameter - the lifetime - from a given data set. This is essentially a statistical or mathematical problem: we are looking for an exact and efficient estimator. The criteria an estimator has to fulfill to be exact and efficient are the following:
  1. On average, the estimator should give the correct or true parameter. If this were not the case one would say the estimator is biased.
  2. The estimator should give the true parameter with high probability or small standard deviation or variance respectively. The variance of the estimator determines its so-called efficiency.

3. Finally there should be a possibility of testing whether the model used for estimating fits the data well.

In statistical literature (Rao 1963), the best estimator in the above sense is known as "bias-corrected maximum-likelihood estimator". The   most important features of this estimator will now be considered.

If one measures $\mathbf{n} = \{n_i, i = 1, 2, ..., k\}$ counts in channels 1 through k, and the true lifetime is τ_{true}, one can assign a probability $P(\mathbf{n}, \tau_{true})$ to this outcome of the experiment. The lifetime is, of course, not known. The maximum-likelihood estimator τ_{ML} for the lifetime τ_{true} is the value that maximizes the expression $P(\mathbf{n}, \tau)$.

The next step is to find an explicit expression for $P(\mathbf{n}, \tau)$. Since single-photon-counting involves distributing N photons, $N = \Sigma n_i$, over k channels, this probability is the multinomial probability

$$P(\mathbf{n}, \tau) = \frac{N!}{n_1! \ ... \ n_k!} \ p_1{}^{n_1} \cdot ... \cdot p_k{}^{n_k}$$

where p_i is the probability that a photon will fall into channel i. The p_i are functions of the parameter τ. This is different from Poisson statistics where N is fitted from the data and not assumed exactly known, which it is - once the measurement is finished.

The three criteria mentioned above for an estimator can now be fulfilled by the bias-corrected maximum-likelihood estimator in the leading terms of a Taylor-expansion in 1/N (Rao 1963):

1. An unbiased estimate of τ_{true} is

$$\tau_{ML} + \frac{1}{N} \frac{1}{2} \sum_i \left(F^{-1}\right)^2 \frac{1}{p_i} \left(\frac{\partial p_i}{\partial \tau}\right) \left(\frac{\partial^2 p_i}{\partial \tau^2}\right) \Bigg|_{\tau \, = \, \tau_{ML}}$$

with $F = \sum_i \frac{1}{p_i} \left(\frac{\partial p_i}{\partial \tau}\right)^2$,   the Fisher information

2. The variance of the estimate is the smallest possible variance and is given by

$$\text{variance} = \frac{1}{N} F^{-1}$$

Therefore the number of photons needed for a certain accuracy of τ is given by

$$N \geq \frac{F^{-1}}{\text{accuracy}^2}$$

which is what we have been looking for.

3. Maximizing $P(\mathbf{n}, \tau)$ is equivalent to minimizing

$$2 \sum_i n_i \ln\left(\frac{n_i}{N p_i}\right) =: 2I^*$$

the Kullback-Leibler minimum discrimination information between the measurement and the model (Kullback 1953), which converges rapidly to the χ^2 distribution with k-2 degrees of freedom. This can be used for testing in the usual way.

3. REMARKS

a) The results can be generalized to the multidimensional case of more than one parameter in a straightforward way giving the Fisher information matrix

$$F_{mn} = \sum_i \frac{1}{p_i} \left(\frac{\partial p_i}{\partial \tau_m} \right) \left(\frac{\partial p_i}{\partial \tau_n} \right)$$

and the variance

$$\text{variance}(\tau_r) = \frac{1}{N} (F^{-1})_{rr}$$

b) Due to the fact that $2I^*$ behaves like χ^2, the well-known algorithm of Marquardt (1963) can be used for the actual estimation process, making the estimation procedure as easy as a least squares estimate.

c) The techniques developed here are not limited to time-correlated single-photon-counting experiments and lifetime estimates, but can be applied to different kinds of single-photon-counting experiments. This latter feature is, however, indispensable since the calculations are based on multinomial statistics.

d) The difference between this maximum-likelihood estimator and a least-squares estimator is a $(1/N)^2$ effect, as shown by Rao (1963), and therefore of no importance. But as Kotze and Gokhale (1980) showed, $2I^*$ provides a better means of testing in the presence of weak signals, i.e. at the limit of sensitivity, than the ordinary χ^2 approach.

e) The difference between the multinomial and the Poisson statistics is not significant either. The advantage of the multinomial distribution chosen here is its mathematical rigour and the fact that the answer to the design question takes a very simple form: $N \geq F^{-1}/\text{accuracy}^2$, as shown above.

f) In general, the bias correction is not very important for the performance of the estimator since it is a $1/N$ effect whereas one is really interested in the standard deviation, i.e. the square root of the variance, which is proportional to the square root of $1/N$. An example will illustrate this.

4. EXAMPLES

Let us consider a background-free monoexponential decay with a lifetime of $\tau = 2.5$ ns. The probabilities p_i are defined by

$$p_i = \int_{\Delta_i} dt \ d(t)$$

where Δ_i is the temporal interval corresponding to channel i and $d(t)$ is the probability density that has the form

$$d(t) = \frac{1}{\tau} e^{-t/\tau} \frac{1}{1 - e^{-T/\tau}}$$

with $T = 8$ ns being the width of the measurement window. When working with multinomial statistics one always has to normalize the functions under consideration; this causes the probability density to be more complicated than a simple exponential. So probabilities p_i are

$$p_i = e^{-i\lambda}\frac{\left(e^\lambda - 1\right)}{\left(1 - e^{-k\lambda}\right)} \quad \text{with } \lambda = \frac{1}{k}\cdot\frac{T}{\tau}$$

$k = 512$ is the number of channels, $i = 1, \ldots, k$. The Fisher information F for this case is 0.087 ns^2. If one aims at an accuracy of 10%, i.e. 250 ps, one would need only 185 photons. This astonishingly low number of photons lets monoexponential lifetime measurements on single molecules in solution seem feasible. The bias in this case is -10 ps and can be neglected.

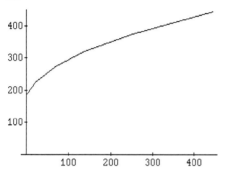

As a third example, figure 1 shows the number of fluorescence photons needed for the same parameters as above, but with a constant background whose strength is Poisson distributed.

Fig. 1. Number of needed fluorescence photons vs. number of background photons

As a third example let us consider a biexponential decay with constant background. The probability density for the multinomial case is

$$d(t) = \frac{b}{T} + (1-b)\left(a\frac{1}{\tau_1}e^{-t/\tau_1}\frac{1}{1 - e^{-T/\tau_1}} + (1-a)\frac{1}{\tau_2}e^{-t/\tau_2}\frac{1}{1 - e^{-T/\tau_2}}\right)$$

with b = portion of background $(0 \le b \le 1)$

          T = width of the measurement window

          $a, 1-a$ = portions of the two fluorescence intensities $(0 \le a \le 1)$

For instance, in a biexponential decay with $\tau_1 = 1$ns and $\tau_2 = 2$ns and a temporal window of 5ns one would need 2.6×10^6 photons to distinguish τ_1 from τ_2 with a certainty of 95 %.

5. FUTURE WORK AND CONCLUSION

Future work will try to incorporate the convolution of the decay curve with the experimentally determined excitation pulse shape. Furthermore, the techniques developed here should allow one to calculate the signal strength one needs to tell whether a decay is e.g. bi- or triexponential (Köllner 1992).

The number of photons needed for parameter estimation based on photon-counting can be predicted with high accuracy.

REFERENCES

Kullback S 1953 Information Theory and Statistics (New York: Wiley) pp 117-119
Köllner M 1992 Chem. Phys. Lett. to be published
Kotze T J v W and Gokhale D V J. Statist. Comput. Simul. 12 1
Marquardt D W 1963 J. Soc. Indust. Appl. Math. 11 431
Rao C R 1963 Sankhya A 25 189

Inst. Phys. Conf. Ser. No 126: Section II
Paper presented at Int. Symp. on Ultrafast Processes in Spectroscopy, Bayreuth, 1991

Three-wave solitons of a new type in Raman scattering of polariton waves

A L Ivanov and G S Vygovskii

Department of Physics, Moscow State University, USSR, 119899

The formation of coupled three-wave solitons of a new type under resonant interaction of two polariton waves by means of LO-phonons in a polar semiconductor is studied theoretically. An adequate approach that consistently considers the polariton effects as well as the relevant exciton-photon-phonon spectra renormalization is developed on the basis of nonlinear wave equations derived from a microscopic approach.

1. INTRODUCTION

The investigation of the coupled three-wave solitons has been intensified since the first experimental observation of the soliton structures in stimulated Raman scattering (SRS) in CO_2-pumped para-H_2 by Druhl et al (1983). In this paper we present an analysis of the three-wave soliton formation under Raman interaction of two excitonic polariton waves in direct-gap semiconductors. The latter are preferable for their large nonlinearities in the vicinity of the excitonic transitions that enable to utilize the electromagnetic pulses of moderate intensity.

The nonlinear propagation of two coherent polariton waves under their strong resonant interaction via LO-phonons may be described (Ivanov and Vygovskii 1991) by the set of macroscopic equations for the positive-frequency components of the two electromagnetic fields $E(\mathbf{r},t)$ and $\mathcal{E}(\mathbf{r},t)$, excitonic polarizations $P(\mathbf{r},t)$ and $\mathcal{P}(\mathbf{r},t)$ and LO-phonon scalar potential $\Phi(\mathbf{r},t)$:

$$\left[\varepsilon_g c^{-2}\frac{\partial^2}{\partial t^2} - \Delta_r\right] E = -4\pi c^{-2}\frac{\partial^2}{\partial t^2} P \quad ;$$

$$\left[\varepsilon_g c^{-2}\frac{\partial^2}{\partial t^2} - \Delta_r\right] \mathcal{E} = -4\pi c^{-2}\frac{\partial^2}{\partial t^2} \mathcal{P} \quad ;$$

$$\left[\frac{\partial^2}{\partial t^2} + \omega_t^2 - (\hbar\omega_t M_{ex}^{-1})\Delta_r\right] P = \omega_t^2 \tilde{\beta} E - \omega_t\rho^{1/2}L\,\mathcal{P}\,\Delta_r\Phi \quad ;$$

$$\left[\frac{\partial^2}{\partial t^2} + \omega_t^2 - (\hbar\omega_t M_{ex}^{-1})\Delta_r\right] \mathcal{P} = \omega_t^2 \tilde{\beta} \mathcal{E} - \omega_t\rho^{1/2}L\,P\,\Delta_r\Phi^* \quad ; \tag{1}$$

$$\left[\frac{\partial^2}{\partial t^2} + \Omega_o^2\right] \Phi = L\rho^{-1/2}(\omega_t\tilde{\beta})^{-1}\,\mathcal{P}^*\,P \quad .$$

Here M_{ex} is the translational exciton mass, ε_g is the background dielectric constant of crystal, ρ is the crystal density, Ω_o is the LO-phonon frequency, ω_t

is the exciton energy, the parameters $\tilde{\beta}$ and L characterize the strength of exciton–photon and Fröhlich exciton–LO–phonon interaction in polar semiconductors, respectively. The most simple configuration will only be considered for a model semiconductor crystal of sulphide cadmium: $\mathbf{e}_p \parallel \mathbf{e}_k$ and $\mathbf{p} \parallel \mathbf{k} \perp$ c-axis, where $\mathbf{p}$ and $\mathbf{k}$ are the carrier wave vectors of the first (E,P) and the second $(\mathcal{E},\mathcal{P})$ polariton pulse, respectively.

2. THE METHOD FOR OBTAINING SOLITON SOLUTIONS

According to set (1), our analysis will concentrate on the case when the carrier frequency ω of the first polariton wave $\mathbf{p}$ lies in the vicinity of the anti-Stokes resonance $\omega_k + \Omega_o$ of the second polariton wave $\mathbf{k}$, treated as a pump nonlinear wave. In this case the traditional approach treating the Raman interaction of electromagnetic waves $E(\mathbf{r},t)$ and $\mathcal{E}(\mathbf{r},t)$ by means of the lowest nonlinear suscep-tibilities $\chi^{(2)}$ and $\chi^{(3)}$ does not work.

Using the slowly varying envelope approximation, one can reduce the set (1) to a more simple set of the first-order ordinary nonlinear differential equations for the envelopes of the corresponding polarizations :

$$\tilde{P}' = -i\, q_1\, \tilde{P} - i\, \rho_1 |\tilde{P}|^2\, \tilde{P} + i\, \beta_1 \tilde{\mathcal{P}}\, \tilde{\Phi} \quad ;$$

$$\tilde{\Phi}' = -i\, a\, \tilde{\Phi} + i\, d\, \tilde{\mathcal{P}}^*\, \tilde{P} \quad ; \tag{2}$$

$$\tilde{\mathcal{P}}' = -i\, q_2 \tilde{\mathcal{P}} - i\, \rho_2 |\tilde{P}|^2\, \tilde{\mathcal{P}} + i\, \beta_2 \tilde{P}\, \tilde{\Phi}^* \quad .$$

where the derivatives are taken over the retarded time $\tau = t - \xi/\upsilon_s$ and the defini-tions are introduced for $i = 1,2$ and $\omega_1 = \omega$, $\omega_2 = \omega_k$, $p_1 = -p$, $p_2 = k$:

$$\rho_i = 2d\omega_t \rho^{1/2} L(p+k) s_i^{-1} \upsilon_s^{-1} \quad ; \qquad a = 0.5\,[\Omega_o^2 - (\omega - \omega_k)^2](\omega - \omega_k)^{-1} \quad ;$$

$$\beta_i = (p+k)\omega_t \rho^{1/2} L(p+k+2a\upsilon_s^{-1}) s_i^{-1} \quad ; \qquad q_i = (a_i - 4\pi\tilde{\beta}\omega_t^2 \omega_i^2 c^{-2} c_i^{-1}) s_i^{-1} \quad ; \tag{3}$$

$$s_i = 2(\omega_i + p_i \hbar\omega_t M_{ex}^{-1} \upsilon_s^{-1}) + 4\pi\tilde{\beta}\omega_t^2 \omega_i (2 + b_i \omega_i c^{-1}) c^{-2} c_i^{-1} \quad ; \qquad c_i = p_i^2 - \varepsilon_g \omega_i^2 c^{-2} \quad ;$$

$$a_i = \omega_t^2 - \omega_i^2 + p_i^2 \hbar\omega_t M_{ex}^{-1} \quad ; \qquad b_i = 2(\varepsilon_g \omega_i c^{-2} + p_i \upsilon_s^{-1}) \quad ; \qquad d^{-1} = 2(\omega - \omega_k)\omega_t \tilde{\beta}\, \rho^{1/2} L^{-1} \quad ;$$

The utilization of polarizations instead of electromagnetic fields allows to explicitly account for the entire series $\chi^{(n)}$ in the framework of the initial microscopic model. Moreover, this procedure is adequate since the polariton waves interact just via polarization components, i.e., exciton components. Introducing the real amplitudes x, y, z and phases φ, χ and ψ for the positive-frequency so-liton envelopes $\tilde{P} = x\, e^{i\varphi}$, $\tilde{\mathcal{P}} = y\, e^{i\chi}$, $\tilde{\Phi} = z\, e^{i\psi}$ one obtains the closed-form set of six nonlinear differential equations:

$$x' = \beta_1\, y\, z\, \sin Q \quad ; \qquad\qquad x\, \varphi' = -(q_1 + \rho_1 y^2)\, x + \beta_1 y\, z\, \cos Q \quad ;$$

$$y' = -\beta_2\, x\, z\, \sin Q \quad ; \qquad\qquad y\, \chi' = -(q_2 + \rho_2 x^2) y + \beta_2 x\, z\, \cos Q \quad ; \tag{4}$$

$$z' = -d\, x\, y\, \sin Q \quad ; \qquad\qquad z\, \psi' = -a\, z + d\, x\, y\, \cos Q \quad ,$$

where $Q = \varphi - \chi - \psi$ is the phase-matching angle for polarization fields. The first three amplitude equations from set (4) have three Manley-Rowe integrals of motion, two of which are independent:

$$\beta_2 x^2 + \beta_1 y^2 = C_1 \quad ; \qquad -dy^2 + \beta_2 z^2 = C_2 \quad ; \qquad dx^2 + \beta_1 z^2 = C_3 \quad . \tag{5}$$

The conventional concept of three-wave solitons assumes that two coupled solitary nonlinear pulses vanishing at $\tau \to \pm \infty$ form a stationary nonlinear perturbation in the cw background of the third wave, which may be referred to as a pump wave. This scheme requires of one of the constant C_i in the integrals of motion (5) to be zero. Only $C_3 = 0$ that corresponds to the case when the second (lower-frequency) polariton wave acts as a pump wave, may lead to stable stationary solutions. Then, one can find the analytical expressions for stationary envelope amplitudes of the polarization fields and phase-matching angle Q. Their type is essentially determined by the sign of parameter β_2. The negative β_2 corres-

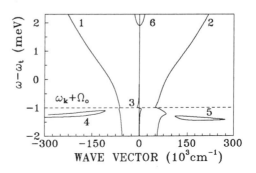

Fig.1. The typical dispersion $p = p(\omega)$ plotted for $\omega_o = 2.513$ eV, $I_o = 500$ MW/cm^2 and $\tau_s = 15$ ps. The parameters are taken for **CdS**: $\omega_t = 2.552$ eV, $\Omega_o = 38$ meV.

ponds to the situation traditionally considered in the theory of three-wave interactions when two nonlinear solitary waves 'burn' a dark soliton in the pump wave, i.e., a dip in the pump intensity profile. When β_2 is positive, the interaction of two solitary waves with the pump leads to the formation of a spike in the intensity profile of pump wave. The characteristic duration of soliton pulses τ_s is given by :

$$\tau_s^{-1} = y_o (-\beta_1 d - a^2)^{1/2} \quad , \tag{6}$$

where y_o is an amplitude of polarization $\tilde{P}$ at $\tau \to \pm \infty$. Assuming vanishing phase modulation of the soliton solutions at $\tau \to \pm \infty$, from the latter three Eqs.(4) and from (5) one can obtain the dispersion equations for the carrier frequencies and wave vectors of the first and the second polariton wave, respectively:

$$q_1 + a = - \rho_1 y_o^2 \quad ; \tag{7}$$
$$q_2 = 0 \quad ,$$

The latter relation represents a dispersion law $k = k(\omega_k)$ for the pump wave in an unperturbed polariton form. The equations (6) and (7) form a complete set of algebraic equations for the soliton velocity $\upsilon_s = \upsilon_s(\omega, \omega_k, I_o, \tau_s)$ and the carrier wave vector $p = p(\omega, \omega_k, I_o, \tau_s)$ of the solitary polariton wave.

3. THE CLASSIFICATION OF SOLITON SOLUTIONS

To classify all the soliton solutions, one should first fix the free parameters of the problem. From a physical point of view it seems reasonable to select the following ones: the carrier frequencies ω and ω_k of polariton pulses determined by the external sources, the intensity of the pump wave I_o, directly related to the quantity y_o, and the common duration τ_s of the soliton pulses. Then, all the soliton solutions may be classified according to the dispersive features $p = p(\omega)$ of the first solitary polariton pulse. The typical dispersion $p = p(\omega)$ for the given values of parameters ω_k, I_o and τ_s is presented in Figure 1. Two different classes of solitons should be considered. The first one deals with the polariton-like dispersion branches *1* and *2* that we denoted as quasi-polariton solitons. A quasi-polariton soliton with the frequency ω sufficiently far from anti-Stokes resonance $\omega + \Omega_o$ propagates with an usual polariton group velocity ω_{pol} while the "wings" of the dispersion branches *1* and *2* transform into an unperturbed polariton dispersion. The second class of solitons, associated with the dispersion branches *3-5*, is of the greatest interest and may be defined as quasi-phonon solitons being "topological fragments" of the phonon dispersion. The branches *4-5* represent closed curves which resemble spectral "droplets" and disappear with decreasing pump intensity. The novel dispersion branches originate from the strong renormalization of the initial LO-phonon term and unperturbed polariton dispersion in the presence of the pump wave (Ivanov and Keldysh 1983). Finally, the branch *6* is related to the upper branch of polariton dispersion.

Unfortunately, in the case of quasi-phonon soliton solutions the approximation of the initial macroscopic Eqs.(1) by the set (2) of the polarization equations becomes invalid, and one should treat a more complicated set of four equations for positive-frequency envelopes $\tilde{P}$, $\tilde{\mathcal{P}}$, $\tilde{\Phi}$ and $\tilde{\mathcal{E}}$ instead of Eqs.(2) . Here we shall only outline the main results. First, the quasi-phonon soliton velocity ω_s, being anomalously small $(10^5 cm/s)$, increases as the pump intensity grows. Such a behavior differs qualitatively from the corresponding dependence $\omega_s = \omega_s(\omega, I_o)$ for the quasi-polariton solitons and directly related to the large phonon component in the internal structure of the considered three-wave quasi-phonon solitons. Second, the stable quasi-phonon soliton solutions exist only for the pump wave intensity I_o that exceeds a

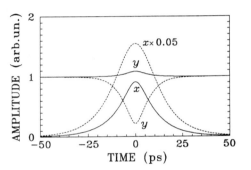

Fig.2. The shapes of soliton components $x(\tau)$, $y(\tau)$ for the quasi-phonon *3* (solid line) and quasi-polariton *1* (dashed line) branches. Pump intensity $I_o = 200 \ MW/cm^2$, $\tau_s = 10 \ ps$.

threshold intensity $I^{th} = I^{th}(\omega, \tau_s, \omega_k)$ ($\approx$ *100-200 MW/cm²* for **CdS**). The frequency band of the stable quasi-phonon solitons is located in the close vicinity of the anti-Stokes resonance and is in fact related to the inverse dispersion region of the branch *3* and the upper sectors of the "droplets" *4-5*. The typical shapes of the envelopes $x(\tau)$, $y(\tau)$ for the stable quasi-phonon *3* and quasi-polariton *1* branches, which could be excited in the forward scattering configuration of wave interaction, are presented in Figure 2.

One of us (A.L.I.) thanks the Alexander von Humboldt Foundation for support.

REFERENCES

Druhl K, Wentzel R G and Carlsten J L   1983 *Phys.Rev.Lett.* **51** 1171
Ivanov A L and Keldysh L V   1983 *Soviet Phys.- J.exper.theor.Phys.* **57** 234
Ivanov A L and Vygovskii G S   1991 *Sol.St.Comm.* **78** 787
Kaup D J, Reiman A and Bers A   *Rev.Mod.Phys.* 1979 **51** 275

Inst. Phys. Conf. Ser. No 126: Section II
Paper presented at Int. Symp. on Ultrafast Processes in Spectroscopy, Bayreuth, 1991

Spatial and temporal distribution of femtosecond pulses after tight focusing

M Kempe, U Stamm, R Gutewort, B Wilhelmi [+], W Rudolph [++]

Friedrich Schiller University, Faculty of Physics and Astronomy, Institute of Optics and Quantum Electronics, Max-Wien-Platz 1, D-6900 Jena, F.R. Germany

[+] Jenoptik GmbH, Carl-Zeiss-Str. 1, D-6900 Jena, F.R. Germany

[++] The University of New Mexico, Department of Physics and Astronomy, Albuquerque, NM 87131, U.S.A.

ABSTRACT: The limits of spatial and temporal resolution achievable in time-resolved laser scanning microscopy are investigated theoretically using Fourier optics. The application of circular and annular lens apertures is compared. First experiments using a CPM-laser are presented.

1. INTRODUCTION

The combination of ultrafast spectroscopy with laser scanning microscopy offers a powerful tool for ultrafast photography of the micro-world with a spatial resolution in the submicron and a temporal resolution in the femtosecond range. Aiming at specific applications knowledge is required about the obtainable spatial and temporal resolution as well as about distortions in the imaging process.

In present the highest spatial resolution can be achieved using microscope objectives made from glass lenses. Investigations about tight focusing of femtosecond pulses by lenses on the basis of geometrical optics (Bor 1988) as well as on the basis of diffraction theory (Bor 1991, Kempe et al. 1991a) have shown that pulse shape and duration is altered considerably if the lenses possess chromatic aberration. Furthermore the spatial distribution of fs-pulses behind lenses can differ from the distribution of a focused monochromatic plane wave that is usually used to describe the imaging properties of a laser scanning microscope (Wilson and Sheppard 1984). These effects are caused by the difference between the phase and group velocity as well as by the group velocity dispersion (GVD) in the lens material. The delay of the pulse front with respect to the phase front has been verified experimentally by Bor et al. (1989).

In Section 2 of the present paper we offer a relatively simple expression for the field amplitude of a Gaussian shaped input pulse near the focal plane of circular lenses. Measurements of the pulse duration and spatial distribution in the focal plane of a microscope objective are presented. The advantages of the use of annular instead of circular apertures are discussed in Section 3.

2. BEHAVIOR OF PULSES AFTER TIGHT FOCUSING

In Fourier optics the propagation of each frequency component of the spectrum of an ultrashort pulse through a lens is described by Kirchhoff's diffraction formula in the Fraunhofer approximation. If one takes into account the frequency dependence of the wave number vector and integrates over all frequency components of the amplitude, this diffraction formula gives the solution of the linear wave equation in the time domain, and, with appropriate initial conditions it describes the propagation of a short light pulse.
In most applications the bandwidth $\Delta\omega$ of the laser pulse is small compared to its carrier frequency ω_0. Therefore it is justified to expand the wave number vector $k(\omega)$ around the center frequency ω_0 of the incoming pulse up to the second order in $\Delta\omega = \omega - \omega_0$.
In this case, assuming a Gaussian shaped input pulse with the envelope

$$A(t) = \exp\left\{-\left(\frac{t}{T}\right)^2\right\} \tag{1}$$

we obtain for the field amplitude behind a circular lens (radius a and thickness d along the optical axis) near the focal plane at $z = f_0 = f(\omega_0)$ the expression

$$U(v, u; t) \propto \int_0^1 dr\, r\, J_0(rv)\, \exp\left\{-j\frac{u}{2}r^2\right\} \sqrt{\frac{1+j\underline{\delta}(r;T)}{1+\underline{\delta}^2(r;T)}} \cdot$$

$$\cdot \exp\left\{-\frac{(t+\tau r^2)^2}{T^2[1+\underline{\delta}^2(r;T)]}[1+j\underline{\delta}(r;T)]\right\}$$

$$with \quad \underline{\delta}(r;T) = \frac{4(\delta'-\delta r^2)}{T^2} \tag{2}$$

(Kempe et al. 1991a). The optical coordinates are given by $v = ak_0 r_2/f_0$ (r_2: radius coordinate in the plane at distance z from the lens) and $u = a^2 k_0(1/f_0-1/z)$, and t is the local time coordinate. Furthermore we use $k_0 = \omega_0/c$ and $n_0 = n(\omega_0)$, where $n(\omega)$ is the frequency-dependent refractive index of the lens material. J_0 denotes the Bessel function of the first kind of order zero.
We have introduced the parameters

$$\delta = \frac{a^2 k_0}{2f_0(n_0-1)}\left(\frac{1}{\omega_0}\frac{dn}{d\omega}\bigg|_{\omega=\omega_0} + \frac{1}{2}\frac{d^2n}{d\omega^2}\bigg|_{\omega=\omega_0}\right) \quad ,$$

$$\delta' = k_0 d\left(\frac{1}{\omega_0}\frac{dn}{d\omega}\bigg|_{\omega=\omega_0} + \frac{1}{2}\frac{d^2n}{d\omega^2}\bigg|_{\omega=\omega_0}\right) \quad and$$

$$\tau(u) = \frac{a^2 k_0}{2 f_0 (n_0 - 1)} \frac{dn}{d\omega}\Bigg|_{\omega=\omega_0} + \frac{u}{2\omega_0} . \tag{3}$$

Equation (2) may be considered as superposition of "partial pulses" with different chirp $\underline{\delta}$, that are delayed with respect to each other, where the delay is described by the parameter τ. The delay and chirp of the partial pulses depend on the radius coordinate in the lens plane r, and therefore they are an inherent result of focusing. It is noteworthy that the focusing gives rise to a specific coupling between the temporal, spectral and spatial properties of the pulse. The strength of this coupling depends on the parameters δ and τ. It can lead to considerable pulse stretching and spreading of energy away from the optical axis in the focal plane of the lens (see for instance Bor 1991). These effects are very sensitive according to the chromatic aberrations of the lens if pulses are tightly focused.
The delay parameter τ is the most important parameter that can be used to characterize the chromatic correction of a lens system. It must be smaller than the pulse duration T to avoid the effects mentioned. Mostly one uses microscope objectives that are well corrected (i.e., $df/d\lambda \approx 0$) around $\lambda_0 = 550$ nm.

In Figure 1 the image of a straight edge and the calculated line spread function (LSF) for a planachromat 100x/0.85 from the company Carl Zeiss Jena is shown. It was obtained by incoherent imaging using 620-nm pulses from a CPM-laser. The simultaneously measured pulse duration (full width at half maximum - FWHM, assuming Gaussian pulses) in the focal plane is 80 fs, whereby the transform-limit is 60 fs. This difference can be explained by the GVD introduced by the glass of the objective. The broadening of the half width of the LSF in comparison with perfect diffraction-limited imaging (200 nm) and with

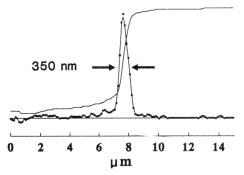

Fig. 1. Image of a straight edge and line spread function, objective: planachromat 100x/0.85, using a CPM-laser.

measurements using a He-Ne-laser (260 nm), respectively, is caused by the relatively bad beam profile of the CPM-laser. In summary it can be seen that though the correction of the objective is not apochromatic the mentioned effects are very small if working at a wavelength of 620 nm. The situation changes completely if other wavelength ranges are used. Thus covering the whole visible spectrum, several very well corrected achromatic lenses or objectives have to be used.
Because this is not very favorable the question arises how the pulse distortion can be avoided in principle.

3. THE USE OF ANNULAR LENS APERTURES

It can be shown that the pulse stretching can be eliminated using annular instead of circular apertures if choosing the inner radius of the annulus according to

$$\epsilon \approx \sqrt{1 - \frac{T}{\tau}} \tag{4}$$

(Kempe et al. 1991b).

The FWHM of the measurable quantities photon flux and energy density that are given by the integral of the absolute square of the field amplitude (2) over v and t, respectively, at the focal plane of a circular and annular lens are shown in Figure 2.

Here the GVD was taken into account for a lens with $f_0 = 25.4$ mm, $a = 12.7$ mm (numerical aperture 0.45), $d = 14.2$ mm made from the Schott glass BK7. The input pulse is chirped in such a way that the partial pulses passing the lens near the aperture edge are transform-limited behind the lens. This chirp δ_{opt}'' which is given by

$$\delta_{opt}'' = \frac{4(\delta' - \delta)}{T^2} \tag{5}$$

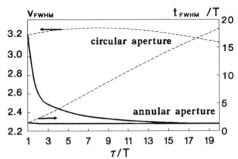

Fig. 2. FWHM of the photon flux and the energy density vs. normalized delay parameter, using circular (dashed lines) and annular apertures (solid lines).

is found to be the optimum using circular lenses if high spatial and temporal resolution is required (Kempe et al. 1991a). More detailed investigations show that nearly transform-limited, undistorted Gaussian shaped pulses are obtained in the focal plane of a lens of annular aperture with an inner radius ϵ due to Equation (4). The spatial distribution in these cases is equal to the distribution of a monochromatic plane wave behind the same aperture. Thus all considerations of annular apertures illuminated by monochromatic light can here be applied too (see, for instance Wilson and Hewlett 1990).

REFERENCES

Bor Z 1988 *J. Mod. Opt.* **35** pp 1907-18
Bor Z 1991 in this issue.
Bor Z, Gogolak Z and Szabo G 1989 *Opt. Lett.* **14** pp 862-4
Kempe M, Stamm U, Wilhelmi B and Rudolph W 1991a submitted to *JOSA B*
Kempe M, Stamm U and Wilhelmi B 1991b submitted to *Opt. Comm.*
Wilson T and Hewlett S J 1990 *J. Mod. Opt.* **37** pp 2025-49
Wilson T and Sheppard C 1984 *Theory and Practice of Scanning Optical Microscopy*
(London: Pergamon)

Inst. Phys. Conf. Ser. No 126: Section II
Paper presented at Int. Symp. on Ultrafast Processes in Spectroscopy, Bayreuth, 1991

Behaviour of femtosecond pulses in lenses

Zs. Bor, Z. L. Horváth

Department of Optics and Quantum Electronics,
József Attila University, Dóm tér 9. H-6720 Szeged, Hungary

ABSTRACT: The diffraction and distortion of femtosecond pulses in lenses is described by using wave optical theory. The propagation of a forerunner pulse along the optical axis is predicted. The forerunner pulse is resulted by the interference of the boundary waves generated by the aperture of the lens. The size of the spot in the focal plane is larger, than that of the Airy pattern. The radius of the diffraction rings in the focal plane increases with time.

1. INTRODUCTION

The spatial profile of a femtosecond pulse front can suffer a considerable distortion upon propagation through a lens [Bor 1988, 1989, Bor et al 1988, 1989 1992a, Gutewort et al 1991, Federico et al 1991, Staerk et al 1988, Szatmari et al 1988]. The effect is due to the difference between the phase and group velocity of the light in the material of the lens. The distortion was successfully described by the geometrical optical theory [Bor 1988, 1989].

The validity of this theory has already been experimentally verified by measuring the pulse front distortion using time of flight interferometry [Bor et al 1989] and autocorrelation technique [Szatmari et al 1988]. It was also shown in [Bor 1988, 1989], that pulse front distortion occurs in any lens, having chromatic aberration. The behaviour of the pulse front is especially interesting at the near vicinity of the focus, where the geometrical optical theory looses it's validity [Bor 1988]. In this paper a wave optical theory is presented.

2. PRINCIPLE OF THE CALCULATIONS

We considered a lens which is homogeneously illuminated by a parallel beam carrying a femtosecond pulse. The input pulse can be represented as a superposition of monochromatic plane waves. The amplitude of each monochromatic wave is given by the Fourier-transform of the input pulse. The lens converts the monochromatic plane waves into spherical-like waves, having a radius of curvature of $f(\lambda)$ converging to the focal point. The focal length was supposed to be a linear function of the wavelength. (i.e. group velocity dispersion in the material of the lens was neglected.) The intensity distribution of one monochromatic component can be calculated by solving the corresponding diffraction integrals [Born et al 1975a]. The pulse front in the vicinity of the focus is calculated as the superposition of the fields of individual spectral components. Mathematical details of the calculations will be given in [Bor et al 1992a].

For comparison, the calculations have been carried out using the parameters of the lens used for the geometrical optical calculations [Bor 1988] (i.e. wavelenght λ=249 nm, focal length f = 150 mm, diameter of the lens 2a = 80 mm, refractive index of the fused silica lens n =1.50799, dispersion of the refractive index λ dn/dλ = -0.1375 (fused silica), longitudinal chromatic aberration of the lens df/dλ = 163.0895 mm/μm). The input pulse was supposed to be Gaussian and 100 fs long (FWHM in intensity).

3. RESULTS OF CALCULATIONS

3.1. Distortion of the pulse front

Figure 1a-e shows the shape of the pulses for moments t=-10, -6, -2, 2, and 6 ps respectively. (Here the origin of time t was chosen so, that the peak of pulse propagating along the optical axis z arrives into the focus at the moment t=0. The coordinate r is the distance from the optical axis.) The insets on the figure show the top view of the contours of equal intensity lines. Comparison of these lines with the pulse fronts calculated from the geometrical optical theory [Bor 1988] (see the broken line inside the contour lines) shows an excellent agreement, proving the validity of the geometrical optical theory concerning the position of the pulse front [Bor et al 1991].

3.2. Boundary wave forerunner pulse on the optical axis

The most unexpected result of the wave optical calculation is that together with the horseshoe-shaped main pulse front an additional narrow pulse is propagating along the optical axis (Fig. 1). This pulse was not predicted by geometrical optics and will be referred to as boundary wave forerunner pulse.

Our simplified interpretation of the boundary wave pulse is the following. According to [Born et al 1975b], in many diffraction problems the Kirchhoff-Fresnel diffraction integral after proper transformations can have such a form, that the electric field is represented as the sum of a direct field and the field of the boundary diffraction wave. The latter may be thought of as arising from the scattering of the incident radiation by the boundary of the aperture [Born 1975b]. The boundary waves in our case have circular symmetry, thus due to interference, significant intensity is resulted only on the optical axis in a spot having a diameter of the order of d$\approx\lambda$f/(2a). Notice, that the boundary wave pulse has completely different physical origin than the Sommerfeld and Brillouin-type of forerunner (precursor) pulse [Brillouin 1966] and therefore they should not be confused.

3.3. Intensity distribution in the focal plane

Figure 2a shows the radial intensity distribution in the focal plane (z=0). The interference rings in the focal plane resemble somewhat the Airy pattern, however, there are two essential differences. Namely, with increasing order number of the maxima of the rings the intensity decays much slower than it is expected from an Airy pattern. Besides, the radius of the rings is increasing with time. The above results are in full agreement with the recent results of Gutewort et al 1991, Federico et al 1991.

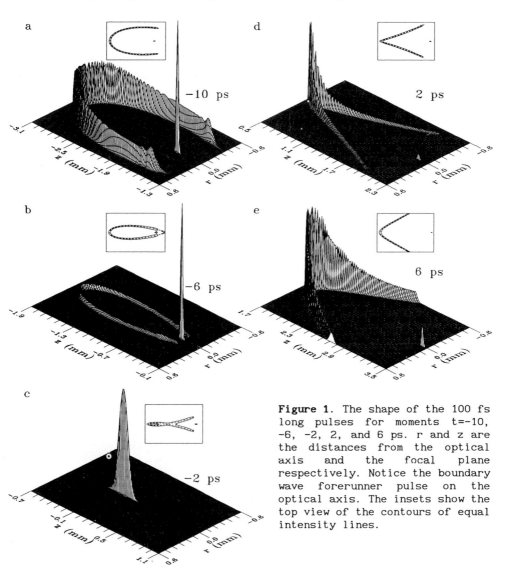

Figure 1. The shape of the 100 fs long pulses for moments t=-10, -6, -2, 2, and 6 ps. r and z are the distances from the optical axis and the focal plane respectively. Notice the boundary wave forerunner pulse on the optical axis. The insets show the top view of the contours of equal intensity lines.

A simplified interpretation of the observed pattern is the following. Due to the pulse front distortion, an observer located in the focal point sees that for the interval of time -4.81 ps to 0 ps a thin bright ring with shrinking diameter appears on the surface of the lens [Bor et al 1992a]. The pattern observed in the focal plane (Figure 2a) is resulted from the diffraction from these bright rings.

3.4. Intensity distribution in a plane neighbouring the focal plane

Figure 2b, c and d show the intensity distribution in planes z= -120 , -200 and -280 μm before the focal pane. (The spatial scale of this figure in the radial direction r is much higher than that used in Figure 1, allowing one to see finer details of the boundary wave pulse.) The temporal separation between the boundary wave pulse and the main pulse is 4.81 ps. The boundary wave pulse has the same duration (100 fs) as the input pulse.

As it was discussed in III.2. the boundary wave pulse may be thought of as arising from the scattering of the incident radiation by the boundary of the aperture [Born et al 1975b]. Consequently, the radial distribution should correspond to the diffraction pattern from a narrow ring having a diameter of the boundary of the aperture. The solution of such diffraction problem might be found for example in [Leith et al 1980]. Applying those results it can be shown [Bor et al 1992a], that the radial distribution should obey

$$\frac{I_b(r)}{I_b(0)} = \left[\frac{J_0\left[\frac{2\pi a}{\lambda f} r \right]}{\frac{2\pi a}{\lambda f}} \right]^2, \tag{1}$$

where J_0 is the 0th order Bessel function. Comparison of this equation with the radial intensity distribution of the forerunner (Figure 2d) shows perfect agreement.

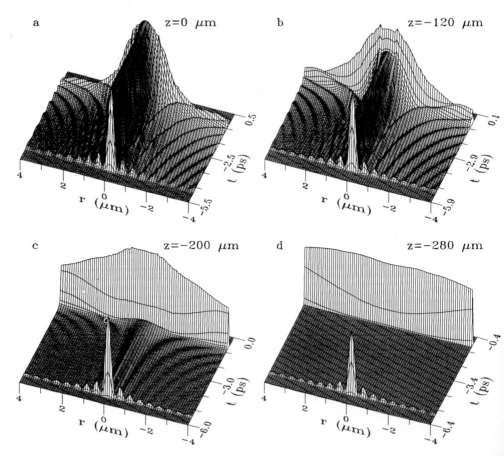

Figure 2. Temporal behaviour of the pulse in the planes z=0, −120, −200 and −280 μm before the focal plane.

Figure 3 shows the shape of the pulse in the moment t=-10 ps for an input pulse of 200 fs. Comparison with Figure 1a shows, that the shape of the pulse front is independent of the input pulse duration. (This is in full agreement with the predictions of the geometrical optical theory [Bor 1988].) The difference is only in the duration of the pulse in a arbitrary point, i.e. the thickness of the "horseshoe" is doubled.

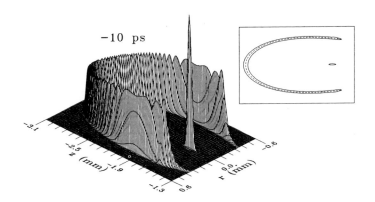

Figure 3. The shape of a 200 fs long pulse for the moment t=-10 fs.

4. CONCLUSIONS

The propagation of a 100 fs long pulse through a lens has been described using wave optical description. The pulse front behind the lens can be calculated as the superposition of the electric fields of the Airy type diffraction patterns of the Fourier components of the input pulse.

The wave optical description supports the validity of the geometrical optical theory, describing the shape of the pulse front.

The intensity distribution in the focal plane is different from the Airy pattern. The diffraction pattern consists of rings, with continuously increasing radius.

The most unexpected result is that in front of the main pulse a boundary wave forerunner pulse also appears. The radial intensity distribution of the boundary wave pulse supports the assumption, that it is resulted from the interference of the boundary waves diffracted by the aperture of the lens. Our preliminary calculations show, that the boundary wave pulse appears also for lenses having no chromatic aberration and for mirrors as well. Apodization of the input beam [Bor et al 1992a,b, Federico et al 1991] leads to the suppression of the forerunner pulse.

5. ACKNOWLEDGEMENTS

This work has been supported by the OTKA Foundation (No 3055 and 3056) and the Hungarian-U.S. Science and Technology Joint Fund in cooperation with National Science Foundation and Hungarian Academy of Sciences under Project (No 061/90).

6. REFERENCES

Bor Z 1988 J. Mod. Opt. **35** 1907
Bor Z and Szabo G 1988 Digest of technical papers of the European Conference on Quantum electronics, Hannover, FRG Sept. 12 1988
Bor Z 1989 Opt. Lett. **14** 119
Bor Z, Gogolak Z and Szabo G 1989 Opt. Lett. **14** 862
Bor Z 1990 Digest of papers of the XVII International Conference on Quantum Electronics, 21-25 May 1990 Anaheim, California, p. 256.
Bor Z, Horváth Z L 1991, Proceeding of the Symposium on Ultrafast Processes in Spectroscopy, Bayreuth, Germany, Oct. 7-10, 1991, paper MO 6-3.
Bor Z, Horváth Z L 1992a , "Dye Laser 25 Years" Ed.: M. Stuke, Topics in Applied Physics, vol. 70 Springer-Verlag (accepted)
Bor Z, Horváth Z L 1992b, submitted to the Conference on Lasers and Electro-Optics, 12-14 May 1992 Anaheim, California
Born M and E. Wolf 1975a, Principles of optics, Pergamon Press, Oxford, New York, Toronto, Sydney, Braunschweig, (1975) p.435
Born M and E. Wolf 1975b, Principles of optics, Pergamon Press, Oxford, New York, Toronto, Sydney, Braunschweig, (1975) p.449
Brillouin L 1960 Wave propagation and group velocity, Academic Press New York, London
Federico A and Martinez O Private Communication
Gutewort R, Kempe M, Rudolf W, Stamm U, Wilhelmi B, 1991, Symposium on Ultrafast Processes in Spectroscopy, Bayreuth, Germany, October 7-10, 1991, paper MO 6-4
Leith E N, Collins G, Khoo I and Wynn T 1980 J. Opt. Soc. Am. **70** 141
Staerk H, Ihlemann J, Helmbold A 1988 Laser und Optoelektronik **20** 6
Szatmari S, Kuhnle G 1988 Opt. Commun. **69** 60

Inst. Phys. Conf. Ser. No 126: Section II
Paper presented at Int. Symp. on Ultrafast Processes in Spectroscopy, Bayreuth, 1991

169

Propagation-time-dispersion in a streak camera lens

K Osvay, Z Bor, B Rácz and G Szabó

Department of Optics and Quantum Electronics, József A University, Dóm tér 9, H-6720 Szeged, Hungary

ABSTRACT: We point out that a time of flight interferometer is suitable to measure the transit time of even complicated optical systems. As the first such application of this method the values of transit time in case of a streak camera lens is presented. The accuracy of the treatment was checked by a control measurement and calculation on a quartz lens and was found to be 10fs.

1. INTRODUCTION

The propagation time of an ultrashort light pulse passing through optical elements depends on the wavelength. This effect can considerable change the temporal and spectral structure of a light pulse (Staerk *et al* 1988). In case of complicated optical systems, however, the calculation of these effects is fairly difficult. Therefore the dispersion of the transit time, which can be extremely large in e.g. streak camera objectives, would rather be measured.

In this paper we demonstrate a method which is suitable to measure the transit time even in lens systems of unknown and complicated construction. The accuracy of this technique exceeds that published by Staerk *et al* (1988).

2. EXPERIMENTAL

Figure 1 shows the time of flight interferometer (Bor *et al* 1989) used for measuring the propagation time of broadband ($\delta\lambda \approx 5$ nm) light pulses. In case of an empty interferometer, interference fringes with unit visibility appear only if the lengths of the two arms are equal. Inserting the lens to be studied into the object arm, it is necessary to increase the length of the reference arm by $\Delta = c \cdot \delta T(\lambda)$ (c is speed of light in vacuum) in order to observe the interference fringes again. Note, that $\delta T(\lambda)$ is the change of the propagation time caused by the lens relative to the vacuum.

There are two effects which can influence the accuracy of measurements such as the pulse front distortion (Bor 1988, Bor and

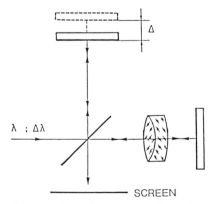

Figure 1. The time of flight interferometer used for measuring the propagation time of a pulse through a lens.

Horvath 1991) and the pulse broadening. Since the duration of laser pulses we used were ≈5ns and the lenses were illuminated along only their optical axis so in the visible range the values of the pulse front distortion caused by the quartz lens are less than 0.3fs. In addition, the calculations carried out according to that of Herrmann, Wilhelmi

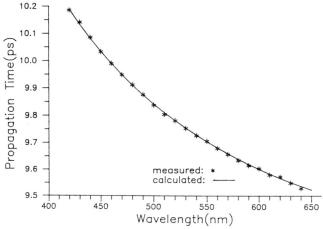

Figure 2. The values of transit time for a quartz lens
(f=100mm, central thickness=6.02mm).

(1987) do not show larger deviation in pulse duration due to the pulse broadening than 0.01fs. Consequently, the accuracy of the measurement does not depend on such phenomena but it is determined by the effects described by Bor *et al* (1990).

In order to test the method first we measured $\delta T(\lambda)$ for a known quartz lens (see figure 2). The appropriate computations were made on the basis of our previously work (Bor *et al* 1990). The accuracy of this technique has been found to be 10fs.

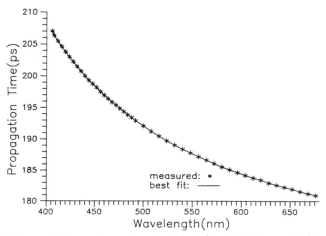

Figure 3. The values of propagation time $\delta T(\lambda)$ in a streak
camera objective.

The values of the transit time measured for the objective of a streak camera (Hamamatsu C979) are seen on figure 3. It occurs an extremely large (26ps) change in the propagation time over the visible range.

This shows, that the streak camera lens can cause severe experimental artifacts by changing the temporal and spectral appearance of broadband ultrafast optical signals such as white continuum, Raman scattering, fluorescence e.t.c.. For instance, it was reported by Chattopadhyay *et al* (1987) that the blue part of the ps-continuum (450nm) lags behind the red part (680nm) about 20ps. We believe, that the observed time lag is mainly due to the transit time dispersion effects in the streak camera objective and other dispersive elements in that experimental set-up.

3. CONCLUDING REMARKS

We have shown that the earlier used technique for group index measurement is also suitable to determine the propagation time in difficult lens system. We have pointed out, in addition, that the used method is not sensitive to neither the pulse broadening nor the pulse front distortion. The transit time measurements carried out for a streak camera lens and the given example prove the importance of proper knowledge of optical elements to be used.

ACKNOWLEDGEMENTS

This publication based on work sponsored by the Hungarian-U.S.Science and Technology Joint Fund in cooperation with National Science Foundation and Hungarian National Academy of Sciences under Project 061/90. This report has been also supported by the OTKA Foundation of the Hungarian National Academy of Sciences under Grant no. 3055.

Bor Z 1988 *J. of Modern Opt.* **35** 1907
Bor Z, Gogolak Z, Szabó G 1989 *Opt. Lett.* **14** 862
Bor Z, Osvay K, Racz B, Szabó G 1990 *Opt. Comm.* **78** 109
Bor Z, Horvath Z L 1991 *Proceedings of VIIth International Symposium on Ultrafast Processes in Spectroscopy*, see in this issue.
Chattopadhyay S K, Craig B B 1987 *J.Phys.Chem.* **91** 323
Herrmann J and Wilhelmi B 1987 *Lasers for Ultrashort Light Pulses* (Amsterdam: Elsevier)
Staerk H, Ihlemann J, Heimbold A 1988 *Laser und Optoelektronik* **20** 34

Inst. Phys. Conf. Ser. No 126: Section II
Paper presented at Int. Symp. on Ultrafast Processes in Spectroscopy, Bayreuth, 1991

Ultrafast cross-phase modulation

P Heist[1], J Krüger[1], W Rudolph[1], T Schröder[1], P Dorn[1], F Seifert[2], B Wilhelmi[3]

[1]   Friedrich-Schiller-Universität Jena, Physikalisch-Astronomische Fakultät, Institut für Optik und Quantenelektronik, O-6900 Jena, Germany

[2]   Zentralinstitut für Optik und Spektroskopie Berlin, Rudower Chaussee 6, O-1199 Berlin, Germany

[3]   Jenoptik Carl Zeiss Jena GmbH, Zentralbereich Forschung und Entwicklung, Carl-Zeiss-Str.1, O-6900 Jena, Germany

ABSTRACT: Optical signals are tuned by more than 10nm within 100fs using cross-phase modulation originating from strong transient changes of refractive index induced by intense femtosecond light pulses in semiconductors and semiconductor microcrystallites in glass matrices.

1.INTRODUCTION

Since the area of communication and information technologies is expected to require huge bandwidths and high parallelity complex systems used for signal processing there is an increasing interest in all-optical devices, which at least in principle, have several advantages compared with optoelectronic approaches, see, e.g., Tamir (1988).

Ultrafast cross-phase modulation induced in optical signals by picosecond or even femtosecond light pulses might be used to

- impress information onto these signals
- switch these signals to various geometrical positions by refraction, diffraction or interference via changes of the signal wavelength and
- change the signal intensity by spectral filtering or the signal polarization direction.

For many applications it is of importance that pump and signal mid frequencies need not coincide. First, it is easier to separate pump and signal after their interaction, when their mid wavelengths differ considerably, e.g., Alfano et al (1986) modulated second harmonic radiation at 0.53μm in glass fibers using the fundamental wave at 1.06μm for pumping. Second, very large effects have been achieved in highly nonlinear semiconductors (Kaschke et al 1988 and Wilhelmi et al 1988) and organic polymers with ultrashort pump pulses being

strongly absorbed and signal pulses passing the sample in loss-free spectral regions. Third
in some applications it is favourable to use signals that are completely absorbed at the
unshifted wavelength and can only be detected as long as the wavelength is shifted into the
transmission range of the spectral filter.

In analogy to self-phase modulation, cross-phase modulation has moreover been applied to
increase the spectral bandwidth of signal pulses in such a manner that the latter could be
compressed in duration down to the new Fourier limit. In this way Yamashita et al (1990)
obtained compressed pulses of 5fs and 12fs duration via cross-phase modulation induced by
pump pulses at the signal wavelength λ_s and at another wavelength $(2\lambda_s)$, respectively.

In general, phase and frequency modulation of optical signals can be achieved by changing
the refractive index at the signal wavelength. The field strength

$$E(t,z) = A(t,z) \cdot \cos[\varphi(t,z)]$$

of the signal as a function of time t and space coordinate z is described by its amplitude
$A(t,z)$ and the phase

$$\varphi(t,z) = \omega_s\left(t-n_s\frac{z}{c}\right) + \varphi_0(t)$$

where ω_s is the carrier frequency, n_s is the refractive index at ω_s and $\varphi_0(t)$ is the input
phase. The instantaneous frequency of the field at the sample exit $z = L$ is given by the
temporal derivative of the phase φ as

$$\omega(t) = \dot{\varphi}(t) = \omega_s - \omega_s \frac{L}{c} \dot{n}_s(t) + \dot{\varphi}_0(t)$$

Here the second term of the sum represents the influence of the refraction index modulation
which may arise from the influence of hf electric fields via the Pockels or Kerr effect or of
intense light fields via self- and cross phase modulation. The latter nonlinear optical pheno-
nomena stem from the third order polarization $P^{(3)} \propto \chi^{(3)}:E\,E\,E$ where two factors in
the field product are related to the pump pulse used for cross-phase modulation.

With irradiation of the sample by pump and signal far off resonance the response is almost
instantaneous, and the refractive index change $\Delta n_s(t)$ and frequency change $\Delta\omega_s(t)$ are
given by

$$\Delta n_s(t) = f(t,\omega_s;I_p,\omega_p) = n_2 I_p$$

$$\Delta\omega_s(t) = -\omega_s \frac{L}{c} n_2 \dot{I}_p(t),$$

where n_2 is a nonlinear optical parameter of the sample and I_p is the pump intensity. Thus
we have a very simple behaviour of the nonlinear optical process, which, for a bell-shaped
temporal profile of the pump pulse, leads to a dispersion-shaped profile of the frequency
change, i.e. with $n_2 > 0$ the frequency change is negative at $t < 0$, zero at $t = 0$ and
positive at $t > 0$. The simple temporal response and the almost linear rise of the frequency
in the vicinity of the pump pulse maximum can advantageously be used in many applications. On the other hand, the optical nonlinearity, described by $\chi^{(3)}$ or n_2, is rather small

or off-resonant processes, and consequently, long interaction paths, e.g. in fibers are required.
The optical nonlinearities become far greater in (or near) resonance where, however, the temporal response becomes more complex in general. Here we restrict the introductory remarks to one limiting case, where the refractive index change originates from level repopulation by n-photon absorption and the life time of the upper level is large compared to the pulse duration. Then we have

$$\Delta n_s(t) \quad \propto \quad \int_{-\infty}^{t} dt' \ [I_p(t')]^n$$

and

$$\Delta \omega_s(t) \quad \propto \quad [I_p(t)]^n$$

i.e. the frequency variation does not change its sign. In reality, when pumping with very short light pulses, the changes of the refractive index arise from several effects at the very beginning.
In semiconductors, e.g., the refractive index is changed in time by

- the dynamic Stark effect,
- exciton-exciton and electron-exciton screening,
- gap shrinkage and
- phase-space filling,

where each individual process is characterized by a specific temporal response (see, e.g., Zimmermann 1988)

2.DETERMINATION OF ULTRAFAST REFRACTIVE INDEX CHANGES

The measurement of fast nonlinear optical changes of absorption and refractive index was performed on pump-and probe spectrometers using picosecond (Kaschke et al 1988) and femtosecond (Lap 1990) light pulses.
Figure 1a represents as an example the spectral transmission of a GaSe platelet with and without irradiation by an intense picosecond pump pulse, where the fringe structure results from the Fabry-Perot interferometer built up by the parallel semiconductor faces.
Strong bleaching occurs near the band edge which is accompanied by a considerable refractive index change in a wide spectral range, including regions of high transmission, as can be seen from the spectral shift of the interference maxima. From the transient transmission spectra the absorption and refractive index changes can be deduced, see Dneprovskij et al (1987). When the absorption changes are known, the refractive index changes can be calculated using the Kramers-Kronig relation. The resulting refractive index changes are depicted in Fig.1b.
Figure 2 shows the absorption and refractive index changes after ps excitation in CdS. Obviously, refractive index changes as high as 0.2 could be obtained.

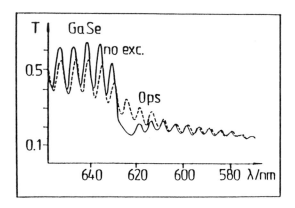

Fig.1a: Transient transmission spectra of a GaSe sample withou excitation (full line) and after excitation (dashed line) by a p pump pulse ($\lambda_P = 527nm$, excitation density $E_P = 10mJ/cm^2$) at delay of 0ps.

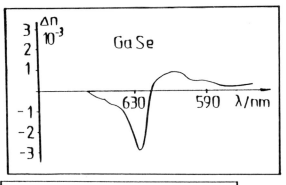

Fig.1b: Nonlinear refractive index changes in GaSe after p excitation calculated from transmission changes (Fig.1a) using the Kramers-Kronig relation.

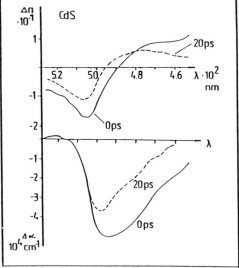

Fig.2: Nonlinear absorption and refractive index changes in CdS after ps excitation ($\lambda_P = 351nm$ energy density $E_P = 6mJ/cm^2$) for two different delays. The change in refractive index Δn have been obtained by Kramers-Kroni transformation of the transient absorption changes.

Colour filters which consist of semiconductor microcrystallites embedded in glass matrices are also attractive candidates for refractive index modulation. Figure 3 represents the change of the refractive index versus wavelength at various delays for the filterglass RG8 (Jenaer Glaswerk GmbH, l=0.5mm excited at λ=620nm and E_P=10mJ/cm^2) which are obtained by the help of the Kramers-Kronig relation from fs pump-and probe transmission spectra (Lap 990).

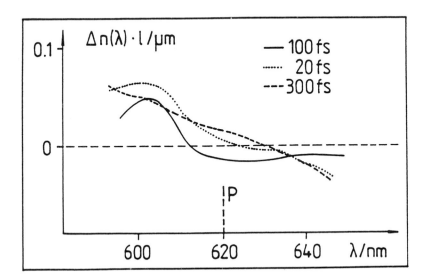

Fig.3: Change of refractive index versus wavelength for various delays between pump and probe pulses. Sample: Semiconductor doped glass (SDG) filter RG8 (Jenaer Glaswerk GmbH, pump: λ=620nm, $E_P \approx$ 10mJ/cm^2).

.MEASUREMENT OF WAVELENGTH SHIFTS

With resonant irradiation by femtosecond light pulses the transient behaviour of absorption and refraction was investigated experimentally in regions of (i) high transmission and (ii) strong absorption of the sample. Let us discuss examples for these two types of interaction phenomena.

A 3μm thin $CdS_{0.52}Se_{0.48}$ sample was excited by two - photon absorption of strong fs pump pulses (λ_P=616nm, T_P=120fs). Thus we impressed strong time dependent phase modulation onto weak signal pulses (λ_S=605nm, bandwidth $\approx$ 1nm) wich passed through the sample in the transparent spectral region. We observed new spectral components in a 20nm broad range on both sides of the mid frequency of the signal pulse (Fig.4). The spectral features on the high-frequency side of the signal can, at least qualitatively, be explained using the simple formula given in the introduction for resonant two-photon excitation. The extent of the modulation on the low-frequency side indicates strong influence of very fast transient processes, which occur before carrier thermalization in the conduction band takes place. The main initial contribution stems from fast gap renormalization by two-photon excited carriers which leads to a decrease of transmission at the very beginning of the pump pulse. Later on the in traband relaxation of these highly excited carriers towards the band edge yields an increase of transmission via band filling. Such an increase and decrease in absorption might be responsible for the almost symmetrical spectral modulation around the signal wavelength. This is confirmed by experimental results of Lap (1990) who measured the time dependence of probe beam transmission at certain wavelengths in thin ($\sim 3\mu m$) $CdS_{0.52}Se_{0.48}$ platelets. In the case of probe pulse photon energy below the band edge he observed an initial drop of transmittance which can be attributed to gap shrinkage and consecutive transmittance increase, which was caused by bandfilling.

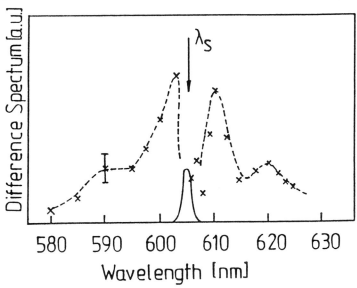

Fig.4: Difference spectrum (dashed line) of the signal pulse ($S_{pump\ on} - S_{pump\ off}$) after inter action with a strong fs pump pulse in a $CdS_{0.52}Se_{0.48}$ platelet. The undisturbed signal pulse spectrum is depicted (full line) for comparison.

ii)   In another measurement the influence of time dependent carrier densities in excited states was investigated for the case of signal pulse propagation through the sample in the absorption region. The sample was a semiconductor doped glass (SDG) filter RG8 with an absorption edge at 680nm. We used a pump pulse at 616nm with a duration of 120fs and a signal pulse at 630nm with a bandwidth of 0.5nm (pulse duration > 1ps). The spectrum of the signal pulse was measured as a function of the delay between pump and signal pulses (Fig.5). The fact, that strong spectral modulation was observed mainly at negative delays, i.e. when the signal pulse was preceding the pump pulse, can be explained in the following way: Only a small part of the signal pulse was modulated by the pump, however the dip near the wavelength of the input wavelength does only appear if the signal is supressed to some extent also after the pump pulse. This has been confirmed by computer simulation. Consequently, the new spectral components can be observed only when the nonmodulated parts of the signal pulse are supressed as, in the case of negative delays, it occurs for a large part of the signal by small signal absorption in the RG8 filter, because the pump pulse is short compared with the signal. For positive delays the nonmodulated signal parts govern the spectra behaviour and spectral modulations could not be observed (though they are present).

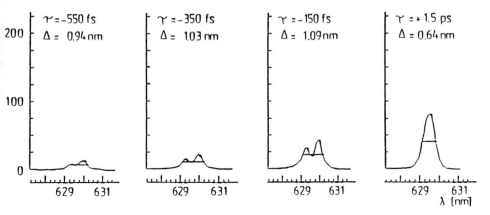

Fig.5: Spectrum of the signal pulse after propagation through a 1mm thick semiconductor doped glass filter RG8 for different delays between pump and signal pulses.

There exist several processes, which can influence the spectral behaviour of the signal pulse. The following processes might contribute to the formation of the measured modulated signal pulse spectrum:

- spectral hole burning (e.g. Gong et al 1991)
- coherent coupling of pump and signal pulses
- nonresonant interaction with the glass matrix of the SDG
- Rabi oscillation of state occupation
- Spatial beam distorsion after refractive index changes

In particular we considered the role of Rabi oscillations by means of a numerical simulation of the experiment described in ii) using the Maxwell-Bloch equations for an equivalent two level system with a phase decay time of about 50fs. The calculated results for the behaviour of the signal pulse spectrum as a function of the delay between pump and signal pulses (Fig.6) are similar to that measured in experiment (Fig.5).

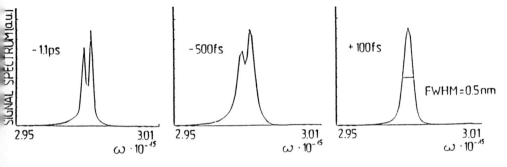

Fig.6: Calculated spectra of the signal pulse for various delays between pump and signal pulses.

The calculated spectral modulation was caused by Rabi oscillations of the occupation number. A sharp spectral dip at the mid frequency of the signal input pulse was obtained for small population of the upper band after propagation of the pump pulse.

Further calculations were performed using the time dependence of the refractive index n and the absorption coefficient α where the latter follows directly from transmission measurements of Lap (1990). Using n(t) and α(t), spectra for the signal pulse were obtained which are very similar to the observed ones.

Though the numerical calculated spectra are in qualitative agreement with the experimental results, other possible explanations given above might be partially responsible for the spectral profile of the signal pulse.

We conclude, that in the case of signal propagation in lossy spectral ranges the changes of both refraction index and absorption coefficient contribute to the spectral modulation.

CONCLUSION

Strong ultrafast spectral modulation of optical signals can be achieved by the use of cross-phase modulation. The signal interacts with short pump pulses which produce ultrafast changes of refraction and absorption in the sample. Resonantly excited semiconductors and semiconductor doped glasses exhibit high optical nonlinearity in loss-free as well as lossy spectral regions.

The experiments were supported by the Deutsche Forschungsgemeinschaft (DFG).

REFERENCES

Alfano R R, Li Q X, Jimbo T, Manassah J T, Ho P P 1986, Opt.Lett.**11**, 636
Dneprovskij V S, Egorov V D, Khechinashvili D S, Nguyen H X, Zimmermann R 1987 in "Ultrafast Phenomena in Spectroscopy IV"
Gong T, Mertz P, Nighan L, Fauchet P M 1991, Appl.Phys.Lett.**59**, pp 721-723
Kaschke M, Wilhelmi B, Egorov V D, Nguyen H X, Zimmermann R, 1988, Appl.Phys.B **45**, pp 71-75
Lap D 1990, PhD Thesis, Jena University
Tamir T (ed) 1988 "Guided-wave optics optoelectronics", Springer Series in elelectronics and photonics **26**, Springer Berlin, Heidelberg, New York, London, Paris, Tokyo
Wilhelmi B, Kaschke M 1988 in "Ultrafast Phenomena in Spectroscopy V", World Scientific, Singapore 1988, 212
Yamashita M, Torizuka K, Uemiya T 1990 in "Ultrafast Phenomena 1990" (Optical Society of America, Washington D.C.), **6**, 175
Zimmermann R 1988 "Many-particle theory of highly excited semiconductors", Teubner Texte zur Physik Bd.**18**, Leipzig

Study of spectral narrowing for femtosecond pulses propagating in single-mode optical fibres

X Zhu and W Sibbett

J. F. Allen Physics Research Laboratories, Department of Physics and Astronomy, University of St. Andrews, North Haugh, St. Andrews, Fife KY16 9SS, Scotland, U. K.

ABSTRACT: Experimental data and the associated analysis of spectral narrowing for femtosecond pulses propagating in anomalously dispersive optical fibres are presented.

1. INTRODUCTION

Propagation of ultrashort optical pulses in single-mode optical fibres has been a subject of much research interest during the past two decades. The significance of the work relates both to the attractive transmission capability of digital optical communication systems (Stolen 1980) and to the potentially exploitable nonlinear effects (Lin 1986) that are readily observed in optical fibres. By using a coupled-cavity mode-locked KCl:Tl colour-centre laser (Zhu *et al* 1989, Zhu and Sibbett 1990) we have been able to perform single-pass studies of femtosecond pulses propagating in several monomode optical fibres. In addition to the reported spectral extensions in several fibre types (Zhu and Sibbett 1991) spectral narrowing of the propagating pulses in two anomalously dispersive fibres (at 1.5 μm) has also been observed. In this paper we present the experimental results of this unusual phenomenon of spectral narrowing. Corresponding discussions on the physical mechanisms involved are also included.

2. EXPERIMENTAL RESULTS

The two fibre types where spectral narrowing was observed are AT & T fibre, which is from the same production run as that used in the first soliton laser (Mollenauer and Stolen 1984), and standard communication fibre. The two fibre types, (one is polarization-perserving and one is not), have anomalous group velocity dispersion D ≈ 15 ps/nm/km at 1.5 μm. The core diameter is 9 μm for the AT & T fibre, and 9.6 μm for the standard communication fibre. A group of spectra for the exiting pulses from a AT & T fibre sample at different coupling powers are shown in Fig. 1. It can be seen that with an increasing intrafibre power the bandwidth of the exiting spectrum decreases. (Note that for all the experimental results reported in this work the duration of the incident optical pulses, which are generated by our mode-locked coupled-cavity KCl:Tl laser, are maintained at ~130 fsec. From such a value and the available average power the maximum intrafibre peak power can be estimated to be around 1 kW). Associated with the spectral narrowing, we observed temporal broadening of the propagating pulses (Fig. 2). Similar results have been obtained for the standard single-mode communication fibre, for which an example of the observed spectral narrowing is shown in Fig. 3. As indicated, the outer trace is the incident spectrum, the bandwidth of which is 20.5 nm, and the inner one is the associated exiting spectrum which has a bandwidth of just under 19 nm. Therefore, without ambiguity, the data in Figs. 1, 3 clearly indicate that after propagating in the two anomalously dispersive fibre samples the spectrum of the pulses became narrower.

From the measured spectral data such as those given in Fig. 1, a plot of the relative change of spectral bandwidth for the exiting pulses as a function of the coupling power can be obtained as shown in Fig. 4. The relative spectral narrowing in figure 4 is defined as the ratio of exiting bandwidth to incident bandwidth. From Fig. 4 it can be seen that at the maximum available optical power the spectrum for the propagating pulses shrinks by 30% for the AT & T fibre and 20% for the standard fibre.

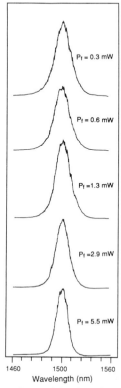

Fig. 1. Spectra of the pulses exiting the 2.7-m-long AT & T fibre as recorded for increasing average power levels.

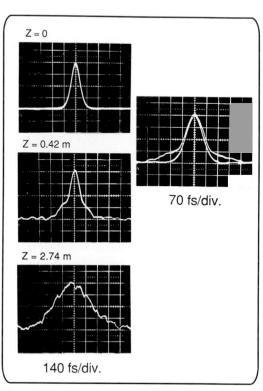

Fig. 2. Autocorrelations for femtosecond pulses propagating in the AT & T fibre. (The trace measured at Z=0 represents the autocorrelation for the incident pulses. The traces shown in the right hand photograph is a direct comparison of the autocorrelation for incident pulses and that for pulses exiting from the 0.42-m-long fibre sample.)

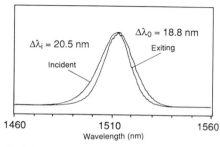

Fig. 3. Spectra for the pulses incident on and exiting a 2.6-m-long standard communication monomode fibre.

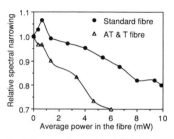

Fig. 4. Decrease of spectral bandwidth ($\Delta\lambda_0/\Delta\lambda_i$) as a function of the increase of coupled pulse power in the two fibre samples indicated in Figs. 1, 3.

3. DISCUSSIONS AND CONCLUSIONS

The spectral narrowing phenomenon evident in Figs. 1, 3 and 4 is rather unusual when compared with the reported data that relate more generally to spectral extensions. Nevertheless, we must recognize the interplay between group-velocity-dispersion (GVD) and self-phase-modulation (SPM) effects in an anomalously dispersive medium. This may lead to the pulses being either temporally compressed or expanded during propagation - depending on the power and the frequency chirping features of the incident pulses - such that it then follows that a "spectral narrowing" can also arise through a solitonic-type influence.

For the fibre where bright optical solitons can be sustained, the dependence of the spectral bandwidth of the propagating pulses on the propagation distance and the input power levels may be described by Fig. 5(a), (b), where P_1, P_2, P_n denote the power levels required for first, second, n_{th} - order solitons and L_0 is the soliton period. Fig. 5(a) illustrates that for an increasing propagation distance the spectral bandwidth $\Delta\lambda$ of the optical pulses increases monotonically for $P < P_1$, remains constant for $P = P_1$ and increases and decreases periodically for $P = P_n$ ($n > 1$). (In the latter case each time the propagation distance is equal to an integral multiple of L_0, $\Delta\lambda$ is restored to the initial value of the incident pulses.)

For a given fibre length, say $L = L_0$, the data of Fig. 5(b) shows that, as the coupling power increases from zero to P_1, the spectral bandwidth first increases and then decreases. At $P = P_1$, the exiting spectrum becomes identical to the incident spectrum ($\Delta\lambda = \Delta\lambda_i$) because in this case the pulses are fundamental solitons. For $P > P_1$ as the launching power becomes progressively greater, $\Delta\lambda$ will increase sharply due to the pulse compression arising from higher-order solitonic effects and it subsequently decreases until $P = 4P_1$, at which the spectral bandwidth again becomes equal to the initial value (i.e. 2nd-order solitons). Such a pattern of extension, narrowing and restoration of the spectra of the propagating pulses is retained for yet higher power levels except that the rise and fall features are expected to be more dramatic.

Although Fig. 5(a), (b) represent only a qualitative description of this spectral behaviour in anomalously dispersive fibres, it can be appreciated that it clearly illustrates the existence of specific power ranges, [shadowed portion in Fig. 5(b)], within which an increase of optical power leads to a decrease in spectral bandwidth. This physical interpretation may be used to provide a satisfactory explanation of the spectral narrowing data in Figs. 1, 3, 4 presented in the previous section. By using the relevant fibre properties it can be estimated that the observed spectral narrowing can be related to an optical power range just less than that required for the establishment of fundamental solitons in the fibre samples involved [see Fig. 5(b)].

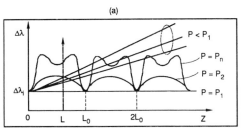
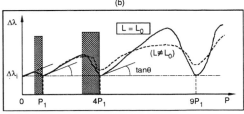

Fig. 5. Illustration of spectral features in anomalously dispersive fibres: (a) variation of the bandwidth for exiting pulses as a function of fibre length for different coupling powers; L_0-soliton period, P_i ($i = 1, 2, n$) - the optical powers corresponding to the soliton orders. (b) The predicted changes of spectral bandwidth of propagating pulses versus coupling powers for the selected fibre lengths (solid line: $L = L_0$; dashed line: $L \neq L_0$). The shadowed area designates the regions where the increase in intrafibre power level leads to the decrease of spectral bandwidth. The slope characterised by $\tan\theta$ is the assumed rate of increase in spectral extension for a pure SPM medium.

In the study of the mechanism for the observed spectral narrowing phenomena we realized that in contrast to the temporal compression of optical pulses a scheme for spectral compression of optical pulses could also be considered. It is known that for temporal compression the pulses are initially propagated in a nonlinear medium such as an optical Kerr material, a phase modulator or a cell of resonant atomic vapour etc. to expand their spectra. The spectrally stretched pulses then pass through an appropriate dispersive delay line, so that the different spectral components are brought together in a compressed timescale. For spectral compression (see Fig. 6) we can simply reverse the above process in that we first send the beam into a dispersive delay line to chirp the pulses. (Note that in this step there is no change in spectrum but just temporal broadening.) When these chirped pulses pass through a selected length of nonlinear medium, which we designate in Fig. 6 as a spectral shrinker, the spectrum of the pulses will then be compressed.Such a scheme can also be appreciated in terms of the analogy between the effect of group velocity dispersion on the temporal features of the pulses and that of self-phase modulation on their spectral features. The key point is that the interplay between GVD and the optical Kerr-type nonlinearity can not only lead to temporal compression of optical pulses but also spectral compression under suitable conditions.

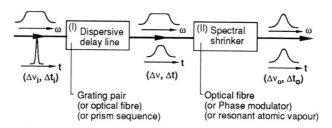

Fig. 6. A scheme for spectral compression for optical pulses.

In summary, we have presented experimental results which show unambiguously the spectral narrowing of femtosecond near-infrared pulses which propagate in anomalously-dispersive fibres. We have given an explanation for this unusual behaviour based on solitonic effects in the fibres. The understanding of this observed phenomenon of spectral narrowing has also led us to suggest a general scheme for spectral compression. It is expected that this method of spectral compression may find some uses in time-domain spectroscopy, where advantage may be taken of an optimally confined spectrum for the excitation/probe ultrashort pulses.

REFERENCES

Lin C 1986 *J. Lightwave Technol.* **LT-4** 1103.
Mollenauer L F and Stolen R H 1984 *Opt. Lett.* **9** 13.
Stolen R H 1980 *Proc. IEEE* **68** 1232.
Zhu X, Kean P N and Sibbett W 1989 *IEEE J. Quantum Electron.* **25** 2445.
Zhu X and Sibbett W 1990 *J. Opt. Soc. Amer. B.* **7** 2187.
Zhu X and Sibbett W 1991 *IEEE J. Quantum Electron.* **27** 101.

Inst. Phys. Conf. Ser. No 126: Section II   185
Paper presented at Int. Symp. on Ultrafast Processes in Spectroscopy, Bayreuth, 1991

Bistable solitons

J.Herrmann, S.Gatz
Central Institute for Optics and Spectroscopy, Rudower Chaussee 6,
O-1199 Berlin

ABSTRACT: The propagation and collision of solitons in doped fibers with a non-Kerr-like nonlinearity has been investigated using numerical and analytical methods. It is demonstrated that bistable (or two-state) solitons exist, which describe pulses with the same duration but different peak powers. At collision solitons of the upper solution branch fuse to a single high-energy soliton. The predicted properties of bistable solitons can be used for novel all-optical switching devices.

Recent research concerning soliton dynamics in single-mode fibers has mainly involved intensity-depending refractive index change with a Kerr-like nonlinearity. But specially in materials with high nonlinear coefficients, e.g. semiconductor doped fibers, organic polymers or others higher order nonlinearities come into play at a not to high intensity. Such higher order effects not only modify the known properties of optical solitons but lead under certain conditions to completely new phenomena.

In this paper we present a study of soliton propagation in dispersive materials with a non-Kerr-like nonlinearity

$$\Delta n = n_2 \cdot f(I), \qquad I = P_0 \cdot |q|^2,$$

where the function $f(I)$ describes the nonlinear refraction index change in highly nonlinear materials, as e.g. semiconductor doped fibers, thin liquid-filled capillaries or others . For the normalized soliton amplitude $q(s, \xi)$ we consider the equation

$$-i \cdot \frac{\partial}{\partial \xi} q + \frac{1}{2} \cdot \frac{\partial^2}{\partial s^2} q + f(|q|^2) q = 0 \tag{1}$$

where dimensionless soliton units were introduced

$$s = 1.76 \cdot t/\tau_0, \quad \xi = z/z_0, \quad z_0 = \tau_0^2/|k_1''|, \quad k_1'' = \frac{d^2}{d\omega^2} k_1,$$

$$\tau_0 \quad \text{pulse duration.}$$

We studied various nonlinear processes as e.g.

(i) $f(|q|^2) = |q|^2 + \alpha \cdot |q|^4,$ (ii) $f(|q|) = |q|^2 \cdot (1 + \gamma \cdot |q|^2)^{-1}$

and   (iii) $f(|q|^2) = (|q|^2 + \alpha \cdot |q|^4)/(1+\gamma \cdot |q|^2)^{-1}$.

For the case (i) an exact analytical soliton solution was found and the cases (ii) and (iii) were solved numerically. It is shown that for all cases for $\alpha < 0$ and $\gamma > 0$ bistable (or two-state) solitons exist, which describe undistorted pulses with the same duration but different soliton peak power $P_0 \cdot B^2$, $B = |q_{max}|$.

In fig.1 for the most general case (iii) the maximum soliton amplitude B is depicted as a function of the parameter α depending on the pulse duration τ_0 ($\alpha \propto \tau_0^2$). For small α one gets B = 1 as in the case of a Kerr-like nonlinearity. As a striking phenomenon one can see that for larger α the soliton amplitude becomes a two-valued function of the pulse duration τ_0. Using a stability criterion we have proven that both branches of the bistable soliton solution are stable against small fluctuations. Besides we have shown that the solitary waves

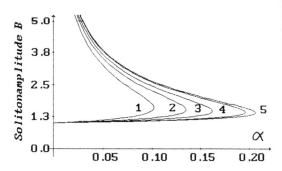

Fig.1: Dependence of the solitonamplitude amplitude B from the parameter α curve 1: $\gamma = 0.5 \cdot \alpha$, curve 2: $\gamma = \alpha$, curve 3: $\gamma = 2 \cdot \alpha$, curve 4: $\gamma = 10 \cdot \alpha$, curve 5: $\gamma = 0$.

on both solution branches can indeed arise from arbitrary initial pulse shapes. The influence of linear loss and higher order solitons were also investigated.

The nonlinear model (i) shows a larger bistability range than the saturable nonlinearity (ii). In order to find realistic conditions for the observation of bistable solitons we propose to use fibers doped with two appropriate materials. One dopand should have a positive sign $n_2^{(a)} > 0$ and a high saturation intensity $I_{sat}^{(a)}$, the other a negative sign $n_2^{(b)} < 0$ with nearly the same magnitude $n_2^{(a)} - |n_2^{(b)}| \lesssim 0.1 \cdot |n_2^{(b)}|$ and a low saturation intensity. The nonlinear refraction indexchange of double doped fibers is described by the nonlinearity (iii). Using some typical parameters in such fibers it is possible to achieve the necessary value $|\alpha| \cong 0.1$ and $\gamma = 0.01$.

For the potential application of such solitons we investigated the collision of two bistable solitons with a small frequency shift $\Delta\omega$ (or a corresponding difference of the group velocities). Solitons of the lower solution branch in fig.1 behave as solitons in the case of a Kerr-nonlinearity and remain their shape after the collision. Contrary the collision of two solitons of the upper solution branch shows a dramatic change of its behavior. As an example in fig.2 the collision of two such solitons with a frequency $\Omega_{in} = 0.5$ ($\Omega = \Delta\omega/\tau_0$) is shown. As one can see both pulses fuse to one soliton of the upper branch with a larger energy and a frequency $\Omega_{out} = 0$. In fig.3 the collision of two solitons of the upper solution branch with the same parameters as in fig.2 but with a larger frequency shift $|\Omega_{in}| = 1.5$ is shown. Now a part of the pulse energy is radiated into two side pulses, which are solitons of the lower solution

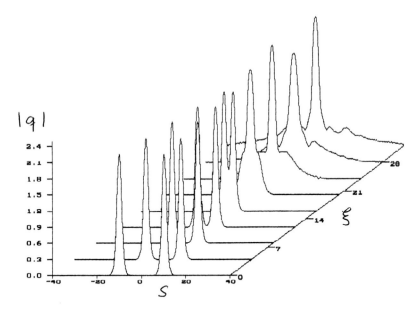

Fig.2: Collision of two solitons of the upper solution branch with
parameters $|\Omega_{in}| = 0.5$, $\alpha = 0.1$, $\gamma = 0.1$.

branch. The central pulse with $\Omega = 0$ is a soliton of the upper solution
branch. Still higher frequency shifts lead to the appearance of four or
more solitons. The behaviour of the solitons at collisions can be explained by a large increase of "bonding" energy after a fusion and has no
analog in optics.

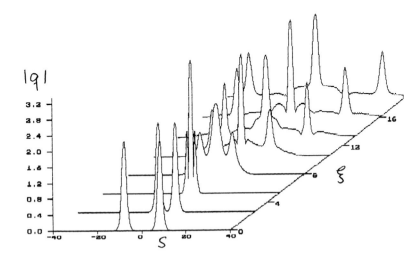

Fig. 3: Collision of two solitons of the upper solution branch with
parameters $|\Omega| = 1.5$, $\alpha = 0.1$, $\gamma = 0.1$.

The shown collision properties of bistable solitons could be applicable for all-optical switching. For a soliton fusion we consider the collision of two solitons of the upper solution branch with a small frequency diffe- rence $2|\Omega| \lesssim 1$ (in physical units this means a frequency difference of both pulses $\Delta \omega \lesssim 1.76/\tau_0$). Both the signal soliton with the frequency shift Ω and the delayed control soliton with the frequency shift $-\Omega$ are coupled into the double doped fiber as described above. Due to the higher group verlocity the control soliton moves faster and collides with the signal soliton. After collision both solitons fuse to a single soliton with nearly the double pulse energy and the middle frequency ($\Omega = 0$). A switched pulse can easily detected due to the shifted frequency, the higher group velocity or the changed pulse shape with a higher pulse ener- gy compared to a non-switched pulse.

The results can also be used for the description of two-state or bistable spatial soliton collision in planar waveguides. In this case two colliding solitary beams show a reflection of light at light. For small input angles both solitary beams fuse to a single beam propagating into the di- rection of axial axis.

Inst. Phys. Conf. Ser. No 126: Section II
Paper presented at Int. Symp. on Ultrafast Processes in Spectroscopy, Bayreuth, 1991

Coherent effects on soliton amplification in a doped fiber

I V Mel'nikov and R F Nabiev

General Physics Institute, 38 Vavilov Str., Moscow SU-117942, USSR

ABSTRACT: The evolution of ultrashort optical soliton in a fiber amplifier is examined numerically. The results of our simulation are found to draw the conditions enabling the amplifier to operate in a quasi-soliton mode when the probing pulse, keeping its well-behaved profile, pools the energy from the inverted dopants. Beyond these conditions, however, amplification has effect of soliton collapse and formation of a multiple-peak pattern.

1. INTRODUCTION

Over the past twenty years of intense technological efforts, predated by the pioneering works of Young (1963), Koester and Snitzer (1964), and Snitzer (1966), rare-earth doped fibers have been recognized as one of the most promising candidate for various lightwave processing systems. Growing applications of these fibers are driven by a number of features such as high gains with moderate pump powers of arbitrary polarization, availability of wide tuning range in the near infrared region of the spectrum, established operation at room temperature, etc. However, there is serious concept of whether it is possible to offer a distortionless scheme of fiber amplifier in a view of optical communications. This has prompted more fundamental consideration on the nonlinear dynamics of solitary waves.

In this respect, recent experiment of Nakazawa et al. (1990) are of extremely interest. There was demonstrated that the pump increasing above a certain threshold value ultimately leads to the femtosecond soliton shortening whereas the pulsewidth remains constant at low pump power. One further motivation may clarify the physical mechanism behind the problem of ultrashort soliton stability in an amplifying medium. Whenever the incoming soliton is chosen to have a duration comparable with or less than the relaxation time of resonant polarization, its interaction with the amplifying medium should be coherent and the response of the active medium depends nonlinearly on the preceding history of the optical pulse. The outcome of the coherent amplification is that initial smooth pulse can be reshaped into a typical ringing. Moreover, the output pulse may be compressed linearly with amplifier length, and its energy grows linearly as well (Manakov 1982). The objective of this report is aimed at a link between retarded coherent response of the resonant dopants and spatiotemporal development of the solitary pulses in the presence of dispersion, Kerr-like and Raman self-interactions. Our numerical simulations seem to indicate very reach dynamical behavior of amplifier for ultrashort solitons.

2. NUMERICAL ANALYSIS

The evolution of pulse envelopes along the fiber length was studied as follows. The electric field of the pulse was governed by the nonlinear Schrödinger equation (NSE) modified by the appropriate inclusion of resonant effects, which obeyed the Bloch equations for a homogeneously broadened two-level system, and the response function of Stolen et al. (1989) was used to describe the self-stimulated Raman scattering. Such an approach is more adequate for femtosecond pulses in comparison with that of our earlier study (Mel'nikov et al. 1990).

This coupled set of equations was solved numerically for various ratios of the input pulsewidth τ to the homogeneous relaxation time of the resonant dopants T. As an input pulse, we used p -soliton solution of NSE with real p that physically means the initial profile takes *sech* - like form. The resonant dopants were supposed to be completely inverted, and, on the application side, the concentration of dopants was equal to 1 ppm, the resonant wavelength was chosen to be 1.5 μm and matrix element was 0.025 D. Figure 1 presents the temporal shape of output pulse; time is given in the units of initial pulsewidth, which is written in the figure caption

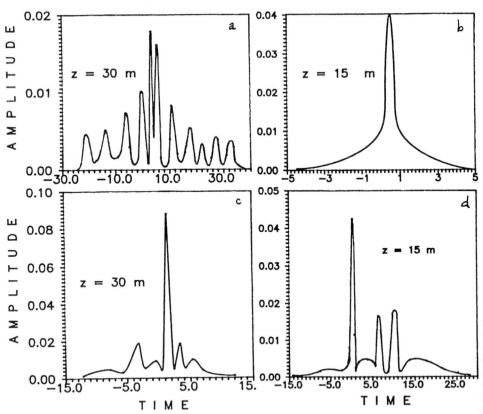

Fig. 1. The output pulses from Er^{3+}-doped fibers. (a) input - 0.3 soliton, τ = T = 0.1 psec; (b) input - single soliton, τ = T = 0.1 psec; (c) input - single soliton, τ = 0.3 psec, T = 0.1 psec; (d) input - single soliton, τ = 0.1 psec, T = 0.3 psec.

along with the initial amplitude and the homogeneous relaxation time. One can note the following features.

First, if the input pulse amplitude is less than that required for a single soliton pulse [Fig. 1a], the pulse widens, is amplified, and the 30 m length fiber yields a pulse train owing to the modulational instability. Figure 1b shows the output pulse from the 15 m length fiber for a specific case $\tau = T = 0.1$ psec. The most notable feature is that the pulse maintains closely the *sech* - like shape, the amplitude of the pulse increases approximately five times, while the pulsewidth decreases by the same factor. Anyone can be easily convinced by Fig. 2a that the product of amplitude and pulsewidth remains almost constant during propagation of the pulse. This case corresponds to the quasi-soliton mode, which is believed to have a potential benefit for optical communications. However, such a distortionless operation is an unstable phenomenon, as one can see from the below.

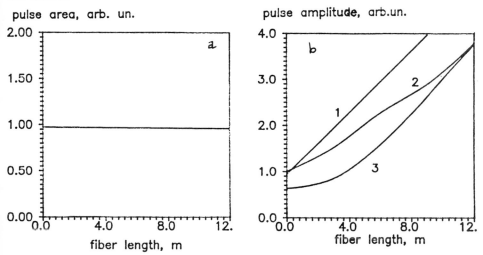

Fig. 2. the pulse area (a) and the pulse amplitude (b) versus fiber length

The next two examples demonstrate the origin of substructure under condition identical to that of Fig. 1b ($p = 1$). In the case $\tau > T$ [Fig.1c] the shape of the output pulse is nearly symmetric, but in the opposite case [Fig. 1d] the shape yielded is seen to disintegrate into asymmetrical multiple-peak pattern in the pulse tail for fiber length 15 m. The former case of amplification is close to that of an N -soliton pulse (see, e.g. Agrawal 1989 and references therein), whereas the latter case is evidence of appearance of coherent effects. Physically, the resonant dopants remember the polarization during the time $\simeq T$ after the pulse passes, thus developing the pulse tail. Later this tail is amplified and breaks up into subpulses. Once adopting the terms of two-level amplifier in the absence Kerr effect and material dispersion (Belenov et al. 1988), one can imagine these subpulses as a sequence of self-similar solitons which interaction eventually forces the leading pulse of sequence to escape the background.

Figure 2b shows an increase of the pulses during the first stage of evolution, where they are smooth and solitonlike. It is is clearly seen the nonlinear amplification of the pulse amplitude for curve (2), which corresponds Fig. 1b, and the nearly linear increase for curves 1 [Fig. 1d] and 3 [Fig. 1c]. Also, it was pointed out, that the product of the amplitude and the pulsewidth (pulse area) remains approximately constant throughout active medium but that the energy increases. The pulse thus drains the energy from the resonant dopants, and its energy grows linearly with distance for curves 1 and 3. This is one of the most striking manifestation of coherent amplification, as indicated above.

The results obtained differ from those of Blow et al. (1988), in which the propagation of solitons in the medium with the gain having finite spectral width is considered. A saturation of the increase of soliton magnitude, obtained there, is connected, as far as we understand, with the parabolic gain spectrum profile used in the referenced paper. Indeed, some part of the pulse of sufficiently short duration has to be absorbed in such a model. In addition, an instantaneous response of the active medium was assumed there, in contradiction to our model.

Finally, our simulations failed to reveal the presence of the Raman self-interaction in the output pulses. This is thought to emanate from our choice of the fiber of ten meters length which is too short to display Raman effects on the amplification of subpicosecond pulses.

3. CONCLUSION

This report has therefore discussed the role of coherent effects on the quality of pulse amplification in the rare-earth doped fibers. The results of our calculations reveal strong dependence on the parameters. Typically an input pulse evolves into a pulse train. Pulse launched close to the single soliton with the pulsewidth about the relaxation time of resonant polarization is amplified nonlinearly and does not break up.

4. ACKNOWLEDGMENT

It is a pleasure to acknowledge the contribution to this work of Dr. A V Nazarkin; we also thank Professors A Yariv and H A Haus for encouraging discussions.

References

Agrawal G P 1989 *Nonlinear Fiber Optics* (New York: Wiley)
Belenov E M, Kryukov P G, Nazarkin A V et al 1988 *J.Opt.Soc.Am.* B5 946
Blow K J, Doran N J and Wood D 1988 *J. Opt. Soc. Am.* B5 1301
Koester C J and Snitzer E 1964 *Appl. Opt.* 3 1182
Manakov S V 1982 *Sov. Phys. JETP* 56 37
Mel'nikov I V, Nabiev R F and Nazarkin A V 1990 *Opt. Lett.* 15 1348
Nakazawa M, Kurokawa K, Kubota H and Yamada E 1990 *Phys.Rev.Lett.* 65 1881
Snitzer E 1966 *Appl. Opt.* 5 1487
Stolen R H, Gordon J P, Tomlinson W J, and Haus H A 1989 *J. Opt. Soc. Am.* B6 1159
Young C G 1963 *Appl. Phys. Lett.* 2 151

Inst. Phys. Conf. Ser. No 126: Section II
Paper presented at Int. Symp. on Ultrafast Processes in Spectroscopy, Bayreuth, 1991

193

Memory effects assisted phase-conjugation of sub-nano and picosecond pulses in optical fibers

N. I. Minkovski and T. P. Mirtchev

Dept. of Quantum Electronics, Faculty of Physics, Sofia University
5 A. Ivanov , Sofia, BG-1126, Bulgaria

Abstract: We have experimentally observed for the first time an efficient Brillouin scattering of a train of short pulses in a single-mode optical fibers.

The process of stimulated Brillouin scattering (SBS) is among the most widely studied third order nonlinear phenomena, because it is characterized by the lowest threshold and do not require special phase-matching conditions (Zel´dovich et al. 1986). It has been observed in various nonlinear media (Maier 1968) and recently is studied extensively in the context of the fiber optics. The reason is that the utilization of SBS effect allows the construction of extremely low threshold fiber Brillouin lasers (Stokes et al. 1982) and efficient phase-conjugators (Petrov et al. 1982), while, on the other hand its onset limits the power in optical fiber communications (Tkach et al. 1986).

In virtually all studies of SBS the relation between the gain bandwidth of the Brillouin medium $\Delta \nu_b$ and spectral width of the pump laser $\Delta \nu_p$ is an important parameter. The slow damping of the hypersound wave results in a narrow bandwidth of the process, which for a Ge-doped fibers is $\Delta \nu_b \approx 40$ MHz (Kurashima et al. 1990), or equivalently in a long phase-relaxation time $T_2 = (1/ \pi \Delta \nu_b) \approx 8$ ns. A well known result of the transient stimulated scattering theory (Zel´dovich et al. 1986) connects the reflected Stokes intensity E_s, the pump intensity E_p and T_2

$$| E_s (z,t) |^2 \approx \exp \left\{ -2(t / T_2) + 2 \left[2gz \int_{-\infty}^{t} | E_p (z,t^{'}) |^2 dt \right]^{1/2} \right\} \quad (1)$$

where g is the gain coefficient, z and t are the spatial and time coordinates respectively. From (1) the relation between the duration of the pump pulses t_p and the threshold pump intensity $\bar{E}_p$ in the case of pulsed excitation can be estimated (Zel´dovich et al. 1986):

$$g | \bar{E}_p (z,t) |^2 L > (15 + 2 \pi \Delta \nu_b t_p)^2 / 4 \pi \Delta \nu_b t_p \quad (2)$$

where L is the interaction length. As it is seen, the transient gain for pump pulses shorter than T_2 , i.e. sub-nano or picosecond pulses, is greatly reduced compared with the steady state gain $G_s = g | \bar{E}_p (z) |^2 L$. In the same time the damage threshold of the fiber or the onset of competing processes (like forward Raman scattering), restricts the possibility to increase the transient gain by directly scaling the pump pulse intensity or the fiber length. In result, as far as we know, up to now SBS in optical fibers have never been obtained with pump pulses shorter than $\approx$ 10 ns (Dianov et al. 1989).

Recently it has been realized that it is possible to overcome the reduction of the Brillouin or Raman transient gain by assuring the interaction of the short pump pulse with an allready coherently excited medium (Greiner-Mothes et al. 1986 and Mirtchev et al. 1990). Thus it is possible to produce gain which tends to the steady-state value (Mirtchev et al. 1990), and is restricted only by group-velocity mismatch that limits the interaction length, not by the slow buildup of the phonon field. The easiest way to have vibrationaly excited medium prior to the pump pulse appearance is to use succession of pulses with repetition rate T_r comparable to T_2 (Greiner-Mothes et al. 1986). Every subsequent pulse interacts with the non-completely relaxed medium producing in turn more initial excitation for the next pulse. Thus gradually a steady state can be reached in which the Stokes gain depends on the pump intensity, T_2, t_p and T_r.

Our implementation of the cited above technique to optical fibers was relieved by the fact that T_2 in fused silica fibers is of the same order of magnitude as the repetition period of a mode-locked laser. Hence we worked directly with the laser output, avoiding the necessity to multiply the repetition rate. In the experimental setup we used acoustooptically Q-switched and mode-locked CW Nd-YAG laser (Minkovski et al. 1991). The mode-locking frequency, obtained from a stabilized synthesizer was 50 MHz, which resulted in round trip time $T_r \approx$ 10 ns.

The duration at FWHM of the Q-switched trains was 300 ns., while their repetition rate was 0.5 KHz. For efficient control of the single pulse duration, which was measured in real time by an autocorelator, the mode-locker RF power was adjustable in the range 0.5 - 10 W. For assuring constant time duration of the single pulses along the train we operated the laser in a regime with a long ($\approx$ 500 μs) prelasing period. After passing through a Glan polarizer and $\lambda/4$ plate the pulses were launched by a microscope objective into the fiber. We used 20 cm. long pieces of four different, GeO_2- doped, single mode fibers with diameters 1.9, 2.2, 2.8, 4.4 μm and GeO_2 concentration 28, 22, 16, 6.5 mol % respectively. The cladding index of the first three fibers was matched to the silica tube index.

Because of the short fiber lengths needed, the losses at the pump wavelength were not an important parameter. The transmitted and reflected radiation were detected by a fast photodiodes and a 100 MHz oscilloscope.

Fig 1. shows typical oscilloscope traces of the transmitted (upper) and reflected (lower curve) signals obtained near the threshold of the discussed effect. It is clearly seen that the first detectable backward

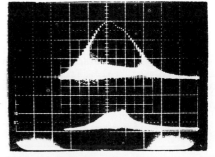

Fig 1. Shapes of the pump and Stokes trains at threshold, pump power 30 mW, duration of the pump pulses 0.5 ns.

scattered pulse was produced by a pump pulse of the second half of the train, rather than by the most powerfull pulse (the symmetric background under the signal is the usual Fresnel reflection from the fiber). This fact indicates that the SBS in our case can not be described as a purely transient phenomenon, produced by isolated pulses. The scattering conditions were most favorable for a pulse from the trailing part of the train, because of the gradual accumulation of vibrational amplitude. The exact position of the maximum reflected Stokes pulse in the train depends on the total energy and duration of the train and on the duration and repetition rate of the single pulses. Fig 1. corresponds to average pump power 30 mW, 0.5 ns. FWHM of the pulses, circular input polarization and the traces were taken with the fourth fiber.

It is noteworthy, that the discussed effect was obtained in all four types of fibers used. Lowest threshold was received with the highest GeO$_2$ concentration fiber, apparently because of its good acoustic waveguiding properties .

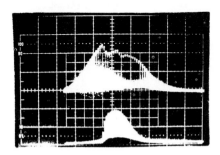

The traces at Fig 2. were taken at input power of 75 mW and other parameters identical to those of Fig 1. They show saturation of the process to the point when efficient scattering of pulses from the leading part of the train is observable also.

Fig 2. Depletion of the pump pulses at pump power of 75 mW.

At that power level, which is approx. two times lower than the damage threshold of the fiber, average power reflectivities of 23% were obtained, which means conversion efficiencies of $\approx$ 55 % for the maximum Stokes pulse. In the same time, according to our calculations, the SBS threshold for an isolated pulse with the same parameters is 90 GW/cm^2, which is above the damage threshold. The importance of the phase memory and the vibrational amplitude buildup for the observed effect is confirmed also by the strong asymmetry between the leading and the trailing slopes of the reflected signal. This asymmetry is illustrated clearly by the hysteresis type curve at Fig 3. which shows the intensity of the maximum Stokes pulse in the backward scattered signal versus maximum intensity of the pump. Every abscissa point represents two pump pulses with equal intensity, belonging to the leading and trailing parts of the train respectively. The higher of the coresponding two reflectivities is produced always by this pump pulse, which enters the fiber later.

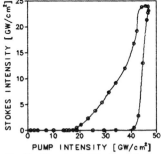

Fig 3. Reflected Stokes intensity versus the intensity of the single pump pulses along the pump train.

Fig.4 shows the calculated threshold intensity for the usual transient SBS (curve a) and the measured threshold intensity for multiple-pulse SBS (curve c), in dependence of the pump pulse duration

t_p. In the latter case the intensity values refer to the maximum pulse in the train, and the duration of the train is the same as for previous figures. In the examined range of t_p curves (a) and (c) show similar behaviour. The reason is that if the repetition rate of the pulses and the FWHM of the train are fixed, then the multiple-pulse threshold should be a fraction of the one for a single pulse, which, according to (1) depends on the single pulse energy. For comparison, Fig 4. (b) shows also the measured damage threshold of the fiber.

Currently our work is oriented towards achieving efficient SBS in optical fibers pumped by even shorter pulses.

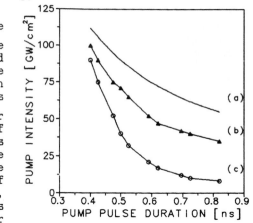

Fig 4. The dependence of the calculated single pulse Brillouin threshold (a), experimentally obtained damage threshold of the fiber (b) and multiple-pulse Brillouin threshold (c) on the pump pulse duration.

For that purpose it is necessary to alter the Brillouin gain bandwidth, andÖor to increase the total pump train energy launched in the fiber. We have therefore modified our laser, obtaining extremely long trains of up to 3 μs FWHM duration (Minkovski et al. 1991), whose intensity is below the damage threshold. Adequate theoretical modeling of the complex dynamics accompanying the process is needed also. In conclusion we note that the discussed effect is connected, for example, with the crosstalk in a high bit-rate soliton based optical communication system, or with the problem of distortionless amplification of short pulses (Zel´dovich et al. 1986).

References

B. Zel´dovich, Pilipetsky N. and Shkunov V. 1986 *Principles of Phase-Conjugation* (Springer-Verlag Berlin Heidelberg) pp 1-100
Maier M. 1968 Phys. Rev. 166 113
Stokes L. F., Chodorow M. and Shaw H. J. 1982 Opt. Lett. 7 509
Petrov M. and Kuzin E. 1982 Sov. Tech. Phys. Lett. 8 316
Tkach R. W. and Chraplyvy A. R. 1986 J. Lightwave Technol. LT-4 1655
Kurashima T., Horiguchi T. and Tateda M. 1990 Photonics Technol. Lett. 2 718
Dianov E.M., Karasik A.Ya., Lutchnikov A.V.and Pilipetski A.N. 1989 Opt. and Quantum Electr. 21 381
Greiner-Mothes M. A. and Witte K. J. 1986 Appl. Phys. Lett. **49** 4
Mirtchev T., Minkovski N. and Tomov I. 1990 Opt. Commun. 80 143
Minkovski N., Mirtchev T. and Tomov I. 1991 Opt. Commun. 81 199

Pressure induced vibrational relaxation in molecular crystals by picosecond coherent Raman spectroscopy in a diamond anvil cell

Eric L. Chronister and Robert A. Crowell

Department of Chemistry, University of California, Riverside, CA USA 92521

ABSTRACT: Pressure dependent vibrational relaxation is observed for crystalline naphthalene and CS_2 using picosecond coherent Raman scattering at low temperature in a high pressure diamond anvil cell. Pressure induced vibrational relaxation is analyzed in terms of shifts in the phonon density of states and pressure induced anharmonic vibrational mode couplings. Time-resolved relaxation measurements as a function of density provide a sensitive probe of the intermolecular interactions which give rise to relaxation phenomena.

1. INTRODUCTION

Conventional Raman spectroscopy under high pressure conditions has been a useful probe of the structure of a wide variety of molecular solids (Ferraro 1984). More recently, time-resolved coherent Raman experiments in high pressure diamond anvil cells have been used as a probe of pressure induced vibrational dephasing in benzene (Baggen *et al* 1987; Della Valle *et al* 1988) and nitromethane (Rice *et al* 1989) at room temperature. However, time-resolved measurements of pressure induced dephasing at room temperature have been complicated by thermally induced dephasing and relaxation processes (Baggen et al 1987; Kroon et al 1989). For this reason, we present high pressure picosecond Coherent anti-Stokes Raman Scattering (psCARS) results for single crystals at low temperature as a means of minimizing inhomogeneous and thermally induced dephasing mechanisms (Chronister et al 1991; Baggen et al 1991)

Due to the reasonably harmonic nature of vibrational motion in molecular crystals, a perturbative anharmonic treatment has proven useful for calculating vibrational lifetimes (Califano et al 1981). Although existing solid state intermolecular potential energy models are consistent with many static observable quantities (e.g. crystal structure, spectra, and dispersion curves), a good quantitative theoretical description of solid state dynamics is lacking. Therefore, high pressure is being used as an additional tool with which to study the anharmonic interactions and density of states effects responsible for vibrational relaxation.

2. EXPERIMENTAL

The picosecond CARS experiments were performed using a mode-locked, Q-switched, Nd^{3+}:YAG laser with two synchronously pumped and cavity dumped dye lasers. The samples were pressurized in miniature Merrill-Bassett type high pressure diamond anvil cells (Merrill and Bassett 1974). Crystalline high pressure samples were grown from the melt within the diamond anvil cell and annealed near the melting point (Chronister et al 1991). Pressure calibration was obtained by the frequency shift of the R_1 Ruby fluorescence line at low temperatures, (Forman et al 1972; Buchsbaum et al 1984) and correlated with observed phonon frequency shifts and the pressure temperature phase diagram for the pure liquids (Bolduan et al 1986). The low temperature measurements were made by immersing the cell in liquid helium (1.1K) or by putting it in thermal contact with the cold tip of a closed cycle helium refrigerator (9-300K). Due to the thermal load of the diamond anvil cell, the closed cycle refrigerator typically yielded sample temperatures of ~15K. Fig. 1 shows a sketch of the experimental apparatus.

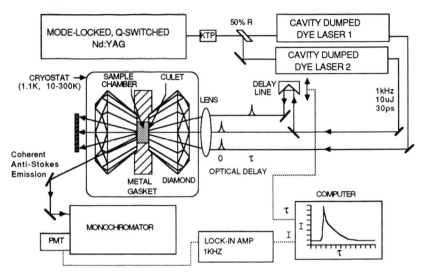

Fig. 1. Schematic drawing of the experimental apparatus for a psCARS experiment in a diamond anvil cell. The three laser pulses are focussed to a 50μm spot within the 300μm sample aperture of the cell.

3. VIBRATIONAL RELAXATION AT INCREASED DENSITY

Phenomenological potentials are often used to account for the repulsive and dispersive interactions in lattice dynamical calculations. In addition, the intermolecular potential for molecular crystals is often calculated as a sum of intermolecular atom-atom interactions. The Lennard-Jones and Buckingham "6-exp" atom-atom potentials, $\Phi_{ij} = 4\varepsilon_{ij}[(\rho_{ij}/r_{ij})^{12} - (\rho_{ij}/r_{ij})^6]$ and $\Phi_{ij} = A_{ij}\exp(-B_{ij}r_{ij}) - C_{ij}r_{ij}^{-6}$, respectively, are two widely used phenomenological forms for van der Waals potentials (Kihara 1977; Buckingham 1980). For a three dimensional lattice of particles, the energy per particle is the sum of all atom-atom interactions, $\Phi = {}^1/_2\Sigma_{i\neq j}\Phi_{ij}$.

3.1   Pressure induced anharmonic coupling

The connection between anharmonicity in the intermolecular potential and the rate of vibrational relaxation motivates our efforts to experimentally correlate measured vibrational relaxation with solid state potential energy functions. Variable density at low temperature is an appropriate probe of the potential energy function since spontaneous relaxation is the only depopulation mechanism at low temperature. For example, the lowest order anharmonic term in an expansion of the intermolecular potential in terms of normal coordinates is the cubic anharmonic term, $\Phi^{(3)} = {}^1/_3!\Sigma_{i\neq j}(\partial^3\Phi/\partial q_i\partial q_j\partial q_k)_0 \, q_iq_jq_k$, where q_i, q_j, and q_k represent the three normal coordinates involved. The resulting "golden rule" vibrational relaxation rate from an initially prepared state Ψ_i to a final state $\Psi_j\Psi_k$ is,

$$\Gamma_{relax} = 2\pi/\hbar \; \Sigma_{jk} \; |<\Psi_i| \; \Phi^{(3)}| \; \Psi_j\Psi_k>|^2 \; \rho(\omega_j) \; \rho(\omega_k) \; \delta(\omega_i-\omega_j-\omega_k) \qquad (1)$$

A reasonable estimate of the magnitude of pressure induced anharmonicity is given by recent calculations on benzene which indicate an increase in the square of the cubic coupling matrix element by only a factor of 2 for a 40kbar pressure increase (Della Valle et al 1988). Qualitatively, a monotonic increase in vibrational relaxation with applied pressure is to be expected since the anharmonicity of the repulsive part of the potential will dominate at high pressure.

3.2 Pressure induced changes in vibrational densities of states

Although the average density of phonon states per volume at a given frequency is relatively constant over the pressure range investigated (due to competition between spectral expansion and volume compression), pressure induced phonon spectral expansion results in higher frequency phonon modes which can become important in the relaxation process. In addition, there is a considerable amount of structure on the phonon density of states (DOS). Thus, even though the *average* DOS is relatively pressure independent, large DOS effects may still dominate the pressure induced dephasing.

As a diagnostic tool we characterize the cubic coupling matrix elements by an average coupling strength, $|\overline{<\Phi^{(3)}>}|^2$, between the initial state and all 2-phonon combination states into which the initial excitation can relax by a cubic fission process.

$$\Gamma_{relax} = 2\pi/\hbar \; \overline{|<\Psi_i| \; \Phi^{(3)}| \; \Psi_j\Psi_k>|^2} \; \Sigma_{jk} \; \rho(\omega_j) \; \rho(\omega_k) \; \delta(\omega_i-\omega_j-\omega_k) \qquad (2a)$$

$$= 2\pi/\hbar \; \overline{|<\Psi_i| \; \Phi^{(3)}| \; \Psi_j\Psi_k>|^2} \; \rho_{2P}(\omega_i) \qquad (2b)$$

where $\rho_{2P}(\omega_i)$ is the total 2-phonon density of all combination states $\Psi_j\Psi_k$ at an energy $\hbar\omega_i$. Cubic relaxation typically occurs by fission into a lower energy vibration (v), and a phonon mode (p). Thus, $\rho_{2P}(\omega_i)$ represents a 2-phonon density of states which includes the sum of all vibration+phonon combination states with a combined energy of $\hbar\omega_i$,

$$\rho_{2P}(\omega_i)=\Sigma_v\rho_p(\omega_i-\omega_v) \qquad (3)$$

where $\rho_p(\omega_i-\omega_v)$ is the phonon density of states at the difference in energy between the initially excited vibration ω_i and the lower frequency vibrations ω_v. The effect of pressure on $\rho_{2P}(\omega_i)$ is three fold; it can cause frequency shifts, it can cause large increases by opening new relaxation channels, and it will cause an increase due to volume compression, given by $P = -(\partial\Phi/\partial V)_{eq}$.

4. RESULTS AND DISCUSSION

The psCARS technique involves the coherent excitation of a Raman active vibration by a pair of laser pulses whose frequency separation corresponds to the vibrational frequency. Following this excitation a third laser pulse of variable delay is used to induce coherent anti-Stokes emission proportional to the remaining coherent intensity. For a single crystal at low temperature the measured dephasing time, T_2, is used as a direct measure of twice the relaxation time, i.e. $T_2 = 2T_1$.

A schematic energy level diagram of the vibrational modes and lattice phonons in crystalline naphthalene and CS_2 is shown in Fig. 2. Although, the vibrational spectra of CS_2 and naphthalene are complicated by naturally occurring isotopic species (90% $^{12}C^{32}S_2$ plus 8% $^{12}C^{32}S^{34}S$ for CS_2 and 10% $^{13}C^{12}C_9H_8$ for naphthalene), the vibrational frequencies of the isotopic species are usually resolvable and selectively excited.

4.1 Relaxation of the $2v_2$ mode of CS_2

The pressure dependence of the vibrational dephasing for the v_1 and the $2v_2$ modes of CS_2 are plotted in fig. 3 and show a monotonic increase in vibrational dephasing with pressure. The vibrational dephasing time ($T_2/2$) of the $2v_2$ mode decreased from 330 ps to 17 ps over a pressure range of 19kbar, while the coherence decay time for the v_1 mode decreased from 100 ps to 39 ps over the same pressure range.

The effect of pressure on the vibrational dephasing of the $2v_2$ mode is much greater than for the v_1 mode. At low pressure, the $2v_2$ mode cannot relax into the v_1 mode by a cubic mechanism since the $130cm^{-1}$ energy gap is larger than the maximum phonon frequency of about $100cm^{-1}$. However, the phonon frequencies increase significantly with pressure, and at 30 kbar the phonon density of states extends to about $150cm^{-1}$. Since the separation between the intramolecular $2v_2$ and v_1 modes remains relatively unchanged over this pressure range, efficient relaxation from $2v_2$ to v_1 occurs when the phonon density of states spans the energy gap

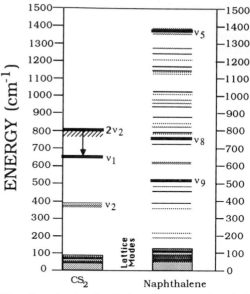

Fig. 2. A sketch of the energy levels of the fundamental vibrational modes and lattice phonons of naphthalene and CS_2. The vertical arrow from $2v_2$ to v_1 represents a pressure induced relaxation channel.

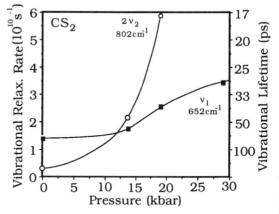

Fig. 3. Vibrational dephasing rate, $2/T_2$, versus applied pressure is shown for the v_1 and the $2v_2$ modes of solid CS_2. The lines are guides for the eye.

between these two modes. The onset of a new relaxation channel at high pressure is indicated in fig. 2 by the vertical arrow connecting $2v_2$ with v_1 at elevated pressures. We conclude that the large pressure induced dephasing observed for the $2v_2$ mode is dominated by the opening up of this new cubic decay channel.

The vibrational dephasing time ($T_2/2$) for the v_1 symmetric stretch (652 cm^{-1}) was measured to be 100 ps at ambient pressure. The next lowest energy vibrational mode is the

v_2 bend, approximately 260cm^{-1} lower in energy, as shown in fig. 2. Since the maximum phonon frequency is much less than this energy gap (even at high pressure) vibrational relaxation for isotopically pure CS_2 can only occur by a high-order multiphonon decay mechanism involving simultaneous fission into more than two other modes.

4.2 Vibrational relaxation of the v_5,

v_8 and v_9 modes of naphthalene

The vibrational relaxation rates for the v_5, v_8 and v_9 modes of naphthalene, shown in fig. 4, all increase dramatically at high pressures. In order to examine the pressure induced density of states effects for naphthalene, eq. 3 is used to calculate the two phonon DOS for naphthalene at two different pressures, the results of which are shown in fig. 5.

Vibrational relaxation by fission into two lower energy modes (i.e. cubic decay) typically requires the emission of a phonon to conserve energy. Thus, we define a 2-phonon DOS, $\rho_{2p}(\omega)$, given by eq. 3. Using an average Gruneisen parameter, γ, for the phonon modes, the *average* 2-phonon DOS per unit volume scales as $\rho_{2p}(\omega)_{ave} \propto (V^{3\gamma-1})(V^{-1}) = V^{3\gamma-2}$, and is relatively pressure independent. Although, on average, the phonon spectral expansion and volume decrease nearly cancel, there is much structure in the 2-phonon DOS, which can give rise to dramatic pressure induced changes, as shown in fig. 5. For example, the 2-phonon density of states for the v_8 (766cm^{-1}) and v_5 (1385cm^{-1}) modes of naphthalene experience large increases due to the pressure induced shifts in the lattice phonon DOS. It is also important to realize that the high pressure 2-phonon DOS in fig. 5 has not been corrected for volume compression, thus the actual DOS per unit volume at 8 kbar will be about 10% higher.

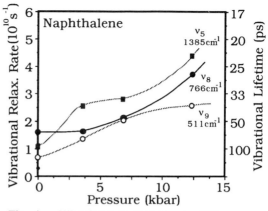

Fig. 4. Vibrational dephasing rate, $2/T_2$, versus applied pressure is shown for three of the symmetric stretching modes of naphthalene, v_5, v_8, v_9. The lines are guides for the eye.

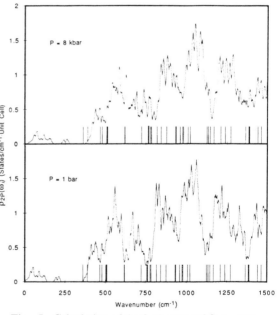

Fig. 5. Calculation of the 2-phonon DOS, $\rho_{2p}(\omega_i) = \Sigma_v \rho_p(\omega_i - \omega_v)$, for naphthalene at a pressure of 0 kbar and at 8 kbar. Also shown is the phonon density of states (≤ 200cm^{-1}) and the intramolecular vibrations used in the calculation of $\rho_{2p}(\omega_i)$. The three intramolecular vibrations investigated with psCARS are v_5 (1385cm^{-1}), v_8 (766cm^{-1}) and v_9 (511cm^{-1}) and are inducated by bold lines.

Since recent calculations on crystalline benzene indicate that a 40kbar pressure increase will only cause a factor of two increase in the anharmonic coupling (Della Valle et al 1988), it is likely that pressure induced DOS changes are a significant factor in pressure induced vibrational relaxation. Fig. 5 indicates that the 2-phonon DOS for the v_5 and v_8 vibrational modes greatly increase as the pressure is changed from 0 to 8kbars, with a much smaller DOS change for the v_9 mode. Experimentally, the observed relaxation rate for the 511cm^{-1} v_9 mode is in fact less than the other two modes at all pressures, as was shown in fig. 4. Although the slower relaxation of the 511cm^{-1} mode at high pressure is consistent with a less dramatic pressure induced 2-phonon DOS change, the pressure induced vibrational relaxation of this mode is still quite significant, and increased anharmonic interactions may be an important contribution to the high pressure vibrational relaxation process in naphthalene.

5. CONCLUSIONS

The observed monotonic pressure induced increase in vibrational relaxation with applied pressure for all of the vibrations studied is qualitatively attributed to pressure induced increases in the density of states. We also conclude that the opening of new relaxation channels due to pressure induced phonon frequency increases is expected to be a general feature in high pressure crystals. However, the relative importance of pressure induced anharmonicity versus density of states effects is still not completely clear.

Density of states effects appear to be responsible for the observed dynamics for the $2v_2$ mode of CS_2 and for two of the three naphthalene vibrations studied, particularly in light of recent calculations on benzene which show relatively mild pressure induced anharmonic couplings. In order to better quantify the relative importance of DOS versus anharmonic effects we are currently studying vibrations for which fig. 5 shows a pressure induced *decrease* in the 2-phonon DOS. These results should more clearly separate the relative importance of density of states versus anharmonic effects.

ACKNOWLEDGMENT

We acknowledge the National Science Foundation (#CHE-9008551), the Los Alamos Center for Nonlinear Studies INCOR Program, and to the donors of The Petroleum Research Fund, administered by the ACS, for financial support of this research.

REFERENCES

Baggen M, van Exter M, Lagendijk A 1987 *J. Chem. Phys.* **86** 2423
Baggen M, Lagendijk A 1991 *Chem. Phys. Lett.* , **177**, 361
Bolduan F, Hochheimer H D, Jodl H J 1986 *J. Chem. Phys.* **84**, 6997
Buchsbaum S, Mills R, Schiferl D 1984 *J. Phys. Chem.* **88** 2522
Buckingham A D 1980 *Vibrational Spectroscopy of Molecular Liquids and Solids*, ed S. Bratos and R. Pick (Plenum Press: New York)
Califano S, Schettino V, Neto N 1981, *Lattice dynamics of Molecular Crystals*, (Springer-Verlag, New York)
Chronister E, Crowell R 1991 *Chem. Phys. Lett.* **182** 27
Chronister E, Crowell R 1991 *J. Phys. Chem.* in press
Della Valle R G, Righini R 1988 *Chem. Phys. Lett.* **148** 45
Della Valle R G, Righini R 1988 *Chem. Phys. Lett.* **148** 45
Ferraro J 1984 *Vibrational Spectroscopy at High External Pressures*, (Acad. Press, Orlando)
Forman R, Piermarini G, Barnet J, Block S 1972 *Science* **176** 284
Kihara T 1977 *Intermolecular Forces* (John Wiley: New York)
Kroon R, Baggen M, Lagendijk A 1989 *J. Chem. Phys.* **91** 74
Merrill L, Bassett W 1974 *Rev. Sci. Instrum.* **45** 290
Rice S, Costantino M 1989 *J. Phys. Chem.* **93** 536

Inst. Phys. Conf. Ser. No 126: Section III
Paper presented at Int. Symp. on Ultrafast Processes in Spectroscopy, Bayreuth, 1991

203

Time-resolved investigation of LO-phonon dynamics in semi-insulating and n-type GaAs

F. Bogani * and F. Vallée

Laboratoire d'Optique Quantique du C.N.R.S.
Ecole Polytechnique, 91128 Palaiseau cedex, France

ABSTRACT : By means of time-resolved Coherent Anti-Stokes Raman Scattering we investigate the LO phonon dynamics in semi-insulating and n-doped GaAs crystals. The processes giving rise to the relaxation of the LO phonon are determined from the measured temperature dependence of its dephasing rate. The influence of an electron plasma on the LO phonon dynamics is investigated in n-doped GaAs samples; we find, in the investigated range of doping, an increasing of the dephasing proportional to the plasma density .

Time resolved coherent non-linear techniques have been extensively used in the study of the vibrational dephasing of liquids and crystals and have greatly inproved our knowledge of the vibrational dynamics. Only recently time resolved four wave mixing spectroscopy has been used in the study of phonon relaxation in semiconductors with a large gap. On the other hand in III-V compounds such as GaAs and InP ,of great both fundamental and technological interest, the understanding of the dephasing processes for the longitudinal optic (LO) phonon is still poor; the knowledge of its dynamics and of the modifications introduced by an electron plasma is of crucial interest since the LO phonon strongly influences the electronic properties of these semiconductors (Lyon 1986).

By means of time resolved Coherent Anti-Stokes Raman Scattering (CARS) we have determined the LO dephasing time in semi-insulating and n-doped GaAs crystals. In order to avoid unwanted plasma excitation and to investigate bulk properties, we have used infrared pulses with photon energy below the GaAs gap (1.52 eV at 4 K). In the experiment the three pulses were produced by an IR version of a system already described and utilised for visible CARS experiments (Vallée 1988) A passively mode-locked Nd:glass laser system provides pulses at 1054 nm with energy $\sim$ 2 mJ and 5 ps duration. This beam is split in three parts : the first ($\sim$ 30 μJ) is used as one of the excitation pulses (ω_L) ,the second one ($\sim$ 10 μJ), after passing through a variable delay line, is used as probe (ω_p)..The remaining part is frequency shifted by Stimulated Raman Scattering in $CHBr_2Cl$ to generate the second excitation beam at ω_s ($\sim$ 3 μJ at 1084 nm). By Coherent Raman Scattering the picosecond pulses at frequencies ω_L and ω_s create, in the crystal, a coherent population of the LO phonon at frequency $\omega_{LO} = \omega_L - \omega_s$ ($\sim$ 295 cm^{-1} in GaAs). The loss of coherence of the excitation is probed by CARS at frequency $\omega_{as} = \omega_p + \omega_{LO}$ of a time delayed probe pulse (ω_p). The measurement of the scattered intensity at ω_{as} as a function of the delay t_D allows the determination of the dephasing time T_{2L} of the LO phonon.

The experiments were performed in GaAs crystals with a (100) orientation placed in a variable temperature cryostat. The thickness ($\sim$ 350 μm) and the polarizations of the beams were choosen in such a way to minimize the effects of multiple reflections from the surfaces on the temporal resolution of the system.The scattered signal at ω_{as} is detected by a S1 photomultiplier after spatial , spectral and polarization filtering.

The measured CARS signal is reported in Fig (1) as a function of the probe delay t_D for a semi-insulating sample at 78 K. The curve a) refers to a low focalization experiment where the photoexcited plasma density is negligible. In this case ,after a coherent artifact due to the system response function, we observe an exponential decay of the coherent signal indicating a lorentian broadening of the line; from the slope of this line we determine an intrinsic dephasing time $T_{2L}/2 = 6.4 \pm .4$ ps The comparison with the depopulation time $T_1 = 7 \pm 1$ ps measured by time resolved Raman Scattering (von der Linde 1980, Kash 1988) shows that the dephasing is due to population relaxation.

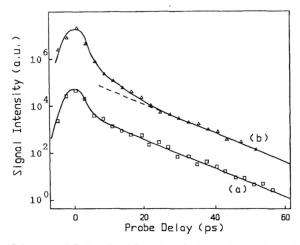

Fig. 1. Coherent anti-Stokes signal from the LO phonon in semi-insulating GaAs at 78K obtained for(a) a weak focalization of the laser beams., (b) a strong focalization (the diameter of the focal spot is decreased by a factor of about 2). The (b) curve is up-shifted for clarity.

At stronger focussing of the laser beams, one photon absorption from impurities and two photon absorption produce a plasma density large enough to modify the LO phonon dynamics.(Abstreiter 1984) . This behaviour has been evidentiated from the measurements made at stronger focussing reported in fig 1 (b): here the spatial inhomogeneity of the plasma photoexcited by the excitation beams leads to the observation of a non-exponential decay of the coherent signal originating in the plasmon-phonon interaction..

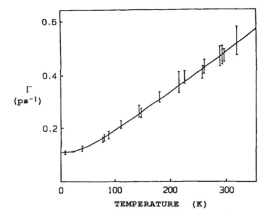

Fig. 2 Observed temperature dependence of the linewidth $\Gamma_{LO} = 1/T_2$ in semi-insulating GaAs. The continous line is the best fit obtained assuming for the phonon relaxation a down conversion process involving a TA and a LO phonon (see text).

The intrinsic relaxation time of the LO phonon has been measured for various temperatures between 10 and 320 K. The results are shown in fig.2. The observed temperature dependence of the dephasing rate corresponds to a down conversion process involving a transverse acoustic phonon (60 cm⁻¹) and a LO phonon (235 cm⁻¹) of opposite wavevectors at the L critical point of the Brillouin zone , in agreement with our preliminary results (Vallée 1991).

The influence of the electrons on the dephasing of the LO phonon has been investigated more quantitatively, using the same CARS technique, in n-doped (Si) GaAs samples. The electric field associated with the LO phonon and the electron plasma couples the two excitations giving rise to a mixed mode. The full dielectric response function of the hybrid mode can be written, in the long wavelength region, as (Abstreiter 1984)

$$\varepsilon = \varepsilon_{\infty}\left(1 + \frac{\omega_{LO}^2 - \omega_{TO}^2}{\omega_{TO}^2 - \omega^2 - 2i\omega\,\Gamma_{LO}} - \frac{\Omega_p^2}{\omega(\omega + i\gamma)}\right) \tag{1}$$

where $\Gamma_{LO} = 1/T_2$, γ is the damping of the electrons and $\Omega^2 = 4\pi n e^2/\varepsilon_{\infty}\, m^*$ is the electron plasma frequency. At low carrier density ($n < 10^{17}$ cm⁻³),as for the samples under study, the LO frequency of the phonon like mode is shifted of a very small amount and its damping results from (1) to be:

$$\frac{1}{T_2} = \Gamma_M = \Gamma_{LO} + \gamma(\omega_{LO} - \omega_{TO})\Omega_p^2 / \omega_{LO}^3 \tag{2}$$

Therefore the plasma affects mainly the relaxation properties of the hybrid mode giving rise to a faster exponential decay of the coherence.The measured relaxations for three crystals with different doping are shown in fig. 3. The dephasing time is identical to that of the semi-insulating crystal for the lower concentration (< 10^{15} cm⁻³) and decreases for higher doping.

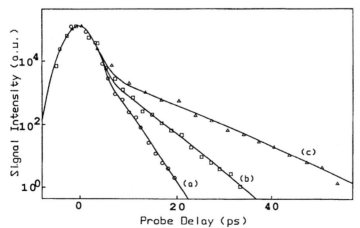

Fig. 3. CARS signal from the phonon-like plasmon-phonon hybrid mode in n-type GaAs at 78K. The electronic plasma density are (a) 1.1 10^{17} cm⁻³,(b) 4.7 10^{16} cm⁻³ and (c) < 10^{15} cm⁻³. The measured dephasing rate are respectively (a) 1.8 ps.,(b) 3.2 ps and (c) 6.4 ps.

The contribution of the plasma to the LO-phonon width, as a function of the carrier concentration, is reported in Fig 4 : The electron collision time $\tau = \gamma^{-1}$ can be evaluated from the measured T ; we find $\tau = 60$ fs for n = 4.7 10^{16} cm^{-3} and $\tau = 50$ fs for n = 1.1 10^{17} cm^{-3} at 78 K. Its linear dependence on the plasma density shows that the electron damping is nearly concentration independent up to n ~ 10^{17} cm^{-3}.

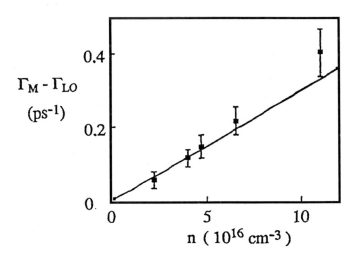

Fig 4. Dependence of the additional damping $\Gamma_M - \Gamma_{LO}$, at 10 K, due to the electron plasma, as a function of the carrier density

The dependence of $\Gamma_M - \Gamma_{LO}$ on temperature for a given sample evidentiates a contribution ,decreasing when the temperature is raised, due to ionized impurity scattering of the electrons and an almost constant contribution coming from neutral impurity scattering.

The LO mixed mode thus constitutes a sensitive probe of the electron plasma properties, allowing the investigation of very fast electronic properties. The comparison of T_2 measured in presence of a photoexcited plasma and T_2 measured in doped samples should give interesting informations about the plasmon and the LO phonon relaxation processes.

We wish to thank Prof. G.Guillot, INSA Lyon, France and Dr. R. Magnanini, MASPEC Parma, Italy,for providing the samples. We are also indebited to C. Gaonach, LCR Thomson, France, for measuring the carrier concentration.

* Permanent address: Dipartimento di Fisica , Universita di Firenze, Largo E. Fermi 2, 50125 Firenze , Italy.

References:

Abstreiter G. , Cardona M. and Pinczuk A. 1984 " Ligth Scattering spectra in solids" vol. IV eds. M. Cardona and G. Guntherodt.(Berlin : Springer Verlag) pp 5-150

Kash J.A. and Tsang J.C. 1988 Solid State Electr. **31** 419

Lyon S.A. 1986 J.of Lum. 35 121

Vallée F , Gale G.M. and Flytzanis C. 1988 Phys. Rev. Lett. 61, 2102 .

Vallée F. and Bogani F. 1991 Phys. Rev. B 43 12049

von der Linde D., Kuhl J. and Klingerberg H. 1980 Phys. Rev. Lett. 44 1505

Inst. Phys. Conf. Ser. No 126: Section III
Paper presented at Int. Symp. on Ultrafast Processes in Spectroscopy, Bayreuth, 1991

Dephasing of excitons in polar semiconductors—The cuprous chloride case

F. Vallée, F. Bogani* and C. Flytzanis

Laboratoire d'Optique Quantique du CNRS, Ecole Polytechnique, 91128 Palaiseau, France

ABSTRACT : We propose a nonlinear optical technique which allows the direct observation of spatiotemporal evolution of coherent short exciton pulses in polar semiconductor. The technique is demonstrated in the case of the transverse and longitudinal components of the Z_3 exciton in Cuprous Chloride (CuCl) and provides a detailed picture of the intrinsic dephasing of these excitations ; this is dominated by exciton-phonon scattering process.

We address here the problem of the coherence of exciton-polaritons in polar semiconductors using a nonlinear time resolved technique that gives [Vallée et al (1991] direct access to their dephasing time. This problem has been addressed in the past with limited success using indirect techniques that proceed via linear polariton excitation within the absorption layer close to the crystal surface where, however, the polariton characteristics and behavior may be different from the bulk. In addition these techniques have other drawbacks ; for instance, the incoherent ones [Masumoto and Shionoya (1982), Askary and Yu (1985), Oka et al (1986)], like the time-resolved luminescence or the induced absorption, can only give global information after a few scattering events, while the time resolved four wave mixing [Masumoto et al (1983), Takagahara (1985), Dagenais and Sharfin (1987)] gives access to the high excitation regime where the polariton dephasing is dominated by polariton-polariton interactions.

The technique [Vallée et al (1991)] we shall discuss here is exempt of these restrictions and allows a direct determination of the intrinsic dephasing time. Besides being very flexible and applicable to essentially any electric dipole allowed transition in crystals without inversion symmetry it can also be complemented to provide spatial resolution as well by exploiting the nonlocal possibilities of the nonlinear interactions. Here we summarize the results of the measurement of the intrinsic dephasing time of the transverse (polariton) and longitudinal components of the Z_3 exciton in cuprous chloride (CuCl).

The principle of the technique was outlined elsewhere and is also schematically depicted in Fig.1. It consists of picosecond second order coherent two-photon excitation and detection of an exciton wave packet ; the two stages can be separated both in time and space which is important when studying propagating excitations like the exciton-polaritons. This was previously demonstrated for the phonon-polaritons by a similar technique for the lower branch [Gale et al (1986)].

The coherent excitation of the exciton polaritons of frequency $\omega_{e\pi}$ and wave vector $k_{e\pi}(\omega_{e\pi})$ is realized in the bulk of the crystal by two photon absorption of two synchronized picosecond

* Dipartimento di Fisica, Universita di Firenze, 50125 Firenze, Italy

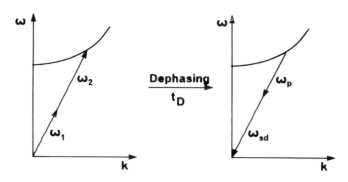

Fig. 1

pulses with frequencies ω_1 and ω_2 and wave vector $\underline{k}_1$ and $\underline{k}_2$ such that $\omega_{e\pi} = \omega_1 + \omega_2$ and $\underline{k}_{e\pi}$ $= \underline{k}_1 + \underline{k}_2$. These conditions introduce certain restrictions in the applicability of the technique ; thus for the case of polaritons they can only be satisfied for the upper-branch [Fröhlich et al (1971 and 1973)]. Close to the exciton resonance and neglecting spatial dispersion, the amplitude of the coherently driven exciton packet is proportional to :

$$d_\lambda = d_{E\lambda}\left(\omega_e^2 - \omega_\lambda^2\right) + d_{M\lambda}$$

(1)

where ω_e is the bare exciton frequency at k=0 ($\omega_e \sim 3.202$eV in CuCl) and $\lambda=\pi,L$ labels the exciton polariton (π) and longitudinal (L) exciton respectively. The coupling parameters $d_{E\lambda}$ and $d_{M\lambda}$ are related, respectively, to two-photon absorption and sum frequency generation close to ω_e. The evolution of the exciton coherence, after the local excitation process has terminated, is followed by phase-matched parametric emission at $\omega_d = \omega_e-\omega_p$ stimulated by a third picosecond pulse of frequency ω_p delayed by t_D and spatially separated with respect to the excitation stage. We wish to stress here the fact that the excitation and probing can be done at will anywhere inside the crystal since all involved frequencies are in the transparency range of the medium : the spatial resolution is fixed by the overlap extension of the interacting beams.

The demonstration of this technique was performed on two upper-branch polaritons in CuCl with energies $\hbar\omega_{e\pi}^b \approx 3.208$eV and $\omega_{e\pi}^f = 3.217$eV corresponding respectively, to a backward,

$\theta=180°$ and a forward $\theta=0°$ excitation geometry and on the longitudinal exciton. As pointed out previously the technique allows the study of the spatiotemporal evolution of the polariton pulse by separating the excitation and probing stages in time and space. However, in the case of CuCl, relaxation was found to occur much faster than propagation and the polariton wave packet was probed only locally and similarly for the longitudinal exciton since it is not a propagating mode.

The experiments were performed using a passively mode locked Nd^3 glass delivering a single 5-ps pulse in the infrared ($\omega_I = 1.054\mu m$) which is frequency converted to create three independent pulses ω_1, ω_2 and ω_p. The one at ω_1 is a small part of the initial infrared pulse while ω_2 is tunable around 611μm and the probe beam frequency ω_p, in the infrared, is adapted to phase matching requirements. The experimental procedure is cursively discussed in Vallée et al (1991). Two samples with different surface orientations were used, one with 110 surface for the study of the transverse (polariton) exciton and another with 111 surface for the study of the longitudinal exciton.

In Fig.2, we reproduce a measurement for the dephasing rate for the two investigated polaritons $\omega_{e\pi}^{f}$ and $\omega_{e\pi}^{b}$ at crystal temperature of 7K. The low intensity ratio of the signals, $I_f/I_b \approx 10^{-4}$, is a consequence of the destructive interference for $\theta=0°$ between the material and electric contributions in d in (1) [Fröhlich et al (1971 and 1973), Kramer and Bloembergen (1976)]. In Fig.3, we report the measured values of the dephasing rates Γ for different temperatures in the range 7-60K and the calculated ones with an analytical expression of this rate based on the assumption that the dephasing is due to the three main exciton-phonon scattering processes namely the Fröhlich (LO), the deformation potential (DP) and the piezoelectric (PE) interaction mechanisms. Their compound effect as depicted in Fig.3 satisfactorily reproduces the observed behavior of Γ.

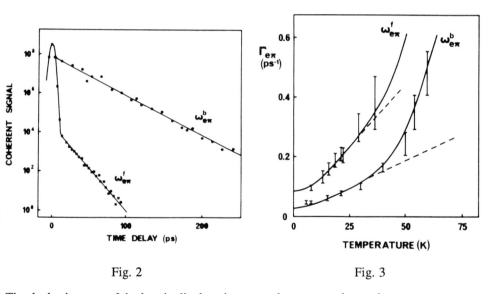

Fig. 2   Fig. 3

The dephasing rate of the longitudinal exciton was also measured over the same temperature range ; the values are within the same range as those of the transverse (polariton) exciton and the overall temperature dependence is similar implying that the same exciton-phonon mechanisms are at work her too. Thus at low temperature (T<40K) the main relaxation mechanism is scattering by acoustic phonons mediated either by the deformation potential (DP) for the LA phonon and by the piezoelectric effect (PE) for the TA phonons ; both up-and down- conversion processes are involved (Fig.4). At higher temperatures (T≥40K) the impact of the LO-phonon assisted scattering process mediated by the Fröhlich interaction [Weisbuch and Ulbrich (1982)] becomes important and in fact dominates the relaxation as the temperature increases ; here too as in the case of the transverse (polariton) exciton, only the up-conversion procession process is relevant because of the very low density of accessible states for the down conversion mechanism. In general the variation of the dephasing rate with frequency both for the transverse and longitudinal excitons can be roughly traced to the variation of the density states over the exciton dispersion curve.

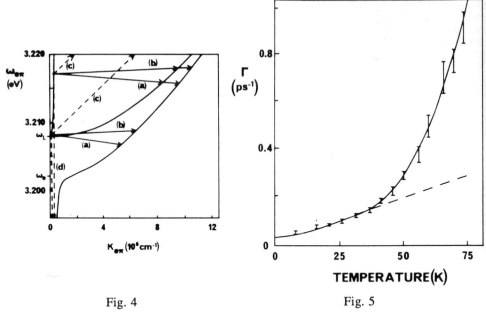

Fig. 4   Fig. 5

Including all three exciton-phonon mechanisms then one can write :

$$\Gamma = \gamma_{LO}\, n\left(\omega_{LO}\right) + \gamma_{DP}\left(1 + n\left(\omega_{LA}^{-}\right) + n\left(\omega_{LA}^{+}\right)\right) + \gamma_{PE}\left(1 + n\left(\omega_{TA}^{-}\right) + n\left(\omega_{TA}^{+}\right)\right) \quad (2)$$

for the intrinsic dephasing rate and obtain a temperature dependance in good agreement with the experimental results. In (2), $n\,(\omega_i)$ is the occupation number for the ω_i-phonon whose frequency is imposed by energy and wave vector conservation [Mita et al (1980)]; γ_{LO}, γ_{DP} and γ_{DP} are frequency-dependent coupling parameters for the three scattering mechanisms LO, DP and PE respectively. These parameters can be estimated by fitting the measured values of the dephasing time with expression (2) ; they can also be independently estimated using known parameters of CuCl [Takagahara (1985)]. The fitting of the experimental values of Γ with (2) as depicted in Figs.3 and 5 gives the following table :

	γ_{LO}	γ_{DP}	γ_{PE}
$\omega_{e\pi}^{f}$	26 µeV	37 µeV	25 µeV
$\omega_{e\pi}^{b}$	26 µeV	14 µeV	3 µeV
ω_{eL}	26 µeV	14 µeV	3 µeV

The broken lines in Figs.3 and 5 represent the temperature dependance of the dephasing rate if the Fröhlich mechanism was altogether neglected ; as can be seen such a neglect would lead to substantial deviations from the measured values as the temperature increases beyond 40K.

In the case of the longitudinal exciton when the phase matching configuration in the probing stage is relaxed a slower relaxation component was observed at low temperature in addition to the one discussed above. This may originate from other exciton states with slower relaxation rates accessible either through spatial dispersion or through localisation by defects. The precision was not sufficient to make precise statements concerning the origin of this relaxation component.

In conclusion, we have demonstrated that the intrinsic dephasing rate of exciton in polar semiconductors can be directly and selectively investigated by use of a time-resolved two-photon nonlinear technique which also provides spatial resolution. The results in CuCl reveal that the dephasing rate is dominated by population relaxation and the variation of density of final states accessible by exciton-phonon scattering over the exciton dispersion curve accounts for the main frequency and temperature dependance of this rate. In concluding we also point out that this technique is quite general and can be extended to high exciton densities where exciton-exciton scattering important or to surface excitons. Combined with spatial separation of the interrogation and probe stages this technique should allow direct investigation of the spatiotemporal evolution of polariton pulses.

We wish to thank Professor D. Fröhlich of the University of Dortmund, Germany, for providing samples and for valuable information ; we also thank J. Godard of the Laboratoire de Physique des Solides, Université de Paris-Sud, France, for preparing the samples.

References

Askary F and Yu P Y 1985 *Phys. Rev.B* **31** 6643
Dagenais M and Sharfin W F 1987 *Phys. Rev. Lett.* **58** 1776
Fröhlich D, Mohler E and Wiesner P 1971 *Phys. Rev. Lett.* **26** 554
Fröhlich D, Mohler E and Uihlein C 1973 *Phys. Stat. Sol.(b)* **55** 175
Gale G M, Vallée F and Flytzanis C 1986 *Phys. Rev. Lett.* **57** 1867
Hönerlage B, Lévy R, Grun J B, Klingshirn C and Bohnert K 1985 *Phys. Rep.* **124** 161
Kramer S D and Bloembergen N 1976 *Phys. Rev.B* **14** 4654
Kuwata M, Kuga T, Akiyama H, Hirano T and Matsuoka M 1988 *Phys. Rev. Lett.* **61** 1226
Masumoto Y and Shionoya S 1982 *J. Phys. Soc. Jap.* **51** 181
Masumoto Y, Shionoya S and Takagahara 1983 *Phys. Rev. Lett.* **51** 923
Mita T, Sôtome K and Ueta M 1980 *Sol. St. Com.* **33** 1135
Oka Y, Nakamura K and Fujisaki H *Phys. Rev. Lett.* **57** 2857
Takagahara T 1985 *Phys Rev. B* **31** 8171
Weisbuch C and Ulbrich R G 1982 *Ligh Scattering in Solids III* Ed M Cardona and G Guntherodt (Springer Verlag) p 207

Inst. Phys. Conf. Ser. No 126: Section III
Paper presented at Int. Symp. on Ultrafast rocesses in Spectroscopy, Bayreuth, 1991

Analysis of experimental data on time-domain spectroscopy of molecular gases

D V Kolomoitsev, S Yu Nikitin

Department of Physics, Moscow State University,
119899 Moscow, USSR

ABSTRACT: Pulse response of time-domain spectroscopy was calculated for ensembles of molecules with inhomogeneously broadened spectra. Curves of pulse response were built using strong collisions model for nitrogen, ammonia and methane under various gas pressures, theoretical and experimental data comparison was fulfilled.

1. INTRODUCTION

The possibility to obtain information about inelastic rotational transitions analyzing the signal of time-domain coherent anti-Stokes Raman spectroscopy of nitrogen was demonstrated earlier by Kolomoitsev and Nikitin (1989). The aim of the present paper is the analysis of time-domain spectroscopy data of molecules with more complicated and dense structure of spectra - ammonia and methane.

2. THE SIGNAL OF THE TIME-DOMAIN SPECTROSCOPY

In the time-domain spectroscopy investigated transition is ex-cited by sufficiently short laser pulse (or pulses). The relaxation process is interrogated by probe pulse, which is delayed in time τ relatively exciting one. The probe pulse monitors the coherent molecular vibrations amplitude $q(t)$, and the time dependence of the time-domain spectroscopy signal energy $W(\tau)$ (pulse response) contains information about mechanisms and rates of the dephasing of vibrations in the matter. Pulse response may be written in the following form (Nikitin 1985 ; Kolomoitsev and Nikitin 1986a, 1991):

$$\begin{cases} W(\tau) = \varkappa \int_{-\infty}^{+\infty} \left| F_P(\theta) \ q(\theta+\tau) \right|^2 \ d\theta, \qquad \varkappa = const \\ q(t) = \int_{0}^{+\infty} F_e(t-\theta) \ h(\theta) \ d\theta \end{cases} \tag{1}$$

Here $F_e(\theta)$ and $F_P(\theta)$ are functions describing the averaged shapes of the exciting and probing pulses, $h(\theta)$ is Green function of the investigated transition, which depends only on properties of the medium. If several statistically independent dephasing mechanisms exist, Green function $h(\theta)$ can be written as product of Green functions describing influence of each mechanism separately. In conditions of considered experiments (temperature T = 295°K, pressure until 7 bar) Q-branch shape is mainly determined by rotational dephasing, caused by the difference of vibrational transition frequencies for the molecules in different rotational states.

3. ROTATIONAL DEPHASING. FREQUENCY EXCHANGE

Under collisions of gas molecules, their rotational states can change. Because of vibrational transition frequency dependence on molecular rotational state, it leads to the change of vibrational transition frequencies with time. Thus, "frequency exchange", caused by collisions, can occur in inhomogeneous set of oscillators. System of kinetic equations describing the time evolution of rotational Green function $h_R(\theta)$ may be presented in the following form (Kolomoitsev and Nikitin 1991; Burshtein et al 1991):

$$h_R = \sum_J h_J, \qquad h_J(0) = P_J, \qquad \frac{d\,h_J}{d\,\theta} + i\omega_J h_J = \sum_{J'} \Gamma_{J'J}\, h_{J'}. \qquad (2)$$

Here P_J denotes the probability to find the molecule in the state with rotational quantum number J, ω_J is the frequency of vibrational transition from given state J, and $\Gamma_{JJ'}$ is the frequency exchange operator, whose elements have sense of the $J \to J'$ transition probabilities per unit time. The operator of such transitions must satisfy the requirements of the detailed balance and conservation of the number of particles:

$$P_J\,\Gamma_{JJ'} = P_{J'}\,\Gamma_{J'J}, \qquad\qquad \Gamma_{JJ} = -\sum_{J \neq J'}\Gamma_{JJ'}\,. \qquad (3)$$

At present a number of frequency exchange operator models is known, and one of the simplest is the model of "strong collisions" (Kolomoitsev and Nikitin 1986b, 1989, 1991), in which the transition probability depends only on the statistical weight of final state:

$$\Gamma_{JJ'} = \frac{1}{\tau_R}\,(P_J - \delta_{JJ'})\,. \qquad\qquad (4)$$

Here τ_R is the mean time between rotationally inelastic collisions. The substitution of (4) into (2) gives:

$$h_R = \sum_J h_J\,, \qquad h_J(0) = P_J, \qquad \frac{d\,h_J}{d\,\theta} + i\omega_J h_J + \frac{h_J}{\tau_R} = \frac{P_J}{\tau_R}\sum_{J'} h_{J'}\,. \qquad (5)$$

So, formulas (5) describe the rotational component of Green function in the "strong collisions" approximation. The time τ_R can be considered as the adjustable parameter, depending on the gas pressure. To calculate the Green function $h(\theta)$ by formulas (5), one must know the spectral structure of the inhomogeneously broadened transition (parameters P_J, ω_J).

Formula (5) is directly usable for diatomic and linear poli-atomic molecules; in this case index "J" coincides with rotational quantum number J. Ammonia (NH_3) and methane (CH_4) molecules belong to the group of symmetrical tops, their energy levels structure is determined not only by vibrational (v) and rotational (J) quantum numbers, but also by independent quantum number of total angular momentum projection on the top axis (K). Elastic ammonia molecule is characterized by the energy levels inversion doubling. We used described above theoretical model for poliatomic molecules understanding index "J" in (5) as the totality of all quantum rotational numbers, i.e. "J" numbers components of investigated spectrum. One can see that in this case we suppose that all collisions, changing any rotational number of molecule, may be characterized by the single parameter τ_R.

4. RESULTS OF CALCULATIONS

Theoretical analysis was fulfilled for time-domain CARS of nitrogen (Akhmanov et al 1985, Tarasevich 1985) and ammonia (Magnitskii 1983), free induction decay (FID) spectroscopy of methane (Bratengeier et al 1989). Frequency exchange , Doppler dephasing (for N_2 and NH_3) and different amount of components of investigated spectra (25 for N_2 (Gerzberg 1949), 91 for NH_3 (Angstl et al 1985) and 65 for CH_4 (Barnes et al 1972)) were taken into account. In calculations parameters of laser pulses (incoherent 50 ps (FWHM) with gaussian profile for CARS and coherent 3 ps with gaussian profile for FID) were used. Experimental points and optimal theoretical curves in Figures 1, 2, 3 demonstrate satisfactory agreement. Using obtained values of τ_R we can estimate the cross-section of rotationally inelastic collisions for the investigated gas: $S_R = (n \langle v \rangle \tau_R)^{-1}$, where n is the number of molecules in cm^3 and $\langle v \rangle = \sqrt{16kT/\pi m}$ is the mean speed of molecular thermal motion. Our estimations of S_R for considered molecules are: N_2- 61 $Å^2$; NH_3- 243 $Å^2$; CH_4- 38 $Å^2$. Note, that S_R values are single for different pressures for NH_3 and for CH_4.

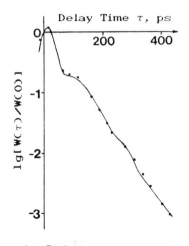

Fig. 1. Pulse response of CARS of Q-branch of N_2 : experimental points and calculated curve $T=295°K$, $p=1.0$ atm; $\tau_R=98$ ps.

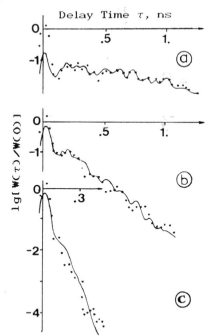

Fig. 2. Pulse response of CARS of ν_1 Q-branch of NH_3 for various pressures p (T=295°K):
(a) p=1.7 torr, $\tau_R=7.8$ ns;
(b) p=45 torr , $\tau_R=300$ ps;
(c) p=150 torr, $\tau_R=90$ ps.

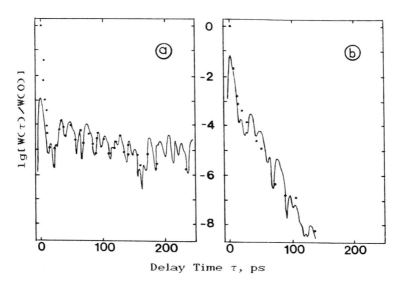

Delay Time τ, ps

Fig. 3. Conversion signal representing FID of the ν_3 Q-band of CH4:   (a) p=0.27 bar, τ_R=450 ps;   (b) p=7 bar; τ_R=17.5 ps.

5. CONCLUSION

We have considered application of the strong collisions model of frequency exchange for describing relaxation process in three various gases and have obtained values of rotationally inelastic collisions cross-sections. Theoretically calculated and experimentally obtained pulse responses satisfactorily agree, and it proves necessity of taking into account frequency exchange with help of more advanced models for obtaining precise numerical parameters of relaxation process.

6. REFERENCES

Akhmanov S A , Koroteev N I , Magnitskii S A , Morozov V B , Tarasevich A P , Tunkin V G 1985 JOSA B 2 640
Angstl R , Finsterholzl H , Frunder H , Illing D , Papousek D, Pracna P , Narahari Rao K , Schrotter H W , Urban S 1985 J.Mol.Spectrosc. 114 454
Barnes W L, Suskind J, Hunt R H, Plyler E K 1972 J.Chem.Phys. 56 5160
Bratengeier K, Purucker H-G, Laubereau A 1989 Opt.Comm. 70 393
Burshtein A I, Kolomoitsev D V, Nikitin S Yu, Storozhev A V 1991 Chem. Phys. 150 231
Gerzberg G 1949 Vibrational and Rotational Spectra of Diatomic Molecules (Moscow: IL)
Kolomoitsev D V, Nikitin S Yu 1986a Opt.Spectr.(Sov.) 60 559
Kolomoitsev D V, Nikitin S Yu 1986b Opt.Spectr.(Sov.) 61 1201
Kolomoitsev D V, Nikitin S Yu 1989 Opt.Spectr.(Sov.) 66 286
Kolomoitsev D V, Nikitin S Yu 1991 Proc.SPIE 1402 11
Magnitskii S A 1983 Ph.D.Thesis (Moscow: MSU)
Nikitin S Yu 1985 Vestn.Mosc.Univ. Ser.3 26 48
Tarasevich A P 1985 Ph.D.Thesis (Moscow: MSU)

Inst. Phys. Conf. Ser. No 126: Section III
Paper presented at Int. Symp. on Ultrafast Processes in Spectroscopy, Bayreuth, 1991

217

Time-domain spectroscopy of molecular vapors with subpicosecond pulses of THz radiation

H. Harde*, D. Grischkowsky

IBM Watson Research Center, P.O. Box 218, Yorktown Heights, New York 10598

A newly developed optoelectronic beam system generating subpicosecond pulses of THz radiation was used for time-domain studies of N_2O vapor. After the excitation the molecules were observed to radiate coherent transients on the free-induction decay originating from the periodic rephasing of coherently excited rotational absorption lines. Numerical simulations of the terahertz pulse propagation through the sample, based on the linear dispersion theory, exactly reproduce the measured pulse structures. Additional calculations of terahertz pulses interacting with OCS and HCN vapor are presented.

1. INTRODUCTION

While in recent years spectroscopy with ultrashort pulses was restricted to optical and infrared frequencies, with the newly developed terahertz beam sources, producing subpicosecond pulses of terahertz radiation (van Exter *et al* 1989a, Zhang *et al* 1990), a new frequency range is becoming accessible for time domain spectroscopy and the studies of fast transients. At IBM an optoelectronic terahertz transmission and detection system has been developed and used for measurements of dielectrics and semiconductors (Grischkowsky *et al* 1990a) as well as high T_c substrates (Grischkowsky *et al* 1990b). Also molecular vapors of H_2O (van Exter *et al* 1989b) and N_2O (Harde *et al* 1991a, Harde and Grischkowsky 1991b) have been investigated with this setup. The teraHz pulses are unique in terms of their pulse duration and corresponding bandwidth. They essentially consist of a single oscillation and are characterized by a transform-limited white spectrum extending to beyond 2 THz. The radiation is well collimated into a low-divergence beam and can be detected with a signal-to-noise ratio of better than 20,000.

In this paper we report an experimental and theoretical study of time-resolved terahertz spectroscopy of N_2O vapor, where we refer to some of the results presented in an earlier publication (Harde and Grischkowsky 1991b), and we add numerical simulations of the pulse propagation through OCS and HCN vapor. These molecules with a similar structure to N_2O but with larger electric dipole moment are interesting candidates for studies of propagation effects through optically dense media. Terahertz pulses passing through one of these vapors excite a multitude of pure rotational transitions in the impact approximation and cause the molecules to reradiate a free-induction-decay (FID) signal that consists of a series of well defined ultrashort pulses (for similar observations in the infrared see, Forster *et al* 1974, Heritage *et al* 1975, Woerner *et al* 1989, and Bratengeier *et al* 1989). These pulses have their origin in a periodic rephasing of the equally spaced rotational lines. For N_2O vapor we have measured trains of coherent transients extending to beyond 600 ps which are compared with theoretical simulations that we derived from calculations based on the linear dispersion theory. Owing to the exceptionally

---

* Permanent address: Universität der Bundeswehr, Holstenhofweg 85, 2 Hamburg 70, Germany

high time resolution of the measurements, a reshaping of the individual pulses in the train can be observed which in the case of N_2O and OCS originates from the anharmonicity in the line spacing. For HCN vapor a strong reshaping and even new formation of the emitted pulse train is calculated which is due to dispersive propagation effects.

2. EXPERIMENTAL SETUP

The optoelectronic terahertz radiation source and detection system is shown in Figure 1 and has been already described by van Exter *et al* (1989a). For the case here, a transmitting antenna with bow-tie geometry was used, fabricated on an intrinsic, high-resistivity gallium arsenide wafer. The antenna is driven by photoconductive shorting the biased $5\mu m$ wide antenna gap with 70 fs pulses coming at a rate of 100 MHz in an excitation beam from a colliding-

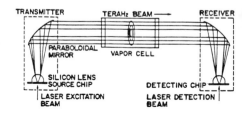

Fig.1. Experimental setup of the terahertz beam system.

pulse mode-locked dye laser. The generated THz radiation is collimated by a silicon lens and paraboloidal mirror and directed towards the receiver, where it is focused onto an ultrafast receiving antenna, which uses a micron-sized dipole structure on an ion-implanted, silicon-on-sapphire wafer. The electric field of the teraHz pulses across the receiving antenna gap is sampled by shorting the gap with the 70 fs optical pulses of a detection beam and monitoring the current versus the time delay between the optical excitation and detection beam. For the measurement with N_2O vapor a gas cell of 38.7 cm length with high resistivity silicon windows is placed into the beam.

3. N_2O-MEASUREMENTS

When the terahertz pulses propagate through the vapor they will be reshaped due to the absorption and dispersion in the sample. These changes can be directly observed by comparing a transmitted pulse with a reference scan, taken without vapor in the cell. Such a measurement is displayed in Figure 2a. The pulses are detected with a time resolution of better than 0.2 ps and a signal-to-noise ratio of more than 20,000. They consist only of a single oscillation and their numerical Fourier transform corresponds to a smooth amplitude spectrum extending up to 1.5 THz (see also dashed curve in Figure 5a).

When the cell is filled with 800 hPa of N_2O vapor, the output changes to that shown in Figure 2b. Within the broad spectral range covered by the terahertz pulses the N_2O molecule has a multitude of equally spaced rotational lines which can be excited simultaneously in the impact approximation. Consequently, a periodic rephasing and dephasing of the entire ensemble of more than 50 transitions occurs during the

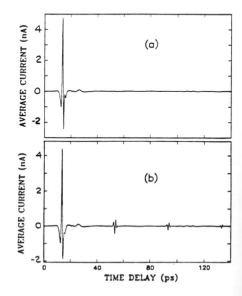

Fig.2. (a) Measured terahertz pulse without vapor in the cell, and (b) with 800 hPa of N_2O vapor.

free-induction-decay, manifest as a train of subpicosecond THz pulses with a repetition rate equal to the frequency separation between adjacent lines and corresponding to a pulse separation of 39.8 ps. Because the rephasing has a close relation to spin or photon echoes, and the frequencies are numerical multiples of a fundamental frequency, equal to the spacing between adjacent lines, we term these periodic pulses "THz commensurate echoes" (Harde *et al* 1991a).

The decay of the echoes is mainly determined by the molecular collisions and hence by the gas pressure, while Doppler dephasing in this spectral range can be completely neglected. So, when the N_2O pressure is further reduced, not only the amplitudes of the echoes become smaller but their damping is reduced as well. Figure 3 shows an example for a measured pulse train over 600 ps with 60 hPa of N_2O vapor in the cell. To observe the echoes more clearly, we have plotted the difference between the measurement and reference scan, mainly suppressing the strong excitation pulse but also reflections from the transmitting antenna chip (at 250 ps)

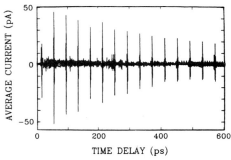

Fig.3. Radiated pulse train shown as difference of measurement and reference scan for 60 hPa N_2O vapor.

and the cell windows (at 470 ps). The amplitude of the first echo in Figure 3 is only 1% that of the exciting pulse. But due to the system's exceptionally high signal-to noise ratio of better than 20,000 with respect to the excitation pulse, the echoes can still be measured on this scale with a S/N of greater than 250. The irregular structure between the emitted transients is not noise but is due to residual water vapor in the beam path.

Compared to studies in the infrared and optical region, where the signals are detected as pulse intensities or as beat notes on a carrier, in our experiment directly the actual electric field and phase of the transmitted pulse as well as that radiated by the sample is measured. A closer look at the individual echoes of the pulse train of Figure 3 shows on an expanded scale a significant reshaping with time, as illustrated when comparing the first and eleventh transient emitted by the sample 40 and 440 ps after the excitation (see Figure 4). This reshaping is caused by small deviations from the rigid-rotator-model of the N_2O molecules. As centrifugal forces increase the moment of inertia of a rotating molecule, the frequency spacing between adjacent rotational lines is not constant but decreases slightly with increasing rotational quantum number J. This anharmonicity causes a gradual dephasing of the individual transitions and manifests itself as a change in the echo-shape. We term this phenomenon centrifugal dephasing which due to the exceptionally high time resolution can be clearly observed in the time domain.

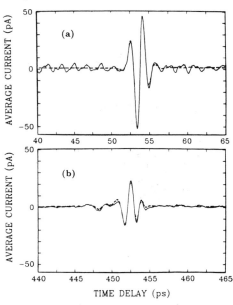

Fig.4. Measurement (—) and calculation (- - -) of (a) the 1st, and (b) the 11th commensurate echo for 60 hPa of N_2O vapor.

4. THEORY

An accurate theoretical description and simulation of the measurements requires the solution of the coupled Maxwell-Bloch equations, which under certain conditions can be solved by introducing the pulse area concept. As the terahertz pulses are characterized by a bandwidth comparable with the frequencies of the spectral distribution, it is not appropriate to describe them by a carrier and a time-dependent envelope as is usually done in the infrared and visible regions. This precludes the use of the pulse area concept for pulse propagation effects, and also the slowly varying envelope approximation cannot be applied for pulses essentially consisting only of a single cycle.

However, under our experimental conditions we can assume a linear response of the sample to the exciting pulses. Then the Maxwell-Bloch equations reduce to that of linear dispersion theory (see, e.g., Allen and Eberly 1975). The interaction of the electric field with the medium then is best described in the frequency domain by means of the dispersion $\Delta k(\omega) L$ and amplitude absorption $\alpha(\omega) L/2$ over the sample length L. A pulse $E(z,t)$ propagating in z-direction through the medium can then be calculated by

$$E(z,t) = \int_{-\infty}^{\infty} E(\omega) e^{-i(\omega t - k_0 z)} e^{i\Delta k(\omega) z} e^{-\frac{1}{2}\alpha(\omega) z} d\omega \tag{1}$$

where $E(\omega)$ is the spectral distribution of the input pulse. The absorption coefficient of three-atomic linear molecules on a single rotational transition from $J \to J+1$ and with Lorentzian lineshape is given by

$$\alpha_J(\omega) = \frac{p f_0 \mu^2 h B_V \omega^2}{6 n c \epsilon_0 (kT)^3} (J+1) e^{-hB_V J(J+1)/kT} \frac{\Delta\omega_J}{(\omega_J - \omega)^2 + (\Delta\omega_J/2)^2} \tag{2}$$

with the transition frequency $\omega_J/2\pi = 2 B_V (J+1) - 4 D_V (J+1)^3$. Here, $\Delta\omega_J$ is the linewidth, B_V the rotational constant of the vibrational state and D_V the centrifugal stretching constant. k_0 and n represent the wave vector and refractive index in the sample far from any material resonance and p is the gas pressure, f_0 the fraction of molecules in the lowest vibrational state, μ the electric dipole moment, kT the thermal energy, h Planck's constant, and c the speed of light. For the change of the wave vector it is found:

$$\Delta k_J(\omega) = \frac{p f_0 \mu^2 h B_V \omega^2}{6 n c \epsilon_0 (kT)^3} (J+1) e^{-hB_V J(J+1)/kT} \frac{\omega_J - \omega}{(\omega_J - \omega)^2 + (\Delta\omega_J/2)^2}. \tag{3}$$

The absorption and dispersion over the whole spectral range of the teraHz pulse is obtained by summing Eqs.(2) and (3) over all transitions within this spectral width.

5. NUMERICAL SIMULATIONS OF THE PULSE PROPAGATION

Our numerical calculations of pulse propagation were performed as follows. The measured reference pulse, shown in Figure 2a, was chosen to be the input pulse to the vapor. Its numerical Fourier transform $E(\omega)$ was multiplied by the amplitude absorption and dispersion which were calculated from Eqs.(2) and (3) using the known molecular constants. The inverse numerical Fourier transform then gives the predicted output pulses. If such a calculated pulse is fitted to the measurement in Figure 2b, where the spectral linewidth, here assumed to be identical for all transitions, is used as the only free parameter, on a double plot the different curves cannot be further distinguished. From the fit the self-pressure broadening of N_2O vapor can be accurately determined (Harde and Grischkowsky 1991b). The calculation reproduces the measurement in pulseshape and absolute amplitude, for the exciting pulse as well as the commensurate echoes remarkably well. This is demonstrated by the comparison in Figure 4 between theory (- - -)

and experiment (—). Here, the change of the echo-shape is well simulated with a stretching constant of $D_V = 5.3\ kHz$, as known from the literature. The background oscillations on the measurement result from the residual water.

Due to the excellent agreement between measurements and calculations for the N_2O we can also expect to obtain reliable results for the simulations of the pulse propagation through OCS and HCN vapor. The OCS molecule is distinguished by a frequency spacing between adjacent lines of $2B_V = 12.2\ GHz$ with an anharmonicity of only $D_V = 1.3\ kHz$, while HCN has a spacing of 88.6 GHz and a stretching constant of about 80 kHz. The calculated absorption spectra for 133 hPa of OCS and HCN are displayed in Figure 5a and 5b respectively. A terahertz pulse propagating through OCS vapor excites the entire manifold of absorption lines with rotational quantum numbers of up to $J = 100$. The Fourier transform spectrum of the input pulses is shown as the dashed curve in Figure 5a. Because the anharmonicity is related to the third order of J, even with the small stretching constant of OCS a significant reshaping of the echoes which is only due to centrifugal dephasing can be calculated. This is illustrated by Figure 6 showing the seventh echo of the emitted train, while the first echo appearing at 82 ps after the excitation pulse has the same shape as displayed in Figure 4a. For this calculation a pressure of 13.3 hPa and a self-pressure broadening of 8.7 MHz/hPa was assumed.

The rotational spectrum of HCN extends to larger frequencies than the spectrum of the input pulses used in our simulations. In this case only the lowest 16 transitions are excited by the terahertz radiation and contribute to the radiated pulse train with a pulse separation of 11.3 ps.

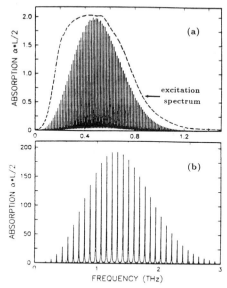

Fig.5. Calculated absorption spectra for 133 hPa of (a) OCS, and (b) HCN vapor.

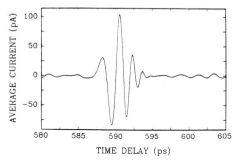

Fig.6. Calculated shape of the 7th echo for 13.3 hPa of OCS vapor.

Since the width of a spectral line is proportional to the pressure, the absorption on line-center is independent of the pressure. This peak absorption (see Figure 5) is for HCN more than two orders of magnitude larger that of OCS and N_2O ($\alpha_{max}L/2 = 0.6$), assuming the same propagation length of 38.7 cm through the cell. Therefore, already at low pressures strong echoes can be observed, as this is demonstrated in Figure 7a for a propagating pulse through 1.33 hPa vapor of HCN.

For a peak absorption greater than unity the vapor behaves as an optically thick sample and propagation effects become noticible. So, at the line center the absorption is already saturated and causes an additional spectral broadening which in the time domain is manifest as a faster decay of the measured pulse train than expected from the coherence relaxation time. In Figure 7a the observed damping is only attributed to this absorptive effect, while that contribution due to collisional dephasing with $T_2 = 2/\Delta\omega = 6.2\ ns$, cannot be observed on this time scale.

Propagation effects due to dispersion become noticible with further increasing pressure when a broader spectral range is affected by the phase shifts, caused by the spectral lines. For a pulse passing through 13.3 hPa of HCN, as shown in Figure 7b, a considerable reshaping from echo to echo is observed as a result of strong dispersion in the vapor. The frequency components of the teraHz pulse are differently delayed and cause this reshaping. At this pressure the first reradiated echo is almost as large as the transmitted excitation pulse. At 133 hPa (see Figure 7c) the phase shifts of the spectral components are great enough to produce an additional beating between the echoes until the original transients cannot be further distinguished. Whether these pulses obey any pulse area theorem, is not clear and still has to be investigated.

6. CONCLUSION

We have demonstrated the performance and features of time-domain spectroscopy with sub-picosecond pulses of terahertz radiation. A newly developed optoelectronic terahertz beam system was used to study propagation effects and coherent transients in N_2O vapor. The molecules were observed to emit a train of sub-picosecond pulses originating from the periodic rephasing of coherently excited absorption lines. We have measured trains extending to beyond 600 ps and compared them with theoretical simulations that we derived from linear dispersion theory analyses. Additionally, calculations of the terahertz pulse propagation through OCS and HCN vapor have been performed, showing significant reshaping of the pulses which in the case of OCS is attributed to centrifugal dephasing and for HCN is caused by dispersive propagation effects.

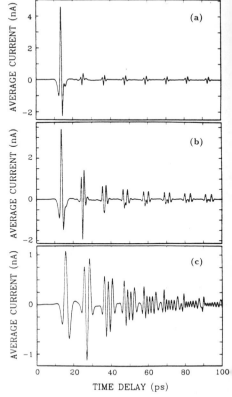

Fig.7. Calculated pulse propagation of terahertz pulses through (a) 1.33 hPa, (b) 13.3 hPa, and (c) 133 hPa of HCN vapor.

REFERENCES

Allen L and Eberly J H 1975 *Optical Resonance and Two-Level-Atom* (New York: Wiley)
Bratengeier K, Purucker H-G and Laubereau A 1989 *Optics Comm.* **70** 393
Exter van M, Fattinger Ch and Grischkowsky D 1989a *Appl. Phys. Lett.* **55** 337; Exter van M and Grischkowsky D 1990 *IEEE Trans. Microwave Theory and Techn.* **38** 1684
Exter van M, Fattinger Ch and Grischkowsky D 1989b *Opt. Lett.* **14** 1128
Foster K L, Stenholm S and Brewer R G 1974 *Phys. Rev. A* **10** 2318
Grischkowsky D, Keiding S, van Exter M and Fattinger Ch 1990a *J. Opt. Soc. Am. B* **7** 2006
Grischkowsky D and Keiding S 1990b *Appl. Phys. Lett.* **57** 1055
Harde H, Keiding S and Grischkowsky D 1991a *Phys. Rev. Lett.* **66** 1834
Harde H and Grischkowsky D 1991b *J. Opt. Soc. Am. B* **8** 1642
Heritage J P, Gustafson T K and Lin C H 1975 *Phys. Rev. Lett.* **21** 1299
Woerner M, Seilmeier A and Kaiser W 1989 *Opt. Lett.* **14** 636
Zhang X C, Hu B B, Darrow J T and Auston D H 1990 *Appl. Phys. Lett.* **56** 1011

Two-photon processes with chirped pulses: spectral diffraction and focussing

B Broers[1], L D Noordam[1], and H B van Linden van den Heuvell[1,2]

[1]FOM-Institute for Atomic and Molecular Physics, Kruislaan 407, 1098 SJ Amsterdam, The Netherlands.
[2]Van der Waals-Zeeman Laboratorium, Valckenierstraat 65, 1018 XE Amsterdam, The Netherlands.

ABSTRACT: It is demonstrated that a small chirp over a pulse can heavily affect the results of a multi-photon excitation process. In addition, an analogy is pointed out between Fresnel diffraction from a slit and a two-photon process driven by a pulse with a quadratic phase chirp. This idea can be used to 'focus' the spectral energy of a chirped pulse onto a small portion of the original bandwidth.

1. INTRODUCTION

In many situations where short laser pulses play a role, these pulses are believed to be well-described by their wavelength, energy and duration. Sometimes, in addition to these, the power spectrum is measured. If the product of bandwidth and pulse duration then turns out to be less then about 1.4 times that of a perfectly bandwidth-limited pulse, the pulse is generally termed 'nearly bandwidth-limited'. In such cases no attention is paid to the exact phase relations between the frequencies within the bandwidth of the pulse. This is often justified for a one-photon process. In the case of multi-photon process, however, small deviations from the bandwidth-limited situation can heavily affect the resulting excitation probabilities. This will be discussed in the first part of this contribution. In the second part it is pointed out that an analogy exists between a two-photon process driven by a pulse with a quadratic phase chirp, and Fresnel diffraction from a slit.

2. EFFECT OF SMALL CHIRP ON TWO-PHOTON ABSORPTION

In this section results of two two-photon experiments will be discussed, showing that a small chirp over a pulse can have a large effect on the transition probabilities for two-photon absorption. Experimental details can be found elsewhere (Broers 1991). In the first experiment two-photon excitation of Rydberg states in Rubidium was studied as a function of (small) chirp of the exciting pulse. The bandwidth was taken large compared to the spacing of the Rydberg levels, so several levels could be excited. After the pulse the distribution of excited levels was measured. The second experiment consisted of a measurement of the power spectrum of second-harmonic light after frequency-doubling in a non-linear crystal, again as a function of chirp of the fundamental pulse. The crystal was thin enough to assure proper phase-matching of all frequencies. In both experiments use was made of a laser system (Noordam 1991) in which a pulse shaper (Weiner 1988) provided tunability of wavelength, bandwidth, and chirp. The resulting pulses had a square-shaped power spectrum and a quadratic phase chirp, so the electric field in the frequency domain, $E(\omega)$ could be written as: $E(\omega) = E_0 e^{i\alpha(\omega-\omega_0)^2}$, for $|\omega - \omega_0| \leq \Delta\omega/2$. Here ω_0 denotes the central frequency, $\Delta\omega$ the bandwidth, and α the amount of chirp of the pulse.

Results of the measurements are summarized in Figure 1. The left column shows the distribution of excited levels, n, in the Rubidium atom, and the middle one the power spectra of the frequency-doubled light. The first row results from chirp-free pulses. The obtained triangular form can easily be explained by looking at the amount of frequencies (ω_1, ω_2) within the bandwidth of the fundamental pulse that add up to one particular frequency $\omega_f = \omega_1 + \omega_2$ at the two-photon level: it is seen that most combinations (ω_1, ω_2) add up to $2\omega_0$, and that the number of possibilities decreases with increasing detuning from $2\omega_0$. The second and third row show the results for pulses with a small chirp; the pulse durations, relative to that of a chirp-free pulse, are 1.10 and 1.27, respectively. Again, the results for both experiments are completely similar. One big difference between the chirp-free case the and chirped case is immediately clear: the triangular shape changes into a symmetric two-peak structure with a large dip in the center. This means that levels at exactly twice the central frequency of the fundamental pulse, $2\omega_0$, are less efficiently populated than levels with a slight detuning from $2\omega_0$.

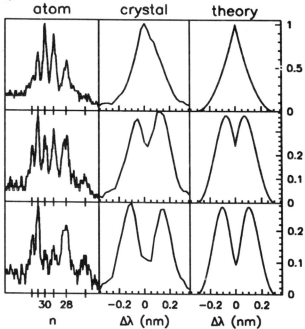

atom   crystal   theory

n $\Delta\lambda$ (nm) $\Delta\lambda$ (nm)

Fig. 1. *Populations of Rydberg levels in the Rubidium atom (left column), and power spectra of frequency-doubled light (middle column), resulting from pulses with varying chirp. Upper figures result from a chirp-free pulse. Middle and lower figures correspond to pulses with a bandwidth-time product, relative to that of a chirp-free pulse, of 1.10 and 1.27, respectively. The right column gives the calculated power spectrum $|E^{(2)}(\omega)|^2$ of $E^2(t)$. The vertical axis is normalized to the top height of the distribution resulting from a chirp-free pulse.*

Apart from this drastic change in form of the distribution of excited levels, there also is an unexpectedly fast drop in the *absolute* excitation probabilities. This can be concluded from the vertical scales in Figure 1, which are normalized to the top height of the chirp-free distribution: 10% lengthening of the pulse by chirp causes the total excited population, integrated over all levels, to decrease by a factor of two, and the population of the level at $2\omega_0$ by a factor of four. This latter value even drops by a factor of ten for a pulse lengthened by 27%. This implies that a simple measurement of energy and duration of a pulse, not considering the possibility of a small chirp, can result in large errors in, for example, cross-sections of multi-photon processes derived from experimental data.

In the right column of Figure 1, calculated excitation spectra for two-photon processes are given, originating from a pulse with a quadratic phase chirp. This 'effective two-photon power spectrum', $|E^{(2)}(\omega)|^2$, is the squared amplitude of the self-convolution of $E(\omega)$:

$$|E^{(2)}(\omega)|^2 = |\int_{-\infty}^{\infty} d\omega' E(\omega')E(\omega - \omega')|^2 \tag{1}$$

Note that the self-convolution $E^{(2)}(\omega)$ is the Fourier transform of $E^2(t)$, the 'effective time evolution of the light field at the two-photon level'.

3. ANALOGY WITH FRESNEL DIFFRACTION

Suppose one calculates the two-photon yield at exactly twice the central frequency of the fundamental pulse, $|E^{(2)}(\omega)|^2$. It is then seen that Eq. (1) takes the form of a simple integration over all frequencies within the bandwidth of the pulse, taking into account the phase of each frequency component. This is schematically depicted in Figure 2a. This picture also gives an idea about the origin of the dip in the profiles in Figure 1: if two 'paths' to $2\omega_0$, i.e. two combinations of frequencies (ω_1, ω_2) adding up to $2\omega_0$, have a phase difference of π, they will interfere destructively and cancel.

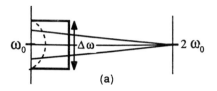 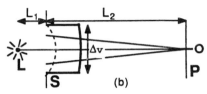

Fig. 2. *Analogy between a two-photon process driven by a pulse with a quadratic phase chirp, and Fresnel diffraction from a slit.*
Fig. 2a. *Situation for the two-photon process. The thick full and dashed curves denote the power spectrum and phase profile of the fundamental pulse, respectively. Each frequency component contributes with its own phase to the spectral energy at $2\omega_0$.*
Fig. 2b. *Situation for Fresnel diffraction. The thick full and dashed curves denote the intensity distribution and phase profile over the slit, respectively. Each part of the slit contributes with its own phase to the intensity at the central part of the screen, O.*

Exactly the same type of 'path integral' is encountered when Fresnel diffraction from a slit is studied, as in Figure 2b. In this figure L denotes a line source, directed perpendicular to the paper, S a slit with width Δv, and P the screen on which the diffracted pattern is recorded. To calculate the diffracted intensity at the central point O on the screen, one has to integrate the contributions from all parts of the slit, again taking into account the phase of each contribution. So we can make the following identifications:

bandwidth of the pulse $(\Delta\omega)$	$\longleftrightarrow$	slit width (Δv)
power spectrum of the pulse $(|E(\omega)|^2)$	$\longleftrightarrow$	intensity distribution over the slit $(I(v))$
phase profile of the pulse (chirp)	$\longleftrightarrow$	phase profile over the slit

This last identification can be made more explicit: $\alpha \longleftrightarrow \frac{1}{\lambda}(\frac{1}{L_1} + \frac{1}{L_2})$, where λ is the wavelength of the light, L_1 the distance from light source to slit, and L_2 the distance from slit to screen (see Figure 2b). This formulation is not surprising, since α measures the curvature of the phase front of the pulse, and $(\frac{1}{L_1} + \frac{1}{L_2})$ is the summed geometric curvature of the wave fronts (=phase fronts) at the slit.

If one looks at all possible paths from L to O through the slit, it is clear that all are longer than the direct straight path from L to O. Some paths, however, have a phase difference of $\approx 2\pi n$ (n integer) with respect to the direct one, so they will interfere constructively with the direct path, while others have a phase difference of $\approx 2\pi(n+1)$, resulting in negative interference. By putting a so-called Fresnel zone plate in the slit, which blocks those zones through which paths go that would contribute negatively, one greatly enhances the intensity in O. Translating this idea to the situation of a two-photon process with a chirped pulse gives the following: put a mask in the pulse shaper, that blocks those frequencies that would contribute negatively to the spectral energy at $2\omega_0$. This will give rise to 'focussing' of spectral energy onto $2\omega_0$. The result of a calculation illustrating this point is given in Figure 3: by introducing the Fresnel zone plate in the shaper, the excited population at the $2\omega_0$-level is increased by a factor of ten at the cost of population at other levels.

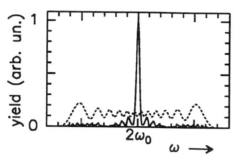

Fig.3.   *Calculated result showing 'focussing' of spectral energy onto $2\omega_0$, at the cost of energy at other frequencies. Full and dotted curves show the effective power spectra at the two-photon level with and without a Fresnel zone plate, respectively.*

4. CONCLUSIONS

In this contribution, we have demonstrated that a small chirp over a pulse can heavily affect the results of a multi-photon process. This is a great problem if one wants to use experimental data to derive absolute cross-sections for multi-photon transitions. Also, we pointed out an analogy between Fresnel diffraction from a slit and a two-photon process driven by a pulse with a quadratic phase chirp. This analogy can be used to 'focus' the spectral energy of a chirped pulse onto a small portion of the original bandwidth.

5. ACKNOWLEDGEMENTS

It is a pleasure to acknowledge many instructive discussions with H.G. Muller and H.J. Bakker. The work in this paper is part of the research program of the 'Stichting voor Fundamenteel Onderzoek van de Materie' (Foundation for Fundamental Research on Matter) and was made possible by financial support from the 'Nederlandse Organisatie voor Wetenschappelijk Onderzoek' (Netherlands Organisation for the Advancement of Research).

6. REFERENCES

Broers B, van Linden van den Heuvell H B and Noordam L D 1991 *to be published*
Noordam L D, Joosen W, Broers B, ten Wolde A, Lagendijk A, van Linden van den Heuvell H B and Muller H G 1991 *Opt. Comm.* **85** 331
Weiner A M, Heritage J P and Kirschner E M 1988 *J. Opt. Soc. Am. B* **5** 1563

Inst. Phys. Conf. Ser. No 126: Section III
Paper presented at Int. Symp. on Ultrafast Processes in Spectroscopy, Bayreuth, 1991

227

Using of broad-bandwidth stimulated Raman scattering in subpicosecond transient spectroscopy of optical Kerr effect

P A Apanasevich, V P Kozich, A I Vodchitz, B L Kontsevoy

The Institute of Physics, Academy of Sciences of Byelarus
Lenin avenue, 70, Minsk, Byelarus, 220602, USSR

ABSTRACT: The technique of ultrafast relaxation measuring of optically induced anisotropy in liquids is presented. It is based on four-wave mixing of broad-bandwidth laser radiations. In the realized Kerr-shutter configuration of interaction the correlated radiations of the nanosecond superluminescent dye laser and its first Stokes component of stimulated Raman scattering were used. The relaxation constants of the Kerr nonlinearity of a number of organic liquids have been measured. The time resolution was limited by the correlation time equal to 130 ± 30 fsec.

1. INTRODUCTION

A number of recent publications, for example by Vasil'eva et al (1980), Morita and Yajima (1984), Beach et al (1985), Kobayashi et al (1988), Nakatsuka et al (1990), Hattori and Kobayashi (1991), are devoted to the transient spectroscopy techniques based on the four-wave mixing (FWM) of broad-bandwidth "noise" laser light. The possibilities of the measuring of electronic and vibrational dephasing times, population relaxation, orientation relaxation of molecules have been demonstrated. In those investigations the broad-bandwidth beam was splitted to get correlated light beams having the same frequency.

The possibility of the technique can be widen if the correlated radiations at different frequencies are used. In the picosecond time scale we have applied the Nd:YAG laser radiation and its second harmonic (see the articles by Apanasevich et al (1988) and Vodchitz et al (1991)).

In the present work the process of stimulated Raman scattering (SRS) of light was used for receiving the broad-band radiation at the shifted frequency. It was demonstrated by Apanasevich et al (1983) and Li et al (1988) that the first Stokes component of SRS radiation has spectral bandwidth and statistical properties similar to those of pump one. We have developed the method of the subpicosecond relaxation measuring of Kerr nonlinearity using the superluminescent nanosecond dye laser radiation and its SRS first Stokes component in a Kerr-shutter scheme of interaction.

2. EXPERIMENT

The amplified radiation of superluminescent Rh6G dye laser pumped by the second harmonic of Nd:YAG laser was as initial one. It had close to Gaussian statistics, spectral width 250cm^{-1} centered at 560nm and correlation time equal to 130±30 fsec. The output beam was splitted into two ones. The 30% part of the radiation was directed into the sample investigated to induce optical anisotropy. The other part served as the pump beam in stimulated Raman scattering process in compressed to 60atm hydrogen. The substantially nonlinear regime of SRS with the conversion efficiency ≥ 7% was realized. The spectral bandwidth of Stokes pulse centered at 730nm was equal to that of the laser pulse. The first Stokes component of scattered light was used as a probe beam in Kerr-shutter scheme of four-wave process realized. The polarizations of interacting beams were crossed at an angle of 45 degrees. In the experiment the energy of the radiation passed through the polarizer crossed at the probe beam polarization versus the time delay between exciting and probe beams was measured. The dye laser and Stokes pulse energies had the values approximately 1 and 0.2mJ in the sample. Their pulse durations were 7 and 4 nsec FWHM, respectively.

3. RESULTS AND DISCUSSION

The correlation functions $G_A(\tau) = \langle I_D^2(t) I_D(t+\tau) \rangle$ and $G_C(\tau) = \langle I_D^2(t) I_{ST}(t+\tau) \rangle$ were measured to study the statistical properties of the dye laser radiation (I_D) and the correlation between Stokes (I_{ST}) and dye laser radiation being used as a pump one in SRS process (Fig.1). They were measured in a glass where the electronic contribution is dominant for a both FWM processes. The coincidence of the curves confirms the assumption used by Li et al (1988) that Stokes wave field can be performed as $E_{ST}(t)$ = $\xi(t) * E_D(t)$, where $\xi(t)$ is slowly varying function.

The peak/pedestal ratio on the experimental curves differs from 3 that has to be for Gaussian statistics of radiation. It may be because of deviation from ideal Gaussian statistics, apparatus effects, and noninstantaneous response of the fastest Kerr nonlinearity in glass – the sample where the FWM process was realized. In the verification to the last supposition we have found that good fitting the experimental and calculated curves has taken place at the ultrafast relaxation time T_F= 40 fsec and Gaussian statistics of radiations.

In the experiment the Kerr nonlinearity relaxation of benzene, toluene, CCl_4, tetrahydrofuran and number of alcohols were studied. The experimental results for benzene and CCl_4 are shown by points in Fig.2.

To get the relaxation constants the fitting of experimental and calculated curves was done. The Gaussian statistics of radiations and maximum two dominant contributions to the Kerr nonlinearity were assumed. The signal field is proportional to the medium polarization

$$P(t,\tau) = E_{ST}(t,\tau) * \int_{-\infty}^{\infty} r(t-t') E_D(t') E_D^*(t') dt',$$

where response function and exciting fields are performed as

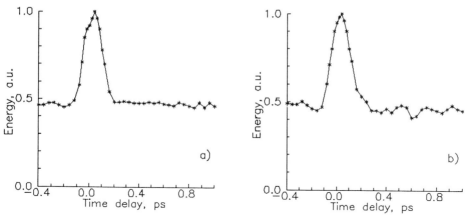

Fig.1. The correlation functions $G_A(\tau) = \langle I_D^2(t)I_D(t+\tau)\rangle$ (a) and $G_C(\tau) = \langle I_D^2(t)I_{ST}(t+\tau)\rangle$ (b) have been written through the FWM a glass.

$$r(t-t') = r(0) * \{\exp[-(t-t')/T_F] + \beta\exp[-(t-t')/T_S]\},$$

$$E_D(t) = \alpha_1(t)R(t), \qquad E_{ST}(t,\tau) = \xi(t)\alpha_2(t)R(t-\tau).$$

Here $\xi(t)$ and $\alpha_{1,2}(t)$ are slow functions of time, $R(t)$ is a complex random function describing the stationary Gaussian process with following properties: $\langle R^*(t)R(t-\tau)\rangle = G(\tau) = \exp(-|\tau|/\tau_c)$.

Theoretically the dependence of the signal energy

$$\varepsilon(\tau) \propto \int_{-\infty}^{\infty} \langle P^2(t,\tau)\rangle dt$$

versus the time delay was calculated. The best coincidence of theoretical and experimental curves has been received at the parameters presented in the table below

Substance	Fast time T_F	Slow time T_S	β
Benzene	160± 50 fsec	4±1 psec	0.09
Toluene	150± 50 fsec	10±2 psec	0.06
CCl$_4$	≤40 fsec	450±80 fsec	0.1

In all cases it is possible to fit experimental and theoretical curves if to take into account two dominant contributions in Kerr nonlinearity. The nature of these contributions may be different (see for example the article by McMorrow et al (1988)). They are, probably, librations and orientations for benzene and may be associated with electron motion and local translation anisotropy for CCl$_4$.

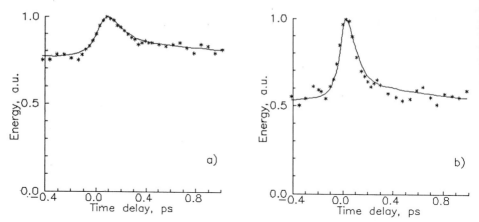

Fig.2. Experimental (*) and calculated model (-) dependences of the signal energy versus time delay between dye laser and Stokes beams for benzene (a) and CCl_4 (b).

4. CONCLUSION

The four-wave mixing of broadband "noise" light was realized to study the femtosecond relaxation times of optically induced anisotropy in transparant liquids. The production of correlated radiations with femtosecond correlation time using stimulated Raman scattering process was demonstrated. The correlated dye laser and SRS radiations were used in Kerr-shutter configuration of interaction. The most liquids investigated have the ultrafast dominant contribution with relaxation time less than 40 fsec. Here the characteristics of the samples with essential more slow contributions are presented. Not high accuracy is because of three fitting parameters for model calculated curves. The polarization suppression of large electronic contribution to the signal detected is a way to more high accuracy. The work is in progress.

REFERENCES

Apanasevich P A, Kozich V P, Vodchitz A I 1988 J.Mod.Opt. **35** 1933

Apanasevich P A, Batische S A, Gandza V A, Grabchikov A S et al 1983 Izv. AN SSSR. Ser. fiz. **47** 1551 (in Russian)

Beach R, Debeer D, Hartmann S R 1985 Phys.Rev.A **32** 3487

Hattori T, Kobayashi T 1991 J.Chem. Phys. **94** 3332

Kobayashi T, Terasaki A, Hattori T, Kurokawa K 1988 Appl.Phys.B **47** 107

Li Z W, Radzewicz C, Raymer M G 1988 JOSA B **5** 2340

McMorrow D, Lotshaw W T, Kenney-Wallace G A 1988 IEEE J.of Quantum Electron. **QE-24** 443

Morita N, Yadjima T 1984 Phys.Rev.A **30** 2525

Nakatsuka H, Katsahima Y, Inouye K 1990 Opt.Commun. **69** 169

Vasil'eva M A, Malyshev V I, Masalov A V 1980 FIAN Short Phys.Commun. **No1** 35 (in Russian)

Vodchitz A I, Kozich V. P, Kontsevoy B L 1991 Sov.J.Opt.and Spectr. **70** 883 (in Russian)

Inst. Phys. Conf. Ser. No 126: Section III
Paper presented at Int. Symp. on Ultrafast Processes in Spectroscopy, Bayreuth, 1991

Third order nonlinear optical properties of substituted poly(p-phenylene vinylene) derivatives

H L Li, S Rentsch and H Bergner

Faculty of Physics and Astronomy, Friedrich Schiller University Jena
Max-Wien-PLatz 1, O-6900 Jena, Germany

ABSTRACT: Through third order harmonic generation measurements on thin films of substituted poly(p-phenylene vinylene) (PPV) derivatives we have determined a third order susceptibility $\chi^{(3)}(-3\omega,\omega,\omega,\omega)$ of 2.6, 2.0, 1.8, 1.1*10^{12} esu for DP-PPV, MP-PPV, DMOP-PPV and DPOP-PPV at 1.06 μm, respectively. The nonlinearity is associated with the delocalization of π-electrons in the conjugated polymer chains.

1. INTRODUCTION

The investigation of optical nonlinear properties of organic polymers aroused great interest recently due to their large third order susceptibility, fast time response as well as versatility of their fabrication and realization. The conjugated polymers show potential future application as nonlinear optical devices, such as optically gated optical switches, optical bistable devices, transient optical memory elements as well as waveguides (Williams 1983, Prasad and Ulrich 1988, Smith 1987, Bartuch 1991 etc.). PPV is a very good candidate for such applications. Kaino et al.(1987) have shown the precourser-route made PPV with a $\chi^{(3)}$ value at 1.85 μm of 7.8*10^{-12} esu.

There are many different measuring methods for evaluating the third order nonlinearities such as degenerated four-wave mixing (Hattori et al. 1987, Bubeck et al. 1989), nonlinear interferometer (Moran et al.1975), nonlinear etalon measurement (Blau 1987), but the third harmonic generation (THG) method is the simplest one. In this work we present experimental results on the third order susceptibility of substituted poly(p-phenylene vinylene) using THG method.

2. EXPERIMENTAL DESCRIPTION

The substituted PPV samples (see figure 1) were prepared by reductive dehalogenation of double geminal dichlorides obtained from diketones. The details of synthesis can be found in the work of Hörhold et al.(1987). The substitutions through phenylene and dimethoxy groups on the side chains make it soluble in many common solvents such as toluene, chlorobenzene, chloroform etc. Thin polymer films with a thickness of about 0.25 μm were deposited on silica glass (d=1 mm) substrates. The films show bright yellow colour and exhibit very good optical qualities.

The third harmonic (TH) signal was generated by exciting the samples with pulses from a passively mode locked Nd:YAG laser with a pulse duration of 25 ps and energy of about

2 mJ at the fundamental wavelength (λ=1.06 μm). The sample was mounted on a rotation stage and the generated TH signal from the polymer film was measured as a function of the angle of incidence. The created TH signal was detected using a photomultiplier connected

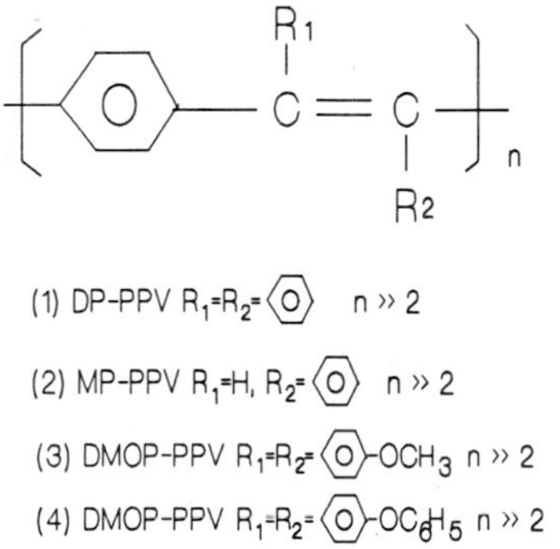

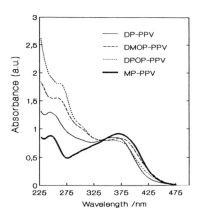

(1) DP-PPV R_1=R_2=⟨O⟩   n ≫ 2

(2) MP-PPV R_1=H, R_2=⟨O⟩   n ≫ 2

(3) DMOP-PPV R_1=R_2=⟨O⟩-OCH_3 n ≫ 2

(4) DMOP-PPV R_1=R_2=⟨O⟩-OC_6H_5 n ≫ 2

Figure 1. The chemical structures of substituted PPV derivatives.

Figure 2. The stationary absorption spectra of PPV derivatives.

with a boxcar integrator. According to Krausz et. al (1989) we have tightly focused the laser beam on the sample (f=7.14 cm) so that the confocal parameter of the fundamental wave is much smaller than the coherence length of the laser pulses in air ($l_{c,air}$=16.3 mm), thus air contribution to the TH signal can be greatly suppressed. To obtain the absolute third harmonic susceptibility we used the TH signal generated in silica glass substrate which has a $\chi_s^{(3)}$ value at 1.06 μm of $3.1*10^{-14}$ esu (Kajzar and Messier 1985) as a standard.

3. EXPERIMENTAL RESULTS AND DISCUSSIONS

The absorption spectra of substituted PPV in the UV and visible range are shown in Figure 2. They exhibt a common feature with a broad absorption band around 370 nm which is associated with π-π^* electron transitions. In figure 3, the TH signal generated in a 1 mm thick silica glass substrate in DP-PPV is shown in dependence on the angle of incidence. Because the optical path varies with the angle, the TH singnal of the silica glass substrate shows the typical Maker fringe pattern due to the coherence length of the phase matching of the nonlinear process. In the polymer films the coherence length of phase matching is greater then the film thickness, thus variation of the consecutive optical path only give rise to a small phase matching mismatch and the TH signal from the polymers show only a monotonic decreasing envelop with increasing angle of incidence. Using the relation (Kaino et al. 1988)

$$\chi^{(3)} = \frac{2}{\pi}\chi_s^{(3)}\frac{\sqrt{I_{3\omega}}/l}{\sqrt{I_{3\omega,s}}/l_{c,s}}$$

where $I_{3\omega}$, $I_{3\omega,s}$ are the created TH signal peak intensities of sample and silica glass, respectively, l is the sample thickness and $l_{c,s}$ the coherence length of the substrate which is equal to 6.69 μm (Krausz et al. 1989). The obtained results are shown in table 1.

The results of different PPV derivatives show similar $\chi^{(3)}$ values within a factor of two and in the order of 10^{-12} esu, typical values of conjugated polymers. It demonstrates that nonlinear optical properties are directly associated with the nonlinear polarization of the π-electrons in the conjugated polymer backbones. The wavelength of TH singnal (λ=355nm) is already within the absorption band of the samples, so the resonance effect through three-photon absorption and reabsorption must be included. The values in table 1 were obtained without this corrections.

Table 1. The measured $\chi^{(3)}$ values of different PPV derivatives. λ_g and λ_{max} indicate the positions of the corresponding band gap and the maximum absorption.

Sample	DP-PPV	MP-PPV	DMOP-PPV	DPOP-PPV
λ_g(nm)	426	442	438	431
λ_{max}(nm)	370	373	375	365
$\chi^{(3)}(10^{-12}$ esu)	2.6	2.0	1.8	1.1

If we include the reabsoption the $\chi^{(3)}$ values in table 1 will be increase 5-6 times. Then these values agree well with the results of Kubodera and Kaino (1989). According to the theoretical calculations (Agrawal et. al 1978, Flytzanis 1987), the bulk $\chi^{(3)}$ of one-dimensional semiconductor should obey the scaling law and is proportional to the sixth power of the electron delocalization length or is proportional to λ_{max}. Thus selecting substitutents to shift the absorption band into long wavelength direction may yield materials of high nonlinear susceptibility.

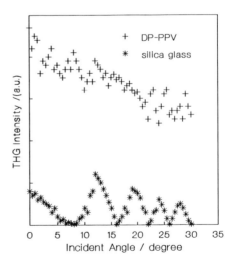

Figure 3. The Maker Fringe patterns of THG signals from PPV samples and silica glass.

In conlusion we have measured the third order susceptibility of different PPV derivatives at 1.06 μm. The obtained values are smaller than the calculated results of PPV by Shuai et al. (1991) who have used the one-electron tight-binding approach theory and get a $\chi^{(3)}$ value of 3.8×10^{-11} esu at 1.06 μm. To compare the experimental results with theory in the whole spectrum, the measurement of the wavelength dependence of $\chi^{(3)}$ is in preparation using an optical parametric oscillator system. It is expected that the samples will exhibit large enhancement of nonlinear properties when the excitation wavelength tunes near to multiphoton resonance.

ACKNOWLEDGMENT

We are grateful to Professor Dr. H. H. Hörhold for supplying us the polymer samples and D. Grebner for preparing polymer films.

REFERENCES

Agrawal G P, Cojan C and Flytzanis C 1978 Phys. Rev. B **17** 776

Bartuch U, Bräuer A and Dannberg P 1991 Inter. J. of Optoelectronics **5** No. 3

Blau W 1987 Opt. Commun. **64** 85

Bubeck C, Kaltbeitzel A, Lenz R W, Stenger-Smith J D and Wegner G 1989 *Nonlinear Optical Efffects in Organic Polymers*, NATO ASI series E: Appl. Sci. vol. **162** pp 143 (H Kuzmany, M Mehring and S Roth ed, Kluwer Acad. Publ., Dordrecht)

Flytzanis C 1987 *Nonlinear Optical Properties of Organic Molecular and Crystals* vol **2** pp 121 (D S Chemla and J Zyss ed, Acadmic Press)

Hattori T and Kobayashi T 1987 Chem. Phys. Lett. **133** 230

Hörhold H H, Helbig M, Raabe D, Opfermann J, Scherf U, Stockmann R and Weiß D 1987 Zeitschrift für Chemie **27** 126

Kaino T, Kubodera K, Kobayashi H, Kuihara T, Saito S, Tsutsui T, Tokito S, and Murata H 1988 Appl. Phys. Lett. **53** 2002

Kaino T, Kubodera K, Tomaru S, Kurihara T, Saito S, Tsutsu T and Tokito S 1987 Electron Lett. **23** 1095

Kajzar F and Messier J 1985 Phys. Rev. A **32** 2352

Krausz F, Wintner E and Leising G 1989 Phys. Rev. B **39** 3701

Kubodera K and Kaino T 1989 *Nonlinear Optics of Organics and semiconductors* pp 163-170 (T Kobayashi ed, Springer-Verlag Berlin, Heidberg)

Moran M J, She C Y and Carman R L 1975 IEEE J. Quantum Electron QE-11 259

Prasad P N and Ulrich D R ed 1988 *Nonlinear Optical and Electroactive Polymers* (Plenum Press, New York)

Shuai Z and Bréads J L 1991 Phys. Rev. B **44** 5962

Smith P W 1987 IEEE Circuit and Devices Magazine (**May**) pp 9-14

Williams D J (ed.) 1983 *Nonlinear Optical Properties of Organic and Polymeric Materials* (Academic, New York)

Inst. Phys. Conf. Ser. No 126: Section III
Paper presented at Int. Symp. on Ultrafast Processes in Spectroscopy, Bayreuth, 1991

Four wave mixing processes in strontium vapour generated by tunable femtosecond pulses

J. Ringling, O. Kittelmann, F. Seifert, J. Herrmann

Zentralinstitut für Optik und Spektroskopie, Rudower Chaussee 6,
O-1199 Berlin, Germany

ABSTRACT: The spectral behaviour of coherent radiation by various third order nonlinear processes in strontium vapour is investigated in the sub-ps time domain. The measurements were performed in a strontium filled heat pipe oven using 500 fs, > 0.1 mJ pulses tunable around the $5s^2-5s5d$ $^1D_2$ two-photon resonance of strontium.

1. INTRODUCTION

Generation of coherent radiation using the third order nonlinearity of metal vapours is a well established tool for nanosecond spectroscopy in the wavelength range below 200 nm. Satisfactory conversion efficiencies of various four wave mixing processes including third harmonic generation (THG) are achieved by resonance enhancement of $\chi^{(3)}$ (Vidal et. al.).

The mechanism of resonance enhancement is considerably altered if applying high intensity sub-ps pulses with a spectral bandwidth large compared to the narrow lines of the atomic transitions. In this strongly coherent regime (Diels et. al.) self induced transparency, dynamic stark shift of the atomic levels and stimulated four wave mixing play an important role. Hence a detailed investigation of the various nonlinear effects during ultrashort pulse irradiation of metal vapours is necessary to obtain valuable information on the competition between these processes, and, in particular, on possible conversion efficiencies and temporal profiles, e.g., of third harmonic and sum frequency generation.

2. EXPERIMENTAL SETUP

Figure 1 shows the experimental setup. The pulses were generated with a Rh6G dye laser synchronously pumped by the second harmonic of a mode-locked cw Nd:YAG laser with fiber-grating pulse compressor (Spectra Physics 3800 S). The average output power of the dye laser was about 50 mW at a repetition rate of 82 MHz. The single pulses had a duration of less than 500 fs.

These low energy pulses were amplified in a three-stage dye amplifier transversally pumped by a frequency-doubled, Q-switched ns-Nd:YAG laser (5 Hz repetition rate). The overall amplification was higher than 10^5 providing pulse energy output as high as 200 µJ with an ASE content less than 5 %. The amplifier design allowed a tunability of a few nm around 576 nm.

For the nonlinear medium we used a modified concentric heat pipe oven which operates at a temperature level of 1100 K corresponding to a Sr vapour pressure of 5 mbar. The amplified pulses were focused into the center of the 5 cm long active vapour zone using a f = 28 cm lens.

The diagnostic system consisted of a solar blind photomultiplier in connection with a boxcar averager and two OMA-system controlled spectrographs for simultaneous accumulation of the spectra of incoming fundamental and generated pulses.

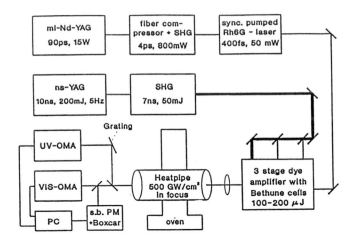

Fig. 1.
Experimental setup

3. RESULTS

The energy level scheme of the system under study is shown in fig. 2. In order to get information about the mechanism of resonance enhancement and linewidth broadening we tuned our pulses around the two photon resonance and measured both the energy of the generated third harmonic signal and the spectra of the fundamental and the third harmonic pulse. Due to the high intensity of our sub-ps pulses the two photon resonance is considerably broadened. The range of resonance enhancement is extended to several nanometers as can be seen in fig. 3. The individual third harmonic pulses had a spectral width of about 70 cm^{-1}.

The conversion decreased faster on the low-frequency side than on the high-frequency side of the two photon resonance. Because of the limited tunability of our amplifier system an investigation of the nonlinear transformation process at higher frequencies was not possible. The maximum conversion efficiency for phasematched THG was about 10^{-4} at a Strontium-Xenon pressure ratio $p_{Xe}/p_{Sr} \approx 25$ (fig. 4). The phase matching curve shows the same asymmetry which can be attributed to the temperature gradient at the ends of the vapour zone, already observed by Vidal et. al.. The third harmonic signal has a maximum energy of about 10 nJ.

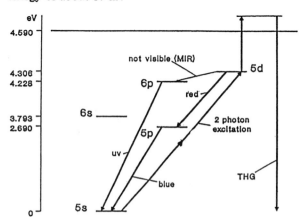

Fig. 2. Energy levels of strontium and observed coherent radiation

We observed that the spectral maximum of the generated third harmonic signal was not exactly at the position of the threefold frequency of the fundamental but was slightly shifted to higher frequencies. Theoretical considerations propose that an intensity dependent shift should occur. Further detailed investigations are therefore necessary and in preparation.

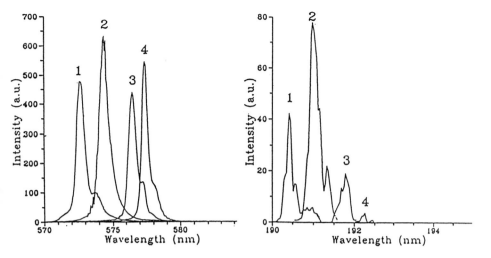

Fig. 3 Single shot spectra of several fundamental pulses (left) and their corresponding third harmonic (right).

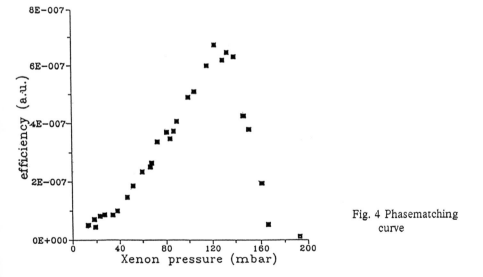

Fig. 4 Phasematching curve

Simultaneously to the THG, due to the strong pumping of the two photon resonance, some other lines at 293.2 nm, around 463 nm and at 767.4 nm were observed corresponding to stimulated four wave mixing processes resonantly enhanced through one photon allowed transitions in strontium (fig. 2).

The generated blue light is emitted on a cone with a broad spectrum considerably affected by phase modulation whereas the spectrum of the red counterpart is a relatively narrow line as well as the radiation at 293.2 nm (fig. 5). The MIR wavelength (fig. 2) which is necessary to fulfill the conservation laws of the parametric process could not be observed because of the limited spectral sensitivity of our diagnostic system.

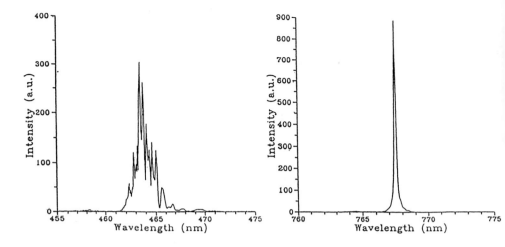

Fig. 5 Spectra of the blue and the red radiation

4. SUMMARY

We presented first measurements of THG and related four wave mixing processes in a Sr-Xe system with intense sub-ps pulses. The experiments showed that the system is able to transform such spectrally broad pulses and that the enhancement of the third harmonic due to the two photon resonance covers a spectral range much larger than known from investigations with ns pulses. Because of the large enhancement range with this nonlinear medium it should be possible to transform pulses much shorter than these used in our experiments.

We achieved phase matching by varying the Xenon pressure and the maximum conversion efficiency was 10^{-4}. To get further information about the dynamics of four wave mixing processes with sub-ps or shorter pulses in Strontium vapour it is necessary to perform time resolved measurements.

5. REFERENCES

Diels, J.C., and Georges, A.T., 1979 Phys. Rev. A **19** 1589

Vidal, C.R., 1987 Four wave frequency mixing in gases, in Topics in applied Physics **59** 57 and references therein

Inst. Phys. Conf. Ser. No 126: Section III
Paper presented at Int. Symp. on Ultrafast Processes in Spectroscopy, Bayreuth, 1991

239

Laser pulses shortening at transient backward SRS and forward scattering suppression

R.G. Zaporozhchenko, I.S. Zakharova, A.V.Kachinskii,
G.G. Kotaev, I.V.Pilipovich

The Institute of Physics, Belorussia Academy of Sciences, Minsk

ABSTRACT: Experimental results and mathematical simulation of the transient SRS process is performed taking into account backward scattering and Stokes - anti-Stokes parametric coupling of the forward waves. It is shown that the parametric coupling has a slight influence on the suppression of forward scattering. A main contribution to the SRS asymmetry is made by pumping exhaustion by the backward Stokes wave.

I. INTRODUCTION

The recent investigations of backward stimulated scattering (Blombergen 1967, D'jakov et al 1982, Gorbunov et al 1984, Sokolovskaja et al 1987, Gakhovich et al 1988, Apanasevich et al 1991) have demonstrated the unique possibilities of this effect for spatial-temporal and spectral-frequency characteristics control of laser radiation. However, the reasons giving rise to the forward and backward scattering competition and the conditions of advantage radiation concentration in the backward direction have not been adequately considered. In this paper we present mathematical simulation of the transient SRS process taking into account the Stokes- anti-Stokes parametric coupling of forward, backward scattering and interacting waves amplitude variation due to focusing and experimental results this process.

2. BASIC EQUATIONS

The SRS process have been analysed taking into account the Stokes -anti-Stokes parametric coupling of forward waves and backward scattering. The set of equations in normalized variables were :

$$\frac{\partial a_1}{\partial t} + h_1 \frac{\partial a_1}{\partial z} = ig_1 \frac{\omega_1}{\omega_s}[x^+ a_s^+ + x^- a_s^- + (x^+)^* a_a e^{-i\Delta kz}], \qquad (1.a)$$

$$\frac{\partial a_s^+}{\partial t} + h_s \frac{\partial a_s^+}{\partial z} = ig_1 (x^+)^* a_1 + g_2 a_1, \qquad (1.b)$$

$$\frac{\partial \bar{a}_s}{\partial t} - h_s \frac{\partial \bar{a}_s}{\partial z} = ig_1 (x^-)^* a_1 + g_2 a_1, \tag{1.c}$$

$$\frac{\partial a_a}{\partial t} + h_a \frac{\partial a_a}{\partial z} = ig_1 \frac{\omega_a}{\omega_s} x^+ a_1 e^{i\Delta kz}, \tag{1.d}$$

$$\frac{\partial x^+}{\partial t} = -T_2' x^+ + ig_3 [(a_s^+)^* a_1 + a_1^* a_a e^{-i\Delta kz}], \tag{1.e}$$

$$\frac{\partial x^-}{\partial t} = -RT_2' x^- + iRg_3 a_1 (a_s^-)^*. \tag{1.f}$$

Here $a_i = E_i/E_{sat}$, where E_{sat} is saturated laser field; ω_1, ω_s ω_a are the frequencies, respectively, of pumping, Stokes ar anti- Stokes waves. x^+, x^- - the polarization of scatterir medium of forward and backward scattering. The coefficients i equations (1) are given by: $\Delta k = (2k_1 - k_s - k_a)L_0$ is the wav detuning, $g_1 = 2\pi \omega_s \beta_{12}(\omega_s)N_0 T_n$, $g_3 = (2\pi \beta_{12}(\omega_s)/h)I_{sat}T$ $g_2 = \sigma_{sp} T_n$, $h = h_1 = cT_n/L_0$, $R = \Delta\bar{\nu}/\Delta\bar{\nu}$, $\Delta\bar{\nu}^-$ - is the width of RS line $\beta_{12}(\omega_s)$ - the RS tensor, N_0 - the scattering particles densit depending on pressure, The pumping wave amplitude was take into account due to focusing of radiation: $a_1 = a_1(z=0)f(z)$ where

$$f(z) = \{S_0[(1-z/f_0)^2 + \pi\theta z^2/4S_0]\}^{-1/2}. \tag{2}$$

Here, f_0 is the lens focal length, θ- the divergence, S_0 is th crossection of beam.

The system of equations (1,2) was solved for inital dat corresponded to scattering of the 2nd harmonic radiation c YAG:Nd laser with the $\tau_0 = 200$ ps, $S_0 = 0.04$ cm^2. A cell wit hydrogen was 18 cm in length; focusing was carried out in th centre of the cell, and the lens focal length in this case wa varying from 9 to 14 cm. Pumping energy and pressure of H_2 als varied.

3. RESULTS AND DISCUSSIONS

Data of the experimental results and calculation are give in Figs. 1. Fig.1 shows the calculated dependencies of th generation efficiency of backward (1), forward Stokes (2) an anti-Stokes (3) waves, as well as those obtained experimentall (*)-for backward, (o)-for forward waves, on the pumping energ W_{in} at hydrogen pressure of 50 Atm. The system of equation (1-2) is written in the plane wave approximation and it numerical solution provides us the dependence of efficiency o the peak intensity of light pulse $\eta(W_{in})$. As far as the experimentally measured light beam spatial distribution wa nearly gaussian $R(r) = (\ln 2/\pi r_0^2)\exp[-\ln 2(r/r_0)^2]$, the obtained

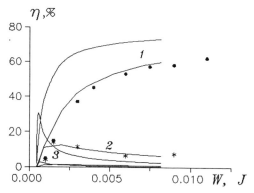

Fig.1. Efficiency of the SRS conversion.

(W_{in})dependence was corrected according to

$$\eta = 2\pi \int_0^\infty R(r)\,\eta[W_{in}R(r)]\,r\,dr.$$ The results of the calculation with

account of pump beam spatial distribution the solid lines are show. The data taking without account of spatial distribution he discontinuous lines are show.

The dynamics of the SRS process in time and space is show hat initial scattered radiation arising near the centre of the ell is amplified with backward SRS pulse shortening.The hreshold of forward scattering at Stokes and anti-Stokes requencies is achieved a head of the cell centre and may be ttributed to the Stokes - anti-Stokes coupling and repumping f the radiation to counter-propagation wave. Trailing edge of ll pulses turned out to be modulated.

The above calculations show that the anti-Stokes wave nvolves only a negligible part of scattering radiation. To lucidate the role of Stokes - anti-Stokes parametric coupling hich is regarded to be responsible for asymmetry of scattering forward - backward" we have eliminated equation (1d) from the et of equations (1). The time envelopes of pulses scattered ear the SRS threshold pumping is shown that the parametric oupling slightly affects on the intensity and pulse duration f the backward SRS (within a few per cent). This is due to the act that only a small portion of radiation is converted into he anti-Stokes component, thus the relation between the nergies is $W_a/W_s^+ \simeq 0.07\text{-}0.1$. We have not measured xperimentally energy at anti-Stokes frequency. The pulse ontrast of the backward SRS varies from 2,7 ($W_{in}= 9*10^{-4}$J to .3 ($W_{in}=4*10^{-3}$J). This may be attributed to the movement of he onset of the forward scattering process along the pumping ulse edge. The threshold conditions of the formation of the orward Stokes pulse are achieved at the trailing of edge umping pulse.In the case the scattered pulse consists of a hort pulse and small spite at the "tail". With increased umping a maximum of the forward SRS shifts first to the center f the pumping pulse and then to its origin.As the maximum hifts at the trailing edge of the forward SRS there appear scillations of radiation which can contain a great energy of

the scattered radiation. In this case the contrast reduces
From this it follows that to obtain high contrast short pulse
it is more advantageous to excite the backward SRS in th
pumping region below the saturating ones so that the pulse o
scattered radiation appeared at the trailing edge of th
pumping radiation. Pulse duration of the backword SRS is 32 p
near the threshold and 12 ps in the region of saturation.

4. CONCLUSION

Thus, numerical simulation of the transient SRS proces
taking into account Stokes-anti-Stokes parametric coupling o
the forward waves and backward scattering permits analyzing th
competition and dynamics of forward and backward SRS and th
effect of anti-Stokes wave on their characteristics. From th
analysis of the modelling results and their correlation wit
the experiment (Apanasevich et al 1991, Kachinskii et al 1989
it follows that the parametric processes slightly affects th
asymmetry only in the near-threshold region. Their action o
dynamics of the process is negligibly small and decreases wit
increasing pumping. However, pumping exhausting by the backwar
Stokes radiation turners out to be quite essential. Th
obtained results show that the origin and development of wav
a^- lead to strong suppression of the forward SRS.

An investigation of the pulse formation dynamics reveal
that at a great excess of the scattering threshold the backwar
Stokes pulse has a broad background of sufficient energy. It
further increase inevitably results in the deformation an
lengthening since the condition of amplification of th
background and the pulse itself differ greatly from each other
according to the calculation results. One way of decreasing th
background is the excitation of scattering at trailing edge o
pumping. In this case the duration of the background radiatio
reduces and the share of the energy contained in it decreases.

References

Apanasevich P A, Kotaev G G, Zaporozhchenko R G et al. 1991
Technical Digest of the third European Quantum Electronics
Conference, Edinburgh NOTuP 36

Bloembergen N 1967 Amer. J. Phys. **35** 988
D'jakov Ju E, Nikitin S Ju 1982 Kvant. Electron. **9** 1258
   (in Russian)
Gakhovith D E, D'jakov Ju E, Zhmakin I N et al. 1990
   Program and book of abstracts of 13 Intern. conference on
   Coherent and Nonlinear Optic Minsk 1988 part 2 189
Gorbunov V A, Ivanov V B, Papernyi S B, Startzev V R 1984,
   Izv. Akad. Nauk SSSR, Phys. **48** 1580 (in Russian)
Kachinskii A V, Kotaev G G, Pilipovich I V 1989 Preprint No555
   Institute of Physics Minsk
Murray I R, Goldhar I, Eimerl D, Szoke A 1979 IEEE J.
   Quant. Electron. **15** 342
Sokolovskaja A I, Brekchovskikh G L, Kudrajvtzeva A D 1987
   IEEE J. Quant. Electron. **23** 1332

Inst. Phys. Conf. Ser. No 126: Section III
Paper presented at Int. Symp. on Ultrafast Processes in Spectroscopy, Bayreuth, 1991

Synchronization of quantum transitions by light pulses: Raman transitions in highly excited hydrogen atom

V A Ulybin

Institute for Laser Physics, Siberian Branch of the USSR
Academy of Sciences, Novosibirsk, USSR-630090.

ABSTRACT: The synchronization of quantum transitions with
natural Raman oscillations by ultrashort light pulses is
discussed. A computer simulated application of the effect
to microwave transitions of the hydrogen atom is presented
as a new possibility for measuring the Rydberg constant.

INTRODUCTION

Precision frequency measurements of quantum transitions lying
in microwave and in optical ranges are currently made by using
methods of resonant interaction between atomic systems and
highly monochromatic electromagnetic fields. The synchronization of quantum transitions (Chebotayev 1989, Chebotayev and
Ulybin 1990) is a spectroscopic method which is based on using
pairs of field pulses whose duration is shorter than a period
of the investigated quantum transition. If the interpulse time
is tuned for different pairs of pulses, the population
probabilities oscillate at the natural atomic frequency. The
method provides direct measurements of time and frequency when
the short pulses are controlled by a time standard.

1. MAIN FEATURES OF THE METHOD

If atoms or molecules interact with pair of perturbation pulses
delayed in time, they make quantum transitions which are
synchronous with natural oscillations. The first of the pulses
mixes atomic states and excites coherence oscillations and the
second one constrains atoms to make the synchronous
transitions.

The phenomenon of synchronization of quantum transitions is not
critically dependent on the nature of the pulse's perturbation.
For mixing quantum states light pulses or pulses of an electric
or a magnetic fields may be used.

An atom (or a molecule) is assumed to be initially in the
ground state or in a metastable state $|1>$. It is clear that a
single light pulse may excite a superposition of the initial
state $|1>$ and an other close quantum state $|2>$ if the duration

τ of the pulse is shorter then the inverse transition frequency ω_{21}. The excitation of the quantum transition $|1\rangle \longrightarrow |2\rangle$ is realized by the two-photon Raman process. As both photons belong to the same wave packet, this Raman process does not depend on the optical phase of the electromagnetic field with carrier frequency $\omega \gg \omega_{21}$.

The excited coherence ρ_{12} oscillates at the transition frequency ω_{21}. If the time delayed light pulse arrives in phase with the atomic oscillations, the atom makes a transition from the mixed state into the pure state $|2\rangle$. Tuning the interpulse time T one may observe oscillations of population probabilities $\rho_{22}(T)$ and $\rho_{11}(T)$ at the frequency ω_{21}.

In the time between the short pulses the atomic system is not perturbed by any measurement fields and thus measurement does not influence the atomic frequencies. The interaction between an atom and a pair of time-separated light pulses depends on the time delay T but not on the difference of theirs optical phases.

2. SYNCHRONIZATION OF RAMAN TRANSITIONS IN THE HYDROGEN

Synchronization of Raman transitions in hydrogen atoms (Chebotayev and Ulybin 1991) is a way of measuring times and frequencies of microwave transitions between the high-lying atomic states by using ultrashort light pulses.

Let a hydrogen atom be excited into the Rydberg state $|n_0\rangle$, with n_0 being the principal quantum number ($n_0 \gg 1$). The frequency intervals between $|n_0\rangle$ and nearby states $|n_0 \pm 1\rangle$ are approximately equal to the frequency $\omega_0 = n_0^{-3} = 2\pi/T_0$, where T_0 is the orbital period, and atomic units ($e = \hbar = m = 1$) are used.

The duration τ of the light pulse is assumed to be shorter than the inverse orbital frequency but longer than inverse ionization energy ($1 \gg \omega_0 \tau \gg n_0^{-1}$). In this case the two-photon Raman excitation of the superposition of Rydberg states is possible. After the light pulse, the state of the atom may be written in the form

$$|\Psi\rangle = \sum_n a_n |n\rangle ,$$

where a_n is the expansion coefficient and $|n - n_0| \ll n_0$.

The time evolution of the superposition of Rydberg states generated by one light pulse manifests quantum beats of states with different quantum numbers n (Parker and Stroud Jr 1986, Alber, Ritsch and Zoller 1986). The light pulse delayed from the first one by the time T generates a new coherent superposition of Rydberg states, and this gives the interference with the superposition of high-lying states excited by the first light pulse. After the second light pulse the total probability of detecting the atom in any of the bound states $|n\rangle$, which are differ from $|n_0\rangle$, is equal to the sum of individual probabilities $W_n(T)$ of the transitions $|n_0\rangle \longrightarrow |n\rangle$

$$W(T) = \sum_{n \neq n_0} W_n(T),$$

where

$$W_n(T) = W_n \exp[-\gamma_{n_0} t_{ex} - \gamma_n(t-T)] \left[\exp(-\gamma_{n_0} T) + \exp(-\gamma_n T) + 2\exp(-\gamma_{n_0} T/2 - \gamma_n T/2) \cos(\omega_{nn_0} T) \right],$$

where W_n is the single-pulse individual probability, t_{ex} is the time interval between excitation of the initial Rydberg state $|n_0\rangle$ and the first ultrashort light pulse, γ_k is the decay rate of the state $|k\rangle$, t is the elapsed time $(t-T \gg \tau)$.

After the interaction between an atom and the two light pulses, Rydberg states of the atom may be detected by the selective Stark ionization (Ducas *et al*. 1975) to independently find the results of the excitation of each Rydberg state $|n\rangle$. If the ionizing electric field is increased, the current observed in an ion detector will be a sequence of pulses. Each of the pulses corresponds to the selective ionization of one of the Rydberg states $|n\rangle$. The total electric charge received by the detector for a time of sweepping (ignoring the pulse related to n_0) is proportional to the probability of the excitation of the Rydberg manifold.

If the second light pulse arrives in phase with beating between one of the states $|n\rangle$ and the initial state $|n_0\rangle$ $(\omega_{nn_0} T = 2\pi k)$, then a maximum of the respective ionization signal may be detected, but in the case of antiphase arriving $(\omega_{nn_0} T = \pi(2k-1))$ the ionization signal will be absent. This individual probability behavior coincides exactly with that occured in the synchronization of quantum transitions for two-level atomic systems. The observation of the behavior as function of the delay time T allows the direct measurements of the times of quantum transitions.

Results of computer calculations for the total probability of the excitation of the Rydberg-state superposition by pairs of light pulses at different delay times T are shown in Fig.1. The case is refers to light pulses of the duration 8.8ps exciting the Rydberg manifold from the initial state with $n_0=90$ and to the observation time equal to 600 periods T_0 of quasiclassical electron motion (T_0=110.8ps). The Fourier transformation was performed for the calculated signal. The two lines in Fig.2 refer to the quantum transitions between the initial state $|n_0>$ and adjacent states $|n_0\pm1>$. The linewidth at the half-maximum is the inverse time of the observation. The accuracy of the determination of resonance frequencies depends on the observation time and on the increment of the delay time. The two resonance frequencies are found with an accuracy of order 10^8.

Fig.1. Probability of the excitation of the Rydberg-state superposition by pairs of light pulses vs. the interpulse time T, with T measured in units of the orbital period T_0 (n_0=90, $\tau\omega_0$=0.5).

Fig.2. Portion of the Fourier spectrum. The values in brackets are theoretical ones. F is frequency in units ω_0: $F(\Delta n)=\omega_{nn_0}/\omega_0$, $\Delta n=n-n_0$.

The upper limit of the observation time T is the lifetime of the Rydberg state. For a Rydberg state with a small angular momentum l=1-10 the radiative lifetime in units of the period T_0 is $T_{nl}/T_0=10^6\div10^8$. Thus, the accuracy in measuring resonance frequencies may be of the order of $10^9\div10^{10}$. For more accurate measurements, atoms in high angular momentum states (l~n) should be used.

The results and analysis given here represent a new method for measuring the Rydberg constant. The method may also be used for frequency and time measurements in molecules.

REFERENCES

Chebotayev V P 1989 *Pisma ZhETF* **49** 429
Chebotayev V P and Ulybin V A 1990 *Appl. Phys. B* **50** 1
Chebotayev V P and Ulybin V A 1991 *Appl. Phys. B* **52** 347
Parker J and Stroud C R Jr 1986 *Phys. Rev. Lett.* **56** 716
Alber G, Ritsch H and Zoller P 1986 *Phys. Rev. A* **34** 1058
Ducas T W, Littman M G, Freeman R R and Kleppner D 1975 *Phys. Rev. Lett.* **35** 366

Inst. Phys. Conf. Ser. No 126: Section III
Paper presented at Int. Symp. on Ultrafast Processes in Spectroscopy, Bayreuth, 1991

Intracavity laser spectroscopy of fast running processes

M V Pyatakhin

P. N. Lebedev Physics Institute, SU-117924 Moscow, Leninsky pr. 53

It is suggested to find the parameters of ultrafast processes in spectral measurements by broadband intracavity laser spectroscopy (ILS). The information on energy transfer phenomena may be obtained from the shape of absorption line. To study the spectrum time evolution and the maximum theoretical sensitivity, the multimode laser dynamics is considered, taking into account stimulated Brillouin scattering (SBS) in the active medium. It is found that SBS in active medium leads to "red" shift of oscillation spectrum and decrease of its stabilization time.

The broad band (intracavity) laser spectroscopy (ILS) [1] is based on high generation spectrum sensitivity to frequency dependent losses of intracavity radiation. The ILS, used more often as a supersensitive method, was put successfully into practice to study the processes with time resolution about 1 μs [2]. In principle it is possible to rise the resolution for ultrafast phenomena investigation, but experimental and theoretical work must be done to determine the sensitivity. Another way may be found that could prove to be more promising, when the parameters of ultrafast processes are found in spectral measurements by ILS. The information on energy transfer intramolecular phenomena, for example, about energy distribution velocity between different vibration modes, may be obtained from the shape of absorption line [3]. For both the first and the second cases the questions about spectrum time evolution in the ILS method and about maximum sensitivity are very important.

In this paper the stimulated Brillouin scattering (SBS) in the active medium was considered. The spectrum of reflected wave is shifted with relation to the spectrum of the falling wave. It may redistribute the energy in wideband laser generation spectrum and decrease the sensitivity of ILS. Let us mention possibilities of mode locking in multimode laser and ultra short laser pulse generation by using the stimulated Brillouin scattering.

As the first step, in the present paper we limit ourselves by qualitative consideration of multimode laser with SBS in active medium time evolution. We will analyze rate equation [1]. To take SBS into account, it is necessary to add corresponding terms into balance equation for a mode photons number. Let's write it, using expression in self-consistent system of equations for the complex amplitudes of the laser modes and acoustic waves [4]. In this approach the rate equations have the form:

$$\frac{dM_q}{dt} = -\alpha\varepsilon_1 (M_q m_{q-1})^{1/2} - \frac{M_q}{T_{q+}} + U_q(t)(M_q+1) + Rm_q$$

$$\frac{dm_q}{dt} = \alpha\varepsilon_1 (m_q M_{q+1})^{1/2} - \frac{m_q}{T_{q-}} + U_q(t)(m_q+1) - Rm_q$$

$$\frac{dU(t)}{dt} = P - \xi U(t)\sum_q (M_q+m_q) - \frac{U(t)}{\tau}$$

Here q is the mode number; M_q and m_q, photons numbers in strong and weak waves; $U_q=U_0(1-\beta q^2)$ describes the shape of gain line; $U(t)U_0$, maximum gain in fixed time; $U_q(t)=U(t)U_q$; P, the pump rate; τ, relaxation time of the upper laser level; $T_{q\pm}$, photons lifetime of q mode in cavity taking into account absorption in the media; R, coefficient of reflection from weak wave to strong one [5]; $\varepsilon_1=\gamma\sum_q (m_q * M_{q+1})^{1/2}$, steady-state solution for sound [4]; 1, number of distances between the modes, corresponding to frequency shift of scattering radiation; ξ, β, α, γ, U_0, the coefficients. The coefficient α equals to the ratio of active medium and cavity length, γ is determined by nonlinear and hydrodynamical properties of matter.

Fig.1. The spectrum dynamics
without absorption

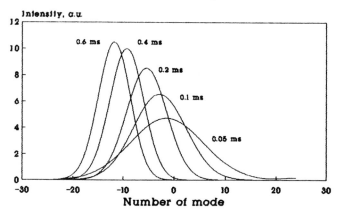

The estimation shows that mode width for typical lifetime of the mode in cavity $T_{q+}=5*10^{-8}s$ is $\simeq 0.7*10^{-3}cm^{-1}$ and is much smaller than intermode distance [6]. Therefore we neglect the mode width. The typical hypersound bandwidth, for example in ethylene glycol, widely used as a solvent for dye, is about 0.18 cm^{-1}, and is not more than intermode shift of nonselective cavity with the length ≤ 35 cm. Additional mode selection usually exist, and hypersound band width is less than intermode shift for a longer cavity. In our consideration we neglect sound bandwidth. Changing the cavity length we may find conditions, when the sound frequency, equal to frequency shift of the scattering wave, is multiple frequency of intermode distance. Without restriction of community we assume that intermode distance is equal to frequency shift at scattering.

The equations have been solved for 50 modes of each of the interacting wave. Parameters were: $T_{q+} = T_{q-} = 5 \times 10^{-8}$s; $U_0 = 1/T_{q+}$; $\tau = 0.2 T_{q+}$; $P = 1.3 \times P_t$; $P_t = 1/\tau$; $\xi = 2 \times 10^{-10}/3\tau$; $\gamma M_0 = 5 \times 10^7 s^{-1}$, M_0, steady-state solution for the mode with $q = 0$ at $R = 0$, $\gamma = 0$; $\alpha = 10^{-2}$; $\beta = 10^{-5}$. The reflection of weak wave with coefficient $R = 0.075/T_{q+}$ provides different losses from strong wave.

First let's discuss the dynamics of the laser without frequency selected losses in cavity. Fig.1 shows the spectrum time evolution of the multimode laser. A "red" shift of the oscillation spectrum with the time was found. At $R = 0$ the shift is absent. In the discussed model weak reflection of the weak wave to the strong one provides a feedback. The spectrum maximum firstly shifts and then the maximum stabilizes on certain mode. It is possible, that this result qualitatively describes the phenomenon experimentally observed [7].

Fig.2. The spectrum dynamics
with absorption on lines -13,-12

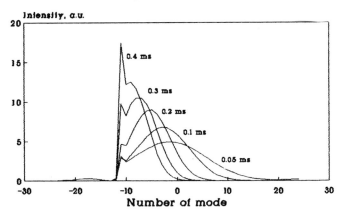

We have also analyzed the case, when the matter with narrow absorption line is put into cavity. Fig.2 shows the spectrum dynamics with absorption on line $q = -13, -12$ ($T\pm = 4.73 \times 10^{-8}$s.). The phenomenon analogous to condensation of the spectrum near the absorption line [3] has been discovered theoretically.

Generation spectra of strong and weak waves concentrate near the "blue side" of the absorption line. A small hole arises near the generation spectrum maximum of strong wave. We associate this with the peculiarity of the spectrum dynamics of weak wave.

Fig.3. The dynamics
of hole depth

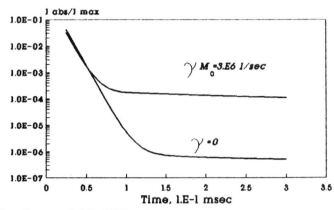

Absorption on q=-7;-6 (Tq=4.97E-8 sec)

The dynamics of the depth of spectrum hole, corresponding to absorption lines, was investigated at different nonlinearity value. Fig. 3 shows temporal dependence of the ratio between line intensities: the line with absorption and the spectrum maximal line. The calculations proved that increase of sound intensity in a wide band laser leads to a decrease of stabilization time of the holes in the laser generation spectrum and to a significant decrease of relative holes depth. This could not be explained in terms of traditional systems of multimode laser rate equations. We found that the rate of the generation spectrum "red shift" increases with nonlinearity. This effect was also observed in [7] under the increase of the pump level above the threshold.

1. A. F. Suchkov 1970 Preprint FIAN Moskow N 126
   L. A. Pahomicheva, E. A. Sviridenkov, A. F. Suchkov, L. V. Titova, S. S. Churilov 1970 ZETF Lett. 12 p. 60 (in Russian)
2. J. P. Reilly, J. H. Clark, C. B. Moor, G. C. Pimentel 1978 J. Chem. Phys. 15 pp. 4381-4395
3. O. M. Sarkisov, E. A. Sviridenkov, A. F. Suchkov 1982 Khimichesk. Phizika 9 p. 1155 (in Russian)
4. M. V. Pyatakhin at al 1991 to be published
5. V. I. Malischev, A. V. Masalov, A. A. Sichev 1970 ZETF Lett. 11 p. 324 (in Russian)
6. V. M. Baev, T. P. Belikova, E. A. Sviridenkov, A. F. Suchkov 1978 ZETP 74 p. 43 (in Russian)
7. Y. M. Aivazajn, V. M. Baev, A. A. Kachanov et al 1986 Kvantovaya Elektronika 13 p. 1723 (in Russian)

Inst. Phys. Conf. Ser. No 126: Section III
Paper presented at Int. Symp. on Ultrafast Processes in Spectroscopy, Bayreuth, 1991

Optical implementation of a Hopfield-type neural network by the use of persistent spectral hole burning media

O Ollikainen and A Rebane

Institute of Physics, Estonian Academy of Sciences, Riia 142, 202400 Tartu, Estonia

ABSTRACT: To demonstrate a possible solution to put to use a unique storage capacity of persistent spectral hole burning (PSHB) media optical   implementations of four-dimensional (4-D), two-dimensional (2-D) and three-dimensional (3-D) interconnection tensors in Hopfield-type and quadratic associative memories are implemented by the use of PSHB media. The capability of present optical schemes to correct errors in the input images are demonstrated.

1. INTRODUCTION

The phenomenon of persistent spectral hole burning firstly observed by Gorokhovskii *et al* (1974) and Kharlamov *et al* (1974) occurs in a variety of low-temperature impurity solids and makes possible writing and reading of optical data in the form of narrow 10^{-2} - 10^{-4} cm^{-1} holes in the inhomogeneously broadened absorption (transmission) spectrum of impurities. In principle, thc density of frequency-domain storage reaches 10^{4} - 10^{6} bits per diffraction-limited size spatial spot.

Holography is a suitable method for obtaining zero-background signal (Renn *et al* 1985) and parallel storage and read-out of optical data in PSHB media. Simultaneous spatial- and frequency domain parallel optical storage has been employed in the method of time- and space domain holography (Rebane A *et al* 1983, Saari *et al* 1986 and Saari *et al* 1989). In these experiments PSHB media storages an analog optical signal with an arbitrary wavefront, a polarization state and temporal structure (the last is Fourier-related to the spectrum of the signal). The possibility to reconstruct full holographic image from partial input (holographic analog of an associative memory) has also been demonstrated by Rebane A (1988).

On the other hand optical neural networks (Farhat *et al* 1985, Jang *et al* 1988, Shariv and Friesem 1989) have received considerable attention for their applications in information storage and processing. There are certain applications such as, e.g., machine vision and optical pattern recognition, which might require more interconnections than can be realized by usual optical storage media. To solve this problem, it is prospective to utilize the unique storage capacity of persistent spectral hole burning media.

In present paper, three experiments are performed to demonstrate potentiality of PSHB media to implement optical multidimensional interconnection tensors (associative memory tensors), which carry out weighted connections of each element of a input image with all elements of a output image.

Recently other related theoretical possibilities of implementing multidimensional optical interconnections based on hole burning holography (Mazurenko 1990, Henshaw 1990 and Weverka *et al* 1991) and on photon echoes (Belov and Manykin 1991a, 1991b) have been discussed.

2. EXPERIMENTAL PROCEDURE

We used a PSHB media prepared from polystyrene doped simultaneously with two types of impurity - molecules of octaethylporphine and protoporphyrine. The PSHB-active absorption bands of the two impurities combine and form in the wavelength interval of 615 - 623 nm a broad band. The dimensions of the PSHB sample are 30x30x4mm. During the experiments the PSHB plate is immersed in liquid Helium. As a laser source, a picosecond dye laser, synchronously pumped by an Ar ion laser is used. In the first experiment, temporal shapes of the recollected signals are measured with a 20 ps time resolution synchroscan streak camera.

In the first and the second experiment a modification of Hopfield memory (Hopfield 1982) which has been described and optically implemented by the use of non-PSHB media (Shariv and Friesem 1989) serves as a rule for associative memory. By this rule two 2-D binary images (1,0) are stored in 4-D memory. To materialize 4-D memory tensor four different physical variables are utilized: two orthogonal spatial coordinates, spectral range of optical signal and time delay between read-out pulse and holographic response pulses. By the use of the method of time-space domain holography spectral dimension is applied twice: at first as frequency of optical signal and secondly as time delay between read-out pulse and holographic response pulses.

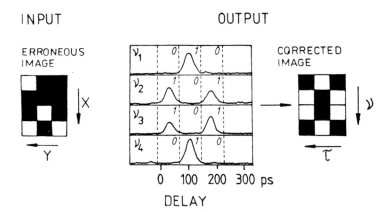

Fig.1. Read-out of the 4-D memory by interrogation with an erroneous 2-D input image. Output signal measurement is carried out by monitoring the time-domain intensity of the diffracted signal at four different read-out laser frequencies.

The read-out of the memory is accomplished by the erroneous input image which resembles one of the stored images, but differs from it by two error bits. At the output a focussing lens collect the diffracted beam onto a single spot at the entrance slit of the streak camera. The temporal structure of the spatially integrated hologram response is measured at four different read-out frequencies and is presented in Fig.1. When arranged into a (4x3) matrix the read-out data clearly reproduces one of the stored images. The fact that the recollected image is inverted with respect to the original follows from the used mathematical algorithm and is not connected with our experimental implementation. It is noteworthy that the input image is coded in two spatial coordinates and the output information in the frequency and temporal domain of an optical signal.

In the second experiment PSHB media is used as 2-D spectral-spatial filter for the implementation of 2-D memory tensor (12 frequencies * 12 spatial stripes). Two 12 bit vectors are stored in memory, after that read-out is accomplished by input vector which has two error bits (Fig.2.). As in the first experiment at the output appears corrected vector, more exactly a negative of one of the stored vectors.

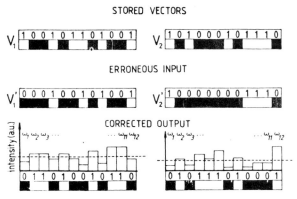

Fig.2. Read-out of the 2-D memory by interrogating with two different 1-D input images. The horizontal dashed line- thresholding level.

In the third experiment by a rule of quadratic associative memory (Chen *et al* 1986, Psaltis and Park 1986) four 32 bits vectors are stored in 3-D memory tensor. The usual, not time-space holograms are written-in at 32 fixed frequencies. Each hologram has a different spatial pattern (32x32 transparent-opaque pixels). The read-out vector contains four error bits and is coded in spatial coordinates (Fig.3). The corrected output vector is coded as in the second example in frequency domain. In present realization, 32 x 32 x 32 = 32768 interconnections between the elements of input and output images are accoplished.

Actually, in read-out processes optical multiplications of 4-D tensor and matrix, matrix and vector, 3-D tensor and matrix are implemented in the first, second and third experiment respectively.

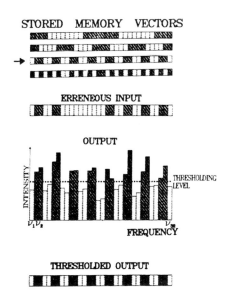

Fig.3. Read-out of the quadratic associative memory.

In conclusion, the potentials of persistent spectral hole burning media for optical implementations of multidimensional interconnection tensors in models of neural networks have been demonstrated.

3. REFERENCES

Belov M N and Manykin E A   1991a   *Opt. Lett.* **16**   327

Belov M N and Manykin E A   1991b   *Opt. Commun.* **84**   1

Chen H H, Lee Y C, Sun G Z, Lee H Y, Maxwell T and Giles C L   1986 *AIP Conf.Proc.* **151** 86

Farhat N, Psaltis D, Prata A and Paek E   1988 *Appl. Opt.* **24** 1469

Gorokhovskii A A, Kaarli R K and Rebane L A 1974 *JEPT Lett.* **20**   474

Henshaw P   1990 in *Digest of Optical Society of America Annual Meeting* (Optical Society of America, Washington, D.C.) p 57

Hopfield J J 1982 *Proc.Natl.Acad.Sci. USA* **79** 2554

Jang J S, Shin S Y and Lee S Y 1988 *Opt. Lett.* **13** 693

Kharlamov B M, Personov R I and Bykovskaya L A 1974 *Opt. Commun.* **12**   191

Mazurenko Yu T 1990 *Opt. Spectrosc.(USSR)* **69** 76

Psaltis D and Park C H   1986 *AIP Conf.Proc.* **151** 370

Rebane A ,Kaarli R, Saari P, Anijalg A and Timpmann K 1983 *Opt. Commun.* **47** 173

Rebane A 1988 *Opt. Commun.* **65** 175

Renn A, Meixner A J, Wild U P and Burkhalter F A 1985 *Chem. Phys.* **93** 157

Saari P, Kaarli R and Rebane A 1986 *J. Opt. Soc. Amer.* **B3** 527

Saari P M, Kaarli R K, Sarapuu R V and Sônajalg H R   1989 *IEEE J.Quantum Electron.* **QE-25**   339

Shariv I and Friesem A A 1989 *Opt. Lett.* **14** 485

Weverka R T, Wagner K and Saffman M 1991 *Opt. Lett.* **16** 826

Inst. Phys. Conf. Ser. No 126: Section IV
Paper presented at Int. Symp. on Ultrafast Processes in Spectroscopy, Bayreuth, 1991

255

Femtosecond generation of coherent optical phonons in condensed media

H. Kurz,

Institute of Semiconductor Electronics, RWTH Aachen, Germany

ABSTRACT: The generation of coherent optical phonons in semiconductors and high temperature superconductors by impulsive excitation with femtosecond laser pulses is reported. Distinct differences in the generation of these coherent phonons are discussed.

I. INTRODUCTION

Coherent optical phonons can be induced by femtosecond laser pulses whose durations are shorter than the period of vibration. They are observed either in reflection or transmission mode as well as in experiments where a vibrational grating scatters off a third laser pulse. In the first two cases a pronounced oscillatory feature is found superimposed on a decaying background, while in the latter method more or less background free detection of coherent oscillation is possible. Usually, the explanation of photon driven phonons is based on the process of stimulated Raman scattering (SRS) where certain vibrational modes are driven by the difference frequency of two optical fields (Blombergen (1965), Shen (1984 and 1986), Laubereau et al. (1978), Penzhofer et al. (1979), Levenson et al. (1974)). Time resolved CARS appears to be the closest analogy in frequency domain (Bron et al. (1986), Zinth et al. (1988)). When the duration of laser pulses is shorter than the period of optical phonons, impulsive stimulated Raman scattering (ISRS) has been invoked in the case of dye molecules (Yan et al. (1985), DeSilvestri et al. (1985)) From the theoretical point of view phase coherent excitation has been treated as a very general feature of femtosecond spectroscopy (Ruhmann et al. (1987and 1988),Chesnoy et al. (1988)).

Since the first discovery of optically induced coherent phonons in semiconductors it became clear that a variety of new phenomena may be able to launch coherent phonons synchroneously (Seibert et. al. (1990), Cho et al. (1990)). In the case of GaAs, LO-phonons are triggered in phase by the impulsive screening of surface space charge fields (Cho et al. (1990)). In GaSe only the highly symmetric A_1'-Raman modes are generated, while E-modes are not observable (Seibert et. al. (1990)). Similar deviations from the SRS-symmetry are reported in the case of metals (Cheng et al. (1990)).

II. GENERATION OF COHERENT PHONONS

The most distinct feature of coherent excitations is the phase relationship between the elementary phonon modes. In a classical frequency domain model the laser pulse drives coherently the Raman mode and imposes a forced vibration on the statistical thermal motion. The driving force $F^Q(t)$ arises from the well known electron–phonon coupling. The collective amplitude of a microscopic ensemble of harmonic oscillators obeys the equation of motion:

$$\mu \left(\frac{\partial^2 Q}{\partial t^2} + 2\Gamma\frac{\partial Q}{\partial t} + \omega_0^2 Q \right) = F^Q(t) \qquad (1)$$

where μ is the reduced lattice mass, Q is the amplitude of coherent vibration, Γ a phenomenological damping rate and ω_0 the phonon frequency considered. In the case of the well known stimulated Raman scattering the driving force is excerted via the Raman coupling tensors a_{kl}:

$$F^Q(\vec{r}, t) = \sum_{k,l} a_{kl} E_k(\vec{r}, t) E_l(\vec{r}, t) , \quad a_{kl} = \frac{\partial \epsilon_{kl}}{\partial Q} \qquad (2)$$

where a_{kl} are the Raman coupling tensors defined by the change in the dielectric function with respect to the normal coordinate of displacement Q. Only the difference frequency part of $E_k E_l$ can excite resonantely the modes or phase the Raman active vibrations. Femtosecond laser pulses cover a sufficient bandwidth to provide the difference in frequencies required within one pulse ($\omega_L - \omega_S = \omega_0$). This means in time domain that the duration of the exciting laser pulse is shorter than the period of vibration.

In polar semiconductors the Raman driving force is composed by a force involving a Raman interaction and by a force associated with the longitudinal component of a nonlinear polarization (Bron et al. (1986)). The amplitude of coherent phonons is driven by ionic Raman interactions reduced by the depolarizing field of nonlinear polarizations P_{NL}. Changes in P_{NL} alone are able to launch coherent phonons in polar media therefore. Thus any nonlinear polarization of sufficient short duration can phase the vibrational modes. In the lowest order the nonlinear polarization arises from $\chi^{(2)}$–type contributions, transforming with the same symmetry as the Raman tensor. The electrooptic coupling is 180^0 out of phase to the ionic contributions, e.g. there is a competition between the two parts in constituting the driving force. The explicit expression for the driving force at excitation across the band gap of semiconductors has to consider resonance phenomena e.g. nonlinearities due to the formation of electron–hole pairs and due to their separation in internal fields (Cho et al. (1990),Kütt et. al. (1991)).

III. TIME RESOLVED OBSERVATION OF COHERENT PHONONS

The coherent lattice vibrations are readily observable in time resolved reflectivity and transmission experiments with sufficient high time resolution and sensitivity. In most cases oscillatory features whose frequencies match exactly certain lattice vibrations are superimposed upon smooth decaying backgrounds. As known from Raman scattering the lattice vibration modulates the electron energy bands through deformation potential coupling as well as by electrooptic coupling. Both contributions exhibit strong resonances at laser frequencies close to the optical energy gaps at interband critical points (Cardona et al. (1982)).

According to the definition of Raman susceptibilities mentioned above they govern the change in dielectric functions as:

$$\Delta\epsilon_{ij}(\vec{r}, t) = \sum_{\alpha} a_{ij}^{\alpha} Q^{\alpha}(t) \tag{3}$$

where all modes are taken into account. Thus the amplitude <u>and</u> phase of the coherent motion are monitored in time resolved reflectivity measurements ($\Delta R \propto \Delta\epsilon$).

The transmission signatures are reproduced by solving the classical wave equation for the probing field (E_l):

$$\Delta\vec{E}_p - \frac{n^2}{c^2}\frac{\partial^2 \vec{E}_p}{\partial t^2} = \frac{4\pi}{c^2}\frac{\partial^2 \vec{P}^{NL}}{\partial t^2} \tag{4}$$

Here the components of nonlinear polarization P^{NL} are composed by Raman terms and nonresonant $\chi^{(3)}$ electronic contributions only present during the excitation:

$$P_i^{NL} = a_{il} Q(t) E_l(t, \Delta t) + \chi_{ijkl}^{(3)} E_j(t) E_k(t) \cdot E_l(t, \Delta t) \tag{5}$$

Unique interference effects are predicted in the case of ultrashort probing pulses. Since the spectral bandwidth of the probing pulse exceeds ω_0, the three emerging fields centered at $\omega_L, \omega_s+\omega_0, \omega_s-\omega_0$ interfere (Yan et al. (1988)) destructively or constructively depending on the phase of coherent vibration. The solution of equ.(5) with expression (6) predicts a periodic sequence of Stokes and Antistokes scattering of the transmitted spectrum. During the excitation with the pump beam ($E_j E_k \neq 0$), cross phase modulation and two photon absorption are expected. These phenomena are generally observed at high excitation levels, e.g. when amplified fs-laser pulses are used for excitation.

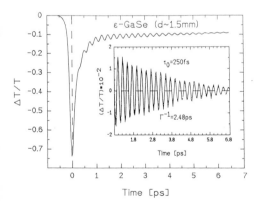

A particularly illustrative example of this high excitation behaviour is shown in fig. 1, where the time resolved transmission signature $\Delta T/T$ of GaSe (d = 1,5 mm) is shown during and after the 50fs excitation with 2eV photons at an intensity level of 100 GW/cm². The transmission probed at 2eV under orthogonal polarization drops during this high power excitation and recovers with a time constant of one picosecond. The superimposed oscillations are clearly visible. Their period of 250 fs matches exactly the inverse phonon frequency of the A_1' mode. The oscillations dephase with a time constant of 2.48 ps.

Fig.1: Time resolved transmission changes in GaSe induced at 100 GW/cm²

Similar results are obtained in InSe and GaS. In the latter material the excitation with 2eV is far below the bandgap (E_g=3eV). Strong two photon absorption features are observed, whereas the baseline of residual induced absorption after the excitation is modulated with a period of 182 fs.

In fig. 2 the time resolved transmission of an extremely thin   InSe sample (E$_g$=1,26eV) is shown at two different wavelengths of the probing pulse (618nm and 635nm). At time delay Δt=0 the relative transmission ΔT/T of the 618nm pulse, filtered out from the blue part of the amplified femtosecond pulse, rises up to value of 12%, while the transmission of the red part (635nm) drops in a complementary way. This complementary behaviour is due to cross phase modulation via nonresonant $\chi^{(3)}$ contributions. After the excitation the transmission signatures are modulated through coherent A$_1'$ phonons with a 180^0 phase shift between the blue (618nm) and red(635nm) part of the transmitted probe spectrum. This implies a periodic blue–and red–shift of the transmitted spectrum as predicted by equs.(4) and (5) and demonstrated more clearly in the insert of fig. 2.

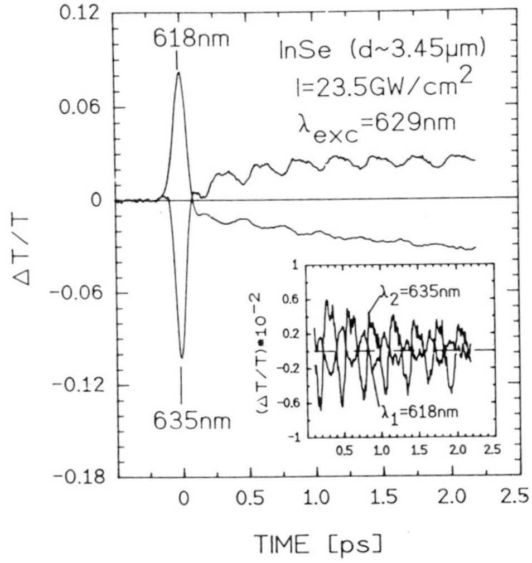

The same antiphased gain/loss behaviour is observed in high temperature superconductors after 10 GW/cm^2 excitation. In YBaCuO–films deposited on SrTiO$_3$ clearly two different modes are detected which decay with a time constant of several picoseconds (Kurz (1991)). Although in spontaneous Raman experiments more frequency modes are observed our fs-data indicate only two different A$_g$-modes (150 and 120 cm^{-1}). These modes are also confirmed in reflectivity measurements.

Fig.2:   Time resolved transmission changes of InSe at two different wavelengths

The optical generation of coherent modes is not bound to the excitation at GW–levels. They are induced and observed with unamplified fs-laser pulses too. In experiments with highly stable CPM-lasers the reflectivity changes can be monitored down to ΔR/R~10^{-7} using a special detection scheme, where the CPM noise is reduced by digital electronic filters. With this improved detection, coherent phonons are revealed in reflectivity signatures of a large variety of semiconductors, metals and high temperature superconductors (Kurz(1991),Chwalek (1991), Cheng (1991)). As an example the reflectivity changes ΔR/R$_0$ of Ge after excitation at 2eV with CPM-pulses is shown in fig. 3. Minute oscillations superimposed on a large electronic back ground are visible, whose period

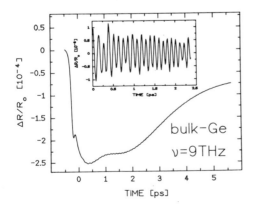

Fig.3:   Time resolved reflectivity changes of Ge at 2eV

matches exactly the Γ_{25}-phonon mode of 9.0 THz (300cm⁻¹). In the case of Ge the appearance of coherent phonon modes is closely related to the selection rule of Raman modes. The dephasing of the modes observed for the first time in an opaque semiconductor takes more time than in III-VI compounds.

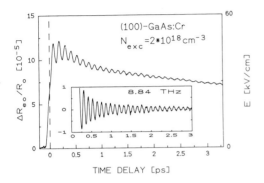

Fig.4: Time resolved REOS-signatures of GaAs:Cu

Completely different excitation mechanisms have been discovered in III-V compound semiconductors. There the sudden generation of electron-hole pairs at the surface of the highly absorbing semiconductors leads to an ultrafast screening of the surface space charge fields (G.C.Cho (1990), W.Kütt et. al. (1991)). The resulting impulsive depolarization field drives optical LO-phonons via electrostatic coupling. As a consequence coherent LO-phonons can be launched for a large range of electron-hole pair densities. In fig. 4 the relative transient reflectivity changes $\Delta R_{E0}/R_0$ arising from electrooptic reactions are plotted versus time. In this reflective electrooptic sampling (REOS) technique the difference between reflectivity along the (O11) axis and that along the (O1̲1) direction is monitored. These differences are determined by changes in the longitudinal electric field normal to the surface of the crystal. The temporal development of $\Delta R_{E0}/R_0$ yields the dynamics of electric field changes in the surface region of the sample. The 8,8 THz oscillations correspond to the frequency of LO-phonons modulating the reflectivity through the electrooptic effect. Detailed investigation of the dephasing have been performed to explore the role of plasmon-phonon coupling in the time domain at different excitation levels (W.Kütt et al. (1991)).

IV. CONCLUSIONS

The superior time resolution and the high sensitivity of femtosecond spectroscopy in detecting very small changes in the optical dielectric functions allows the direct visualization of synchroneously launched optical phonons in time. Both the amplitude and phase of phonons can be determined in reflectivity measurements, whereas in transmission specific interference process lead to a temporal sequence of blue-and redshifts of the transmitted spectrum, whose period matches exactly the period of the coherently driven phonon mode. In a large variety of semiconductors, metals and high temperature superconductors the details of excitation cannot be reconciled with the simple extrapolation of SRS down to time scales of femtoseconds. The direct visualization of dephasing open the way to quantitative investigation of dephasing due to carrier-phonon interaction in time domain.

ACKNOWLEDGEMENTS

The skillful experimental work of G.C. Cho, W. Kütt, K. Seibert, A. Esser, H. Heesel, Th. Pfeifer and T. Dekorsy is greatfullly acknowledged. This work has been entirely supported by the Alfried-Krupp Foundation.

REFERENCES

Blombergen N., 1965, Nonlinear Optics, Benjamin, New York
Bron W.E., Kuhl J. and Rhee B.K., 1986, Phys. Rev. B 34, 6961
Cardona M. and Güntherodt G. (eds.), 1982, Light Scattering in Solids II, Topics in Applied Physics 50, Springer-Verlag
Cheng T.K., Brorson S.D., Kazeroonian A.S., Moodera J.S. Dresselhaus G., Dresselhaus M.S. and Ippen E.P., 1990, Appl. Phys. Lett. 57, 1004
Chesnoy J. and Mokhtari A., 1988, Phys. Rev. B A, 38, 3566
Cho G.C., Kütt W. and Kurz H., 1990, Phys. Rev. Lett. 65, 764
Chwalek S.M., Uher G., Wittacker J.F., Mourou G., Agostinelli J. and Lelenthal M., 1990, Appl. Phys. Lett., 57, 1696
DeSilvestri S., Fujimoto J.G., Gamble E.B., Williams L.R. and Nelson K.A., 1985, Chem. Phys. Lett., 116, 146
Kütt W., Cho G.C., Pfeifer T. Strahnen M. and Kurz H., 1991, Digest of CLEO, Optical Society of America, Washington DC
Kütt W., Cho G.C., Pfeifer T. and Kurz H., 1991, Proceedings of HCIS 7, Nara Japan, to be published in Semiconductor Science and Technology
Kurz H., 1991, Digest of the Quantum Electronics and Laser Society Conference, Optical Society of America, Washington DC
Laubereau A. and Kaiser W., 1978, Rev. Mod. Phys., 50, 607
Levenson M.D. and Blombergen N., 1978, Phys. Rev. B 10, 4447
Penzhofer A., Laubereau A. and Kaiser W., 1979, Prog. Quant. Electr. 6,55
Ruhmann S., Joly A.G. and Nelson K.A., 1987, J. Chem. Phys. 86, 6563
Ruhmann S., Köhler B., Joly A.G. and Nelson K.A., 1988, IEEE-QE 24, 470
Seibert K., Heesel H., Albrecht W., Geurts J., Allakhverdiev K. and Kurz H., 1990, 20th International Conference on "The Physics of Semiconductor 3", eds. Anastassakis E.M., Joannopoulos J.D., World Scientific
Shen Y.R., 1984, The Principles of Nonlinear Optics, Wiley, New York
Shen Y.R., 1986, IEEE-QE, 1196
Yan Y.X., Gamble E.B. and Nelson K.A., 1985, J. Chem. Phys., 83, 5591
Zinth W., Leonhardt R., Holzapfel W. and Kaiser W., 1988, IEEE-QE 24, 455

Inst. Phys. Conf. Ser. No 126: Section IV
Paper presented at Int. Symp. on Ultrafast Processes in Spectroscopy, Bayreuth, 1991

Femtosecond spectroscopy of YBa$_2$Cu$_3$O$_{7-\delta}$: electron−phonon interaction measurement and energy gap observation

S.V.Chekalin, V.M.Farztdinov, V.V.Golovlev, V.S.Letokhov,

Yu.E.Lozovik, Yu.A.Matveets, and A.G.Stepanov

Institute of Spectroscopy, USSR Academy of Sciences, 142092
Troitsk, Moscow Region, USSR

ABSTRACT: The femtosecond dynamics of the difference reflection and transmission spectra of an YBa$_2$Cu$_3$O$_{7-\delta}$ film have been studied in the spectral range 620-680 nm at the initial temperatures T_0 above and below the critical temperature T_c. Based on the experimental data, the electron-phonon interaction parameter $\lambda \langle \omega^2 \rangle$ has been estimated at $(4 \pm 2) \cdot 10^2 (\text{meV})^2$.

Real-time studies into the relaxation of charge carriers in oxide superconductors can help to obtain unique information about the nature of superconductivity in such materials, particularly the role of electron-phonon interaction in them. In the works reported by Brorson et al (1990), Kaseroonian et al (1991), Chwalec et al (1990), Han et al (1990) the femtosecond charge-carrier dynamics in cuprates using a single wavelength for both pump and probe pulses have been investigated. In our work, we have studied the femtosecond dynamics of the difference reflection and transmission spectra of an YBa$_2$Cu$_3$O$_{7-\delta}$ film in the spectral range 620-680 nm at the initial temperatures T_0 above and below the critical temperature T_c. Studying the time history of electron-phonon relaxation throughout the spectral range has allowed us to estimate the electron-phonon interaction parameter (Eliasberg 1960, Allen 1987) $\lambda \langle \omega^2 \rangle$ at $(4 \pm 2) \cdot 10^2 (\text{meV})^2$.

The sample under study was a 150 nm thick YBa$_2$Cu$_3$O$_{7-\delta}$ film deposited on a SrTiO$_3$ substrate and coated with a protective . MgO layer 5 nm in thickness. The superconducting and protective layers were grown by high-frequency magnetron sputtering in a single production cycle. The crystallites in the superconducting layer were around 300 nm across, their c-axis being oriented normal to the substrate surface. The critical temperature T_c was equal to 80 K.

The sample was excited with a 150-fs laser pulse at 612 nm. The pump pulse intensity on the sample was some 10^{11} W/cm^2. The excited sample area was probed with a weak wide-band continuum pulse of the same duration. The probe pulse that passed through (or reflected from) the sample was detected with a multichannel optical analyzer built around two CCD arrays. In our experiment, we measured the difference spectrum $\Delta A(\lambda, \iota)$ at various delay times τ between pump and probe pulses.

In Figure 1 showing some of the experimental difference transmission spectra, one can see the drastic difference between the results for $T_0 > T_c$ and $T_0 < T_c$. While transmission in the former case is observed to decrease during the pump pulse all over the spectral range under study (see Figure 1(a)), it grows on the region 1.93-2.01 eV in the latter case (for $T_0 < T_c$), there being only a slight drop in transmission in the range 1.89-1.93 eV (see Figure 1(b)). With the delay time between the pump and probe pulses amounting to around a picosecond, a quasiequilibrium sets in, so that fast (~ 100 fs) changes in the spectra cease. The only difference in observation conditions between the spectra of Figures 1(a) and 1(b) was that between the initial temperatures of the samples, and so it is reasonable to

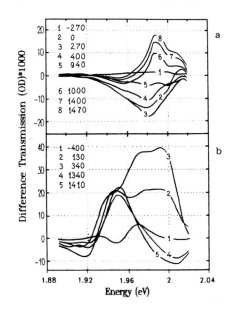

FIGURE 1. Difference transmission spectra of the sample for various delay times ι (fs). a) T = 92 K, b) T = 70 K.

suppose that the changes the spectra are seen to undergo at ι < 1 ps (Figure 1(b)) contain information on the dynamics of destruction of the energy gap as a result of the sample being heated by the powerful femtosecond laser pulse.

Figure 2 shows the relationship between the optical density of the difference transmission spectra of ΔA and the delay time ι for the most typical regions. At T_0 = 92 K (Figure 2(a)), the shape of the curves remains qualitatively the same throughout the spectral range of interest: ΔA is seen to decrease for a characteristic time of the order of 100 fs and then relax during $\lesssim$ 1 ps. The difference between the curves for various regions of the spectrum is that asymptotic value of ΔA is less than zero at $\hbar\omega$ < 1.96 eV and greater than zero at $\hbar\omega > 1.96$ eV. Similar curves, though relaxing to $\Delta A \simeq$ 0, were observed at $T_0 = 70$ K for $\hbar\omega < 1.93$ eV (Figure 2(b)). The curves for $\hbar\omega > 1.93$ eV at this temperature reversed sign (Figure 2(c)), i.e. ΔA was first observed to increase and then relax to various asymptotic values, depending on wavelength. The relaxation of ΔA in the vicinity of $\hbar\omega = 1.96$ eV was nonmonotonic: two to three ΔA oscillations were observed to occur on a 300-500-fs scale. The nature of this nonmonotonicity is as yet vague. An essential feature of the curves obtained at $T_0 < T_c$ is the delay of the response for ~ 200 fs in the region of 1.98 eV (Figure 2 (c)), compared to that in the region of 1.91 eV (Figure 2(b)) and the response at $T_0 > T_c$ (Figure 2(a)). The magnitude of this delay is roughly three times the relative lag of the various spectral components of probe pulse, observed to occur in the experiment as a result of dispersion.

To recover $\Delta\varepsilon_2$ from the difference transmission and

reflection spectra, use was made of Bjorneklett et al (1989) ellipsometric data on the equilibrium dielectric constant, ε_1 and ε_2, of $YBa_2Cu_3O_{7-\delta}$. The difference spectra of $\Delta\varepsilon_2$ are shown in Figures 3(a) and (b). As can be seen from the figures, the spectra for $T_0 > T_c$ also greatly differ from those for $T_0 < T_c$. At $t \geq 700$ fs, the spectral dependence of the response for $T_0 = 92$ K becomes alternating: $\Delta\varepsilon_2 < 0$ for $\hbar\omega < 1.99$ eV and $\Delta\varepsilon_2 > 0$

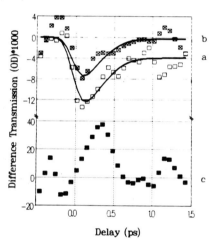

for $\hbar\omega > 1.99$ eV. Subsequently, starting from $t \simeq 1$ ps, the shape of the spectrum ceases to change -it remains alternating, with $\Delta\varepsilon_2 = 0$ at $\hbar\omega = 2$ eV (Figure 3(a)). An alternating response is characteristic of the spectral range corresponding to transitions either into or out of the Fermi level region (Eesley.1986, Gershenson et.al.1990) and is associated

FIGURE 2. Difference transmission of the sample as a function of time for various energy of the probe quanta $\hbar\omega$(eV). a)T = 92 K, $\hbar\omega = 1.95$ eV, b)T = 70 K, $\hbar\omega = 1.91$ eV, c) T = 70 K, $\hbar\omega = 1.98$ eV. The solid curves are the numerical fits to the experimental data for $\lambda \langle\omega^2\rangle = 450$ $(meV)^2$, $C_e = 16$ mJ/mole K^2, and $I_p = 2\cdot10^{11}$ W/cm^2.

with the temperature smearing of the charge carries distribution function in the vicinity of the Fermi level. The decrease of transmission (ε_2 rise) for short delay times over a wide spectral range (Figure 3(a)) is to all appearance due to the shift of the Fermi level as a result of strong heating of the charge carriers. The absence of any substantial changes in the $\Delta\varepsilon$ spectrum at $\iota \gtrsim 1$ ps points to the establishment of a quasiequilibrium. By this moment, the temperature of the sample reached 500 K, and the system loses its memory of the initial conditions, so that the difference of a mere 30 K between the initial temperatures cannot bring about any significant dissimilarities in the spectra of the excited sample. It can therefore be argued that the diversities observed in the difference spectra at $\iota \gtrsim 1$ ps (see Figures 3(a) and 3(b)) can only result from the differences between the spectra of the sample in the initial (unexcited) state. This can be explained by the presence of an energy gap in the vicinity of the Fermi level for $T_0 = 63$ K and the

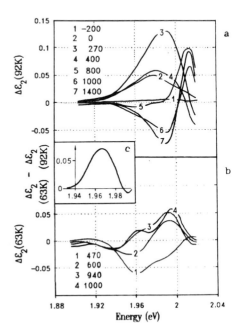

shift of this level because of the difference between the T_0 values.

Subtracting the difference spectrum $\Delta\varepsilon$ for $T_0 = 92$ K at $\iota = 1$ ps from that for $T_0 = 63$ K, we obtain a peak around 30 meV wide in the neighborhood of some 1.97 eV (Figure 3(c)). Relating the width of this peak to

FIGURE 3. Difference spectra of the imaginary part of the dielectric constant ε for various delay times ι(fs). a) $T_0 = 92$ K, b)T= 63 K, c) the result of subtracting the spectra of $\Delta\varepsilon$ for $T_0=63$ K and $T_0=92$ K ($\iota=1000$ fs).

the width Δ of the energy gap, we get $2\Delta_0/T_c = 8\pm3$, which agrees well with the data reported by Imer et al (1989), Shutzmann et al (1989), Collins et al (1990), Person et al (1990).

The presence of an energy gap in the initial state can also explain the lag of the optical response maximum observed to occur for $T_0 < T_c$ in the vicinity of the Fermi level. The magnitude of this lag (~ 200 fs) matches the time it takes for superconductivity to be destructed.

The experimental curves of Figure 2 were used to estimate the electron-phonon interaction parameter $\lambda\langle\omega^2\rangle$ as follows. The dielectric constant of the sample varies as a result of the heating of electrons and the shifting of energy bands due to the heating of the lattice and thermoelasticity effects. Within the framework of linear response, this variation (the change of transmission in our case) can be represented in the form $\Delta A \simeq a \cdot \Delta T_e(t) + b \cdot \Delta T_L(t)$, where T_e is the electron temperature and T_L the lattice temperature. To calculate the dynamics of ΔA, the temporal variations of T_e and T_L were determined on the base of Allen's (1987) theory.

The calculated (with due regard for the Gaussian shape of the pump and probe pulses) functions $\Delta A(t)$, were compared to the experimental curves (see Figure 2). Comparison was made for the spectral range 1.92-1.95 eV far from the Fermi level in order to avoid the possible nonlinear effects associated with the shift of the level as a result of strong heating of electrons. The parameters a and b were selected so as make the calculated curves approximate the experimental data for the maximum ($\iota \sim 100$ fs) and asymptotic ($\tau \geq 1$ ps) values, the temporal behavior of the experimental curves being approximated by means of the parameter $\lambda\langle\omega^2\rangle$. The resultant estimate is $\lambda\langle\omega^2\rangle = (4\pm2)\cdot10^2$ (meV)^2. Using for $\langle\omega^2\rangle$ the estimate $\langle\omega^2\rangle = \theta_D^2/2$ and taking the Debye temperature θ_D at 350 K (Kvavadze et al 1990), we get the following estimate for the electron-phonon coupling constant: $\lambda = 0.9\pm0.4$. This estimate agrees with Brorson et al (1990) experimental measurements and with the value of $\lambda\sim1$ found by Rodrigues et al (1990) with the help of local density functional method.

We would like to thank M.E.Gershenzon and M.I.Falei for kindly providing the samples, I.B.Kedich and A.P.Yartsev for their assistance in the experiment, and also L.B.Kislichenko for her aid in calculations.

REFERENCES

Allen P B 1987 *Phys.Rev.Lett.* <u>59,</u> 1460.

Bjorneklett A, Borg A, and Hunderi O 1989 *Physica* (Amsterdam) A<u>157</u>, 164.

Brorson S D , Kazeroonian A , Face D W, Cheng T K , Doll G L , Dresselhaus M S , Dresselhaus G , Ippen E P , Venkatesan T , Wu X D , and Inam A 1990 *Solid State Comm.* 74, 1305.

Chwalec J M, Uher C, Whitaker J F , Mourou G A, Agostinelli J, and Lelental M., 1990, *Appl.Phys.Lett.* <u>57,</u> 1696.

Collins R T, Schlesinger Z, Holzberg F, and Freidl C 1989 *Phys.Rev.Lett.* <u>63</u>, 422.

Eesley G L 1986 *Phys.Rev.* B<u>33</u>, 2114.

Eliashberg G M   1960   Zh.Eksp.Teor.Fiz. 38, 966; 39, 1437.

Gershenson M E , Golovlev V V , Kedich I B , Letokhov V S ,
    Lozovik Yu E , Matveets Yu A , Sil'kis E G , Stepanov A G ,
    Titov V D , Faley M I , Farztdinov V M , Chekalin S V , and
    Yartsev A P   1990   Pis'ma v Zh.Eksp.Teor.Fiz. 52, 1189.

Han S G , Vardeny Z V , Wong K S , Symko O G , Koren   G
    1990   Phys.Rev.Lett. 65, 2708.

Imer Y M , Pattney F , Dardel B , Schneider W - D, Baer Y ,
    Petroff Y , and Zettl A   1989   Phys.Rev.Lett. 62, 336.

Kazeroonian A , Cheng T K , Bronson S D , Li Q , Ippen E P ,
    Wu X D , Venkatesan T , Etemad S , Dresselhaus M S ,
    Dresselhaus G 1991   Solid State Comm.78,95.2.

Kvavadze K A , Igitkhanishvili D D , Nadareishvili M M ,
    Takhnishvili L A , Zinzadze G A , and Chubabriya M Y ,
    1990   Sverkhprovodimost' 3, 1628 (in Russian).

Person B N J , Demuth J E   1990   Phys.Rev.B42,8057.

Rodriguez C O , Liechtenstein A I , Mazin I I , Jepsen O ,
    Andersen O K , and Methfessel M   1990   Phys.Rev.B42, 2692.

Shutzmann J, Ose W, Keller J, Ronk K F, Roas B , Shulz L,
    and Saeman-Ischenko G   1989   Europhys.Lett. 8, 679.

Inst. Phys. Conf. Ser. No 126: Section IV
Paper presented at Int. Symp. on Ultrafast Processes in Spectroscopy, Bayreuth, 1991

267

Investigation of melting processes in GaAs in the femtosecond time domain

T Schröder, W Rudolph, S Govorkov[*], I Shumay[*]

Friedrich-Schiller-University Jena, Institute of Optics and Quantum Electronics, Max-Wien-Platz 1, O-6900 Jena, Germany
[*]Moscow State University, Institute of Nonlinear Optics, 119899 Moscow, USSR

Abstract: The dynamics of symmetry changes in a GaAs sample were investigated with the help of intensive femtosecond light pulses ($\lambda = 620$ nm), generating the Second Harmonic in reflection from the surface of the GaAs (110) wafer. Using different experimental techniques we observed a symmetry loss after optical excitation in less than 100 fs.

1.Introduction

Laser-induced phase transitions at the surface of semiconductors (i.e. melting, evaporation, amorphization etc.) have been of great interest since the discovery of pulsed-laser annealing (ed. by Bieglsen et al 1985). So far a number of experiments of this kind have been performed. Shank et al (1983a) measured laser-induced modification of Si reflectivity. Kash et al (1985) used the spontaneous Raman spectroscopy to estimate the electron-phonon energy transfer time in GaAs. Shank et al (1983b) studied the onset of melting of Si surface layers after femtosecond pulse excitation through second harmonic generation (SHG) in reflection. The latter experiment on SHG is particularly important since it has demonstrated that the nonlinear optical response of the centrosymmetric material can have strong anisotropy reflecting the bulk symmetry. Here we report on subpicosecond dynamics of laser-induced melting of GaAs monocrystals studied by SHG in reflection as well as by linear reflectivity measurements. The experiments were provided using the pump-and-probe as well as the "self-action" techniques proposed by Bloembergen et al (1984).

GaAs is a noncentrosymmetric material and exhibits a large second-order nonlinear susceptibility. This allows strong SHG in the bulk in electric dipole approximation. In contrast, melted material is an isotropic liquid where SHG is forbidden in electric dipole approximation due to the symmetry selection rules. Thus the deviation of the SH intensity as a function of pump-laser intensity from the square law unambiguously indicates a modification of the surface layer spatial symmetry, which can be attributed to surface melting. By means of pump-and-probe experiments the dynamics of SH decay and reflectivity changes can be monitored.

2. "Self-action" Experiments

In a "self-action" experiment (Schroder et al 1990) we used 50µJ of energy of the pulses from a CPM dye laser amplified in four gain stages. The pulses were coupled out before recompression and had a duration of 260fs. These pulses were focused onto a monocrystalline GaAs (110) wafer. The incident pulse energy could be varied by a set of calibrated neutral filters. The reflected SH signal was detected by a photomultiplier.

The sample was moved after each laser shot to ensure equal excitation conditions for each pulse. For each set of neutral filters the incident laser pulse energy and the SH energy were averaged over 100 shots. The measured dependence of SH pulse energy versus incident pulse energy is shown in Fig.1. The arrow indicates the onset of amorphization.

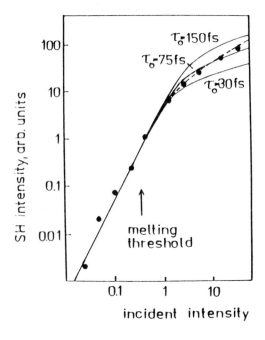

Fig.1:
Second harmonic signal (experimental results: ●) versus pump laser intensity. The arrow indicates the melting threshold which is at energy densities of about 50mJ/cm². Results of a numerical modelling are shown for cosh⁻² pulses (solid curves) and for Gaussian pulses (dashed curve, $\tau_0 = 75$fs).

As can be clearly seen the slope of the curve shows a quadratic behaviour at low laser intensities. The slope decreases at an intensity corresponding to the threshold for laser-induced surface amorphization which we attribute to surface melting. This means that the laser fluence in the central part of the laser spot is sufficient to melt the surface layer of the GaAs sample during the pulse duration. In this manner the melted area gives no contribution to the SH signal. A numerical treatment yields a time constant for laser-induced melting of $\tau_0 = 75$fs. This time constant is shorter than the time which was measured for complete energy transfer to LO-phonons by Kash et al (1985). Assuming that in polar semiconductors electrons and holes are coupled with LO phonons near the center of the Brillouin zone (Compaan 1985) our result gives some more evidence to the hypothesis of "cold melting" introduced by Tom et al (1988).

3.Pump-and-probe Experiments

Our experiments (Shumay et al 1991) were carried out by means of strong fs light pulses (λ=616nm, T_L=100fs), which were divided into pump and probe pulses. The total laser fluence on the surface was estimated to be about 0.15J/cm² which is three times the threshold value for GaAs surface amorphization. Probe pulse SH intensity was detected by a photomultiplier (PMT) and a part of the reflected probe pulse was detected by a photodiode for linear reflectivity measurements.

   In our experiments we have measured the difference in SH intensity ΔI_{SH} and linear reflectivity ΔI_R of the probe pulse with pump pulse on and off as a function of probe pulse delay τ with respect to the pump pulse. Obtained dependencies are shown in Fig.2. Intensity cross correlation function of the pump and probe pulses with peak position corresponding to τ=0 is shown in the insert.

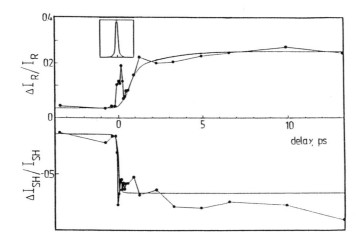

Fig.2: Second harmonic intensity and linear reflectivity from GaAs (110) surface versus probe time delay. Dots: experimental data (each point corresponds to average over 60 laser pulses with proper energy selection). Smooth curves: numerical fit. Broken line is drawn through experimental data to guide an eye.

It is clearly seen, that SH intensity drops by 70% during the 100 fs long pump pulse. On the other hand, linear reflectivity increases more gradually and after approximately 1ps obtains the characteristic value of molten GaAs. The reflectivity spike in the vicinity of zero delay can be attributed to coherent interaction of the pump and probe pulses leading to pump pulse scattering in the direction of probe pulse. However, we can not exclude state filling effect in valence and conduction bands coupled by laser excitation. Since relaxation time of carriers from their initially excited states was found to be of the order 100 fs [4] this effect can be observed only during laser excitation. The same effect leads to "bleaching " of the second-order susceptibility and can be the reason for the appeerence of negative 200 fs long spike in SH intensity at zero delay.

The characteristic time constants of SH decay $\tau_{SH} \sim 100$fs and reflectivity changes $\tau_R \sim 1$ps were found by fitting the experimental data.

To attribute the observed fast drop of SH intensity to atomic disordering one should rule out several possible reasons of $\chi^{(2)}$ decrease. First, band-filling effect seems to be not in the case because its relaxation time is gouverned by interband transitions on the time scale of picoseconds, thus it should manifest itself in linear reflectivity in delay time range $\tau = 0.1 \div 10$ ps. Second possible reason is that the high carrier density screens the ionic potential (Tom 1988), but this effect is unlikely to cause a considerable decrease of nonlinear susceptibility of polar semiconductor.

It is worth to noting that SH probes the long-range crystal lattice order at a distance comperable with wavelength, while linear reflectivity is more sensitive to the short range order. Thus, both self-action and pump-and-probe experiments show that GaAs lattice seems to lose its long-range crystal order as revealed by sharp decrease of second harmonic intensity on a time scale less than 100fs. This time appeare to be shorter than the characteristic electron-phonon energy transfer time as measured by Kash (1985). At the same time linear reflectivity obtains the characteristic value of molten GaAs in several picoseconds, the time consistent with thermal model of laser-induced melting. Consequently, we conclude that the onset of melting of GaAs monocrystal following strong excitation by an ultrashort laser pulse is preceded by a semiconductor-like nonequilibrium phase lacking long range order or by a centrosymmetric crystalline phase, which is realized with the lattice remaining essentially cold.

References

Bieglsen D K, Rosgonyi G A, Shank C V (editors) 1985 Pittsburgh, "Energy-Beam Solids Interactions and Transient Thermal processing"

Bloembergen N, Malvezzi A M, Liu J M 1984 Appl.Phys.Lett.**45**, 1019

Compaan A 1985 J.of Luminesc.**30**, 425

Kash J A, Tsang J C, Hvam J M 1985 Phys.Rev.Lett.**54**, 2151

Shank C V, Yen R, Hirliman C 1983 Phys.Rev.Lett.**50**, 454

Shank C V, Yen R, Hirliman C 1983 Phys.Rev.Lett.**51**, 900

Schröder T, Rudolph W, Govorkov S, Shumay I 1990
    Appl.Phys.A **51**, 49

Shumay I, Govorkov S, Rudolph W, Schröder T 1991
    Opt.Lett.**16**, 1013

Tom H W K, Aumiller G D, Brito Cruz C H 1988 Phys.Rev.Lett.**60**, 1438

Inst. Phys. Conf. Ser. No 126: Section IV
Paper presented at Int. Symp. on Ultrafast Processes in Spectroscopy, Bayreuth, 1991

271

Time-resolved spectroscopic investigations of Ti doped YAlO$_3$ and pure YAlO$_3$ crystals

H Chosrovian, S Rentsch and U -W Grummt[1]

Faculty of Physics and Astronomy, 1) Faculty of Chemistry
Friedrich - Schiller - University Jena, Max - Wien - Platz 1
O - 6900, F R G

ABSTRACT: The difference of optical density after light pulse excitation for Ti doped and pure YAlO$_3$ crystals was measured within 6, and 1000 ps and 10 μs and 10 s. Ti^{3+}:YAlO$_3$ crystals show a broad light induced transient absorption (LIA) between 350 to 750 nm, which consists of some independent bands. The bands are formed within a few hundreds ps and have about 10^3 to 10^5 times longer lifetimes as the fluorescence of Ti^{3+} (14 μs). In order to explain the experimental results the model of bound small polaron is used.

1. INTRODUCTION

Ti^{3+} doped YAlO$_3$ with the fluorescence range 540 - 800 nm, Figure 1, is an attractive candidate for a tunable solid-state laser in the visible region (Kaminski 1975). This material exhibits the crystal structure of a distorted perovskite. Various research groups have tried to get this material lasing, but without success (Khattak et al. (1989), Wegner et al. (1989), Wall et al. (1989)). Only Kvapil et al. (1988) reported lasing with an efficiency of only about 0.5*10^{-3}. This extremely low value of the lasing efficiency was temptatively attributed to parasitic absorptions at the emission wavelengths by Kvapil et al.(1989). Petermann et al. (1991) explained the experimental results within a simple configurational model assuming that a Ti^{3+}/Ti^{4+} complex center was formed, which diminishes the lasing efficiency of Ti:YAlO$_3$ strongly. Thereby, divalent impurities like Fe^{2+} may stabilize the Ti^{4+} concentration and thus also the number of Ti^{3+}/Ti^{4+} pairs.

The aim of this investigation was to study the spectral and dynamical properties of Ti:YAlO$_3$ crystals in order to get the reasons of the extremly low value of the lasing efficiency.

2. EXPERIMENTAL RESULTS

The pump-probe double beam arrangement for ps time resolved absorption spectroscopy is described in detail by Damm et al. (1989). For the excitation of the crystals at room temperature the second harmonic of the Nd phosphate glass laser at 527 nm was used with a pulse energy of 1 to 3 mJ, pulse duration of 6 $\pm$ 1 ps and a beam cross section of about 1 mm^2. The spectral changes in the visible region are obtained by measuring the optical density difference with and without excitation using the picosecond continuum at various delay times after excitation. 15 pulses were averaged to give one ΔD spectrum at a certain delay time with an accuracy of about 5%.

The temporal development of transient absorption, as shown in Figure 3, is different fo
samples with different annealing and codoping conditions.

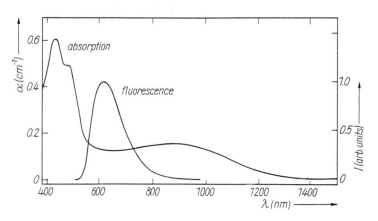

Figure 1. Absorption and fluorescence spectra of Ti:YAlO₃ crystal

In the cases a), b), c) and d) the Ti^{3+} fluorescence maxima (negative ΔD values) due t
induced amplification of the probe beam will be achieved within a) 25 ps, b) 10 ps, c) <1C
ps and d) 60 ps, respectively. The gain process is diminished within a) 300 ps, b) 40 ps, c
100 ps and a broad absorption is formed (positive ΔD values). The suppression of Ti^{3+}
fluorescence is clearly shown. In case d) the structure of the fluorescence bands gives hin
for the superposition of absorption bands. In all samples a long (730 nm), a medium (65

nm) and a short (600 nm)
wavelength absorption band
could be recorded. The lifetime
of the recorded bands is rather
long, no decay was detectable
within the nanosecond time
range. The difference of optical
density $\Delta D(\lambda,t)$ with time
resolution of about 10 μs were
carried out using a flashlamp
arrangement. The crystals were
excited perpendiculary to the
axis of the testing beam and
tested parallel to it.   The
temporal decay of ΔD as well
as the transient absorption
spectra integrated over 10 μs
were recorded for fixed time
delays. A broad light induced

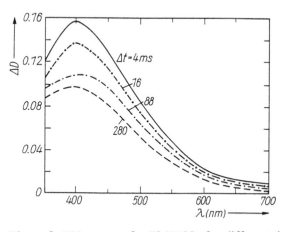

Figure 2. LIA spectra for Ti:YAlO₃ for different time
delays integrated over 10 μs

absorption could be found for all samples with a maximum at 400 nm. The observe
transient absorption is located in a spectral region where the pump laser of Ti^{3+}:YAlC
works. Moreover, the observed absorption superimposes the intended laser region (560
750 nm) and reduces the lasing efficiency strongly.

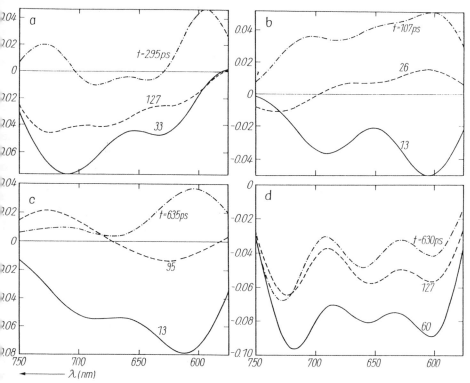

Figure 3. The $\Delta D(\lambda, t)$ spectra of Ti:YAlO$_3$ for three characteristic delays after excitation with pulses of 6 ps and 527 nm a) Ti:YAlO$_3$ annealed in air, b) Zr codoped Ti:YAlO$_3$, c) Ti:YAlO$_3$ unknown annealing, d) Ti:YAlO$_3$ annealed in H$_2$ atmosphere

For comparision, measurements on pure (undoped) YAlO$_3$ crystals have been performed under the same excitation conditions. The decay of LIA was fitted using a double exponential function. The results let us to believe that in both Ti-doped and pure YAlO$_3$ crystals the effects of LIA have a similar nature.

DISCUSSION

In summery, we have observed strong light induced absorption in Ti-doped and even pure YAlO$_3$ crystals. In Ti-doped crystals there is a weak portion of the Ti^{3+} excited state absorption arising from charge transfer transitions to energy levels of defects. For explanation of the transient absorption of the YAlO$_3$ the model of small polarons has been developed by Schirmer *et al.* (1975) and by Akkermann *et al.* (1989). The bound small polaron model is based on the assumption that the hole trapping is accompanied by a lattice distortion around the trapping site. The hole together with the lattice relaxation can be understood as a small polaron bound to a defect. In YAlO$_3$, depending on the irradiation and thermal history of the crystal, two types of independent centers O$_I^-$ and O$_{II}^-$ in correlation with ESR spectra were obtained and discussed by Schirmer *et al.* (1975). The induced optical absorption in pure YAlO$_3$ was explained by a light induced transfer of the hole from one to another equivalent O^{2-} site near the Y^{3+} vacances. The hole which is coupled to the lattice can be understood as a small polaron bound to a defect. Flashlamp - spectroscopic investigations in times from μs to seconds of both Ti doped and pure YAlO$_3$ crystals show similar spectral and temporal behaviour of absorption differences. We believe, however, tha the polaronic absorption not only leads to coloration of nominally undoped $YAlO_3$, but mos probably also contributes to the absorption in the doped material. We do not exclude othe processes as hole trapping at Ti^{3+} site, i.e. the following charge transfer process may b possible: $Ti^{3+} + e^+ \to Ti^{4+}$. In this way the number of Ti^{3+} ions can be diminished, an this process harmful influenced on the laser generation. The temporal relationship of bot processes, i.e. Ti^{3+}/Ti^{4+} pair building and holes trapping at O^{2-} ions near defects of lattice depending on the grown and annealing conditions could be anderstand now. After ligh excitation the hole trapping occurs in all samples and causes the LIA with a typical lifetime from some milliseconds up to some seconds. However, the Ti^{3+}/Ti^{4+} pair building depend on the annealing conditions and the LIA due to this mechanism with a typical lifetime o about 15 μs could be diminished by conditions of annealing. Our results show that in th reducing atmosphere this second mechanism is negligibly. The opinion about practica impossibility to get laser action on this material one has to refer to the hole trappin mechanism near the defects (poor optical quality of used crystals) and not to "useless" of T ions. Intensive efforts have to be done to diminish such harmful defects.

ACKNOWLEDGEMENTS

The authors are indebted to J Kvapil for supplying of $YAlO_3$ crystals.

REFERENCES

Akkermann B A, Bylka G R, Beinstein D I, Winokyrov B M, Winokyrova B B, Galeev A A, Garmash B M, Ermakov G A, Markelov A A, Nisamytdinov N M and Hasanova N M 1989 Soviet Phys.- Solid State **31** 214
Damm T, Kaschke M, Kresser M, Noack F, Rentsch S and Triebel W 1989 Exper. Tech Phys. **33** 409
Kaminski A A 1975 Lasernii kristali Izd. Nauka Moskva
Khattak C P, Schmid F, Wall K F and Aggarval R L 1989 SPIE Conference Orlando Florid
Kvapil J, Koselja M, Kvapil J, Perner B, Skoda V, Kubelka J, Hamel K and Kubecak V 1988 Czech. J. Phys. **B38** 237
Kvapil J, Koselja M, Kvapil J and Hamel K 1989 CLEO'89 Paper MC2 Baltimore
Petermann K, Danger T and Seedorf E 1991 Adv. Solid State Lasers to be published
Schirmer O F, Blazey K W, Berlinger W and Diehl R 1975 Phys. Rev. **B11** 4201
Wall K F and Aggarwal R L 1989 Int. Conf. Tunab. Sol. State Lasers Cape Cod ME3 -
Wegner T and Petermann K 1989 Appl. Phys. **B49** 275

Inst. Phys. Conf. Ser. No 126: Section IV
Paper presented at Int. Symp. on Ultrafast Processes in Spectroscopy, Bayreuth, 1991

Pulsed propagation of luminescence in anthracene

T Reinot, J Aaviksoo

Institute of Physics, Estonian Acad. Sci. , Riia 142, 202400 Tartu, Estonia

ABSTRACT: Time-resolved low-temperature luminescence spectra of thin crystal flakes of anthracene are studied. Depending on the distance between the excitation and the detection spots the luminescence kinetics has two maxima, the second one corresponding to luminescence polaritons propagating ballistically over several millimeters. The propagation time may exceed tenfold the previously reported polariton life-times in anthracene and depends on the emission wavelength, distance and orientation between crystal axes and the propagation direction. CCD camera imaging of the luminescent crystal allows one to distinguish the regions of diffusive and ballistical energy propagation.

1. INTRODUCTION

Polariton features of excitonic resonances have been found in a number of organic and semiconductor crystals (anthracene, naphtalene, CdSe, GaAs, CuCl etc.) (Rashba and Sturge 1982, Aaviksoo 1991, and refs. therein). The polariton luminescence has many specific features, amongst them are spatial effects: dependence of the spectrum on the dimensions of the crystal (Ferguson 1975) and the location of registration (Nishimura *et al* 1985), as well as the propagation of luminescence pulses through the crystal plate (Kuwata *et al* 1988). Differently from the latter work we have studied the propagation of luminescence pulses in the crystal plate plane and the distances envolved are several millimeters. Owing to the strong dipole transition the low-temperature luminescence of anthracene crystals is described in the polariton framework and the corresponding kinetic phenomena have been related to energy relaxation processes at the excitation spot (Aaviksoo *et al* 1987). In the present report the spatially resolved luminescence kinetics is shown to reflect the energy transfer in the crystal (Reinot and Aaviksoo 1991).

2. EXPERIMENTAL

The sublimation-grown anthracene monocrystal, which is a 5-50 μm thick flake in **ab**-plane, was immersed into liquid helium at T = 4.2 K. Mode-locked cavity-dumped Oxazine-750 dye laser was used as the luminescence excitation source. The frequency-doubled pulsetrain of ~1ps pulses, with the repetition rate of 4 MHz had the average power of ~0.5 mW. The excitation energy was ~1400 cm^{-1} above the exciton resonance at $\nu_{oo} = 25097$ cm^{-1}. Low-temperature time-resolved luminescence spectra of anthracene from different spatial locations of the crystal were studied in the forward scattering

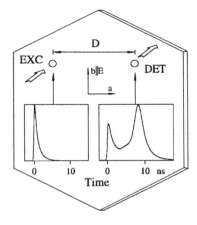

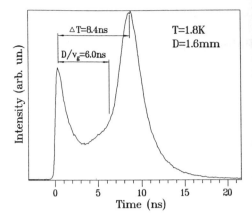

Figure 1. Geometry of the off-spot luminescence experiment with temporal resolution. Arrows EXC and DET indicate the excitation and detection directions, correspondingly. Distance D, detection wavelength λ and orientation of D is varied in the experiment.

Figure 2. Luminescence kinetics of the ν_{00} - 46 cm^{-1} line detected D = 1.6 mm away from the excitation spot. ΔT indicates the delay of the second luminescence pulse and D/v$_g$ marks the delay calculated from the group velocity of b-polarized polaritons at the detection wavelength.

geometry (Figure 1.). Time-correlated photon counting technique was used with resolutions $\Delta t = 200$ ps and $\Delta\nu = 2$ cm^{-1}. The spatial distribution of integral luminescence intensity was studied by means of a CCD camera, which had a dynamical range of ~300.

3. RESULTS AND DISCUSSION

Polariton luminescence decay kinetics was detected in the forward scattering off-spot geometry (Figure 1.). In a general case, these curves have two maxima, with delay ΔT between them. A characteristic kinetic curve of off-spot luminescence for the most intense line in the anthracene spectrum (ν_{00} - 46 cm^{-1}) is given in Figure 2. The first component has a fast rise time, $\tau_R < 50$ ps, and an almost exponential decay time $\tau_D = 2$ ns. As there is no dependence of the first component on distance D, its direction and orientation (within our temporal resolution), and the kinetics coincides with that of the excitation spot luminescence, this component is due to excitations which are created "instantly" at the detection spot, relax there through the bottleneck and contribute to the luminescence at $\nu_L < \nu_{00}$. The delay ΔT between the pulses can be described by the following formula: $\Delta T = T_0 + D/v_g(\nu_L,\alpha)$, where $v_g(\nu_L,\alpha)$ is polariton group velocity at the luminescence frequency ν_L and for the propagation angle α with respect the crystal b-axis.

The anisotropy of the luminescence kinetics with respect to crystal axis is due to different oscillator strenghts and resonance frequencies of the a- and b-polarized polaritons, which results in different group velocities, refraction indices and TIR angles (the calculated values for the most intensive lines in anthracene spectrum are given in Table 1.).

The dependencies of the time delay ΔT on D, α and ν_L suggest that the second maximum is due to polaritons, which, after relaxation through the bottleneck near the excitation spot, propagated afterwards ballistically to the detection spot at group velocities v$_g$ at the detection frequency ν_L, which is orders of magnitude less than in vacuum (see Table 1.).

Correspondence between the theoretical calculations proceeding from the standard polariton dispersion expression and experiment is good provided a constant initial delay $T_o = 2$ - 3 ns is subtracted from the observed delay of luminescence pulses. This delay reflects the formation period of the ballistic luminescence pulse during which the initially created polaritons (excitons) undergo several scattering/relaxation steps before the mean free path L of the final polariton is comparable with the crystal dimensions. This is the diffusive propagation regime. The temporal width of the second maximum contains

Frequency $\nu_0 - \nu$ (cm^{-1}) E ∥ b	Refraction index E ∥ b	Group velocity E ∥ b E ∥ a cm/s		TIR angle E ∥ b (deg)
2	21	$2.3 \cdot 10^5$	$6.1 \cdot 10^8$	2.7
22	6.5	$8.6 \cdot 10^6$	$7.1 \cdot 10^8$	8.9
46	4.6	$2.7 \cdot 10^7$	$8.4 \cdot 10^8$	12.6
22+46	3.9	$4.9 \cdot 10^7$	$9.6 \cdot 10^8$	14.9
394	2.2	$8.0 \cdot 10^8$	$3.4 \cdot 10^9$	27.0
1403	1.77	$6.3 \cdot 10^9$	$6.3 \cdot 10^9$	34.4

Table 1. Refraction indices, group velocities, and angles of total internal reflection (TIR) for the intense bands in anthracene luminescence spectrum.

the broadening due to the finite spatial extent of the diffusive propagation regime and different pathlengths of the quasi-waveguide polariton modes.

Due to small total internal reflection (TIR) angles $\theta_{TIR} = 2$-$35°$ for $\nu_L < \nu_{oo}$ - $1500\ cm^{-1}$ the majority of polaritons created near the excitation spot travels ballistically in quasiplanar waveguide modes along the crystal. These polaritons form the off-spot luminescence, they can escape the crystal if they meet crystal edges, scatter from some surface defect (impurity, phonon) and change their wavevector. The minor part of polaritons at the excitation spot, travelling at $\theta < \theta_{TIR}$, escape the crystal after a few reflections from the surfaces. This emission constitutes the excitation spot (in-spot) luminescence.

We have observed group delays as large as 20 ns. At the same time lifetimes measured from the excitation spot are ~ 1-3 ns (Aaviksoo *et al* 1987). Therefore the luminescence trapping phenomenon occurs in a flake crystal.

The CCD camera image of a luminescing anthracene flake is given in Figure 3. The black circle denotes the excitation spot with a diameter of .1 mm; the bright area around it shows the spatial distribution of time and frequency integrated luminescence. Let us note the strongly emitting crystal edges as well as emission due to some imperfections.

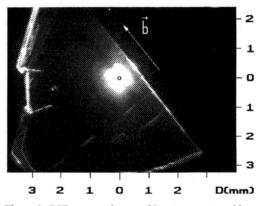

The total luminescence flux from the excitation spot constitutes around 3% of the total luminescence output from the crystal. This indicates the importance of ballistic (radiative) energy transfer in anthracene crystals.

The cross-section of the integral intensity distribution in **b**-axis direction is given in Figure 4.

Figure 3. CCD camera image of low-temperature (time and wavelength integrated) luminescence of a thin anthracene flake. The circle marks the excitation spot, the arrow shows crystallographic **b**-axis.

As can be seen, two different regions can be singled out. First, the Gaussian region, which is 2-3 times broader than the excitation spot, and second, inversely linear region at the edges of the distribution. We relate these regions, correspondingly, to diffusive and ballistic propagation regimes. From the above kinetic results we conclude that the diffusive stage lasts 2 - 3 ns and further ballistically propagating polaritons form the final distribution.

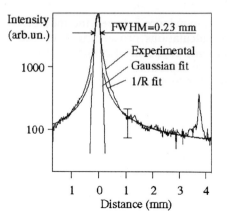

Figure 4. The time and wavelength integrated luminescence intensity cross-section in b-direction and corresponding fits. The errorbar notes the variations from crystal to crystal. FWHM of the excitation spot is .1 mm.

The cross-sections of the luminescence intensity in **a** and **b** directions almost coincide, the "butterfly" shape of the luminescence spot (Figure 3.), however, is observable in all crystals and arises from "excess" broadening in the directions 20-25° from the **b**-axis. This phenomenon, due to crystal anisotropy, is presently under study. The darker axis of the luminescence spot is always in the **b**-direction.

4. CONCLUSIONS

Polariton luminescence pulses of a few ns duration propagate over several millimeters through the anthracene flake. Taking into account the wavelenght, distance and direction dependencies we conclude that the second, delayed maximum in the off-spot luminescence kinetics is a clear evidence of ballistic propagation of luminescence pulses in the crystal. This pulsed off-spot luminescence is not a small effect; it prevails in the integral luminescence from a crystal flake and reflects the importance of the polariton effect in the anthracene crystal. The propagation of luminescence pulses is highly anisotropic, polariton group velocities along **a**-axis are hundreds of times slower than along **b**-axis and the structure in luminescence distribution reveals the crystal symmetry. Gaussian and inverse linear dependencies of the luminescence intensity distribution on distance allows one to distinguish between diffusive and ballistic energy transfer.

5. REFERENCES

Aaviksoo A, Freiberg A, Lippmaa J and Reinot T 1987 *J. Lumin.* **37** 313

Aaviksoo J 1991 *J. Lumin.* **48&49** 57

Ferguson J 1975 *Chem. Phys. Lett.* **36** 316

Kuwata M, Kuga T, Akigawa H, Hirano T and Matsuoka M 1988 *Phys. Rev. Lett.* **61** 1226

Nishimura H, Yamaoka T, Hattori K, Matsui A and Mizuno K 1985 *J. Phys. Soc. Japan* **54** 4370

Rashba E I and Sturge M D eds. 1982 *Excitons* (Amsterdam: North-Holland)

Reinot T and Aaviksoo J 1991 *J. Lumin.* **51** 00

Inst. Phys. Conf. Ser. No 126: Section IV
Paper presented at Int. Symp. on Ultrafast Processes in Spectroscopy, Bayreuth, 1991

Photoinjected charge carrier trapping and recombination in one-dimensional conducting polymers

Igor Zozulenko

Institute for Theoretical Physics, Academy of Sciences of Ukraine
Kiev, 252143, Ukraine

ABSTRACT: The effect of impurities on the recombination kinetics of photoinjected solitons in quasi-one-dimensional conducting polymers is studied. It is shown that soliton trapping by defects causes an acceleration of the photocurrent decay (in comparison with the conventional one-dimensional geminate recombination theory) and a deceleration of the photoinduced bleaching signal decay. The comparison of the results obtained with the experimental data concerning the photoexcitation dynamics in *trans*-polyacetylene is performed and restrictions of the model are discussed.

1.INTRODUCTION

The existence of the soliton-like excitations in conjugated polymers, in particular, in *trans*-polyacetylene, causes the number of novel electrooptical properties of these materials. In the pioneer work of Su and Shriffer (1980) it has been shown that the photoinjection of an *e-h* pair leads to its fast thermalization (of the order of the inverse optical phonon frequency, or 10^{-12} sec.) with the following conversion to the pair of charged topological defects, i.e. to the soliton-antisoliton $(S^+\text{-}S^-)$ pair. After thermalization solitons diffuse in polymer chains until they decay due to recombination or trapping on impurities.

One of the most powerful tool to study a soliton dynamics is a usage of a time-resolved technique in experiments on photoconductivity (PC) or photoinduced absorption spectroscopy (PA). Initial (picosecond) stages of soliton relaxation kinetics are shown by the numerous experiments to describe in the framework of a geminate recombination of a pair of quasiparticles diffusing in an ideal infinite chain, the quasiparticles decay obeying the $t^{-1/2}$ dependence. In the subnanosecond region an acceleration of the photocurrent decay (Phillips and Heeger 1988, Yu *et al.* 1990) and a deceleration of photoinduced bleaching (Vardeny *et al.* 1982, Vardeny *et al.* 1990) and reflectivity (Weidman and Fitchen 1987) were observed, the differences being associated with the charge carrier trapping by impurities. Indeed, being trapped, the charge carriers do not contribute to the overall current but they do contribute to the photoinduced absorption signal and the probability for a pair with one immobile carrier to recombine is obviously lower than that for a pair in which both carriers are mobile.

In our communication the soliton dynamics in polymer crystals with impurities playing the role of traps is investigated and the results obtained are associated with the observed characteristics of photoexcitations in conducting polymers - transient photocurrent and photoinduced changes in optical absorption.

2. MODEL

We assume the infinitely deep traps (impurities) to divide the polymer chains in segments where a soliton motion is an independent and is a strictly one-dimensional, the distribution of impurities in the chains being random. A light pulse creates $S^+ - S^-$ pairs, in the initial moment solitons being separated by the distance n_0; $n_0 \ll c^{-1}$, c is the impurity concentration. The position of solitons in a chain of length n at the moment t is described by the distribution function $g_n(x, y, t)$ obeying the diffusion equation

$$\frac{\partial g_n(x,y,t)}{\partial t} = W \frac{\partial^2 g_n(x,y,t)}{\partial x^2} W + \frac{\partial^2 g_n(x,y,t)}{\partial y^2}, \tag{1}$$

where x (y) is the coordinate of the first (second) soliton counted off from the left (right) end of a chain, W is the diffusion rate. The boundary conditions for Eq.(1) correspond to the soliton trapping on the chain ends $g_n(0, y, t) = g_n(x, 0, t) = 0$ and soliton recombination on the same cell $g_n(x, y, t) = 0$, $x + y = 0$. The details of the calculations of $g_n(x, y, t)$ are given by Zozulenko (1990a). Here we write down the survival probability of a recombining pair averaged over random distribution of the chain length which determines the soliton concentration in polymer

$$\rho(t) = c^2 \int_{n_0}^{\infty} dn\, n\, e^{-cn} \int_0^n dx \int_0^{n-x} dy\, g_n(x,y,t) \tag{2}$$

$$= \left(\frac{2}{\pi}\right)^3 c^2 \sum_{m=1}^{\infty} \sum_{l=1}^{\infty} \frac{(1-(-1)^l)(1+(-1)^m)}{(m^2-l^2)^2} \frac{m}{l} \int_{n_0}^{\infty} dn\, n$$

$$\times\ e^{-\frac{\pi^2}{n^2}(m^2+l^2)Wt - cn}(l \sin \pi m \frac{n_0}{n} + m \sin \pi l \frac{n_0}{n}) \tag{3}$$

$$= \operatorname{erf}\left(\frac{n_0}{2\sqrt{2Wt}}\right),\ \sqrt{Wt}\, c \ll 1, \tag{4}$$

$$= \exp\left(-\frac{3}{2}(10\pi^2 c^2 Wt)^{\frac{1}{3}}\right),\ 10\pi^2 c^2 Wt \gg 1. \tag{5}$$

3. PHOTOCONDUCTIVITY AND PHOTOINDUCED CHANGES IN OPTICAL ABSORPTION.

The transient photocurrent at a small bias field is proportional to $\rho(t)$, thus, Eq.(3) can be used to describe the photoconductivity decay caused by the soliton recombination in conjugated polymers with impurities. As one can see from Eq.(3), at small times, $c\sqrt{Wt} \ll 1$, the charge carrier concentration decay follows the power-law dependence which is characteristic for the case of geminate recombination in an infinite ideal chain; in other words, at these times the solitons do not "feel" the chain ends. The effects, connected with the restriction of soliton motion in segments between two traps manifest itself in the decay kinetics at times $c\sqrt{Wt} \leq 1$ ($W \sim 10^{12} \sec^{-1}$, $c \sim 10^{-3}$, therefore, $t \sim 1$ns, i.e. in subnanosecond region, as in the experiments of Phillips and Heeger (1988) and Yu et al. (1990)), at which the power asymptotics (4) is replaced by the faster exponential-type dependence (5) (see Fig.1).

Soliton decay kinetics in conducting polymers can be retraced also upon photoinduced changes in optical absorption. These changes are associated with the bleaching of the interband absorption or with changes in photoinduced absorption (or reflectivity) at so called "high energy peak" at ~ 1.4 eV. The intensity of PA signal is determined by the relation

$$IP_A(t) = \rho(t) + T(t), \tag{6}$$

where $\rho(t)$ corresponds to the part of charge carriers survived in the polymer chains (Eq.(3)), and $T(t)$ to immobile excitations trapped by impurities by the moment t. The expression for $T(t)$ has the form (details of calculation are given by Zozulenko (to be published))

$$T(t) = \left(\frac{2}{\pi}\right)^3 c^2 \sum_{m=1}^{\infty} \sum_{l=1}^{\infty} \frac{(1 - (-1)^l)(1 + (-1)^m)\, m}{(m^2 + l^2)(m^2 - l^2)} \frac{m}{l} \int_{n_0}^{\infty} dn\, n$$

$$\times \ e^{-\frac{\pi^2}{n^2}(m^2+l^2)Wt - cn} (l \sin \pi m \frac{n_0}{n} + m \sin \pi l \frac{n_0}{n}). \tag{7}$$

In Fig.2 the kinetic curves of PA signal are presented. The deviations from the power-law dependence $\propto t^{-1/2}$ are seen to happen on times $\sim (10\pi^2 c^2 W)^{-1} (\sim 100$ ps if $c=0.01$, $W = 10^{-12} \text{sec}^{-1})$ which is in agreement with the results of Vardeny *et al.* (1982) and Vardeny *et al.* (1989).

To conclude, we emphasize that the results obtained are applicable for a description of the initial subnanosecond stages of PC and PA kinetics when the concentration drop depth does not exceed two or three orders. With greater times, the effects produced by the violation of one-dimensionality of quasiparticle motion (interchain hopping), as well as by the recombination of quasiparticles on neighboring chains have to be taken into account.

References

Phillips S D and Heeger A D 1988 *Phys.Rev.* **B38** 6211
Su W P and Schrieffer J R 1980 *Proc.Natl.Acad.Sci.USA* **77** 5626
Vardeny Z, Strait J, Moses D, Chang T C and Heeger A G 1982 *Phys.Rev.Lett.* **49** 1657
Vardeny Z, Chang H T, Chen L and Leising G 1989 *Synth.Met.* **28** D167
Weidman D L and Fitchen D B 1987 *Synth.Met* **17** 355
Yu G, Phillips S D, Tomozava H and Heeger A J 1990 *Phys.Rev.* **B42** 3044
Zozulenko I V 1990a *Solid State Comm.* 1990 **76** 1035
Zozulenko I V b *Phys.Rev.B*, to be published

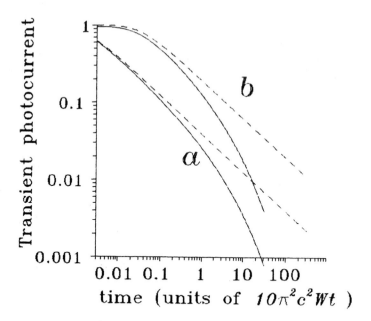

Figure 1: Photocurrent kinetics in one-dimensional polymer crystals with traps, Eq.(3) (solid lines). Dashed lines represent the recombination kinetics in an infinite chain without traps; $c = 0.01$, $n_o = 1$ (a), 5 (b)

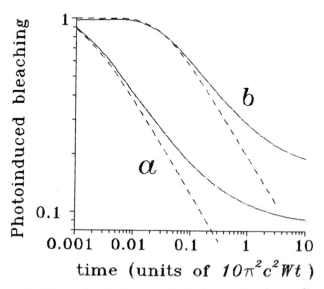

Figure 2: Photoinduced changes in optical absorption in one-dimensional polymer crystals with traps (see Eq.(6)) (solid lines). Dashed lines represent the recombination kinetics in an infinite chain without traps; $c = 0.01$, $n_o = 1$ (a), 5 (b)

Inst. Phys. Conf. Ser. No 126: Section IV
Paper presented at Int. Symp. on Ultrafast Processes in Spectroscopy, Bayreuth, 1991

Ultrafast relaxation of plasma oscillations in Ag islands studied by second-harmonic generation

D. Steinmüller-Nethl, R. A. Höpfel, E. Gornik*, A. Leitner **, F. R. Aussenegg**

Institut für Experimentalphysik, Universität Innsbruck, A-6020 Innsbruck, Austria
* Walter Schottky Institut, Technische Universität München, D-8046 Garching, Germany
** Institut für Experimentalphysik, Universität Graz, A-8010 Graz, Austria

ABSTRACT: The relaxation of surface plasma oscillation in Ag islands is investigated experimentally via the second-harmonic autocorrelation of femtosecond laser pulses. Decay times of 40 ± 7 fs were obtained.

1. INTRODUCTION

Several theoretical (Kawabata et al. 1966, Lushnikov et al. 1969, Genzel et al. 1975, Sarid 1981, Thoai et al. 1982) and experimental (Doremus 1964, Smithard 1973, Chen et al. 1976, Abe et al. 1979, Exeter et al. 1988) works concerning the determination of lifetimes and damping mechanisms of surface plasma oscillations have been reported. Collective oscillations of conduction electrons are either localized or nonlocalized surface plasmons (SP) (Agranovich 1982). Lifetimes of localized SP in small systems of Fermions have been obtained by the linewidth of the optical absorption spectra. The linewidth of plasmon resonances, however, do not yield unambiguous lifetimes of the SP since the nonuniformity of the particle sizes with different resonance frequencies gives rise to an inhomogeneous broadening of the absorption. In the literature the full width at half maximum (FWHM) varies between 0.1 to 1 eV for Ag depending on the particle size (Smithard 1973, Abe et al. 1979). Quantum-mechanical and semiclassical theoretical calculations for the lifetime of SP differ strongly from each other (Thoai et al. 1982 and references therein). Recently Exeter and Lagendijk (1988) were successful in measuring the short-range SP propagation in the prism configuration with a picosecond pump-and-probe technique via heat distribution. To our knowledge, until now, <u>no direct</u> observation of the decay of SP has been reported.

2. EXPERIMENT

The principle of our experiment is the following: A reference second-harmonic (SH) autocorrelation signal of femtosecond pulses was first generated with nonabsorbing KDP crystal, in which the SH generation occurs quasi instantaneously. Then the KDP was replaced by an Ag island film sample and a broadening of the generated SH autocorrelation signal on the island film in comparison to the autocorrelation of KDP was observed. The broadening is caused by the resonant excitation of electron plasma oscillation with finite lifetimes τ_{sp}.

The measurements were carried out at room temperature using a balanced colliding pulse mode locked dye laser (CPM) (λ = 610 nm, FWHM: 80 - 200 fs, average output power: 20 mW, repetition rate: 100 MHz). The laser beam was focussed to a diameter of 4 μm; the power density was in the order of $1 \cdot 10^{14}$ W/m². In the inset of Figure 1 the experimental setup is shown. The two divided laser beams were recombined, after temporal delay τ of one pulse relative to the other in steps of 8.3 fs, and focussed collinearly onto the surface of the sample using a microscope objective. Both beams were polarized parallel to the incident plane to obtain more efficient excitation of longitudinal surface plasma oscillations. The angle of excitation was chosen to be 45° in order to minimize the compensation of the induced depolarization field and to optimize the nonlinear polarization. To separate the fundamental reflected wave from the reflected SH signal we used a UG11 filter before entering the single 1/4 m spectrometer with slit width of 2 mm. The detector is a cooled photomultiplier with a GaAs photocathode. With the help of a polarization filter in the reflected beam the polarization of the 2ω-wave was determined to be p-polarized too.

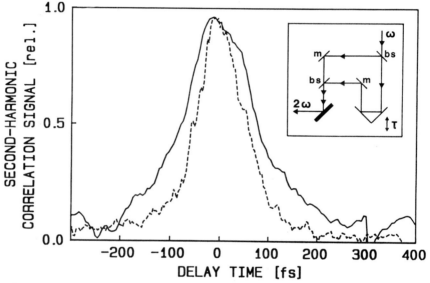

Fig.1:   Second-Harmonic autocorrelation signal of the laser pulses on an Ag island film (solid) and KDP (dashed). The inset shows a schematic of the experimental setup (m - mirror, bs - beam splitter).

The Ag island films were prepared by low rate evaporation (0.05 nm/s) on optically transparent indium-tin-oxide (ITO), which has a thickness of 200 nm, on silicate glass substrates and annealing afterwards (140° C, 2 min.) in Argon atmosphere. The typical diameter of the Ag islands is 20 nm, as determined by transmission electron microscopy. ITO was chosen as substrat since island films on this material produce a larger SH signal as on glass under the same conditions of excitation. The ITO substrat, however, does not show any detectable SH signal.

We use island films in order to prevent the influence of propagation effects. Island films illuminated in an ATR configuration show no narrow angle reflection minimum as it is observed with compact thin metal films. An estimation for the interaction of the particles which each other can be determined by the shifts and broadening of the plasmon resonance from p- and s-polarized light (Meier and Wokaun 1985). Using the parameters of our samples we get a values of about 20 nm. From both it can be concluded that the propagation can be neglected, the surface plasmons are *localized*. A spreading of the plasma wave packet due to the propagation does not have to be considered and the measured broadening of the autocorrelation signal obtained by SH generation gives direct information on the duration of the plasma oscillation.

In Figure 1 it is obvious that the width of the autocorrelation signal of Ag particle films (solid curve) obtained by SH generation is considerably broader than the width of the autocorrelation from KDP (dashed curve). As we will show later *the broadening gives direct information on the duration of the plasma oscillation since the SH signal is generated mainly by the localized plasmons and is present for the duration of the plasma oscillation and not only for the duration of the light pulse.* Measurements for various pulse widths of the CPM laser are shown in Figure 2, where the FWHM of the autocorrelation signals from silver islands are plotted in comparison to the FWHM of the autocorrelations from KDP. Again we can see that the FWHM of the Ag correlation signal is always broader than the KDP autocorrelation signal.

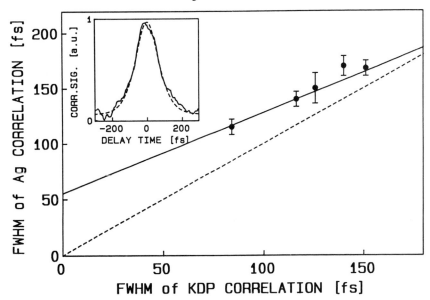

Fig. 2: FWHM of the autocorrelation signal measured on Ag islands versus FWHM of the autocorrelation signal of KDP. The solid line gives the ratio of the calculated FWHM of the island with τ_{sp} = 40 fs to the measured FWHM of KDP autocorrelation. The dashed line shows a 1:1 relation. Inset: Experimental correlation of silver particles (solid) in comparison with the calculated curve (dashed) with a lifetime τ_{sp} of 40 fs.

3. INTERPRETATION

The linear polarization $P(t,\omega)$ of one metal particle in the presence of an incident laser field $E_{in}(t,\omega)$ can be described by

$$P(t,\omega) = (\varepsilon_{met}(\omega) - \varepsilon_m) \{E_{in}(t,\omega) + E_{dep}(t,\omega)\} = (\varepsilon_{met}(\omega) - \varepsilon_m) E_{loc}(t,\omega)$$

where $E_{dep}(t,\omega)$ is the depolarization field arising from polarized matter in the particle and

$E_{loc}(t,\omega)$ is the total local electric field. $\varepsilon_{met}(\omega)$ is the complex Drude dielectric function and ε_m the dielectric function of the surrounding medium. $E_{dep}(t,\omega)$ depends on the geometry of the metal islands described with the ´depolarization factor´ L (Bohren and Huffman 1983). The total local field $|E_{loc}(t,\omega)|$ is much larger than the incident field $|E_{in}(t,\omega)|$. Therefore the total linear polarization can be well approximated by the plasma oscillation field $E_{dep}(t,\omega)$ alone. The field enhancement is expressed by

$$E_{loc}(t,\omega) = f(\omega) \, E_{in}(t,\omega) = [1 + \{\varepsilon_{met}(\omega) - \varepsilon_m\} \, L]^{-1} \, E_{in}(t,\omega) \approx E_{dep}(t,\omega)$$

with $f(\omega) \gg 1$. L depends on the ratio of the semiaxis of the metal ellipsoid and on $\varepsilon_{met}(\omega)$ (Bohren, Huffman 1983). The resonance condition for the single island is

$$1 + [\mathrm{Re}\{\varepsilon_{met}(\omega_{res})\} - \varepsilon_m] \, L = 0.$$

When the particle is excited at this frequency ω_{res} by a laser, it absorbs maximum power and we can describe the process as a resonantly driven damped harmonic oscillator on which our model of evaluation is based.

The depolarization factor L changes depending on the particle size and the dipole interaction. Accordingly the resonance frequency changes too. The statistic size distribution of the film limits the theoretical enhancement since only a small part of the surface, where the particle size corresponds to the resonance condition, gives a contribution to $f(\omega)$.

As explained the fields due to the presence of surface plasma resonances are dramatically enhanced leading to nonlinear effects e. g. second-harmonic generation and Raman scattering. The second-order polarization $P^{(2)}(t,2\omega)$ is proportional to the following expression $P^{(2)}(t,2\omega) \propto \chi^{(2)}_{eff,loc}(2\omega):E_{loc}(t,\omega)E_{loc}(t,\omega)$, where the local effective second-order susceptibility $\chi^{(2)}_{eff,loc}$ of the island-substrate-system takes into account the processes arising from bulk and surface effects. The presence of the surface and the inhomogeneity of the surrounding medium breaks the inversion symmetry of silver (Leitner 1990). Therefore $\chi^{(2)}_{eff,loc} \neq 0$, which allows the generation of the 2ω-wave. The intensity of the second-harmonic wave is proportional to $|P^{(2)}(t,2\omega)|^2 \propto |E_{loc}(t,\omega)|^4$. The frequency of the 2ω-wave is higher than the plasma frequency of silver. Thus we are not dealing with a 2ω-plasmon but with the sources of a free light wave at 2ω, which is emitted as long as the fundamental localized SP with frequency ω is excited. Since we have collinear excitation the intensity of the SH sum field is measured, given by the fourth power of $E_{loc}(t,\omega)$. Since in the experiment we average over the interference fringes of the second-order interferometric autocorrelation as described by Diels (1985) the measured signal corresponds to

$$G^{(2)}(\tau,\tau_{sp}) \propto \int_{-\infty}^{\infty} I_{loc}(t,\omega) I_{loc}(t+\tau,\omega) dt$$

which is equal to the intensity autocorrelation of $I_{loc}(t,\omega) = |E_{loc}(t,\omega)|^2$. In order to describe the time-dependence of $E_{loc}(t,\omega)$, driven by the external field, we describe the plasma oscillations by a damped harmonic oscillator model driven by $|E_{in}(t,\omega)| \propto$

sech^2 (1.76 t /Δ) with Δ being the FWHM of the laser pulse. The corresponding one-dimensional equation of motion for $x(\omega,t)$ with $P(\omega,t) = Nex(\omega,t)$ is solved numerically. In Figure 3(a) the normalized driving laser field and in Figure 3(b) the corresponding depolarization field $E_{dep}(t,\omega)$ of the surface plasmon in dependence of the time is shown.

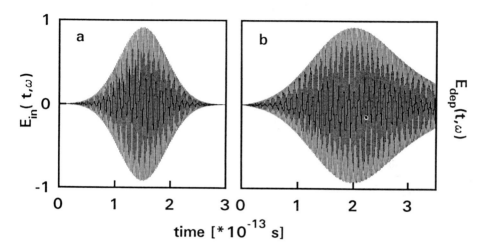

Fig. 3: (a): driving laser field $E_{in}(t,\omega)$ (b): corresponding depolarization field of the surface plasma oscillation.

With $I_{loc}(t,\omega)$ being proportional to $\int |P(t,\omega)|^2 \, dA$, where the integration is executed over the laser focus, we can fit our experimental correlation curves. *The lifetime of the plasma oscillation τ_{sp} is the only parameter used to fit the experimental data.*

4. RESULTS AND DISCUSSION

The inset in Figure 2 shows an experimental Ag correlation curve (solid) in comparison to the calculated ones (dashed). Best agreement is obtained for values of τ_{sp} = 40 fs. The error bars indicate the reproducibility of the experiment. From the whole set of experiments we obtain a value for the lifetime of the plasma excitation of 40 ± 7 fs, independent of the laser pulse width but depending on the laser frequency. We calculated the homogeneous linewidth of the electron plasma oscillation of a single silver particle (following Bohren and Huffman) using the dielectric function of bulk silver (Johnson and Christy 1972) and our particle size conditions. The homogeneous linewidth corresponds to a relaxation lifetime of 40 fs in best agreement with our experimentally measured value. In addition this result agrees well with the relaxation time of propagating SP indirectly measured by Exeter et al. (1988). In comparison with the halfwidth of absorption measurements giving ΔE between 0.1 and 1 eV (τ_{sp} between 1.8 and 18 fs). Theoretical lifetimes determined by Thoai et al. (1982) vary for the considered range of our particles (R = 10 nm) between 5.6 to 18 fs.

In addition we have performed the same experiments on different substrates, e.g. GaAs, giving the same result of 40 ± 7 fs. From this we conclude that the decay is due to intrinsic scattering mechanisms.

For large particles, where the diameter is in the order of the wavelength, retardation effects have to be taken into account. An estimation for the validity of the Rayleigh theory (electrostatic) (Kreibig et al. 1969) gives a value for the diameter (d) of an Ag sphere of d ≤ 25 nm. Since our particle size is typically 20 nm the broadening due to the size-effects and dipole interaction in comparison to the lifetime of the surface plasmons can be neglected.

5. CONCLUSION

We have shown the first direct observation of energy relaxation of surface plasma oscillations in small metal particles. The experiments were performed by comparing the autocorrelation of femtosecond laser pulses in nonabsorbing KDP with the autocorrelation of Ag islands using the nonlinearity of the surface plasmon polarization. We obtain a decay time of 40 ± 7 fs depending on the laser frequency, but independent of laser pulse widths. Since this method leads to a direct determination of lifetimes of localized SP further investigations for nonlocalized SP are feasible too. Taking into account the propagation of these nonlocalized SPs the time delayed laser beam must be displaced in space. In addition, information on coherence, phase relaxation and interaction of surface plasmons with each other can be obtained from similar experiments.

This work was supported by the "Fonds zur Förderung der wissenschaftlichen Forschung" (P7558) and by the "Gesellschaft für Mikroelektronik, Wien". We acknowledge cooperation with R. Christanell, S. Juen and K. F. Lamprecht at the femtosecond laser system and H. Brunner for sample preparation.

REFERENCES

Abe H, Schulze W, Tesche B, Chem. Phys. 47, 95 (1979)
Agranovich V M, Mills D L (ed.): *Surface Polaritons*, (North Holland Publishing Company, Amsterdam, New York, Oxford, 1982)
Bohren C G, Huffman D R (ed.): *Absorption and Scattering of Light by Small Particles*, (John Wiley & Sons, 1983)
Chen W P, Ritchie G, Burstein E, Phys. Rev. Lett. 37, 993 (1976)
Diels C-J M, Fontaine J J, McMichael I C, Simoni F, Appl. Opt. 24, 1270 (1985)
Doremus R H, J. Chem. Phys. 40, 2389 (1964)
Exeter E, Lagendijk A, Phys. Rev. Lett. 60, 49 (1988)
Genzel L, Martin T P, Kreibig U, Z. Phys. B21, 339 (1975)
Johson P B, Christy R W, Phys. Rev. B6, 4370 (1972)
Kawabata A, Kubo R, J. Phys. Soc. Japan 21, 1765 (1966)
Kreibig U, Fragstein C v, Z. Phys. 224, 307 (1969)
Leitner A, Mol. Phys. 70, 197 (1990
Lushnikov A A, Simonov A J, Z. Phys. 270, 17 (1974)
Meier M, Wokau A, J. opt. Soc. Am. B 2, 931 (1985)
Sarid D, Phys. Rev. Lett. 47, 1927 (1981)
Smithard M A, Solid State Commun. 13, 153 (1973)
Thoai D B, Ekhardt W, Solid State Commun. 41, 687 (1982)

Laser induced electron emission processes of an Au-surface irradiated by single picosecond pulses at $\lambda = 2.94$ μm in the intermediate region between multiphoton and tunneling effects

Cs Tóth[1], Gy Farkas[1], K L Vodopyanov[2]*

Central Research Institute for Physics, P.O.Box 49,
H-1125 Budapest, Hungary

General Physics Institute, Vavilov str. 38,
SU-117942 Moscow, USSR

ABSTRACT: Electron emission from the surface of a gold target irradiated by single picosecond pulses of an erbium laser was investigated. The laser intensity (5-120 GW/cm^2) corresponded to the intermediate interaction region between the pure multiphoton and optical tunnel effects, where the Keldysh-γ is in the range $1 < \gamma < 12 = n_0$. At $I_L < 80$ GW/cm^2 laser intensities, where the surface heating was excluded, with increasing I_L the slope of the measured logarithmic intensity dependence of the photocurrent decreases from the $n_0 = 12$ perturbative value down to $n \cong 5$. The experiment shows that the Keldysh-type theories are also valid in the case of photoeffect of metals.

1. INTRODUCTION

Photoelectron emission from metals at high light intensities (Farkas 1978, Farkas 1987) - similarly to the multiphoton ionization of atoms - manifests itself in two different forms distinguished by the Keldysh (or perturbation) parameter, γ (Keldysh 1964, Bunkin and Fedorov 1965). The Keldysh-parameter is defined as $\gamma = \omega (2mW)^{1/2}/eE$, where W denotes the depth of the potential well ("work function"). The first limiting case, $\gamma \gg 1$, for low light intensities and high frequencies represents the pure multiphoton mechanism, when the interaction may be considered perturbatively. Here the order of nonlinearity is $n_0 = [W/\hbar\omega + 1]_{int}$, which is equal to the minimum number of interacting photons required to produce one free electron. The other limit, $\gamma \ll 1$, is the optical tunneling, when the electrons escape from the potential well by quantum mechanical tunneling through the barrier, which is periodically broken by the oscillating electromagnetic field. This limiting case occurs for high laser intensities and low frequencies.

*Present address: Experimental Physik III, Univ. Bayreuth,
                  D-8580 Bayreuth, Germany

2. PREVIOUS EXPERIMENTS

Description of detailed experimental studies have been summarized by Farkas (1978, and references therein) for the pure multiphoton case and for the intermediate region ($\gamma \sim 10$) at $\lambda = 1.06$ μm, as well as for the tunneling case (CO_2 -laser, $\lambda = 10$ μm) by Farkas and Chin (1985). Vodopyanov *et al.* (1989) have reported on a preliminary experiment with regard to the intermediate region at $\lambda = 2.8$ μm, when the value of the Keldysh-parameter γ was 6. In this latter experiment the heating of the metal surface by laser radiation - caused by the accumulation of successive laser pulses of the applied mode-locked train - significantly influenced the electron emission processes and led to thermally assisted photoelectron emission.

3. NEW RESULTS IN THE TRANSITION REGION

In the present contribution we report on our new experimental results on laser induced electron emission of metal surfaces: an Au surface was investigated as a cathode using single, selected picosecond pulses of an actively mode-locked Er^{3+}-YAG laser at $\lambda = 2.94$ μm wavelength. The applied laser intensity (5-100 GW/cm^2) corresponded to the intermediate interaction region between the pure multiphoton and tunnel effects, where the decisive Keldysh-parameter, γ, was in the range $1 < \gamma < 12 = n_o$.

The main differences between the previous (Vodopyanov *et al.* 1989) and the present investigations are the following: (i), the use of single pulses instead of a laser pulse train to avoid cumulative heating; (ii), the application of grazing incidence, $\Theta \sim 89°$, to further decrease the possible heating effects; (iii), the laser intensities are $\sim 2-5$ times higher, in order to get closer to the $\gamma = 1$ value.

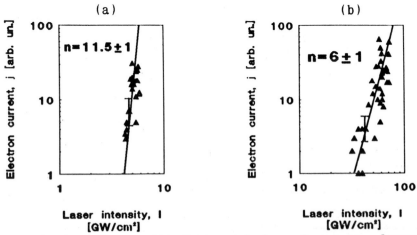

Fig.1   Photoemitted current dependence on laser intensity around (a): $I_1 = 5$ GW/cm^2, and (b): $I_2 = 63$ GW/cm^2, with the fitted logarithmic slope values.

When determining the experimental $n = \partial \log j / \partial \log I$ slope values of the laser intensity (I) dependences of the photoemitted current (j), at low laser intensities ($I_1 = 5$ GW/cm^2, $\gamma = 12$) we observed a pure perturbative multiphoton photoeffect with $n_0 = 11.5 \pm 1$ (see Figure 1a). In contrast, at higher laser intensities (above I=10 GW/cm^2) the observed slope value starts to decrease (Figure 1b). After reaching an $n \cong 5$ minimum at I~80 GW/cm^2, the slope value starts to increase again, reaching n=23 at 110 GW/cm^2 (see experimental points on Figure 2).

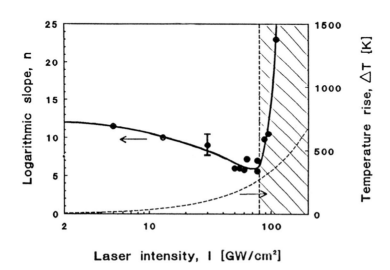

Laser intensity, I [GW/cm²]

Fig.2 Logarithmic slope values of light intensity dependences of photocurrent at various intensity regions. The circles are experimental points, the continuous line is a smooth fit to show the main trends. The shaded area represents the intensity region of the thermally assisted processes (see also dashed curve at the bottom-right part of the figure showing the increase of the surface temperature in the laser spot region).

4. DISCUSSIONS

If one estimates the increase in temperature of the surface under the effect of single laser pulses one can determine the light intensity region (I> 80 GW/cm^2) in which the thermally assisted processes are certainly dominant (the shaded region of Figure 2). On the other hand, comparison with the tunneling theories (Keldysh 1964, Bunkin and Fedorov 1965, Silin 1970) - fitting the experimental points with $\gamma' = \gamma/3$ and correspondingly with a W'=W/9 effective work function value (see Figure 3) - leads to the conclusion that the observed decrease of the slope values below 80 GW/cm² may be attributed to the dominant tunnel type character of the electron emission (Tóth *et al.* 1991). A similar feature of the ionization of atoms has recently been proved by Perry *et al.* (1988) and by Gibson *et al.* (1990) in the intermediate region (1<γ<4) of the Keldysh-γ. Further experimental efforts are necessary to study the γ<1 case for metals, where interesting new properties of the emitted electrons (angular and energy distributions, coherence properties) can be expected.

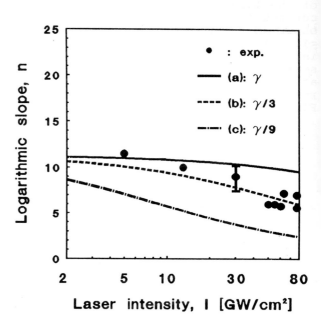

Fig.3   As for Fig. 2 without the thermally influenced laser intensity region above 80 GW/cm². The circles are experimental points, the fitted theoretical curves correspond to various γ values: (a) $\gamma=\gamma$;   (b) $\gamma'=\gamma/3$; (c) $\gamma''=\gamma/9$.

REFERENCES

Bunkin F V and Fedorov M V 1965   *Zh.Eksp.Teor.Fiz.* **48** 1341 [1965 *Sov.Phys.JETP* **21** 896]

Farkas Gy 1978 *Multiphoton Processes* ed J H Eberly and P Lambropoulos (New York: Wiley) pp 81-100

Farkas Gy and Chin S L 1985 *Appl.Phys.* B37 141

Farkas Gy 1987 *Photons and Continuum States of Atoms and Molecules* ed N K Rahman, G Guidotti and M Allegrini (Berlin: Springer) pp 36-46

Gibson G, Luk T S and Rhodes C K 1990 *Phys. Rev.* A41 5049

Keldysh L V 1964 *Zh.Eksp.Teor.Fiz.* **47** 1945 [1965 *Sov.Phys. JETP* **20** 1307]

Perry M D, Landen O L, Szöke A and Campbell E M, 1988 *Phys. Rev.* A37 747

Silin A P 1970 *Fiz. Tverd.Tela* **12** 3553

Tóth Cs, Farkas Gy and Vodopyanov K L 1991 *Appl.Phys.* B53 2163

Vodopyanov K L, Kulevskii L A, Tóth Cs, Farkas Gy and Horváth Z. Gy 1989 *Appl.Phys.* B48 485

Inst. Phys. Conf. Ser. No 126: Section V
Paper presented at Int. Symp. on Ultrafast Processes in Spectroscopy, Bayreuth, 1991

Ultrafast electronic and thermal processes in hydrogenated amorphous silicon

D. Hulin[1], P.M. Fauchet[2,] R. Vanderhaghen[3], A. Mourchid[1,3],
W.L. Nighan Jr.[4], J. Paye[1] and A. Antonetti[1]

1 - Laboratoire d'Optique Appliquée, ENSTA - Ecole Polytechnique, 91120 Palaiseau, France

2 - Laboratory for Laser Energetics, University of Rochester, Rochester NY 14623, USA

3 - Laboratoire de Physique des Interfaces et Couches Minces, Ecole Polytechnique, 91128 Palaiseau, France

4 - Department of Electrical Engineering, Princeton University, Princeton NJ, USA

ABSTRACT: The properties of free carriers photogenerated in the extended states of hydrogenated amorphous silicon have been investigated using femtosecond time-resolved spectroscopy. The carriers recombine non-radiatively in a time that can be as short as 1 ps. The carrier thermalization rate is found to be around 1 eV/ps. The decay time of "optic" phonons into "acoustic" phonons is less than 200 fs. In general, the characteristic times for all these processes are much shorter than in crystalline semiconductor.

I. INTRODUCTION

Hydrogenated amorphous silicon (a-Si:H) presents the general features of a semiconductor such as valence and conduction bands separated by an energy gap, with extended states in the bands. However, there are also localized states with a total number of the order of $10^{20} cm^{-3}$. These localized states specifically arise from disorder and are usually found at an energy below the bottom of the conduction band or above the top of the valence band. Alloying silicon with hydrogen has already removed most of the localized states in the middle of the gap.

In contrast to the case of crystalline semiconductors, the special problem one encounters with amorphous semiconductors is that the total number of localized states is very large (Tauc 1990). Thus, a carrier injected in the extended states will be trapped very quickly and a DC or even an AC electrical measurement does not provide much useful information on the free carrier. These difficulties can be overcome by studying a large density of photogenerated carriers on a time scale shorter than the trapping time.

Most of the optical investigations usually performed in crystalline materials are not possible in amorphous semiconductors, because momentum is not a good quantum number (is not conserved). Fortunately, we have been able to devise ways to obtain useful information on free carrier dynamics from femtosecond pump and probe experiments. In this paper, we present different experimental results obtained on electronic and thermal processes in hydrogenated amorphous silicon.

II. EXPERIMENTAL PROCEDURE

The samples are device-grade undoped hydrogenated amorphous silicon deposited as a thin film on fused silica and are prepared by RF glow discharge. All samples had a Tauc gap of 1.7 eV and a thickness between 0.25 and 0.50 μm to insure quasi-uniform excitation at 2 eV. All experiments reported here are performed at room temperature.

The experimental scheme follows the classical pump and probe scheme. The ultrashort optical pulses (less than 100 fs duration, 620 nm central wavelength) are generated from a colliding pulse mode-locked dye laser. Pulse amplification is achieved with a 20 Hz frequency-doubled Nd:YAG laser (Quantel) or a 6 kHz copper vapor laser. A white light continuum of the same duration is generated by focusing part of the beam into a water cell or an ethylene glycol jet used as a probe. The probe light wavelength is then selected using appropriate interference filters. The reflectivity and the transmission of the probe beam are measured simultaneously on a shot-to-shot basis or using phase-sensitive lock-in techniques. The real and imaginary parts of the refractive index are obtained by inversion of the Fabry-Perot formulae.

Some experiments required the use of a second pump pulse, intense and tuned in the near infrared. To achieve this, part of the probe beam was selected through an interferential filter and sent to an infrared-dye amplifier pumped by the Nd:YAG laser (Migus et al., 1985). Pulses of 200 fs duration and 50 μJ of energy were obtained at 870 nm.

III. PROPERTIES OF FREE CARRIERS

Electron-hole pairs are photoinjected with an excess energy of 0.3 eV with respect to the band gap. When measuring their optical properties just after excitation, the majority of carriers are still in the extended states. Furthermore, the energy distribution profile is mainly determined by the pump parameters; therefore, when we vary the pump intensity, it will remain identical (except for a multiplication factor), no matter the decay processes are. Lattice heating is also negligible. Thus Δn and Δk, the changes of the refractive index and the extinction coefficient ($\Delta k = \lambda \Delta \alpha / 4\pi$), recorded at $t=0^+$, are a direct measure of the free carrier susceptibility.

Usually, the carrier response to an electromagnetic field is well described by the Drude model. This model predicts that Δn and Δk vary linearly with the e-h pair density N. Their ratio is independent of N (if all the other parameters do not vary with the carrier density) and takes a very simple expression if we assume that the scattering time τ is identical for electrons and holes:

$$\Delta k / \Delta n = -1/\omega\tau$$

We have verified experimentally that both Δk and Δn measured at $t=0^+$ vary linearly with the carrier density and that their ratio is proportional to the probe wavelength (Mourchid 1990a). This ratio is free of any incertitude related to the determination of the carrier density. For a fixed density, Δk can be well fitted over our spectral range of investigation by the expression deduced from the Drude model.

The good agreement between the predictions of the Drude model and the experimental observations lead us to deduce an effective scattering time of the order of one femtosecond ($\tau = 0.7$ fs). This time is very short and is at the limit of the Drude model validity if τ is considered as an usual scattering time. In fact, we have shown that τ is independent of the carrier density and is therefore more related to the fluctuations of the lattice due to disorder. The values of the mobility in the extended states (6 cm^2/Vs) which can be inferred from the scattering time (Mourchid 1990a), are in the range of the lowest values used to explain conventional time-of-flight measurements.

IV. CARRIER RECOMBINATION

The absorption measured with a probe photon energy significantly smaller than the band gap is produced by intraband transitions. It is strongly related to the presence of free carriers and is therefore a good tool to study their evolution after excitation. At high carrier densities (10^{19} to 10^{21} cm^{-3}), we observe, on a picosecond time-scale, a rapid decrease of the probe absorption (Tanguy 1988).

The intraband photoinduced absorption is a product of the carrier density N and of the optical absorption cross-section σ:

$$\Delta k_{intraband}(t) = a \int N(t,E) \ \sigma(E) \ dE$$

The decrease of the intraband absorption can originate from the carrier disappearance, from a variation of σ due to trapping in localized states (Vardeny 1988) or both.

The strong reduction of the photoinduced absorption after a few picoseconds can be due either to the carrier recombination or to trapping in deep states where their optical absorption cross-section could be strongly reduced. In the first case, the energy released towards the lattice is 2 eV per e-h pair if the recombination is non-radiative and less if the recombination is radiative or if trapping occurs. We have observed that, in our experimental conditions, the radiative recombination is negligible. We have measured the energy received by the lattice (Mourchid 1990b) and found (1.85 ± 0.2) eV per disappearing e-h pair, a value very close to the initial 2 eV value. Therefore we can conclude that after a few tens of picoseconds, almost all carriers have recombined non-radiatively.

We define an effective lifetime τ_{eff} given by $d(\Delta k)/dt = -\Delta k/\tau_{eff}$. The determination of the slope of Δk at $t=0^+$ is a signature of free carrier disappearance only, since just after photogeneration, carriers are in the extended states and σ does not depends on the carrier density. Therefore τ_{eff} measures the recombination time of electrons and holes in the extended states. In contrast, the same measurement at later time can be affected also by a variation of σ.

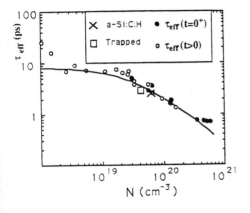

FIGURE 1
Carrier lifetime for various samples and densities for A-Si:H and alloys. Carriers can be free or in the tails

Figure 1 shows the measured τ_{eff} obtained as defined above, over one order of magnitude of N at $t=0^+$, and over more than two orders of magnitude at longer times. N is estimated from Δk. τ_{eff} exhibits a rather complex variation with N. It is almost constant for the highest densities, starts increasing with decreasing N and reaches a roughly constant value for low densities (as also seen by Esser et al., 1990) before increasing again. For intermediate densities, τ_{eff} is fitted by a $1/(A+BN)$ where $1/A=8$ps and $B=3\ 10^{-9}$cm^{3}s^{-1}. The curve represents this fit.

The same type of lifetime variation with Δk is observed when it is obtained from the slope of $\Delta k(t)$. This implies that the most important factor in the temporal evolution of the intraband absorption is related to the decrease of the carrier density and not to a strong change of σ when the carriers cool down in the band tail states (weakly localized states). In fact we have measured a change of σ by only a factor 2 for carriers directly created in traps by a 1.4 eV pump pulse. We can also note that, if the above statement were not correct, Δk at $t=0^+$ should not be linear with the pump intensity since, at high pump fluence, we saturate the number of carriers directly created in the localized states. We can conclude that σ (measured at 1 μm) for free carriers in the extended states or for trapped carriers near the band edge, is constant within less than 20%. The variation of Δk, for times of the order of a few picoseconds, is therefore representative of recombination and not of trapping.

Our experimental data can be fitted with the following model. The saturation value of τ_{eff} at very high carrier density excludes a recombination process between hot carriers. At least one carrier needs to be cold or trapped at the top of the band tail. There is a limited number of recombining states near the band edge, so that there is a minimum value of τ_{eff}, as observed (0.8 ps). We have searched for a possible delay between generation and recombination: the upper limit for this possible delay is less than 200 fs. Therefore, we can exclude a recombination via traps if the trapping time is larger than 200 fs. After photogeneration, the carrier thermalize and their recombination is then bimolecular. We attribute this bimolecular recombination to multiphonon mechanism. The recombination efficiency is the same for free carriers and for trapped carriers up to 100 meV below the band edge. It is reduced for carriers in deep traps.

V. ENERGY TRANSFER

Photogenerated carriers at high densities disappear rapidly through non-radiative recombination. Simultaneously, there is an increase of the interband absorption that can be seen by using probe wavelengths shorter than the energy gap (Tanguy 1988). In this case, the induced absorption shows a long-lived component, the magnitude of which increases for shorter probe wavelengths. This increase of the interband absorption is due to the band gap shrinkage produced by the heating of the lattice.

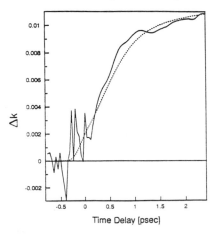

FIGURE 2
Lattice temperature increase measured through the band gap shrinkage (solid line) and energy released by the carriers (dotted line). The two curves have been normalized.

We compare the temporal behavior of the carrier non-radiative recombination and the lattice temperature rise. This gives us an estimate of the delay between optic phonon emission (pair recombination) and their decay into acoustic phonons (lattice heating). The carrier density is known through the induced intraband absorption measured with a probe in the spectral transparency region. The band gap shrinkage due to the lattice heating is obtained from $\Delta k(t)$ measured at 720 nm, after subtraction of the electronic part. This probe wavelength has been selected since the electronic contribution can still easily be characterized (zero delay and magnitude) although the thermal component has already a sizeable contribution. Figure 2 shows the band gap shrinkage (continuous line) and the calculated amount of energy released by the disappearing carriers (dashed line). This last curve is obtained assuming that carriers lose instantaneously 0.3 eV due to thermalization in the extended states and then 1.7 eV when they recombine. The two curves match quite well, indicating that the delay between carrier non-radiative recombination and lattice heating, is experimentally at most 200 fs.

It is unlikely that, on this short time scale, carriers emit directly long-range (acoustic) phonons. Thus this delay may be interpreted as the average time for an "optic" (localized) phonon to decay into "acoustic" phonons. This has to be contrasted with the situation in crystalline materials where a time of 7 ps has been measured in GaAs (Von der Linde 1980) for the decay of optic phonons into acoustic phonons. The major difference in amorphous materials is the absence of the momentum conservation rule; this reduces strongly the slowing down due to possible bottle-neck in the transformation of optic phonons in acoustic phonons in crystalline material.

VI. CARRIER THERMALIZATION

We have performed a different experiment in order to measure the carrier thermalization rate in the extended states of a-Si:H. It relies on the use of two pump beams in the otherwise classical pump-probe scheme. The first pump beam with a 2 eV photon energy creates a large density of electron-hole pairs through interband absorption. The second pump beam at 1.4 eV heats the previously created carriers by intraband absorption. The heated carriers release their added excess energy towards the lattice by phonon emission. By monitoring the resulting lattice temperature rise, we deduce the carrier energy relaxation rate.

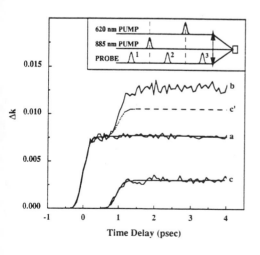

FIGURE 3
Photoinduced absorption of the probe beam at 520 nm with and without the different pumps: (a) the 2 eV pump pulse alone, (c) the 1.4 eV pump pulse alone, (b) the two pump pulses delayed by 1 ps, (c') the sum of curves (a) and (c).

Figure 3 shows the photoinduced absorption of a probe beam at 520 nm, with and without the different pumps. Curve (a) corresponds to the 2 eV pump pulse alone, curve (b) to the addition of the 1.4 eV pump pulse at a delay of 1 ps after the first excitation and curve (c) to the 1.4 eV pump alone. The signal in this last situation is due to the carriers which have been generated directly in the localized states and then participate also to the intraband absorption. Their contribution has to be taken into account when analyzing the effects of the infrared pump beam. We take advantage of this signal to obtain a full characterization of our infrared pump pulse, both in delay and in duration. Curve (c') represents the sum of curves (a) and (c).

A careful analysis relying on pump intensity variations and on experiments at other wavelengths gives a thermalization rate dE/dt around 1 eV/ps. Although there is some uncertainty on this result, it cannot be two times smaller and it cannot be much larger. Our experimental result shows that the carrier thermalization process in a-Si:H is very rapid, more rapid than in crystalline material where it is less than 0.5 eV/ps.

Our model is based on the assumption that the excess energy is shared by all the free carriers: part of the population is photoexcited by the IR pulse and collisions redistribute rapidly the energy before phonons can be emitted. This assumption is reasonable in view of the large density of photogenerated carriers. When we do not use it, we obtain a much more rapid thermalization rate since now carriers have to lose 1.4 eV in the same amount of time.

VII. CONCLUSION

Using femtosecond time-resolved spectroscopy, we have measured electronic and thermal processes in hydrogenated amorphous silicon; they all appear to occur on an ultrashort time-scale.

The effective scattering time is in the range of one femtosecond. Non-radiative recombination takes place in a few picoseconds for our densities of photoexcited carriers. This is interpreted as a bimolecular multiphonon recombination of carriers in the extended states or in the band tail states.

We have studied different thermalization processes and found that they are more rapid than in crystalline materials. The carrier-phonon and phonon-lattice interactions have an increased efficiency in amorphous materials, most likely because of the relaxation of the momentum conservation rule due to disorder.

One of the authors (P.M.F.) acknowledges support by ONR (NOOO14-91-J-1139) and NSF (ECS-9196000).

REFERENCES

Esser A, Seibert K, Kurz H, Parsons G N, Wang C, Davidson B N, Lucovski G and Nemanich R J 1991 *Phys. Rev. B 41 2879*

Mourchid A, Hulin D, Vanderhaghen R, Nighan W L Jr, Gzara K and Fauchet P M 1990a *Solid State Comm. 74 1197*

Mourchid A, Vanderhaghen R, Hulin D and Fauchet P M 1990b *Phys. Rev. B 42 7667*

Tanguy C, Hulin D, Mourchid A, Fauchet P M and Wagner S W 1988 *Appl. Phys. Lett. 53 880*

Tauc J and Vardeny Z 1990 *Critical Reviews in Solid States and Material Sciences 16 403*

Vardeny Z, Thomsen C and Tauc J 1988 *Solid State Comm. 65 601*

von der Linde D, Kuhl J and Klingenberg H 1980 *Phys. Rev. Lett. 44 1505*

Inst. Phys. Conf. Ser. No 126: Section V
Paper presented at Int. Symp. on Ultrafast Processes in Spectroscopy, Bayreuth, 1991

Femtosecond spectroscopic study of free carrier induced optical non-linearities in crystalline silicon

A. Esser, A. Ewertz, T. Zettler, W. Kütt, and H. Kurz
Intitute of Semiconductor Electronics, RWTH Aachen,
W-5100 Aachen, Germany

Abstract: Time resolved spectroscopy of optical nonlinearities in c–Silicon introduced by free carriers is performed with fs-time resolution. The results are interpreted in the framework of a Drude formalism as a consequence of the optical conductivity concept.

1) Introduction

Femtosecond spectroscopy is a standard pump/probe technique where reflectivity and transmission changes induced by an intense pump pulse are measured by a continuum probe pulse in a broad range of photon energies. In this study we apply femtosecond spectroscopy to probe optical nonlinearities induced by free carriers in crystalline Silicon. At an excitation energy of 2eV free carriers are generated via indirect transitions between the central valley of the valence band and the X-valley of the conduction band. In the following, we will treat the temporal and the spectral dependence of the free carrier induced optical nonlinearities, in the frame of optical conductivity concepts.

2) Experimental

Femtosecond time resolved spectroscopy is performed on a SOS(Silicon on Saphire)-film using a standard pump/probe set-up to measure induced reflectivity and transmission changes. The SOS-film is excited by 50fs-pulses from a mode locked ring-dye laser (CPM) amplified at a repetition rate of 7kHz. A part of the amplified pulse train is focussed into an ethylen glycol cell to generate chirped broadband continuum pulses for probing. Differential transmission and reflectivity changes between 1.2 and 1.9eV are detected seperately in an optical multichannel analyzer for a sequence of time delays between pump and probe beam. In order to compare the influence of the inherent experimental chirp, a numerical recalibration of transmission and reflectivity spectra is accomplished using the chirp data of the continuum pulse measured immediately before or after the actual experiment via two photon absorption in GaP [Albrecht et al., 1991]. Finally the measured differential reflectivity and transmission spectra are transformed into the corresponding change of the complex dielectric function $\Delta\varepsilon(\omega)$, using the standard thin film optical equations.

3) Results and Discussion

Fig.1 shows the induced changes in the real $(\Delta\varepsilon_1)$ and imaginary $(\Delta\varepsilon_2)$ part of the dielectric function in a 0.5µm thick SOS-film observed at two different time delays Δt, <u>after</u> the end of the excitation pulse. The optical changes are dominated by the response of free carriers, e.g., $\Delta\varepsilon_1$ and $\Delta\varepsilon_2$ are mainly determined by the the optical conductivity of free electrons and free holes. We fit the change of the dielectric function by model calculations with the Drude Ansatz using two frequency independent fit parameters τ_j and m^*, which are usually identified as a current relaxation time and the combined optical conductivity mass of electrons and holes, respectively [Grosse, 1979].

$$\Delta\varepsilon_1(\omega) = -\varepsilon_1 * \omega_p^2/(\omega^2 + \tau^{-2}) \quad ; \quad \Delta\varepsilon_2(\omega) = \varepsilon_1 * \omega_p^2/(\omega\tau*(\omega^2 + \tau^{-2}))$$

$$\text{with} \quad \omega_p^2 = N_{ex}*e^2/(\varepsilon_1\varepsilon_0 m^*)$$

(1)

The model calculations represented by the solid lines in fig.1 fit surprisingly well the experimental data. This result clearly demonstrates, that the optical response of free carriers is dominated by intraband transitions. Possible interband contributions due to the direct E_1 transition can be excluded [Kütt et al.,1990] .

However at each time delay a separate set of fitting parameters is required. To analyze the required temporal changes of the fit parameters in more detail, we applied the same fit-procedure on a set of experimental data taken for a sequence of time delays between 0.2ps and 2ps.

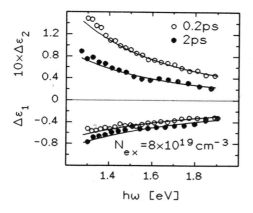

Fig.1 Free carrier induced change of the complex dielectric function for two different time delays after the end of the excitation pulse.

For m^* the results of this fit procedure is shown in fig.2. For the smallest time delay ($\Delta t = 0.2$ps), m^* starts at a value of about $0.17 \times m_0$. It then decreases monotonically with increasing time delay and levels off at $m^* = 0.14 \times m_0$ after 1ps. The relaxation time τ, however, increases with increasing time delay. Since any type of recombination or diffusion can be neglected in crystalline Silicon on a subpicosecond timescale [Kütt et al. , 1990], energy relaxation is argued as modifying these parameters in time. Extremely hot carriers are excited at an excess energy of 0.9eV across the indirect bandgap. These hot carriers loose their excess energy by carrier phonon scattering on a subpicosecond timescale and the carrier temperature approaches the lattice temperature. The optical conductivity mass drops from $0.17 m_0$ to $0.14 m_0$ during the whole process. According to Yang et al. [Yang et al., 1986], the temperature dependence of the effective mass can be explained by the nonparabolicity of the conduction band and the valence band. This temperature dependence can be approximated by:

$$m^*(T_C) = m^*(300K) + 2.85 * 10^{-5} K^{-1} * (T_C - 300K) * m_0 \qquad (2)$$

where T_C is the carrier temperature and m_0 is the free electron mass. The solid line in fig.2 describes the transient behaviour of m^* through equ.2 with $m^*(300K) = 0.14 m_0$ assuming a phenomenological energy relaxation time of $\tau_R \approx 200$fs. In detail, this corresponds to a carrier temperature of 1800K at 0.2ps time delay.

As demonstrated in fig.2, the relaxation time τ decreases from 2 fs at 0.2 ps time delay to 5 fs at 2 ps delay. For these values $\omega\tau \gg 1$ holds. Thus the relaxation time approximation is still valid. The transient behaviour of the relaxation time can be related to the temporal evolution of the carrier temperature. The temperature dependence of the scattering time can be understood within a simple carrier-carrier scattering approach. In this model, the scattering time is proportional to the thermal velocity of the carriers, which scales with the square root of the carrier temperature. The ratio of the relaxation times at 0.2ps and 2 ps matches exactly the square root ratio of the carrier temperatures, e.g. $(1800K/300K)^{1/2} \approx 5fs/2$fs. According to simple statistical considerations one would further expect an inverse linear dependence between the relaxation time and the carrier density. This behaviour is cosistent with our experimental findings.

The carrier-carrier scattering approach discussed so far is based on a single particle description of carrier motion. The energy of a single particle, which is scattered at an average rate τ^{-1}, can only be defined with an uncertainty ΔE given by $2\pi\Delta E * \tau \approx h$. The experimentally derived values for the parameter τ range between 2fs and 5fs, corresponding to an uncertainty of 0.3eV and 0.1eV, respectively. For these uncertainty values, the definition of a single particle energy within the bandstructure of crystalline Silicon becomes problematic.

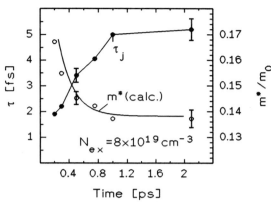

Fig.2 Temporal evolution of m^* and τ derived for the fitting procedure

According to Bohm and Pines [Pines et al. 1952] however, long range Coulomb interactions between carriers can not be neglected. The long range interactions lead to a collective motion of carriers, which can be viewed as a new degree of freedom with respect to the single particle motion. In addition to the energy spectrum of single particles a new spectrum of energies corresponding to plasma oscillations appears.

The extremely small values of τ are not related to single particle scattering. They describe the damping of plasma oscillations. In our case the scattering rate is always higher than the plasma frequency $(2\pi\omega_p\tau\approx0.25)$. Since the plasma frequency defines the Eigen-frequency of the collective motion, this situation corresponds to an overdamped collective motion of free carriers due to carrier-carrier collisions.

References

T.F. Albrecht, K. Seibert and H. Kurz, Optics Communications <u>84</u>, 223 (1991)

P. Grosse, "Freie Elektronen in Festkörpern" Springer-Verlag
Berlin Heidelberg New York, (1979)

W. Kütt, A. Esser, K. Seibert, U. Lemmer, and H. Kurz, Proceedings of ECO , Editor: A. Antonetti, SPIE <u>1268</u>, 154 (1990)

D. Pines and D. Bohm, Physical Review <u>85,2</u> 338 (1952)

G.Z. Yang and N. Bloembergen, IEEE Journal of Quantum Electronics <u>QE22</u>, 195 (1986).

Inst. Phys. Conf. Ser. No 126: Section V
Paper presented at Int. Symp. on Ultrafast Processes in Spectroscopy, Bayreuth, 1991

Picosecond infrared studies of hot holes in germanium

W. Kaiser, M. Woerner, and T. Elsaesser

Physik Department E 11, Technische Universität München,
D-8046 Garching, Federal Republic of Germany

Hot holes in p-type germanium are investigated via picosecond changes of the inter-valence band absorption in the mid-infrared. The transient absorption from the heavy to the light hole band studied in spectrally and temporally resolved measurements gives evidence of the subpicosecond carrier-carrier and inter-valence band scattering of the excited holes. These processes are followed by the cooling of heavy holes within several tens of picoseconds. Nonequilibrium split-off holes created by intense picosecond excitation give rise to a short-lived population inversion between the split-off and the light hole band. A transient gain of 5 cm^{-1} is observed around 260 meV. The amplification is mainly due to the high generation rate of split-off holes and the strong radiative transition probability.

The dynamics of hot charge carriers has been of interest to physicists and electronic engineers since the early fifties. After early transport studies with short electric pulses, recent investigations with ultrashort optical pulses have provided new insight in the interaction and cooling processes of hot carriers. In most optical experiments, hot electron-hole plasmas were generated by excitation of electrons across the band gap to the conduction band. The cooling of electrons and holes and the interaction between them is complicated to treat quantitatively. In this paper we report on experimental investigations of hot holes only.

The valence band structure of various semiconductors, e.g. silicon, germanium, and III-V compounds, consists of three bands : the heavy hole, the light hole, and the split-off band. Detailed parameters of the three bands (nonparabolicity and warping) are available for germanium (Kane 1956, Fawcett 1965). The investigations reported here are concerned with the transient inter-valence band absorption in p-type germanium (Woerner 1990, 1991). Spectrally and temporally resolved experiments with picosecond pulses in the mid-infrared give direct insight in the time evolution of hot hole distributions. The data are well suited for a quantitative comparison with theory, giving the strength of the optical deformation potential.

Our experimental system consists of a Nd:glass laser generating pulses of a duration of 4 ps which pump two traveling wave dye lasers tunable in the near infrared. Parametric difference frequency generation between the Nd-laser and the dye lasers in two nonlinear crystals of AgGaS$_2$ provides laser pulses of approximately 2 ps in the wavelength range between 3.5 μm (350 meV) and 12 μm (105 meV) (Elsaesser 1985). In this way, we are able to excite and probe the samples independently over an important range of the infrared spectrum. Measurements were made at sample temperatures between 10 K and 300 K. Germanium crystals with Ga-doping varying between 10^{16} and 7×10^{17} cm^{-3} were investigated.

In Fig. 1, the infrared absorption of p-type Ge is depicted (Kaiser 1953). The well known spectrum serves to illustrate the studies reported here. At long wavelengths, one sees a strong absorption which is due to transitions from the heavy to the light hole band (hh→lh). We note that at low carrier temperature the absorption decreases substantially. In the first part of this article, we discuss picosecond absorption changes of the hh→lh transition. In a second type of experiment, carriers are excited from the heavy to the split-off band (hh→so) and the transition from the light hole to the split-off band (lh→so) is monitored. We point to the fact that at low temperature the absorption of the lh→so transition disappears, i.e. the sample becomes transparent in this wavelength range.

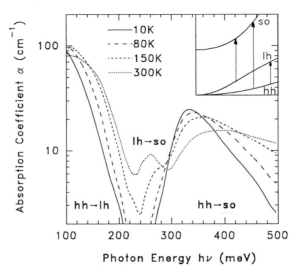

Fig. 1 Stationary absorption spectra of p-type Ge for four different lattice temperatures (hole density 7×10^{16} cm^{-3}). The valence band structure (E vs k) of Ge is depicted in the inset. The three inter-valence band transitions are indicated by arrows.

With intense pulses around 10 μm, holes are excited from the heavy hole to the light hole band. At carrier concentrations exceeding 10^{16} holes per cm^3, carrier-carrier and inter-valence band scattering occur on a subpicosecond time scale establishing a quasi equilibrium distribution of hot heavy holes within our time resolution of 1 ps. The elevated temperature of the hole distribution leads to a substantial change of the population density of the initial hole states and as a result to a strong variation of the hh→lh absorption. In Fig. 2, the hole absorption is plotted as a function of the hole temperature. There exists a distinct maximum around room temperature. Heating of holes which are initially at 300 K (B) leads to a reduced absorption (C) due to the smaller population of the initial states with rising temperature. The situation is completely different when starting at low temperature (A). The steep Fermi tail at low temperature, i.e. the carrier freeze out, gives rise to a smaller initial absorption but leads to larger absorption values as a result of carrier heating. For very high temperatures, the

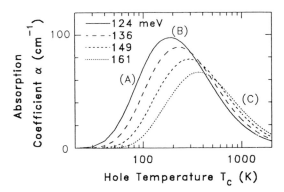

Fig. 2 Calculated inter-valence band absorption of the hh→lh transition for four different photon energies used in the time resolved experiments (hole density 7×10^{16} m^{-3}). The absorption coefficient α is plotted as a function of the carrier temperature T_c.

substantial thermal broadening of the hole distribution gives rise to an absorption even lower than the stationary value.

The predictions of Fig. 2 are born out by our experimental data. In Fig. 3 a, one readily sees the fast bleaching of a 300 K sample while at a lower lattice temperature of

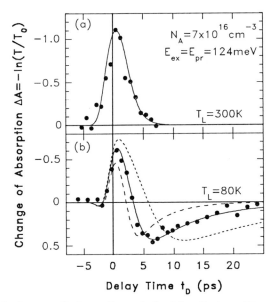

Fig. 3 Transient change of absorption of the hh→lh transition after excitation at $E_{ex}=E_{pr}=$ 124 meV. The absorption change $\Delta A = -\ln(T/T_0)$ is plotted vs the time delay between pump and probe (T_0, T : transmission before and after excitation). The data (points) were taken at lattice temperatures of (a) $T_L=$ 300 K and (b) 80 K (hole concentration 7×10^{16} cm^{-3}). The solid lines are the result of a model calculation with deformation potential $D_0=6.3\times10^8$ eV/cm. The short and long-dashed lines give the result for $D_0 = 5 \times 10^8$ and 9×10^8 eV/cm, respectively.

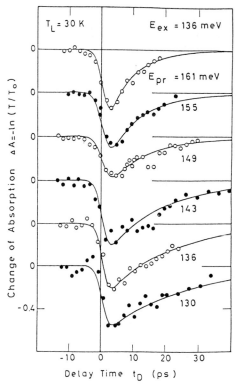

Fig. 4 Transient change of absorption of the hh→lh transition at six different probe frequencies 130 meV$\leq$ E$_{pr}$ $\leq$161 meV after excitation at E$_{ex}$= 136 meV (lattice temperature T$_L$ = 30 K). The absorption change is plotted vs the time delay between pump and probe pulses.

80 K the initial decrease of absorption is followed by a strong induced absorption (Fig. 3 b). This behavior directly reflects the time evolution of the hole temperature during carrier cooling. Cooling of the hot holes is studied in more detail in Fig. 4, where the enhanced absorption is monitored at six different photon energies. At higher photon energies one probes transitions starting from higher heavy hole states. These states are faster depopulated during cooling than the lower states as indicated by the different decay times in Fig. 4. The solid lines drawn through the experimental points of Figs. 3 and 4 are calculated for carrier cooling via LO-phonon emission (optical deformation potential scattering). The agreement between experiment and theory is excellent. The only adjustable parameter, the deformation potential of $D_0 = 6.3 \times 10^8$ eV/cm is in fair accord with values deduced from transport measurements (Reggiani 1976). The dashed lines in Fig. 3 b which deviate substantially from the data points give the time dependent absorption change calculated with values of $D_0 = 5 \times 10^8$ eV/cm and $D_0 = 9 \times 10^8$ eV/cm. Thus our data define the absolute value of the deformation potential with an accuracy of ±20 percent.

Next, transitions involving the split-off band will be discussed (Woerner 1991). On the right hand side of Fig. 5 part of the valence bands of Ge is depicted and the transitions for excitation and probing are indicated. In Fig. 6, the experimental data for pump photons of E$_{ex}$= 363 meV ($\simeq$ 3μm) and probing photons of E$_{pr}$= 263 meV ($\lambda \simeq$ 4.7μm)

re presented. During the excitation of the sample the transmission increases but de-
reases quite rapidly at the end of the pumping pulse. We recall that the sample is
ransparent at the probing wavelength prior to excitation. The enhanced transmission
.t early times corresponds amplification of the probing pulse. For $\Delta T/T_0 = 0.1$ and a
ample thickness of 0.02 cm one derives a substantial gain coefficient of 5 cm^{-1} from the
data. It should be noted that gain in this wavelength range has never been anticipated
r seen before.

The optical gain in Ge at 4.7 μm may be rationalized as follows. The transitions from
he split-off to the light hole band are favored by two facts : first, the dipole matrix
element is large (six times larger than for transitions between the heavy and light hole
bands) and, second, a pecularity of the band structure at our probing wavelength. As
een in Fig. 5, the split-off band and the light hole band have identical slope at E_{pr}
and thus the joint density of states has a singularity in this part of the Brillouin zone.
This singularity is lifted on account of the fast scattering processes, i.e. carrier-carrier
and LO-deformation potential scattering. The first scattering process is expected to
establish very rapidly (< 100 fs) a quasi-equilibrium distribution in the split-off band
with an approximate temperature of 700 K (see Fig. 5 left). Via the second scattering
mechanism the holes in the split-off band return very rapidly to the two lower valence
bands with a time constant of roughly 200 fs. The high excess energy of the light holes
gives rise to a higher temperature of approximately 1000 K which establishes a smaller
occupation probability in the terminal states of the so$\rightarrow$lh transition. In other words, for
a short time during pumping a small inversion (of approximately 0.01) between split-off
and light hole band is realized explaining the observed gain at 4.7 μm. Detailed

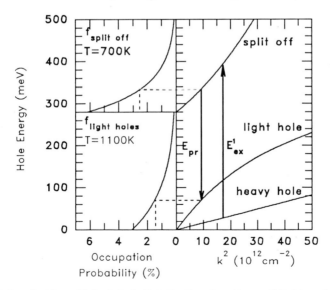

Fig. 5 Left hand side : Calculated distribution functions of light and split-off holes
during picosecond excitation at $E_{ex}^1 = 363$ meV. At delay zero, a transient population
inversion, i.e. optical gain, is found for the optically coupled states around 260 meV.
Right hand side : Valence band structure of Ge. The hole energy is plotted versus k^2
(k : wavevector). The arrows indicate the relevant optical transitions.

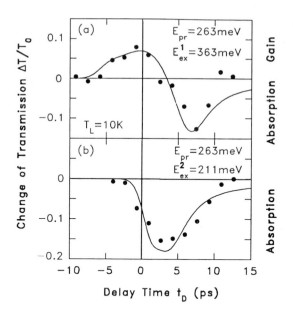

Fig. 6 Picosecond transmission changes of the lh→so transition at $E_{pr}=$ 263 meV after inter-valence band excitation. The transmission change is plotted as a function of delay time. (a) Heavy holes excited to the split-off band at $E_{ex}^1 =$363 meV give rise to optical gain. (b) Excitation of the hh→lh transition at $E_{ex}^2=$ 211 meV results exclusively in an absorption increase.

measurements showed that the gain exists only over a narrow spectral range of 20 meV around 263 meV. When the split-off band is depleted of excited holes, the transient population of the light hole band leads to an absorption at 263 meV. The latter is clearly seen in Fig. 6, several picoseconds after delay zero. Cooling of the hot hole distribution reestablishes the original transparency of the sample in this wavelength range.

As pointed out at the beginning of this article, a series of semiconductors exhibits a similar structure of the valence band. Experimental studies of the type reported here are certainly not restricted to germanium. In fact, recent studies of p-type GaAs gave detailed information on the relaxation of hot holes in this III-V compound. These results will be published elsewhere.

References

Elsaesser T., Lobentanzer H., Seilmeier A. 1985 *Opt. Commun.* **53** 355
Fawcett W. 1965 *Proc. Phys. Soc.* **85** 931
Kaiser W., Collins R.J., Fan H.Y. 1953 *Phys. Rev.* **91** 1380
Kane E.O. 1956 *J. Phys. Chem. Solids* **1** 82
Reggiani L. 1976 *J. Phys. Chem. Solids* **37** 293
Woerner M., Elsaesser T., Kaiser W. 1990 *Phys. Rev.* **B 41** 5463
Woerner M., Elsaesser T., Kaiser W. 1991 *Appl. Phys. Lett.* **59** 2004

Inst. Phys. Conf. Ser. No 126: Section V
Paper presented at Int. Symp. on Ultrafast Processes in Spectroscopy, Bayreuth, 1991

Nonthermalized electron distribution at low density in GaAs

D.W. Snoke, W.W. Rühle, Y.-C. Lu, and E. Bauser

Max-Planck-Institut für Festkörperforschung, 7000 Stuttgart 80, Germany

ABSTRACT: We have observed the energy distribution of electrons in GaAs on a time scale of 10 ps following near-resonant creation by short (5 ps) laser pulses. A streak camera is used to detect the band-to-acceptor luminescence in GaAs samples with low p-type germanium doping. At excitation densities above $10^{14}/cm^3$ the electron energy distribution is a Maxwell-Boltzmann at all times, but at lower density the initial distribution is a non-Maxwellian broad peak. This peak evolves as time progresses until it becomes a Maxwell-Boltzmann distribution. We model the thermalization via electron-electron Coulomb scattering and find that the timescale for thermalization is consistent with our data.

1.INTRODUCTION

Much recent work has been aimed at observation of nonthermal electrons in GaAs. Transients in the differential absorption spectrum associated with nonthermalized carriers have been observed in bulk GaAs and GaAs quantum wells (Oudar 1985, Knox 1988) and in AlGaAs, (Bradley 1989, Nunnenkamp 1991) on femtosecond timescales, although the electron distribution had to be deduced indirectly. Elsaesser et al. (Elsaesser 1991) have observed the carrier energy distribution via luminescence up-conversion on timescales of 100 fs, but the high densities necessary for luminescence up-conversion led to the observation of a thermalized distribution at all times.

Several authors have used the electron-acceptor recombination process in p-type GaAs, however, to observe the conduction-electron energy distribution in a manner independent of carrier density in the low density regime (e.g. Mirlin 1980, Fasol 1986, Ulbrich 1989). Because an acceptor-bound hole is very localized in space, it spans a wide range of k-states, allowing it to recombine with conduction electrons having nearly any momentum, with a well-defined matrix element (Dumke 1963).

We report direct observation of the conduction electron kinetic-energy distribution on timescales of 10 ps using a streak camera to observe the electron-acceptor luminescence in p-type GaAs. We see a clear transition between Maxwellian statistics at high density and non-Maxwellian statistics at low carrier density at early times.

2. EXPERIMENTAL RESULTS

We use p-type GaAs:Ge samples grown by liquid-phase epitaxy on semi-insulating GaAs with acceptor density in the range of $2-8 \cdot 10^{16}$ /cm³. Both single epitaxial layers of GaAs:Ge and $Al_7Ga_3As/GaAs : Ge/Al_7Ga_3As$ heterostructures with 2 μm GaAs layers are used, with the same results. Carriers are generated near the 1.519 eV band gap in the Γ-valley using a synchronously pumped dye laser with Styryl-8 or Styryl-9 dye, with pulse full width at half maximum of roughly 5 ps and repetition rate of 80

MHz. The time and spectral resolution of the streak camera measurements are 15 ps and 1 meV, respectively.

Figure 1 shows typical spectra of the band-to-acceptor luminescence at two delays relative to the maximum of an exciting laser pulse at E=1.536 eV (17 meV excess energy) using two different laser powers, on GaAs:Ge with doping density $5 \cdot 10^{16}$ /cm³ in helium vapor at roughly 10 K. The cw component of the luminescence due to donor-acceptor recombination with very long lifetime has been subtracted. In the high power case, the initial carrier density is estimated as $4 \cdot 10^{14}$ /cm³. In the low-power case, initial density is ten times less, or roughly $4 \cdot 10^{13}$ /cm³.

As seen in Figure 1(a), at the earliest times the electron distribution at low laser power (solid line) is much broader than the electron distribution at higher power (dotted line). The electron distribution in the high-power excitation case fits a Maxwell-Boltzmann distribution already by t=10 ps, while in the low-density case, the initial distribution is clearly non-Maxwellian. At times greater than 50 ps, both the high-power and the low-power spectra fit Maxwell-Boltzmann distributions with the same density of states D(E), as seen in Fig. 1(b). The non-thermalized distribution at low density evolves into a Maxwellian distribution.

Fig. 1. Time-resolved energy distribution of the conduction electrons in p-type GaAs:Ge in helium vapor at 10 K, as seen in the band-to-acceptor luminescence following a 5-ps laser pulse at photon energy of 1.536 eV, for two laser pulse powers at two different delays with respect to the laser pulse. Dots: approximately $4 \cdot 10^{14}$ /cm³ pair density. Solid line: ten times lower density. Dashed lines: fits to the Maxwell-Boltzmann distribution, convolved with a homogenous broadening function with full width at half maximum of 2 meV. a) T=53 K Maxwell-Boltzmann, fit to the *high*-density distribution. b) T=50 K Maxwell-Boltzmann, fit to the *low*-density data.

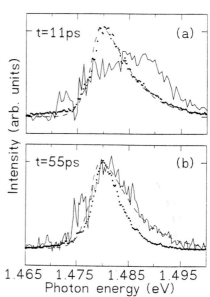

The Maxwell-Boltzmann fits to the spectra are given by

$$N(E) = e^{-\frac{E}{k_B T}} D(E). \tag{1}$$

Here D(E) is the density of states, proportional to $(E - E_0)^{\frac{1}{2}}$, where E_0, the gap energy minus the acceptor binding energy, corresponds to the point of zero kinetic energy for the electrons. The distribution N(E) of Eq. 1 is also multiplied by the matrix element for electron-acceptor recombination,(Dumke 1963) which is nearly constant over this energy range.

Although the distribution at early times in the case of low laser power is non-Maxwellian, it is still quite broad compared to the spectral width of the laser light.

A narrower distribution is never seen. Therefore even within the time period of our time resolution, substantial scattering of the electrons must occur. This scattering rate is independent of density, however; we see the same broad distribution at densities as low as $4 \cdot 10^{12}$ /cm^3. We find that the rate of *inelastic* scattering processes for the electrons such as electron-phonon scattering or electron-donor scattering occur much too slowly to provide the broad distribution seen within 10 ps. *Elastic* scattering, which occurs much more quickly, will not broaden the distribution. In modeling to be presented elsewhere,(Snoke 1991) we have found that assuming that the electron scatters primarily at short range with the hole created by the same photon, which is possible in the case of rapid momentum dephasing, gives good fits to the low-density distribution at early time.

The transition at higher carrier density to a Maxwell-Boltzmann distribution at early time comes about via electron-electron scattering. Since the early-time distribution is non-Maxwellian at slightly lower densities, we can infer that electron-electron scattering establishes the Maxwell-Boltzmann distribution at a density of $4 \cdot 10^{14}$ /cm^3 within approximately 15 ps and not faster.

This result is consistent with the results of Elsaesser et al.,(Elsaesser 1991) who found for a carrier density 500 times higher than in this experiment that a significant degree of thermalization of electrons in bulk GaAs occurred in less than 100 fs. From our measurements, linear scaling of the scattering rate with density would imply that a Maxwell-Boltzmann distribution would be reached within a time of 20 fs at density of $2 \cdot 10^{17}$ /cm^3, with significant thermalization on timescales even less than that. Electron screening in fact results in a sublinear dependence on density above 10^{16} /cm^3, but the broad distibution at early time observed by Elsaesser et al. is certainly not surprising.

3. MODEL OF ELECTRON-ELECTRON SCATTERING

The primary experimental result of this work is that electron-electron scattering causes the electrons to obtain a Maxwell-Boltzmann distribution within about 15 ps at a density of $4 \cdot 10^{14}$ /cm^3, and not faster. Can we see this thermalization time as arising from the simple Coulomb scattering of the electrons?

Previous work (Snoke 1989) has shown that two-body scattering of a fermion or boson gas far from equilibrium can be modeled via a numerical solution of the Boltzmann equation for scattering rate. We have performed numerical simulations of the scattering of the electrons for our experimental conditions. In essence, we simply calculate the rate of scattering given by Fermi's Golden Rule over all possible two-body scattering events in momentum space. The distribution $N(E)$ is then updated for some small time step dt based on the calculated $dN(E)/dt$. Details will presented elsewhere (Snoke 1991). The matrix element for the carrier-carrier scattering at low density is the screened Coulomb-potential matrix element, (Ridley 1988)

$$M = \frac{e^2}{\epsilon V} \frac{4\pi}{(\underline{k}^2 + q_0^2)}. \tag{2}$$

Here ϵ is the optical dielectric constant, and q_0 is a screening parameter.

Fig. 2 shows the calculated evolution of a distribution of electrons interacting via a Coulomb potential, with very low q_0. The timescale on the figure gives the calculated elapsed time in picoseconds for electrons at a density of $4 \cdot 10^{14}$ /cm^3 and average energy of 17.5 meV, the experimental conditions at high density of Figure 1.

For q_0 much less than the thermal wavelength, the thermalization depends only very weakly of the exact value of q_0 used. We find that for $q_0 = 0.2 \cdot 10^5$ /cm^{-1} the gas distribution evolves to within 10% root-mean-squared deviation from a Maxwell-Boltzmann distribution within 20 ps.

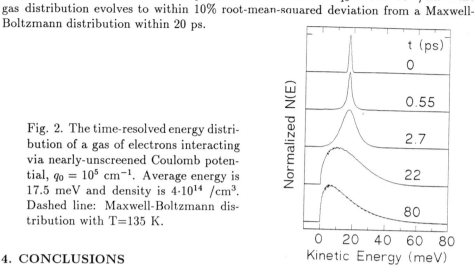

Fig. 2. The time-resolved energy distribution of a gas of electrons interacting via nearly-unscreened Coulomb potential, $q_0 = 10^5$ cm^{-1}. Average energy is 17.5 meV and density is $4 \cdot 10^{14}$ /cm^3. Dashed line: Maxwell-Boltzmann distribution with T=135 K.

4. CONCLUSIONS

Previous experiments have concentrated on very short-timescale observations at relatively high density (10^{17} /cm^3). Necessary assumptions about screening in each case have made the results difficult to interpret. In these experiments, screening is extremely weak and therefore these results indicate the "bare" scattering rate of electrons in the regime where scattering rate is linear with density. We have found that the timescale for thermalization is in very good agreement with a model for weakly-screened Coulomb scattering of electrons.

5. REFERENCES

C.W.W. Bradley, R.A. Taylor, and J.F. Ryan, Sol. State Electronics **32**, No. 12, 1173 (1989).

W.P. Dumke, Phys. Rev. **132**, 1998 (1963).

T. Elsaesser, J. Shah, L. Rota, and P. Lugli, Phys. Rev. Lett. **66**, 1757 (1991).

G. Fasol and H.P. Hughes, Phys. Rev. B **33**, 2953 (1986).

W.H. Knox, D.S. Chemla, G. Livescu, J.E. Cunningham, and J.E. Henry, Phys. Rev. Lett. **61**, 1290 (1988).

D.N. Mirlin, I. Ya. Karlik, L.P. Nikitin, I.I. Reshina, and V.F. Sapega, Sol. State Comm. **37**, 757 (1980).

J. Nunnenkamp, J.H. Collet, J. Klebniczki, J. Kuhl, and K. Ploog, Phys. Rev. B **43**, 14047 (1991).

J.L. Oudar, D. Hulin, A. Migus, A. Antonetti, and F. Alexandre, Phys. Rev. Lett. **55**, 2074 (1985).

B.K. Ridley, **Quantum Processes in Semiconductors** (Clarendon, Oxford, 1988).

D.W. Snoke and J.P. Wolfe, Phys. Rev. B **39**, 4030 (1989).

D.W. Snoke, W.W. Rühle, Y.-C. Lu, and E. Bauser, to be published (1991).

R.G. Ulbrich, J.A. Kash, and J.C. Tsang, Phys. Rev. Lett. *62*, 949 (1989).

Inst. Phys. Conf. Ser. No 126: Section V
Paper presented at Int. Symp. on Ultrafast Processes in Spectroscopy, Bayreuth, 1991

Carrier thermalization in GaAs and InP studied by femtosecond luminescence spectroscopy

T. Elsaesser[1,2], J. Shah[1], L. Rota[3], and P. Lugli[4]

[1] AT&T Bell Laboratories, Holmdel, NJ 07733, USA
[2] Permanent address : Physik Dept. E 11, Technische Universität München, D-8046 Garching, Fed. Rep. Germany
[3] Dipartimento di Fisica, Università degli Studi di Modena, I-41100 Modena, Italy
[4] Dipartimento di Ingegneria Meccanica, II. Università degli Studi di Roma, I-00173 Roma, Italy

Thermalization of a nonequilibrium distribution of electrons and holes generated by femtosecond photoexcitation is studied in GaAs and InP via spectrally and temporally resolved luminescence. In both materials, a rapid onset of luminescence is observed over a broad spectral range from the bandgap up to 1.7 eV. The data demonstrate the redistribution of both electrons and holes over a wide energy range within 100 fs, even for excitation densities as low as 10^{17} cm^{-3}. Equilibration is is dominated by carrier-carrier collisions with scattering rates higher than predicted by theoretical simulations using static screening of the interaction potential. Our Monte-Carlo calculations including dynamical screening via a molecular dynamics scheme account for the experimental results.

The nonequilibrium properties of electrons and holes in semiconductors are important for understanding the fundamental scattering processes as well as for device applications. Optical spectroscopy with femtosecond pulses gives direct experimental access to the carrier dynamics (Erskine 1984, Shah 1987, Lin 1988, Zhou 1989,1991), whereas Monte Carlo simulations allow a theoretical modelling of the relevant scattering events (Stanton 1990, Lugli 1989, Zhou 1990). In polar semiconductors like GaAs and InP, non-thermal carrier distributions created by femtosecond photoexcitation equilibrate on a subpicosecond time scale to hot Fermi distributions by carrier-carrier and longitudinal-optical (LO) phonon scattering. For excitation with pulses around 2 eV, the initial distribution of electrons consists of three narrow peaks generated by transitions from the heavy hole, light hole, and split-off valence bands. Equilibration of this distribution has mainly been studied by monitoring transient changes of interband **absorption** (Erskine 1984, Lin 1988). Electrons and holes in different regions of k-space contribute to bleaching at a specific probe wavelength and - consequently - unambiguous information is difficult to extract from the data. In contrast, the **luminescence** spectrum of GaAs and InP at photon energies below 1.7 eV is dominated by recombination of electrons with heavy holes, requiring the presence of both electrons and holes at the same k-vector. Thus luminescence spectroscopy gives new independent information on carrier thermalization (Elsaesser 1991). In this paper, we present luminescence studies of carrier thermalization in GaAs and InP with a time resolution of 100 fs. The time evolution of the luminescence spectra of GaAs and InP within the first picosecond after excitation at 1.93 eV is monitored for electron-hole densities between 1.7×10^{17} and 7×10^{17} cm^{-3}. The broad structureless emission spectra at very early delay times demonstrate the redistribution of carriers over a wide energy range within 100 fs.

Thin GaAs and InP layers grown by molecular beam epitaxy were excited by 120 fs pulses from a dye laser emitting at 1.93 eV. The sum frequency of the luminescence and of a second laser pulse was generated in a 0.5 mm thick LiIO$_3$ crystal, spectrally dispersed in a double monochromator (spectral resolution 5 meV) and detected by single photon counting. The time resolution of the experiment is approximately 100 fs.

Luminescence spectra of GaAs and InP measured at very early delay times are presented in Figures 1 and 2, respectively. For GaAs, the excitation densities were approximately 7×10^{17} cm^{-3} (solid lines) and 1.7×10^{17} cm^{-3} (dashed lines). In InP, the carrier density was 5×10^{17} cm^{-3}. The distinct temporal evolution of the spectra in Figures 1 and 2 demonstrates the high temporal resolution of our experiments. At all time delays, luminescence is emitted over a wide range of energies from the respective band gap (GaAs : 1.42 eV, InP : 1.34 eV) up to 1.7 eV. This region is considerably broader than

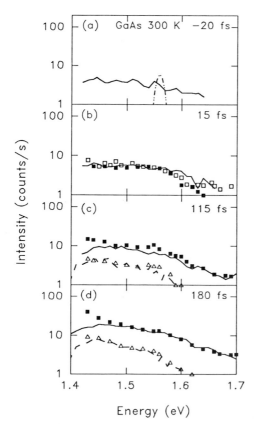

Fig. 1   Luminescence spectra of GaAs after femtosecond excitation at 1.93 eV. The excitation density was 7×10^{17} cm^{-3} (solid lines) and 1.7×10^{17} cm^{-3} (dashed lines). The emission intensity is plotted as a function of photon energy for delay times of (a) -20 fs, (b) 15 fs, (c) 115 fs, and (d) 180 fs. The solid squares and the triangles represent the spectra calculated from Monte Carlo simulations using static screening. The open squares in (b) give give the result from the molecular dynamics model.

the three individual peaks of the initial nonequilibrium distribution. As depicted in Fig. 1 (a) (dash-dotted line), luminescence due to non-thermalized electrons excited from the split-off band in GaAs would be centered at 1.57 eV with a spectral width of roughly 20 meV. Note that broad structureless emission spectra are also observed with the lower carrier concentration of 1.7×10^{17} cm^{-3}.

The onset of luminescence was studied for a number of fixed photon energies between the bandgap and 1.6 eV (Elsaesser 1991). In all cases, the emission intensity exhibits a first femtosecond rise within the time resolution of the experiment, giving direct information on carrier thermalization. This fast kinetics is followed by a picosecond rise of intensity. In GaAs, the latter contribution is due to the slow backscattering of electrons from the L and X valleys to the Γ valley and subsequent cooling, resulting in an accumulation of electrons in states close to the bandgap. In InP, no transfer to the upper valleys occurs and only carrier cooling contributes to the somewhat faster picosecond build-up of emission intensity. At even longer delay times, the luminescence decays with the lifetime of the electron-hole plasma.

The large spectral width of luminescence emitted at very early delay times (c.f. Fig. 1), the absence of any narrow structure in the spectra, and the rapid onset of emission at all photon energies give evidence that electrons and holes are redistributed over a wide energy range within the time resolution of our experiments of 100 fs. Carrier-carrier scattering and interaction with longitudinal-optical phonons represent the main mechanisms of thermalization. The very rapid onset of luminescence within 100 fs and the absence of any structure with optical phonon spacing in the spectra gives evidence, that the first mechanism dominates at the carrier densities studied in our experiments. We now discuss Monte-Carlo (MC) simulations of the femtosecond carrier redistribution. The dominant carrier-carrier scattering is treated in two alternative ways : (i) In the k-space approach, we assume a screened interaction potential with a screening length that is updated every 50 fs according to the time evolution of the carrier distribution (time dependent static screening) (Osman 1987). The total scattering rate is obtained from Fermi's golden rule adding up all contributions from different types of carriers.

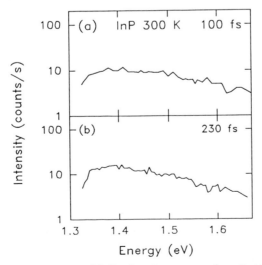

Fig. 2 Luminescence spectra of InP after femtosecond excitation at 1.93 eV for delay times of (a) 100 fs and (b) 230 fs. The carrier density was 5×10^{17} cm^{-3}.

The main limitation of this model comes from the assumption of a static screened interaction.

(ii) An alternative method for the treatment of the carrier-carrier interaction is based on a molecular dynamics simulation for the real-space trajectories of the ensemble of carriers which interact via a bare Coulomb potential (Lugli 1986). This method does not require any assumptions on the screening between carriers and thus fully accounts for dynamical screening.

The luminescence spectra obtained from a time dependent static screening formulation are plotted in Fig. 1 (solid squares). Good agreement between theory and experiment is found for delay times longer than 200 fs. The static screening model does not reproduce the data at shorter times. In particular, during the laser pulse the MC spectra (Fig. 1 b) drop sharply above 1.57 eV indicating an insufficient energy exchange between electrons excited from the split-off band and those at higher energies. This finding is due to the intrinsic failure of the static screening approach that underestimates the carrier-carrier interaction at the earliest times.

The molecular dynamics simulation gives much better agreement with the experimental data. In this model, enhanced scattering rates between the different electron populations are found resulting in smoother distribution functions and much better agreement with the luminescence spectra measured at early times (Fig. 1 b, open squares). These findings demonstrate the important role of dynamical screening for carrier equilibration on a time scale of 100 fs. Simulations for InP are currently being performed. In the present theoretical simulation of the data, the coherent coupling of the initial polarization in the semiconductor to the femtosecond excitation pulse is not explicitly taken into account. Investigation of the influence of coherent phenomena on the analysis of the data would be interesting.

In conclusion, our results give direct evidence that photoexcited electrons and holes in GaAs and InP are redistributed over a wide energy range within the first 100 fs after excitation. The ultrafast carrier equilibration is dominated by carrier-carrier collisions with scattering rates higher than predicted from a statically screened interaction potential. Our Monte Carlo simulations including dynamical screening via a molecular dynamics formalism are in good agreement with the experimental findings.

References

Elsaesser T., Shah J., Rota L., Lugli P. 1991 *Phys. Rev. Lett.* **66** 1757

Erskine D.J., Taylor A.J., Tang C.L. 1984 *Appl. Phys. Lett.* **45** 54

Lin W.Z., Schoenlein R.W., Fujimoto J.G., Ippen E.P. 1988 *IEEE J. Quant. Electron.* **24** 267

Lugli P., Ferry D.K. 1986 *Phys. Rev. Lett.* **56** 1295

Lugli P., Bordone P., Reggiani L., Rieger M., Kocevar P. Goodnick S.M. 1989 *Phys. Rev.* **B 39** 7852

Osman M. A., Ferry D.K. 1987 *Phys. Rev.* **B 36** 6018

Shah J., Deveaud B., Damen T.C., Tsang W.T., Gossard A.C., Lugli P. 1987 *Phys. Rev. Lett.* **59** 2222

Stanton C.J., Bailey D.W., Hess K. 1990 *Phys. Rev. Lett.* **65** 231

Zhou X.Q. Cho G.C., Lemmer U., Kütt W., Wolter K., Kurz H. 1989 *Solid State Electron.* **32** 1591

Zhou X.Q., Kurz H. 1991 *This conference; paper $Th - P10$*

Zhou X., Hsiang J. 1990 *J. Appl. Phys.* **67** 7399

Femtosecond nonlinearities and hot-carrier dynamics in GaAs

T. Gong and P. M. Fauchet

Laboratory for Laser Energetics and Department of Electrical Engineering, University of Rochester, Rochester, NY 14623, U.S.A.

ABSTRACT: Pump-probe techniques are used to perform a series of measurements on intrinsic GaAs samples at room temperature with a temporal resolution of 75–100 fs. Changes of both absorption coefficient and refractive index are measured over a wide spectral region (550–950 nm) for various carrier densities ($\sim 10^{17}$–10^{19} cm^{-3}) injected at 2 eV. These measurements provide insight on the fundamental properties of nonequilibrium carriers, including electron-electron scattering, electron-hole scattering, electron-phonon intervalley scattering, band-gap renormalization, plasma screening of Coulomb interactions, and free-carrier absorption.

1.   INTRODUCTION

The properties of hot carriers are determined primarily by the electronic band structure of the host material and carrier-lattice and carrier-carrier interactions—topics that have historically been of sustained interest in solid-state physics. Because of the continuing size reduction of conventional semiconductor devices and the emergence of new devices based upon advanced growth techniques, the study of hot-carrier dynamics remains of importance. The advent of ultrashort laser pulses provides a very direct means of investigating hot-carrier relaxation and cooling processes with excellent temporal resolution.

Extensive investigations of hot-carrier dynamics in GaAs have been performed using ultrafast spectroscopy [1–14]. In this paper, we present a new series of studies of hot-carrier dynamics. The transient absorptive and refractive nonlinearities of GaAs have been measured with a temporal resolution of less than 100 fs. Various hot-carrier processes and their influences on the ultrafast optical response are discussed.

2.   TECHNIQUES

We have used a copper vapor laser amplified CPM laser to perform pump-probe measurements on GaAs at room temperature. About 3% of the 75–100 fs pulses at 620 nm is used as the pump beam. The remainder is focused onto a jet of ethylene glycol, producing a white-light continuum, which is used as a probe pulse. The probe wavelength is selected with interference filters over a spectral region from 550 nm to 950 nm. The polarization of the pump and probe pulses is orthogonal to reduce coherent artifacts. The intensity of the pump pulse can be varied over nearly two orders of magnitude.

The time-resolved transmission T(t) and reflection R(t) are measured simultaneously on a thin (<0.3 μm) intrinsic GaAs film obtained using a lift-off technique [15] and attached to a sapphire window. The absorption coefficient $\alpha(t)$ and refractive index n(t) are than deduced from the measured T and R by inversion of the Fabry-Perot formulae. The changes of refractive index $\Delta n(t)$ are also obtained by measuring the time-resolved differential reflection $\Delta R/R$ on thick GaAs samples [13].

3.   RESULTS

3.1   Refractive Index Spectral Hole Burning

For excitation of GaAs at 2 eV (620 nm), the electrons are initially injected into the Γ valley at the three distinct excess energies of 0.50 eV, 0.43 eV, and 0.15 eV due to transitions from the heavy-hole, light-hole, and split-off valence bands with a relative strength of 42%, 42%, and 16% respectively. The transient decrease of absorption, caused by the nonthermal carrier distribution generated after femtosecond excitation, is characterized by an absorption spectral hole near 2 eV [6,16]. According to the Kramers-Kronig relation, the existence of a spectral hole in the imaginary part of the refractive index (absorption) will cause a spectral resonance in the change of a real part of the refractive index. We have indeed observed this resonance around 2 eV, which we call refractive index spectral hole burning [13]. Figure 1 shows our measurements of the spectral dependence of the refractive-index change for $N \sim 2.5 \times 10^{18}$ cm^{-3}. As shown in Figure 1(b), the spectral hole disappears quickly, because a large fraction of the electrons scatter to the X and L valleys and the rest thermalize within the Γ valley on a sub-100-fs time scale.

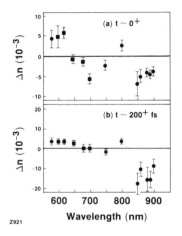

Fig. 1   Measured Δn spectrum for $N \sim 2.5 \pm 1.2 \times 10^{18}$ cm^{-3} at time delays: (a) $t \sim 0^+$ and (b) $t \sim 200^+$ fs. ■ are the measured average values for both a film and a bulk sample; ● are the measured average values for a thin film. The error bars are obtained from the standard deviation of the measured values and a conservative estimate of the uncertainty in carrier density. A clear spectral resonance around 620 nm appears in (a).

Z921

If the absorption spectral hole burned at $t \sim 0^+$ followed the symmetric pump spectrum around 620 nm (2 eV), we expect that the change of the refractive index would be zero at 620 nm. However, our data clearly show that the zero change of the refractive index is near 650 nm, and not at 620 nm. Furthermore, measurements on the absorption changes also confirm that at $t \sim 0^+$, $|\Delta\alpha|$ is larger near 650 nm than at 620 nm, a result than can also be inferred from absorption spectral hole burning measurements made around 620 nm by others [16]. Therefore, on a time scale shorter than our temporal resolution, the initial excited carrier "distribution" is strongly deformed and becomes spectrally asymmetric, and the peak of the distribution appears to be red-shifted. The Γ-valley electrons excited from the heavy-hole valence band by 2 eV photons can scatter to both X and L valleys; by contrast, those electrons excited from the light-hole valence band can only scatter to the L valley. Therefore the electrons excited from the heavy-hole valence band are more likely to be transferred to the satellite valleys than those excited from the light-hole valence band. This "preferred" scattering produces a deformation of the excited-carrier distribution. This process also appears to be instantaneous within our temporal resolution because carriers excited by the earlier part of the pump pulse have already started undergoing this effect. Another effect, which was recently observed and termed "resonant intervalley scattering" by Bigot *et al.* [17], can also contribute to the ultrafast deformation (shift) of the absorption spectral hole. Because of the vanishing density of $|X\rangle$ states at the Γ-X transition point, the return of electrons from the bottom of the X valley to the Γ valley is more probable, thereby enhancing the concentration of electrons in the Γ valley near the transition point. We note that the peak of the "trapped" electron distribution (~1.92 eV or 646 nm, 30 meV below the X-valley minimum) is very close to the zero change of the refractive-index spectral resonance. It still not clear why this resonant process takes place in ~50 fs [17], because an electron has to emit or absorb a phonon twice in order to complete this process. Nevertheless, this explanation is supported by the following facts: the shift has little carrier-density dependence; it does not appear [3] or is very small [18] when the spectral hole is burned below the L-valley minimum; it does not appear in InP [19], a material in which there is no intervalley scattering with 620 nm-excitation; and Δn is nearly zero at the pump wavelength (620 nm) at t ~ 0+ for GaAs at low temperature (2 K) [20] when the Γ-X transition is suppressed.

3.2 Studies of the Initial Scattering Time

If pump-probe measurements are performed at the same wavelength, the recovery of the bleaching measures the scattering rate of carriers from the initially optically coupled states. Since $\Delta n(t)$ measures the spectrally integrated population change near the as-excited states, it is best to use $\Delta\alpha(t)$ to deduce the initial scattering time. Figure 2 shows $\Delta\alpha(t)$ at 620 nm for various injected carrier densities.

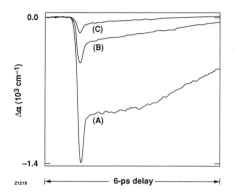

Fig. 2 Time evolution of $\Delta\alpha$ at 620 nm for different N ~: (A) 3.3×10^{18} cm^{-3}, (B) 9.5×10^{17} cm^{-3}, (C) 3×10^{17} cm^{-3}.

We have developed a simplified model in which the initial fast bleaching is modeled by two equations:

$$\frac{dN}{dt} = \frac{N_o I_{pump}(t)}{\int_{-\infty}^{+\infty} I_{pump}(t)\,dt} - \frac{N}{\tau_T} \tag{1}$$

$$S(t) \propto \int_{-\infty}^{+\infty} I_{probe}(t'-t)N(t')\,dt' \quad, \tag{2}$$

where N_0 is the total injected carrier density, N is the time-dependent carrier density in the excited states, $1/\tau_T$ is an effective total scattering rate from the initial excited states, $I_{pump}(t) = I_{probe}(t) = sech^2(1.763\ t/\tau_p)$, and S(t) is the detected signal. If $\tau_T \gg \tau_p$, the decay part of N(t) can be simply described by an exponential with time constant τ_T. However, when $\tau_T \sim \tau_p$, as in our measurements, N(t) does not have a simple form.

From Figure 2, it is clear that the change of the signal is not governed by a single decay process, but rather that a fast process is superimposed on a much slower variation. It is important to note that the slow variation is rather different at each carrier density. This result prevents the traces obtained for different carrier densities from being normalized by the amplitudes of their slow variations, a method used by others [6]. However, at least for low carrier densities (N < 5×10^{17} cm^{-3}), the slow variation of $\Delta\alpha(t)$ is well described by an exponential decay. The solid line shown in Figure 3(a) is a fit of the slow decay of $\Delta\alpha(t)$ for N ~ 3×10^{17} cm^{-3} using Equations (1) and (2) with τ_p = 100 fs and τ_{T2} = 2 ps. This slow component is subtracted from the experimental trace to obtain the "effective" fast bleaching component, as shown by a dotted line in Figure 3(b). We then use Equations (1) and (2) again to fit the fast bleaching. The solid line shown in Figure 3(b) is the best fit using τ_p = 100 fs and τ_{T1} = 50 fs.

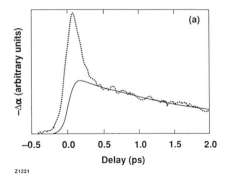
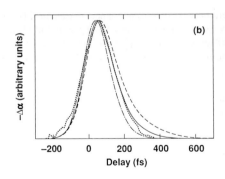

Z1221

Fig. 3   (a) Measured time-resolved $\Delta\alpha$ at 620 nm for N $\sim 3 \times 10^{17}$ cm^{-3} (dotted line) and a fit (solid line) for a slow decay with $\tau_p = 100$ fs and $\tau_T = 2$ ps; (b) The fast component (dotted line), obtained by subtracting a slow component [solid line in (a)] from the experimental curve [dotted line in (a)], is compared to different fits with $\tau_p = 100$ fs and $\tau_T = 50$ fs (solid line), $\tau_T = 70$ fs (dashed line), and $\tau_T = 30$ fs (dashed-dotted line). The best fit is obtained using $\tau_T = 50$ fs.

It is worth mentioning that only less than 35% of the *total* excited electrons ever accumulate in the "as-excited" states; most of them have already scattered away within the pulse width since $\tau_{T1} \sim 0.5\tau_p$. After ~200 fs, the split-off probe makes a major contribution to $\Delta\alpha$; it monitors the band filling and the cooling of the Γ-valley electron distribution ~150 meV above the conduction band edge. The fact that the split-off probe samples the Boltzmann "tail" for low carrier densities ($<5 \times 10^{17}$ cm^{-3}) is probably responsible for the exponential decay of the slow component.

At higher carrier densities, the slow component is rather complicated and the simple fit described above does not work well. However, information on the initial "effective" scattering rate can be obtained by monitoring the peak value of the transient spike of $\Delta\alpha(t)$ as a function of carrier density. Figure 4 plots $\Delta\alpha_{max}$ versus N. The weak sublinearity of the data indicates that τ_T decreases little at higher carrier densities: τ_T is reduced from ~50 fs to ~40 fs when N increases from 3×10^{17} cm^{-3} to 7×10^{18} cm^{-3}. The value of τ_T at high N (~40 fs) is very close to those obtained by other groups with a three-time-component fit [2,9] or a similar two-time-component fit [21].

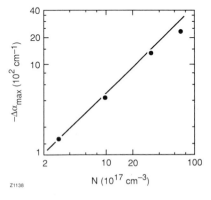

Z1138

Fig. 4 $\Delta\alpha_{max}$ (t ~ 0$^+$) as a function of N at 620 nm. The straight line represents linear relationship between $\Delta\alpha_{max}$ and N.

Let us now discuss the microscopic origin of the ultrafast initial scattering time. The contributions by electron excited from the heavy-hole, light-hole, and split-off valence bands all need to be considered. Intervalley scattering is a very important process, and its rate is currently under investigation [22]. Recently Zollner *et al.* [23] introduced the concept of an "effective" intervalley deformation potential (IDP) in which the contribution of the TA phonons is included in a temperature-dependent IDP for LO-phonon scattering. If we use values of Zollner *et al.* [23] ($D_{\Gamma L} \sim 5.8 \times 10^8$ eV/cm and $D_{\Gamma X} \sim 9.4 \times 10^8$ eV/cm at $T_L = 300$ K) to calculate the intervalley scattering times at different excited energies, we obtain $\tau_{\Gamma X}(\sim 0.5$ eV$) \sim 130$ fs, $\tau_{\Gamma L}(\sim 0.5$ eV$) \sim 120$ fs, and $\tau_{\Gamma L}(\sim 0.43$ eV$) \sim 150$ fs. τ_{LO} (the unscreened LO-phonon emission time) is taken to be ~ 180 fs [4,24]. The effective initial scattering time is then given by

$$\frac{1}{\tau_T} = 0.42 \left(\frac{1}{\tau_{\Gamma X}(0.5\ \text{eV})} + \frac{1}{\tau_{\Gamma L}(0.5\ \text{eV})} + \frac{1}{\tau_{LO}} \right) + 0.42 \left(\frac{1}{\tau_{\Gamma L}(0.43\ \text{eV})} + \frac{1}{\tau_{LO}} \right) + 0.16 \left(\frac{1}{\tau_{LO}} \right) \quad (3)$$

The calculated value of 65 fs is close to the measured $\tau_T \sim 50$ fs at $N \sim 3 \times 10^{17}$ cm^{-3}. The 30% difference can be attributed to carrier-carrier scattering. The fact that τ_T only decreases to ~ 40 fs at high N, points out that carrier-carrier scattering does not strongly affect the initial scattering time, a conclusion also reached by others [2,9].

3.3 Band-Edge Nonlinearities

The interactions between hot carriers strongly affect absorptive and refractive nonlinearities around the band edge. Figure 5 shows the time-resolved changes of absorption for the same $N \sim 1.5 \pm 0.7 \times 10^{18}$ cm^{-3} at probe wavelengths of 880, 890, 900, and 920 nm, which are below the band edge, and of 860 nm, which is slightly above the band edge. The short-lived increase of absorption observed at 880, 890, and 900 nm right after excitation is attributed to band-gap renormalization accompanied by plasma screening of Coulomb interactions. Coulomb screening is often neglected in the interpretation of many experiments. The decrease of absorption slightly above the band edge (850 and 860 nm) immediately after excitation, which is caused by the reduction of Coulomb (Sommerfeld) enhancement factor, reveals the importance of this effect. Plasma screening of Coulomb interactions, along with band-gap renormalization, cause a clear spectral resonance in $\Delta\alpha$ around the band edge, which takes place instantaneously when most carriers are still "hot" [11]. The subsequent broadband decrease of absorption indicates that the states near the perturbed band edge become filled.

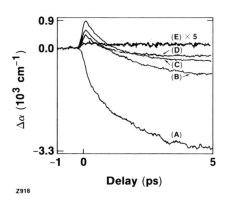

Fig. 5 Measured time-resolved $\Delta\alpha$ for $N \sim 1.5 \pm 0.7 \times 10^{18}$ cm^{-3}. The probe wavelengths are: (A) 860 nm, (B) 880 nm, (C) 890 nm, (D) 900 nm, and (E) 920 nm. Curve (E) has been multiplied by five.

Z918

At 920 nm (~ 75 meV below the original band edge), a small, long-lived induced absorption is observed. This effect is attributed to intra-band (free-carrier) absorption. We investigate it by probing at 950 nm, which is ~ 120 meV below the original band edge. Figure 6(a) and (b) show $\Delta\alpha(t)$ and $\Delta n(t)$ for various (high) carrier densities. Both $|\Delta\alpha_{max}|$ and $|\Delta n_{max}|$ scale approximately linearly with N, consistent with free-carrier absorption (FCA) being the main contribution to the nonlinearities at this wavelength. The cross section for FCA, defined as $\sigma_{eh} = \Delta\alpha/N$ is $\sim 2.6 \pm 1.0 \times 10^{-17}$ cm^2 at 950 nm. Free-carrier absorption in n-type GaAs has been systematically studied earlier [25], and σ_e (solely due to electrons) deduced from those data is $\sim 8 \times 10^{-18}$ cm^2. Other measurements [25,26] also show that σ_h (solely due to holes) is more than a factor of two larger than σ_e. Our estimated σ_{eh} is close to $\sigma_e + \sigma_h$. It is interesting to note that ~ 5 ps after excitation, when the carriers are cooled down, $\Delta\alpha$ is only $\sim 20\%$ smaller than after <0.5 ps, when the carriers are still hot [27] and many of them reside in the satellite valleys. Also, 950 nm |Δn| is quite large because band filling causes a large negative Δn, which extends far below the band edge [28]. Such large refractive nonlinearities should be useful for designing optoelectronic devices such as phase modulators.

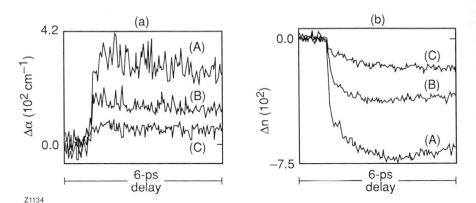

Z1134

Fig. 6   Measured time-resolved (a) Δα and (b) Δn at 950 nm for various N ~: (A) 1.3×10^{19} cm^{-3}, (B) 5×10^{18} cm^{-3}, (C) 2×10^{18} cm^{-3}.

3.4   Band-Edge Gain Dynamics

For high injected carrier densities (N > 10^{18} cm^{-3}), the combination of nearly instantaneous intra-Γ-valley redistribution of electrons [12,23] with the rapid scattering of the high-energy Γ-valley electrons to the X and L valleys [10] should make it possible to observe band-edge gain on subpicosecond time scales even for an excitation at 2 eV. Figure 7 shows the absorption coefficient measured at 880 nm. It is clear that α becomes negative on a subpicosecond time scale for high carrier densities. In fact, gain is observed in a wide spectral region (850–900 nm) on subpicosecond and picosecond time scales for N > 3×10^{18} cm^{-3} [14]. The time delay for gain to occur is ~280±80 fs at 880 nm, 450±150 fs at 860 nm, and 650±200 fs at 850 nm for the highest carrier density (N ~ 8–10 × 10^{18} cm^{-3}) used in these measurements. This time increases to ~800 fs±300 fs at 880 nm and >3000±800 fs at 860 nm for lower carrier density (N ~ 3.3 × 10^{18} cm^{-3}).

At a probe frequency ω, gain should be observed at any given time when the condition

$$\hbar\omega < \mu_{eff} + E_g \tag{4}$$

with

$$\mu_{eff} = \frac{m_e + m_h}{m_h T_h + m_e T_e} \left(\mu_e T_h + \mu_h T_e \right) \tag{5}$$

is satisfied. Here $\mu_{e(h)}$ is the quasi-chemical potential of the Γ-valley electrons (holes) with respect to the conduction (valence) band edge, and E_g is the normalized band gap. In our model the Coulomb enhancement factor is neglected due to the strong plasma screening at these high carrier densities [11,29]. The calculation of μ_{eff} involves solving a set of kinetic equations for the density and kinetic energy of the Γ-valley electrons and holes [10,14]. Electrons and holes are assumed to equilibrate instantaneously. The initial temperature of the electrons and holes are ~3000 K and ~600 K respectively, based on the kinetic energies of carriers with 2-eV excitation. Scattering to the X and L valleys is incorporated into the kinetic model as an energy-dependent sink for the high-energy, Γ-valley electrons, again using the effective IDP summarized by Zollner *et al.* [23]. Zone-center LO-phonon emission is also neglected because only a very small amount of energy will be lost by electrons on a subpicosecond time scale when screening is included at high carrier densities [30]. We have compared the calculated delay-times for gain to occur using Equations (4) and (5) with those deduced from measurements. The hole temperature is treated as a parameter. Good agreement is obtained between model and measurements [14].

In addition to revealing the importance of ultrafast equilibration of carriers near the Γ point of the Brillouin zone and the efficient cooling mechanism provided by intervalley scattering, these results also give some insights on the hot-hole dynamics. Figure 8 displays the hole temperature which gives a best fit to the data. It appears that the hole distribution is heated to ~800 K within 300 fs, and then cools down to ~300 K within <1 ps. A heated hole distribution indicates that Coulomb-mediated electron-hole scattering is important on this time scale, in particular when the electron distribution is hotter and for high carrier densities. This result is supported by theoretical calculations including contributions from intervalance and intravalence band scattering processes [14,31]. The rapid "cooling time" of holes that we obtain also agrees well with those measured in n-type GaAs using 100-fs time-resolved luminescence [32].

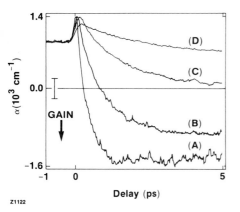

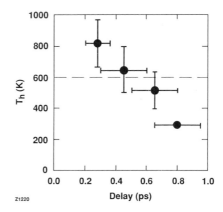

Fig. 7 Measured temporal evolution of the absorption coefficient $\alpha(t)$ at 880 nm for different nominal N~: (A) 9.0×10^{18} cm^{-3}, (B) 3.3×10^{18} cm^{-3}, (C) 1.0×10^{18} cm^{-3}, (D) 3.3×10^{17} cm^{-3}. The error bar associated with the uncertainty on the absolute values of α is indicated.

Fig. 8 The hole temperature (T_h) obtained from the "best fit" between the experiment and the model at different time delays. The dashed line indicates the initial T_h, based on the kinetic energy of the holes with 2-eV excitation.

4. CONCLUSIONS

In conclusion, we have performed a series of measurements on time-resolved absorptive and refractive nonlinearities induced by hot carriers injected at 2 eV on GaAs. Measurements near the initial excited states yield the first observation of refractive-index spectral hole burning. Studies of spectral hole burning and initial scattering times reveal the importance of intervalley scattering and carrier-carrier interactions. Measurements near the band edge show that optical nonlinearities on femtosecond and picosecond time scales are governed by various carrier effects, such as band-gap renormalization, plasma screening, band filling, and free-carrier absorption. Subpicosecond gain near the band edge for high injected carrier densities demonstrates that the rapid cooling of Γ-valley electrons is provided by ultrafast intra-Γ-valley equilibration and intervalley scattering. Evidence of a transient heated hole distribution provides insights for the study of electron-hole and hole-lattice interactions.

5. ACKNOWLEDGEMENTS

The authors thank Jeff F. Young and P. J. Kelly at the National Research Council of Canada for performing calculations used in Section 3.4 on band-edge gain dynamics, and also for many valuable discussions. Contributions from W. L. Nighan Jr., P. Mertz, C. Peng, M. Shayegan, and G. W. Wicks are also appreciated. This work was supported by the U.S. Army Research Office under Contract DAAL03-91-G-0173, the U.S. Office of Naval Research under Grant N00014-91-J-1488, and the National Science Foundation under Contract ECS-9196000.

6.　REFERENCES

1.　C. V. Shank, R. L. Fork, R. L. Leheny, and J. Shah, Phys. Rev. Lett. **42**, 112 (1979).
2.　M. J. Rosker, F. W. Wise, and C. L. Tang, Appl. Phys. Lett. **49**, 1726 (1986).
3.　J. L. Oudar, D. Hulin, A. Migus, A. Antonetti, and F. Alexandre, Phys. Rev. Lett. **55**, 2074 (1985).
4.　J. A. Kash, J. C. Tsang, and J. M. Hvam, Phys. Rev. Lett. **54**, 2151 (1985).
5.　J. Shah, B. Deveaud, T. C. Damen, W. T. Tsang, A. C. Gossard, and P. Lugli, Phys. Rev. Lett. **59**, 2222 (1987).
6.　W. Z. Lin, R. W. Schoenlein, J. G. Fujimoto, and E. P. Ippen, IEEE J. Quantum Electron. **QE-24**, 267 (1988), and references therein.
7.　P. C. Becker, H. L. Fragnito, C. H. Brito Cruz, R. L. Fork, J. E. Cunningham, J. E. Henry, and C. V. Shank, Phys. Rev. Lett. **61**, 1647 (1988).
8.　J. A. Kash, Phys. Rev. B **40**, 3455 (1989).
9.　X. Q. Zhou, G. C. Cho, U. Lemmer, W. Kütt, K. Wolter, and H. Kurz, Solid State Electron. **32**, 1591 (1989).
10.　D. Kim and P. Y. Yu, Phys. Rev. Lett. **64**, 946 (1990).
11.　T. Gong, W. L. Nighan, Jr., and P. M. Fauchet, Appl. Phys. Lett. **57**, 2713 (1990).
12.　T. Elsaesser, J. Shah, L. Rota, and P. Lugli, Phys. Rev. Lett. **66**, 1757 (1991).
13.　T. Gong, P. Mertz. W. L. Nighan, Jr., and P. M. Fauchet, Appl. Phys. Lett. **59**, 721 (1991).
14.　T. Gong, P. M. Fauchet, J. F. Young, and P. J. Kelly, Phys. Rev. B **44**, 6542 (1991).
15.　E. Yablonovitch, T. Gmitter, J. P. Harbison, and R. Bhat, Appl. Phys. Lett. **51**, 2222 (1987).
16.　C. H. Brito-Cruz, R. L. Fork, and C. V. Shank, in the *XV International Conference on Quantum Electronics Technical Digest Series 1987, Vol. 21* (Optical Society of America, Washington, DC, 1987), p. 82.
17.　J.-Y. Bigot, M. T. Portella, R. W. Schoenlein, J. E. Cunningham, and C. V. Shank, Phys. Rev. Lett. **65**, 3429 (1990).
18.　D. Hulin (private communication).
19.　T. Gong and P. M. Fauchet (unpublished).
20.　G. Bohne, S. Freundt, S. Lehmann, and R. G. Ulbrich (paper WE 4-4), in the *Technical Digest of the VIIth International Symposium of Ultrafast Processes in Spectroscopy*, Bayreuth, Germany, 7–10 October 1991.
21.　P. C. Becker, H. L. Fragnito, C. H. Brito Cruz, J. Shah, R. L. Fork, J. E. Cunningham, J. E. Henry, and C. V. Shank, Appl. Phys. Lett. **53**, 2089 (1988).
22.　M. A. Alekseev and D. N. Merlin, Phys. Rev. Lett. **65**, 274 (1990); J. A. Kash, J. C. Tsang, and R. G. Ulbrich, Phys. Rev. Lett. **65**, 275 (1990).
23.　S. Zollner, S. Gopalan, and M. Cardona, Solid State Commun. **76**, 877 (1990).
24.　J. C. Tsang and J. A. Kash, Phys. Rev. B **34**, 6003 (1986).
25.　J. S. Blakemore, J. Appl. Phys. **53**, R123 (1982), and references therein.
26.　H. C. Casey, Jr. and M. B. Panish, *Heterostructure Lasers, Part A: Fundamental Principles* (Academic Press, San Diego, 1978), p. 175, and references therein.
27.　K. Seeger, *Semiconductor Physics* (Springer-Verlag, New York, 1973), p. 379.
28.　B. R. Bennett, R. A. Soref, and J. A. Del Alamo, IEEE J. Quantum Electron. **QE-26**, 113 (1990).
29.　H. Haug and S. W. Koch, Phys. Rev. A **39**, 1887 (1989).
30.　M. A. Osman and D. K. Ferry, Phys. Rev. B **36**, 6018 (1987).
31.　J. F. Young, P. Kelly, and N. L. Henry, Phys. Rev. B **36**, 4535 (1987).
32.　X. Q. Zhou and H. Kurz (paper TH P10), in the *Technical Digest of the VIIth International Symposium of Ultrafast Processes in Spectroscopy*, Bayreuth, Germany, 7–10 October 1991; and G. M. Gale, A. Chebira, J. Chesnoy, and E. Fazio (paper TU 6-1), *ibid.*

Inst. Phys. Conf. Ser. No 126: Section V
Paper presented at Int. Symp. on Ultrafast Processes in Spectroscopy, Bayreuth, 1991

Optical nonlinearities at the bandedge of GaAs at room temperature

H. Heesel, T. Zettler*, A. Ewertz and H. Kurz

Institute of Semiconductor Electronics II, RWTH Aachen, W-5100 Germany

*Central Institute of Optics and Spectroscopy, Rudower Chaussee 6,

O-1199 Berlin, Germany

ABSTRACT

We report on femtosecond measurements of carrier-induced changes in the dielectric function of GaAs films at room temperature. In the vicinity of the fundamental gap, optical nonlinearities are predominant caused by bandfilling and plasma screening of electron-hole interactions. We present a detailed study of absorptive and dispersive nonlinearities by Kramers-Kronig consistent evaluation of differential transmission and reflection spectra.

INTRODUCTION

Laser excitation of III-V compound semiconductors with photon energies exceeding the energy of the fundamental gap creates electrons and holes in the respective bands. The interaction processes of this many-body system determines the optical properties of the excited semiconductor. Optical nonlinearities are introduced through many-body effects[1,2] like plasma screening of the Coulomb potential, band-gap shrinkage and state- as well as bandfilling. The resulting changes of the semiconductor absorption and dispersion become quite large in the spectral vicinity of resonances as the fundamental absorption edge. Such optical nonlinearities are of special interest for potential applications in all-optical and electro-optical devices[3].

EXPERIMENTAL

Our copper vapor laser amplified CPM system generates pulses of 60 fs duration at 2eV at a repetition rate of 7kHz. Aa attenuated part of the amplified pulse is used as a pump beam. The remainder is focused into a ethylene glycol cell to generate a white-light probe continuum between 1.3 and 1.9eV. The experimental configuration is a conventional pump-probe scheme. The pump induced changes in transmission and reflectivity of the probe beam are recorded by an optical multichannal analyzer (OMAIII). The obtained differential transmission/reflection spectra (DTS,DRS) are numerically chirp corrected[4] to remove their inherent temporal distortion.

The measurement of transmission and reflection permits the separation of dispersive and absorptive contributions by inversion of the optical equations including the Kramers–Kronig relation. In a first step, we determine the thickness and the static dielectric function of our samples by inversion of linear transmission and reflection measurements (Fig.1). The inversion of DTS and DRS-spectra finally determines the transient changes $\Delta\varepsilon_1$, $\Delta\varepsilon_2$ of the static dielectric function. This procedere yields a consistent set of $\Delta\varepsilon_1$, $\Delta\varepsilon_2$ - spectra.

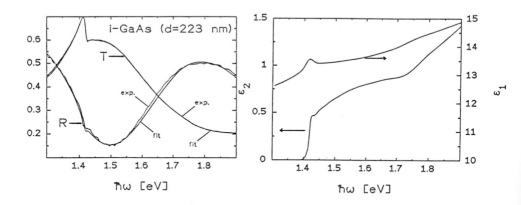

Fig.1: a) Comparison of measured and calculated transmission- and reflection spectra of a thin GaAs-film (d=223nm) on BK7-glas-substrate

b) Static dielectric function calculated from the data in a)

RESULTS AND DISCUSSION

We have studied the excitation density range from 1×10^{16} to 1×10^{19} cm^{-3}. As an example Figure 2 shows as an example the calculated changes in ε_1 and ε_2 of a 223nm thick GaAs-film excited with $N_{exc} = 1 \times 10^{18}$ cm^{-3} at different time delays. For negative time delay, an oscillatory feature is observed close to the bandedge. Similar features have been observed in GaAs at low temperatures and can be explained by coherent effects arising from exciton bleaching[5].

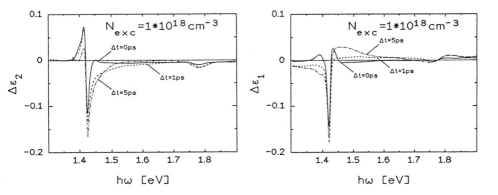

Fig.2 : $\Delta\varepsilon_1$, $\Delta\varepsilon_{||}$ - spectra of a thin GaAs layer (d=223nm) at three different time delays: $\Delta t=0$, $\Delta t=1$ps, $\Delta t=5$ps ($N_{exc} = 1 \times 10^{18}$ cm^{-3})

At zero time delay, a reduced absorption due to bandfilling and reduction of Coulomb enhancement is observed above the bandgap over a wide spectral range. This indicates an instantaneous scattering of carriers out of their initial states leading to an *extremly broad nonthermal distribution* from the bandedge to to 1.9eV. For positive time delays, we observe a fast plasma thermalization in less than 200fs and a subsequent cooling of the thermalized distribution on a picosecond timescale. The bleaching peak at 1.78eV is a replica of the carrier distribution emerging from probing the split-off to conduction band transition.

ACKNOWLEDGEMENT

This work has been supported by the DFG (Ku540/6-1) and the Alfred Krupp Stiftung. The authors are grateful to A. Förster and H. Lüth (ISI-HL, KFA Jülich) for supplying the samples.

REFERENCES

1) H.Haug, S.Schmitt-Rink, J.Opt.Soc.Am.B 1135, (1985).

2) H.Haug, S.W.Koch, Phys.Rev.A, 39, 1887 (1989).

3) S.W.Koch, N.Peyghambarian, and H.M.Gibbs, J.Appl.Phys. 63, R1 (1988).

4) T.F.Albrecht, K.Seibert, and H.Kurz, Opt.Commun. 84, 223 (1991).

5) M.Joffre, D.Hulin, A.Migus, and A.Antonetti, Opt.Lett. 13, 276 (1988).

Transient reflectivity and transmission spectra of gallium arsenide: dependence on doping and excitation density

G Böhne, S Freundt, S Lehmann, R G Ulbrich

IV. Physikalisches Institut der Universität Göttingen,
Bunsenstr. 13-15, W-3400 Göttingen, Germany

ABSTRACT: Time resolved reflectivity at 2 eV and transmission spectra at the band edge of GaAs were measured after electron-hole-pair excitation with 2 eV, 70 fs laser pulses. Variation of sample doping, excitation density and temperature (300 K, 2 K) provided information about relaxation processes in GaAs on the 100 fs...10 ps time scale. We discuss the change in optical properties in the framework of a dielectric function determined by time-dependent carrier distributions.

1. INTRODUCTION

After laser pulse excitation of electron-hole-pairs in semiconductors different relaxation processes take place, which can be detected in classical excite- and probe-experiments by the temporal change in optical properties of the sample (Zhou 1989). We report two classes of measurements: (i) the transient reflectivity of intrinsic and doped GaAs-samples with 2 eV incident light pulses of 70 fs duration. The dependence on type and level of doping, excitation density, temperature and polarization directions show the role of Coulomb scattering, intervalley scattering, LO phonon emission and coherent interaction of excite- and probe-pulses inside the sample. (ii) Investigation of the transient absorption in the region of the direct band gap around 1.5 eV after optical excitation at 2 eV. These measurements are more sensitive to phase space filling at the band edges and to the screening of the Coulomb interaction at frequencies corresponding to the exciton Rydberg R^*.

2. TRANSIENT REFLECTIVITY OF GaAs

All measurements were performed with 70 fs-laser pulses from a classical CPM-Laser (Rh 6G, DODCI; 4 prisms). We reached excitation densities up to $7 \cdot 10^{17}$ cm^{-3}. One of the two laser beams was divided into strong pump pulses and time-delayed, weaker (1:20) test pulses, which detect the change of reflectivity after excitation of the sample.

The relative change of reflectivity $\Delta R/R_0$ of GaAs at 2 eV as a function of the change of the real part $\Delta \epsilon_1$ and of the imaginary part $\Delta \epsilon_2$ can be calculated from Fresnel's formulas with the known optical constants at 2 eV: $\Delta R/R_0 = 36 \cdot 10^{-3} \cdot \Delta \epsilon_1 + 6 \cdot 10^{-3} \cdot \Delta \epsilon_2$. The occupation of electron and hole states after the absorption of excite beam photons results in two contributions to the change of the dielectric function: (i) interband transitions, which connect the upper three valence bands with the conduction band and (ii) intraband transitions of free carriers. If the probe beam photon energy lies above the centre of the carrier distribution, the first contribution will increase the reflectivity and decrease otherwise. The second contribution is always nonpositive, but usually it has a smaller value than the interband contributions. Additionally a contribution from the electro-optic effect may occur. The screening of the ubiquitous surface space charge electric field leads –via the electro-optic effect– to a change of refractive index and of reflectivity, respectively. Because of the zincblende type electro-optic tensor ($r_{41} = r_{52} = r_{63}$, all others $r_{ij} = 0$) this contribution depends on the investigated sample surface and on the probe beam polarization. Figure 1 shows the time dependent change of reflectivity for two different polarization directions. The measurements have fast oscillations, which agree with the frequency of LO-phonons (Cho 1990). In our following experiments we suppressed this surface effect by a suitable choice of the probe beam polarization:

– On a {110}-surface (cleaved surface) the electro-optic effect is zero in general.
– On a {100}-surface (wafer surface) this contribution has no influence in the case of polarization directions along a main crystal axis.

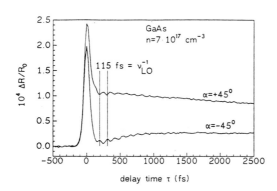

Fig. 1: Polarization-dependent time resolved reflectivity of (100)-GaAs with LO-Phonon oscillations (α: angle of probe beam polarization to the (010)-crystal axis).

The measurements on undoped GaAs-samples at room temperature show a fast increase (see Figure 2) of reflectivity, which follows closely the shape of the laser pulses. For $\tau > 200$ fs the signal decreases slowly.

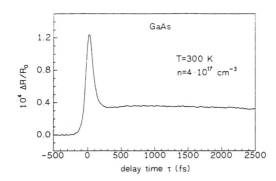

Fig. 2: Transient change of reflectivity of an intrinsic GaAs-sample at room temperature with perpendicular excite–probe–polarization.

There are different possible origins for the spike around zero delay time. If one or more fast relaxation processes exist, they will lead to a decrease of reflectivity determined by the corresponding relaxation times. Additionally, the interference of excite- and probe-beam will lead to a transient grating inside the sample, which will diffract a part of the excite beam intensity into the direction of the reflected probe beam. This will lead to a peak of reflectivity in the case of coincidence of the two laser pulses (Böhne 1990). The measurements with parallel polarization of excite- and probe-beam on GaAs show a higher spike than with perpendicular polarization. Measurements on InP have only a small spike in perpendicular polarization. These two facts suggest that in GaAs the coherent contribution is important only for parallel polarization. The fast depopulation of the initial excited electronic states by intervalley scattering, which is in InP less important than in GaAs at photon energies of 2 eV, induces a fast decrease in reflectivity of GaAs for both polarization configurations (Schäfer 1990). We could confirm this conclusion with measurements at low temperature (2 K) and low excitation density. The low temperature causes a higher band gap, and the lower excitation density decreases the Coulomb scattering rate. Under these circumstances the intervalley scattering rate decreases strongly. The experiment confirms this: the spike around zero delay time disappears at perpendicular polarization. Only for parallel polarization the coherent interaction of excite- and probe-beams still leads to a spike.

Figure 3 shows a significant dependence of the transient reflectivity signal on type and level of doping. All measurements were performed at room temperature with excitation densities of about $4 \cdot 10^{17}$ cm^{-3}. The very different height of $\Delta R/R_0$ for delay times > 1 ps can be explained with Coulomb scattering between the initial cold –doping dependent– carriers and the optically excited hot carriers. After some picoseconds they reach thermal equilibrium at a temperature depending on the ratio of the optically excited carriers and the doping density.

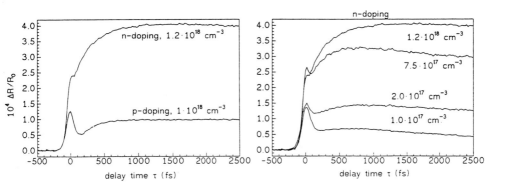

Fig. 3: Time resolved reflectivity of GaAs for different type (Fig. 3a) and level (Fig. 3b) of doping ($n_{ex} = 4 \cdot 10^{17}$ cm^{-3}).

3. TIME RESOLVED ABSORPTION SPECTRA AT THE BAND EDGE OF GaAs

We repeated the classical experiment of Shank et al. (Shank 1979) with somewhat higher time resolution and measured the transient absorption at the band edge of GaAs at 2 K and 77 K. In order to get synchronized short laser pulses at 1.5 eV we amplified the CPM-laser pulses with a copper vapour laser pumped six–pass dye laser. The amplified pulses were focused on fused silica to generate white light continuum pulses. The sample was excited at 2 eV and we measured time delayed spectra with a liquid nitrogen cooled CCD-camera. The sample was immersed in liquid helium or nitrogen. Figure 4 shows the absorption spectra for different time delays between excite and probe pulses at 2 eV. The band gap energy of the unexcited sample is 1.52 eV, the 1s-exciton lies 4.2 meV below. Carrier scattering in the first few hundred femtoseconds broadens the exciton line and the height of absorption decreases. We observed no shift of the peak position. If there were any shift of the 1s-exciton energy, its value must be small compared with the broadening of the absorption line. For $\tau > 1$ ps the exciton peak disappears completely and the absorption decreases due to Pauli-blocking of electron–hole–pair states near the band edge.

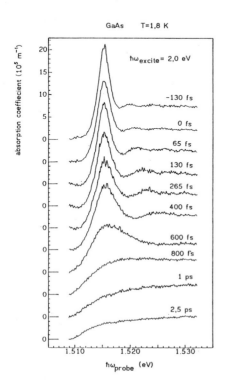

Fig. 4: Transient absorption spectra at the band edge of GaAs.

4. CONCLUSION

We measured the time resolved reflectivity at 2 eV and transmission spectra around 1.52 eV on GaAs after excitation with short pulses from a CPM-laser. The reflectivity measurements show the influence of phase space filling, intervalley scattering, doping of the samples and in parallel polarization the coherent interaction of the two laser beams inside the sample. The time resolved absorption spectra show strong bleaching of the 1s-exciton due to the scattering by hot carriers and a subsequent decrease of absorption due to the filling of electron- and hole-states near the band edges.

Böhne G, Sure T, Ulbrich R G, Schäfer W 1990 *Phys. Rev. B* **41** 7549
Cho G C, Kütt W, Kurz H 1990 *Phys. Rev. Lett.* **65** 764
Schäfer W, Böhne G, Sure T, Ulbrich R G 1990 *J. of Lum.* **45** 211
Shank C V, Fork R L, Leheny R F, Shah J 1979 *Phys. Rev. Lett.* **42** 112
Zhou X Q, Cho G C, Lemmer U, Kütt W, Wolter K, Kurz H 1989 *Solid-State Electr.* **32** 1591

Inst. Phys. Conf. Ser. No 126: Section V
Paper presented at Int. Symp. on Ultrafast Processes in Spectroscopy, Bayreuth, 1991

333

Femtosecond relaxation of photoexcited single-species carriers in bulk doped semiconductors

G. M. Gale, A. Chébira and E. Fazio

Laboratoire d'Optique Quantique du CNRS, Ecole Polytechnique, 91128 Palaiseau, France

J. Chesnoy

Alcatel Alsthom Recherche, 91460 Marcoussis, France

Time-resolved studies of the luminescence of weakly photoexcited doped semiconductors by femtosecond up-conversion techniques allow the selective investigation of the dynamic behavior of the minority species. This assertion is illustrated by experiments on the relaxation of heavy holes to the top of the valence band in n-doped bulk gallium arsenide and on the decay of electrons to the bottom of the conduction band in p-doped bulk indium phosphide. For n-doped gallium arsenide at 77 K luminescence rise-times $\approx$ 350 fs are observed and simply interpreted in terms of the interaction of the holes with the electron bath and the lattice. In the case of indium phosphide time-resolved luminescence behavior is more complex and requires a more detailed interpretation.

INTRODUCTION

Picosecond and sub-picosecond laser spectroscopy experiments in semiconductors have recently allowed the probing of the ultrafast relaxation dynamics of photo-generated electron-hole plasmas (see for example Shah et al 1987, Elsaesser et al 1991). These experiments provide information germane to the temporal evolution of the hot-carrier distributions. This knowledge is important, both for the understanding of the fundamental physics of the relaxation processes involved and for the technology of device development.

It now seems clear that carrier-carrier interaction, free-carrier screening and phonon emission are important features of the thermalisation and relaxation process of photoexcited semiconductors. However, experimental observables (for example time-resolved luminescence) contain averaged data concerning the electron and hole distribution functions and it is difficult to estimate the relative contribution of each species to the overall relaxation process. The theoretical description of the electron-hole system is complex because carrier temperature cannot be defined at short times and because of the difficulties of a correct description of dynamic screening effects. A further complication in semiconductors such as GaAs or InP may be provoked by intervalley transfer which occurs when the electron energy exceeds the Γ/satellite-valley gap.

Many of the above problems may be simplified by studying the band-gap luminescence of weakly photoexcited doped semiconductors where the number of photo-generated electron-hole pairs is small with respect to the number of ionized donors or acceptors. In this case the time-resolved band-gap luminescence is directly related to the to the rate of arrival of minority carriers at the band edge, as the majority carrier distribution remains essentially unchanged. The theoretical treatment of this situation should be considerably facilitated with respect to that of the electron-hole plasma in undoped semiconductors because in the present case the majority carriers retain thermal equilibrium at the lattice temperature.

EXPERIMENTAL

The experiment is performed with a 65 fs duration, 50 MHz repetition rate, 30 mW average power synchronously-pumped dye laser operating at 620 nm. A small fraction of the laser beam is employed to excite an electron-hole plasma in the doped semiconductor crystal, which is cooled to 77 K, and the emitted luminescence is collected in a back-scattering geometry by a double Cassegrain telescope and refocussed into a 100 μm thick nonlinear urea crystal. A second laser beam (probe), suitably time-delayed with respect to the first, is also focussed into the urea crystal to perform nonlinear sum frequency generation with the luminescence and the UV signal produced by this up-conversion process is detected by a photon counting system after spectral selection to eliminate light at 310 nm. This system allows time sampling of the luminescence with a time-resolution down to 90 fs and a band-width of more than 100 meV. The system time-resolution and the position of zero delay may be determined by detecting the sum at 310 nm of laser light back-scattered from the crystal and the probe beam.

N-DOPED GALLIUM ARSENIDE

We have investigated the band-gap luminescence of n-doped gallium arsenide for three silicon doping concentrations of 6×10^{17}, 1.5×10^{18} and 2.4×10^{18} cm^{-3}. At 77 K the donors are fully ionized as determined by Hall-effect measurements between 5 and 300 K. The band-gap of undoped gallium arsenide at 77 K is 1.51 eV. Figure 1 shows the result obtained in the sample of GaAs containing 2.4×10^{18} free electrons at 77 K. The number density of electron-hole pairs created in the conditions of this experiment, with a laser excitation energy of 2.0 eV, is estimated to be $\approx 10^{17}$ cm^{-3}. The luminescence signal rises from zero to a plateau level within about 1.3 ps. Subsequent decay of the luminescence signal is very slow (about 10% over 100 ps) and this decay can be attributed to recombination. At higher laser excitation levels, approaching 2×10^{18}, we observe quite similar behavior at short times but at longer times the luminescence signal continues to increase over ≈ 20 ps before reaching a maximum (not shown in figure). This continued luminescence increase can be attributed to the slow increase of the number of electrons in the Γ valley due to the return of photo-excited electrons from the lateral valleys after rapid initial intervalley transfer. When the number of photo-excited electrons is negligible compared to the number of donors this effect is no longer observable. The inset of fig. 1 shows the initial rise of the band-gap luminescence. We attribute this initial rise to the decay of holes to the top of the valence band. Luminescence rises fairly exponentially as can be seen from the fit in the inset (smooth curve) which is obtained by convolving $[1 - \exp(-t/\tau)]$ with the system response function. From the fit we obtain τ = 330 fs ($\pm$ 30 fs) at a sample temperature of 77 K for this doping concentration of 2.4×10^{18} cm^{-3}. Luminescence rise-times τ of 340 ± 40 fs and 460 ± 90 fs are obtained at the lower doping concentrations of 1.5×10^{18} and 6×10^{17} cm^{-3} respectively. For the lowest donor concentration it proved impossible to decrease the excitation sufficiently to obtain a perfect plateau at short times (we note that the signal observed at low excitation is proportional to the product of hole and donor concentrations and thus varies approximately as donor concentration squared) and hence the longest time has been corrected for the effect of the slow increase of electron number density in this case. We therefore observe hole relaxation times in n-doped gallium arsenide $\approx$ 350 fs which are fairly insensitive to donor concentration.

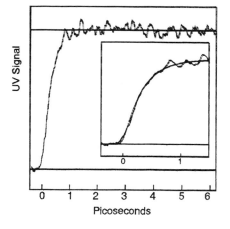

Fig. 1. Up-converted luminescence signal in n-doped gallium arsenide at 77 K.

We may interpret these results in terms of a simple theoretical model which, although somewhat crude, appears to provide some insight into the mechanisms governing relaxation of the holes.

As the 'light' holes in gallium arsenide do not contribute significantly to band-edge luminescence due to their small density of states at small wave-vector and have a dispersion equivalent to that of the heavy holes at moderate to large wave-vector, we will consider that both the 'light' and the heavy holes may be represented by a single isotropic and parabolic band with the mass of the heavy hole. We will initially neglect the approximately 20% of holes excited into the split-off band. Because we are considering a situation where the hole number density is small with respect to the electron number density we will neglect hole-hole interactions compared to hole-electron interactions (this hypothesis can always be justified at sufficiently low relative hole/electron density). The hole distribution hence thermalizes uniquely though collision with the cold electron bath.

Spatial diffusion of the holes is negligible in the times considered here and the only factors affecting hole population are laser excitation, coupling with the electron bath and interaction with the lattice. Because of the wide spread of hole/TO phonon coupling constants cited in the literature we will opt for the lowest value, which is negligible compared to the two remaining interactions with LO phonons and the electrons.

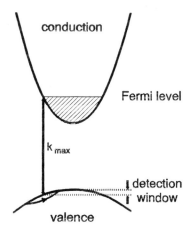

Fig. 2. Hole detection-window and electron band Fermi-level.

We assume that the holes interact with the equilibrium electron bath via a static screened Coulomb potential (with a screening parameter ß), which is a good approximation here due to the large hole/electron mass ratio. At the electron densities considered in this experiment the electron distribution is always degenerate and one must use the Fermi-Dirac screening parameter ß which is smaller than the corresponding Maxwell-Boltzmann parameter. In contrast, because of the relatively large hole mass, at *thermal equilibrium*, the hole occupation function is given by a Maxwell-Boltzmann distribution. In these conditions the transition rate T^e_{ij} of a hole of energy E_i to a state of energy E_j due to interaction with the electron bath can be calculated semi-analytically. Similarly, the transition rate involving LO phonons T^p_{ij} of a hole of energy E_i to $E_i - E_{LO}$ can also

be calculated semi-analytically. This simplification of the calculation of time-independent hole transition rates to a semi-analytic form allows the calculation of a total 120 x 120 T_{ij} transition matrix, with 1 meV steps, in a few minutes of microcomputer time. The evolution of an arbitrary hole distribution may then be rapidly evaluated in a time-dependent simulation.

The results of such simulations at a temperature of 77 K show that an initial laser-excited hole distribution centered near 70 meV thermalizes in ≈ 300 fs and relaxes completely to a Maxwell-Boltzmann distribution at the lattice temperature in ≈ 1.3 ps. Further, even though the hole/electron bath interaction is dominant, the hole distribution evolution rate is very insensitive to donor (electron) concentration. This insensitivity is due to the nature of the Fermi-Dirac distribution. As the hole mass is large, the average energy loss per collision with electrons is small for the holes. This means that only electrons near the top of the Fermi-Dirac distribution can accept hole

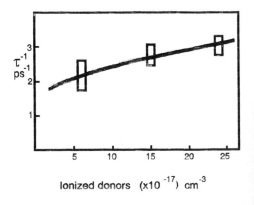

Fig. 3. Three inverse luminescence rise-times in n-doped GaAs at 77 K. The full curve is calculated (see text).

energy because of the degeneracy of the Fermi-Dirac distribution. Hence the number of electrons which can interact with holes is smaller than the total number of electrons and it turns out that the variation of this number with total electron density is almost perfectly compensated by screening as borne out by the simulations. Nevertheless, the luminescence rise-time can vary with electron density because the detection window width depends on the position of the Fermi-Dirac level in the conduction band as illustrated in fig. 2, assuming k-selection. Holes can only recombine and produce luminescence when their k-vector is less than k_{max}. The value of k_{max} increases with rising donor concentration, leading to a corresponding increase in detection window width.

Figure 3 shows the calculated luminescence rise-times (full curve) compared to experimental values (open rectangles). Agreement is good, especially considering that the calculation contains no adjustable parameters. However this agreement may be partially fortuitous as we have replaced the three valence bands of GaAs by a single band in our calculation.

P-DOPED INDIUM PHOSPHIDE

The behavior of electrons in indium phosphide may be isolated in an analogous fashion by studying the time-resolved luminescence of p-doped samples. Figure 4 shows the temporal variation of the band-edge luminescence excited with a 2 ev laser in a bulk sample of InP at 77 K, doped with 2×10^{18} atoms/cm^3 of Zn. The luminescence first rises rapidly (see inset), in about 600 fs, then continues to rise much more slowly over times $\approx$ 100 ps. This slow secondary rise, which we observe in all samples at 77 K, is reminiscent of the intervalley transfer effect observed in gallium arsenide and indeed we can fit the curves in fig. 4 using the phenomenological expression Lum(t) = $[1 - \exp(-t/\tau_1)] \times [1 - f \exp(-t/\tau_2)]$ with τ_1 = 600 fs and τ_2 = 45 ps and where f, $\approx$ 0.7, is the fraction of electrons

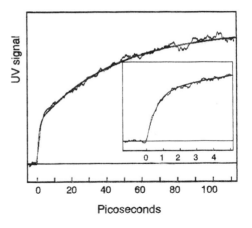

Fig. 4. Up-converted luminescence signal in p-doped (2E18 Zn) indium phosphide at 77 K over 110 and 5 ps (inset).

escaping from the Γ valley and returning slowly with the time-constant τ_2. However, the accepted Γ/L-valley separation in InP appears to be somewhat larger than the initial electron energy excess in the conduction band which should normally rule out intervalley transfer. Nevertheless, we note that intervalley transfer could occur, even in these conditions, if the electron thermalisation time is small compared to the electron relaxation time, as appears to be the case in gallium arsenide (Elsaesser et al 1991).

At a constant laser excitation level of $\approx 10^{18}$cm^{-3}, the fast-rising component, τ_1, exhibits a strong dependence on doping concentration N_D. τ_1 varies from ≈ 2 ps in undoped samples, through ≈ 600 fs at $N_D = 2 \times 10^{18}$ to 250 fs at $N_D = 10^{19}$. The latter value may be limited by the system detection window of ≈ 100 meV. The variation of electron relaxation time with donor concentration could reflect the change in the electron/hole-bath interaction in InP with doping density, given that the hole distribution is not strongly degenerate in this case, in contrast to the situation for electrons in gallium arsenide. However, we note a possible role for LO-phonon saturation effects here (these effects were absent in n-doped GaAs due to the low excitation intensity and the small number of phonons emitted by the holes). If LO-phonon saturation increases electron decay times at low doping levels then an augmentation of doping concentration could lead to a decrease in electron relaxation time due to a possible decrease in LO-phonon lifetime similar to that observed in doped GaAs (Bogani and Vallée, 1992).

CONCLUSION

The relaxation to thermal equilibrium of holes excited by 2 eV photons in n-doped gallium arsenide at 77 K can be characterized by a decay time ≈ 350 fs, almost independent of doping concentration in the range 6×10^{17} to 2.4×10^{18} cm^{-3}. The results are well described by a preliminary model which includes only hole/LO-phonon and hole/electron-bath interactions.

Electrons excited by 2 eV photons in p-doped indium phosphide in contrast exhibit two-component relaxation behavior with the fast component showing a pronounced dependence on doping concentration which may either be due to a direct or an indirect influence of the hole bath. A very slow component with a ≈ 45 ps risetime is also observed. Further investigation of InP is required to elucidate the relaxation pathways of hot electrons in this material.

REFERENCES

Bogani F and Vallée F 1992 *ibid*
Elsaesser T, Shah J, Rota L and Lugli P 1991 *Phys. Rev. Lett.* **66** 1751
Shah J, Deveaud B, Damen T L, Tsung WW T, Gossard A C and Lugli P 1987 *Phys. Rev. Lett.* **59** 2222

Inst. Phys. Conf. Ser. No 126: Section V
Paper presented at Int. Symp. on Ultrafast Processes in Spectroscopy, Bayreuth, 1991

Hot carrier relaxation in doped III–V compounds studied by femtosecond luminescence

X.Q. Zhou*, and H. Kurz

Institute of Semiconductor Electronics, Technical University of Aachen,
W-5100 Aachen, Germany

*: Present address: Max-Planck-Institut für Festkörperforschung,
Heisenbergstrasse 1, 7000 Stuttgart 80, Germany

ABSTRACT:

The use of doped samples allows us to separate the electron and hole dynamics in time-resolved luminescence. Femtosecond thermalization and subpicosecond energy relaxation of electrons and holes are demonstrated in GaAs and InP. Different relaxation channels such as electron-electron, electron-hole, as well as hole-TO-phonon scattering are discussed.

1. INTRODUCTION

In recent years ultrafast relaxation processes of hot carriers in intrinsic semiconductors have been extensively studied with time-resolved optical spectroscopy. Carrier-carrier-scattering (Leheny et al 1979, Knox et al 1984, 1988, Zhou et al 1989, 1990, Elsaesser 1991), carrier-LO-phonon interaction and intervalley transfer (Taylor et al 1985, Shah et al 1987) have been found to be important processes. However, a detailed analysis of experimental data is often difficult, because the usual optical techniques such as transmission and luminescence probe only the sum or the product of the electron and hole distribution functions. An experimental separation of electron and hole dynamics is then practically not possible. In previous studies of electron relaxation in GaAs, hole relaxation has been assumed to be completed within 300 fs (Shah et al 1987). The ultrafast relaxation of hot holes is less understood because of experimental difficulties. In this report, we attempt to separately monitor the relaxation of electron and hole plasmas by using doped samples. Our data also provide insight into the influence of doping on energy relaxation of hot carriers.

2. EXPERIMENTALS

With a CPM-Laser, optical pulses (50fs, 25 mW per beam) at 2 eV are fed into a luminescence up-conversion setup. The temporal resolution of the up-conversion system is $\leq$ 100 fs in the whole luminescence energy range, and the energy resolution is about 30 meV (Zhou 1991). The samples used are epitaxial layers grown by MOCVD. All experiments are performed at room temperature.

3. RESULTS AND DISCUSSIONS

3.1. Electron internal thermalization and energy relaxation in p-doped samples

Fig. 1 shows the luminescence spectra of p-doped GaAs and InP at a delay time of 100 fs after optical excitation. Since the excitation density of $10^{18}cm^{-3}$ is kept smaller than the doping density of $p = 2\times10^{18}cm^{-3}$, the spectra are determined mainly by the dynamics of hot electrons. In both materials the spectra exhibt Maxwellian behaviour even at energies as high as 1.7 eV. These features, induced by strong electron-electron scattering, give evidence for an internal thermalization of the hot electron subsystem within 100 fs. The effective carrier temperature (Tc) extracted is 1100 K for p-GaAs and 2150 K for p-InP, respectively. Assuming that the hole temperature does not deviate largely from the lattice temperature because of p-doping, the electron temperature can be evaluated from Tc. In p-GaAs the evaluated electron temperature is about 1700 K, which is much lower than 3400 K deduced from the excess energy of excitation. Similar large differences in temperature have also been observed between p-doped and intrinsic GaAs. Obviously, a dramatic energy loss has to be anticipated during the optical excitation, which results essentially from the enhanced inelastic interaction with majority holes. In contrast, the measured temperature in p-InP does not differ very much from the excess energy of excitation as well as from the temperature in intrinsic InP. This indicates a relative weaker influence of doping on the initial energy relaxation of electrons in InP, which is not inconsistent with the stronger electron-LO-phonon coupling in InP.

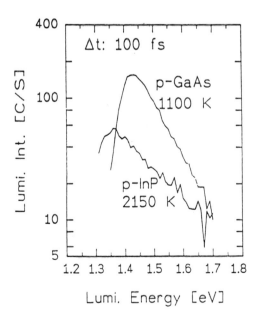

Fig.1: Luminescence spectra of p-GaAs and p-InP with $p = 2\times10^{18}cm^{-3}$ at a delay time of 100fs. Excitation density : $10^{18}cm^{-3}$.

In a complementary fashion the band-edge luminescence (BEL) monitors the relaxation of electrons to the central valley minimum. Fig. 2 shows the temporal evolution of BEL of GaAs at different p-doping levels. Compared to the intrinsic case, the rise time is drastically shortened with increasing p-doping. Since the slow return time of electrons from the L-valley controls the BEL in undoped GaAs (Shah et al 1987, Zhou et al 1990), the reduction of rise time in heavily p-doped GaAs indicates a significant increase of intravalley scattering process for excited electrons, consistent with the low initial electron temperature measured here. As a consequence, the phonon mediated intervalley transfer must be bypassed for the highest doping level.

3.2. Energy relaxation of holes in n-doped samples

Fig. 3 illustrates the temporal evolution of BEL of n-doped GaAs in an energy window of 30 meV at the band-edge. The luminescence transients are normalized to the signal maximum. At an excitation density of $5x10^{16} cm^{-3}$ and doping levels above $2x10^{17} cm^{-3}$, the BEL is controlled by the hole dynamics. Due to the degenerated doping, electron heating at the band-edge is not important. The measured BEL rises rapidly to its maximum and remains there for a couple of picoseconds before diffusive carrier transport and recombination. At both doping levels, the signal has already reached 30% of the maximum intensity during the optical excitation, implying a fast broadening of hole distributions down to the band-edge in less than 50 fs. The signal maximum is reached 700 fs after excitation. Since any significant relaxation of holes would lead to a change in the band-edge signal, we conclude that the thermal equilibrium between excited holes and lattice has to be established in about 700 fs in GaAs at a hole density of $5x10^{16} cm^{-3}$. The same rise times for different doping densities imply that the energy

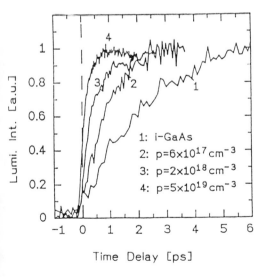

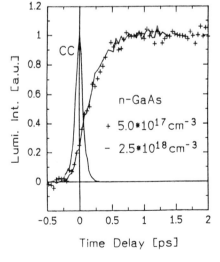

Fig.2: Band-edge Luminescence of p-GaAs at an excitation density of $5x10^{16} cm^{-3}$ and different doping levels.

Fig.3: Band-edge Luminescence of n-GaAs at an excitation density of $5x10^{16} cm^{-3}$ and different doping levels.

relaxation of holes does not depend on the concentration of electrons. This indicates that intrinsic relaxation process in the valence band dominates the electron-hole interaction in the energy relaxation of holes. An extrapolation of this result to the case of undoped material is therefore allowed. Similar results are obtained in n-InP. The thermalization time of holes with the lattice is measured to be about 400 fs, consistent with the stronger hole-phonon coupling in this material.

In order to explain the relaxation mechanism of hot holes from BEL data, a theoretical model including hole - LO- and TO-phonon scattering has been carried out. An instantanous internal thermalization of holes among the three valence bands is assumed. Such an ultrafast thermalization of holes has previously been reported by Knox et al (1988). Our results reveal the dominance of hole-TO-phonon scattering in hole relaxation. The optical deformation potential coupling constant is further estimated to be 2. (+10,-5) ev for GaAs and 30 (+/-10) eV for InP, respectively.

4. SUMMARY

Thermalization and energy relaxation of photoexcited carriers in doped GaAs and InP have been studied by femtosecond luminescence spectroscopy. For an excitation density of $10^{18} cm^{-3}$ the electron internal thermalization time is measured to be ≤ 10 fs. A significant energy loss during the optical excitation is observed in p-GaAs. The enhanced electron-hole interaction in the central valley due to heavy p-doping leads to a bypass of electron intervalley transfer. In the n-doped case, relaxation of holes i dominated by TO-phonon emission. Thermal equilibrium between holes and the lattice is established on a subpicosecond time scale.

5. ACKNOWLEDGEMENT

The authors thank H.M. van Driel for a critical reading of the manuscript and K. Lee for a fruitful discussion. The work has been supported by the Deutsche Forschungsgemeinschaft (Ku 540/3) and by the Alfried Krupp Stiftung.

6. REFERENCES

Elsaesser T., Shah J., Rota., and Lugli P. 1991, Phys. Rev. Lett. **66**, 1757

Knox W.H., Downer M.C., Fork R.L., and Shank C.V. 1984 Opt. Lett. **9** 150

Knox W.H., Chemla D.C., Livescu G., Cunningham J., and Henry J.E. 1988 Phys. Rev Lett. **61**, 1290

Leheny R.F., Shah J., Fork R.L., and Shank C.V. 1979 Solid State Commun. **31**, 809

Shah J., Deveaud B., Damen T.C., Tsang W.T., Gossard A.C., and Lugli P. 1987 Phys. Rev. Lett. **59**, 2222

Taylor A.J., Erskine D.J., and Tang C.L. 1985 J. Opt. Soc. Am. **B2**, 663

Zhou X.Q., Cho G.C., Lemmer U., Kütt W., Wolter K., and Kurz H. 1989 Solid State Electron. **32**, 1591

Zhou X.Q., Lemmer U., Seibert K., Cho G.C., Kütt W, Wolter K., and Kurz H. 1990 Proceedings of 20th ICPS Thessaloniki, World Scientific, 2522

Zhou X.Q. 1991 Ph.D.thesis, Technical University of Aachen

Inst. Phys. Conf. Ser. No 126: Section V
Paper presented at Int. Symp. on Ultrafast Processes in Spectroscopy, Bayreuth, 1991

343

Reversible gallium arsenide bleaching under the action of picosecond light pulses

N.N.Ageeva, I.L.Bronevoi, S.E.Kumekov and V.A.Mironov

Institute of Radio Engineering & Electronics, USSR Academy of Sciences, USSR, 103907, Moscow, GSP-3, Marx Avenue, 18

ABSTRACT: A reversible threshold increment to the bleaching of gallium arsenide under picosecond light at $\hbar\omega \approx E_g$ illumination has been discovered and studied. The increment is a result of additional photogeneration of carriers (under intraband carrier heating by the light) and recombination superluminescence. On completion of the pump pulse the bleaching spectra acquire a universal form.

An almost reversible change in the transparency spectrum of an epitaxial layer of GaAs under interband absorption of picosecond light pulses with photon energies $\hbar\omega_{ex}$ slightly above the width of the forbidden band E_g has been discovered by Bronevoi *at al* (1985). The duration of exciting pulses was 30 ps to 40 ps, spontaneous recombination time for the samples studied was $\tau_R \gtrsim 500$ ps. All measurements were at room temperature. Changes in transparency were measured with the aid of a probing beam with wavelength λ_p tunable from 871 nm to 620 nm and a variable optical delay.

At exciting pulse energies W_e above a certain threshold value the sample's transparency was found to rise and decay almost synchronously with variations of the pulse intensity. A certain residual brightening was observed to be present after the pulse (at times substantially shorter than τ_R). The residual brightening spectrum has been shown (Bronevoi *at al* 1985, Ageeva *at al* 1989) to be independent both from the exciting pulse energy W_e and from the exciting photon energy $\hbar\omega_{ex}$. This poses two problems: 1) what determines the state responsible for residual brightening, and 2) what mechanism governs reversible brightening.The state of carriers after a pulse is determined by their temperature T_e and position of Fermi quasi-levels μ_e and μ_h. In the time scale used T_e may be taken to coincide with the lattice temperature, as confirmed by the residual brightening spectrum pattern. Equal concentrations of electrons and holes $n(T_e,\mu_e)=p(T_h,\mu_h)$ provide one interrelation between μ_e and μ_h. Saturation condition $\mu_e-\mu_h=\hbar\omega_{ex}$ could be the second possible interrelation, but this, however, is contradicted by the independence of the residual brightening spectrum from $\hbar\omega_{ex}$ (Ageeva *at al* 1989). Experiment and analysis (Ageeva *at al* 1989, Bronevoi *at al* 1986) show the main factor determining the state of charge carriers after the pulse to be an abrupt shedding of redundant concentration via superluminescence to a level, at which the maximal concentration of electron – hole pairs yet prevents population

E_g of the unexcited sample, and found to correlate with the exciting pulse in the picosecond range, with the same energy threshold W_e as that of reversible brightening.

The quasi–equilibrium state of charge carriers may be considered to be present also during the exciting pulse, as confirmed by comparing experimental and analitical spectra. At $\hbar\omega_{ex}$ slightly over E_g this comparison shows $\mu_e-\mu_h=E_g$, which is undiscernable from $\mu_e-\mu_h\approx\hbar\omega_{ex}$ in this case. The comparison also disclosed the reversible share in brightening to be related to carrier heating during excitation. This heating may be assumed to be due to intraband absorption. This assumption is confirmed both by analysis (Bronevoi *at al* 1989, Kumekov and Perel' 1988) and direct experiments (Ageeva *at al* 1990), in which the sample was heated by a pulse with photon energies below E_g simultaneously with the exciting pulse. This heating pulse was found to substantially increase the reversible share in brightening.

It should be noted, that carrier heating may also be caused in part by superluminescence, since carriers extinction is at lower energies than their generation.

1. Bronevoi I L, Gadonas R A, Krasauskas V V, Lifshits T M, Piskarskas A S, Sinitsin M A and Yavich B S 1985 *JETP Lett.* **42** 395
2. Ageeva N N, Bronevoi I L, Dyadyushkin E G, Mironov V A, Kumekov S E and Perel' V I 1989 *Solid State Commun.* **72** 625
3. Bronevoi I L, Kumekov S E and Perel' V I 1986 *JETP Lett.* **43** 473
4. Ageeva N N, Bronevoi I L, Dyadyushkin E G and Yavich B S 1988 *JETP Lett.* **48** 276
5. Kumekov S E and Perel' V I 1988 *JETP* **67** 193
6. Ageeva N N, Borisov V B, Bronevoi I L, Mironov V A, Kumekov S E, Perel' V I, Yavich B S and Gadonas R 1990 *Solid State Commun.* **75** 167

Inst. Phys. Conf. Ser. No 126: Section V
Paper presented at Int. Symp. on Ultrafast Processes in Spectroscopy, Bayreuth, 1991

Theory of hot LO–phonon decay in laser excited direct-gap semiconductors

U.Wenschuh , E.Heiner† and K.W.Becker

Max-Planck-Institut für Festkörperforschung, Heisenbergstr.1, 7000 Stuttgart 80, Germany
†JINR Dubna, Head Post Office, P.O.B.79, 101000 Moscow, USSR

1. Introduction

We consider the relaxation of hot longitudinal optical (LO) phonons during the evolution of a highly excited electron–hole plasma of a direct-gap semiconductor. This problem has been considered recently by several authors in connection with the evaluation of energy loss rates of a hot electron-hole plasma in semiconductors and in quantum wells [1-6]. There it was always assumed that (i) the hot LO–phonons decay exponentially and (ii) a temperature rise in the acoustical phonon system can be neglected. Whereas the first assumption is not very satisfying at least from a theoretical point of view the latter assumption seems also not to be fulfilled experimentally for instance in GaAs, where the carrier relaxation times are in the order of some ps. There, the sound velocity is s=35.7 nm/ps [7] and the laser focus on the sample in general has a diameter of about $15\mu m$ [8]. It is the aim of this paper to avoid these shortcomings. We derive equations of motion which describe the hot LO-phonon decay in a coupled phonon system by taking into account anharmonic phonon interactions. Thereby use is made of a time dependent projection technique [9]. The theory is developed for excitation densities of about $10^{17} \sim 10^{18} cm^3$ and initial temperatures of about 60K.

2. Model Hamiltonian

We start from the following Hamiltionian

$$H = H_0 + H_1(t) + H_2 + H_3 + H_4 \tag{1}$$

where

$$H_0 = \sum_{k,\sigma,\nu=\pm 1} \epsilon(k,\nu)a^+_{k\nu\sigma}a_{k\nu\sigma} + \sum_{q\mu=\mu_1,\mu_2} \omega(q,\mu)(b^+_{q\mu}b_{q\mu} + 1/2) \tag{2}$$

$$H_1(t) = f(t)\sum_{k\sigma}(\varphi_k e^{-i\omega t}a^+_{k+1\sigma}a_{k-1\sigma} + h.c.) \tag{3}$$

$$H_2 = \frac{1}{2}\sum_{q}\sum_{kk'\nu\nu'\sigma\sigma'} V(q)a^+_{k+q\nu\sigma}a^+_{k'-q\nu'\sigma'}a_{k'\nu'\sigma'}a_{k\nu\sigma} \tag{4}$$

$$H_3 = \sum_{q\mu}\sum_{k\nu\sigma} M_\mu(q)a^+_{k\nu\sigma}a_{k+q\nu\sigma}(b^+_{q\mu} + b_{-q\mu}) \tag{5}$$

$$H_4 = \sum_{q_1 q_2 \mu_1 \mu_2} c(q_1\mu_2, q_2\mu_1, q_1 - q_2\mu_1)b_{q_1\mu_2}b^+_{q_2\mu_1}b^+_{q_1 - q_2\mu_1} + h.c.. \tag{6}$$

Here we have introduced the following abbreviations

$$\varphi_k = \left(\frac{\hbar e^2 I(1-R)}{m\omega\epsilon_0 c}b_{k\nu\nu'}\right)^{1/2} \tag{7}$$

$$V(q) = e^2/4\pi\epsilon_0 q^2 \tag{8}$$

$$M_\mu(q) = \frac{2\pi}{i}\left(\frac{e^2\hbar\omega}{2\pi V}\left(\frac{1}{\epsilon_\infty}-\frac{1}{\epsilon_0}\right)\right)^{1/2}\frac{1}{q^2} \tag{9}$$

$$c(q_1\mu_2, q_2\mu_1, q_1 - q_2\mu_1) = -\frac{i}{G^{3/2}}\gamma\frac{2M}{\sqrt{3}}\frac{1}{s}w(q_1\mu_2)\,w(q_2\mu_1)\,w(q_1 - q_2\mu_1). \tag{10}$$

The first part of the unperturbed Hamitonian H_0 describes free electrons ($\nu = +1$) and holes ($\nu = -1$) and the second part unperturbed phonons. Here, the sum over μ runs over all longitudinal optical branches μ_1 and over all accustical (LA, TA) branches μ_2. H_2 is the interaction with the laser light, where $f(t)$ is the envelope of the laser pulse. The parameter φ_k, defined in (7), is the transfer matrix element in dipole and rotating wave approximation where use has been made of the so-called slowly varying amplitude approximation (SVAA) [10]. I is the intensity of the laser pulse, R is the reflectivity of the sample and $b_{k\nu\nu'}$ is the oscillator strength [11]. Finally, the Hamiltonian H_3 describes the coupling between the charge carriers and the phonons and H_4 anharmonic phonon interactions where umklapp processes [12] have been neglected. Note that in H_4 only a coupling between optical and accoustical phonons is taken into account but no couplings among optical and accoustical phonons. The parameter γ in (10) is the Grüneisen constant.

3. Projection Technique

By applying a modern version [9] of Robertsons time dependent projection technique [13], we derive equations of motion for the nonequilibrium phonon system. We introduce the following set of relevant observables:

$$\{O_i\} = \{1, H_{ac}, H_{op}(q)\}. \tag{11}$$

Here H_{ac} and $H_{op}(q)$ are the unperturbed acoustical and optical phonon energies where the latter quantity is decomposed with respect to the wave vector q

$$H_{ac} = \sum_{q\mu_1} \omega(q, \mu_1)(b^+_{q\mu_1}b_{q\mu_1} + 1/2) \tag{12}$$

$$H_{op}(q) = \sum_{\mu_2} \omega(q, \mu_2)(b^+_{q\mu_2}b_{q\mu_2} + 1/2). \tag{13}$$

As is shown below the decomposition (13) into wave vectors q leads to a more detailed description of the time evolution of q–dependent optical phonon occupations $< N_{q\mu_2} > (t)$. In the framework of the Robertson theory we start from the so–called relevant statistical operator for the set (11)

$$\sigma_0(t) = exp\{\lambda_0 1 - \sum_i \lambda_i O_i(t)\} \tag{14}$$

where the parameters λ_i are a set of relevant Lagrange multipliers which corresponds to (11)

$$\{\lambda_i\} = \{\lambda_0, \beta_{ac}, \beta_{op}(q)\}. \tag{15}$$

For $\sigma_0(t)$ the following master equation can be derived by use of a short memory approximation over a coarse grained time axis ($\Delta t \approx 200 fs$)

$$(i\partial_t - \mathcal{PL})\sigma_0 = (-i)\mathcal{P}(t)\,\mathcal{L}(t)\int d\omega E(\omega)\,\delta(\omega - \mathcal{L}_0(t))\,\mathcal{S}(\omega, \mathcal{L}_1(t))\,\mathcal{Q}(t)\,\mathcal{L}(t)\,\sigma_0(t). \tag{16}$$

Here $\mathcal{P}$ is a projector on the set of relevant observables (11), $\mathcal{Q} = 1 - \mathcal{P}$, and $\mathcal{L} = \mathcal{L}_0 + \mathcal{L}_1$ is the Liouville operator defined by $\mathcal{L}A = [H, A]$ for any operator A of the unitary space. The quantity $\mathcal{S}(\omega, \mathcal{L}_1(t))$ is the so-called 'screening superoperator', introduced in ref. [14]. Finally $\mathcal{E}(\omega)$ is defined by $E(\omega) = \pi\delta(\omega) - iP\left(\frac{1}{\omega}\right)$. By multiplying the master equation (16) from the left with the relevant observables (11) and taking the trace we obtain the following evolution equations which describe the dynamics in the nonequilibrium phonon system:

$$\partial_t < H_{op}(q) >= 2\pi \sum_{k\nu\mu_2} |M^s_{\mu_2}(q)|^2 \omega(q,\mu_2) *$$

$$\{ \delta(\epsilon(k,\nu) - \epsilon(k+q,\nu) + \omega(q,\mu_2)) *$$
$$[< \overline{N_{q\mu_2}} >< n_{k,\nu} >< \overline{n_{k+q,\nu}} > - < N_{q\mu_2} >< \overline{n_{k,\nu}} >< n_{k+q,\nu} >]$$
$$- \delta(\epsilon(k,\nu)\epsilon(k-q,\nu) - \omega(q,\mu_2)) *$$
$$[< \overline{N_{q\mu_2}} >< n_{k,\nu} >< \overline{n_{k-q,\nu}} > - < N_{q\mu_2} >< \overline{n_{k,\nu}} >< n_{k-q,\nu} >]\} -$$

$$-4\pi \sum_{q_2\mu_1\mu_2} |c(q\mu_2, q_2\mu_1, q - q_2\mu_1)|^2 \omega(q,\mu_2)\delta(\omega(q,\mu_2) - \omega(q_2,\mu_1) - \omega(q - q_2,\mu_1)) *$$
$$\{< \overline{N_{q\mu_2}} >< \overline{N_{q_2\mu_1}} > + < N_{q-q_2\mu_1} >\} \tag{17}$$

$$\partial_t < H_{ac} >= 4\pi \sum_{q,q_2,\mu_1,\mu_2} |c(q\mu_1, q_2\mu_1, q_2 + q\mu_2)|^2 \omega(q,\mu_1) *$$

$$\delta(\omega(q,\mu_1) - \omega(q_2,\mu_1) - \omega(q_2 + q,\mu_2))\{< \overline{N_{q_2+q\mu_2}} > (< N_{q\mu_1} > + < \overline{N_{q_2\mu_1}} >)\} +$$

$$+4\pi \sum_{q,q_1,\mu_1,\mu_2} |c(q\mu_1, q_1 - q\mu_1, q_1\mu_2)|^2 \omega(q,\mu_1)\delta(\omega(q\mu_1) - \omega(q - q_1\mu_1) - \omega(q_1\mu_2)) *$$
$$\{< \overline{N_{q_1\mu_2}} > (< N_{q\mu_1} > + < \overline{N_{q_1-q\mu_1}} >)\}. \tag{18}$$

The quantity $M^s_\mu(q)$ in (15), defined by $M^s_\mu(q) = M(q)/\epsilon(q,\omega)$, is the screened electron-phonon coupling matrix element [15]. Moreover we have introduced the following abbreviations

$$< \overline{n_{k,\nu}} > = 1- < n_{k,\nu} > \tag{19}$$
$$< \overline{N_{q\mu}} > = 1+ < N_{q\mu} > . \tag{20}$$

4. Method of Solution

Note that the expectation values $< H_{op}(q) > (t)$ and $< H_{ac} > (t)$ on the left hand sides of (17) and (18) can be expressed by time-dependent expectation values $< N_{q\mu} > (t)$ of the phonon occupation number operators. Note also that the right hand sides depend on time dependent expectation values $< n_{k,\nu} > (t)$ ($\nu = \pm 1 = e, h$) of the electron and hole occupation number operators. In the following we assume that the phonons as well as the electrons are in a quasi-equilibrium. This assumption means that the time dependences of $< N_{q\mu} > (t)$ and $< n_{k,\nu} > (t)$ can be expressed by Bose and Fermi functions, respectively

$$< N_{q\mu_1} > (t) = \{\exp[\omega(q,\mu_1)/T_{ac}(t)] - 1\}^{-1} \tag{21}$$
$$< N_{q\mu_2} > (t) = \{\exp[\omega(q,\mu_2)/T_{op}(q,t)] - 1\}^{-1} \tag{22}$$
$$< n_{k,\nu} > (t) = \{\exp[(\epsilon(k,\nu) - \nu\mu_\nu(t))/T_\nu(t)] + 1\}^{-1}. \tag{23}$$

where the Bose distributions depend on time dependent temperatures $T_{ac}(t) = 1/\beta_{ac}(t)$ and $T_{op}(q,t) = 1/\beta_{op}(q,t)$ for the acoustical and optical phonons, respectively. Similarly, the Fermi distributions depend on time dependent temperatures $T_{e,h}(t)$ and on chemical potentials $\mu_{e,h}(t)$. By use of (19), (20) and of

$$
\begin{pmatrix} \partial_{T_{ac}} < H_{ac} > & \partial_{T_{op}(q_i)} < H_{ac} > & \cdots \\ \partial_{T_{ac}} < H_{op}(q_i) > & \partial_{T_{op}(q_i)} < H_{op}(q_i) > & \cdots \\ \vdots & \vdots & \ddots \end{pmatrix} \begin{pmatrix} \partial_t T_{ac} \\ \partial_t T_{op}(q_i) \\ \vdots \end{pmatrix} = \begin{pmatrix} \partial_t < H_{ac} > \\ \partial_t < H_{op}(q_i) > \\ \vdots \end{pmatrix}
$$

(24)

the equations of motions (17), (18) can be rewritten as a system of differential equations for the quasi–equilibrium temperatures $T_{ac}(t), T_{op}(q,t)$. This system of equations is our final result. I is nonlinear and of dimension $(1+N)$, where N is the number of different q–vectors of the optical phonons. Obviously, the system can only be solved if the time dependent carrier temperature $T_{e,h}(t)$ and chemical potentials $\mu_{e,h}(t)$ are known. These can be obtained by two different ways (i) experimentally: The time dependent quantities T_e, T_h, μ_e, μ_h can be deduced from time resolved pump-probe spectroscopy [16] without the assumption of an exponential decay of the phonon occupation. (ii) theoretically: The obervation level (11) has to be extended by introducing observables H_e, H_h, N_e, N_h for the electron system. This leads to a $(5+N)$ dimensional system of differential equations [17] which contains the whole information about the dynamic of the system. In this way it is also possible to investigate the influence of an acoustical phonon temperature which increases in time on the dynamical decay of the LO-occupation.

5. References

[1]   W.Pötz and P.Kocevar, Phys.Rev.B28, 7040 (1983)
[2]   M.Asche and O.G.Sarbei, Phys.Stat.Sol.B141, 487 (1987)
[3]   J.R.Senna and S.Das Sarma, Solid State Commun.64, 1397 (1987)
[4]   A.C.S.Algarte, Phys.Rev.B32, 2388 (1985)
[5]   J.Collet, J.L.Oudar and T.Amand, Phys.Rev.B34, 5443 (1986)
[6]   K.Leo, Ph.D.thesis, University of Stuttgart 1988
[7]   T.B.Bateman, H.J.McSkimin and J.M.Whelan, J.Appl.Phys.30, 544 (1959)
[8]   W.W.Rühle, private communication
[9]   R.Balian,Y.Alhassid and H.Reinhardt, Phys.Rep.131, 1 (1986)
[10] U.Wenschuh and E.Heiner, Phys.Stat.Sol.B162, 303 (1990)
[11] H.Ehrenreich and M.H.Cohen, Phys.Rev.115. 786 (1959)
[12] P.G.Klemens, Phys.Rev.148, 845 (1966)
[13] B.Robertson, Phys.Rev.144, 152 (1966)
[14] E.Heiner, Phys.Stat.Sol.153, 295 (1989)
[15] U.Wenschuh and E.Heiner, to be published
[16] C.W.W.Bradley, R.A.Taylor and J.F.Ryan, Soled State Electronics32, 1173 (1989)
[17] U.Wenschuh, unpublished

Dynamic Burstein–Moss effect in InAs at room temperature and passive mode-locking of a 3 μm Er laser

K.L.Vodopyanov *, H.Graener
    EP III Uni.Bayreuth, Postfach 101251, D-8580 Bayreuth, FRG
C.C.Phillips, I.T.Ferguson
    Imperial College, London SW7 2AZ, UK
* Permanent address: *General Physics Inst, Vavilov str 38, 117942 Moscow*

ABSTRACT: Picosecond pump-probe experiments performed in ultrathin MBE InAs epilayers near the absorption edge (2700 - 3900 cm^{-1}) have shown large Burstein-Moss shifts and significant absorption decreases near the excitation frequency with recovery times in the range 20-3000 ps. Nonthermal photoexcited carrier distributions were observed shortly after excitation, and they are responsible for much faster initial component of the bleaching process. Using these effects we have achieved passive mode-locking of a 3 μm Er^{3+} laser for the first time.

1. INTRODUCTION

The rapid progress in the research into 3 μm band erbium solid state lasers (Er^{3+}:YAG,YSGG,YLF,CaF2..) operating at room temperature requires the creation of appropriate nonlinear elements for passive resonator Q-switching and mode-locking to produce giant and ultrashort light pulses.

To date electro-optical methods have mainly been used for Q-switching and mode-locking these types of lasers (Bagdasarov et al 1980, Vodopyanov et al 1987), because at present bleachable passive shutters based on solutions of organic dyes or color center crystals (as used in visible and near-infrared lasers) do not exist for the 3 μm band. An appropriate passive Q-switching material would make it possible to create a simple 3 μm high repetition rate giant pulse laser for medical applications. Alternatively a passive shutter with a small enough absorption recovery time could produce passive mode-locking of erbium lasers, giving pulse durations down to 3 ps, corresponding to the inverse width of erbium laser's gain profile (9-12 cm^{-1}).

2. OPTICAL PROPERTIES OF InAs

Molecular Beam Epitaxy-grown ultrathin InAs layers at room temperature seem to be extremely promising as bleachable absorbers for the 3 μm radiation (Phillips et al 1991) because of the following features: InAs is a direct band gap semiconductor; its bandgap of 0.35 eV (corresponding to the frequency of 2800 cm^{-1} or λ=3.54 μm) is slightly lower than the laser photon energy (Fig.1) and a bleaching effect is expected due to conduction band filling (dynamic Burstein-Moss effect). The small electron effective mass produces large dynamic Burstein-Moss shifts in the absorption edge at moderate carrier densities, and an electron concentration of $5 \cdot 10^{17}$ cm^{-3} is sufficient to shift the band edge to 0.44 eV, corresponding to λ =2.8 μm laser photon energy.

3. PUMP-PROBE EXPERIMENTS AT 2700-3900 CM^{-1}

In the first stage of our work we used a pump probe technique. A double-resonance IR picosecond spectrometer was used, based on LiNbO3 parametric superradiant generators pumped by a Nd:YAG laser (Graener et al 1987). The pump and probe pulses had a duration of 12 ps and were independently tunable in the range 2700- 3900 cm^{-1}. The pump beam intensity was $5 \cdot 10^7$ W/cm^2. InAs epilayers of different thicknesses grown on GaAs substrates were studied at room temperature.

A pump photon whose energy is higher than the bandgap creates carriers with excess energy which are thermalized to Fermi distribution on a subpicosecond time scale; the Fermi distribution then cools to the lattice temperature on a picosecond time scale (using data obtained experimentally for the bulk GaAs by Shank et al (1979)). The conduction band filling by the electrons causes the dynamic absorption edge shift to the blue, which recovers on the time scale of order of the carrier recombination time.   Fig.2 shows the

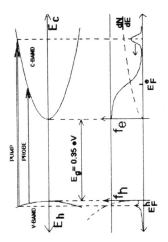

InAs 300 K

Fig.1 InAs at 300 K.
Band scheme

result of a spectral resolved experiment, where the frequency of the pump pulse and the time delay between pump and probe beams are fixed, and the frequency of the probe pulse is changed. Curve 1 is a small signal absorption $A_0(\nu)$ of InAs; curve 2 represents a dynamic spectrum $A(\nu)$ taken 20 ps after the excitation. One can see a dramatic change in the absorption spectrum due to the band filling effect. The electron occupation probability f_e (curve 3) in this state is calculated from the known dynamic spectra using the relation

$$A(\nu) = A_0(\nu) \{1-f_e-f_h\},$$

where f_e and f_h are occupation probabilities for the electrons in the c-band and holes in the v-band; here we neglect the term f_h compared to f_e because of 18 -fold difference in the effective masses (and densities of states in E-space) between electrons and heavy holes in InAs. One can see that at ν(PU)=3600 cm^{-1} the electron distribution closely follows a Fermi function with a quasi-Fermi energy of 1090 cm^{-1} (0.14 eV) and temperature of 400 K - a little bit higher than room temperature. The temperature rise is explained by the lattice heating effect of the pump pulse. Excitation at lower frequencies causes weaker shifts to the blue . Also apparent is a small negative absorption (amplification) region below 3100 cm^{-1} (Fig.2,curve 2) taking place due to population inversion. Curve 4 shows absorption edge.

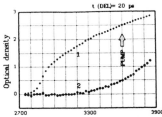

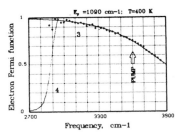

Fig.2 Dynamic spectrum
at t(DEL)=20 ps

Now we turn to time resolved experiments. In this case the frequencies ν(PU)=ν(PR) are fixed and the time delay between the pulses is changed. Fig 3 shows a strong dependence

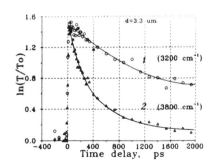

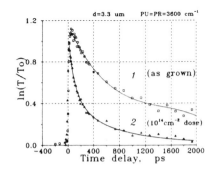

Fig.3,4 Recovery time dependence on frequency and proton dose

of the absorption recovery time on the excitation frequency: $\tau \cong 1500$ ps (3200 cm^{-1}, curve 1) and $\tau \cong 300$ ps (3800 cm^{-1}, curve 2). The absorption recovery is non-exponential and τ is defined here as the time for a 2-fold decrease of the optical density change. When the samples were irradiated with high energy protons this led to a strong decrease of recovery time. For a 3.3 μm thick InAs sample irradiated with a dose of 10^{14} cm^{-2} at 200 keV, $\tau \cong 130$ ps at 3600 cm^{-1}(Fig.4, curve 2) while for the as grown sample at the same conditions $\tau \cong 670$ ps (curve 1). The absorption recovery time is also very sensitive to the thickness of the InAs layer. This is illustrated on Fig.5, where bleaching dynamics for the different thicknesses of InAs are shown. At pump and probe frequencies of 3600 cm^{-1}, τ changes from 670 ps (3.3 μm thickness, Fig.5, a) to <20 ps (0.1 μm, Fig.5, d). These results can be explained in terms of the influence of lattice defects in decreasing the carrier lifetime. It is clear that the role of the defects at the interface between InAs and GaAs due to the lattice constant mismatch is larger when the InAs is thinner; the proton bombardment creates additional volume defects thereby decreasing the carrier lifetime.

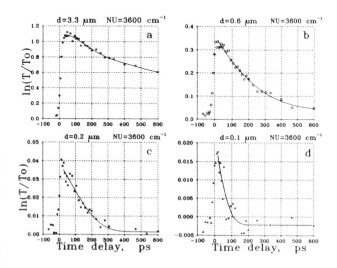

Fig.5 Recovery time dependence on sample thickness

Fig.6 represents a dynamic spectrum of the sample taken at a very early stage, before the excitation pulse reached its maximum (time delay= -7 ps, ν(PU)=3600 cm^{-1}). One can see that a spectral hole is burned (~ 100 cm^{-1} width) in the absorption continuum of InAs, centered at the position of the pump frequency (Fig.6, curve 2), due to generation of nonthermal photoexcited (hot) carriers. Due to the very fast thermalization time, the dynamic distribution of hot carriers follows the instantaneous value of pump intensity, that is why it disappears at later moments. The hot carrier thermalization time is beyond the temporal resolution of our present experimental setup.

4. PASSIVE MODE LOCKING OF AN Er:YSGG LASER ($\lambda=2.8$ μm)

For passive mode-locking the $\lambda=2.8$ μm flashlamp pumped erbium laser we used a 0.27 μm thick InAs epilayer grown by MBE on a 0.25 mm thick GaAs substrate whose sides were parallel to $\cong 0.2°$ (Vodopyanov et al 1991). The sample was proton bombarded at 15 keV with a dose of 10^{13}ions/cm^2 to reduce the absorption recovery time to 100 ps (at an excitation frequency of 3600 cm^{-1} corresponding to $\lambda=2.8$ μm). The sample had an initial absorption for the laser wavelength of 10-15% and was put at an angle close to Brewsters angle within the resonator (Fig.7,a). The laser output (Fig.7,b) was in the form of a train of about 20 pulses (estimated duration 20-30 ps) with the total energy of 5-10 mJ. The laser repetition rate was 1-2 Hz.

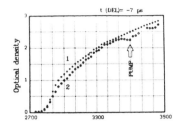

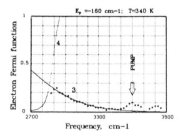

Fig.6 Dynamic spectrum at t(DEL)=-7 ps

5. CONCLUSION

A significant bleaching has been observed in InAs in the 3 μm spectral region which occurs due to the dynamic Burstein-Moss effect. Its measured recovery time was in the range of 3 ns-20 ps and strongly depended on the excitation frequency, sample thickness and proton implantation dose. Spectral measurements showed an absorption shift to the blue of up to 600 cm^{-1}. A dynamic short-lived hot carrier distribution due to intense excitation was directly observed in InAs. Passive mode locking of the 3 μm erbium laser has been demonstrated for the first time.

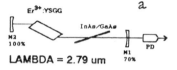

LAMBDA = 2.79 um

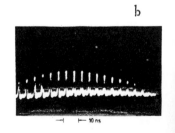

Fig.7 Passive mode-locking of a 2.8 um Er laser

Bagdasarov Kh S, Zhekov V I et al 1980 Sov.J.Quantum. Electron. **10** 1127
Graener H, Dohlus R and Laubereau A 1987 Chem. Phys. Letters **140** 306
Phillips C C, Li Y B, Stradling R A and Vodopyanov K L 1991 J.Phys. D: Appl.Phys. **24** 437
Shank C V, Fork R L, Leheny R F and Shah J 1979 Phys.Rev.Letters **42** 112
Vodopyanov K L, Kulevskii L A et al 1987 Sov.J.Quantum. Electron. **17** 776
Vodopyanov K L, Lukashev A V, Phillips C C, Ferguson I T 1991 Appl.Phys. Lett. **59** · 1658

Kinetics and relaxation of nonequilibrium electron−hole system in direct gap semiconductors

V. Dneprovskii, V. Klimov, M. Novikov*

Moscow State University, Department of Physics, 117234, Moscow, USSR
*Institute of Spectroscopy, USSR Academy of Sciences, 142092, Troitzk, USSR.

ABSTRACT: The drastic modification of the electron-hole plasma emission spectra arising from the significant reduction of the energy relaxation rate was observed under intensive picosecond interband excitation. The registered difference in the stimulated emission kinetics at low and high excitation levels evidences for the two different gain (recombination) mechanisms (of the plasma and excitonic origin), although no significant changes were observed in the luminescence spectra.

1. TRANSFORMATION OF THE EMISSION SPECTRA OF THE CdSe ELECTRON-HOLE PLASMA DUE TO THE STRONG REDUCTION OF THE INTRABAND RELAXATION RATE

The energy losses of the electron-hole plasma (EHP) in the direct gap semiconductors such as CdS and CdSe are governed by the polar interaction with the LO-phonons. This process occurs on the femtosecond time scale. So the time of the complete energy relaxation of the EHP (even for the carrier excess energy of about 1 eV) is expected to be of the order 0,1-1 ps. But as evidenced from the experiments (Saito and Göbel 1985, Junnarkar and Alfano 1986, Baltrameynas et al 1987) the cooling is much slower in the case of the high density electron-hole system and its characteristic times may exceed tens of picoseconds. In the present paper we will show that the cooling time of the dense EHP in CdSe may be comparable with its recombination time. We studied CdSe single crystals (T=80K) excited by the ultrashort second harmonic pulses from the Nd:glass laser (photon energy 2,339 eV) with a duration of about 5 ps. The spectra of the luminescence from the excited volume and those of the stimulated emission scattered at the crystal boundaries were measured simultaneously.

The emission spectra of CdSe (80K) corresponding to the different energies (W) of the pumping pulses are presented in Fig.1. At W=13 μJ we observed a Q band (λ=690 nm) of the spontaneous luminescence and R band of the stimulated emission. The spectral position and the shape of these bands are well described by the emission spectra of EHP (concentration $n_e = 1,2.10^{18} cm^{-3}$, electron temperature T_e=150K) calculated under the assumption of the direct and indirect (accompanied by simultaneous emission of LO-phonon) recombination. The lines of the direct (R_d; λ=690 nm) and indirect (R_i; λ=700 nm) recombination are well resolved in the stimulated emission (the spectral distance between them (25 ± 3) meV corresponds to the LO-phonon energy $\hbar\omega_{LO}$ in CdSe). At W=30 μJ spectral properties of the luminescence were drastically changed. A gigantic short-wave tail extending to the pumping wave length and a new B-band (λ=644,6 nm) appeared in the emission spectra. This spectrum could no longer be explained by the assumption of the quasi-equilibrium distribution of the carriers with a certain electron temperature, but rather is indicative of strong changes of the plasma temperature during the recombination (the initial EHP temperature may exceed 10^3K because of the large excess energy of the carriers). The B-band probably arised from the strong nonequilibrium population of the B valence subband in the case of the "hot" EHP. So the observed emission of the strong over heated plasma shows that the rate of EHP cooling was reduced to that of recombination.

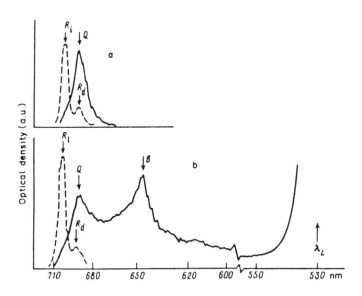

Fig.1 The spectrum of CdSe (80K) interband luminescence. The energy of the pumping pulses - a) 13 μJ; b) 30 μJ. Solid lines - the emission from the excited volume, dashed lines - from the facet of the crystal.

Two processes may lead to slowing down of the intraband energy relaxation: the nonequilibrium filling of phonon modes and the screening of electron(hole)-phonon interaction.

The EHP energy losses are determined by the filling of the phonon modes which are most strongly coupled to the carriers (Kocevar 1985). Corresponding phonon wave numbers q_o^h (for electrons) and q_o^h (for holes) are given by $q_o^{e,h} = \omega_{LO}\sqrt{m_{e,h}/3kT_e}$ (m_e and m_h are the effective carriers masses). When the phonon temperature for the states q_o^e and q_o^h is close to the electron temperature of EHP LO-phonon subsystem looses its cooling efficiency and energy losses are governed by the "slow" decay of LO-phonons into the acoustic phonons. The characteristic time of LO-acoustic phonons interaction τ_a for CdSe (80K) is about 4 ps (Baltrameynas et al 1987). So the time of complete energy relaxation under the experimental conditions may be as long as 100 ps being comparable with characteristic recombination times (0,3-0,8 ns (Dneprovskii et al 1990)).

The critical concentration $N_f^{e,h}$ for the filling is determined by the condition $R(q_o^{e,h})\tau_a = exp(-\hbar\omega_{LO}/kT_e)$, where $R(q_o^{e,h})$ is the emission rate of LO-phonons with wave numbers q_o^e and q_o^h . On taking into account that $R(q_o^{e,h}) = (\Omega q^2/(2\hbar e^2))|V_q|^2 Im(1/\epsilon(\omega_{LO}, q_o^{e,h}))$ (V_q - the matrix element of the polar electron-phonon coupling, $\epsilon = \epsilon_1 + i\epsilon_2$ - the longitudinal dielectric constant, Ω - volume) and on using ϵ in random-phase approximation one gets

$$N_f^{e,h} = 0,055\hbar\omega_{LO}^2\epsilon_\infty^2 m_{e,h}/(\pi e^2\epsilon_o\epsilon_\infty/(\epsilon_\infty - \epsilon_o)\tau_a kT_e),$$

where ϵ_o and ϵ_∞ are the static and high frequency dielectric constants respectively the above relation shows that $N_f^{e,h} \propto m_{e,h}$ and hence, $N_f^e < N_f^h$. Thus an increase in carrier concentration lead at first to the blocking of the electron channel of energy relaxation and then - of the hole channal. So the effective slowing-down of the cooling due to the filling occurs if $n_e > N_f^h$. This condition was satisfied under experimental conditions (at $T_e > 80K$; $N_f^h < 10^{18}cm^{-3}$).

The influence of the screening is determined by the $|\epsilon|^2$ value at $\omega = \omega_{LO}$ and $q = q_o^{e,h}$. The corresponding critical concentration $N_s^{e,h}$ may be found with the aid of the criterion $\epsilon_2(\omega_{LO}, q_o^{e,h})/\epsilon_\infty = 1$ which gives

$$N_s^{e,h} = 0,055\hbar\omega_{LO}^3\epsilon_\infty m_{a,h}/(e^2kT_e)$$

The ration $N_s^{e,h}/N_f^{e,h} = \pi\omega_{LO}\tau_a(\epsilon_o - \epsilon_\infty)/(\epsilon_o\epsilon_\infty^2)$ does not depend upon temperature or effective carriers masses and is about 5 for CdSe parameters. The screening comes into play therefore at higher EHP densities than that at which the filling occurs. In the range $n_e > N_s^{e,h}$ the screening suppresses the phonon mode filling, since in this case $|\epsilon(\omega_{LO}, q_o^{e,h})|^2 \propto n^2$ and $R(q_o^{e,h}) \propto 1/n_e$. thus the rate of the phonon emission no longer increases but rather decreases with the concentration n_e.

2. THE DYNAMICS OF THE STIMULATED EMISSION AND THE MOTT TRANSITION IN DIRECT GAP SEMICONDUCTORS

A strong screening of electron-hole interaction arising under intensive laser excitation may result in the ionization of the excitons and the transition (Mott transition) to the EHP. The attempts to observe this transition in the luminescence, transmission or reflection spectra failed because the emission lines of the EHP and dense exciton gas overlaps with each other and the exciton structure is present in the transmission (refection) spectra of EHP due to the effect of exciton enhancement.

In the present section we report the results of the time resolved luminescence (spontaneous and stimulated) study which enables to distinguish between two state of EH system in CdSe (80K): the exciton state and plasma one.

We used the same experimental technique as in section 1, except for the registration set up. The pulses of spontaneous (from the excited volume) and stimulated (from the crystal boundaries) emission along with the reflected from the sample pumping pulses, were registered with the aid of the high speed streak camera (time resolution 5 ps). The usual spectral measurements of the luminescence was performed simultaneously.

In the whole range of the excitation levels (0,05 - 1 μJ) a broad Q-band of the spontaneous emission ($\lambda = 690$ ns) dominated in the luminescence spectra. At $W > 0,2\mu$J an intensive R - line of the stimulated emission ($\lambda = 702$ nm) arised at the long wave wing of the Q-band. The spectral properties of the highly excited CdSe emission give no indication that the changes in the EH system take place.

On the contrary the temporal properties of the luminescence (stimulated in particular) strongly differed at the low and high intensities of the excitation (Fig.2). At $W < 0,2\mu$J we observed the pulses of the spontaneous emission with a build up time of about 0,6 ns and a characteristic relaxation time τ_I=4 ns. Just above the threshold for the induced processes (W=0,2 - 0,3 μJ) we registered the relatively short pulse of the stimulated emission strongly delayed (Δt about 1 ns) in respect to the pumping pulse. The pulses of spontaneous luminescence in this case had nearly the same features as in the region W< 0,2μJ. At $W > 0,4\mu$J the delay time of the stimulated emission pulse drastically shortened down to the values of about 30-40 ps.

The observed peculiarities in the stimulated emission kinetic we believe are indicative of the transformation of the EH system state. The pumping intensities used in the work correspond to the EH pair densities being well above the critical value for the Mott transition in CdSe ($n_m = 10^{17}cm^{-3}$ at 80 K). So the initial state of the photoexcited system is the plasma one. At the high excitation level (W> 0,4μJ) the plasma density is high enough for the induced processes to be effective. The stimulated emission develops in this case. The delay time of the induced pulse is linked with the characteristic cooling time (the EHP must be cold enough to provide a negative absorption, i.e. gain). The relatively large delay (30-40 ps) is in a good agreement with the results of the previous section.

At the lower pumping intensities the plasma density is not high enough to provide a gain. So it decays via relatively "slow" spontaneous processes. After plasma reaching the critical Mott density the formation of the excitons becomes possible. This process is strongly suppressed by the screening of the electron-hole interaction so it occurs on the subnanosecond time scale. The observed long delay of the stimulated emission in this case may be attributed to the time which is necessary for the accumulation of the exciton density being sufficient for the induced processes to develop in the exciton system.

In conclusion we have studied the spectra and kinetics (with picosecond time resolution) of the spontaneous and stimulated emission of CdSe crystals excited by the powerful picosecond pulses. The transformation of the CdSe luminescence spectra observed at the intensive pumping evidences for the strong reduction of the EHP cooling rate down to the values being comparable with the rate of the recombination. The registered difference in kinetics of the stimulated emission at the low and high excitation level allows to distinguish between the exciton and plasma state of the EH system.

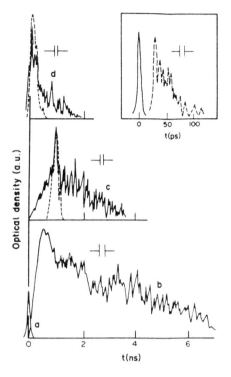

Fig.2 The pulses of luminescence of CdSe (b,c,d) and the pumping picosecond pulse (a). The energies of the pumping pulses: b. - 0.08 μJ; c. - 0.28 μJ; d. - 0.44 μJ; insert. - 0.8 μJ. Solid lines - the emission from the excited volume, dashed lines - from the facet of the crystal.

REFERENCES

Saito H. and Göbel 1985 Phys.Rev. **B31** 2360
Junnarkar M.R. and Alfano R.R. 1986 Physs.Rev. **B34** 7045
Baltrameynas R., Zhukanskas A., Latinis V. and Yurshenas S. 1987 JETP Lett. **46** 481
Kocevar P. 1985 Physica **134B** 155
Dneprovskii V.S., Klimov V.I. and Novikov M.G. 1990 Solid State Commun. **73** 669

Inst. Phys. Conf. Ser. No 126: Section V
Paper presented at Int. Symp. on Ultrafast Processes in Spectroscopy, Bayreuth, 1991

Investigation of carrier temperature relaxation with femtosecond transient grating experiments in CdS_xSe_{1-x}-semiconductors

Dao van Lap , U Peschel , H E Ponath and W Rudolph

Friedrich–Schiller–Universität, Max–Wien–Platz 1, O-6900 Jena, Germany

ABSTRACT: The relaxation of the electron temperature in CdS_XSe_{1-X} was measured by transmission and transient grating experiments using fs light pulses. As compared with ordinary ambipolar diffusion, at high excitation a pronounced delayed rise and a faster decrease of the diffraction signal was observed. A theoretical model which includes one– and two–photon absorption, gap–shrinkage, diffusion and temperature relaxation explains these experimental findings as a result of temperature relaxation with a characteristic time constant of 1 ± 0.2 ps.

1. INTRODUCTION

The dynamics of nonequilibrium carriers in semiconductors such as their relaxation and trans- port mechanism can be studied favorably by means of ultrashort light pulses . With the pro- duction of carriers with energies larger than the gap energy the first processes after excitation include thermalisation of electrons and holes on a sub ps time scale due to carrier–carrier in- teraction which leads us to a carrier–temperature and a following slower relaxation of this temperature down to the lattice one due to electron–phonon interaction (Osman et al 1987).

2. EXPERIMENTAL

We derive pulses with a duration of 100 fs, an energy of $100\mu J$, at a wavelength of 618nm from a CPM dye laser followed by a 4–stage amplifier pumped by an excimer laser (Rudolph et al 1990). These pulses are split into two parts (1),(2) serving as probe and pump channel, respectivly, see Fig. 1. The pump pulse is then divided into two partial pulses (2.1), (2.2) which are incident on the sample under a certain angle. Mirror S_4 mounted on a translation stage is adjusted for their exact temporal overlap. The delayed probe pulse is focused by means of lens L_1 onto the excited area where its spot size is about half of the pump spot. Whith the detectors 3, 1, 2, 4 the the probe light diffracted into the diffraction orders −1, 0, 1, 2 can be measured simultaneously. By blocking one of the pump beams the signal recorded with D_1 corresponds to the transient sample transmission. The samples were two thin platelets of $CdS_{0.52}Se_{0.48}$ and $CdS_{0.43}Se_{0.57}$ with the c–axes parallel to the polarisation direction of the light. Since the mean photon energy is below the gap energy ($CdS_{0.52}Se_{0.48}$: 134meV, $CdS_{0.43}Se_{0.57}$: 68meV) the absorption is due to single– and two–photon absorption (SPA and TPA). Typical results of our experiments are sketched in Figure 2.

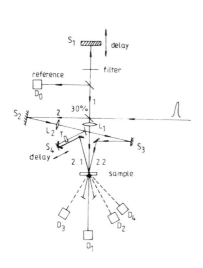

Fig. 1. Experimental setup

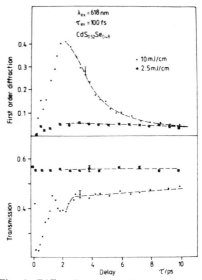

Fig. 2. Diffraction and transmission transients
in for two different excitation energies

3.RESULTS AND DISKUSSIONS

If we look at the dynamics of transmission, depicted in Figure 2b we find after a rapid increase
caused by thermalisation a very fast decrease of absorption due to temperature relaxation. To
describe the optical behaviour of our sample we need the susceptility χ

$$\chi \sim \int_0^\infty dE \sqrt{E} \frac{1 - f_n(E) - f_p(E)}{E_G(n) + E - \hbar\omega + \frac{i\hbar}{\tau_2}} \tag{1}$$

where τ_2 is the phase relaxation time with a value of 75fs and E_G the carrier-density dependent
bandgap energy which is determined by (Taylor et al 1987)

$$E_G(n) = E_g^0 - \beta \sqrt[3]{n} \tag{2}$$

If we assume thermal equilibrium the distribution functions $f_{n/p}$ of carriers depend on the
actual density (n) and temperature (T) of electrons only. In the case of the transmission
expertiment the density remains constant and the temperature is assumed to fall exponentially
with a characteristic time τ_w

$$T(t) = (T_0 - T_{lattice}) e^{-\frac{t - t_0}{\tau_w}} + T_{lattice} \tag{3}$$

If the mean photon energy $(\hbar\omega)$ is below the gap energy the initial temperature T_0 can be
approximated by the ratio of SPA and TPA

$$T_0 \approx \frac{2}{3K_B} \frac{(2\hbar\omega - E_G(n)) n_{TPA}}{n_{SPA} + n_{TPA}} \tag{4}$$

Comparing the results of our simulations (Figure 3) with the experimental data (Figure 2b) we
determine a temperature relaxation time $\tau_w \approx (1 \pm 0.2)$ps and a gap shrinkage of 90 meV for
an electron density of $10^{18} cm^{-3}$.

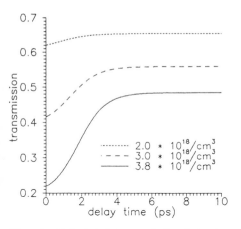

Fig. 3. Calculated tramsmission behaviour of $CdS_{0.52}Se_{0.48}$ for three different electron densities

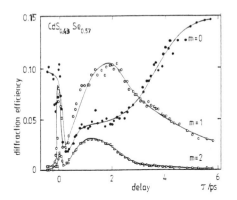

Fig. 4. Measurement of zeroth (m=0), first (m=1) and second (m=2) order of diffraction (data from Rudolph et al 1990)

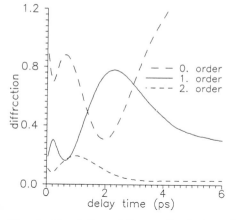

Fig. 5. Simulation of the zeroth–,first– and second–order diffraction signal for the experimental situation belonging to Fig.4

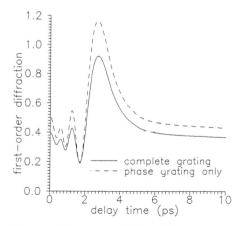

Fig. 6. Simulation of the diffraction at a complete and a pure phase grating for the experimental situation belonging to Fig.2a

Comparing the different diffraction orders depicted in Figure 4 we find the intensity in the first order to be some times larger than the transmitted one, a fact which suggests the dominant influence of a phase grating. Fortunately the influence of the initial conditions, which we can't exactly determine, on the dynamics of the diffraction is relatively weak. The best fit we achieved with a step like distribution of carriers generated by SPA. Further experiments–depicted in Figure 7–showed that the decay time of the first diffraction order depends linearly on the squared grating constant λ what is typical for a diffusion effect which is even much greater as one would expect. If we assume the carrier distribution to be Boltzmann–like and the diffusion to be ambipolar where μ is the corresponding mobility we can simulate the dynamics of the spatial distributions of carriers (n) and temperature (T) or inner energy U_n of electrons by eq.5 and 6.

$$\frac{\partial}{\partial t}n - \frac{2}{3}\frac{\mu}{e}\frac{\partial^2}{\partial x^2}U_n = 0 \tag{5}$$

$$\frac{\partial}{\partial t}U_n - \frac{1}{\tau_w}\left(U_n - \frac{3}{2}K_BT_{lattice}n\right) - \frac{10}{9}\frac{\mu}{e}\frac{\partial^2}{\partial x^2}\left(\frac{U_n^2}{n}\right) = 0 \tag{6}$$

Figure 5 and 6 show the calculations for the two different samples using the same parameters as by the simulation of transmission. As we expected the pure phase grating plays an important role (see Figure 6) what explains the fast changes of diffraction during a constant transmission (see Fig 2). The reasons for the calculated oszillations shortly after the excitation are phase changes greater than two π. We can't decide whether these fluctuations really occur or not. May be that they are disguised by thermalisation effects we neglected in our simulations. The calculations for different grating constants depicted in Figure 8 show the same strong influence of the grating spacing on the dynamics of diffraction as we measured experimentally. But notice that the amount of change of diffraction efficiency must not correspond to the same rates of alteration of the carrier distributions. We have partly a strong amplification of the visibility of diffusion effects, e.g. by gap shrinkage in originally unexcited regions of the grating.

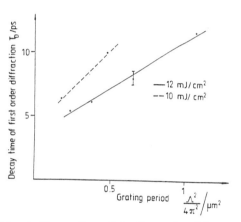

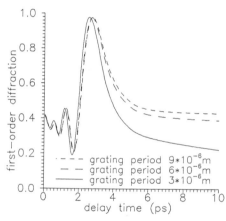

Fig. 7. Decay time of the first order diffraction versus the grating period squared for two excitation energies (sample $CdS_{0.52}Se_{0.48}$)

Fig. 8.Calculation of the transient first order diffraction for different grating spacings in $CdS_{0.52}Se_{0.48}$

4. SUMMARY

The fast dynamics measured in transient transmission experiments can be explained as a result of temperature relaxation with a time constant of (1 ± 0.2)ps. The optical induced grating is above all a phase grating which is quickly changed by temperature relaxation and diffusion. The optical visibility of diffusion effects is much amplified by nonlinearities of the susceptility.

5. REFERENCES

Osman M A and Ferry D K 1987 Phys. Rev. B36 6018
Rudolph W, Puls J, Henneberger F and Lap D 1990 phys.stat.sol (b) 159 49
Taylor A S and Wiesenfeld 1987 Phys. Rev. B35 2321

Inst. Phys. Conf. Ser. No 126: Section V
Paper presented at Int. Symp. on Ultrafast Processes in Spectroscopy, Bayreuth, 1991

361

Nonlinearity near half-gap in bulk and quantum well GaAs/AlGaAs waveguides

M N Islam,[a] C E Soccolich,[a] R E Slusher,[b] W S Hobson,[b] and A F J Levi [b]

a) AT&T Bell Laboratories, Holmdel, NJ 07733, USA
b) AT&T Bell Laboratories, Murray Hill, NJ 07974, USA

ABSTRACT: We report the nonlinear index and two-photon absorption coefficient near half-gap in bulk and multiple quantum well (MQW) GaAs/AlGaAs waveguides. In the bulk material we measure a nonlinear index $n_2=+3.6\times10^{-14}cm^2/W$ and a two-photon absorption coefficient $\beta=0.33\times10^{-4}cm/MW$ with a resulting figure-of-merit $F_m = (2n_2/\beta\lambda) = 13$, which shows the appropriateness of this material for all-optical switching. Although we measure an n_2 up to 1.7 times larger in the MQW, we obtain nearly the same value for the figure-of-merit because of increased two photon absorption. Furthermore, time resolved pump-probe measurements show an intriguing exchange of energy between orthogonal axes that can be explained by a low frequency Raman process.

All-optical switching and quantum optics applications require materials that can provide a π-phase shift due to the nonlinear index with less than 3dB of absorption. The third-order nonlinear optical properties of semiconductors are of interest for making compact, integrable devices where many devices can be grown simultaneously on the same wafer. We present measurements of the nonlinear index, n_2, and two-photon absorption coefficient, β, as well as time resolved pump-probe data for both bulk AlGaAs and GaAs/AlGaAs multiple quantum well (MQW) waveguides near half-gap. The GaAs/AlGaAs material system studied here is important for several reasons: (1) by varying the alloy composition the half bandgap energy can cover the 1.3 to 1.6μm infrared window that is important for optical communications; (2) the material is almost perfectly lattice matched; and (3) already a very mature fabrication technology exists for these alloys.

In general terms, the goal is to implement an all-optical, three terminal switch or logic module with two inputs (control and signal) and a cascadable output (an output that can serve as the signal input to an identical device). What are the material considerations for such an ultrafast device that is based on virtual transitions? First, we need to know the strength of the interaction, which for refractive index devices is proportional to n_2: this determines the switching energy, which is typically inversely proportional to the product of n_2 and the length of the device. Second, we need to know if the pulse is distorted while propagating through the device. Analogous to propagation in optical fibers, sources of pulse distortion include nonlinear absorption and low frequency Raman effects. A commonly used figure of merit for nonlinear materials is $F_m = (2n_2/\beta\lambda)$ (Mizrahi 1989), which must be greater than one for all-optical switching and quantum optics applications. Therefore, to characterize the semiconductor below half of the

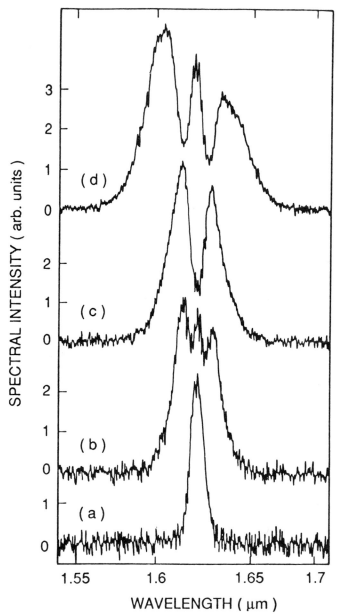

Figure 1   Experimentally measured self-phase-modulation spectra at the output of the bulk waveguide as a function of increasing pump intensity. The series corresponds to : (a) the input spectrum, (b) $\Delta\Phi \approx \pi$, (c) $\Delta\Phi \approx 1.5\pi$, and (d) $\Delta\Phi \approx 2.5\pi$.

:nergy gap we measure n_2, β, and the low frequency Raman coefficient.

The laser source used in the experiments is a passively modelocked NaCl color center aser that operates between 1.6 and 1.7 μm and produces 300 to 500 fsec pulses separated y 11.4 nsec (Islam 1989). The laser output is separated into orthogonally polarized ump and probe beams, and the probe is frequency shifted by 80 MHz by passing hrough an acousto-optic modulator. A stepper-motor controlled delay line is used to ary the time separation between the two pulses, which are recombined at a polarizing eam splitter and then sent to the waveguides. In the bulk samples a ridge waveguide is ormed in a 2.55μm thick layer of $Al_{0.2}Ga_{0.8}As$, and guiding is provided by a 2.55μm nderlying layer of $Al_{0.5}Ga_{0.5}As$. The ridge height is 0.7μm and its width is 3.6μm. The MQW waveguide core consists of 200 periods of 40 Å GaAs wells and 70 Å $Al_{0.3}Ga_{0.7}As$ barriers. We assure guiding in the vertical dimension by using a 3 μm hick layer of lower index $Al_{0.5}Ga_{0.5}As$ below the MQW guide, and lateral confinement s provided by etching a 3.8 μm wide ridge with a 0.6μm height. The MQW and bulk vaveguides are grown in the same MOCVD reactor and special care was taken to ninimize oxygen in the chamber.

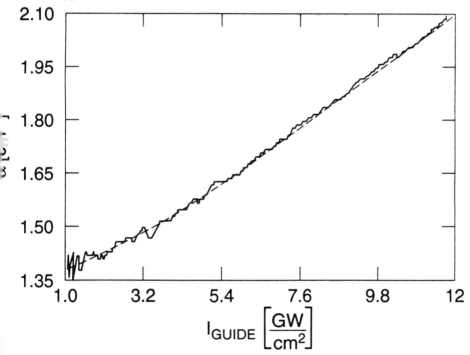

igure 2 Nonlinear absorption coefficient in the MQW for $\hat{e}\perp\hat{z}$ as a function of pump tensity. The dashed curve is a cubic least square error fit to the data, and from the low tensity slope we can obtain the two-photon absorption coefficient.

As in the case for optical fibers, we use the simple self-phase-modulation spectral technique to measure n_2 in our semiconductor waveguides (Stolen 1978). Self-phase-modulation dramatically broadens the spectrum and forms a series of spectral peaks at phase shifts of π and larger for an instantaneous n_2 with negligible dispersion. For example, Figure 1 shows the experimental spectra at the output of the bulk waveguide as a function of power. In the bulk material the intensity required for a π phase shift is 3.1 ± 0.5 GW/cm^2, which yields a value of $n_2=+3.6(\pm0.5)\times10^{-14}$ cm^2/W if we assume that the effective mode cross-sectional area A_{eff} is one-half of the geometric area A_{geom}. We deduce the positive sign of n_2 in our waveguides from the observed pulse broadening at high intensities. In the MQW for the electric field polarized in the plane of the quantum wells $\hat{e}\perp\hat{z}$ we obtain a value of $n_2 = +6(\pm0.8)\times10^{-14}$ cm^2/W $\approx 1.7\times n_2^{(bulk)}$, and for the polarization perpendicular to the quantum wells $\hat{e}\|\hat{z}$ we find a value of $n_2 = +4.8(\pm0.7)\times10^{-14}$ cm^2/W $\approx 1.33\times n_2^{(bulk)}$. The enhancement in the MQW may be due to excitonic effects in the quasi-two-dimensional material (Islam 1991).

We measure the transmission of the pump pulse as a function of intensity to deduce the nonlinear absorption coefficients. The logarithm of the transmission function is proportional the the absorption coefficient, and β is obtained from the slope at low intensity. Figure 2 shows exemplary data for the MQW sample, and the dashed curve is a least square error fit to the data using a cubic function. After correcting for the Gaussian temporal profile of the pulse (Mizrahi 1989) and assuming $A_{eff}\approx\frac{1}{2}A_{geom}$, we obtain for the bulk material $\beta=0.33(\pm.05)\times10^{-4}$ cm/MW. For the MQW material we obtain for $\hat{e}\perp\hat{z}$ $\beta=0.53(\pm.08)\times10^{-4}$ cm/MW $\approx 1.6\times\beta^{(bulk)}$ and for $\hat{e}\|\hat{z}$ $\beta=0.38(\pm.06)\times10^{-4}$ cm/MW $\approx 1.15\times\beta^{(bulk)}$. The resulting figures-of-merit for the materials below half-gap are: (a) for the bulk $F_m \approx 13$; (b) for the MQW with $\hat{e}\perp\hat{z}$ $F_m \approx 13.5$; and (c) for the MQW with $\hat{e}\|\hat{z}$ $F_m \approx 15$.

We also perform time-resolved pump-probe measurements with orthogonally polarized pump and probe pulses. Only the probe beam is modulated, and the signal from the detector at the waveguide output is fed to a lock-in that is referenced to the same modulation frequency. In Figure 3 we show typical pump-probe data at a pump intensity of 3.85 GW/cm^2 for the bulk sample (similar behavior is observed in the MQW). We include three sets of data that are collected as a function of pump power: (a) no polarizer to see the complete change in output due both to the pump and probe; (b) polarizer at output along probe axis to see the change in probe due to the pump beam; and (c) polarizer at output along pump axis to see the change in pump due to the probe. Note that although the three curves in Figure 3 are drawn on the same scale, the various data are displaced for ease of display. With no polarizer at the waveguide output we observe the expected behavior for multi-photon absorption. For example, in Figure 4 we plot the normalized peak change in transmission ($\Delta T/T$) as a function of peak pump intensity at the bulk waveguide input and find contributions from both two and three photon absorption. When we add the polarizer to the waveguide output, the behavior becomes much more complicated in both the bulk and MQW. The probe pulse is attenuated when overlapped with the leading edge of the pump pulse and amplified in the

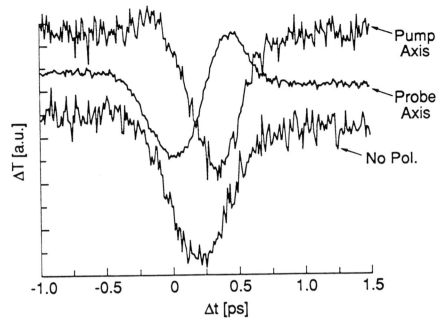

Figure 3 Time-resolved pump-probe data using 285 fs pulses at 1.67μm with no polarizer, polarizer at output along probe axis, and polarizer at output along pump axis.

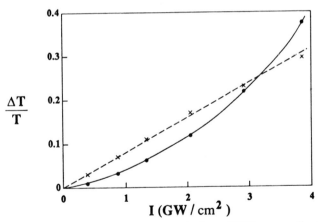

Figure 4 Peak-to-peak change in transmission ΔT/T normalized to the probe transmission in the bulk waveguide as a function of output pump intensity. The solid dots correspond to no polarizer, and the ×'s correspond to a polarizer at the output along the probe axis.

trailing edge. In Figure 4 we also include the peak-to-peak change $\Delta T/T$ of the probe and see a linear behavior, thus confirming that the derivative-like signal is due to a $\chi^{(3)}$ process . Note that when the probe rises above its background level ($\delta t < 0$) that the probe experiences significant gain (as large as 15%), indicating a transfer of energy between the two axes.

One possible explanation for the exchange of energy between the two axes is a low frequency Raman gain. Self-phase-modulation chirps the pump pulse (red-shifts the leading edge and blue-shifts the trailing edge), while Raman gain transfers energy from the higher to the lower frequency pulse. The data implies a Raman coefficient between orthogonally polarized pulses of $R_\perp \sim 5.5 \times 10^{-11}$ cm/W for $\Delta v \leq 30$ cm^{-1}, which is more than two orders of magnitude larger than in fused silica fibers for comparable Δv. A low frequency Raman gain of this magnitude may arise in the AlGaAs alloy, where the random occupation of sites relaxes the Raman selection rule $\vec{k}=0$ and allows all modes of vibration to participate in the scattering (Shuker 1971). In addition, even in the perfect crystal there are second order Brillouin processes that involve two phonons and that are Raman-like; i.e. wave vector conservation is satisfied by the difference between two acoustic phonon wave vectors.

In summary, we have measured the real and imaginary components of $\chi^{(3)}$ near the half-gap in bulk AlGaAs and GaAs/AlGaAs MQW waveguides. For the bulk material we find $n_2 = +3.6 \times 10^{-14}$ cm^2 /W and $\beta = 0.33 \times 10^{-4}$ cm/MW, which imply a figure of merit $F_m = 13$. In the MQW we find that both n_2 and β are enhanced up to 1.7 times over the value in bulk, so the figure-of-merit is approximately the same as in the bulk. Finally, time-resolved pump-probe measurements provide evidence for a low frequency Raman gain that leads to an exchange of energy between the two orthogonal axes.

Islam M N, Soccolich C E, Slusher R E, Levi A F J, Hobson W S and Young M G 1991 *to be published in J Appl Phys* "Nonlinear Spectroscopy Near Half-Gap in Bulk and Quantum Well GaAs/AlGaAs Waveguides"

Islam M N, Sunderman E R, Soccolich C E, Bar-Joseph I, Sauer N and Chang T Y 1989 *IEEE J Quantum Electron* **25** 2454

Mizrahi V, DeLong K W, Stegeman G I, Saifi M A and Andrejco M J 1989 *Opt Lett* **14** 1140

Shuker R and Gamon R W 1971 *Scattering in Solids* ed M Balkanski (Paris: Flammarion Sciences) pp 334-8

Stolen R H and Lin C 1978 *Phys Rev A* **17** 1448

Inst. Phys. Conf. Ser. No 126: Section V
Paper presented at Int. Symp. on Ultrafast Processes in Spectroscopy, Bayreuth, 1991

Resonant electron and hole tunneling between GaAs quantum wells

A.P. Heberle, W.W. Rühle, K. Köhler[+]

Max-Planck-Institut für Festkörperforschung,
Heisenbergstr. 1, 7000 Stuttgart 80, Germany

[+]Fraunhofer-Institut für Angewandte Festkörperphysik,
Eckerstr. 4, 7800 Freiburg, Germany

ABSTRACT: Time-resolved photoluminescence measurements in the picosecond regime are reported for a $GaAs/Al_{0.35}Ga_{0.65}As$ asymmetric double quantum well structure biased by electric fields of positive or negative polarity. A large number of electron and hole resonances is detected. A splitting of the ground state resonances reveals the importance of excitonic effects. Longitudinal optical phonon assisted tunneling plays a minor role for narrow quantum wells in comparison with impurity or interface roughness assisted transfer.

1. INTRODUCTION

The dynamics of carrier transfer through thin barriers in semiconductor heterostructures has been intensively investigated in the last years since this process is not only of basic physical interest but also of fundamental importance for future electronic, optoelectronic, or all-optical devices. Optical spectroscopy on an asymmetric double quantum well (ADQW) structure allows the direct study of transfer through a single barrier. Electron and hole transfer from a narrow (QW_n) to a wide quantum well (QW_w) and vice versa through a thin barrier can directly be traced with high temporal resolution luminescence spectroscopy. Resonant tunneling is observed when quantized levels in the two QWs are energetically aligned e.g. by an electric field in growth direction (Oberli et al. 1989, Alexander et al. 1990, Matsusue et al. 1990, Leo et al. 1990, Nido et al. 1991, Norris et al. 1991) or by sample design (Deveaud et al. 1990, Roussignol et al. 1991).

For the first time, we observe a large variety of electron and hole tunneling resonances in *one* ADQW by investigating the decay times of *both* QW_w *and* QW_n luminescence over a wide range of electric fields applied for *both* polarities. These results provide clear evidence for the importance of excitonic effects on the resonance position, for the strength of longitudinal optical (LO) phonon-assisted and LO-phonon-free transfer, and for a strong decrease of the transfer time at resonance with increasing subband number of the target QW.

2. EXPERIMENTAL

The experiments are based on a sample grown by molecular beam epitaxy on a n^+-doped GaAs:Si substrate. After a buffer structure follows 100 nm $Al_{0.35}Ga_{0.65}As$, a 10 nm wide GaAs QW, a 6 nm $Al_{0.35}Ga_{0.65}As$ barrier, a 5 nm wide GaAs QW, a 100 nm $Al_{0.35}Ga_{0.65}As$ layer, and a 5 nm GaAs layer on top. The ADQW structure is repeated 10 times with 20 nm $Al_{0.35}Ga_{0.65}As$ in between. Ohmic contacts on the substrate and semitransparent Schottky contacts of 1 mm diameter on top allow application of electric fields up to 100 kV/cm. The application of fields in both polarities is possible due to the $Al_{0.35}Ga_{0.65}As$ blocking layers.

The sample temperature is kept at 10 K. The ADQWs are excited with pulses from a tunable Styryl 8 dye laser which is synchronously pumped by an Ar^+ laser at a repetition rate of 80 MHz. The carrier excitation density is below $10^{10} cm^{-2}$. The photoluminescence is dispersed by a 0.32 m spectrometer and detected by a two dimensional streak camera with a spectral and overall temporal resolution of 0.5 nm and 7 ps, respectively.

3. RESULTS AND DISCUSSION

3.1. Resonant Electron Tunneling

Figure 1 shows the electric field dependence of the QW_w ($\square$) and QW_n (+) luminescence decay times of the sample. The diagrams on top show schematically the tilting of the valence and conduction band due to the applied electric field. The field strength is determined from the Stark shift of the QW_w luminescence. A large number of resonances is visible. The assignment of the resonances to level crossings is much better if one takes into account the fact that electrons and holes form excitons at these low temperatures and excitation densities. Figure 1 shows that we get good agreement with the calculated positions where electron and hole levels of a direct exciton cross the free carrier levels in the neighbouring well as indicated by arrows at the observed resonances. The label $e_{1n}^X \rightarrow e_{1w}$ corresponds to an electron bound in an exciton tunneling from the first subband of QW_n to the first subband of QW_w.

The decay times of the QW_n luminescence show two sharp minima at -23 kV/cm and +43 kV/cm, which we assign to the tunneling resonances $e_{1n}^X \rightarrow e_{1w}$ and $e_{1n}^X \rightarrow e_{2w}$. The decay of QW_w luminescence exhibits a sharp decrease at -47 kV/cm where there is a tunneling resonance from $e_{1w}^X \rightarrow e_{1n}$. Note, the different positions of the $e_{1n}^X \rightarrow e_{1w}$ and $e_{1w}^X \rightarrow e_{1n}$ resonances clearly indicate the importance of excitonic effects: the onsets of the two ground state tunneling resonances differ by two times the exciton binding energy of 9 meV. The small shifts of the resonances due to the binding energy of spatially indirect cross excitons and to inhomogeneous broadening is neglected since both effects partially compensate. A comparable field dependence of the interwell transfer due to scattering at charged impurities and interface defects was calculated and detected in dc photoluminescence for a similar structure by Delalande et al. 1990.

Also indicated are the electric fields where a transition of the electron of an exciton to the neighbouring free electron level with emission of an LO phonon becomes possible, as first shown by Oberli et al. 1990 (label $e_{1w}^X \rightarrow e_{1n} + LO$, etc.). The LO phonon resonances are very weak if detectable at all. The $e_{1n}^X \rightarrow e_{1w}$ and $e_{1w}^X \rightarrow e_{1n}$ resonances are impossible via LO phonon intersubband scattering. LO phonon assisted tunneling is therefore not significant in this structure.

A weak, longlived cross luminescence between e_{1n} and h_{1w} (first heavy hole subband level of QW_w) appears for electric fields below -38 kV/cm due to the spatial separation of electrons and holes. The spectral position of the cross luminescence confirms the electric field calibration by the Stark effect.

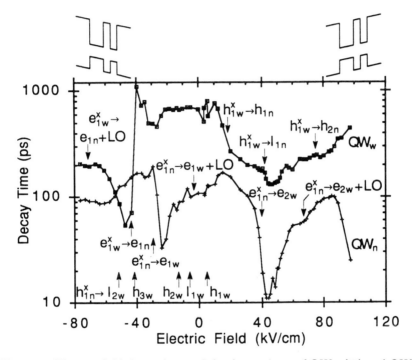

Figure 1: Electric field dependence of the decay times of QW_w ($\square$) and QW_n (+) luminescence. The diagrams on top indicate how the bands are tilted by the field. The positions of various calculated resonances are indicated by arrows. The labels for the resonances are explained in the text.

3.2. Resonant Hole Tunneling

Hole tunneling resonances are detected in the decay times of the QW_w luminescence for positive electric field as shown in Figure 1. Three resonances are observed: $h_{1w}^X \rightarrow h_{1n}$, $h_{1w}^X \rightarrow l_{1n}$, and $h_{1w}^X \rightarrow h_{2n}$, where h denotes a heavy and l a light hole level. The calculated positions include the exciton binding energy. The decay times rise again to the purely radiative value for the highest fields. Calculations by Delalande et al. 1990 showed that tunneling between heavy and light hole levels is possible via impurity scattering due to band mixing effects as described by the Luttinger-Kohn formalism. Experimentally, the carrier transfer in the hole resonances is an order of magnitude slower than in the electron resonances whereas the theory of Delalande et al. 1990 predicts comparable transfer times. However, in the experiment the sharpness of the theoretical resonances is strongly reduced by inhomogeneous broadening.

Hole tunneling from $QW_n \rightarrow QW_w$ is also observed. The resonances are indicated by arrows at the bottom of Figure 1, but they are too closely spaced to be resolvable. Note that these resonances near 0 kV/cm may cause the decay time of the QW_n luminescence at zero field to not actually reflect the nonresonant electron transfer times since it is strongly influenced by resonant hole tunneling in agreement with the work of Roussignol et al. 1991.

4. SUMMARY

The occurence of fast ground state transfer times is a direct proof of impurity or interface roughness assisted tunneling. A heavy to light hole resonance supports this result. LO phonon assisted tunneling plays only a minor role for narrow wells. Excitonic effects must be included in order to understand the splitting of the ground state resonance. Resonant electron tunneling is an order of magnitude faster than resonant hole tunneling.

5. ACKNOWLEDGEMENT

The authors thank K. Rother and H. Klann for technical assistance and J. Kuhl for a critical reading of the manuscript. The work has been partly supported by the Bundesminister für Forschung und Technologie.

6. REFERENCES

Alexander M.G.W., Nido M., Rühle W.W., and K. Köhler 1990 Phys. Rev. **B41**, 12295

Deveaud B., Chomette A., Clerot F., Auvray P., Regreny A., Ferreira R., and Bastard G. 1990 Phys. Rev. **B42**, 7021

Delalande C. and Rolland P. 1990 Superlattices Microstruct. **8**, 7

Leo K., Shah J., Gordon J.P., Damen T.C., Miller D.A.B., Tu C.W., Cunningham J.E., and Henry J. E. 1990 Phys. Rev. **B42**, 7065

Matsusue T., Tsuchiya M., Schulman J.N., and Sakaki H. 1990 Phys. Rev. **B42**, 5719

Nido M., Alexander M.G.W., Rühle W.W., and Köhler K. 1991 Phys. Rev. **B43**, 1839

Norris B.T., Vodjdani N., Vinter B., Costard E., and Böckenhoff E. 1991 Phys. Rev. **B43**, 1867

Oberli D.Y., Shah J., Damen T.C., Tu C.W., Chang T.Y., Miller D.A.B., Henry J.E., Kopf R.F., Sauer N., and DiGiovanni A.E. 1989 Phys. Rev. **B40**, 3028

Oberli D.Y., Shah J., Damen T.C., Kuo J.M., Henry J.E., Lary J., and Goodnick S.M. 1990 Appl. Phys. Lett. **56**, 1239

Roussignol P., Gurioli M., Carreresi L., Colocci M., Vinattieri A., Deparis C., Massies J., and Neu G. 1991 Superlattices Microstruct. **9**, 151

Inst. Phys. Conf. Ser. No 126: Section V
Paper presented at Int. Symp. on Ultrafast Processes in Spectroscopy, Bayreuth, 1991

Dynamical aspects of pressure induced Γ−X electron transfer in a (GaAs)₁₅/(AlAs)₅ type-I superlattice

J. Nunnenkamp, K. Reimann, J. Kuhl, and K. Ploog

Max-Planck-Institut für Festkörperforschung,
Heisenbergstr.1, D-7000 Stuttgart 80, FRG

Abstract: Application of hydrostatic pressure induces a transition of a type-I $(GaAs)_{15}/(AlAs)_5$ superlattice (SL) to type-II character at $P_c=1.2$ GPa. In the case of type-II character (> 1.2 GPa) the electrons exhibit a rapid Γ-X transfer from the GaAs Γ-states to the AlAs X-states. Time-resolved femtosecond pump/probe measurements of this Γ-X electron transfer show a square-root dependence on the energy separation $\Delta_{\Gamma X}$ between Γ and X states : $\tau_{\Gamma-X}^{-1} \propto \Delta_{\Gamma X}^{1/2}$. Additionally we observe a shift of P_c with carrier density to <u>lower</u> pressures which is explained in terms of different renormalization of the Γ and X states.

1.Introduction

GaAs/AlAs type-II SL's are characterized by the fact that the lowest conduction band state is located in the AlAs X-state whereas the holes are located in the GaAs Γ-state. The X-Γ electron-hole transition is therfore indirect in k <u>and</u> real space.

By means of molecular-beam epitaxy (MBE) it is possible to produce either type-I $(GaAs)_m/(AlAs)_n$ (m,n number of monolayers) SL's as well as type-II $(GaAs)_m/$ $(AlAs)_n$ SL's just by choosing proper thicknesses of the constituent layers. According to the work of Dawson et al. (1990) we observe type-II character in the case of m≤ 12. Electrons and holes, both created in the GaAs layer, are spatially separated due to the possible Γ-X electron transfer from the GaAs to the AlAs layer. Recently, we could show (Feldmann et al. 1990) that this Γ-X transfer depends on the overlap of the involved Kronig-Penney wavefunctions Ψ_1^Γ (initial state), $\Psi_i^{X_z}$ (all possible final states) as

$$\tau_{\Gamma-X}^{-1} \propto \sum_i |< \Psi_1^\Gamma|\Psi_i^{X_z} >|^2. \tag{1}$$

In these experiments on 12 different samples we have varied the overlap $| < \Psi_1^\Gamma|\Psi_i^{X_z} > |$ as well as the energy separation $\Delta_{\Gamma X}$ of the Γ- and X-state (Feldmann et al. 1990). Assuming that the wavefunction overlap is constant, we now apply pressure on <u>one</u> type-I GaAs/AlAs sample in order to study the dependence of $\tau_{\Gamma-X}$ on $\Delta_{\Gamma X}$ alone. This method allows switching from type-I to type-II character as well as continous change of $\Delta_{\Gamma X}$. Due to the elastic nature of the transfer process for samples with thin GaAs layers (Feldmann et al. 1990) and taking into account the large miniband width of short-period SL's we can rewrite Eq.(1):

$$\tau_{\Gamma-X}^{-1} \propto | < \Psi_1^{\Gamma} | \Psi_1^{X_z} > |^2 \cdot \Delta_{\Gamma X}^{1/2} \tag{2}$$

with $D(E)=\Delta_{\Gamma X}^{1/2}$ being the 3-dimensional density of states in the frequency band where the wavefunction overlap is assumed to be constant. Realizing that we shift a broad Γ miniband upwards with respect to the X-states by applying pressure may justify this approximation.

2.Cw and time-resolved measurements

In cw photoluminescence measurements we have determined the type-I/type-II crossover pressure ($P_c=1.2$ GPa) with type-I character below and type-II behavior above P_c and the pressure coefficients of the direct and indirect transitions (Fig.1).

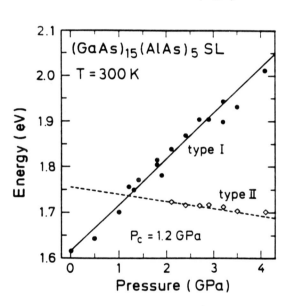

Fig.1: *Pressure depen-dent photoluminescence peaks of the Γ-hh (full dots) and X-hh (open diamonds) transitions in the (GaAs)₁₅/(AlAs)₅ SL together with theo-retically estimated pres-sure coefficients of 100.1 meV/GPa (solid line) in the case of the direct and -15.9 meV/GPa (dashed line) in the case of the in-direct transition.*

Futhermore, we performed time-resolved measurements to prove the validity of Eq.(2), using 2 eV femtosecond pulses from a CPM laser/copper-vapor laser pumped amplifier system to excite the sample and white-light continuum pulses to probe the transmission as a function of delay between pump and probe pulse. The overall time resolution of this setup was $\leq$ 150 fs. A more detailed description of the experiment is presented elsewhere (Nunnenkamp et al. (1991)).

Fig. 2 shows four differential transmission (DT) measurements on the Γ-hh transition taken at a constant carrier density ($I_0 \Leftrightarrow n = 1.5 \cdot 10^{12} cm^{-2}$) but different pressures. The type-I DT (Fig.2, 0.2 GPa) demonstrates a rapid absorption bleaching similar to the findings of Knox et al (1985). Coulomb screening causes the initial bleaching at t ≈ 0 fs whereas at longer times (≥ 200 fs) phase-space-filling (PSF) becomes the dominant process.

When switching to type-II character (Fig.2, 1.5 GPa and 2.3 GPa) the Γ-X transfer of the electrons to the AlAs layer leads to a decrease of PSF and therefore to a decay of the DT with a time constant which decreases with increasing pressure. The increase of the DT at longer times mainly monitors the cooling of holes left in the GaAs layer. At 3.0 GPa (Fig.2) a new scattering channel (Γ-X GaAs intralayer scattering) is opened thus decreasing the DT decay time to ≤ 50 fs.

At 1.1 GPa (below P_c) we observe a carrier-induced type-I/type-II switching, indicating that P_c is shifted to <u>lower</u> pressures with increasing carrier density. This fact explains the significant shift of the crossover-pressure P_{cdyn} observed at fs pulse excitation with an excitation density of $1.5 \cdot 10^{12} cm^{-2}$ compared to P_{cstat} (determined in the cw experiments) in Figure 3 where the type-II DT decay times of the Γ-hh transition are shown versus pressure. From 0.9 GPa to 2.6 GPa the DT decay times are explained by:

$$\tau_{\Gamma-X}^{-1} \propto (\Delta_{\Gamma X}(P, n = 0) + \Delta E(n))^{1/2} \tag{3}$$

where $\Delta E(n)$ corresponds to a carrier dependent shift of P_c (solid line).

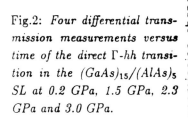

Fig.2: *Four differential transmission measurements versus time of the direct Γ-hh transition in the* $(GaAs)_{15}/(AlAs)_5$ *SL at 0.2 GPa, 1.5 GPa, 2.3 GPa and 3.0 GPa.*

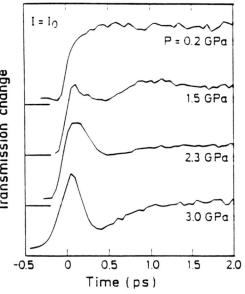

At even higher pressure (> 2.6 GPa) the opening of a new scattering channel (Γ-X GaAs intralayer scattering) results in a further reduction of $\tau_{\Gamma-X}$ (dashed line).

The deduced shift $\Delta E(n = 1.5 \cdot 10^{12} cm^{-2}) = 34$ meV (see Eq. 3) results in a new crossover-pressure of $P_c \approx 0.9$ GPa (P_{cdyn}).

3. Bandgap-Renormalization

Calculation of the expected bandgap-renormalization (BGR) of the direct transition according to Cingolani et al. (1990) gives $\Delta E_{BGR}^{\Gamma} = -40$ meV. Hence, if only the direct transition is renormalized we would expect the full line (type-I) in Fig. 1 to be shifted by -40 meV leading to a new $P_c \approx 1.7$ GPa. The deduced value of 0.9 GPa ($n = 1.5 \cdot 10^{12} cm^{-2}$) can only be explained by considering also the renormalization of the indirect transition yielding $\Delta E(n) = \Delta E_{BGR}^{\Gamma}(n) - \Delta E_{BGR}^{X}(n)$.

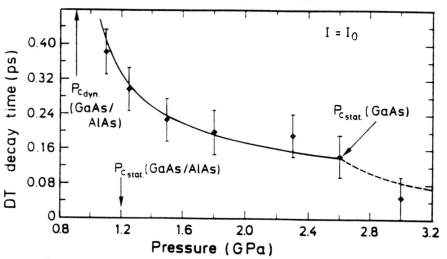

Fig.3: *Decay times of the DT at the Γ-hh transition in the GaAs layers versus pressure (compare with Figure 2). The solid line corresponds to Eq. 3 wheras the dashed line also considers Γ-X GaAs intralayer scattering, i.e. $\tau_{\Gamma-X}^{-1} = (\tau_{\Gamma-X}^{GaAs/AlAs})^{-1} + (\tau_{\Gamma-X}^{GaAs})^{-1}$.*

From this relation we deduce $\Delta E_{BGR}^{X}(1.5 \cdot 10^{12} cm^{-2}) = -74$ meV to be approximately twice as large as the renormalization of the direct transition $\Delta E_{BGR}^{\Gamma}(1.5 \cdot 10^{12} cm^{-2}) = -40$ meV.

We would like to thank A. Schulz for excellent technical assistance.

References:

Cingolani R., Kalt H., and Ploog K. 1990 Phys. Rev. **B42**, 7655

Dawson P., Foxon C.T., and van Kesteren H.W. 1990 Semicond. Sci. Technol.5, 54

Feldmann J., Nunnenkamp J., Peter G., Göbel E.O., Kuhl J., Ploog K., Dawson P., and Foxon C.T. 1990 Phys. Rev. **B42**, 5809

Knox W.H., Fork R.L., Downer M.C., Miller D.A.B., Chemla D.S., Shank C.V., Gossard A.C., and Wiegmann W. 1985 Phys. Rev. Lett. **54**, 1306

Nunnenkamp J., Reimann K., Kuhl J., and Ploog K. 1991 Phys. Rev. **B44**, 8129

Inst. Phys. Conf. Ser. No 126: Section V
Paper presented at Int. Symp. on Ultrafast Processes in Spectroscopy, Bayreuth, 1991

Hot luminescence and nonlinear effects in shortperiod superlattices under picosecond excitation

E A Vinogradov, A V Zayats, D N Nikogosyan, Yu A Repeyev

Institute of Spectroscopy, USSR Academy of Sciences,
142092, Troitsk, Moscow region, USSR

ABSTRACT: Optical properties (luminescence and it excitation) of short-period (≤ 20Å) amorphous Si/SiO_2 superlattices is investigated under picosecond excitation. Three types of radiative transitions (intersubband luminescence in $a - Si$ layers, cross-luminescence between states of well and barrier layers and impurity luminescence in $a - SiO_2$ layers) were observed in ps excited SL contrary to continuous excitation when only impurity recombination in $a - SiO_2$ layers took place. Photoluminescence intensity dependencies were investigated on pumping intensity. Carrier transfer between states of adjacent layers influences nonlinear effects in optical properties of superlattices.

1. INTRODUCTION

Semiconductor superlattices (SLs) consisting of tunneling thin layers are perspective structure for application in ultrafast optoelectronics. Quantum size effects which depend on the SL period give rise to appreciable changes in the spectra of electron states in superlattices. The photoluminescence (PL) spectrum, luminescence intensity and the lifetime of excited carriers are determined by the quantum confinement as well as by the interaction between the electron states of adjacent layers which form the SL (Murayama 1986). The influence of quantum size effects in amorphous SLs results in the appearance of k subbands ($k = 0$, 1, 2...) of the conduction and valence bands, and electron transitions between various subbands of valence and conduction bands obey the rule $\Delta k = 0$ (Hattori et al 1988). This structure of electron states of $a - Si$ layers is revealed in modulated absorption spectra of $a - Si/SiO_2$ SLs (Vinogradov et al 1990 and 1991).

In contrast to widely investigated SLs based on III-V and others compounds, the large difference (about 7 eV) between $a - Si$ and $a - SiO_2$ band gaps lead to the increase of the influence of impurity states of $a - SiO_2$ layers on optical properties of $a - Si/SiO_2$ SLs. Investigation of luminescence in these structures under continuous excitation shows that the PL due to electron transitions between impurity states of fused quarts layers is revealed only and no luminescence related to $a - Si$ layers of SLs is observed (Vinogradov et al 1990).

Excess carriers excited close to resonance with band gap alter the refractive index of SL and, therefore, cause the nonlinear effects. Since these nonlinearities depend on the density of carriers excited, relaxation processes and carriers lifetimes will affect the nonlinear behavior (Poole and Garmire 1985).

This paper presents the results of experimental study of luminescence properties of amorphous Si/SiO_2 superlattices under picosecond excitation.

2. EXPERIMENTAL

The SLs to be investigated were prepared by RF magnetron sputtering of, in turn, amorphous SiO_2 and Si in spectrally pure argon onto substrates of silicon single crystals with (100) or (111) orientation (Plotnikov et al 1987). There is Ni interlayer between substrate and SL sample. The thickness of $a-SiO_2$ layers is 10 Å and thickness of $a-Si$ layers is 2.7 Å for SL-1 sample, 5.5 Å for SL-2 sample and 8.2 Å for SL-3 sample. Total thickness of SLs is about 1000 Å.

Photoluminescence and PL excitation spectra of SLs were recorded at room temperature using for excitation 20 ps- light pulses of tunable parametric oscillator pumped by second harmonic of Nd:YAG laser radiation. The output radiation wavelength was continuously tunable from 370 to 1890 nm. The energy of pulses at most often used wavelengths (450 nm and 580 nm) was about 0.1 mJ.

3. RESULTS AND DISCUSSION

The possibility to observe PL bands in SL investigated strongly depends on the energy of excitation quanta, whereas in bulk semiconductors under interband excitation all PL lines are seen. A number of PL bands (I_0) is observed (Fig.1) under longwave (580 and 600 nm) excitation whose spectral positions correspond well to the energy distance between the nearest subbands (k=0). PL bands (C_0) were also observed inside the first subband for SL-2 and SL-3 where no peculiarities in the absorption spectrum were detected. No other PL bands were observed in the spectra under this excitation. A change over to excitation wavelength 450 nm leads to disappearance of all previously observed bands. Besides luminescence band of fused quartz ($\propto 2.1$ eV) which are visible for all SLs investigated, for the SL-2 sample PL bands were observed near the edge (I_1) and inside (C_1) the second (k=1) subband (Fig.1). In the PL spectrum of SLs 1 and 3 the a-quartz PL band at 2.1 eV has a shortwave branch. At still shorter-wave (370 nm) excitation (Fig.1) in the spectrum of SL-1 there is only one broad PL band near the edge of the second subband of $a-Si$, the sample SL-2 has no luminescence and for SL-3 a broad new line near 2.9 eV is observed.

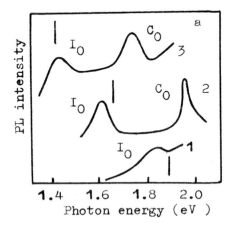
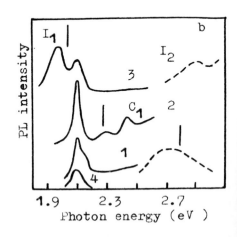

Fig.1. Photoluminescence spectra of SLs under picosecond excitation. (a) for SL-1 at λ_{ex} =580 nm (1), SL-2 at λ_{ex} =600 nm (2), SL-3 at λ_{ex} =580 nm (3). The arrows show the position of k=0 subband in SLs; (b) for λ_{ex} = 450 nm (solid line) and 370 nm (dash line) for SL-1 (1), SL-2 (2), SL-3 (3) and $a-SiO_2$ (4, intensity x10). T=300 K. The arrows show the position of k=1 subband in SLs.

It turns out that I_0 luminescence lines, whose spectral positions correlate with those of the edges of the nearest $a - Si$ subbands, are excited by the light absorbed at electron transitions between these $k=0$ subbands (Fig.2). As the energy of exciting photon increases the absorption due to transitions between the next subbands ($k=1$) begins and I_0 luminescence bands disappears but, simultaneously, I_1 bands appear whose positions are determined by the edge of subbands k=1. Further decrease in the excitation wavelength leads to repetition of the situation when the edge of the next subband $k=2$ is reached.

Usually, due to thermalization of excited carriers, only lower states of the electronic spectrum show up in luminescence. in our case, if carriers are excited to other than k-th the I_k subband-to-subband luminescence is not observed. These facts evidence the slow relaxation between different subband within conduction and valence bands of $a-Si$ in the SLs investigated. The slow intersubband relaxation results in observation of hot luminescence due to higher subbands of well layers. Both hot electrons and hot holes simultaneously take part in the recombination.

Second type of hot luminescence observed is related to the transitions between higher subbands of conduction band of $a - Si$ layers and impurity states of neighbour $a - SiO_2$ layers (Zayats et al 1991). Corresponding PL lines are excited with similar intensity both in Stockes and anti-Stockes ranges of conduction band subbands from which the transitions take place (Fig.2).

The electrons excited in $a - Si$ layers in SL can transfer to adjacent $a - SiO_2$ layers either with photon emission (cross-luminescence) or by nonradiative processes (resonant transfer to impurity states which are close in energy to electron states in $a-Si$ or Auger-recombination with subsequent leaving the well layer).Excitation of impurity transition in $a - SiO_2$ layers via the states of the subbands of $a - Si$ layers in SL leads to increased intensity of this SL luminescence by aabout 10^4 times compared to separate $a - SiO_2$ layers (Vinogradov et al 1990).

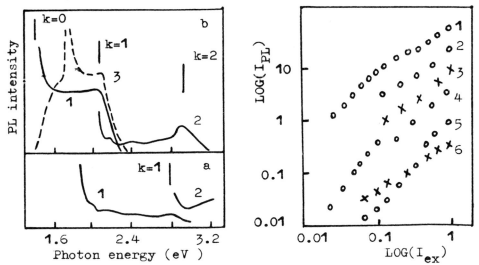

Fig.2. Photoluminescence excitation spectra: (a) for I_0 (1) and I_1 (2) PL bands in SL-1; (b) for I_0 (1), I_1 (2) and C_0 (3) in SL-3. The arrows show the positions of the subband edges.
Fig.3. PL intensity dependencies on effective mean internal pumping intensity: for I_0 band in SL-3 under excitation 860 nm (1) and 580 nm (2); for I_0 band in SL-1 λ_{ex} =580 nm (3); for C_0 band in SL-3 λ_{ex} =540 nm (4); for impurity luminescence bands in SL-3 (5) and SL-1 (6) under excitation 450 nm.

Low intersubband relaxation results in the increase of influence of excess carriers accumulated near subbands edges on dispersive nonlinearity that reveals in wide spectral range in this conditions. The dispersive nonlinearity (intensity dependent refractive index) is observed in Fabry-Perot cavity formed by SL on metal layer in structures investigated. (Nonlinearity caused by saturated absorption in these SLs is not observed down to 30 ps time delay.) The refractive index dependence on pumping intensity is caused by excess carriers (ΔN) generated by excitation. The models of dispersive nonlinearity give the opportunity to consider linear dependence between Δn and ΔN (Burt 1990). Excess carrier concentration dependence on pumping intensity can be nonlinear due to radiative and Auger recombination because recombination rate depends on carrier concentration. This effect reveals in nonlinear dependent PL intensity (Miller and Parry 1984).

Figure 3 shows the dependencies of PL intensity (I_{PL}) on effective mean intensity (I_{ex}) inside SL for different types of luminescence transition. These dependencies $I_{PL} = CI_{ex}^{m}$ can be determined by the ratio between recombination times (and rise time of pulse). The intensity of I_0 luminescence band in SL-3 has bimolecular recombination dependence in the presence of Shockley-Read recombination under excitation near $k=0$ subband edge: $m=1.9$ (theoretical calculation gives $m=2$) for low excitation intensity and $m=1.1$ for high I_{ex} (theory gives $m=1$). Such dependence is resulted in the competition between $k=0$ subband-to-subband and cross- luminescence. If the excitation quanta are slightly less than $k=1$ subband edge the intensity dependence is described by $m=1.22$ (analogous dependence is observed for I_0 band in SL-1 also). THe new rate is probably determined by Auger recombination that limits carrier lifetime in subband. In this case $m=1.33$ is predicted that is close to $m=1.22$ obtained in experiment. The intensity dependencies observed for other type of luminescence confirm the electron transfer between SL layers due to Auger recombination.

4. CONCLUSION

Subband-to-subband and subband-to-impurity luminescence that is absent under cw excitation was observed in shortperiod amorphous Si/SiO_2 superlattices under picosecond excitation. In contrast to well investigated SLs based on III-V and other compounds, in $a - Si/SiO_2$ SLs relaxation of excited carriers between various subbands of c- and v- bands is low. This allows observation of two type of hot PL caused by higher subbands in $a - Si$ layers: subband-to-subband hot luminescence with participation both hot electrons and hot holes, and hot cross-luminescence in which only hot electrons take part. The intensity dependent carrier lifetime influences dispersive nonlinearity which are observed in density dependent PL intensity. Types of carriers relaxation is dependent on pumping intensity.

ACKNOWLEDGEMENTS. Authors grateful to F.A.Pudonin for a preparation of the SLs samples.

REFERENCES

Burt M G 1990 Semicond. Sci. Technol. **5** 1215
Miller A and Parry G 1984 Opt. Quantum Electron. **16** 339
Murayama Y 1986 Phys. Rev. B36 2500
Plotnikov A F, Pudonin F A and Stopachinsky V B 1987
    Sov.Phys.:JETP Letters **46** 443
Poole C D and Garmire E 1985 IEEE J. Quantum Electron. **QE-21** 1370
Vinogradov E A, Denisov V N, Zayats A V, Mavrin B N, Makarov G I, Oktyabrsky S R
    and Pudonin F A 1990 Spectroscopy and Optoelectronics in Semiconductors and
    Related Materials ed S C Shen et al (Singapore: World Scientific) pp 121-131
Vinogradov E A, Zayats A V and Pudonin F A 1991 Sov. Phys.:Solids **33** 197
Zayats A V, Repeyev Yu A, Nikogosyan D N, Vinogradov E A 1991
    Phys. Lett. A **155** 65

Coulomb relaxation kinetics in a dense quasi-twodimensional electron plasma

K. El Sayed, T. Wicht, H. Haug, and L. Bányai

Institut für Theoretische Physik, J. W. Goethe Universität Frankfurt,
Robert-Mayer-Strasse 8, D-6000 Frankfurt am Main, Germany

ABSTRACT: For carrier-carrier scattering the relaxation kinetics of a quasi-twodimensional degenerate electron gas is studied in the framework of the Boltzmann equation using a static RPA screening. The relaxation of a given initial nonequilibrium distribution is calculated both by ensemble Monte Carlo (MC) simulations and by expansions in terms of eigenfunctions of the linearized collision operator. A systematic comparison of the results of both methods yields a rather detailed understanding of the Coulomb relaxation kinetics and of the strengths and limitations of both methods.

In recent experiments on direct semiconductors (Oudar 1985, Knox 1986, Becker 1988) a degenerate nonequilibrium carrier distribution has been excited by a femtosecond (fs) laserpulse and its relaxation has been studied on a fs timescale. In these situations the most important scattering processes are the carrier-carrier collisions. The understanding of ultrafast optical spectroscopy in semiconductors requires a detailed treatment of the kinetics due to this interaction beyond a phenomenological lifetime approximation. In order to start with a simple model we choose a quasi-twodimensional homogeneous electron gas in a parabolic band. For all numerical results the GaAs conduction band mass is used. For this model the full nonlinear Boltzmann equation is given by

$$\frac{\partial f_{\mathbf{k}}}{\partial t} = 2 \sum_{\mathbf{p},\mathbf{k}',\mathbf{p}'} W_{\mathbf{k},\mathbf{p},\mathbf{k}',\mathbf{p}'}\left[- f_{\mathbf{k}} f_{\mathbf{p}}(1-f_{\mathbf{k}'})(1-f_{\mathbf{p}'}) + (1-f_{\mathbf{k}})(1-f_{\mathbf{p}}) \, f_{\mathbf{k}'} f_{\mathbf{p}'} \right] , \tag{1}$$

with

$$W_{\mathbf{k},\mathbf{p},\mathbf{k}',\mathbf{p}'} = \frac{2\pi}{\hbar} \left| V_s(|\mathbf{k}'-\mathbf{k}|) \right|^2 \delta_{\mathbf{k}+\mathbf{p}-\mathbf{k}'-\mathbf{p}'} \, \delta(e_{\mathbf{k}}+e_{\mathbf{p}}-e_{\mathbf{k}'}-e_{\mathbf{p}'}) .$$

$f_{\mathbf{k}}$ is the electron distribution function and $\hbar\mathbf{k}$ the 2d momentum. The matrix element W contains the statically screened Coulomb potential (static long-wavelength limit of the RPA, see e.g. Haug 1990.) The first term of the collision integral describes the scattering rate out of the momentum state $\mathbf{k}$ and the second one the rate into this state. Using this equation, we investigate the evolution of a distribution which consists of a Fermi distribution and an additional Gaussian peak with a width σ, centered at the energy e_0 with an amplitude C.
In order to obtain a symmetric linear collision operator L we linearize eq. (1) with respect to a renormalized deviation $\phi_{\mathbf{k}}$ instead of the distribution deviation $\delta f_{\mathbf{k}}$

$$f_{\mathbf{k}} = \frac{1}{e^{\beta(e_{\mathbf{k}} -\mu)-\phi_{\mathbf{k}}} +1} = f_{\mathbf{k}}^0 + \delta f_{\mathbf{k}} = f_{\mathbf{k}}^0 + \phi_{\mathbf{k}} f_{\mathbf{k}}^0 (1 - f_{\mathbf{k}}^0). \tag{2}$$

The linearized Boltzmann equation reads

$$\dot{\phi}_{\mathbf{k}} = - (L\phi)_{\mathbf{k}} . \tag{3}$$

For an initial deviation $\phi(t=0)$ a solution of eq. (3) can be found by expanding $\phi(t=0)$ in terms of the eigenfunctions $\phi^{(\lambda)}$ of L which are determined by the eigenvalue equation $L\phi^{(\lambda)} = \lambda\phi^{(\lambda)}$. The eigenvalue spectrum of L is continuous. The density of states is shown in Figure 1 for an electron density of $0.64 \cdot 10^{12}/cm^2$ at a temperatur of 250K. We find eigenvalues close to zero (not shown) belonging to the conserved quantites (density, momentum, energy) and a broad continuum of eigenvalues. The spectrum has an upper bound which gives the highest relaxation rate in this system ($\lambda_{max} \simeq (40fs)^{-1}$). The upper part of the spectrum is independent of the cut-off wavenumber taken as $6k_F$. The cut-off dependent lower part of the spectrum is unimportant for our investigations. Obviously, it is impossible to approximate this broad and featureless spectrum by a single relaxation rate.

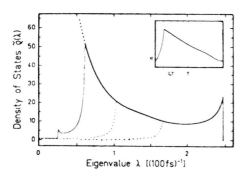

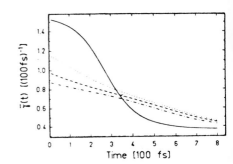

Fig. 1. Density of states of the collision eigenfrequencies for the plasma density $n=.64 \cdot 10^{12} cm^{-2}$, and three different values of the cut-off wavenumber k_N: $k_N = 2k_F$ (dotted line), $k_N = 4k_F$ (dashed line), $k_N = 7k_F$ (full line). The inset gives a double logarithmic plot.

Fig. 2. Relaxation rate $\bar{\gamma}(t)$ versus t for the initial distribution deviation with $e_0 = .8\mu$ and various width values: $\sigma = 0$ (full line); $\sigma = .2\mu$ (dashed line); $\sigma = .3\mu$ (dashed-dotted line).

The eigenfunctions $\phi^{(\lambda)}_{\mathbf{k}}$ are oscillating functions which range from k=0 up to a λ-dependent maximum wavenumber. Therefore several eigenfunctions (each of them decays exponentially) will contribute to the expansion of an localized initial deviation $\phi(t=0)$ causing a nonexponential decay of the disturbance. Such an nonexponential decay can be described by a time-dependent relaxation rate $\bar{\gamma}$ defined by

$$S(t) = S(t=0) \, e^{-g(t)} \quad , \quad \bar{\gamma}(t) = dg(t)/dt , \tag{4}$$

where S(t) is the signal under consideration. A constant $\bar{\gamma}$ describes an exponential decay. Figure 2 shows $\bar{\gamma}(t)$ for various peak widths. Even for a strongly localized deviation (the width of which given by the grid spacing $\Delta k = k_F/50$) the decay slows down considerably for larger times. On the scale of a typical relaxation time of about 100fs ($\bar{\gamma} = 1$), however, the rate is relatively constant. MC simulations performed for initial conditions *without* equilibrium background show significant variations even on this scale.

In addition Figure 2 shows that the initial relaxation time $1/\bar{\gamma}(t=0)$ increases with the width of the disturbance. This can be understood by considering the relaxation as a drift and diffusion process in momentum space (Haug 1991). In Figure 3 the initial relaxation time is displayed as a function of the peak center e_0 for a fixed width. A deviation centered around the chemical potential will decay fastest because there free states as well as scattering partners can be found. These facts indicate that the effective relaxation time depends on the form and the position of the disturbance.

In order to check the validity of the linearization, we perform MC simulations of the full nonlinear Boltzmann equation for the same initial conditions and find good agreement. The solid triangels in Figure 3 show the MC results.

The evolutions of an initial distribution deviation δf evaluated with the eigenvalue expansion and with the MC simulation are shown in Figures 4a and b, respectively. The MC data are averaged over five runs. This shows that in certain situations the linearization can be used even for quite large deviations. Detailed conditions are given in El Sayed 1991.

In the MC method (Jacoboni 1983) one considers a particular population n_k of electrons on the k-grid. Each k state is either occupied or empty ($n_k=1$ or 0). The distribution function f_k is the ensemble average of n_k. (In an isotropic situation, the ensemble average

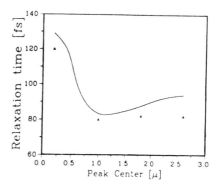

Fig. 3. Relaxation time versus peak position e_0 for the peak width $\sigma=.4\mu$ and a density of $n=.64\cdot10^{12}\mathrm{cm}^{-2}$. The solid triangles show the MC results.

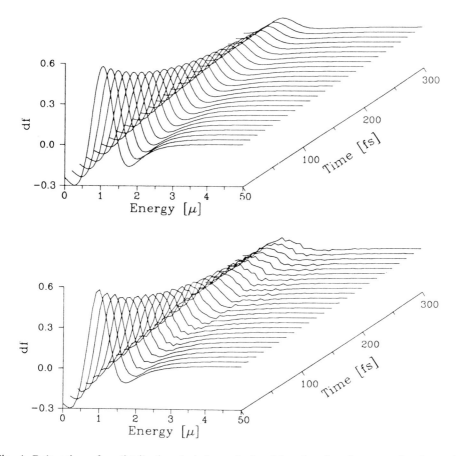

Fig. 4. Relaxation of a distribution deviation calculated by eigenfunction expansion (upper) and by MC simulation (lower) averaged over five runs.

can be replaced by an angle average.) After calculating the transition probalilities between the electron states the stochastic evolution of the system can be simulated. (For details of the method see El Sayed 1991.) Because we use the simple static screening, the algortithm is very fast. A typical run takes just 10 minutes on a 80386-PC.

In Figure 5 the density dependence of the initial relaxation time is shown. The initial distribution was a Gaussian one centered at $e_0=60$meV with a width of $\sigma=60$meV and a density-dependent amplitude. (The temperatur of the final Fermi distribution is given by the labels.) A power-law fit to the MC data yields

$$\tau = 185 \text{ fs} \left(\frac{n}{10^{11}\text{cm}^{-2}} \right)^{-0.30} .$$

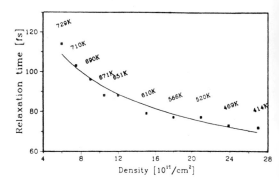

Fig. 5. Relaxation time versus density for an initial Gaussian distribution with $e_0=60$meV and $\sigma=60$meV. The lables show the temperature of the final equilibrium Fermi distribution. The MC data (solid sqares) is fited by a power-law (solid line).

If we vary amplitude, position and width of the initial Gaussian in order to keep the temperature of the final distrubution fixed we find nearly the same density dependence.

References

Becker P C, Fragnito H, Brito-Cruz C, Shaw J, Fork R L, Cunningham J E, Henry J E, and Shank C V 1988 Phys. Rev. Lett. **61** 1647
El Sayed K, Wicht T, Haug H, and Bányai L Z. Phys. B to be published
Haug H and Koch S W 1990 *Quantum Theory of the Optical and Electronic Properties of Semiconductors* (Singapore: World Scientific)
Haug H and Henneberger K 1991 Z. Phys. **B 83** 447
Jacobini C and Reggiani L 1983 Rev. Mod. Phys. **55** 645
Knox W H, Hirlimann C, Miller D A B, Shah J, Chemla D S, and Shank C V 1986 Phys. Rev. Lett. **56** 1191
Oudar J L, Hulin D, Migus A, Antonetti A and Alexandre F 1985 Phys. Rev. Lett. **55** 2074

Inst. Phys. Conf. Ser. No 126: Section V
Paper presented at Int. Symp. on Ultrafast Processes in Spectroscopy, Bayreuth, 1991

An infrared spectrometer for time-resolved intersubband spectroscopy

A. Seilmeier, T. Dahinten, U. Plödereder, G. Weimann *

Phys. Institut, Universität Bayreuth, D-8580 Bayreuth, Germany
* Walter Schottky-Institut, Technische Universität München,
D-8046 Garching, Germany

ABSTRACT: Picosecond light pulses tunable between 4 μm and 18 μm are generated by difference frequency mixing in $AgGaS_2$ or GaSe crystals. The pulses are applied to study intersubband relaxation in n-modulation doped GaAs/Al_xGa_{1-x}As multiple quantum well structures by an infrared bleaching technique. A rise of the time constants from 4 ps to 7.5 ps is observed for increasing doping concentrations between 3.2×10^{11} cm^{-2} and 1.4×10^{12} cm^{-2}.

1. INTRODUCTION

In recent years intersubband relaxation has been addressed in several experimental and theoretical papers. Direct information is available from time resolved experiments with ultrashort light pulses. The dynamics of subbands has been studied by several techniques, e.g. by a Raman probing technique (Oberli, 1987 and Tatham, 1989), and by an infrared bleaching technique (Seilmeier, 1987).

In undoped samples time constants in the order of a picosecond are observed, which are consistent with theoretical calculations taking into account electron – LO phonon interaction as the relevant relaxation mechanism (Tatham, 1989). In this paper the intersubband scattering in n-type modulation doped GaAs/Al_xGa_{1-x}As quantum well structures is studied by infrared bleaching experiments. In modulation doped structures time constants up to 14 ps are found (Seilmeier, 1987), which are clearly longer than the values estimated from a simple electron-LO phonon interaction model. This fact is a strong indication for a more complex relaxation process which may be attributed to a transfer of electrons to the potential minimum in the barrier generated by the ionized donors.

2. EXPERIMENTAL

The samples investigated here are grown by molecular beam epitaxy on (100) semi-insulating GaAs substrates of 350 μm thickness and consist of 25 or 50 thin, undoped GaAs layers. They are embedded in approximately 400 Å thick Al_xGa_{1-x}As layers in which the central 80 Å to 100 Å are doped with Si. The total multi-quantum well structures are clad between 0.2 μm thick Al_xGa_{1-x}As layers to avoid surface depletion and substrate effects.

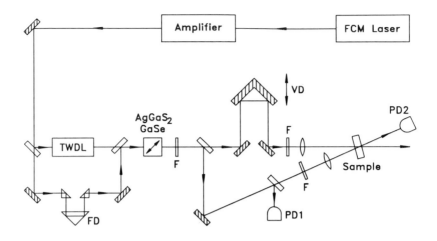

Fig. 1. Experimental system for time resolved infrared bleaching experiments in the wavelength range 4–18 μm.

Intersubband scattering is investigated by an infrared bleaching technique, which requires ultrashort light pulses tunable at wavelengths longer than 7 μm. An intense infrared pulse resonantly excites a considerable number of electrons from the lowest subband to the first excited subband of the well resulting in a bleaching of the intersubband transition. A second weaker light pulse of the same frequency monitors the absorption change as a function of time. The absorption recovery reflects the repopulation of the lowest subband.

The infrared picosecond light pulses are generated by a parametric amplification process (Fig. 1). We start with a feedback controlled modelocked (FCM) Nd:glass laser system which generates single pulses of 2.3 ps duration (Heinz, 1989). The pulses pump a traveling wave infrared laser (TWDL) producing tunable pulses between 1.15 μm and 1.40 μm (dye No. 5) (Elsässer, 1984) or between 1.1 μm and 1.16 μm (dye A 9860), respectively. A second part of the glass laser pulses and the dye laser pulses are mixed in an AgGaS$_2$ or a GaSe crystal. In this way pulses at the difference frequency tunable between 4 μm and 18 μm are generated (Elsässer, 1985). The pulses exhibit an energy in the order of a microjoule, a spectral width of about 12 cm^{-1} and a pulse duration of 1.8 ps as determined by background–free cross correlation measurements. Details will be published elsewhere.

The tunable infrared pulses are divided into pump and weaker probe pulses which are focused on the sample. A noncolinear geometry with an angle of ˜ 30° between pump and probe beam is used. The change of transmission of the probe beam is studied as a function of delay time.

3. EXPERIMENTAL RESULTS

In the following time–resolved data are presented for thin multiple GaAs/Al$_x$Ga$_{1-x}$As quantum well structures which are n–type modulation doped. The investigations are performed at room temperature. Three samples W 1757, W 1756 and W 1755 are investigated with a well thickness of 59 Å

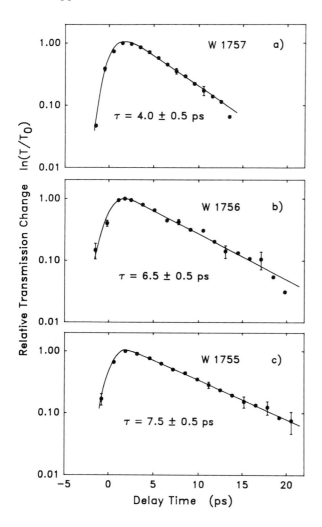

Fig. 2: Transmission change of the probe pulse as a function of the delay time for samples of a well thickness of $d_{GaAs} \sim$ 59 Å and carrier densities/well a) of 3.2 $\times$ 10^{11} cm^{-2}, b) of 7.5 $\times$ 10^{11} cm^{-2}, and c) of 1.4 $\times$ 10^{12} cm^{-2}. T_0 is the transmission of the sample without excitation.

and different doping concentrations of 3.2 $\times$ 10^{11} cm^{-2} (50 wells), 7.5 $\times$ 10^{11} cm^{-2} (50 wells), and 1.4 $\times$ 10^{12} cm^{-2} (25 wells), respectively. In a first approximation the intersubband transition frequencies do not depend on the doping concentration. The spectra show only a small shift of the absorption bands to lower frequencies with rising carrier concentration (W 1757: 1250 cm^{-1}, W 1756: 1210 cm^{-1}, W 1755: 1200 cm^{-1}; Plödereder, 1992). The maximum infrared absorption observed in a special prism geometry (Seilmeier, 1989) exceeds A = 1.0 (T < 10 %).

In Figure 2 time resolved results are shown for the three samples. The signal plotted in Fig. 2 $\ln(T/T_0) = \sigma \cdot N (n_2 - n_1 + n_0)$ depends on the instantaneous carrier density/well in the upper state n_2 and on the depletion of the lower state ($n_0 - n_1$). (n_0 and n_1 are the total carrier density/well and the instantaneous carrier density/well in the lower state, respectively; σ is the absorption cross section, N is the number of quantum wells.) We observe a rapid bleaching of the absorption and an absorption recovery on a longer time scale. Decay times of 4 ps, 6.5 ps and 7.5 ps are found for

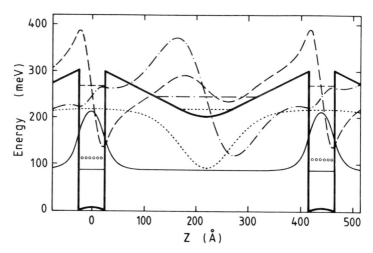

Fig. 3: Potential, eigenstates and wavefunctions for a thin n–modulation doped GaAs/Al$_x$Ga$_{1-x}$As quantum well structure obtained from self–consistent calculations. Relevant parameters: well width 50 Å, doping concentration 6 × 10^{11} cm^{-2}.

the samples W 1757, W 1756, and W 1755, respectively, representing a surprisingly strong rise of τ with increasing doping concentrations from n = 3.2 × 10^{11} cm^{-2} to n = 1.4 × 10^{12} cm^{-2}.

4. DISCUSSION

The mechanism of intersubband relaxation and the origin of the relatively long time constants obtained in n–modulation doped samples are of interest. Electron–polar LO phonon interaction is generally accepted as the dominant relaxation mechanism in thin quantum wells (subband separation > $\hbar\omega_{LO}$). A rough estimate of the scattering time τ_{21} of an intersubband transition between the first excited subband and the lowest subband gives a value of 2 ps for a well thickness of 50 Å. Time constants consistent with the estimated numbers have been observed in undoped samples (Tatham, 1989).

The clearly longer time constants observed in n–modulation doped structures have to be attributed to specific properties of their band structures (Plödereder, 1992). First we have to discuss the band structure of these samples which is believed to be of particular importance. The ionized impurities in the center of the barrier give rise to a shallow potential minimum containing also several bound subbands (see Fig. 3). Generally, the eigenstates and wavefunctions of the entire heterostructure potential consisting of quantum wells and potential minima in the center of the barriers have to be determined by a self– consistent band structure calculation. As an example the self–consistent potential, calculated eigenstates and the corresponding envelope wavefunctions are shown in Fig. 3 for a structure with a well width of 50 Å and a doping concentration of 6 × 10^{11} cm^{-2} (Abstreiter, 1991).

The wavefunctions of the two bands of lowest energy (solid and dotted line in Fig. 3) indicate a strong confinement of the carriers in the well and in the barrier, respectively. The higher subbands are made up from the first excited subband of the wells and excited barrier subbands which are close to resonance. The corresponding wavefunctions exhibit a large probability of finding electrons both in the well and in the barrier. An efficient spatial transfer of electrons from the wells to the potential minima of the barriers and vice versa is expected. In Fig. 3 the energy levels are drawn in the well and in the barrier minimum, depending where they are arising from.

A strong mixing between the upper well subband and barrier subbands only takes place, if they are located at approximately the same energy. Such a situation occurs in very thin quantum wells where the excited subband of the well is located high enough, or in highly doped samples with a pronounced potential minimum in the barrier.

In the time resolved experiment electrons are excited to the subband with the largest transition moment which arise from the first excited subband of the quantum wells. In Fig. 3 the wavefunction of this state (broken line) is partially localized in the doping region of the barrier. Electrons spatially transferred to the barrier may relax to the lower confined bands of the potential minimum. The electrons can only return to the well via excited states with wavefunctions exhibiting again a large amplitude in both the barrier and the well (e.g. the dash-dotted band). Such a relaxation path requires absorption of a phonon. Taking into account the electron transfer from the well to the barrier and the way back via excited states a delay in the relaxation is expected resulting in longer time constants. This effect is believed to be responsible for the long time constants particularly in the samples W 1756 and W 1755. In sample W 1757 with the lowest carrier concentration the barrier states are located at considerably higher energies compared to the excited upper subband of the wells. In this case only weak interaction takes place, manifested by a shorter time constant of 4 ps. This value is already close to the values estimated for undoped samples.

5. CONCLUSIONS

Progress has been made in the understanding of intersubband scattering processes. There is strong indication that a transfer of excited electrons from the wells to the doping region of the barrier delays the repopulation of the lowest subband. This effect is more pronounced in samples with high doping concentrations. In this contribution only a qualitative discussion of the mechanisms is presented; a detailed quantitative model is still lacking.

6. ACKNOWLEDGMENT

The work was done in close collaboration with G. Abstreiter, Walter-Schottky-Institut, Technische Universität München. The GaSe crystal was made available by K.R. Allakhverdiev and T.A. Kerimov, Acad. Sci. Azerbaijan SSR, Baku. The project is supported by the Deutsche Forschungsgemeinschaft.

REFERENCES

Abstreiter, G., Besson, M., Heinrich, K., Köck, A., Weimann, G. and Zachai, R., 1991, in: Proceedings of the NATO Workshop on Resonant Tunneling in Semiconductors: Physics and Applications, L.L. Chang, ed., Plenum Press, to be published

Elsässer, T., Polland, H.-J., Seilmeier, A. and Kaiser, W., 1984, **IEEE J. Quant. Electron.**, 20, 191

Elsässer, T., Lobentanzer, H. and Seilmeier, A., 1985, **Opt. Commun.**, 52, 355

Heinz, P. and Laubereau, A., 1989, **J. Opt. Soc. Am.**, B 6, 1574

Oberli, D.J., Wake, D.R., Klein, M.V., Klem, J., Henderson, T. and Morkoc, H., 1987, **Phys. Rev. Lett.**, 59, 696

Plödereder, U., Dahinten, T., Seilmeier, A., Weimann, G., 1992, in: Proceedings of the NATO Workshop on Intersubband Transitions in Quantum Wells, E. Rosencher, B. Vinter and B. Levine ed., Plenum Press, to be published

Seilmeier, A., Hübner, H.-J., Abstreiter, G., Weimann, G. and Schlapp, W., 1987, **Phys. Rev. Lett.**, 59, 1345

Seilmeier, A., Wörner, M., Abstreiter, G., Weimann, G. and Schlapp, W., 1989, **Superlattices and Microstructures**, 5, 5569

Tatham, M.C., Ryan, J.F. and Foxon, C.T., 1989, **Phys. Rev. Lett.**, 63, 1637

Femtosecond relaxation of excited carriers in microcrystallites in a glassy matrix at excitation intensity 10^{10}–10^{13} W/cm^2

S.V.Chekalin,   V.M.Farztdinov,   V.V.Golovlev,   Yu.E.Lozovik,
Yu.A.Matveets,A.G.Stepanov, and A.P.Yartsev

Institute of Spectroscopy, USSR Academy of Sciences,142092
Troitsk, Moscow Region, USSR

ABSTRACT: The carriers relaxation time has been found to be shorter than 30 fs. The role of different processes in nonmonotonic temporal behaviour of absorption spectra is discussed.

Methods of femtosecond laser spectroscopy have recently been used actively to study the relaxation of excited carriers in semiconductors (Williams et al 1988). Semiconductor CdS$_x$Se$_{1-x}$ doped glasses (SDG) have been of particular interest (Williams et al 1988). This readily available material has a very large nonlinear susceptibility and a short response time and thus holds promise for use in optoelectronics. The dimensions of the microcrystallites in commercial glass samples are of the order of 100 A, and their concentration is $\sim 10^{15}$ cm^{-3}. It was shown by Nuss et al (1986) that with the excited-carrier density N below 10^{18}cm^{-3}, the primary mechanism for intraband relaxation in these glasses is scattering of excited carriers by optical phonons with a time scale of 200 - 500 fs, while the interband recombination time is measured in tens of picoseconds. In this paper we are reporting a study of the relaxation of excited carriers in RG - 8 commercial filters at much higher densities, at which the intensity of the femtosecond exciting pulse is close to the damage threshold.

A test sample was excited with 120-fs pulse at a wavelength of 612 nm (200 meV above the bottom of the conduction band). The excitation intensity I could be varied over the range 10^{10}- 10^{13} W/cm^2. The optical density of the sample at the excitation wavelength was $\sim$ 0.3. To study the time evolution of the absorption by the sample, we probed the excitation region with a wide band (continuum) femtosecond pulse, which was delayed with respect to the exciting pulse (Lozovik et al 1990). The intensity of the probe pulse did not exceed 10^9 W/cm^2. Using a multichannel optical analyser connected to a computer, we measured the spectra of the

excited sample and of the unexcited one in the spectral
interval 500 - 720 nm.The delay of the probe pulse was varied.

   Glass photodarkening - the usual effect when exposing a
SDG by pico- and nanosecond optical pulses (see for example
(Roussignol et al 1987)) - was not observed despite the fact
that we have used rather high intensities . Such behaviour of
SDG   may   be   the   consequence   of   twostageness   of   the
photodarkening process. At the first stage the trapping of
photoexcited carriers on the surface states (low energy traps)
occures with a characteristic time of 800 fs to 8 ps according
to different estimations. The second stage is the escape of
charged carriers from the surface states to deep electron
traps in the host glass due to the absorption of the second
quantum.   The   last   process   results   in   glass   darkening
(Grabovskis et al 1989). In our case there is not enough time
for charged carriers to be trapped on the surface states
during a pump pulse, so the second stage can't be realized.
That is why photodarkening is not observed.

   Almost   complete   bleaching   was   observed   for   the
wavelengths $\lambda > \lambda_{pump}$ near zero delay. Appreciable   bleaching
at this time delay is also observed for the wavelengths $\lambda <$
λ_{pump}. From this it follows that the energy states higher than
the excitation level have been populated in the course of the
excitation and that the carrier - carrier scattering plays a
basic role in the relaxation at the shortest ($< 10^{-13}$s) time
scale.From   the   fact   that   instant   bleaching   of   sample
absorption in the wide spectral interval 500 - 700 nm at zero
delay of the probe pulse it follows that the relaxation time
of excited carriers   does   not   exceed   the   time   resolution
(10^{-13}s). The comparison of the observed bleaching value at
the excitation wavelength with the number of absorbed photons
yields an upper estimate $\tau < 30$ fs on the intraband relaxation
time ; this estimate is much shorter than the time observed at
10^9 W/cm^2 and agrees with our previous result (Lozovik et
al 1990) , obtained from SDG investigation at the wavelengths
$\lambda > \lambda_{pump}$.

   Figure 1 shows the time evolution of difference optical
density for some wavelengths. The relaxation of the absorption
changes observed after the maximum of the exciting pulse had
characteristic times from some hundreds fs to some hundreds
ps.Partial restoring of optical density with a characteristic
time of ˜ 2 ps seems to be connected with excited carriers
cooling, which occurs due to the interaction with optical
phonons. The relaxation with 10 - 100 ps time is due to the
interband recombination of charge carriers. The nonmonotonic
temporal behaviour of bleaching was observed on the leading
front of the pump pulse in a number of experiments, its time
scale being of the order of our experimental resolution.   The
sharp (< 500 fs) decrease of bleaching is observed just after
the maximum of the excitation pulse and its recovery observed
in some hundreds femtoseconds after relaxation at long delays
is wavelength dependent.

Figure 2 shows the result of our computer simulations of the time dependence of optical density spectra of SDG after the excitation by femtosecond pump pulse. In our model we accounted for the following effects : nonradiative Auger recombination , charged carriers relaxation on optical phonons, band gap renormalization due to exchange and correlation effects , two - photon absorption and some other effects. Values of the most of material constants were taken from (Pugnet et al 1981). We also supposed that the carrier - carrier scattering is the fastest process and occurs on a time scale of $< 10^{-14}$ s. Auger recombination coefficient was taken with due account of its strong temperature dependence. Its room temperature value was varied in the region $\sim 10^{-30} cm^6 s^{-1}$.

According to our calculations, the decrease of bleaching at zero delay in Figure 2 is caused by two-photon absorption.The decrease of bleaching at picosecond delay is due to strong heating of charged carriers caused by Auger recombination.The sharp increase in the rate of Auger recombination as the pulse passes is a consequence of a rapid increase in N and a high carrier temperature, because the frequency of the exciting pulse is about 200 meV greater than the band gap and due to the heating by two-photon absorption. After the pump pulse has passed, Auger processes are ceased in a time about a picosecond, because of rapid carrier cooling.

The calculated time dependences of spectra are in a qualititative agreement with the experimental ones.

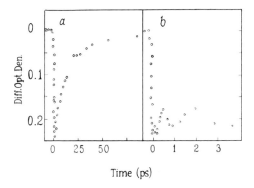

Figure 1. The time evolution of difference optical density for intensity $I = 5 \cdot 10^{11}$ W/cm^2 ; $a - \lambda = 550$ nm, $b - \lambda = 650$ nm.

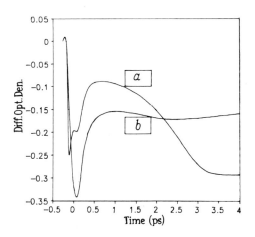

Figure 2. The result of computer simulations of the time dependence of optical density spectra of SDG after the excitation by femtosecond pump pulse with $I = 5 \cdot 10^{11}$ W/cm^2 ; $a - \lambda=650$ nm, $b - \lambda=550$ nm.

CONCLUSIONS

We have demonstrated that there are three characteristic time scales where the different processes are playing a basic role at intensities higher than 10^{10} W/cm^2. The smallest time scale is of the order of 10^{-14} s and it is the carrier - carrier scattering that plays a main role on this time scale. The second time scale is of the order of 10^{-12} s and here Auger recombination and carrier - phonon scattering play a main role. The third time scale is of the order of 10^{-11}s and here radiative interband recombination plays a main role. The importance of Auger processes and two-photon absorption is increased with increasing intensity and becomes to play a prominent role at intensities $\geq 10^{11}$W/cm^2.

REFERENCES

Grabovskis V Ya et al 1989 *Fiz. Tverd. Tela.* 31,272.
Lozovik Yu E , Matveets Yu A et al 1990   *JETP Lett.*52,221.
Nuss M C , Zinth W and Kaizer W 1986   *Appl. Phys. Lett.*, 49,171
Williams S V, Olbright G R et al   1988   *J. Mod. Opt.* 35, 1979.
Pugnet M, Collet J, and Cornet A   1981 *Solid State Comm.*38,531.
Roussignol P, Ricard D et al   1987 *J.Opt.Soc.Am.* B4,5

Inst. Phys. Conf. Ser. No 126: Section V
Paper presented at Int. Symp. on Ultrafast Processes in Spectroscopy, Bayreuth, 1991

Time-resolved studies on the carrier dynamics in quantum dots

V Jungnickel, J Puls and F Henneberger

Humboldt-Universität, Fachbereich Physik, Invalidenstr. 110, 1040 Berlin, Germany

ABSTRACT: The paper compares data of recent time-resolved studies on luminescence and nonlinear absorption of specially prepared CdSe nanocrystals embedded in glass.

1. INTRODUCTION

At present much attention is paid to semiconductor quantum dots as quasi zero-dimensional systems. A recent review of the field and a compilation of relevant papers is given by Bawendi et al (1990). There are expectations that those structures exhibit large optical nonlinearities and fast relaxation times. However, recent luminescence studies (Bawendi et al 1990, Bugayev et al 1991, Esch et al 1991, Hache et al 1991, Henneberger et al 1991, Misawa et al 1991) have revealed slow decay components related to long-living states. Based on the comparison of time-resolved luminescence and nonlinear absorption data we present in this paper a model for the carrier dynamics of quantum dots embedded in glass.

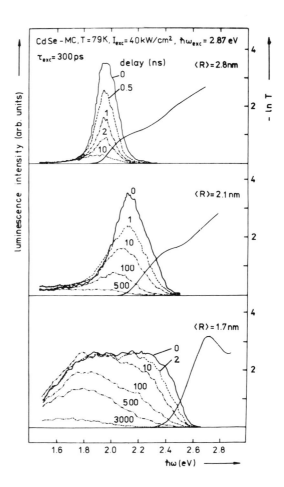

2. SAMPLES

The studies were made on specially prepared dots of pure CdSe embedded in a borosilicate glass. Two kinds of samples grown in different ways were investigated. One containes (apart from the usual components) Na, the other K, which was checked by SIMS. The linear absorption edge of the samples (compare fig. 1) is shifted by hundreds of meV's relative to bulk CdSe. The Bohr radius of bulk CdSe is 4.8 nm which is 2 to 3 times larger than <R> in our samples (1.7... 2.8 nm). This situation is refered to as the strong - confinement regime (Efros and Efros 1982). Hole-burning firstly studied by

Fig. 1 Time-resolved luminescence spectra of three different averaged-sized CdSe-quantum dots. Time resolution: 500 ps. For comparison the linear absorption spectra are given (right hand scale).

Fluegel *et al* (1990) and later in more detail by Spiegelberg *et al* (1991) has yielded a substructure of the hole state not resolved in the absorption spectrum. Three intrinsic, nearly equidistantly spaced hole levels are found which are radiatively coupled to the 1s electron (compare fig. 3). They are related to hole-hole-coupling in a quantum dot. Details are discussed in a previous paper (Henneberger *et al* 1991).

3. TIME-RESOLVED LUMINESCENCE

The luminescence was studied by pulsed dye laser excitation at 2.87 eV. This is much below the glass mobility edge so that the dots are directly excited. The intensity of the 300 ps pulses was intentionally kept low, typically less than 100 kW/cm², to avoid any kind of saturation in the system. This is an essential difference to other luminescence studies at 100 or 1000 times larger excitation intensities (see also Uhrig *et al* 1990, Dnjeprovskii *et al* 1990). The luminescence was passed through a monochromator and then detected with a MCP photo-multiplier of 500 ps time resolution. Gating and signal averaging was performed by a 150 ps boxcar-integrator. By careful deconvolution with the apparatus function a final time-resolution of 150 ps is yielded.

Fig. 1 summarizes the luminescence spectra of three differently aver-age-sized samples at low temperature. It is clearly seen how the emission spreads out to lower photon energies when the dot size decreases. A detailed analysis of the data by a deconvolution procedure described elsewhere (Jung-nickel 1991) reveals that the luminescence is composed of three different contributions each of which decays single-exponentially in time. In Fig. 2 the extracted decay times and the time-integrated contribution of the re-spective components are plotted versus photon energy. The total yield is dominated by the most low-energy shifted compo-nent which de-cays in the 1 μs range. It is followed by a component of about 100 ns life-time and a third most high-energy one, which decays on the sub-ns time scale.The slight change of the decay time across the emis-sion bands may be due to the size distribu-tion. This de-composition of the luminescence not obvious in the plots of fig. 1 occurs independently of the average dot size.

So far we have only consi-dered "fresh" samples. Under

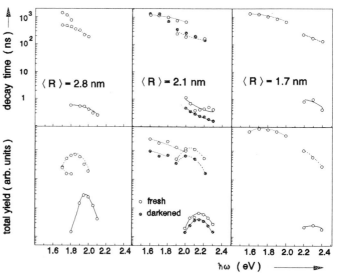

Fig. 2 Decay time and total yield for three different radiative recombination channels observed versus photon energy. The data are obtained by deconvolution of the decay curves measured at thecorresponding photon ener-gies.

reiterative excitation over longer times quenching of the emission is obser-ved for all three components. However, only the decay of the ps-component becomes faster, whereas the life-times of the other two components remain unchanged. As a short-hand notation we use "darkening" for these long-term emission changes although we do not see a remarkable increase of the linear absorption. The data in fig.1 and 2 were obtained on samples containing Na.

n samples with K the 100 ns-component is absent, otherwise they behave
imilar.
ith rising temperature the emission is quenched again in all three
omponents. The Arrhenius plots reveal a common activation energy ≈20 meV of
he quenching process. Above 200 K the decay of the two slower components
tarts to become faster and the life-times at 300 K are shorter by factor of
oughly 2.

. NONLINEAR ABSORPTION

The concept of the experiment is based on the substructure in the
ole energy spectrum of the dots discussed above (see insert of fig. 3). The
nergy of the exciting photons - provided by the same dye laser as used in
he luminescence study - was chosen resonant to the third hole transition of
he 1.85 nm sample
elected. The absorp-
ion change was pro-
ed by the Ar⁺ laser
hich makes available
arious lines at the
bsorption edge of
his sample. In par-
icular, the 496 and
58 nm line are in
lose resonance to
he first and second
ole transition,
espectively. The
ransmitted probe
ulses were detected
y a Si-APD and re-
orded in a digital
torage oscilloscope
ith 3 ns time reso-
ution in a single
hot mode. In this
ay long-term changes
n the nonlinear ab-
orption could be
easured shot by
hot. In fig.3 the
ifferential trans-
ission signal
TS = (T - T₀)/T₀
t the two wave-
engths is plotted
ersus pump-probe
elay for different
urations of laser
xposure. In accor-
ance with measure-
ents using ns-probe
ulses (Spiegelberg
t al 1991), absorp-
ion bleaching is
bserved at both
esonances. However,
he recovery behaves
ifferent. At the
irst hole transition
faster process is
uperimposed to a
low μs-background.
ith an increasing
umber of shots the
ackground disappears
nd only a reso-
ution-limited needle

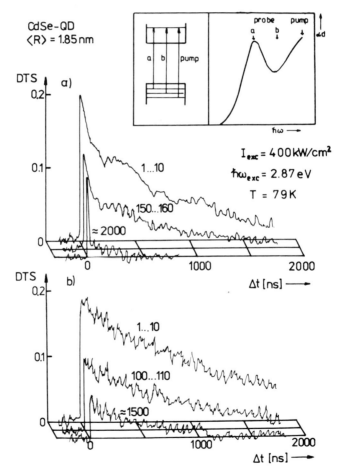

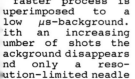

Fig. 3 Differential transmission versus pump-probe
delay for probing at (a) 496 and (b) 458 nm. The
numbers at the plots are the excitation shots applied. The energy scheme is depicted in the insert.

survives. At the second hole transition μs-recovery is only seen. No speed-up occurs, but the nonlinear change disappears at all when the excitation is repeated for a longer time.

5. DISCUSSION

The slow absorption recovery at *both* hole resonances on fresh samples shows that the electron survives for ≈1 μs in the 1s level, while the hole is strongly localized on a much shorter time scale. This is consistent with the μs-component seen in the emission. The nature of this trap may be surface related. Na is known to be easily built into CdSe as an acceptor. K is too large to be stable. Therefore, the ns-component is related to an acceptor-like hole trap. Long-term optical excitation creates (or opens the path to) a non-radiative electron trap so that the emission is quenched. The absorption recovery at the first hole transition is than controlled by the localization of the hole. No bleaching is observed at the second hole transition as this state is depopulated very shortly after excitation. The activation energy for the temperature quenching of the emission is close to the LO phonon of CdSe. Probably tunneling through a potential barrier occurs to a non-radiative state. This state might be related to the electron trap of the darkening process. The origin of the ps-emission is not fully clear at the moment. Experiments with improved time-resolution are under way.

ACKNOWLEDGEMENTS

The authors are very much indepted to Dr.A.I.Ekimov and Dr.M. Müller for providing us with the nanocrystals. We are also grateful to Dr.W.Frentrup for the SIMS analysis. This work was partly supported by the Deutsche Forschungsgemeinschaft.

REFERENCES

Bawendi M, Steigerwald M L and Brus L E 1990 Ann. Rev. Phys. Chem. 41 477

Bugayev A, Kalt H, Kuhl J and Rinker M 1991 Appl. Phys. A 53 75

Dneprovskij V S, Efros Al L, Ekimov A I, Klimov V I, Kudryavtsev A I and Novikov M G 1990 Solid State Commun. 74 555

Efros Al and Efros A 1982 Sov. Phys. Semicond. 16 1209

Esch V, Klang K, Fluegel B, Hu Y Z, Khitrova G, Gibbs H M, Koch S W and Peygambarian N 1992 J. Nonlin. Opt. Phys. 1 1

Fluegel B, Joffre M, Park S H, Morgan R, Hu Y Z, Lindberg M, Koch S W, Hulin D, Migus A, Antonetti A and Peyghambarian N 1990 J. Crystal Growth 101 643

Hache F, Klein M C, Ricard D and Flytzanis C 1991 J. Opt. Soc. Amer. 8 1802

Henneberger F, Puls J, Spiegelberg Ch, Schülzgen A, Rossmann H, Jungnickel V and Ekimov A.I 1991 Semicond. Sci. Technol. 6 A41

Jungnickel V 1991 diploma thesis Humboldt-Universität Berlin

Misawa K, Yao H, Hayashi T and Kobayashi T 1991 J. Crystal Growth in press

Spiegelberg Ch, Henneberger F and Puls J 1991 Superlattices and Microstructures 9 487

Uhrig A, Banyai L, Hu Y Z, Koch S W, Gaponenko S, Neuroth N and Klingshirn C 1990 J. Noncryst. Solids 112 277

Nonlinear picosecond spectroscopy of zero-dimensional semiconductor

V.S.Dneprovskii, V.I.Klimov, Ju.V.Vandyshev

Department of Physics,Moscow State University
119899 Moscow USSR

Bleaching bands arising from the filling of space quantization levels were observed in the time-resolved differential transmission spectra of quasi-zero-dimensional CdSe microcrystals under picosecond laser excitation. The spectra of nonlinear susceptibility $\chi^{(3)}$ and the relaxation times of the nonlinearity are determined for CdSe microcrystals of different size. The amplification regime and laser generation in the resonator made of the glass doped with CdSe microcrystals were achieved.

The quantum size effect results in the drastic modification of the energy spectrum of quasi-zero-dimensional semiconductor microcrystals (MCs). As shown by Efros and Efros (1982) the quasi-continuous spectrum of the bulk semiconductor is replaced by the discrete one. However this discrete structure is usually unresolved by the methods of linear spectroscopy because of the significant broadening due to the size distribution of MCs. The dominant processes leading to optical nonlinearities in semiconductor MCs are expected to differ from those in bulk materials (Scmitt-Rink *et al* 1987, Hanamura 1988). In this contribution we investigated the recovery of the transmission spectra of CdSe MCs by picosecond pump-and-probe technique. The experimental results allowed to clarify the dominant mechanism and to determine the spectral and temporal characteristics of optical nonlinearities in quasi-zero-dimensional CdSe MCs.

The investigated samples of semiconductor doped glass (SDG) were prepared by the method of secondary heat treatment.They contained CdSe microcrystals of average radius R≈6nm (sample 1) and R≈3.5nm (sample 2) at concentration corresponding to approximately 0.1% semiconductor volume fraction. The samples of SDG cooled to 80 K were excited by powerful picosecond (20-25 ps) pulses of the second harmonic of Nd:YAG mode-locked laser. The transmission of photoexcited samples was probed by the delayed pulses of picosecond continuum. Optical multichannel analyzer was used to measure differential transmission (DT) spectra $DT(\lambda)=[T-T_0]/T_0$, where $T(\lambda)$ and $T_0(\lambda)$ are the transmission spectra of excited and unexcited sample respectively.

The measured DT spectra of sample 1 (excitation energy $w=2mJ/cm^2$) for various time delays (Δt) between the pump and the probe pulses are shown in Figure 1. Excitation resulted in significant bleaching above the absorption edge. Three bleaching bands with maxima at $\lambda_1=(1)$ 649 nm,(2) 616 nm,(3) 563 nm were observed for sample 1, while these bands were not pronounced in the linear absorption spectrum. The kinetics of the absorption recovery differed significantly for λ_1,λ_2 and

λ_3 bands . Bleaching at λ_3 band completely relaxed within 70 ps after excitation. The slower recovery was characteristic of λ_2 band. And λ_1 band could be seen in DT spectra up to the maximum delay of 3 ns.

Spectral position of bleaching bands and their kinetics allowed to attribute the observed bleaching to filling of the energy levels of spatially confined electrons and holes by photoexcited charge carriers. The bleaching bands λ_1 and λ_2 correspond to the 1S-1S and 1P-1P transitions in the spherical quantum well of radius about 6 nm. The transition energies E_{01} and E_{02}, calculated within effective mass approximation, are shown in Figure 1. λ_3 band is in the spectral region of the transitions with energies E_{21}, E_{02}, E_{31}.

The dynamics of the nonlinear transmission spectra of MCs was numerically simulated. The initial stage of the DT spectra kinetics can be attributed to the relatively fast cooling of photoexcited carriers accompanied by intraband transitions of carriers from higher to lower levels. Due to this process λ_3 – band relaxes rapidly while the amplitudes of λ_1 and λ_2 – bands increase slightly.   The final stage of the relaxation is governed by the recombination of carriers which results in the recovery of absorption at λ_1 and λ_2 – bands.

Significant bleaching was observed in DT spectra for sample 2 with smaller size microcrystals. The spectral position of the bleaching band (λ_1'=589nm) corresponds to the average size of microcrystals about 3.5 nm. The recovery time of the transmission was about 60-70 ps. The significant shortening of relaxation for smaller microcrystals is probably due to the enhancement of non-radiative recombination. Auger processes and surface recombination may be involved (Rossignol *et al* 1987, Dneprovskii *et al* 1990).

At room temperature the transmission recovery time did not exceed 30 ps for both samples.

The DT spectra reported in the present work can be used to evaluate the real and imaginary parts of the third-order Kerr-type susceptibility $\chi^{(3)}$. The changes of the absorption coefficient $\Delta\alpha(\omega)$ and refraction coefficient $\Delta n(\omega)$ induced by the resonant monochromatic field of intensity $I(\omega)$ are connected with $Im\chi^{(3)}$ and $Re\chi^{(3)}$ respectively. For small Δn and $\Delta\alpha$:

$$Re\chi^{(3)}(\omega)=\frac{cn_0^2}{4\pi^2}\frac{\Delta n(\omega)}{I(\omega)} \quad , \qquad (1a)$$

$$Im\chi^{(3)}(\omega)=\frac{c^2n_0^2}{8\pi^2\omega}\frac{\Delta\alpha(\omega)}{I(\omega)}, \qquad (1b)$$

where n_0 is linear refraction index ($\Delta n(\omega)$ spectra can be calculated using $\Delta\alpha(\omega)$ and the Kramers-Kronig equation). For quasistationary conditions the radiation-intensity $I(\omega)$ is connected with N – the number of EH pairs per MC:

$$N(\omega)=\frac{(1-T(\omega)-r)\ I(\omega)\ \tau_e}{d\ \hbar\omega\ N_m} \quad , \qquad (2)$$

where N_m denotes the concentration of microcrystals in the

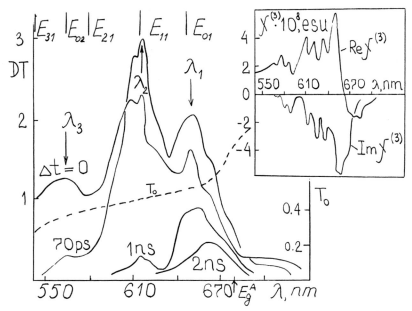

Fig.1 Time-resolved differential transmission spectra of CdSe MCs. Linear transmission spectrum is shown by the dashed line. Inset: $\chi^{(3)}$ spectra of CdSe MCs.

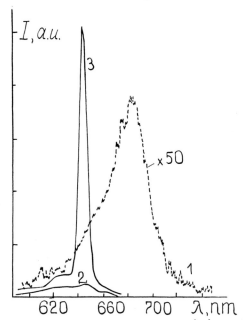

Fig.2 Spectra of spontaneous (1) and stimulated (2,3) emission of sample 1 with reflective coatings: w=0.08 (1), 0.5 (2), 0.8 (3) mJ/cm^2.

glass matrix, τ_e – recombination time of photoexcited carriers, $T(\omega)$ and r –the transmission and reflectivity of the sample respectively, d – its thickness.

The resultant spectra of Re and Im $\chi^{(3)}$ (for sample 1) are presented in Figure 1(inset). The amplitudes of Re and Im$\chi^{(3)}$ are of the same order – about $5*10^{-8}$ esu for larger microcrystals (R≈6nm) and 10^{-9} esu for the smaller ones (R≈3.5nm). Strong reduction of the nonlinearity for MC of smaller size (instead of its enhancement, predicted by Schmitt-Rink *et al* 1987) is caused mainly by the significant shortening of the relaxation time (from 1.5-2 ns in sample 1 to 60-80ps in sample 2).

The amplitude of the differential transmission in the spectral region of λ_1 and λ_2 bands (sample 1) indicates that not only the bleaching of absorption but gain ($\alpha<0$) was achieved under the experimental conditions. The maxima of the gain spectra correspond to the energies of the two lowest transitions 1S–1S and 1P–1P.

To obtain laser generation we used a 1 mm thick Fabry-Perot resonator: dielectric coatings with reflection coefficients of 95% and 100% at 650 nm were deposited on the parallel faces of sample 1. The resonator was cooled to liquid nitrogen temperature. At low excitation levels a broad band of spontaneous emission (λ=682 nm) was observed in the luminescence spectrum (Figure 2). When the energy of excitation was increased to about 0.2 mJ/cm^2 stimulated emission, directed perpendicularly to the mirrors of the resonator, arose at the high energy wing of the spontaneous band. At w≈0.5 mJ/cm^2 intensity of the stimulated emission increased steeply and its linewidth narrowed to about 7 nm, indicating the onset of laser generation. The spectral position of the laser generation line coincided with the energy of the lowest 1S–1S transition in microcrystals.

In conclusion, we have studied the recovery of the transmission of CdSe MCs after intense picosecond excitation. Bleaching bands corresponding to the transitions between the levels of size quantization were clearly observed in differential transmission spectra. The induced decrease of absorption was explained by filling of states in MCs by photoexcited carriers. Spectra of the third – order nonlinear susceptibility were calculated for MCs of different size. The observed decrease of nonlinearity in MCs of smaller size was attributed to the strong shortening of the carriers lifetime. Laser generation in glass doped with CdSe microcrystals was achieved.

References

Dneprovskii V S, Efros Al L, Ekimov A I *et al* 1990 *Solid State Commun.* 74 p 555
Efros Al L, Efros A L 1982 *Fizika i Tekhnika Poluprovodnikov* 16 p 1209
Hanamura E 1988 *Phys.Rev.B* 37 p 1273
Rossignol P, Kull M, Ricard D, Flytzanis C 1987 *Appl.Phys.Lett.* 51 p 1382
Schmitt-Rink S, Miller D A B, Chemla D S 1987 *Phys.Rev.B* 35 p 8113

Absorption saturation and nonlinear diffraction of laser pulses in the commercial CdS$_x$Se$_{1-x}$-doped glasses

V.A.Zaporozhchenko, R.G.Zaporozhchenko, I.V.Pilipovich, A.E.Kazachenko, V.A.Zyulkov*

Institute of Physics, Byelorussia Acad.Sci., Leninskii prospekt 70, 220602, Minsk, USSR

* STA "Polytekhnik", Partizanskii prospekt 77,220602, Minsk, USSR

ABSTRACT:Experimental and theoretical results are presented on the nonlinear absorption and light-induced diffraction in the commercial semiconductor-doped glasses. The validity of the application of the bulk-semiconductor theory to these glasses are discussed.

1. INTRODUCTION

In recent years, a number of researchers studied the nonlinear optical properties of a small semiconductor particles imbedded in a dielectric host. Examples of such materials are the silicate glasses doped with CdS$_x$Se$_{1-x}$ microcrystallites (see for example the review by Williams *et al* 1988). They have been proved to exhibit large nonlinearities, moreover, at a small microstructure dimensions they show quantum-size effects. Nonlinear transmission and four-wave mixing were the main methods investigating semiconductor-doped glasses. Different authors tried to explain their experiments by several mechanisms of nonequilibrium carriers relaxation. Those are the induced absorption by free carriers, Auger-recombination processes and their dependence on the level of excitation (de Rougemont *et al* 1988), the role of deep traps (Tomita and Matsuoka 1990) and others.

In present work, the experimental results have been obtained on the nonlinear transmission and light-induced diffraction in CdS$_x$Se$_{1-x}$-doped glasses and theoretical calculations are proposed to explain these processes within the frames of the model accounting for band-filling, linear- and Auger- recombination. The frames of the bulk semiconductor theory applicability will be discussed too.

2. SAMPLES CHARACTERISTICS AND EXPERIMENT

Two samples of commercial orange glass filters OS-13 were studied. The linear transmission spectra of the samples are presented in Fig.1. Characteristic "shoulders" due to the quantum confinement are clearly seen for both samples. Pump laser wavelength is indicated in Fig.1 by arrow. The average radius $\bar{a}$ and concentration of semiconductor inclusions were determined from the small-angle X-ray scattering. The $\bar{a}\approx30\mathring{A}$ value was the same for both samples, they differed in the concentration of microcrystallites and sulphur fraction x.

Second harmonic of the actively mode-locked YAG:Nd laser at the wavelength λ=532nm was used as a pump radiation. The pulses duration was about 100ps with the energy 5÷10mJ. In the nonlinear transmission experiment pump radiation was focused onto a sample by a lens. Input and transmitted through a sample radiation was measured by a photo-diodes. In the self-induced diffraction experiment the laser beam was split into two beams with approximately equal energies. After the delay line equalizing the optical paths these beams were converged in a sample. The angle between two pumps was kept small, such that 'thin' grating conditions should prevail. Input and diffracted to the first and second order energies were using calibrated photodiodes and a

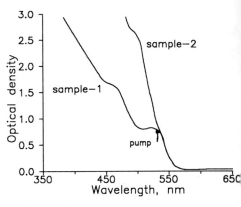

Fig.1. Samples linear transmission spectra.

digital storage scope. In both experiments the pump radiation transverse distribution was thoroughly measured to calculate the pump intensities. Note that the samples were exposed to a smaller possible amount of laser shots to avoid photodarkening.

Experimentally obtained dependencies of the samples transmission versus pump fluence are plotted in Fig.2. The measured diffraction efficiency is shown in Fig.3. In both figures the results on sample-1 and sample-2 are depicted by circles and squares correspondingly.

3. THEORETICAL MODEL.

To describe the experiments, we used the theoretical model that is typical for bulk direct-gap semiconductors and accounts for band-filling effect, linear- and Auger- recombination. Equation for carriers density N is as follows:

$$\frac{dN}{dt} = -\gamma_1 N - \gamma_2 N^2 - \gamma_3 N^3 + \frac{\alpha(N)I}{\hbar\omega\, p} , \qquad (1)$$

where γ_1, γ_2, γ_3 are the probabilities of linear, nongeminate radiative and Auger recombination processes, respectively; $\hbar\omega$ is the laser quantum energy; p- the volume fraction of microcrystallites in glass; I- light intensity. The absorption coefficient $\alpha(N)$ is given as

$$\alpha(N) = \alpha_0 \sqrt{\hbar\omega - E_g}\ (1 - f_e - f_h) ,$$

where $\alpha_0 = \alpha(I=0)$ and E_g is the band-gap energy. f_e and f_h where assumed to be quasi-equilibrium Fermi distribution functions. Low efficiency of the samples luminescence enabled us to consider $\gamma_2 = 0$.

The calculations of the nonlinear transmission were based on the equation for the light intensity

$$\frac{\partial I}{\partial z} + \frac{1}{\omega}\frac{\partial I}{\partial t} = -\alpha(N)I , \qquad (2)$$

where ω is the group velocity of light in the medium. Field equation, which was used to calculate the induced diffraction, were as follows:

$$\frac{\partial \mathcal{E}}{\partial z} + \frac{1}{v}\frac{\partial \mathcal{E}}{\partial z} = \frac{2\pi i}{\lambda}(\Delta n + \frac{i\lambda}{4\pi}\alpha(N))\mathcal{E} \qquad (3)$$

Field $\mathcal{E}$ in (3) was expanded into a series of spatial harmonics, corresponding to the diffracted orders (Apanasevich *et al* 1989). Light-induced change of the refractive index Δn was obtained from the absorption coefficient change through the Kramers- Kronig relationship:

$$\Delta n = \frac{\hbar c \alpha_0}{\pi k T \sqrt{a_0}} \int_0^\infty \frac{\sqrt{\eta}}{\eta^2 - a_0^2}\left[f_e(\mu\eta/m_e^*) + f_h(\mu\eta/m_h^*)\right]d\eta , \qquad (4)$$

where μ is the reduced mass, m_e^*, m_h^* - effective masses of the densities of electron and hole states and $a_0 = (\hbar\omega - E_g)/kT$.

4. DISCUSSION

The calculation results of the nonlinear transmission for both samples are shown in Fig.2 by solid lines. The best fit of theoretical curves was obtained with the following parameters of linear- an Auger recombination:

$$\text{sample-1} - 1/\gamma_1 = 10^{-10}s ; \quad \gamma_3 = 4.2 \cdot 10^{-28}cm^6 s^{-1}$$
$$\text{sample-2} - 1/\gamma_1 = 10^{-10}s ; \quad \gamma_3 = 3.9 \cdot 10^{-27}cm^6 s^{-1}$$

The obtained values of γ_3 correspond

to the Auger recombination processes that occurs through the capture of carriers on deep traps (Ridley 1982). It is seen, that the samples strongly differs in γ_3 value. This difference can be attributed to the larger amount of deep traps in sample-2. The results of the luminescence study of these samples also support this hypothesis (Gribkovskii *et al* 1991). To describe the experiment at the fluences exceeding $100mJ/cm^2$ it was necessary to increase the Auger recombination constant. Such a behavior can be explained either by nonthermal distribution of charge carriers or by considerable rise of the relaxation rate due to the photodarkening.

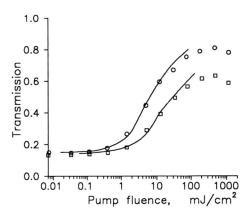

Fig.2. The nonlinear transmission of the samples.

Calculations of light-induced diffraction, the results of which are depicted in Fig.3a,b by solid lines, were performed at the same values of γ_1, γ_3. The calculations data analysis has shown that with the rise of pump fluence the role of phase grating, as compared with the amplitude one, also increases, and at high input intensities diffraction on phase grating dominates completely. Though the used theoretical model describes qualitatively the behavior of the diffraction efficiency for sample-2, the difference between calculated and experimental curves for sample-1 is of the order of magnitude. The main assumption in the calculations of Δn in (4) is the

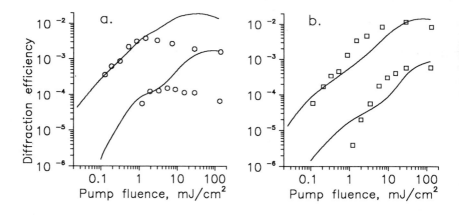

Fig.3. Self-induced diffraction efficiency vs pump fluence
for sample-1 (a) and sample-2 (b).

proportionality of the absorption coefficient to $\sqrt{\hbar\omega - E_g}$, as is the case
for bulk direct-gap semiconductors (Miller. *et al* 1981). As seen from
Fig1., for sample-2 the pump radiation falls in the "red" wing of the
absorption spectrum which is determined greatly by microcrystallites with
radii larger than average. In this case the $\alpha \sqrt{\hbar\omega - E_g}$ dependence may be
rather satisfactorily. For sample-1 pumping is in the vicinity of the first
quantum-size maximum and the above assumption is too rough. The discrepancy
between experimental data and theory is more pronounced at high pumping
levels, where the diffraction on phase grating prevails. Note also that
"thin" grating condition can be violated due to the large phase shifts at
high pump fluencies for the second diffracted order.

5.CONCLUSION

In conclusion, the above investigations have shown that for the correct
description of the nonlinear properties of commercial $CdS_x Se_{1-x}$ glasses
with average radius of microcrystallites about 30Å due account of quantum-
size effects is needed. One should also seek for the relaxation mechanisms,
which lead to a decrease in the transmission and diffraction efficiency at
high pump levels Such investigations are under way now.

REFERENCES

Apanasevich P A, Asimova V D, Zaporozhchenko V A, Zaporozhchenko R G,
   Pilipovich I V 1989 Phys.Stat.Sol.B, **152** 347
Gribkovskii V P, Zaporozhchenko R G, Zuylkov V A, Kazachenko A E 1991
   (to be published) in Superlatt. and Microstr.
Miller D A, Seaton G T, Prise M E, Smith S D 1981 Phys.Rev.Lett. **47** 197
Ridley B K 1982 Quantum processes in semiconductors (Claredon Press
   Oxford)
de Rougemont F, Frey R, Roussignol P, Ricard D and Flytzanis C 1988
   Appl.Phys.Lett. **50** 1619
Tomita M and Matsuoka M 1990 J.Opt.Soc.Am.B, **7** 1198
Williams V S Olbright G R, Fluegel B D, Koch S W and Peyghambarian N 1979
   Journ.of Modern Optics **35** 1979

Inst. Phys. Conf. Ser. No 126: Section V
Paper presented at Int. Symp. on Ultrafast Processes in Spectroscopy, Bayreuth, 1991

Phase relaxation of excitons in semiconductor mixed crystals

Uwe Siegner and Ernst O. Göbel

Philipps–Universität Marburg, Fachbereich Physik und Zentrum für
Materialwissenschaften, F. R. Germany

Optical phase relaxation of excitons in the semiconductor mixed crystals CdSSe
and AlGaAs is studied by spontaneous and stimulated photon echo experiments in
order to investigate the influence of disorder and localization on the dynamics of
optical excitations. The dephasing time of excitons in CdSSe amounts to several
100 ps while phase relaxation of excitons in AlGaAs is much faster. The different
dephasing behaviour is attributed to the difference in the disorder potential and
the degree of localization.

1. INTRODUCTION

Phase relaxation experiments reveal the initial scattering processes of an optical
excitation and thus its interaction with its environment. In ordered semiconductor
crystals optical dephasing of excitons is exclusively due to quasiparticle interaction like
exciton–exciton and exciton–phonon scattering. In semiconductor mixed crystals,
however, the dynamics of excitons is affected by the presence of disorder and by
disorder–induced localization. As compared to an ordered crystal the presence of
disorder has two implications on optical dephasing of excitons: (i) disorder–induced
localization of excitons modifies quasiparticle interaction (ii) elastic scattering at the
static disorder potential might be an additional scattering mechanism, however, it has
to be clarified whether it contributes to optical dephasing.
In this paper we report spontaneous and stimulated photon echo experiments on the
semiconductor mixed crystals CdSSe and AlGaAs. Both types of semicondutor mixed
crystals show a strong inhomogeneous broadening of the exciton transition as the result
of alloy disorder (Cohen 1982, Permogorov 1982, Schubert 1984). In linear optical
experiments the inhomogeneous broadening of the exciton line is the common
characteristic feature of disorder in CdSSe and AlGaAs. In contrast to the linear
optical properties, we show the nonlinear optical properties of both mixed crystals to
be different, depending on the detailed nature of the disorder potential. Consequently,
we suggest nonlinear optical experiments as an appropriate tool to characterize
disorder and localization.

2. EXPERIMENTAL

Optical dephasing can be studied in time–resolved four–wave–mixing experiments,
where three laser pulses delayed in time with respect to each other excite a sample to
produce two coherent signal pulses. For wave vectors k_1, k_2, and k_3 of the incoming
excitation pulses the first signal pulse is emitted into the phase–matching direction
$2k_2-k_1$ as a result of two–pulse interaction. The second signal pulse is generated by
three–pulse interaction and emitted into the direction $k_3+k_2-k_1$. If the excited
transition is inhomogeneously broadened both signal pulses are emitted time–delayed

as a spontaneous (two–pulse signal) or stimulated (three–pulse signal) photon echo. The intensity of both photon echo pulses decays as a function of time delay t_{21} between pulse no.1 and pulse no.2 due to scattering processes, which irreversibly destroy the macroscopic polarization. The decay of the macroscopic polarization is described by the dephasing time T_2 (Yajima 1979). In the stimulated photon echo case the echo intensity additionally depends on the time delay t_{23} between pulse no.2 and pulse no.3 as a result of population decay processes, taking place in this time interval. For t_{21} larger than the pulse duration a phased array is set up, which decays due to recombination, diffusion, and energy relaxation (Weiner 1985). If pulse no.1 and pulse no.2 coincide in time $(t_{21} = 0)$ an ordinary population grating is generated, which decays only due to recombination and diffusion. A comparision of the phased array decay to the population grating decay thus allows to determine the energy relaxation time T_3.

We have studied a $CdS_{0.40}Se_{0.60}$ and a MBE–grown $Al_{0.38}Ga_{0.62}As$ sample at low temperatures. In photoluminescence experiments on both samples an inhomogeneous broadening of the exciton transition of 10–15 meV is observed, indicating similar linear optical properties of both semiconductor mixed crystals.

The photon echo experiments have been performed with picosecond pulses from a cavity–dumped, synchronously mode–locked dye laser. The excitation photon energy is always resonant to localized exciton transitions. In case of the CdSSe sample the excitation field is perpendicular to the optical axis of the crystal, i.e. exciton transitions involving the upper valence band are excited.

3. RESULTS AND DISCUSSION

Figure 1 shows the result of a stimulated photon echo experiment on CdSSe. Original streak camera traces are depicted for different time delays t_{21} between pulse no.1 and pulse no.2. If pulse no.2 precedes pulse no.1, i.e. $t_{21} < 0$, no echo pulse is emitted into the direction of the camera. However, the stimulated photon echo pulse can clearly be observed for positive time delays t_{21} as large as 130 ps. These data unambiguiously reveal that phase relaxation of resonantly excited localized excitons is very slow in CdSSe. In fact, the dephasing time amounts to 400 ps under these experimental conditions.

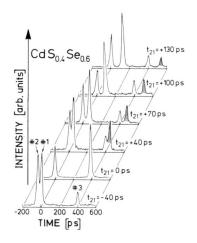

Fig.1. Time–resolved streak camera traces of stimulated photon echo pulses (shaded) from localized exciton transitions in a CdSSe mixed crystal. The time delay t_{21} between excitation pulse no.1 and no.2 increases from the bottom to the top. The dephasing time amounts to 400 ps.

The dephasing times of localized excitons in CdSSe have to be compared to the dephasing time of free excitons in the corresponding binary compound CdSe, where disorder and localization are absent. Dörnfeld and Hvam (1989) report dephasing times

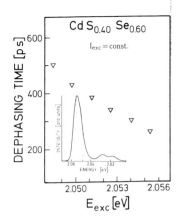

Fig.2. Dephasing time of localized excitons in CdSSe vs. excitation photon energy (triangles). The inset shows the photo‑luminescence spectrum of the sample. The energy range between the dashed lines corresponds to the energy range where the dephasing times are determined.

of free excitons in CdSe which are about one order of magnitude shorter than the dephasing times found for localized excitons in the mixed crystal. Consequently, we conclude that localization strongly reduces exciton–exciton and exciton–phonon scattering in CdSSe.

The influence of localization on the dynamics of excitons in CdSSe becomes even more obvious if the dependence of the dephasing time on exciton density is determined. We have shown that the dephasing time does not depend on exciton density up to densities of at least 10^{16} cm^{-3} (Noll 1990). Therefore exciton–exciton scattering does not contribute to dephasing in this density range. In contrast, free excitons in the ordered semiconductor CdSe do interact even at low densities (Dörnfeld 1989). The different behaviour demonstrates the high degree of exciton localization in CdSSe.

In Fig.2 the dephasing time of localized excitons is depicted as a function of excitation photon energy. The inset shows the photoluminescence spectrum of the sample in order to characterize the spectral range where the data are taken. The photoluminescence spectrum is dominated by emission from localized excitons. At lower energies the phonon replicas are also observed. The dephasing experiments have been performed at energies between the dashed lines, i.e. well below the "mobility edge" of the sample, which coincides with the high–energy edge of the exciton emission band. The dephasing time increases from 260 ps to 500 ps with decreasing excitation energy. The energy dependence of the dephasing time is a result of stronger localization at lower energies, where quasiparticle interaction is more strongly suppressed.

The same energy dependence is reported for the intraband energy relaxation time T_3 (Shevel 1987). T_3 describes hopping within the distribution of the localized states. The energy relaxation time T_3 and the dephasing time T_2 are connected by the fundamental relation $T_2 \leq 2T_3$ if the recombination time is much longer than T_3 and T_2. In CdSSe where the recombination time is in the nanosecond range (Shevel 1987) this assumption holds. Thus T_3 sets an upper limit for T_2. This relation allows to determine the dephasing mechanism in CdSSe. The actual dephasing times well agree with the upper limit of the dephasing time as given by the energy relaxation times reported by Shevel et al. (1987), i.e. we have the relation $T_2 \approx 2T_3$. As a consequence, dephasing of localized excitons in CdSSe is mainly due to exciton–phonon interaction, i.e. hopping within the localized states. In particular, dephasing due to elastic scattering at the disorder potential is negligible.

On the base of this experimental finding we can characterize the disorder potential and the mechanism of localization referring to the theory developed by Bennhardt et al. (1991). The theory treats a strongly localized hole in a disordered valence band while the conduction band is ordered. Localization of the exciton is thus a result of Coulomb attraction between electron and hole. Within this model Bennhardt et al. (1991) show that elastic scattering at the disorder potential does not lead to phase relaxation. This

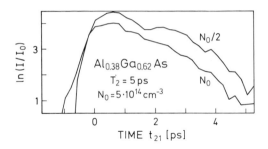

Fig.3. Half–logarithmic plot of the time–integrated spontaneous photon echo intensity vs. time delay t_{21}. The data are taken at the exciton densities N_0 and $N_0/2$.

theoretical prediction agrees with our experimental finding. The underlying model of the localization mechanism is particularly reasonable in the case of CdSSe as the valence band states are mainly formed by the group–VI–ions, where the disorder is present, while the conduction band states are mainly formed by the Cadmium.

The dephasing response of the AlGaAs sample is shown in Fig.3, where the time–integrated spontaneous echo intensity is plotted as a function of time delay t_{21} for two different exciton densities $N_0 = 5 \cdot 10^{14}$ cm^{-3} and $N_0/2$. The excitation photon energy corresponds to the low–energy side of the photoluminescence spectrum. The curves are arbitrarily shifted along the y–axis. The dephasing time obtained from these data amounts to 5 ps. Thus, the dephasing time of localized excitons in AlGaAs is considerably shorter than in CdSSe, demonstrating that the nonlinear optical response of semiconductor mixed crystals sensitively depends on the detailed nature of the disorder although their linear optical properties may be similar.

The data depicted in Fig.3 also reveal that the dephasing time does not increase if the exciton density is decreased from N_0 to $N_0/2$. As a consequence, exciton–exciton scattering does not show at the density N_0. At the density $2N_0$, however, exciton–exciton scattering sets in and results in a decrease of the dephasing time. The onset of exciton–exciton scattering at the density $2N_0$ is demonstrated by the data depicted in Fig.4, where the intensity of the stimulated photon echo is plotted versus time delay t_{21} at the densities N_0 and $2N_0$. The dephasing time decreases from 5 ps to 4 ps if the excitation density is increased from N_0 to $2N_0$. A continuous decrease of the dephasing time is observed with further increase of the exciton density. We have calculated the mean distance between two excitons at the density $2N_0$, where exciton–exciton scattering sets in, and normalized it to the exciton Bohr radius. The normalized exciton distance at the density $2N_0$ in AlGaAs is larger than the normalized exciton distance in CdSSe at a density where exciton–exciton interaction is not found. This result reveals exciton localization in AlGaAs to be weaker than in CdSSe.

At the density N_0 in AlGaAs, however, exciton–exciton scattering does not contribute to phase relaxation. Thus, dephasing at this density can only be due to the population decay and to elastic scattering at the disorder potential. In order to determine the

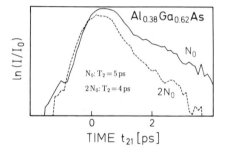

Fig.4. Half–logarithmic plot of the time–integrated stimulated photon echo intensity vs. time delay t_{21} at the exciton densities N_0 and $2N_0$.

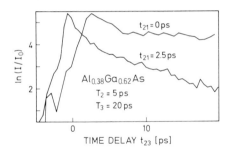

Fig.5. Decay of a phased array ($t_{21} = 2.5$ ps) and decay of an ordinary population grating ($t_{21} = 0$ ps) in a half–logarithmic representation.

contributions to dephasing from these relaxation processes separately we have investigated the population dynamics at the density N_0. In Fig.5 the intensity of the stimulated photon echo is shown as a function of time delay t_{23} between pulse no.2 and pulse no.3 at a constant time delay t_{21} of 2.5 ps. At negative time delays t_{23} the signal originates from a grating set up by pulse no.1 and pulse no.3, which is probed by pulse no.2. This part of the signal is not evaluated further. At positive time delays t_{23}, however, the decay of a phased array is monitored, which is set up by pulse no.1 and pulse no.2 and probed by pulse no.3. This decay is a result of recombination, diffusion, and energy relaxation. The second curve ($t_{21} = 0$ ps) shows the decay of an ordinary population grating, which reflects only recombination and diffusion. The population grating decays with a time constant of 700 ps. Thus, recombination can be neglected in terms of phase relaxation. In addition, the decay of the phased array signal can be corrected for recombination and diffusion as the corresponding time constant is known from the population grating experiment. We then are able to determine the energy relaxation time T_3. The energy relaxation time amounts to 20 ps whereas the dephasing time of 5 ps is much shorter. Therefore we are far off the limit where phase relaxation is determined only by energy relaxation, i.e. the relation $T_2 \approx 2T_3$ does not hold in the case of AlGaAs, in contrast to the result obtained for CdSSe. In the case of AlGaAs elastic scattering at the disorder potential yields an important contribution to the dephasing rate. Comparing this experimental result to the theoretical predictions of Bennhardt et al. (1991) we expect short–range correlated disorder in the valence band and in the conduction band. In this situation the localized exciton state strongly resembles a multi–level system, which is known for rapid dephasing.

4. CONCLUSION

We have investigated dephasing and energy relaxation of localized excitons in the semiconductor mixed crystals CdSSe and AlGaAs. In CdSSe the dephasing time of localized excitons amounts to several 100 ps, corresponding to a homogeneous linewidth of 3–4 μeV. Dephasing of excitons is much faster in AlGaAs, where the dephasing time amounts only to 5 ps at low densities, from which a homogeneous linewidth of 260 μeV is calculated. At low densities, where exciton–exciton scattering does not contribute to dephasing, the homogeneous linewidth of excitons in AlGaAs has two contributions: (i) exciton–phonon scattering yields a contribution of 35 μeV as calculated from the experimentally determined energy relaxation time T_3 (ii) elastic scattering at the disorder potential results in a homogeneous broadening of the exciton line of 225 μeV. Thus, in AlGaAs scattering at the disorder potential is an important scattering mechanism. In contrast, dephasing of localized excitons in CdSSe is mainly due to exciton–phonon interaction, i.e. hopping within the distribution of the localized states. Elastic scattering at the disorder potential does not contribute significantly to phase relaxation in CdSSe. The different dephasing behaviour of excitons in terms of scattering at the disorder is related to the respective properties of disorder in CdSSe and AlGaAs.

Different degrees of exciton localization are revealed by the efficiency of exciton–exciton scattering in CdSSe and AlGaAs. Exciton–exciton scattering does not contribute to dephasing at low and at intermediate densities in CdSSe, demonstrating rather strong exciton localization. In AlGaAs, localization of excitons is weaker, as revealed by the fact that exciton–exciton scattering does not contribute to dephasing only at very low densities whereas it shows up at intermediate densities.

In conclusion, the nonlinear optical response of excitons in semiconductor mixed crystals critically depends on the detailed nature of the disorder and thus offers the possibility to characterize disorder and localization.

We wish to thank D. Bennhardt and P. Thomas, Marburg, for valuable theoretical contributions to this work as well as S. Shevel, Kiev, and K. Ploog, Stuttgart, for providing the samples.

5. REFERENCES

Bennhardt D, Thomas P, Weller A, Lindberg M and Koch S W 1991 *Phys. Rev. B* **43** 8934

Cohen E and Sturge M D 1982 *Phys. Rev. B* **25** 3828

Dörnfeld C and Hvam J M 1989 *IEEE J. Quant. Electron.* **QE–25** 904

Noll G, Siegner U, Göbel E O, Schwab H, Renner R and Klingshirn C 1990 *Journ. Crystal Growth* **101** 731

Permogorov S, Reznitskii A, Verbin S, Müller G O, Flögel P and Nikiforova M 1982 *phys. stat. sol. b* **113** 589

Schubert E F, Göbel E O, Horikoshi Y, Ploog K and Queisser H J 1984 *Phys. Rev. B* **30** 813

Shevel S, Fischer R, Göbel E O, Noll G, Thomas P and Klingshirn C 1987 *Journ. Lumin.* **37** 45

Weiner A M, De Silvestri S and Ippen E P 1985 *J. Opt. Soc. Am. B* **2** 654

Yajima T and Taira Y 1979 *J. Phys. Soc. Jap.* **47** 1620

Inst. Phys. Conf. Ser. No 126: Section V

Paper presented at Int. Symp. on Ultrafast Processes in Spectroscopy, Bayreuth, 1991

Subpicosecond spectroscopy of excitons in GaAs/AlGaAs heterostructures

Karl Leo[a] and Jagdeep Shah

AT&T Bell Laboratories, Holmdel, NJ 07733, U.S.A.

Stefan Schmitt-Rink[b]

AT&T Bell Laboratories, Murray Hill, NJ 07974, U.S.A.

Klaus Köhler

Fraunhofer-Institut für angewandte Festkörperphysik, W-7800 Freiburg, Germany

Abstract. We study the dynamics of excitons in GaAs/AlGaAs quantum wells using time-resolved self-diffracted four-wave-mixing. We discuss two topics: First, we show that the decay of the diffracted signal is different for parallel and perpendicular polarization of the incident laser pulses, and second, we study the dephasing dynamics of a wavepacket in a coupled quantum well structure.

1. Dependence of four-wave mixing decay on polarization

Self-diffracted time-resolved four-wave mixing (FWM) is frequently used to study the dynamics of coherently created excitations in semiconductors. The decay of the diffracted signal as function of the delay time between the two incident laser pulses is expected to be exponential with a time constant equal to half (homogeneous broadening) or a quarter (inhomogeneous broadening) of the phase relaxation time T_2 [1]. In Ref.1, optical excitations were described by non-interacting two-level systems. A large body of experimental studies has been interpreted using this two-level interpretation (see, e.g.,

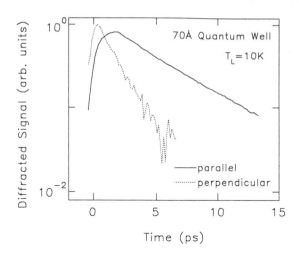

Figure 1. Decay of the diffracted signal for parallel (solid) and perpendicular (dotted) polarization of the incident laser pulses.

[2,3,4]). Theoretical studies based on a more detailed model of excitons [5,6] have predicted deviations from the lineshape expected for two-level systems. These predictions were confirmed by experimental studies [6].

An important parameter in self-diffracted FWM experiments is the polarization of the laser pulses. In the simple two-level picture, the decay of the diffracted signal should be independent of the polarization of the incident beams. For parallel linearly polarized beams, a population grating is formed; for perpendicularly polarized beams, an orientational grating of excitons is created. However, the relevant step for the decay of the signal is in both cases the polarization decay of the exciton ensemble created by the first pulse.

We have studied the polarization dependence of the FWM decay in a number of samples. Figure 1 shows as an example signals for a 70Å $GaAs/Al_{0.3}Ga_{0.7}As$ quantum well (QW). The n=1 heavy-hole exciton is resonantly excited with pulses of 500 fs duration. Lattice temperature and excitation density (6K and $5 \times 10^{-8} cm^{-2}$, respectively) are kept low to minimize the influence of exciton-phonon and exciton-exciton scattering. The diffracted signal for parallel polarizations decays much slower than for perpendicular polarizations: In this case, the decay times (5ps and 1.5ps, respectively) differ by about a factor of three, in qualitative agreement with other observations [7]. The absolute intensity of the signal at T=0 is about a factor of six larger for parallel polarization (note that the traces in Fig. 1 are normalized). In the sample studied here, the linewidth of about 3.5meV is mainly caused by inhomogeneous broadening. We have observed a similar dependence on polarization in all other samples studied, including samples where the excitonic transitions were mainly homogeneously broadened. Preliminary studies of the density dependence show that the differences between the two polarizations remain for increased density and temperature. Rotation of the sample around the growth axis (normal to the QW planes) does not indicate a dependence of the decay on sample orientation. At present, we do not have a concise explanation of the experimental ob-

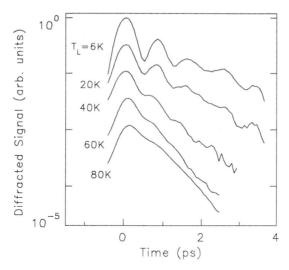

Figure 2. Coherent oscillations in coupled quantum wells for various lattice temperatures.

servations. We are presently performing experiments with circularly polarized light: In this case, states of definite spin are created, which will greatly simplify the theoretical interpretation.

2. Damping of coherent oscillations in coupled quantum wells

The simultaneous excitation of the two delocalized electronic states in an asymmetric coupled quantum well (CQW) leads to the creation of a wavepacket in one well. Subsequently, the wavepacket will oscillate coherently between the wells with a frequency proportional to the inverse of the energy splitting between the two states. We have recently reported the first observation of such coherent oscillations in GaAs/AlGaAs quantum well structures [8]. The coupled quantum wells realized in semiconductor structures are a model system for many other double well potentials in other areas of physics. Leggett et al. [9] have discussed the dissipative dynamics of such systems in great detail. The investigation of wavepacket dynamics in semiconductor CQW might be helpful for a systematic study of these systems.

As a first example, we have studied the influence of lattice temperature on the wavepacket dynamics in a CQW. Figure 2 shows the FWM traces for various lattice temperatures in a $150/25/100\text{Å}$ GaAs/Al$_{0.2}$Ga$_{0.8}$As CQW. It is obvious that the coherent oscillations are strongly damped with increasing temperature. A preliminary analysis of the oscillation frequency as a function of temperature indicates a slight increase. This seems to disagree with the expected slower "hopping rate" for a two-level system coupled to a phonon bath [9]. However, other effects like the coupling to phonons in the barrier or the bandgap reduction due to the increased lattice temperature might obscure the expected effect. Detailed theoretical investigations are in progress.

Acknowledgments

We thank T.C. Damen, E.O. Göbel, and W. Schäfer for collaboration and many stimulating discussions and P. Ganser for help with the crystal growth. The work of K.L. was partially supported by the Max-Planck-Gesellschaft zur Förderung der Wissenschaften e.V.

a) now: Institute for Semiconductor Electronics, Techn. Univ. of Aachen, W-5100 Aachen, Germany
b) now: Physics Department, University of Marburg, W-3500 Marburg, Germany

3. References

[1] T. Yajima, and Y. Taira, J. Phys. Soc. Jpn. **47**, 1620 (1979).

[2] L. Schultheis, M.D. Sturge, and J. Hegarty, Appl. Phys. Lett. **47**, 995 (1985); L. Schultheis, A. Honold, J. Kuhl, K. Köhler, and C.W. Tu, Phys. Rev. B **34**, 9027 (1986).

[3] G. Noll, U. Siegner, S. Shevel, and E.O. Göbel, Phys. Rev. Lett. **64**, 792 (1990).

[4] P.C. Becker, H.L. Fragnito, C.H. Brito-Cruz, R.L. Fork, J.E. Cunningham, J.E. Henry, and C.V. Shank, Phys. Rev. Lett. **61**, 1647 (1988).

[5] K. Leo, M. Wegener, J. Shah, D.S. Chemla, E.O. Göbel, T.C. Damen, S. Schmitt-Rink, and W. Schäfer, Phys. Rev. Lett. **65**, 1340 (1990).

[6] C. Stafford, S. Schmitt-Rink, and W. Schäfer, Phys. Rev. **B41**, 10000 (1990).

[7] H.H. Yaffe, Y. Prior, J.P. Harbison, and L.T. Florez, in it Quantum Electronics Laser Science, 1991 Technical Digest Series (Optical Society, Washington, D.C. 1991) p.196.

[8] K. Leo, J. Shah, E.O. Göbel, T.C. Damen, S. Schmitt-Rink, W. Schäfer, and K. Köhler, Phys. Rev. Lett. **66**, 201 (1991); also in *Proceedings of the Picosecond Electronics and Optoelectronics Conference*, J. Shah and T.C.L.G. Sollner, eds. (Optical Society of America, Washington D.C. 1991) p.56.

[8] A.J. Leggett, S. Chakravarty, A.T. Dorsey, P.A. Fisher, A. Garg, and W. Zwerger, Rev. Mod. Phys. **59**, 1 (1987).

Inst. Phys. Conf. Ser. No 126: Section V   415
Paper presented at Int. Symp. on Ultrafast Processes in Spectroscopy, Bayreuth, 1991

Tunneling versus exciton formation of photo-induced carriers in asymmetric double quantum wells

R.Strobel, R.Eccleston, and J.Kuhl
Max-Planck-Institut für Festkörperforschung
Heisenberstraße 1, 7000 Stuttgart 80, Germany

K.Köhler
Fraunhofer Institut für Angewandte Festkörperphysik,
7800 Freiburg, Germany

ABSTRACT: The dynamics of exciton formation and the exciton formation time are determined in a GaAs asymmetric double quantum well tunneling structure using the non-linear photoluminescence (PL) cross-correlation technique. The electron and hole tunneling times are measured simultaneously.

1. INTRODUCTION

GaAs-$Al_{0.3}Ga_{0.7}As$ asymmetric double quantum well (ADQW) structures consist of a wide and narrow quantum well (10nm and 5nm thick respectively in these experiments) separated by a thin barrier through which tunneling of carriers out of the narrow well may occur, and have been the subject of many investigations (Nido (1990)) of resonant and non-resonant tunneling times by means of time-resolved photoluminescence (TRPL) . We report here strong changes in the time-integrated PL efficiency of the narrow well of such structures as the photo-excited free electron-hole pair density, n_{ex}, is varied. Fig. 1 shows the time and spectrally integrated PL intensity at 8K emitted from the n=1 heavy-hole exciton of the narrow well of two ADQW's as a function of n_{ex}. For a 20nm barrier sample (crosses), through which negligible tunneling from narrow to wide well can occur, a completely linear dependence on density is observed. However, for a 4nm barrier sample (open circles), where TRPL measurements indicate an electron tunneling time of 10ps, the intensity increases with n_{ex}^2. A linear dependence is again observed in the wide well of the same 4nm barrier sample (dots in Fig.1) where tunneling is also absent. We attribute the dramatically different dependence in the narrow well of the tunneling structure to competition between n_{ex}-independent tunneling of free carriers out of the well and n_{ex}-dependent formation of excitons from free carriers in a bimolecular process (ie. at a rate proportional to the product of free electron and free hole densities) . At low n_{ex}, most electrons tunnel before excitons can be formed and so relatively little excitonic PL from the narrow well is observed. At higher density, the tunneling probability is unchanged, but the exciton formation probability is increased. Proportionately more excitonic PL is then observed, leading to the measured non-linear dependence. The role of the exciton formation has been rarely considered in the dynamics of photoexcited carriers in quantum well systems, and the influence on conventional experiments is often weak. For example, the risetime of excitonic TRPL transients is usually dominated by exciton cooling rather than formation (Damen 1990). In ADQW's however, the PL efficiency is strongly sensitive to exciton formation. In this work, this is exploited to measure the exciton formation rate using the time-resolved non-linear PL cross-correlation technique.

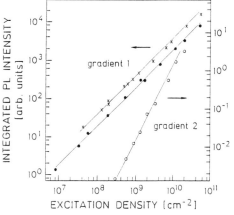

Fig.1 Integrated PL (arb. units) vs. n_{ex} at 8K for: 4nm barrier/narrow QW (circles), 4nm barrier/wide QW (dots) and 20nm barrier/narrow QW (crosses).

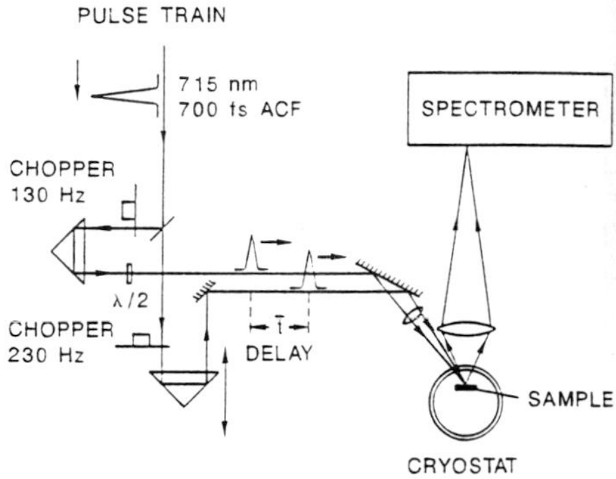

Fig.2   The   non-linear   PL cross-correlation measurement.(ACF - autocorrelation function).

2. EXPERIMENTAL

In the non-linear PL cross-correlation measurement (Fig.2), two 700fs pulses from a hybrid mode-locked Pyridine 1 dye laser operating at 715nm are overlapped on the ADQW sample which is maintained at 8K in a helium cryostat. The time-integrated narrow well exciton PL originating from the photocreated free carriers is detected using a cooled photomultiplier. Beam 1 is chopped at 230Hz and beam 2 at 130Hz and the photoluminescence signal is detected at the sum frequency, 360Hz. A signal is generated at the sum frequency only because the PL emission is a non-linear function of n_{ex}. As discussed above, this arises from the enhanced probability of exciton formation due to the mutual interaction of the two carrier populations created by the two pulses. By varying the delay between the two pulses, the decay of the non-linear change in PL output may be measured. The decay occurs due to depletion of the carrier population created by the leading pulse via electron and hole tunneling or exciton formation. The times associated with these 3 processes can be extracted by modeling the cross-correlation signal using a system of rate equations to describe the carrier dynamics in the narrow well:

$$\frac{dn}{dt} = -Cnp - \frac{n}{\tau_e} + \frac{N}{\tau_h} + G(t) + G(t - \bar{t}) \tag{1}$$

$$\frac{dp}{dt} = -Cnp - \frac{p}{\tau_h} + \frac{N}{\tau_e} + G(t) + G(t - \bar{t}) \tag{2}$$

$$\frac{dN}{dt} = Cnp - \frac{N}{\tau_e} - \frac{N}{\tau_h} - \frac{N}{\tau_{LT}} \tag{3}$$

where n, p, and N are the density of free electrons, free holes, and excitons, respectively; τ_e and τ_h are the tunneling times for electrons and holes; τ_{LT} is the exciton radiative lifetime; $G(t)$ and $G(t-\bar{t})$ are carrier generation terms given by two Gaussian pulses of appropriate FWHM, delayed by $\bar{t}$. The bimolecular generation rate of excitons is given by the term Cnp, where C is the bimolecular exciton formation coefficient; n/τ_e, p/τ_h are the tunneling rates for free electrons and holes. Tunneling of an electron or hole bound in an exciton ('excitonic' electrons or holes) is given by N/τ_e and N/τ_h. The excitonic tunneling terms N/τ_e and N/τ_h not only decrease the density of excitons, but also give a positive contribution to the free hole and free electron densities, respectively. The radiative recombination term, N/τ_{LT}, determines the photoluminescence intensity emitted from the sample. Time integration of $N(t,\bar{t})/\tau_{LT}$ gives the total time-integrated PL as a function of $\bar{t}$. Because exciton formation is a bimolecular process, the growth of the exciton population is non-exponential and has no single characteristic time. But, from (1)(2)(3), we can define a density-dependent exciton formation time, τ_{ex}, equal to the $1/e$ time for a photoexcited population of electrons and holes to form excitons in the absence of tunneling, given by $\tau_{ex} = (e-1)/Cn_{ex}$. Note that calculation of the total PL intensity vs. n_{ex} in the limit of $\tau_{ex} \gg \tau_e$ using (1)(2)(3) gives the n_{ex}^2 dependence measured in Fig. 1.

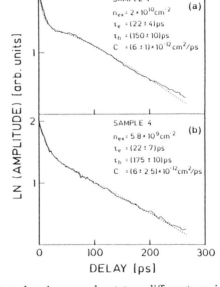

Figure 3. PL cross-correlation signal vs. delay for 4nm barrier/narrow QW at (a) $n_{ex} = 2 \times 10^{10} cm^{-2}$ and (b) $n_{ex} = 5.8 \times 10^9 cm^{-2}$. Also shown is the modeled fit.

3. RESULTS AND DISCUSSION

Figure 3 shows non-linear PL curves from a 4nm barrier sample at two different excitation densities and the best fit using Equs. (1)(2)(3). In the low density case (curve (b)), the signal closely approximates to two exponential decays given by τ_e (=22ps) and τ_h (=160ps). The same result is obtained from both experiment and model if n_{ex} is reduced by a further factor of 10. However, at higher n_{ex} (curve (a)), the faster decay becomes even faster and a non-exponential 'shoulder' region is observed. The faster decay corresponds to the case where the carriers from the first pulse are depleted faster by exciton formation than by electron tunneling. The shoulder arises because as excitonic electrons tunnel, free holes are released which contribute further to the non-linearity. This provides direct evidence that the exciton is broken apart during the tunneling process. The value of C is determined most accurately in curve (a), because the fast decay is then given mainly by the exciton formation process, and corresponds to $\tau_{ex} = 14$ps at this value of n_{ex}, consistent with the upper limit of 20ps inferred from TRPL measurements (Damen (1990)).

Shorter tunneling times ($\tau_e = 7ps$ and $\tau_h = 66ps$) are obtained from a 3nm barrier sample, as expected from a thinner barrier. The non-exponential shoulder is also absent at $n_{ex} = 2 \times 10^{10} cm^{-2}$ as expected because τ_e is now less than τ_{ex}. Note that a similar ratio, $\tau_e/\tau_h \approx 0.1$, is obtained for both the 3nm and 4nm samples. This ratio is in both cases much less than expected from semi-classical tunneling theory (for further details see: Strobel (1991)).

The good agreement between experiment and model as a function of barrier thickness and n_{ex} confirms that the non-linear PL originates from competition between exciton formation and tunneling, and demonstrates the bimolecular nature of exciton formation. It should be stressed that although the quantitative values are obtained from a three parameter fit, the accuracy of the values remains quite high since τ_h is relatively independent of the other parameters, and C and τ_e may be distinguished by varying n_{ex}. The results imply that the correlation of the photocreated 'geminate' electron-hole pair is rapidly broken and that excitons are subsequently formed purely from non-geminate pairs. This is consistent with the large initial excess energy of the carriers. Several LO-phonons are immediately emitted and separation of the geminate pair is likely to be very efficient. We find the same results for excess photon energies between 130meV (715nm) and 20meV above the exciton line. We therefore conclude that the exciton formation and tunneling dynamics are largely unaltered by different excess carrier energy within this regime.

4. CONCLUSIONS

The PL output of the narrow well of ADQW structures varies non-linearly with excitation density. This is due to the bimolecular nature of the exciton formation process, which makes the carrier dynamics in such structures strongly carrier density-dependent. Using PL cross-correlation measurements we determine a value of $6 \times 10^{-12} cm^2/ps$ for the bimolecular exciton formation coefficient, and also the electron and hole non-resonant tunneling times.

5. REFERENCES

T.C. Damen, J. Shah, D.Y. Oberli, D.S. Chemla, J.E. Cunningham, and J.M. Kuo, J.Luminescence **45**, 181 (1990).

D. von der Linde, J. Kuhl, and E. Rosengart, J. Luminescence **24/25**, 675 (1981).

M. Nido, M.G.W. Alexander, W.W. Rühle, T. Schweizer, and K. Köhler, Appl. Phys. Lett. **56**, 355 (1990), and the references therein.

R.Strobel, R.Eccleston, J.Kuhl, and K.Köhler, Phys. Rev. **B 43**, 12564, (1991).

Inst. Phys. Conf. Ser. No 126: Section V   419
Paper presented at Int. Symp. on Ultrafast Processes in Spectroscopy, Bayreuth, 1991

Exciton relaxation dynamics in GaAs/AlGaAs single quantum well heterostructures

Ph.Roussignol^*, A.Vinattieri^, L.Carraresi^, M.Colocci^ and C.Delalande*

^Dipartimento di Fisica dell"universita' di Firenze and LENS, Largo E.Fermi 2, 50125 Firenze-Italy
*Ecole Normale Supérieure, 24 rue Lhomond, 75005 Paris-France

ABSTRACT: We present an experimental study of the photoluminescence risetime τ_R of single GaAs/AlGaAs quantum well structures. The dependence of τ_R on the excitation energy is investigated for different temperatures and excitation intensities. A very slow relaxation process is found at 4K when the excitation energy is resonant with the light-hole exciton. Both temperature or excitation intensity increase makes the relaxation process faster. A tentative explanation in terms of a reduced free carrier-exciton scattering efficiency as a consequence of the resonant excitation is given.

The ultrafast dynamics and the non-linear optical properties of excitons in quantum confined structures are among the most active fields in the present semiconductor research. In particular, the exciton relaxation dynamics in GaAs/AlGaAs quantum wells (QW) has been subject of increasing interest in recent times for the promising potential applications of these structures as ultrafast photonic devices. Nevertheless, the main mechanisms underlying exciton relaxation under different conditions of excitation density and temperature are not yet completely clarified. In fact, the complex interplay between extrinsic effects such as exciton localization at crystal defects and/or exciton transfer between growth islands and the intrinsic ones related to exciton cooling and thermalization makes indeed difficult to draw a clear picture of exciton relaxation in quantum wells.

We will present time-resolved measurements of the excitonic luminescence in GaAs/Al$_{0.3}$Ga$_{0.7}$As single quantum wells showing that the accurate measurement of the PL risetime at the heavy-hole exciton energy can provide valuable information on the exciton cooling and wave-vector relaxation under given conditions of the excitation photon energy, excitation intensity and temperature.

High quality GaAs/Al$_{0.3}$Ga$_{0.7}$As single QWs with thickness ranging between 30 and 120Å were grown by MBE on a semiinsulating (001) oriented GaAs substrate. Al$_{0.3}$Ga$_{0.7}$As barriers, 500Å thick, have been grown between adjacent wells in order to decouple the GaAs QWs from each other. No significant Stokes shift between the PL and PLE peaks has been observed under CW excitation, thus excluding exciton localization and trapping at crystal defects. The high quality of the sample was confirmed by the narrow PL linewidth observed at 4K ($\approx$ 1 meV for the 90Å well at 1W cm^{-2} excitation intensity). The PL dynamics was investigated using a picosecond Nd:YAG synchronously pumped dye laser (pulse duration: 5ps; rep.rate: 76 MHz) for excitation; the detection was performed by using a synchroscan streak camera, after dispersion of the luminescence through a 0.22 m double monochromator with 1 meV energy resolution, providing a time resolution of 20 ps. A helium cryostat has been used for varying the sample temperature between 4 and 50 K.

We report in Fig. 1a the PL time decay of the 90Å well after resonant excitation of the heavy hole exciton, for an excitation sheet density $n_{ex} \approx 10^8$ cm^{-2}. A fast monoexponential decay with a time constant of $\tau_L = 120$ ps is observed, together with a PL rise time τ_R within the experimental time resolution, in good agreement with Damen et al (1990).

We have extracted the PL risetime τ_R and decay τ_L at the heavy hole exciton E_{HH} as a function of the excitation excess energy $\Delta E = E_{exct} - E_{HH}$, by using a simple phenomenological three-level model. In Fig 1b-1d typical decay curves are reported for different excitation photon energy E_{exct}; while the decay constant is found to be independent of E_{exct}, strong variations in the PL risetime are observed for increasing the excitation photon energy.

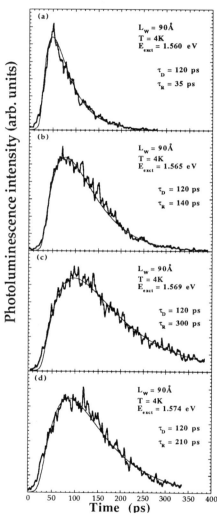

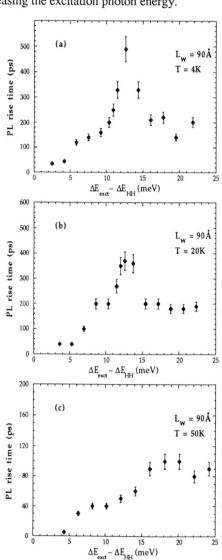

Fig.1 : PL time decay at the heavy-hole exciton energy for different excitation energies E_{exct} : 1.560eV, (a) ; 1.565eV, (b) ; 1.569eV, (c) ; 1.574eV, (d) - The full lines are the fits to the experimental curves within the model discussed in the text ; the reported values of τ_D and τ_R for each excitation energy are the best-fit values for the decay constant and the risetime, respectively.

Fig.2 : PL risetime at the heavy-hole exciton energy as a function of ΔE, for three different temperatures : 4K, (a) ; 20K, (b) ; 50K, (c) - Note the change of scale in (c).

The values of τ_R thus obtained are plotted in Fig.2 as a function of ΔE at low excitation ($n_{ex} = 10^8$ cm^{-2}). We see that, apart from a gradual increase of τ_R with increasing ΔE, a strong increase of the PL risetime is observed for resonant excitation at the light hole exciton energy E_{LH}. While an increase of τ_R has already been reported (Damen et al. (1990), Eccleston et al.(1991),Deveaud et al. (1991)), the peak in τ_R at E_{LH} has not been previously observed, to our knowledge, and seems to disagree with recent measurements by Damen et al (1990), where no strong dependence of the exciton PL risetime has been found up to 100 meV above the heavy-hole exciton energy. We also find that a smoothing of the peak at E_{LH} is observed whenever the excitation density is increased above 1-2 10^9 cm^{-2} (Fig.3) or the sample temperature is raised to a few tens of degrees K, where the peak disappears and an agreement is retrieved with the findings of Damen et al.(1990).

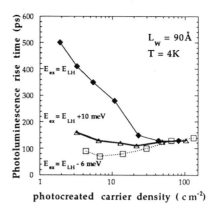

Fig.3 : PL risetime of the 90Å well at 4K as a function of the excitation density for three photon energies : $E_{exct} = E_{LH}$, ($\blacklozenge$) ; $E_{LH} +$ 10 meV, (Δ) ; E_{LH} - 6meV, ($\square$) - Note the independence of τ_R on E_{exct} when n_{ex} is larger than 3-4 10^9 cm^{-2}.

We believe that some significant insight on the dynamics of exciton relaxation in GaAs QWs are obtained by the data presented.

In fact, as well known in the case of excitation excess energies ΔE smaller than the LO-phonon energy, the relaxation of the excited carriers into K=0 radiative excitons can only proceed via exciton acoustic phonon, exciton-exciton and exciton-free carrier interactions or exciton-defect collisions. We will disregard the last mechanism because of the high quality of the sample. At low excitation, nex $\leq 10^9$ cm^{-2}, where the strong slowing down of the exciton relaxation is observed for a resonant excitation of light-hole excitons, exciton-free carrier and exciton-acoustic phonon scatterings seem to provide the main relaxation mechanism. Since exciton formation under resonant excitation is believed to proceed without significant free carriers generation, at least at low temperature, the strong increase in τ_R observed at E_{LH} reflects the less efficient exciton scattering probabilities as compared to the case where free carriers are generated under non resonant excitation.

The smoothing of the structures in τ_R as a function of ΔE for increasing temperature and excitation density is therefore expected as a consequence of the increased scattering rates connected to free carrier generation.

In conclusion, excitonic effects have been found to introduce strong modifications in the relaxation mechanisms in GaAs QWs as compared to the situations where free carriers are mainly involved in the relaxation processes. Further work, both theoretical and experimental is definitely needed in order to achieve a complete understanding of the relaxation mechanisms in these structures.

Acknowledgements

We want to acknowledge F. Bogani, M. Gurioli, J. Martinez-Pastor and R. Ferreira for fruitful discussions and B.Etienne for providing the samples. The Physics Department of the University of Florence is affiliated with the Gruppo Nazionale di Struttura della Materia, the Centro Interuniversitario di Struttura della Materia and the Consorzio Interuniversitario di Fisica della Materia. The Laboratoire de Physique de la Matière Condensée is "Laboratoire associé à l'Université Paris VI et au CNRS".

References.

Damen T.C., Shah J., Oberli D.Y., Chemla D.S, Cunningham J.E. and Kuo J.M. 1990, *Phys.Rev.***B42,** 7434.
Deveaud B., Clérot F., Roy N., Satzke K., Sermage B., Katzer D.S. 1991, *Phys.Rev.Lett.* **67,** 2355.
Eccleston R., Strobel R., Rühle W.W., Kuhl J., Feuerbacher B.F.and Ploog K. 1991, *Phys.Rev.* **B44,** 1395.

Inst. Phys. Conf. Ser. No 126: Section V
Paper presented at Int. Symp. on Ultrafast Processes in Spectroscopy, Bayreuth, 1991

The exciton interband scattering and LA−phonon interaction dynamics in GaAs quantum wells

R.Eccleston, J.Kuhl, R.Strobel, W.W.Rühle, and K.Ploog.

Max-Planck-Institut für Festkörperforschung, 7000 Stuttgart 80, Germany

ABSTRACT: The risetime of the heavy-hole (hh) exciton time-resolved photoluminescence (TRPL) transient in a high-quality 27nm GaAs quantum well (QW) is analysed for both light-hole (lh) exciton and free-carrier photoexcitation to obtain the exciton-LA phonon energy loss rate and the lh-hh exciton transfer time. The results show excitons are formed from free carriers almost elastically.

1.INTRODUCTION

TRPL risetime measurements are reported in an high quality GaAs-Al$_{0.3}$Ga$_{0.7}$As 27nm single QW with an extremely narrow hh exciton PL linewidth (0.2meV) (Ploog (1991)). This narrow linewidth permits selective photo-excitation of lh exciton or free carrier states very close to the hh exciton PL line without the unwanted direct resonant creation of hh excitons that occurs in strongly inhomogeneously broadened samples. For a sufficiently low excess photon energy, rapid LO-phonon emission by free carriers is forbidden. The excess energy of the photoexcitation is then transferred into kinetic energy of the hh exciton population which is subsequently formed. This population then cools mainly by slow LA phonon emission. The finite exciton cooling time causes a finite risetime of the emitted hh exciton PL because the effective exciton oscillator strength increases as the temperature of the exciton population falls (Feldmann (1987)). In this paper, we analyse such risetimes and extract the exciton LA-phonon energy loss rate, and by comparison of results for excitation of lh excitons (which in a 27nm QW also lie in the bandgap) and free carriers, assess the contribution of exciton formation to the energy loss process. We also investigate the time-scale and mechanism for exciton interband transfer from lh exciton to hh exciton states. The sample was cooled in a helium flow cryostat and excited with pulses from a synchronously pumped mode-locked Styryl 8 laser. The PL was spectrally dispersed in a 32cm spectrometer and temporally resolved in a 2D synchroscan streak camera. The temporal resolution of the system was between 15ps and 25ps and the spectral resolution was 0.5meV.

2.LH-EXCITON TO HH-EXCITON INTERBAND TRANSFER

The inset of Fig.1 shows TRPL transients at two initial excited carrier densities, n_0, for excitation at the lh exciton line and detection at the hh exciton PL line at 15K. For large n_0 (curve a), the rise in the transient exhibits two components. First, an immediate fast rise of the hh exciton PL to 80% of the peak value is observed within the time resolution of the streak camera. This indicates that (at this n_0) transfer between lh exciton $\mathbf{K} \approx 0$ states (where the lh excitons are photocreated) and hh exciton $\mathbf{K} \approx 0$ states (where the hh-exciton PL is emitted) occurs within 20ps. The second, slower 20% component of the risetime originates from the finite rise in excitonic PL due to hot exciton cooling described earlier. Transfer between the hh and lh exciton bands therefore establishes an Maxwell-Boltzmann (M-B) hh exciton distribution which is initially hotter than the lattice. However, at lower n_0, the fast rise in the PL due to lh-hh transfer disappears (curve b). Fig. 1 shows the variation of total risetime with n_0. Since cooling by LA-phonon emission is independent of density, this n_0 dependence indicates that the transfer between the optically active states of the two exciton bands is exciton density dependent, i.e. it occurs via exciton-exciton scattering. This result is consistent with either the lh-hh exciton interband scattering process itself occuring via exciton-exciton scattering, and/or that the time required to populate hh exciton $\mathbf{K} \approx 0$ states (i.e. to form a M-B distribution from an initially non-thermal hh exciton distribution) becomes significant at low n_0. Further work to distinguish the relative importance of these two processes is in progress.

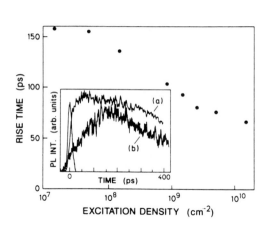

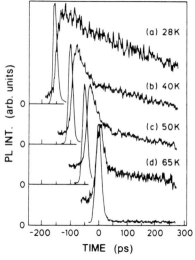

Fig. 1. PL risetime vs. n_0 for lh exciton excitation and hh exciton PL detection at 15K. The inset shows TRPL profiles at (a) $n_0 = 1.5 \times 10^{10} cm^{-2}$, (b) $n_0 = 1.5 \times 10^8 cm^{-2}$.

Fig.2. TRPL profiles for lh exciton excitation and hh exciton PL detection at different T_L and $n_0 = 1.5 \times 10^{10} cm^{-2}$. For clarity, the curves are shifted arbitrarily by 50ps.

The lattice temperature, T_L, was also varied (at large n_o) to determine the initial temperature of the hh exciton population. As T_L was increased, the cooling component of the risetime decreased and near 28K (Fig.2a) was close to the time-resolution limit, indicating that the hh exciton distribution is formed at an initial temperature which is close to 28K, the value expected if all the lh-hh exciton splitting energy (2.4meV) is transferred into hh exciton kinetic energy (i.e. 2.4meV/k $\approx$28K). Scattering into the hh exciton band therefore occurs without significant energy loss from the exciton distribution. Finally we note that for further increase of T_L (Figs.2b-d), an initial fast

TRPL decay appears. This fast decay is the exact inverse of the finite rise of the TRPL signal due to hot exciton cooling observed at 15K. The exciton population internally thermalises to a temperature near 28K and is then heated to the lattice temperature with a corresponding *fall* in the radiative recombination rate.

3. EXCITON LA-PHONON EMISSION

Fig. 3 shows the change in TRPL profile (at T_L=15K) as the excess photon energy, ΔE, above the hh exciton PL line is varied. Fig.3a is the previous resonant lh exciton pumping result (i.e. ΔE=2.4meV). Fig.3b is obtained at ΔE =6.9meV, i.e. only slightly above the n=1 electron-hh subband edge (located at ΔE=5.7meV). In Figs.3a-d, a progressive increase in the PL risetime is observed, indicating a higher initial exciton temperature. However, beyond $\Delta E \approx 10$meV the TRPL profile no longer varies strongly with ΔE (compare Figs.3d and 3e). This is consistent with the onset of very rapid energy loss via free carrier LO-phonon emission prior to exciton formation and within the streak camera time-resolution. No further increase in the TRPL risetime, and therefore initial exciton temperature, with ΔE is therefore apparant.

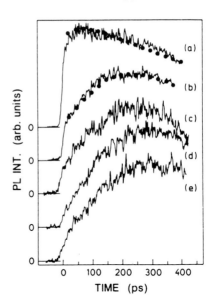

Fig. 3. TRPL profiles at 15K for detection at the hh exciton PL line for different excess photon energy, ΔE, at $n_0 = 1.5 \times 10^{10} cm^{-2}$. (a) ΔE =2.4meV, (b) ΔE =6.9meV, (c) ΔE =8meV, (d) ΔE =12.9meV, and (e) ΔE =20.8meV. Curves (a) and (b) also show the theoretical fit (dots).

The 2D LA phonon deformation potential energy loss rate in eV/s for a 2D M-B distribution of excitons is (Hess (1974)):

$$\langle\frac{dE}{dt}\rangle_{LA} = -2.14 \times 10^4 (D_c - D_v)^2 \left(\frac{M_{ex}}{m_0}\right)^2 (T_{ex} - T_L) \tag{1}$$

where T_{ex} is the exciton temperature, M_{ex} is the total exciton mass ($0.67m_0$ used here), D_c and D_v are the deformation potentials of the conduction and valence band of GaAs in eV (10.7eV and -3.4eV, respectively). Note that Equation (1) implies an exponential decay of T_{ex} to the lattice temperature, and that the free carrier deformation potentials and masses add coherently, giving a larger energy loss rate than that expected from the sum of that for free electrons and holes. The luminescence intensity, $I(t)$, is proportional to the instantaneous exciton recombination rate, given in a QW by (Feldmann (1987)): $I(t) \propto -dN/dt = (1 - exp(-\Gamma/kT_{ex}(t)))N/K$, where Γ is the homogeneous linewidth (0.2meV,(Feuerbacher (1990))), and N is the exciton density. K is a constant determined from the exponential tail of the TRPL where $T_{ex}=T_L$. By evaluating $T_{ex}(t)$ from Equation (1), we can model the TRPL transients and by comparison with experiment obtain the exciton-LA phonon energy loss rate. The fit parameter is essentially the exciton mass. Note however that the use of a single exciton mass is always an approximation in carrier cooling studies of GaAs QW's due to the non-parabolicity of the valence bands, the required value always exceeding the in-plane gamma point masses (Leo(1988)). We consider here only the lh exciton and near-bandedge measurements (Figs.3a and 3b). Any influence of LO phonon energy loss can then be neglected. We assume excitons are formed quasi-elastically, i.e. the entire exciton binding energy is transfered into extra kinetic energy of the bound electron-hole pair. Further details are given elsewhere (Eccleston (1991)). Figs.3a and 3b show (dots) the calculated TRPL transients, giving reasonable agreement with experiment for $M_{ex} = 0.67m_0$. We can conclude that for $\Delta E \leq 10$meV, the exciton cooling rate determined from TRPL risetimes is quantitively consistent with LA-phonon emission using $M_{ex} = 0.67m_0$. The good agreement with theory may also be used to support the assumption that excitons are formed elastically. If significant energy was lost, e.g. by emission of a large energy phonon in the formation process, a much smaller difference in risetime would be obtained between Figs. 3a and 3b (irrespective of the M_{ex} value used). We find that an energy loss of no more than 1meV can be included without generating significant differences from the observed transients.

4. CONCLUSIONS

TRPL risetime measurements can be used to obtain important information about exciton dynamics in quantum wells. We find that bandgap lh **K** ≈ 0 excitons scatter to hh **K** ≈ 0 exciton states via exciton-exciton scattering. No significant energy is lost in the initial lh-hh transfer process and the resultant hh exciton temperature corresponds to a value of kT close to the lh-hh exciton splitting energy. For photoexcitation within 10meV of the hh exciton line, the PL risetime increases with excess photon energy due to a larger initial exciton temperature. Above $\Delta E \approx 10$meV, no further increase in risetime is observed due to the onset of rapid LO phonon emission. The magnitude of the TRPL risetime is quantitatively consistent LA phonon emission rate assuming quasi-elastic exciton formation.

5. REFERENCES

R.Eccleston, R.Strobel, W.W.Rühle, J.Kuhl, B.Feuerbacher, and K.Ploog, Phys. Rev. **B44**, 1395 (1991).
J.Feldmann, G.Peter, E.O.Göbel, P.Dawson, K.Moore, C.T.Foxon, and R.J.Elliot, Phys. Rev. Lett. **59**, 2337 (1987).
B.Feuerbacher, J.Kuhl, R.Eccleston and K.Ploog, Solid State Comm.**74**,1279 (1990).
K.Hess and C.T.Sah, J.Appl.Phys. **45**, 1254 (1974).
K.Ploog, A.Fischer, L.Tapfer, and B.F.Feuerbacher, Appl.Phys. **A52**, 135 (1991).
K.Leo, W.W.Rühle, and K.Ploog, Phys. Rev. **B 38**, 1947 (1988).

Inst. Phys. Conf. Ser. No 126: Section V

Paper presented at Int. Symp. on Ultrafast Processes in Spectroscopy, Bayreuth, 1991

Carrier-induced shifts of exciton energies in CdTe/CdZnTe superlattices

M.K. Jackson, D. Hulin, and J.-P. Foing

Laboratoire d'Optique Appliquée, Ecole Nationale Supérieure de Techniques Avancées-Ecole Polytechnique, 91120 Palaiseau, France

N. Magnea and H. Mariette

Laboratoire de Physique des Semiconducteurs, Centre d'Etudes Nucléaires de Grenoble, 38041 Grenoble Cedex, France, and Laboratoire de Spectrométrie Physique, Université Joseph Fourier, Grenoble, 38402 St. Martin d'Hères Cedex, France.

ABSTRACT: We report the first carrier-induced red shift of excitonic absorption in a type-I superlattice. The red shift lasts for several picoseconds after excitation, and is not observed when the pump is resonant with the excitonic absorption; a spatial separation of electrons and holes is proposed to explain this observation.

1. INTRODUCTION

Progress in the epitaxial growth of wide-bandgap II-VI semiconductors now makes possible the fabrication of heterostructures with quality rivalling that of III-V devices. The CdTe/CdZnTe system is interesting because it has a small valence band offset, and the confinement of holes is largely determined by strain shifts of the valence band edges. This allows the study of systems where electrons are confined to the CdTe layers, but holes can be confined in either layer, or not well confined at all. In this paper we report ultrafast pump-probe measurements of absorption in CdTe/CdZnTe superlattices, and investigate the effects of the weak hole confinement on the excitonic optical response in the presence of free carriers.

2. EXPERIMENT

We have studied two CdTe/CdZnTe superlattices grown by molecular beam epitaxy on (100) $Cd_{0.96}Zn_{0.04}Te$ substrates (Ponchet 1990). The first, sample Z339, consists of 10 periods of 12.8nm CdTe wells and 12.6nm $Cd_{0.90}Zn_{0.10}Te$ barriers. The second, Z322, is similar: 36 periods of 14.0nm CdTe wells and 13.8nm $Cd_{0.91}Zn_{0.09}Te$ barriers. Pump-probe absorption experiments were performed at approximately 13K, using a colliding pulse mode-locked laser, amplified at 6.5kHz using the green line of a copper vapour laser (CVL). The amplified pulses are focused on an ethylene glycol jet to form a spectral continuum, part of which is selected with interference filters, reamplified in a dye jet also pumped by the green line of the CVL, and used as a pump. The pump bandwidth is 8nm and the duration is approximately 150fs. A portion of the continuum is used as a probe pulse, and has a duration of less than 100fs.

In Fig. 1 we show the time-resolved absorption spectra αl for sample Z339, for several pump-probe delays, as indicated. For comparison, the dotted line shows the spectrum for t=-1ps, corresponding to the unperturbed, or linear, absorption. The zero of delay corresponds to pump-probe coincidence, within approximately 250fs. From the spectrum for t=-1ps shown as a dotted line in Fig. 1, we see that the linear absorption spectrum is dominated by the excitonic absorption at approximately 773nm, with substrate absorption and band-to-band absorption of the superlattice contributing to a rising absorption for short wavelengths. Note that although the substrate interferes with transmission of the probe, pump absorption is unaffected since

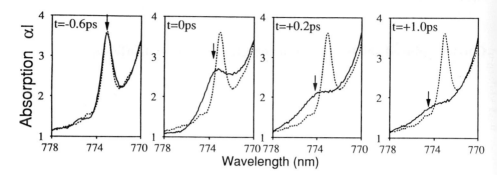

Fig. 1. Time-resolved absorption spectra αl for sample Z339, with a pump at 760nm, for several pump-probe delays, as indicated. The spectrum for t=-1ps is shown as a dotted line for comparison. The wavelength corresponding to peak excitonic absorption is indicated by an arrow in each spectrum.

the pump is incident from the superlattice side. The pump spectrum for Fig. 1 was centered on 760nm. In each spectrum, the wavelength of peak excitonic absorption is indicated by an arrow. The exciton undergoes a significant red shift, an absorption bleaching, and a broadening, due to the free carriers created by the pump.

We repeated the experiment for various pump wavelengths, using the thicker sample, Z322. In Fig. 2 we show the center wavelength of the exciton absorption, as a function of time, for two pump wavelengths. The solid line corresponds to the pump at 760nm. The red shift (positive peak shift) is seen to appear and then disappear, on a timescale of approximately 1ps; for longer times, a blue shift is observed, which then decreases slowly. The dashed curve shown in Fig. 2 is for a pump centered at 772nm. The peak shift behavior with time is strikingly different in this case, which corresponds to direct creation of excitons. No red shift is observed; the spectra are dominated by a blue shift that quickly reaches a peak, and then decreases monotonically. It appears that the red shift is only observed when carriers are created with an excess kinetic energy. Note that the pump intensity for the two pump wavelengths was chosen to give similar absorption bleaching at long times, corresponding to excitation of roughly the same density of carriers.

3.   DISCUSSION

We first discuss the possibility that the excitonic shifts we observe are due to the optical Stark effect (OSE), as it is well known that the OSE can lead to blue (Joffre 1989) or red (Hulin 1990, Chemla 1991) shifts of the exciton. The OSE is characterized by a rapid disappearance of the shift after pump-probe coincidence, occurring on the time scale of the pump duration (Joffre 1989). It can be seen from Figs. 1 and 2 that the shift persists much longer than the 150fs pump pulse duration, and is therefore *not* due to the OSE. We attribute the shift to the presence of carriers created by the pump.

In our samples the excitonic bandgap is type-I, the exciton being formed from the lowest-lying electron and heavy-hole states, both of which are localized in the CdTe layers (Tuffigo 1991). The excitonic absorption seen in Fig. 1 is due to this spatially-direct excitonic absorption. Our experiment consists of measuring the energy of this transition in the presence of carriers created by the pump. The pump-created carriers can have various effects upon absorption depending upon their energy and wavevector, as has been studied in many previous experiments. Another important factor is the distribution of these carriers in the different layers of the sample, and it is this point that we will consider in detail.

When pumping at 772nm, electron-hole pairs bound as excitons are directly created in the CdTe layers, where they stay until they recombine. The situation when pumping at 760nm is

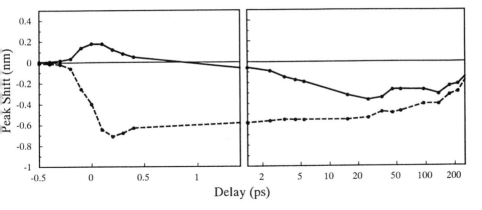

Fig. 2. The peak shift of the excitonic absorption in sample Z322 as a function of time, for a pump at 760nm (solid line), and 772nm (dashed line).

quite different: the pump creates free electrons and holes. Since the pump energy is 27meV greater than the excitonic bandgap, and the exciton binding energy is approximately 14meV (Tuffigo 1991), a kinetic energy of 13meV is shared between the free electrons and holes. The initial distribution of this excess energy between the electrons and holes is unknown, and depends upon the detailed band structure of the superlattice. However, due to carrier-carrier scattering, the kinetic energy will rapidly be shared between the electrons and holes. This kinetic energy will have a very different effect upon the behavior of the free electrons and holes. Using a valence band offset of zero in the absence of strain, and calculating the strain shifts of the band edges, we find that the depth of the potential confining electrons in the CdTe layers is 36meV. The strength of this confining potential is so great that the electrons would remain localized in the CdTe layers even if they had all of the 13meV kinetic energy. For heavy holes the situation is very different. The depth of the confining potential is 9meV; even more importantly, the lowest-lying light-hole level, which is located in the CdZnTe layers, is very close in energy. Therefore, even a small kinetic energy is sufficient for the heavy hole to overcome the localization in the CdTe layer. The exact mechanism for the delocalization, which could be due to band mixing between the heavy- and light-hole states for nonzero parallel wavevector, or scattering of the heavy hole by collision, is beyond the scope of this article. However, we suggest that if they have significant kinetic energy, the heavy holes will rapidly become delocalized in both the CdTe and the CdZnTe layers. This situation is transient and cannot be maintained as the carriers cool and lose their kinetic energy. As the heavy-hole level in the CdTe layer is the lowest-energy state, the holes will be trapped in the CdTe layers, where they will eventually form excitons.

We now discuss the curves of Fig. 2 in terms of the above picture. An excitonic blue-shift in type-I GaAs/AlGaAs multiple quantum wells has been observed (Peyghambarian 1984, Masumoto 1985, Hulin 1986), due to many-body repulsion between the excitons (Schmitt-Rink 1985). We believe that the blue shift we see is of the same origin; the presence of a large density of excitons in the CdTe layers increases the exciton energy, which is seen in the probe absorption. This occurs immediately when pumping at 772nm, as excitons are created directly in the CdTe layers. When pumping at 760nm, some time is required for hole trapping, and exciton formation. In both cases, the blue shift slowly disappears as the excitons recombine.

We now consider the red shift seen in Figs. 1 and 2. We first consider bandgap renormalization (BGR), as it is known that the presence of free carriers red-shifts the *free-carrier* energies. Zimmermann (1978) and Schmitt-Rink (1989) have shown that this does not lead to an *excitonic* red shift because the BGR is offset by a reduction in the exciton binding energy, with the excitonic transition energy remaining nearly constant. Recently Olbright (1991) has considered

BGR specifically in the case of spatially-separated carriers, and shown that a blue shift of the spatially-direct exciton is obtained. We conclude from these studies that it is unlikely that BGR can be used to explain the red shift we see. A small blue shift from BGR in the presence of free carriers may be present in our experiments, but is masked by a much larger red shift.

The spatial separation of electrons and holes will lead to electric fields, which will change the energy of the direct exciton. Sasaki and coworkers (1990) have considered a similar case in their study of type-II AlGaAs/AlAs superlattices with long periods. They developed a model neglecting the excitonic binding energy and showed that the electric fields induced by charge separation lead to a red-shift of the free-carrier bandgap. We have adapted Sasaki's model to our situation, assuming a hole charge uniformly distributed throughout both CdTe and CdZnTe layers, and an electron charge uniformly distributed throughout the CdTe layers. We numerically solve for the lowest bound-state energy for electrons and heavy holes, and find that a carrier density of $4 \times 10^{11} cm^{-2}$ gives a red shift of 1meV (0.5nm). This red shift is approximately what is observed experimentally. Several effects have been neglected in this very simple model, including self-consistency and the change in the exciton binding energy, which will be required in a more rigorous calculation. However, this model does provide a simple explanation for the red shift, and suggests a similarity to long-period type-II superlattices.

4.   CONCLUSIONS

We have observed the first carrier-induced red shift of exciton absorption in type-I CdTe/CdZnTe superlattices. We attribute this red shift to transient delocalization of hot holes, which leads to band bending.

ACKNOWLEDGMENTS

The authors would like to acknowledge helpful discussions with A. Mysyrowicz, R. Romestain, Y. Merle d'Aubigné, and C. Tanguy. One of us (M.K.J.) would like to acknowledge financial support in the form of a NATO Science Fellowship administered by the Natural Science and Engineering Research Council of Canada. The Laboratoire d'Optique Appliquée is CNRS URA 1406, and the Laboratoire de Spectrométrie Physique is CNRS URA 8.

REFERENCES

D.S. Chemla, in *Proceedings of the International Meeting on the Optics of Excitons in Confined Systems*, A. D'Andrea, R. Del Sole, R. Girlanda, and A. Quattropani, eds., (The Institute of Physics, 1991), in press.

D. Hulin, A. Mysyrowicz, A. Antonetti, A. Migus, W.T. Masselink, H. Morkoç, H.M. Gibbs, and N. Peyghambarian, Phys. Rev. B **33**, 4389(1986).

D. Hulin and M. Joffre, Phys. Rev. Lett. **65**, 3425(1990).

M. Joffre, D. Hulin, J.-P. Foing, J.-P. Chambaret, A. Migus, and A. Antonetti, I.E.E.E. J. Quantum Electron. **25**, 2505(1989).

Y. Masumoto, S. Shionoya, and H. Okamoto, Opt. Commun. **53**, 385(1985).

N. Peyghambarian, H.M Gibbs, J.L. Jewell, A. Antonetti, A. Migus, D. Hulin, and A. Mysyrowicz, Phys. Rev. Lett. **53**, 2433(1984).

G.R. Olbright, W.S. Fu, A. Owyoung, J.F. Klem, R. Binder, I. Galbraith, and S.W. Koch, Phys. Rev. Lett. **66**, 1358(1991).

A. Ponchet, G. Lentz, H. Tuffigo, N. Magnea, H. Mariette, and P. Gentile, J. Appl. Phys. **68**, 6229(1990).

F. Sasaki, T. Mishina, and Y. Masumoto, Phys. Rev. B. **42**, 11426(1990).

S. Schmitt-Rink, D.S. Chemla, and D.A.B. Miller, Phys. Rev. B. **32**, 6601(1985).

S. Schmitt-Rink, D.S. Chemla, and D.A.B. Miller, Adv. in, Phys. **38**, 89(1989).

H. Tuffigo, N. Magnea, H. Mariette, A. Wasiela, and Y. Merle d'Aubigné, Phys. Rev. B. **43**, 14629(1991).

R. Zimmermann, Phys. Stat. Sol. B. **90**, 175(1978).

Inst. Phys. Conf. Ser. No 126: Section V
Paper presented at Int. Symp. on Ultrafast Processes in Spectroscopy, Bayreuth, 1991

431

Nonlinear optical response of interacting excitons

A L Ivanov[*], L V Keldysh, V V Panashchenko

Department of Physics, Moscow State University, Leninskii Gory, Moscow, GSP 119899, USSR
[*]*Present address: Institut für Theoretische Physik, Universität Frankfurt am Main, Robert-Mayer-Str. 8-10 , 6000 Frankfurt am Main 11, Germany*

ABSTRACT: The low-intensity effect of the dynamical exciton level shift due to "large" electron-hole complexes (biexcitons) creation is investigated. The mechanism of such exciton-biexciton optical Stark effect is attributable to the direct virtual coupling of two excitons forming a biexciton by Coulomb-like interaction rather than to "giant" oscillator strength of the exciton-biexciton optical transition. The effective red shift of the exciton level may be realized in special "pump-probe" scattering geometry. The traditional conception of the two-photon absorption due to excitonic molecule is critically analysed.

1. INTRODUCTION

The exciton optical Stark effect in semiconductors is receiving extensive theoretical and experimental attention currently (Keldysh 1972, Mysyrowicz et al 1986, Schmitt-Rink and Chemla 1986, Haug and Koch 1990). This effect is manifested as a blue-shift of the exciton line in the presence of a high intensity polariton wave. The corresponding intensity is determined by the parameter $Na_{ex}^3 = |P_k|^2 a_{ex}^3 \simeq 1$, where a_{ex} is the Bohr radius of the exciton in the ground state, while $N=|P_k|^2$ is the exciton component concentration of the pump wave k of frequency ω_k .

The proposed exciton-biexciton optical Stark effect has the same fundamental nature as the exciton optical Stark effect : an exciton-exciton interaction in the semiconductor excited by the pump wave. On the other hand, the specific nature of the exciton-biexciton Stark effect, particulary, its low observation threshold, arisen because under moderate ($Na_{biex}^3 \simeq 1$, a_{biex} is the radius of the biexciton in the ground state) semiconductor excitation, in addition to the single characteristic dimensions a_{ex} in the exciton system, another natural scale, a_{biex}, appears. So it seems possible to realize optical Stark effect without "deformation" of an excitons, i.e. without perturbation of an exciton interior structure. Of course, such statement is valid only for the case where the binding energy of the biexciton is substantially lower than the binding energy of its constituent excitons.

2. LOW-INTENSITY EXCITON-BIEXCITON OPTICAL STARK EFFECT

A consistent microscopic theory of the exciton-biexciton Stark effect can be formulated on the basis of the analysis of the proper photon-electron-hole Hamiltonian taking into account the electron-hole structure of the exciton and biexciton excitations (Ivanov, Keldysh and Panashchenko 1991). The principal problem is connected with the true introduction of these quasiparticles. To put it roughly, in the presence of an intensive polariton wave exciton and biexciton excitations are not more the two- and four-electron (hole) complexes as it is for low-excited semiconductor because effective coupling of them with pump wave is realized. However it is possible to treat "semiconductor + pump wave" as weak-excited system if the proper canonical transformation of the quantum state and Hamiltonian is made. In fact the such construction of the new ground vacuum state of the transient excited semiconductor system permits to introduce the true exciton and biexciton independent boson excitations. Further analysis of diagram (or Heisenberg) equations establishes the dispersion properties of the semiconductor system in the presence of a polariton pump wave :

$$\frac{p^2 c^2}{\omega^2} = \epsilon(p,\omega,|P_\mathbf{k}|) = \epsilon_g + \epsilon_g \frac{\Omega_c^2}{2\omega_t} \frac{\nu^{biex}(\omega,\omega_\mathbf{k})}{\nu^{ex}(\omega)\nu^{biex}(\omega,\omega_\mathbf{k}) - \left| M_2(p,k)P_\mathbf{k} \right|^2} \; ; \tag{1}$$

$$\nu^{ex}(\omega) = \omega_\mathbf{p}^{ex} + \delta_\mathbf{p}^{ex} - \omega - i\gamma^{ex}; \quad \nu^{biex}(\omega,\omega_\mathbf{k}) = \Omega_{\mathbf{p+k}}^{biex} + \delta_{\mathbf{p+k}}^{biex} - \omega_\mathbf{k} - \omega - i\gamma^{biex} \, ,$$

where $\omega_\mathbf{p}^{ex}$, $\Omega_{\mathbf{p+k}}^{biex}$ are the exciton and biexciton unperturbative dispersions, γ^{ex} and γ^{biex} are the inverse exciton and biexciton lifetimes, $\delta_\mathbf{p}^{ex}$ and $\delta_{\mathbf{p+k}}^{biex}$ are non-resonant dynamical shifts of the exciton and biexciton levels, ϵ_g is the background permittivity of the crystal, ω_t is the energy of the transverse exciton state, whereas the polariton parameter Ω_c is defined in terms of the longitudinal - transverse polariton splitting ω_{lt} or in terms of the dimensionless parameter β :

$$\Omega_c^2 = 2\omega_{lt}\omega_t = \left(\frac{4\pi\beta}{\epsilon_g}\right)\omega_t^2 \; ;$$

The Eq.(1) describes the dispersion properties of "matter waves" $p = p_{i=1,2,3}(\omega)$ of the semiconductor and reflects the nature of unification with subsequent splitting of the exciton-photon-biexciton terms of the semiconductor in the presence of the polariton pump wave. The matrix element $M_2(p,k)$ characterizing direct resonance coupling of exciton $\mathbf{k}$ of the pump wave with exciton $\mathbf{p}$ of the probe wave to form a biexciton $\mathbf{p+k}$ is given by

$$M_2(p,k) = \epsilon_{\mathbf{p+k}} \; \Psi\!\left(\frac{\mathbf{p-k}}{2}\right) \simeq \epsilon_{\mathbf{p+k}} \, a_{biex}^{3/2} \, , \tag{2}$$

where $\epsilon_{\mathbf{p+k}} < 0$ is the biexciton potential, i.e. the average potential energy of the exciton-exciton interaction in a biexciton, while $\Psi\!\left(\dfrac{\mathbf{p-k}}{2}\right)$ is the Fourier transform of the wave function of the excitons relative motion in the biexciton. Such coupling of the excitons forming a biexciton has Coulomb nature and can be due to the presence in the system of the third quasiparticles, photons, which resonate with the excitons. In other words the polariton effect itself may be responsible for satisfaction of the law of conservation of energy in the resonance pairing of excitons that forms a biexciton.

The following expression can be obtained for the effective dynamical shift $\tilde{\delta}_\mathbf{p}^{ex}$ of the exciton level from the dispersion Eq.(1) :

$$\tilde{\delta}_\mathbf{p}^{ex} = \delta_\mathbf{p}^{ex} - \frac{\left| M_2(p,k)P_\mathbf{k} \right|^2}{\Omega_{\mathbf{p+k}}^{biex} + \delta_{\mathbf{p+k}}^{biex} - \omega_\mathbf{k} - \omega_\mathbf{p}^{ex}} \; ; \tag{3}$$

Therefore the effective dynamical shift of the exciton level is determined by two components. The first one

$$\delta_\mathbf{p}^{ex} = |P_0|^2 \, V \, W(0) \tag{4}$$

represents from the formal point of view the dynamical shift from renormalization of the vacuum state of the semiconductor in the presence of the pump wave. In fact it resulted from non-resonant interaction of the probe exciton $\mathbf{p}$ with the excitons $\mathbf{k}$ of the pump wave. The potential $W(0)$ of the exciton-exciton interaction has Coulomb nature and drastically depends on the spin structures of the interacting excitons. It is generally well-known that interexciton repulsion predominates in a uniformly excited system in thermodynamic equilibrium. However it is possible to realize the effective exciton-exciton attraction $W(0) < 0$ in a "probe radiation - pump wave" experiment. For example such an optical experiment can be formulated for CdS semiconductor in a geometry with $\mathbf{p} \parallel \mathbf{k} \parallel \mathbf{c}$ -axis using a pump wave and probe radiation of opposite circular polarizations.

But the nature of the low-intensity exciton-biexciton optical Stark effect is directly connected with the second term in (3). It characterizes the resonant part of the exciton-exciton interaction resulting in virtual $\mathbf{p+k}$ biexciton formation from probe $\mathbf{p}$ and pump $\mathbf{k}$ excitons. The efficiency of such process is connected with matrix element $M_2(\mathbf{p},\mathbf{k})$ and as result the new length parameter a_{biex} is naturally appeared. The second term in (3) as usually dominates due to resonant denominator and also results in effective exciton level red shift for the most interesting experimental case $\Omega^{biex}_{\mathbf{p+k}} - \omega_{\mathbf{k}} - \omega^{ex}_{\mathbf{p}} > 0$.

The threshold of the observation of the dynamical exciton level red-shift and the corresponding modifications of the spectra of elementary semiconductor excitations is given by the conditions ($|\omega - \omega_t| > \omega_{lt}$) :

$$\frac{\omega_{lt}\left|M_2(\mathbf{p},\mathbf{k})P_{\mathbf{k}}\right|^2}{(\omega - \omega_t)^2} > \gamma^{biex} > \gamma^{ex} \tag{5}$$

and directly resulted from the proper analysis of the dispersion Eq.(1) .

Studies of the dynamical exciton-photon-biexciton spectra renormalization are well-known after the works of May *et al* (1979) and Haug *et al* (1980). A traditional theoretical description of this effect has customarily been based on an analysis of a system of independent boson excitations: photons $\alpha_{\mathbf{p}}$, excitons $B_{\mathbf{q}}$ and biexcitons A_1 with ordinary polariton exciton-photon mixing and a phenomenological incorporation of three-particle interaction of the form $\frac{1}{\sqrt{V}}M_1(\mathbf{p}, \mathbf{q}) A^{\dagger}_{\mathbf{p+q}}B_{\mathbf{q}}\alpha_{\mathbf{p}}$ (V is the crystal volume). In this case, the matrix element $M_1(\mathbf{p}, \mathbf{q})$, was attributed to the "giant" oscillator strength of the exciton-biexciton optical transition (Rashba 1974) :

$$M_1(\mathbf{p}, \mathbf{q}) = -i\,\frac{\Omega_c}{2}\,\Psi\left[\frac{\mathbf{p-q}}{2}\right] ; \tag{6}$$

The subsequent introduction of the polariton pump wave $\mathbf{k}$ into the quasiparticle system resulted in the spectra renormalization, which is determined by the combination $|M_1(\mathbf{p},\mathbf{k})P_{\mathbf{k}}|^2$. Afterwards the similar results have been obtained by Combescot and Combescot (1988) by means of microscopic approach.

Nevertheless, as was mentioned before, a consistent analysis makes it possible to identify a new spectral modification mechanism determined by the direct virtual coupling of two excitons that form a biexciton by their Coulomb-like interaction. This therefore refers to a process described by terms of the form $\frac{1}{\sqrt{V}} M_2(\mathbf{p}, \mathbf{q}) A^{\dagger}_{\mathbf{p+q}}B_{\mathbf{q}}B_{\mathbf{p}}$ in the corresponding quasiparticle Hamiltonian. And the considered exciton-biexciton optical Stark effect is not determined by the giant oscillator strength (5) of the exciton-biexciton optical transition, but rather by matrix element $M_2(\mathbf{p}, \mathbf{q})$.

In connection with recent experiment of Hulin and Joffre (1990) devoted to observation of the exciton-biexciton optical Stark effect it is necessary to point out that the specific ABC-problem has to be considered for such phenomenon (Ivanov and Panashchenko 1991).

3. BIEXCITON TWO-PHOTON ABSORPTION

Hanamura (1973) has introduced the conception of the giant two-photon absorption due to excitonic molecule. In fact, the first process of the creation of an exciton by means of one photon absorption and the second process of the optical conversion of this exciton to the biexciton state have been considered as the basic processes. The second transition has been connected with the matrix element $M_1(\mathbf{p},\mathbf{k})$ and the strong enhancement of the two-photon absorption arisen both from the resonance character of the first process and giant oscillator force of the exciton-biexciton optical transition.

Also it is possible and important to investigate the biexciton two-photon absorption starting from the considered phenomena of the exciton-photon-biexciton spectra renormalizations. Indeed, we can obtain the absorption coefficient $2\mathrm{Im}p(\omega)$ of the probe electromagnetic wave of frequency $\omega \simeq \Omega_{p+k}^{biex} - \omega_k$ from the corresponding dispersion equation (in our case - from Eq.(1)) of the exciton-photon-biexciton system of the semiconductor in the presence of polariton pump wave. Just this absorption coefficient in the formal limiting case of the diminishing intensity $I_1 \to 0$ of the pump wave describes the usual two-photon ω and ω_k absorption due to excitonic molecule creation.

So let's define the coefficient $K^{(2)}{}_{biex}(\omega, \omega_k)$ of the biexciton two-photon absorption by the usual manner :

$$\frac{d}{dz} I_{i=1,2} = - K^{(2)}{}_{biex}(\omega, \omega_k) \sqrt{I_1} \sqrt{I_2} \, , \qquad (7)$$

where I_1 and I_2 are the intensities of the electromagnetic waves penetrating into the crystal through a surface and absorbed under the condition $\omega + \omega_k \simeq \Omega_{p+k}^{biex}$. Then, using the dispersion Eq.(1), we find the following expression for the two-photon absorption coefficient :

$$K^{(2)}{}_{biex}(\omega, \omega_k) = \frac{\epsilon_g}{c^2} \frac{\left|\Psi\left(\frac{p-k}{2}\right)\right|^2 \epsilon_{p+k}^2 \Omega_c^4}{4\hbar(\omega_k - \omega_t)^2(\omega - \omega_t)^2} \sqrt{\frac{I_1}{\omega_k}} \sqrt{\frac{I_2}{\omega}} \frac{\gamma^{biex}}{(\Omega_{p+k}^{biex} - \omega_k - \omega)^2 + (\gamma^{biex})^2} , \qquad (8)$$

while starting from the traditional approach to the exciton-photon-biexciton spectra renormalization phenomenon we obtain the other result :

$$K^{(2)}{}_{biex}(\omega, \omega_k) = \frac{\epsilon_g}{c^2} \frac{\left|\Psi\left(\frac{p-k}{2}\right)\right|^2 \Omega_c^2}{4\hbar(\omega_k - \omega_t)^2(\omega - \omega_t)^2} \left[(\omega_k - \omega_t)^2 + (\omega - \omega_t)^2\right] \sqrt{\frac{I_1}{\omega_k}} \sqrt{\frac{I_2}{\omega}} \cdot$$

$$\cdot \frac{\gamma^{biex}}{(\Omega_{p+k}^{biex} - \omega_k - \omega)^2 + (\gamma^{biex})^2} ; \qquad (9)$$

The expressions (8) and (9) are derived under the conditions $|\omega - \omega_t| > \omega_{lt}$, $|\omega_k - \omega_t| > \omega_{lt}$. Moreover for the real experimental situation the proper corrections to the intensities $I_{i=1,2}$ of the electromagnetic waves inside the crystal are necessary to make. Namely, for the case of the normal incidence of the waves on the surface of a crystal, the external electromagnetic intensities $I_{0i=1,2}$ are connected with intensities $I_{i=1,2}$ by means of relations :

$$I_{i=1,2} = \frac{4 n_i}{(n_i + 1)^2} I_{0i=1,2} \, ,$$

where the corresponding polariton refractive index $n_{i=1,2}$ is given by

$$n_i = \left[\epsilon_g \left(1 + \frac{\omega_{lt}}{\omega_t - \omega_i}\right)\right]^{\frac{1}{2}} \simeq \sqrt{\epsilon_g}$$

and $\omega_{i=1,2} = \omega, \omega_k$.

It is important to point out that in our calculations of the two-photon absorption coefficient $K^{(2)}{}_{biex}(\omega, \omega_k)$ (see Eq.(8) and Eq.(9)) in contrast with the traditional approach (Hanamura 1973) the polariton character of the propagation of the both electromagnetic waves inside the crystal is taken into account explicitly. To put it roughly, the principal corrections arising from the polariton effect are connected with exciton-photon components structure of a polariton wave as well as with large difference between polariton group velocity and "background velocity" $\frac{c}{\sqrt{\epsilon_g}}$.

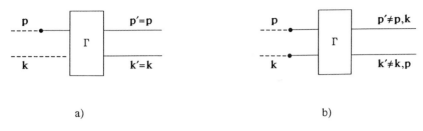

a)   b)

Fig. 1

To identify the nature of the principal differences of two results obtained, i.e. Eq.(8) and Eq.(9), let's return to the analysis of the basic elementary processes of the two-photon absorption due to biexciton formation. The schematic representation of the two different models of two-photon absorption is given in Fig.1 . In these graph the solid and dashed lines denote exciton and photon propagation, respectively, bold dot denotes exciton-photon polariton transition, whereas the square Γ represents the biexciton state. According to the traditional conception (see graph a)) the two photon absorption is determined by matrix element $M_1(p, k)$, which, in turn, is given by

$$\frac{1}{\sqrt{V}} M_1(p, k) = \sum_q \langle 0 | A_{p+k} \left[-i \frac{\Omega_c}{2} B_q^\dagger \alpha_q \right] B_p^\dagger \alpha_k^\dagger | 0 \rangle , \qquad (10)$$

where $|0\rangle$ denotes the ground state of a semiconductor. Treating the biexciton excitation as a two-exciton complex

$$A_1 = \frac{1}{\sqrt{2V}} \sum_q \Psi(q) B_{-q+\frac{1}{2}} B_{q+\frac{1}{2}}$$

and fulfiling the proper convolution of the pairs of operators in the right side of Eq.(10) we immediately find the expression (6) for matrix element $M_1(p, k)$. But it is easily to see that such procedure implies that as the result of the corresponding two-photon absorption process the two constituent excitons of the created biexciton have the same wave vectors p and k as initial photons. According to the well-known Hopfield conception of the crystal absorption it is not act of the true absorption! For the true absorption in such picture it is necessary to introduce the real scattering process. Roughly speaking, just the process of the exciton-exciton interaction is responsible for the real two-photon absorption in our approach to the problem (see graph b)).

In accordance with formulas (8) and (9) the two-photon absorption coefficient $K^{(2)}_{\text{biex}}(\omega, \omega_k)$ has the proper symmetric form concerning the frequencies ω and ω_k . Moreover, the frequency band of the biexciton two-photon absorption represents the usual Lorentz counter with the central frequency $\Omega = \Omega^{\text{biex}}_{p+k} = \omega + \omega_k$ and the frequency width determined by biexciton scattering parameter γ^{biex} . So it is very interesting question to fulfil the experimental comparison of the two different expressions (8) and (9) of the biexciton two-photon absorption coefficient. The main difference between these expressions lies in the different dependences on the resonance factors $(\omega_k - \omega_t)^2$ and $(\omega - \omega_t)^2$. Moreover, the relation $R(\omega, \omega_k)$ of two coefficients (8) and (9) has the simple form :

$$R(\omega, \omega_{\mathbf{k}}) = \frac{(\omega_{\mathbf{k}} - \omega_t)^2 + (\omega - \omega_t)^2}{\epsilon^2_{\mathbf{p+k}}} \; ; \qquad (11)$$

It seems, CuBr and CuCl semiconductors are the best candidates for such possible experiments.

The authors are grateful to H. Haug and A. Mysyrowicz for helpful discussions. One of us (A.L.I.) would like to thank the Alexander von Humboldt Foundation for support.

References

Combescot M. and Combescot R. 1988 *Phys. Rev. Lett.* **61** 117.

Haug H., März R. and Schmitt-Rink S. 1980 *Phys. Lett. A* **77** 287.

Haug H. and Koch S.W. 1990 *Quantum Theory of the Optical and Electronic Properties of Semiconductors* (World Scientific) pp 272-294.

Hanamura E. 1973 *Solid. St. Comm.* **12** 951.

Hulin D. and Joffre M. 1990 *Phys. Rev. Lett.* **65** 3425.

Ivanov A.L., Keldysh L.V. and Panashchenko V.V. 1991 *Zh. Eksper. Teor. Fiz.* **99** 625.

Ivanov A.L. and Panashchenko V.V. 1991 *Zh. Eksper. Teor. Fiz.* **99** 1579.

Keldysh L.V. 1972 *Problems in Theoretical Physics* (Nauka Press) p 433.

May V., Henneberger K. and Henneberger F. 1979 *Phys. Stat. Sol. (b)* **94** 611.

Mysyrowicz A., Hulin D., Antonetti A., Migus A., Massilink W.T. and Markoc H. 1986 *Phys. Rev. Lett.* **56** 2748.

Rashba E.I. 1974 *Fiz. Tekh. Poluprovodn.* **8** 1241.

Schmitt-Rink S. and Chemla D.S. 1986 *Phys. Rev. Lett.* **57** 2752.

Inst. Phys. Conf. Ser. No 126: Section V
Paper presented at Int. Symp. on Ultrafast Processes in Spectroscopy, Bayreuth, 1991

Exciton–exciton inelastic collision dynamics in CuBr microcrystals

G.Tamulaitis, R.Baltramiejūnas, S.Pakalnis

Institute of Physics, Lithuanian Academy of Sciences,
Goštauto 12, 2600 Vilnius, Lithuania

A.I.Ekimov

Physical-Technical Institute, Academy of Sciences of the USSR,
Sankt-Peterburg

ABSTRACT: Time evolution of optical nonlinearity due to exciton-exciton interaction was investigated in CuBr microcrystals after a band-to-band photoexcitation. Fast decaying damping of the $Z_{1,2}$ exciton line was observed and interpreted as a result of an elastic exciton-exciton interaction. Using the relationship between the exciton density dynamics and the time evolution of transient absorption characteristics, the constant of inelastic exciton-exciton interaction was evaluated, the latter effect being proved as prevailing in the decay of photoexcited excitons at appropriate excitation conditions.

1. INTRODUCTION

Recently the optical nonlinearities of semiconductor-doped glasses attract considerable attension mainly due to short relaxation times. The physical origin of these nonlinearities depends on microcrystal radius, temperature and excitation level. Side by side with unique effects caused by quantum confinement of photoexcited quasiparticles in small microcrystals (Ekimov et al 1985), new aspects of the nonlinearities due to excitation of bulk-like electron-hole plasma or exciton gas were observed when the microcrystal radius significantly exceeds the Bohr radius of exciton. At moderate excitation intensity and sufficiently low temperatures, the nonlinearities are caused by an exciton system. At resonant exciton excitation (Henneberger et al 1988, Masumoto et al 1988), the saturation effect of exciton absorption band prevailes. In this paper we report the observation of time behaviour of the excitonic nonlinearities at band-to-band excitation.

2. EXPERIMENTAL

The experiments were carried out on 0.3 mm thick samples of a glass matrix with CuBr microcrystals, their average radius (a=140 Å) being sufficiently large to excite many-exciton system. Excite-and-probe method was employed. The third harmonic of the Stokes component of $KGd(WO_4)_2:Nd^{3+}$ laser radiation (quantum energy 3.05 eV, FWHM duration 5 ps) was used for excitation, and the samples were probed by the radiation of parametric generator, pumped by the second harmonic of the same laser. Our experiments were carried out on the samples when exposed to a high fluence of laser radiation, after permanent photodarkening and the related processes (Roussignol et al 1986) are over.

3. TRANSIENT EXCITONIC ABSORPTION

An absorption edge of an unexcited sample is built up by two bands evidently caused by the $Z_{1,2}$ and Z_3 exciton absorption. The $Z_{1,2}$ exciton with binding energy of 107 meV represents the state with the lowest electron-hole pair energy in CuBr crystals, and manifests itself in the linear absorption spectrum of our sample as a 100 meV wide band. The excitation-induced transformation of the

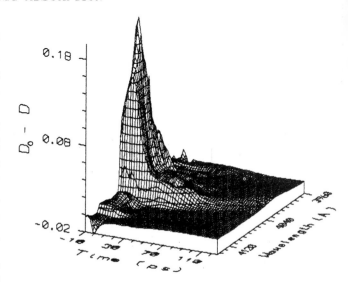

Fig. 1. Time behaviour of transient optical density in the region of $Z_{1,2}$ exciton.

band in time domain is displayed in Fig.1, where the difference between optical densities of the unexcited sample (D_o) and the excited one (D) is presented ($D=\lg(I_o/I)$, I_o and I the intensities of incident and transmitted light, correspondingly). The absorption coefficient of such a composite material as small semiconductor microcrystals in a glass matrix is mainly governed by the absorption of the semiconductor. Therefore, Fig.1 presents the time behaviour of the $Z_{1,2}$ exciton band of CuBr. The bleaching in the vicinity of the band peak and the increasing absorption in the remote spectral regions have to be emphasized as the most striking spectral features, proving the damping character of the band transformation. Assuming the Lorentzian shape of the band, transient absorption can be expressed as

$$\Delta\alpha = const \frac{(\omega_o-\omega)^2 - \Gamma^2 - \Gamma\cdot\Delta\Gamma}{((\omega_o - \omega)^2 + (\Gamma + \Delta\Gamma)^2)((\omega_o - \omega)^2 + \Gamma^2)} \Delta\Gamma, \quad (1)$$

The expression determines the relationship between the

excitation-induced absorption (transmission) and the changes of damping parameter $\Delta\Gamma$ for peak position ω_o and damping parameter Γ available from linear-absorption spectrum.

The spectrum of nonlinear absorption in the region of the $Z_{1,2}$ exciton peak is illustrated in Fig.2, where the experimental points at zero-delay and the calculation results with $\Delta\Gamma=20$ meV are presented. A good coincidence of both spectra confirm the damping model of exciton band transformation being appropriate.

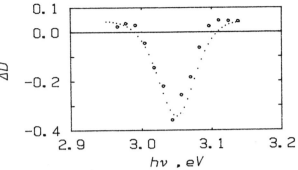

Fig 2. Spectrum of zero-delay transient absorption (o), and the results of the corresponding calculation (·).

4. EXCITON DENSITY DYNAMICS

The most probable mechanism of the damping in our experimental conditions is an elastic exciton-exciton interaction. The latter depends on exciton density n_x. We treated the $\Delta\Gamma$ versus n_e dependence in terms of the classical collision theory, since the criterion for the classical approximation (Thomas et al 1976)

$$\left[\frac{3}{2}\frac{M_x}{\mu_x}\frac{k_B T}{E_x}\right]^{1/2} \cdot N \gg 1 \qquad (2)$$

is satisfied. For the values of $M_x = 1.4 \cdot m_o$ and $\mu_x = 0.23 \cdot m_o$ (m_o is an electron mass) for translational and reduced masses of the exciton with binding energy $E_x = 107$ meV, the coefficient before N at room temperature equals nearly 2. Here N represents a typical impact parameter in units of the Bohr radius, so for N=2 the inequation is satisfactorily fulfilled. Then, in accordance with classical theory of gasses, the direct proportionality

$$\Delta\Gamma \approx 2h\ a_B^2 (k_B T/M_x)^{1/2} n_x \qquad (3)$$

is simply derived from. The coefficient of the proportionality is rather rough and shows only an order of magnitude of exciton density, however, the proportionality itself allows to judge about the n_x decay from absorption dynamics.

Two recombination mechanisms of a dense exciton system have been proposed by Benoit a la Guillaume et al (1967) and used afterwards to interpret the luminescence spectra by many other authors: i) inelastic exciton-exciton interaction, when one of the interacting excitons annihilates emitting a photon, while the other changes its energy and quasimomentum according to conservation laws, ii) exciton-electron collision with radiative annihilation of the exciton. The invariability of

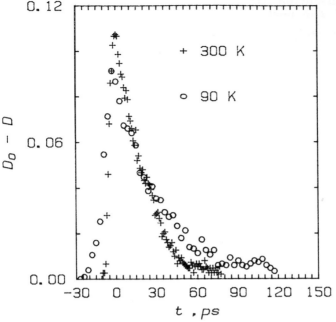

Fig. 3. Time evolution of the maximum optical density.

decay characteristics with temperature decrease from 300 K, when the exciton concentration exceeds the electron one by less than an order of magnitude, down to 90 K, when practically all electron-hole pairs in CuBr are bound, proves the first mechanism as more probable (see Fig.3). So the rate equation for exciton density at our experimental conditions can be expressed as

$$\frac{dn_x}{dt} = G - \frac{n_x}{\tau} - C \cdot n_x^2$$

where G is the generation term, τ linear relaxation time, and C reaction constant of inelastic exciton-exciton collision. Just after the excitation, the inelastic exciton-exciton collisions being the dominating recombination mechanism, the rate equation reduces to a simple form: $dn_x/dt = -C \cdot n_x^2$. Thus, constant C becomes the only fitting parameter when describing the time evolution of exciton concentration. Therefore, following the procedure proposed by Ugumori et al (1982), we evaluate C extracting n_x versus time characteristic from the time evolution of transient optical density as described above. The value of $5 \cdot 10^{-8}$ cm^{3}s^{-1} for C is in agreement with the results obtained by few other authors in different materials, that are dispersed in the range from 10^{-10} to 10^{-6} cm^{3}s^{-1}. Such a wide range is caused more by the indirect methods than by the investigated materials, so far. So the further investigation of the reaction constant of inelastic exciton-exciton interaction seems to be useful for better understanting of the behaviour of dense exciton gas.

Benoit a la Guillaume et al (1967) Phys. Rev. **177** 567
Ekimov et al (1985) Solid State Commun. **56** 921.
Henneberger et al (1988) Appl. Phys. B **46** 19
Masumoto et al (1988) Appl. Phys Lett. **53** 1527
Roussignol et al (1986) J. Opt. Soc Am. B. **4** 5
Thomas et al (1976) Phys. Rev. B. **13** 1692.
Ugumori et al (1982) Jpn. J. Appl. Phys. **21** 1588

Ultrafast polarization switching in ridge-waveguide laser diodes

A.Klehr, A.Bärwolff, R.Müller, G.Berger,
Abt. Halbleiterlaserphysik, Zentralinstitut für Optik und Spektroskopie,
Rudower Chaussee 5-6, D-1199 Berlin, FRG
J.Sacher, W.Elsässer, E.O.Göbel
Fachbereich Physik der Philipps Universität Marburg,
Renthof 5 ,D-3550 Marburg, FRG

ABSTRACT : Time resolved polarization characteristics of 1.3 μm GaInAsP/InP ridge-
waveguide (RW) lasers are presented, showing switching behaviour between the TE-
and TM-mode as well as polarization bistability. A switching time of about 50 ps is
observed for the transition between TE- and TM-polarization under 900 MHz current
modulation. The influence of the lateral waveguide structure on the emission properties
of the RW-laser is theoretically analysed with a beam propagation method.

In recent years polarization switching and bistability have been demonstrated in solitary
lasers. Chen and Liu (1985) observed at low temperature a switching time of 200 ps
between the two polarization modes of a BH - laser driven by fast rising current pulses of
high amplitude. Linton et al (1988) found a room temperature switching time of 250 ps for
the transition between the TE- and TM-polarization modes in a twin - stripe laser with
separate current injection into the two channels.
Klehr et al (1990) demonstrated polarization switching and bistability in a solitary 1.3 μm
ridge waveguide (RW) -laser at room temperature. A transition time of 700 ps between TE
and TM were observed.
This letter reports bistable ultrafast polarization switching in InGaAsP/InP RW-laser diodes
at room temperature.
The 1.3 μm double heterostructure RW-lasers (Figure 1) employed were grown by the conventional liquid-phase epitaxy technique. Basically they consist of a 0.15 μm thick quarternary active and a 0.2 μm thick quarternary etch-stop layer of composition $In_{1-x}Ga_xAs_yP_{1-y}$
($x=0.12$; $y=0.27$). Whereas commonly used RW-lasers have ridges of rectangular shape,
see inset (a) in Figure 1, the cross section of the 3.5μm wide ridges of the lasers employed
in our experiments exhibit a "waist structure" composed of a trapezoidal bottom part and
a rectangular upper part,inset (b). The lateral effective index-profile (along the y-direction)
was created by covering the parts outside the ridge structure with a SiO_2 insulator and an
overall contact metallization.
The lattice composition of the quarternary active layer was chosen to provide a substrate-
related lattice misfit $\Delta a/a = -5 \times 10^{-4} \ldots -1 \times 10^{-4}$ perpendicular to the layer plane (along the
x-direction), which corresponds to a biaxial tension in the order of 10 MPa. This tension,
however, is not sufficient to compensate for the resonator TM-losses by tension-induced
TM-gain enhancement. Figure 2 shows the light power-current characteristics (P-I) for the
total output power per facet (curve 1) and for the output power of the TE (full line) and
TM (dashed line) modes (curves 2), respectively, measured at T = 25°C. Lasing starts at
about 39 mA in the TE-mode. The total power and the TE-mode power depend linearly on

current up to about 65 mA. With a further increase of the current from 65 mA to 92 mA the emission efficiency of the TE-mode decreases accompanied by a slow rise of the TM-mode power. Simultaneously, a broadening of the longitudinal modes of the TE- spectrum is observed in the respective current regime indicating that TE-mode emission tends to become unstable. Finally, the polarization changes abruptly from TE to TM at the switching current I_{S1} connected with an increase of the total output power by about 0.7 mW. Just

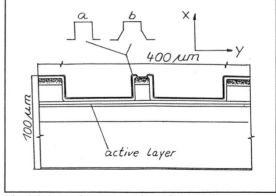

Fig.1 Schematic cross section of a RW-laser perpendicular to the direction of wave propagation

below the switching point I_{S1} oscillations of the TE- and TM-mode with a frequency of 4 GHz were observed as has been reported in a paper by Klehr et al (1990). Above I_{S1} stable TM-emission has been found. With decreasing injection current the polarization flipped back from TM to TE at $I_{S2} = 82$ mA. Thus, the TE-hysteresis loop (curve B) is of the clockwise type and that of the TM-emission (curve A) of the counter-clockwise type. The widths of both hysteresis cycles are almost 10 mA in the case of Figure 2. Generally,

widths of the hysteresis loops have been found to vary in the range from 2 to 20 mA.

In order to determine the switching time for the TE/TM transition a sinusoidal current modulation of 900 MHz with a modulation amplitude of 75 mA was applied to the laser diode biased with 23 mA DC (below threshold). The laser output beam was split into two parts to separate TE- from TM-polarization and afterwards the two beams were focused with the same optical path length onto a streak-camera with 10ps resolution time. The time resolved output of the polarization modes is depicted in Figure 3, showing a complete transition from TE to TM with a switching time, from peak to peak, of about 50ps

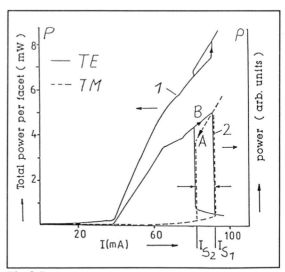

Fig.2 Power-current curves

without any pronounced relaxation oscillation. It is expected that the transition time could be even shorter for higher modulation frequencies or a faster rise time of the switching current. According to these results it seems possible that the polarization of laser emission can be modulated in a multigigahertz range with a high extinction rate. We would like to point out, that only weakly index-guided lasers with the waist ridge and the parameters mentioned above provide reproducible polarization bistability whereas lasers with a commonly used rectangular ridge do not exhibit this effect. In order to understand the reason for their different properties the refractive index profiles, responsible for the guiding of the lightwaves, as well as the stationary intensi-

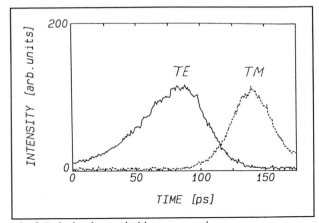

Fig.3 Polarization switching versus time

ties of the TE- and TM-modes have been calculated for both types of lasers using the parameter values given in Table 1. Figure 4 shows the lateral effective-index profiles (left: rectangular ridge; right: waist ridge) which depend on the lateral carrier distribution, along the y-axis, according to an injection current of 80 mA. Current spreading, spontaneous carrier recombination and carrier-induced antiguiding have been considered in the numerical calculations. Gain saturation has not been taken into account, however. It may be deduced from Figure 4 that waveguiding of TE-polarized light is worse in lasers with a

waist ridge due to a smaller index step, i.e. the difference between the maximum and the minimum on the left and right of the index profile, compared to lasers with the rectangular ridge. For TM polarization a decrease in the index step by almost the same amount is found. Nevertheless, TM-waveguiding may be improved in lasers with a waist ridge because the maximum of the corresponding index profile is well above the index level outside the ridge (marked Δn in Figure 4).

emission wavelength	1.3 μm
stripe width	3.5 μm
eff. diffusion coeff.	33.0 cm^{2}s^{-1}
nonradiat. recomb. rate	5.0 ns
radiat. recomb. coeff.	6.3x10^{-9} cm^{3}s^{-1}
Auger recomb. coeff.	4.7x10^{-29} cm^{6}s^{-1}
eff. linear recomb.time	3.0 ns
internal loss	50.0 cm^{-1}
laser length	200.0 μm
antiguiding factor	-6.0
linear gain coeff.	2x10^{-16} cm^2
carrier dens.at transp.	1x10^{18} cm^{-3}
transv.confin.factor TE	0.394
transv.confin.factor TM	0.373
reflectivity TE	0.386
reflectivity TM	0.211

Table 1 Parameters used for calculation

Thus, the TM-mode should experience a higher net gain in lasers with a waist ridge. This is also supported by Figure 5 (top: rectangular ridge; bottom: waist ridge) representing the behaviour of stationary output light intensities along the lateral y-axis for TE- or TM-lasing operation, respectively, where the same injection current as in Figure 4 has been assumed. The results have been obtained by the use of the beam-propagation method (BPM) including gain saturation. All parameter values have been taken to be the same for TE- and TM-mode emission except for the facet reflectivities and the transverse confinement factors.

From this numerical example it becomes evident that the special structure of the ridge used in the experiments is more suitable to achieve TM-mode emission than a ridge of rectangular shape. Therefore, one may suggest that a small increase of the TM-gain coefficient relative to the TE-gain coefficient, as caused by internal stress, is sufficient to obtain polarization switching TE/TM in a laser with a waist ridge. However, further experimental and theoretical investigations are needed to understand in detail the physical mechanism behind polarization switching and bistability observed in the aforementioned experiments.

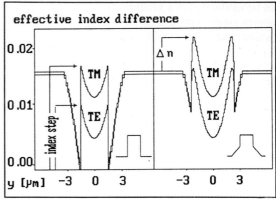

Fig.4 Effective-index profiles along the lateral axis

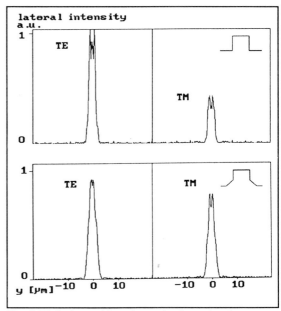

Fig.5 Light intensities along the lateral axis

References

Chen Y.C. and Liu J.M. 1985 Appl.Phys.Lett. **46** pp 16-18
Klehr A., Rheinländer B., Ziemann O. 1990 Int.Journ.of Optoelectr. **5** pp 513-22
Linton R.S., White J.H., Carroll J.E., Singh J., Adams M.J., Henning J.D. 1988
 Electr.Lett. **24** pp.1232-33

Inst. Phys. Conf. Ser. No 126: Section V
Paper presented at Int. Symp. on Ultrafast Processes in Spectroscopy, Bayreuth, 1991

Generation of picosecond pulses by optically pumped short- and ultra-short-cavity semiconductor lasers (theory)

Yu D Kalafati, V A Kokin

Institute of Radioengineering & Electronics of USSR Academy of
Sciences. 18 Marx avenue, Moscow 103907, USSR.

The electron-hole plasma reheating effect on picosecond dynamics of
semiconductor lasers is studied theoretically. It is shown that
reheating effect leads to shoulders on the long wavelength pulses, and
that pulse duration decreases as the wavelength decreases. The analytical expressions for pulse duration and plasma cooling time are obtained.

Over the past 10 years there has been a continued interest in studying the
generation of picosecond pulses from optically pumped semiconductor laser.
During the optical pumping high temperature electron-hole (e-h) plasma is
generated. Koch et al (1982) and Wiesenfeld and Stone (1986) have shown
that under this conditions laser action in ultrashort- and short-cavity
semiconductor lasers has some peculiarities:
1. Generation of λ - shaped short wavelength pulses.
2. Shoulders of the long wavelength pulses.
3. Time-varying wavelength (chirp) in the generated pulses.
Koch et al (1982) carrying out the numerical modeling of picosecond carrier
and lasing dynamics in optically pumped semiconductor laser came to a
conclusion that reheating of e-h plasma due to a preferential elimination
of lower energy carriers could lead to shoulders on the long wavelength
pulses. However, to the best of our knowledge, there is no self-consistent
theoretical explanation of all the effects mentioned above. In the present
paper we suggest theoretical approach that enable to obtain the analytical
expression for pulse duration and to explain the experimental results.

We have investigated the reheating effect numerically and analytically
taking also into account the reheating of plasma due to intraband
absorption of laser emission. To describe the picosecond dynamics of lasing
and of hot e-h plasma we have considered the equations which are a
bookkeeping of the rate of supply, annihilation and creation of e-h pairs
n, their energy W, and photons N inside the laser cavity. In general this
equations are integro-differential due to the broadband nature of the
semiconductor laser output:

$$\frac{d}{dt} n = c \sum_i \alpha(\Omega_i) N_i \, , \tag{1}$$

$$\frac{d}{dt} W = \alpha' B + c \sum_i \alpha(\Omega_i)(\hbar\Omega_i - E'_g) N_i - nJ, \tag{2}$$

$$\frac{d}{dt} N_i = (c\,\alpha(\Omega_i) + \tau_p^{-1}) N_i + g_i, \tag{3}$$

where c - the velocity of light in semiconductor, $\alpha(\Omega)$ - the interband gain (absorption) coefficient of the light in semiconductor, Ω_i- the frequency of the i-th longitudinal mode, α' - intraband absorption coefficient, N_i- the number of photons in the mode, $B = c \sum_i \hbar\Omega_i N_i$ - the intensity of light in optical cavity, $E'_g = E_g - n \, \partial E_g/\partial n$, E_g - the semiconductor band gap; $J(n,T,T_L)$ - the rate of plasma energy loss (per e-h pair) due to interaction of electrons and holes with optical phonones, T is the temperature of e-h plasma and T_L - crystal lattice temperature, τ_p - photon lifetime in optical cavity, g_i -

the rate of spontaneous emission into the mode. In the case of ultrashort cavity the equations (1)-(3) become more simple due to the fact that only one mode must be taken into account.

We have solved this equations numerically with initial conditions $n = 5 \cdot 10^{18}$ cm^{-3}, $T = 1200$ K, $T_L = 300$ K, $N = 0$. This initial conditions occur in experiments as a result of optical pumping of semiconductor laser by intense subpicosecond pulse. For numerical solution we have used $\tau_p = 1$ ps, $\alpha' = \sigma n$, with $\sigma = 10^{-17}$ cm^2. Other parameters have been chosen similar to those in GaAs. The results of numerical solution are shown in the Figure 1.

As it follows from the Figure 1 there are two characteristic time scales in the dynamics of lasing and plasma cooling – "ultrafast" and "fast" (we don't discuss here the "slow" period of laser dynamics that begins when the plasma temperature became equal to the lattice one). Kalafati and Kokin (1991) carried out the analytical consideration of the equations (1)-(3) based on the assumption that $\tau_p \ll \tau_T$ (where τ_T - the characteristic plasma cooling time). Following this approach it is possible to show that during the "fast" period the laser dynamics can be described by the only equation:

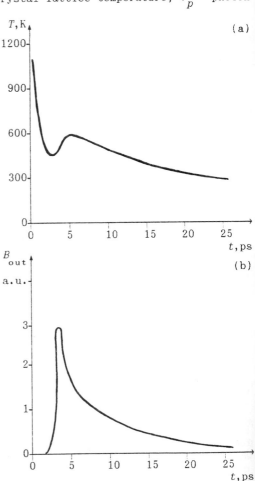

Figure 1. The dependence of plasma temperature 1(a) and laser output 1(b) on time, obtained from numerical solution of equation (1)-(3).

$$\frac{3}{2} k \frac{d}{dt} T = - \frac{J}{\beta} \qquad (4)$$

where $\beta = \beta_1 + \beta_2$ - the reheating dynamic factor; β_1 describes reheating effect that occur due to elimination of lower energy carriers and β_2 - due to intraband absorption of laser emission:

$$\beta_1 = \frac{5}{2} \left[\frac{\mathcal{F}_{3/2}(\eta_G^e)}{\mathcal{F}_{1/2}(\eta_G^e)} + \frac{\mathcal{F}_{3/2}(\eta_G^h)}{\mathcal{F}_{1/2}(\eta_G^e)} \right] + (\hbar\Omega^* - E_g) \frac{dn_G}{dT} \qquad (5)$$

$$\beta_2 = \alpha' c \tau_p \hbar\Omega^*/kT \qquad (6)$$

$\mathcal{F}_j(\eta)$ - Fermi integrals, $\eta_G^{e,h} = \mu_G^{e,h}/kT$; $\mu^{e,h}$ - Fermi energies of electrons and holes; Ω^* - emission frequency which for ultrashort-cavity laser coincides with the frequency of the only mode Ω_1 and for short-cavity laser corresponds to the frequency at which the gain coefficient is maximum; subscript G denotes values corresponding to the threshold. During "fast" period the concentration of carriers and the intensity of laser output B_{out} become the functions of plasma temperature:

$$n = n_G(T); \quad B_{out} = \frac{1}{2} c\tau_p \frac{\hbar\Omega^*}{kT} n_G(T) \ln(\frac{1}{R}) \frac{J}{\beta} \qquad (7)$$

where R - mirror reflectivity. The n_G can be found from equations:

$$- \alpha(\Omega^*) c \tau_p = 1; \quad n_G(T) = N_C \mathcal{F}_{1/2}(\eta_G^e) = N_V \mathcal{F}_{1/2}(\eta_G^h), \qquad (8)$$

where N_C, N_V - densities of states in conductive and valence bands. In the case $m_e \ll m_h$, $\tau_e \gg \tau_h$ (m_e, m_h - electron and hole effective masses, τ_e, τ_h - carrier-lattice energy relaxation times for electron and hole subsystems) and for high plasma temperature when the distribution function of holes remains nondegenerate the equation (4) for T reads:

$$\frac{d}{dt} T = - \frac{T - T_L}{\tau_T} \qquad (9)$$

and

$$B_{out} = \frac{3}{4} c E_g \frac{\tau_p}{\tau_T} n_G(T) \frac{T - T_L}{T} \ln(\frac{1}{R}),$$

where

$$\tau_T = (\beta_1 + \beta_2) \tau_h, \qquad (11)$$

As it follows from (9)-(11) and Figure 1 the τ_T determines the rate of e-h plasma cooling and the decay of lasing. In the present paper we consider the dependence of τ_T or, to be more precise, of β on plasma temperature, emission frequency and semiconductor laser parameters. In the case of short-cavity GaAs laser formula (11) reads:

$$\tau_T \approx (6.2 + 0.34 \tau_p \sqrt{T} E_g) \tau_h, \qquad (12)$$

where τ_T, τ_p and τ_h are measured in ps, T in K and E_g in eV. For $\tau_p = 1$ ps and $T = 500$ K we can find that $\tau_T \approx 16.6\tau_h$. According to results of Kumekov and Perel (1988) for $T = 500$ K, $T_L = 300$ K and concentration of carriers in

GaAs $3 \cdot 10^{15}$ cm^{-3} we have $\tau_h \approx 0.8$ ps and consequently $\tau_T \approx 13.2$ ps. This value is in good agreement with experimental one. It must be noted that in this case the assumption $\tau_T \gg \tau_p$ is fulfilled. The fulfillment of the second inequality $\tau_T \gg \tau_h$ means the considerable slowing down of e-h plasma cooling rate in the presence of stimulated emission.

Up to now we discuss the high temperature limit in calculation of the dynamic reheating factor. The numerical solution of algebraic equations (8) make it possible to find β for the arbitrary value of plasma temperature. The results of this solution for short-cavity GaAs laser with τ_p = 1 ps are shown in the Figure 2. The analytical formula (14) is in good agreement with numerical results down to 100 K.

In the case of ultrashort-cavity semiconductor laser the dynamic reheating factor β depends on temperature and on the frequency of the laser mode. For high temperatures and long wavelength modes ($\hbar\Omega_1 \approx E_g$) the analytical formula (12) for β is valid. The results of numerical calculation of β for ultrashort-cavity GaAs laser with $L = 0.2$ μ and $R = 0.98$ (L - cavity length) are shown in the Figure 3. As it follows from Figure 3 the pulse duration decrease as the frequency of the laser mode increases. This dependence is in good agreement with the experimental results obtained by Wiesenfeld and Stone (1986).

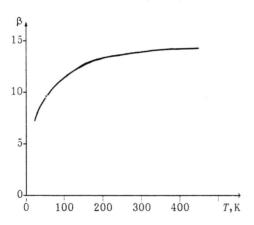

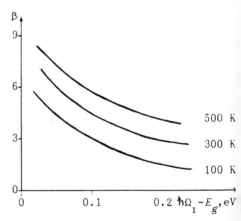

Figure 2. The dependence of β on plasma temperature for short-cavity

Figure 3. The dependence of β on laser mode frequency for ultrashort-cavity semiconductor laser.

It should be added that chirp in short-cavity semiconductor lasers occur due to the change of maximum of gain coefficient with time and in ultra-short-cavity semiconductor laser due to the change of refractive index.

REFERENCES

Kalafati Yu D, Kokin V A 1991 Zh. Exsp. Teor. Fiz. **99** 1793 [Sov.Phys.JETP **72** N6]
Koch T L, Chin L C, Harder Ch and Yariv A 1982 Appl.Phys.Lett. **41** 6
Kumekov S E, Perel V I 1988 Sov.Phys.JETP **67** 193
Wiesenfeld J M, Stone J 1986 IEEE J. Quantum Electron. **22** 119

Inst. Phys. Conf. Ser. No 126: Section V
Paper presented at Int. Symp. on Ultrafast Processes in Spectroscopy, Bayreuth, 1991

449

Picosecond photodetectors fabricated on low temperature GaAs

M. Klingenstein, J. Kuhl, R. Nötzel, K. Ploog,
Max-Planck-Institut für Festkörperforschung, Heisenbergstr.1, 7000 Stuttgart 80, Germany.
J. Rosenzweig, C. Moglestue, Jo. Schneider, A. Hülsmann, K.Köhler,
Fraunhoferinstitut für Angewandte Festkörperphysik, Tullastr.72, 7800 Freiburg, Germany.

ABSTRACT: GaAs metal-semiconductor-metal photodiodes fabricated on GaAs grown at low substrate temperatures (200 °C) have been investigated in the time domain by electro-optic sampling. It could be shown, that these diodes have a faster response and a considerably reduced long time tail. They can be used at larger bias than comparable diodes produced on GaAs grown at 700 °C. Temperature dependent measurements show, that the tail can be described by hopping conductivity and disappears below 50 K.

1. INTRODUCTION

Interdigitated Metal-Semiconductor-Metal (MSM) diodes fabricated on GaAs are attractive photodetectors for use in lightwave communication at 0.8 µm (Sze et al. 1971, Roth et al. 1985). Their most promising advantages are a broad bandwidth, a high quantum efficiency, a small leakage current, and compatibility with GaAs field effect transistors. The key features providing the broad bandwidth of these devices are the low capacitance associated with the small active area and the rapid carrier sweep out obtained by narrow electrode spacing. Recently, we demonstrated that the diode response of MSM-diodes with an area of 10 µm x 10 µm and an electrode spacing of 0.75 µm is limited by the transit time of electrons and holes (Klingenstein et al. 1991). Two dimensional Monte-Carlo (MC) calculations (Moglestue et al. 1991, Rosenzweig et al. 1991) provide a quantitative description of the output current pulse shape except for a long tail of approximately 20% of the peak amplitude lasting 200-300 ps which we attribute to low frequency gain. Two possibilities remain to decrease the response time of these diodes. The first is decreasing the finger distance, which reduces the transit time of the carriers to the contacts (Chou et al. 1991) but increases the capacitance. The second possibility is fast trapping of the photogenerated carriers before they can reach the electrode (Chen et al. 1991) with the disadvantage of reduced sensitivity. Moreover the trapping should reduce the tail mentioned above, because carriers, which are injected from the contacts, are trapped almost instantaneously and can not reach the other contact. GaAs, grown at very low substrate temperatures of 200 °C (LT-GaAs) has been proven to provide a high density of deep traps. Therefore this material is a high resistivity photoconductor (Campbell et al 1990) with carrier recombination times of approximately 1 ps (Frankel et al. 1990).

In this paper we present response measurements of MSM-photodiodes fabricated on LT-GaAs and integrated in a coplanar transmission line. The experimental data are compared to photocurrent pulse shapes obtained from similar diodes on material grown at high temperature 700 °C (HT-GaAs). As we expected, the FWHM of the photocurrent pulse is reduced by a factor of almost 2 and the long tail in the LT-diode photocurrent is significantly smaller. Furthermore, we compare LT-GaAs diode signals to signals received from direct excitation between the contacts of a coplanar stripline on the same material to investigate whether the particle or the displacement current yields the dominant contribution to the photocurent output. Thereby we could also determine the influence of the diode capacitance and structure on the signal. Temperature dependent measurements between 10 K and 300 K identify hopping conductivity to be responsible for the tail of LT-GaAs MSM diodes.

2. EXPERIMENTAL

The diodes have been fabricated by means of electron beam lithography on LT-GaAs using Ti/Pt/Au as contacts. The 2.5µm thick LT-GaAs layer was grown at 200 °C substrate temperature and annealed for 10 min at 600 °C. The diodes under investigation had an active area of 10 µm x 10 µm and 0.75 µm wide fingers with 0.75 µm distance beween them. Electro-optic sampling with a small LiTaO$_3$ tip (Valdmanis 1987) was used to measure the diode response. For the low temperature measurements we used diodes with a finger distance of 1.5 µm and 1 µm finger width mounted in a He-gas-flow cryostate. In this case the signal was sampled photoconductively (Auston 1984) using the LT-GaAs itself as a fast photoconductor. Excitation and sampling were performed using the 100 fs optical pulses derived from a dispersion-compensated colliding-pulse-mode locked (CPM) dye laser (wavelength 620 nm, repetition rate 120 MHz).

3.MEASUREMENTS AND DISCUSSION

Figure 1 shows the response of a MSM-diode on LT-GaAs (solid line) illuminated by 2.4 pJ pulse energy at 4 V bias. We found a risetime of 1.3 ps, a 1/e decaytime of 2.5 ps for the first rapid decay and a full width at half maximum (FWHM) of 3.3 ps corresponding to 70 GHz bandwidth (residual tail neglected). These characteristic times remain unchanged for a variation of the pulse energy from 0.6 pJ-24 pJ and of the voltage from 1 V-6 V. The signal amplitude is approximately

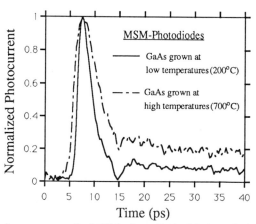

Fig.1: Normalized photocurrent of a MSM-photodiode fabricated on LT-GaAs (solid line) (0.75µm fingerdistance / 4V bias / 2.4pJ pulse energy) and on HT-GaAs (dotted line) (0.5µm fingerdistance / 3V bias / 0.3pJ pulse energy) after excitation by a 100fs pulse at 620nm vs. time.

proportional to the voltage and the intensity in the ranges given above. The dip, which is observed at t=15 ps, originates from the electro-optic sampling and is probably due to internal reflections in the tip. The fastest response obtained from a MSM-diode fabricated on HT-GaAs is depicted by the dotted line in Fig.1. The latter diode had a finger distance of 0.5 μm and was excited with a pulse energy of 0.3 pJ at 3 V bias. The risetime was 2.2 ps, the 1/e time for the first decay 4.8 ps and the FWHM 6 ps. The curves in Fig.1 reveal the faster response time and the lower long time tail of the diode fabricated on LT-GaAs.

The reduction of the FWHM of the pulse from the LT-GaAs diode compared to that of the HT-GaAs diode originates from efficient trapping of the carriers in the LT-GaAs on a timescale shorter than the carrier transit time. Therefore only part of them can reach the electrodes resulting in a reduced sensitivity of LT-GaAs diodes. For optimum operation conditions (3 V bias / pulse energy < 0.6 pJ) the diode on HT-GaAs produces a signal amplitude of 200 mV/pJ pulse energy (for optical pulses short compared to the FWHM of the electrical response) which already decreases to 115 mV/pJ at 2.4 pJ energy because of field screening effects (Moglestue et al. 1991, Rosenzweig et al. 1991). The LT-GaAs diode has a sensitivity of 85 mV/pJ at 0.6 pJ energy, which goes down slightly to 65 mV/pJ at 2.4 pJ and 50 mV/pJ at 24 pJ.

The externally measured photocurrent can be divided in two parts.

$$I(t) = \int_{C.A.} d\vec{f} \left(\vec{j(t)} + \varepsilon_0 \frac{\partial \overrightarrow{E(t)}}{\partial t} \right)$$

Here C.A. is the contact area, j and E are the current density and the electric field, respectively. The first term in the integrand is the particle current, which is caused by the carriers recombining at the electrodes. The second term is the displacement current caused by the change of the internal electric field. To investigate which part provides the main contribution to the response of the LT-GaAs diode, the photodiode pulse was compared to the shape of a photocurrent pulse excited in the gap between the conductors of a coplanar stripline with 10 μm distance and 5 μm width (sliding contact excitation) (Frankel et al. 1990). The bias between the transmission lines of 80 V has been chosen so that the average electric field is the same as in the diode (0.75 μm / 6 V) and the electric signal was again analysed by electro-optic sampling. The result of this

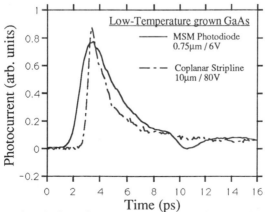

Fig.2: Photocurrent signal of a MSM-diode (solid line) (0.75μm fingerdistance / 6V bias) and a coplanar stripline (dotted line) (10μm / 80V bias) vs. time. The elements were fabricated on LT-GaAs and illuminated by 24 pJ pulse energy.

measurement is shown in Fig.2 (dotted line). The risetime of 600 fs is almost limited by the time resolution of the electro-optic sampling and the optical pulse width. The FWHM of this photocurrent signal is 1.4 ps.

If the particle current dominates the photocurrent, only the particles at the contacts produce the signal. Therefore the response amplitude should be proportional to the contact length. Theoretical Monte Carlo calculations (Moglestue et al., 1991) show that most of the particles recombine near the edges of the contacts because of the high electric fields. Thus the width of the contact has almost no influence on the particle current and the only important measure is the length of the contact edges. Figure 3 depicts the two compared structures. Since the diode has a 6 times larger contact length than the striplines, the diode response amplitude should be 6 times larger than the amplitude of the stripline signal for dominating particle current.

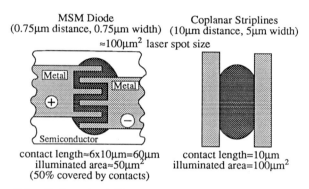

MSM Diode   Coplanar Striplines
(0.75μm distance, 0.75μm width) (10μm distance, 5μm width)
≈100μm² laser spot size

contact length≈6x10μm=60μm   contact length=10μm
illuminated area≈50μm²   illuminated area=100μm²
(50% covered by contacts)

Fig.3: Comparison of the diode and stripline structures

The change of the internal electric field is caused by the displacement of the photogenerated charges inside the device. Therefore, for dominating displacement current the signal amplitude originates from all carriers moving in the electric field. As half the area of the diode is covered by the metallic fingers, its response amplitude should be half the amplitude produced by the stripline detector.

The comparison of the photocurrent pulses from a MSM-diode (solid line) and a coplanar stripline (dotted line) both illuminated with 24 pJ and biased with 6 V and 80 V, respectively, are shown in Fig.2. The fact that both signal amplitudes are almost identical supports the assumption, that mainly the displacement current contributes to the photocurrent response of LT-GaAs diodes. This result implies that the differences between the shapes of the LT-GaAs diode and stripline pulses are mostly due to electrical parasitics and the propagation time of the electrical pulse on the electrodes of the interdigitated diode structure. The deconvolution of both signals gives an electrical diode response function having a FWHM of 1.5 ps. This result agrees fairly well with the estimated capacitance (6-10 fF) and the signal propagation time on the interdigitated electrode structure of the diode.

In order to investigate the origin of the tail for t>10-15 ps, we performed temperature dependent measurements between 10 K and 300 K. Figure 4 presents the photocurrent signals of the MSM-diode at 10 K, 150 K and 250 K using 24 pJ pulse energy. No tail is observed at 10 K. For higher temperatures the tail raises to 7.2% of the peak amplitude at 150 K and to 9.4% at 250 K. The

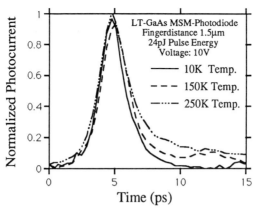

Fig.4: Normalized photocurrent of MSM-photodiode fabricated on LT-GaAs (1.5μm fingerdistance / 10V bias) for 10K (solid line), 150K (dashed line) and 250K(dotted line).

tail height is measured at t=14ps, because the slowly decaying component is clearly seperated from the pulse, there. The temperature dependence of the photocurrent can be ascribed to hopping. The change of the hopping conductivity σ_h, which is proportional to the current, with temperature is given by

$$\sigma_h=\sigma_0\times e^{-\Delta E/kT}$$

with ΔE being the energy difference between adjacent hopping states (Maden and Shaw 1988). σ_0 depends on the density of states at the Fermi energy, the phonon spectrum and the spatial overlap of the wavefunctions of the hopping states. The analysis of the photocurrent shows an activation energy of $\Delta E \approx 7$ meV over the whole temperature range down to 50 K. This value is much smaller than the activation energies obtained from dark current measurements (Kaminska and Weber 1990) which we could reproduce with our samples. We believe, that this discrepancy is due to the strong and highly transient non-equilibrium occupation of deep traps during the first picoseconds after the excitation by ultrashort optical pulses. Therefore hopping processes of photoexcited carriers from occupied trap states which are energetically closer to the empty states are very likely in this transient regime. The hopping conductivity continously decreases when the carriers relax to deeper traps.

4. CONCLUSIONS

We have fabricated interdigitated MSM photodiodes on GaAs grown at low temperatures of 200 °C. These diodes have been compared to similar detectors produced on HT-GaAs. LT-GaAs diodes have a faster response (3.3 ps FWHM) and a lower long time tail but a smaller sensitivity compared to diodes on HT-GaAs (6 ps FWHM). We have shown that mainly the displacement current contributes to the diode signal in the LT-GaAs diodes in accordance with the fact that the carrier trapping times are shorter than the carrier transit times in these structures. In contrast to a simple stripline photodetector the interdigitated diode structure offers the advantage of a significantly lower bias voltage. Our experiments demonstrate that 4-6 V are sufficient to achieve a sensitivity of 85 mV/pJ instead of 80 V needed in a stripline. The successful growth of device quality GaAs on top of a LT-GaAs layer (Smith et al. 1988) and the low bias permit an easy integration of LT-GaAs diodes with high speed electronic circuits.

The authors would like to thank H.Stützler, A.Schulz, S.Wührl, A.Axmann, K.Glorer, M.Lang and R.Osorio for experimental help and sample preparation and H.Grahn for the critical reading of the manuscript. The financial support of the Bundesminister für Forschung und Technologie is gratefully acknowledged.

REFERENCES

Auston D.H.1984 in "Picosecond Optoelectronic Devices", ed. C.H.Lee, Academic, Orlando, FL, 1984, pp.73- 117

Campbell A.C., Crook G.E., Rogers T.J. and Streetman B.G. 1990, J. Vac. Sci. Technol. B8 (2), 305

Chen Y., Williamson S. and Brock T.1991, CLEO Proceedings, C-PDP-10, p.590, Baltimore/Maryland 1991

Chou S.Y., Liu Y. and Fischer P.B. 1991, SPIE Proceedings Vol.1474, Optical Technol. for Signal Processing Systems, Orlando/Florida, April 1991

Frankel M.Y., Whitaker J.F., Mourou G.A., Smith F.W. and Calawa A.R. 1990, IEEE Trans. Electr. Dev. Vol.37, 2493

Kaminska M. and Weber E.R. 1990, Proc. 20th Int. Conf. on the Physics of Semiconductors (ICPS), Vol.1, 473

Klingenstein M., Kuhl J., Rosenzweig J., Moglestue C. and Axmann A. 1991, Appl. Phys. Lett. 58, 2503

Maden A. and Shaw M.P. 1988, The Physics and Applications of Amorphous Semiconductors, Academic Press , pp. 72-97

Moglestue C., Rosenzweig J., Kuhl J., Klingenstein M., Lambsdorff M., Axmann A., Schneider Jo. and Hülsmann A. 1991, J. Appl. Phys., 70 , 2435

Rosenzweig J., Moglestue C., Axmann A., Schneider Jo., Hülsmann A., Lambsdorff M., Kuhl J., Klingenstein M., Leier H. and Forchel A. 1991, SPIE Proceedings Vol.1362, 168

Roth W., Schumacher H., Kluge J., Geelen H.J. and Beneking H. 1985,IEEE Trans. on Electr. Dev. 32,1034

Smith F.W., Calawa A.R., Chen C.-L., Manfra M.J. and Mahoney L.J. 1988, IEEE Electr. Dev. Lett. vol.9, 77

Sze S.M., Coleman jr. D.J. and Loya A., 1971, Solid State Electronics, Pergamon Press (New York) , Vol.14, 1209

Valdmanis J.A. 1987, Electronics Letters 23, 1308

Inst. Phys. Conf. Ser. No 126: Section V
Paper presented at Int. Symp. on Ultrafast Processes in Spectroscopy, Bayreuth, 1991

Picosecond optical diagnostic of semiconductors and elements of integrated circuits

V V Simanovich   G I Onishchukov   A A Fomichev

Laser Center of the Moscow Institute of Physics and Technology, Institutsky per. 9, Dolgoprudny, Moscow reg., 141700 USSR

ABSTRACT: Picosecond electrooptical sampling is used for testing the dynamic of electrical potentials in elements of GaAs integrated circuits. It is shown that under the electrical pulses exitation in the element active zone by laser pulses, in the case of sampling in the exitation zone the nonlinear gyrotropy effect can severely change the response pulse shape.

Progress in development of integrated circuits on the basis of GaAs allows to produce integrated elements with a micron-size active zone and an operating frequency up to several GHz. Nowadays development of non disturbing methods of such elements diagnostic with a high temporal resolution is in progress. One of them is the electrooptical sampling in pico- and subpicosecond range based on the Pokkels effect in GaAs (Valdmanis and Mourou 1986).

In our experimental setup the picosecond laser system ARGO consisted of the CW-pumped Nd:YAG laser (λ=1.06 µm) with active mode locking and the compact optical fiber-grating compressor with the second harmonic generator is used. This system provides two optical beams with the average power of 1 W (λ=1.06 µm) and 100 mW (λ=0.53 µm) and 4 ps pulse duration. The first is used as the probe beam and the second is used for the electrical pulses generation in the tested structure. Using locking-amplifier technique with modulation of the exiting beam we can carry out measurements of statical as well as dynamical electrical potentials in various GaAs integrated elements with the amplitude sensitivity being better than 30 mV and the time resolution being better than 5 ps. The measurements were carried out on microstrip lines, integrated photodiodes and transistors with active zone of 2 to 4 µm for various values of the applied dc voltage.

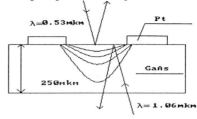

Fig. 1. "Backside" sampling geometry.

The probe radiation came through the back side of the semiconductor plate (so called "backside" geometry, see Fig.1). In this case due to the complexity of the electrical field distribution the depolarization of the probe radiation beam is determined by both longitudinal and transverse electrooptical effects as it is shown in Fig.2 (1-transverse electrooptical effect, 2-longitudinal electrooptical effect, 3-net electrooptical effect).

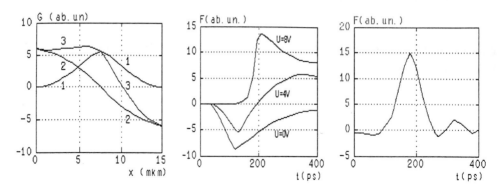

Fig.2. Dependence of the response amplitude on the probe beam position.

Fig.3. Response shapes for different value of the "pulling" field.

Fig.4. Response shape without influence of the nonlinear gyrotropy.

Measurements in the absence of special testing elements as well as structures are possible only via the electrooptical diagnostic with the electrical pulses exitation in the active zone of the tested element. It turned out that in this case when the probe beam was in the exited by light pulses semiconductor region the shape of the time response is determined not only by the electrooptical effect but also by the nonlinear surface girotropy caused by the effective mass anisotropy for the high-level exitation. In Fig.3 shapes of response pulses obtained in such way for the microstrip line are shown for various values of the applied dc voltage (external "pulling" field). If U=0 the change in the probe beam polarization is determined only by the gyrotropy induced by the exitation pulse, while for U=8V the electrooptical effect plays the key role. The light induced gyrotropy response is comparable with that one of the tested electrical pulse. They can be of the same or opposite sign depending on the probe beam position and the crystallographic axes orientation with respect to the exiting light polarization. But we should mentioned that it can be avoided by an appropriate beam adjustment. The purely electrooptical response obtained in such way is shown in Fig.4.

Linear deformation of the refractive index ellipsoid was confirmed experimentally as well as by a numerical simulation, though only the qualitative agreement between experimental and theoretical data is available because of the nonuniform rate relating the external field and the depolarization. Amplitude measurements are possible for the reevaluated rate with the account for the probe beam displacement. The amplitude of the electrical pulse has to be measured with the correction to the double electrooptical effect and the induced nonlinear gyrotropy in the active zone. The results obtained are in a good agreement with estimated values for our experimental conditions.

In addition, the method can provide measurements of such semiconductors electrophysical parameters as field distribution in active zone, carrier lifetime, transfer time, etc.

References:

Valdmanis J A and Mourou G S 1986 IEEE J. of Quantum Electr. QE-22 69

Inst. Phys. Conf. Ser. No 126: Section V
Paper presented at Int. Symp. on Ultrafast Processes in Spectroscopy, Bayreuth, 1991

Investigation of electrical pulse propagation in a n-MOS device by time-resolved laser scanning microscopy

A Krause, H Bergner, K Hempel, U Stamm

Friedrich Schiller University Jena, Faculty of Physics and Astronomy,
Max-Wien-Platz 1, O-6900 Jena, Germany

ABSTRACT: The propagation of an optically generated electrical pulse in a chain of n-MOS inverters is investigated by means of a laser scanning microscope with an actively modelocked argon-ion laser as light source and an OBIC-stage. The pulse duration of the generated pulse and the transition times between two inverters are determined. A simple approach of the mechanism of optical switching of a blocked n-MOS inverter is presented.

1. INTRODUCTION

The development of highly integrated high-speed semiconductor devices requires test procedures with a spatial resolution in the sub-μm region and high frequency characteristic up to the GHz-range. These requirements can be fulfilled by contactless testing methods (Halbout 1988, Wolfgang 1979).
One realization for such a method is a laser scanning microscope (LSM) with high temporal resolution based on the time resolved optical beam induced current method (OBIC) (Bergner 1989, Bergner 1990). In this paper we studied the application of such a LSM to characterize a chain of n-MOS inverters.

2. EXPERIMENTAL ARRANGEMENT AND DEVICE UNDER TEST

In our experimental setup a LSM is combined with an actively modelocked argon-ion laser (pulse duration 100 ps, pulse repetition frequency 123 MHz, average power 600 mW) as light source and an OBIC-stage. A more detailed description of the experimental arrangement is to be found in this issue (Stamm 1991).
The experiment is carried out

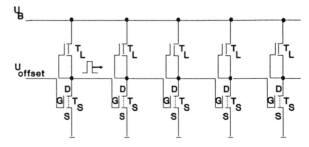

Fig. 1 Electrical scheme of the chain of n-MOS inverters

as a pump and probe experiment. The laser beam is splitted up into one intense part for generating the electrical pulse and a weak part for probing. The pump pulse is focused onto a desired blocked inverter of the inverter chain and generates a short electrical pulse by switching the illuminated inverter. The probe beam is scanned over the device under test (DUT). Because of the used generation method the electrical pulses are synchronized with the probe pulses. The OBIC-current induced by the probe pulse is taken as signal to build up the image of the electrical states of the DUT at the time of incidence of the probe pulse. A series of images is taken for different time delays between pump and probe pulses. The time dependence of the OBIC-signal for several points of the sample is obtained by recording the amount of OBIC-signal at these points vs. delay time.

The DUT is a chain of inverters of an especially prepared test chip. The inverters are enhancement-depletion inverters in n-MOS technology. The switching transistor T_S has a width-length ratio of the inversion channel from 8.0 μm / 2.4 μm, for the load transistor T_L this ratio is 6.4 μm / 6.4 μm. Fig.1 shows an electrical scheme of a part of the chain.

3. OPTICAL SWITCHING OF A BLOCKED n-MOS INVERTER

The idea of our investigation was to switch optically a blocked inverter at the beginning of the chain and to study the response of each inverter to the generated short electrical pulse. A cross-section of one inverter is depicted in fig. 2.

The switching behaviour can be explained by the following simple model. The drain current I_d of a blocked MOS-transistor is determined by the diffusion current given by the formula (Sze 1981),

$$I_d = qAD_n \frac{(n(0)-n(L))}{L} \quad,$$

where $n(0)$ and $n(L)$ are the concentrations of the minority carriers in the gate region near source and drain, respectively. D_n is the coefficient of diffusion of electrons, q the charge of an electron, L the

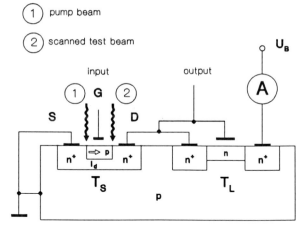

Fig. 2 Scheme of an illuminated MOS inverter

length of the inversion channel and A is in our case the cross-section area of photoexcited carriers. The pump pulse is focused onto the pn-junction gate source. The illumination leads to an increasing number of minority charge carriers in the source region and therefore to an increased source drain current. This corresponds to a decreasing gate voltage. If the power of the incident beam becomes so high that the threshold voltage is reached, the electrical state of the MOS inverter changes from the blocked into the none-blocked one. We have calculated the drain current of a blocked MOSFET in dependence on the laser excitation power taking into account typical device parameters and a δ-function like optical excitation. The result is depicted in fig. 3 showing the possibility of optical switching at moderate optical powers. Experimentally we have found a threshold power of 350 μW. After the illumination the MOS-transistor returns to its original electrical state within a characteristic time. This relaxation time rises with the beam power caused by increased capacity due to the large number of photoexcited carriers.

The time dependence of the optically generated electrical pulse is shown in fig. 4. The high OBIC-signal corresponds to the conducting state of the switching transistor, the low current to the non-conducting state. The inverter is switched from the non-conducting state to the conducting state within 1.9 ns. In this state the inverter stays only about 1 ns and at the delay time 8.3 ns equivalent to 0 ns it returns to the non-conducting state. Hence, the duration of the generated pulse is approximately 2.5 ns.

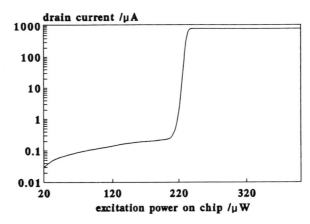

Fig. 3 Drain current of the switching transistor vs. optical power

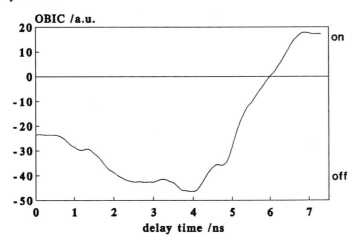

Fig. 4 OBIC-current of the optically switched inverter

4. PULSE PROPAGATION IN THE CHAIN OF INVERTERS

The propagation of the optically generated electrical pulse is shown in fig. 5. The curves show the OBIC-signal generated at the drain gate region of four adjacent inverters. When considering this figure we have to remember that a train of electrical pulses is generated. Because the DUT is relatively slow we have to distinguish on one microscope image several electrical pulses arising from succeeding excitation pulses. Regarding the transistor T6 we see that in the time interval 1.6 to 2.6 ns the transistor T6 switches from the non-conducting state to the conducting state. It remains about 5 ns in the conducting state and returns then to the conducting state. The transistor T7 following the transistor T6 begins to switch at the delay time 3.6 ns. This corresponds a transition time of about 2.5 ns for the trailing edge of the electrical pulse from one inverter to the next. From the time dependent OBIC-signal of transistor T5 a electrical pulse duration 2.5 ns follows.

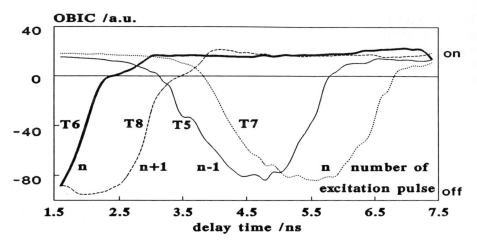

Fig. 5 OBIC-current of four adjacent inverters at different delay times between pump and probe pulses

5. CONCLUSIONS

By means of a LSM coupled with a modelocked laser and an OBIC-stage it is possible to study the pulse propagation of optically generated electrical pulses in a n-MOS device. The duration of the generated pulse was estimated to 2.5 ns. The transition time for the trailing edge from one inverter to the next is 2.5 ns. The rise time of a transistor is about 1.5 ns. The time resolution of this method is only limited by the pulse duration of the optical pulses and the electrical properties of the device under test. The method should be applicable to devices with higher cut-off frequencies.

6. REFERENCES

Bergner H, Damm T, Stamm U and Stolberg K P 1989 Int.Journ.Optoelectr. 4 p 583
Bergner H, Damm T, Stamm U, Müller M and Stolberg K P 1990 Conference
    Proceedings "Frontiers in Electro-Optics", 20-22 March 1990, Birmingham, published
    by Reed Exhibition Companies Ltd., 171.
Halbout J M, May P and Chiu G 1988 J.Mod.Opt. 35 p 1995
Stamm U, Bergner H,Hempel K and Krause A 1991 , in this issue
Sze S M 1981 Physics of Semiconductor Devices, John Wiley & Sons, New York
Wolfgang E et al 1979 IEEE Trans. Electr. Devices ED-26 p 549

Inst. Phys. Conf. Ser. No 126: Section V
Paper presented at Int. Symp. on Ultrafast Processes in Spectroscopy, Bayreuth, 1991

Spectral and temporal holography of ultrashort light pulses and its possible applications in opto-electronic systems and devices

Yu T Mazurenko

S I Vavilov State Optical Institute, SU-199034 St-Petersburg USSR

ABSTRACT: The methods of spectral-and-temporal holography of ultrashort light pulses based on using spectrally nonselective recording media are considered. These methods can be applied in opto-electronics for the processing of fast signals. The possibilities of time-division multiplexing of data streams using dynamic spectral holography are considered in detail.

1. INTRODUCTION

In the previous papers of the author the principles of holographic storage and reconstruction of ultrashort pulses using the spatial spectral decomposition of light were proposed. The review of these works was published by the author (Mazurenko 1990a). The methods considered can be used in opto-electronics for the processing of fast signals. The possible specific applications are recording, reconstruction, and shaping of femto, pico, and nanosecond optical signals, optical implementation of matched temporal filters for optimal detection and recognition of short temporal signals, optimal compression of chirped ultrashort pulses. Probably, one of the most promising application of these methods is the time-, frequency-, and code-division multiplexing of data streams. The term multiplexing means combining the various messages at a common channel. This enables the transmission of multiple messages through a single channel. We here consider the time-division multiplexing based on the dynamic spectral holography in detail.

2. SPECTRAL DECOMPOSITION WAVES

Consider a double spectral instrument that is shown in the figure. The instrument contains the dispersive elements G_1 and G_2, which perform the angular spectral decomposition of radiation, and lenses L_1 and L_2 with the focal length f. In our example G_1 and G_2 are the diffraction gratings. S is the plane of spectrum of the instrument. Two dispersive elements G_1 and G_2 mutually cancel their dispersion.

One can describe the angular spectral decomposition of light by a dispersive element as a result of the passage of plane waves through a transparency which has the frequency dependent amplitude transmission (Mazurenko 1990a). Taking into account the directions of the axes x_1 and x_2, we can represent the transparencies $G_1(\Omega,x_1)$ and $G_2(\Omega,x_2)$, corresponding to G_1 and G_2, in the following form:

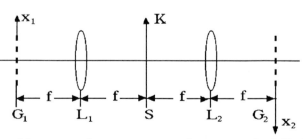

$$G_1(\Omega,x_1) = p(x_1)\exp(i\Omega x_1/v), \tag{1a}$$

$$G_2(\Omega,x_2) = p(-x_2)\exp(-i\Omega x_2/v). \tag{1b}$$

Here x is the transverse coordinate in the plane of transparency, $\Omega = \omega - \omega_0$, ω is the frequency of light radiation, ω_0 is the value of ω for a wave propagating along the optical axis of the instrument, v is the coefficient characterizing the angular dispersion of spectral decomposition, p(x) is the term describing the shape of the dispersive element aperture.

Lenses L_1 and L_2 form in the S plane the spatial frequencies spectra of the coherent images placed in the planes G_1 and G_2 (Goodman 1968). We consider that the transverse coordinate in the S plane is graduated in the units of spatial frequency K. This allows to describe the action of the lenses L_1 and L_2 as the Fourier transform of the optical spatial signal a(x) to the spatial frequencies spectrum A(K) or the inverse transform (Goodman 1968).

Let optical radiation with the plane wave front and time-dependent amplitude is incident from the left onto the entrance of the instrument. The radiation can be characterized by the carrier frequency ω_0, complex envelope (magnitude) r(t), and the Fourier spectrum of the complex envelope $R(\Omega)$. After the radiation is transmitted through the transparency $G_1(\Omega,x)$ and Fourier transform lens L_1, we obtain the following field distribution $U(K,\Omega)$ in the spectrum plane S (Mazurenko 1990a):

$$U(K,\Omega) = \text{const } R(\Omega)W(\Omega-Kv). \tag{2}$$

Here $W(\Omega)$ is the instrument function of the spectral dispersion whose width determines the frequency resolution $\Delta\Omega$. This function is the Fourier transform of the temporal instrument function w(t) which, in turn, is determined by the aperture function p(x) as w(t) = p(vt). The duration of w(t) is equal to $T \approx 2\pi/\Delta\Omega$.

Changing from the frequency-domain representation (2) to the time-domain representation we find the following field distribution in the spectrum plane u(K,t) depending on time t and coordinate K:

$$u(K,t) = \text{const } \exp[-i(\omega_o + Kv)t] \int r(\tau)w(t-\tau) \exp(iKv\,\tau)d\tau. \tag{3}$$

Note that the integral in (3) corresponds to the known definition of the "instantaneous" spectrum of a signal, related to the instant t, and calculated using the temporal window $w(t-\tau)$ whose width equals T.

If r(t) is a pulse with the sufficiently short duration, then, within the limits of the frequency interval which we are interested in, it is possible to consider that in (2)

$$R(\Omega) \approx \text{const } \exp(-i\Omega t_j),$$

where t_j is the center of a pulse. We shall think of such a pulse as an ultimately short or delta-shaped and adopt $r(t) = \delta(t-t_j)$. Therefore, for such pulses, according to (3),

$$u(K,t) = \text{const } \exp[-i(\omega_0 + Kv)t]w(t-t_j). \tag{4}$$

The waves described by (3) and (4) can be named spectral decomposition waves.

3. OPTICAL TIME-DIVISION MULTIPLEXING

Consider several information channels characterized by the same bandwidth. The bandwidth determines the characteristic time of the signals change, which we denote by T_c. If the considered channels are optical, their outputs can be placed tightly along the some common straight line. In so doing the coherent 1-D image is formed. This image depends on time. We denote obtained spatial and temporal distribution as a(x,t), where the coordinate x actually determines the order number of a channel.

Using the normal optical Fourier transform (Goodman 1968), we can obtain the spatial Fourier image (spatial frequencies spectrum) of the a(x,t). We denote this Fourier image as A(K,t). The field distribution A(K,t) also changes during the characteristic time T_c. The considered spatial spectrum can be projected onto the plane S of the double spectral instrument. We may select the properties of the instrument in such a way that the condition $T_c > T$ is satisfied.

Using the monochromatic background we can record in the S plane the time-dependent (dynamic) hologram of the distribution A(K,t). Evidently, the recording medium must be dynamic (reversible) and must be characterized by a reaction time which is $\sim T_c$. We can consider this hologram as the dynamic filter placed in the S plane, the amplitude transmission of this filter being proportional to A(K,t).

Let us direct onto the entrance of the instrument the periodic sequence of delta-shaped pulses with the repeating frequency 1/T. We denote the complete temporal signal corresponding to this sequence as $r_b(t)$:

$$r_b(t) = \sum \delta(t-nT). \tag{5}$$

Spectral decomposition waves of the individual pulses can be described by the expr. (4). The length of each of these waves equals T.

We consider the passage of the n-th pulse from the series (5) through the spectral instrument. Taking into account the condition $T_c > T$, we can assume that, during this pulse passage, the approximately stationary transparency occurs in the spectrum plane, the transmittance of this transparency being proportional to A(K,nT). Taking into account (2), we obtain the following field distribution on the rear surface of the transparency:

$$U'_n(K,\Omega) = \text{const } A(K,nT)W(\Omega-Kv)\exp(-i\Omega nT). \tag{6}$$

We transform this field distribution by the right-hand part of the instrument (lens L_2 and dispersive element G_2). Then we integrate the obtained amplitude distribution over the surface of exit aperture. Changing from the frequency-domain representation to the time-domain representation, we obtain the following output signal:

$$r_n(t) = \text{const } m(t-nT) a[-v(t-nT),nT]. \tag{7}$$

The function m(t) is defined as $m(t) = \int w(\tau)w(\tau-t)d\tau$. It is the relatively slowly varying function whose duration is about T.

It is not difficult to see that the temporal signal (7) is practically the analog of the spatial signal a(x,nT). If the size of the image a(x,nT) equals vT then the length of the signal rn(t) equals T and the adjacent pulses (7) are not superimposed.

Taking into account the input signal (5) completely, we obtain the following complete signal at the instrument exit:

$$r(t) = \text{const } \sum m(t-nT)a[-v(t-nT),nT]. \tag{8}$$

Thus, as a result of the transforms considered, the series of pulses is obtained which contain in a single channel. Each of these pulses actually contains the scan in the time scale of the set of signals corresponding to the single temporal section of the multichannel message. Therefore, the messages corresponding the different initial channels are separated in the time scale of the united signal. This means that we have obtained the time-division multiplexing.

4. DEMULTIPLEXING

Now we will consider the reverse process of demultiplexing. We direct onto the entrance of the spectral instrument a signal r(t) which, particularly, may be the multiplexed message (8). Besides, we direct onto the entrance of the instrument the sequence $r_b(t)$ of the short pulses (5). According to (3), (4), the instantaneous spectra of the signals r(t) and $r_b(t)$ are formed in the S plane.

As two signals considered are actually transformed into the continuous radiations in the S plane, these radiations can interfere, the interference pattern changing with the characteristic time T. The instantaneous spectrum of the short pulses sequence $r_b(t)$ can be considered as the reference radiation. Then the recording of the interference pattern in the dynamic medium which have the suitable response time can be considered as the obtaining of the nonstationary (dynamic) spectral hologram of the continuous radiation r(t). Such hologram is the sequence of holograms repeating after the period T. As this takes place, each of holograms o the sequence actually contains the recording of the instantaneous spectrum of the signal r(t) in the length spar of time $\{nT\text{-}T/2, nT + T/2\}$.

It is possible to separate spatially the radiations r(t) and $r_b(t)$ in the plane perpendicular to the figure plane (Mazurenko 1990a). Using this possibility we can extract the component of the spectral hologram we are interested in. This component is of the form:

$$H(K,t) = \text{const} \int r(\tau)w(t-\tau)\exp(iKv\,\tau)d\tau \sum \exp(-iKvnT)w(t-nT). \qquad (9)$$

We may consider the spectral hologram (9) as the Fourier transform hologram of some 1-D image changing with time with the characteristic parameter T. Reconstructing this hologram with the plane monochromatic wave and producing the optical Fourier transform of the monochromatic radiation distribution thus formed we obtain the following nonstationary amplitude image:

$$a(x,t) = \text{const} \sum r(nT + x/v)w(t\text{-}nT\text{-}x/v)w(t\text{-}nT). \qquad (10)$$

For the fixed value of x the amplitude a(x,t) is the sequence of pulses with the repeat period T. It is not difficul to see that the amplitude of the nth pulse is proportional to the reading of the signal r(t) at the instant $t = nT + x/v$.

Let us distribute the spatial and temporal signal (10) among some number of information channels, allotting a short cut of the x axis to each of them. If the signal r(t) is a multiplexed message then, when the distribution of the signal (10) is correct, the temporal signal which is sent to each channel contains the sequence of reading of the one-channel message. As it does so, the value of x essentially determines the output channel number. And this actually means the demultiplexing of the signal with time division.

5. CONCLUSIONS

We will cite some estimate characteristics of the real application of the proposed method. The minima bandwidth of initial messages is determined by the accessible in optics value of the frequency resolution during spatial spectral decomposition. In use the Fabry-Perot etalon of a special type (Mazurenko 1990b) this value may be as small as 10^6 Hz. Because of this the bandwidth of initial messages can be, say, 10^6 to 10^9 Hz. It also should be noted that the bandwidth of initial messages must be ensured by the the sufficiently fast response o the dynamic medium. The number of initial messages is actually determined by the maximum number of the elements of resolution which can be realized in an optical system in the diffraction limit. This number may be say, 100 to 1000. Correspondingly, the bandwidth of the multiplexed message may be 10^8 to 10^{12} Hz. These values must be provided by the sufficiently small length of duration of the laser radiation pulses which is quite accessible.

6. REFERENCES

Goodman J W 1968 *Introduction to Fourier optics* (McGrow-Hill, San Francisco)
Mazurenko Yu T 1990a *Appl. Phys.* **B 50** 101
Mazurenko Yu T 1990b *Opt. Spektr.* (Russ.) **68** 241

Inst. Phys. Conf. Ser. No 126: Section VI
Paper presented at Int. Symp. on Ultrafast Processes in Spectroscopy, Bayreuth, 1991

465

Measurement and conjecture concerning IVR acceleration by methyl group. CH_3 vs CD_3

David B. Moss

Department of Physics, Lynchburg College, Lynchburg, VA 24501, USA

Charles S. Parmenter, Trent A. Peterson, Christopher J. Pursell and Zhong-Quan Zhao

Department of Chemistry, Indiana University, Bloomington, IN 47405, USA

ABSTRACT: Intramolecular vibrational redistribution (IVR) lifetimes have been measured for the $3^1 5^1$ level of S_1 p-fluorotoluene containing either a CH_3 or CD_3 methyl rotor. The lifetimes are 3.4 and 1.5 psec, respectively. They are two orders of magnitude shorter than that from the same ring level in p-fluorobenzene. The magnitude of the methyl IVR acceleration is accounted for by an increase in state density deriving principally from interactions between internal rotation and ring vibrations. The apparent enigma of nearly equivalent CH_3 and CD_3 IVR rates is also solved by this account.

1. INTRODUCTION

We learned long ago that placement of a CH_3 internal rotor on an aromatic ring has a profound effect on intramolecular vibrational redistribution (IVR) (Parmenter and Stone 1986). The discovery came from comparisons of fluorescence from the S_1 states of p-difluorobenzene (pDFB) and p-fluorotoluene (pFT), molecules that differ by replacement of a fluorine in pDFB with a methyl group to make pFT.

Two types of fluorescence spectroscopy revealed the large IVR differences. The first involved comparisons of single vibronic level fluorescence (SVLF) spectra as excitation climbed the S_1 manifold. The spectra showed that the onset and the dominance of congested fluorescence structure, a hallmark of IVR, occurs at much lower energies in pFT. In a different fluorescence approach, the actual IVR rates were derived from chemical timing, a technique whereby fluorescence time resolution is achieved by using O_2 quenching to restrict fluorescence lifetimes. While the pFT measurements were preliminary estimates, IVR in pFT was always seen to be faster. The difference for some levels was as much as two orders of magnitude. Thus both approaches showed that methyl is an accelerating functional group for IVR.

Much subsequent work has been directed towards uncovering the causes of the phenomenon (Longfellow and Parmenter 1988, Moss et al. 1987, Zhao and Parmenter 1991, Zhao and Parmenter 1992, Martens and Reinhardt 1990), an effect that so far has been observed only in 300 K experiments (Baskin et al. 1988). Some molecular aspects can be securely eliminated. For example, IVR in pFT is observed at S_1 vibrational energies much too low for any involvement of the eight CH_3 modes. The IVR response also seems far greater than could be produced by the small differences in the aromatic ring frequencies. Thus the IVR acceleration is almost certainly due to the methyl internal rotation. It is, in effect, a consequence of adding a single new degree of freedom. This focus on internal rotation makes the effect even more surprising because the methyl interaction with the ring is seemingly small. The internal rotation barrier is only 32 cm^{-1} so that methyl rotation in S_1 pFT is almost free (Okuyama et al. 1985).

We report here a further experimental step in our probe of the factors underlying the IVR acceleration. We have revisited the time-resolved fluorescence from a pFT-h_3 S_1 level to obtain an IVR lifetime that replaces the upper limit from our previous work. We have also measured the IVR lifetime from this level in pFT-d_3 to reveal the consequences of transforming the CH_3 rotor to CD_3. This information is combined with the findings of our cold jet fluorescence studies (Zhao and Parmenter 1991) of the rotor-vibration interactions to conjecture about the principal causes of IVR acceleration by the methyl internal rotation.

2. IVR RATES IN pFT-h_3 vs pFT-d_3

Chemical timing (Coveleskie et al. 1985) has been used to determine the IVR rate after initially pumping the S_1 level $3^1 5^1$ lying at 2030 cm^{-1} in pFT-h_3 and at 2013 cm^{-1} in pFT-d_3. The experimental details and data analyses are similar to those described for pDFB (Holtzclaw and Parmenter 1986) and benzene (Longfellow et al. 1988). Fig. 1 contains the crucial fluorescence data together with a fit from the kinetic model used to extract the IVR rate. Only small differences occur between the sets of the fluorescence data, and accordingly we can expect the IVR rates to be similar in the two molecules.

The modelling (Holtzclaw and Parmenter 1986) treats IVR within the standard framework of radiationless transition theory. The lifetime τ_{IVR} or the rate constant k_{IVR} for decay of the initially created vibrational state by IVR is given by the Golden Rule expression

$$\tau_{IVR}^{-1} = k_{IVR} = \frac{2\pi}{\hbar} V^2 \rho_{eff} \quad (1)$$

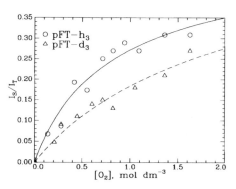

Fig. 1. Fluorescence data presented in the standard form for analysis by chemical timing (Holtzclaw and Parmenter 1986). The solid (pFT-h_3) and dashed lines (pFT-d_3) show the fit of the model to the data using the τ_{IVR} values quoted in the text.

where V is the average interaction matrix element for coupling of the initial state with the field of zero-order states and ρ_{eff} is the density of these coupled states. The analysis presented in Fig. 1 corresponds to IVR lifetimes of 3.4 psec (pFT-h_3) and 1.5 psec (pFT-d_3). These values will change slightly as new data are obtained concerning assumptions of the analysis. It is unlikely, however, that the lifetimes are off by much more than a factor of two. In particular, while the rates for pFT-h_3 and pFT-d_3 appear distinctive, cumulative errors from our kinetic assumptions could conceivably account for much of the difference.

Three facts about relative IVR rates in the species pDFB, pFT-h_3 and pFT-d_3 are now established.

i) Acceleration of IVR by two orders of magnitude upon methyl replacement of a pDFB fluorine atom was inferred from the upper lifetime limits of our earlier pFT measurements. The acceleration is now confirmed for the $3^1 5^1$ level. The IVR lifetime of the $3^1 5^1$ pDFB level is 97 psec (Holtzclaw and Parmenter 1986). In contrast, the pFT lifetime is ~3 psec.

ii) The density of coupled states, ρ_{eff}, has been substantially increased by the methyl substitution. This conclusion follows from the IVR kinetics. In pDFB, the IVR must be modelled as an intermediate case radiationless transition, implying a relatively small ρ_{eff}, known to be 11 per cm^{-1}. On the other hand, statistical limit kinetics must be used to describe IVR in pFT, a certain indicator of substantially increased ρ_{eff}.

iii) Deuteration of the rotor, CD_3 vs CH_3, has a relatively small, if any, effect on the IVR rate in pFT.

3. A RATIONALE FOR THE IVR ACCELERATION

Enough is now known about the interaction between internal rotation and vibration in pFT-h_3 and pFT-d_3 to support conjecture about the principal causes of IVR acceleration. It is convenient and appropriate to use the Golden Rule expression, Eq. 1, as the basis of discussion. Attention centers on how V^2 and ρ_{eff} change upon methyl substitution. We emphasize that ρ_{eff} is not necessarily the full level density of the system, ρ_{vib}, but the density of only those levels that actually participate in the coupling.

We begin by noting that IVR as observed in these experiments is a consequence of coupling among nearly degenerate levels of different vibrational character. That character is built only of ring modes, and with one exception, the ring fundamentals are almost the same in pFT and pDFB. The exception is a surprise that arose in our level structure study of the region 0-400 cm^{-1} in S_1 pFT. Amidst all the similarities between pDFB and pFT, an additional low frequency mode, given the transient assignment ν_x, appeared in pFT (Zhao and Parmenter 1991). Calculations have more recently led to its identity. It is probable that the lowest frequency mode ν_{30} occurs in pFT with a double well potential. The so-called $\nu_x = 136$ cm^{-1} is actually one of the components of the split ν_{30} fundamental. The other is at 110 cm^{-1}. We estimate that this change in ν_{30} will increase the total level density ρ_{vib} over that for pDFB by somewhere between a factor of two or three. Since ν_{30} is active in level coupling, the increase also appears in ρ_{eff}.

Symmetry requirements for level coupling limit the fraction of ρ_{vib} that contribute to IVR. If we ignore rotational mechanisms (probably inappropriate to do so) and stick only with anharmonic interactions, symmetry restrictions limit ρ_{eff} to $(1/8)\rho_{vib}$ in pDFB (Holtzclaw and Parmenter 1986). pFT on the other hand has lower symmetry (molecular symmetry group G_{12} containing four symmetry species), and this change increases the fraction of ring levels that can couple. ρ_{eff} for pFT is $(1/4)\rho_{vib}$. Thus here we gain a factor of two over pDFB.

We have yet to consider internal rotation energy levels, the most obvious source of increased state density in pFT. We are interested in combination levels built from ring vibrations plus internal rotation. Before worrying about their contributions to ρ_{eff}, we need to know whether significant coupling occurs between combination levels containing different quanta of internal rotation, m. For example, the interactions that could contribute to observable IVR would be like those of the level pair $30^3m^3 \dots 8^1m^0$, two combination levels with different vibrational (ν_{30} and ν_8) and different rotor components (m = 3 and m = 0).

A theory of internal rotation-vibration interaction has treated this type of coupling in pFT (Moss et al. 1987). Effective interactions are indicated with coupling matrix elements quite comparable to those of favorable anharmonic interactions. Additionally our spectroscopic probes have characterized more than thirty such interactions in pFT-h_3 and pFT-d_3. Big matrix elements, on the order of cm^{-1}, are common. The example above, in fact, occurs in pFT-h_3 with a 2 cm^{-1} matrix element.

The possible boost in ρ_{eff} by the presence of these levels containing rotor quanta is surprisingly easy to estimate if we use the theoretical selection rules for coupling between internal rotation and vibration, an interaction that we designate as i•v. The rules require that $\Delta m = \pm 3$ for some levels and $\Delta m = \pm 6$ for others.

We consider two sets of nearly pFT levels. The first lies 171 cm^{-1} below the pumped level and is comprised of all the m = 0 levels in that region. It will have approximately the same level density as the m = 0 levels in the pump region since it is not far away. If we now add six quanta of internal rotation to each level of this set, a second set comprised entirely of levels with the component m = 6 is created. At what energy does this set lie? Since the energy of six internal rotation quanta is 171 cm^{-1} (Okuyama et al. 1985), this m = 6 set lies exactly on top of the pumped level. We now have a set of levels that can interact with the isoenergetic m = 0 pumped level by the $\Delta m = \pm 6$ i•v selection rule. Approximately half the set can couple. Thus by this i•v interaction, the pumped level gains access to a new set of levels with an effective density of about (1/2) ρ_{vib}.

The scheme is repeated for the set of levels with m = 3 that can couple by the i•v interaction with $\Delta m = \pm 3$. Additionally the m = 3 and m = 6 internal rotation levels occur with two components so that there are two distinct sets each of m = 3 and m = 6 levels. Each individually will add an effective density of about (1/2) ρ_{vib} to the levels accessible by the pumped level.

The cumulative result of all these changes brought about by methyl substitution is to increase ρ_{eff} by about a factor of 50 over the pDFB value. These discussions of anharmonic and i•v interactions are highly simplified, and we can identify an additional source of level interaction that will further increase ρ_{eff}. Overall molecular rotation, r, is known to be involved with coupling of rotor-vibration states through interactions of three motions, i•v•r (Zhao and Parmenter 1991). Such coupling brings additional sets of levels into interaction with the pumped level. Thus one does not have to work hard to increase the IVR rate by two orders of magnitude through the ρ_{eff} factor alone.

In contrast, it is difficult to find compelling reasons why V^2 should be enormously different in the two molecules. The question centers on whether the $\Delta m \neq 0$ coupling in pFT occurs with matrix elements generically different from those of anharmonic interactions. The theoretical modelling suggests that the matrix elements are similar. The experimental characterizations of level interactions may lend support. We observe many large matrix elements (those couplings that are seen in pFT-h_3 and pFT-d_3) but also many that are small (the allowed interactions that do not appear). Perhaps the average is not greatly different than that of an analogous mix of anharmonic cases.

4. A RATIONALE FOR THE SIMILAR CH_3 VS CD_3 RATES

The preceding analysis identifies increased state densities as the crucial factor leading to faster IVR on account of methyl substitution. The increase is a consequence of introducing a low frequency motion, namely internal rotation. Accordingly, when we introduce an even lower energy set of levels by using deuteration to cut the rotor energies in half, intuition might suggest a big increase in the IVR rate. The observed CH_3 vs CD_3 rates, however, generate an enigma. The rates turn out to be about the same.

The enigma disappears when we extend the analysis of the previous section to the issue of CH_3 vs CD_3 rates. The $\Delta m = \pm 3, \pm 6$ selection rules governing i•v interactions play the central roll. They keep ρ_{eff} essentially the same in pFT whether the rotor is CH_3 or CD_3. In either case, coupling of the initially pumped m = 0 level can occur to only four sets of levels, two using $\Delta m = \pm 3$ and two using $\Delta m = \pm 6$. The additional level densities introduced by those sets are dependent only on ring mode frequencies. Those parameters are almost identical in pFT-h_3 and pFT-d_3, and hence ρ_{eff} will be about the same in each.

The factor V^2, on the other hand, does not appear to change greatly as the rotor transforms from CH_3 to CD_3. No change occurs in the theoretical description of i•v interactions. Also, none is seen in the experimental probes of state-to-state couplings within the two molecules. The average of many observed matrix elements for each molecule is about the same.

By these arguments, our analysis leads to the conclusion that CH_3 and CD_3 internal rotation will influence IVR rates equivalently. The pFT-h_3 vs pFT-d_3 measurements provide a key test. The observed similarities of rates give a strong support for the central concepts of our model.

5. SUMMARY

New measurements on the $3^1 5^1$ state ($\epsilon_{vib} \approx 2000$ cm^{-1}) show explicitly that a 100 fold increase occurs in the IVR rate when a fluorine in pDFB is replaced by a methyl group with nearly free internal rotation to make pFT. The IVR lifetime is changed from 97 to 3 psec. An additional measurement in pFT-d_3 allows for the first time a comparison of the IVR rates with a CH_3 vs CD_3 rotor. While deuteration cuts the internal rotation energies in half, almost no change occurs in the IVR rate. An account of these data is developed in terms of the Golden Rule expression for IVR. As one might expect, the accelerated IVR rate by methyl substitution is largely a consequence of the increased state density introduced by the internal rotation as opposed to changes in average coupling strengths. A special technique is used to estimate the state density increase. The increase leads to a good account of the relative pDFB vs pFT IVR rates as well as the equivalence of IVR rates in pFT with CH_3 vs CD_3 rotors.

6. REFERENCES

Baskin J S, Rose T S and Zewail A H 1988 *J. Chem. Phys.* **88** 1458
Coveleskie R A, Dolson D A, and Parmenter C S 1985 *J. Phys. Chem.* **89** 645
Holtzclaw K W and Parmenter C S 1986 *J. Chem. Phys.* **84** 1099
Longfellow R J, Moss D M, and Parmenter C S 1988 *J. Phys. Chem.* **92** 5438
Longfellow R J and Parmenter C S 1988 *J. Chem. Soc. Faraday Trans II* **84** 1499

Martens C C and Reinhardt W P 1990 *J. Phys. Chem.* **93** 5621

Moss D B, Parmenter C S and Ewing G E 1987 *J. Chem. Phys.* **86** 51

Okuyama K, Mikami N and Ito M 1990 *J. Phys. Chem.* **89** 5617

Parmenter C S and Stone B M 1986 *J. Chem. Phys.* **84** 4710

Zhao Z Q and Parmenter C S 1991 *Proceedings of the 24th Jerusalem Conference in Quantum Chemistry and Biochemistry,* B. Pullman and J. Jortner, editors. (Kluwer Academic Publishers, Dordrecht), (in press).

Zhao Z Q and Parmenter C S 1992 in *Time Resolved Vibrational Spectroscopy,* H. Takahashi, Editor. (Springer-Verlag Co., Heidelburg) (in press)

Infrared transient hole burning with picosecond pulses

H. Graener

Physics Department, University Bayreuth, D-8580 Bayreuth,
PO 101251, Germany

ABSTRACT: Transient hole burning experiments with picosecond pulses provide detailed insight in the structural and dynamical properties of hydrogen bonded systems. Experimental data of polymers and liquid systems will be presented.

Spectral Hole burning is a well established technique on a wide range of time-scales and in various spectroscopic fields (Moerner 1987, Shimoda (1976). Its ultrafast counterpart in the infrared however was lacking until recenty although highly desirable dynamical information can be provided by this powerful method. We have, for the first time, observed transient spectral holes in the infrared on the picosecond time-scale studying the OH stretching vibration in condensed matter. Our results demonstrate that this vibrational mode represents a fast spectroscopic probe of hydrogen bonds and its molecular environment.

In this report experimental data for a polymer system (poly-vinylbutanal, PVB) and a liquid (water) will be discussed. Pump - probe experiments are performed using two independent tunable pulses (duration 12-14 ps, bandwidth 10-17cm^{-1}, tuning range 2500-7000cm^{-1}) with linear polarization and excitation intensities of several 10^{10} Wcm^{-2}. The transient transmission changes are measured as function of probe frequency, delay time and probe polarization and for different frequency settings of the excitation pulse (Graener et al 1990a).

The investigated sample display a broad (200 - 300 cm^{-1}) and strong OH absorption band between 3300 and 3500 cm^{-1} in the conventional IR spectrum. These rather unusual features for a vibrational absorption are believed to be due to hydrogen bonding. The shapes of the observed transmission changes in the transient spectra clearly deviate from the conventional band contour and provide direct evidence for the inhomogeneity of the absorption band. The analysis of the transient spectra gives detailed information about the "quasi"-homogeneous substructure.

In the simplest approach the absorption coefficient α_0 for an inhomogeneous absorption band can be written as

$$\alpha_0(\nu) = \int_{-\infty}^{\infty} N(\nu_0)\ \sigma(\nu_0)\ \mathcal{L}\left(\frac{\nu - \nu_0}{\Gamma(\nu_0)}\right)\ d\nu_0 \qquad (1)$$

$\sigma(\nu_0)$ is the absorption cross section, $\mathcal{L}(y)$ describes the "quasi"-homogeneous lineshape function, which is assumed to be Lorentzian in the following. $N(\nu_0)$ is a frequency distribution function, which is assumed to be time-independant at least on the time scale of the picosecond experiments. The function has to obey the normalisation condition

$$N_{tot} = \int_{-\infty}^{\infty} N(\nu_0)\ d\nu_0 \qquad (2)$$

where N_{tot} is the total number density of absorbing OH groups.

A strong pump pulse with the intensity I_{pu} and the frequency ν_{pu} within the absorption band interacts with a sample of length ℓ and will excite some percent of the molecules. The reduced population between the ground- and the excited state creates a transmission change, which can be described by the following equation:

$$\ln\left(\frac{T}{T_0}\right)\ (\nu, t_D) = \frac{4\ \ell}{\hbar\ \nu_{pu}}$$

$$\times \int_{-\infty}^{\infty} d\nu_0\ N(\nu_0)\ \sigma(\nu_0)\ \mathcal{L}\left(\frac{\nu - \nu_0}{\Gamma(\nu_0)}\right)\ \sigma(\nu_0)\ \mathcal{L}\left(\frac{\nu_0 - \nu_{pu}}{\Gamma(\nu_0)}\right)$$

$$\times \int_{-\infty}^{t_D} dt\ I_{pu}(t)\ \exp\left(\frac{t - t_D}{T_1}\right) \qquad (3)$$

This equation is valid for pulse durations $t_p \gg (\pi c \Gamma)^{-1}$ and in the low saturation limit (T not much larger than T_0). T_1 is the lifetime of the vibrational excited state (assumed to be independent of ν_0) and t_D denotes the delay time between pump and probe pulse. For the interpretation of persistent holeburning experiments (where the time integral will give a constant factor) oftentimes the situation is discussed where σ and Γ are independent of ν_0 and the width of $N(\nu_0)$ much broader than Γ. Under these conditions the above equation reduces to a convolution integral of two Lorentzian functions, resulting in a Lorentzian with a width of 2Γ, which is assumed as first approximation in the following. For a more realistic description of a transient holeburning experiment the excited state absorption has also to be considered; i.e. one has to add on the r.h.s of Eq. 3 a second frequency integral, where the first product $\sigma(\nu)\mathcal{L}(y)$ is replaced by a corresponding term for the $v=1 \rightarrow v=2$ transition.

Equation 3 nicely describes the two principal types of experiments: In a time resolved experiment the frequency integral will give a constant, so the vibrational lifetime T_1 can be determined from the t_D-dependance of the data. In a frequency resolved experiment the time integral will be constant, so that the parameters σ and Γ can be readily deduced from the widths and the amplitudes of the spectral holes. With these data and

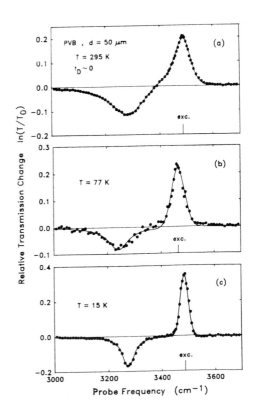

Fig 1.: Transient probe spectra in the OH region of PVB taken during the excitation process ($t_D \sim 0$); pump frequency near the center of the OH stretch absorption band around 3480 cm^{-1};

a) T = 295 K;

b) T = 77 K;

c) T = 15 K;

the solid lines are calculated curves with two lorentzian components convoluted with the spectral shape of the pulses equation 1 the spectral distribution function $N(\nu)$ can be derived.

Fig.1 shows three examples of transient spectra. The sample was a 50μm film of PVB at different temperatures. The pump frequency was ~3480 cm^{-1} near the center of the conventional IR absorption band. The delay time between pump and probe pulse was set near $t_D = 0$ (maximum temporal overlap). The principal shape of the three transmission changes is similar: a transmission increase (positive $\ln(T/T_0)$, spectral hole) around the pump frequency, and an absorption increase (inverse hole) due to the excited state absorption, which is red shifted by ~220 cm^{-1}. Clearly visible is the strong narrowing of the hole with decreasing temperature. At a temperature of 15 K the observed holewidth is limited by the spectral resolution of the experimental system. In this temperature range only an upper limit for the "quasi"-homogeneous linewidth of $\Gamma < 3$ cm^{-1} can be given. For the higher temperatures (T > 77 K) an exponential increase with T^{-1} was found, which can be explained with an extended Anderson - Kubo model (Graener et al 1990b).

In another series of experiments we have investigated the time evolution of the spectral holes. Two examples are shown in Fig. 2 for T = 295 K (a) and T = 15 K (b). The observed transmission change is plotted versus delay time. Pump and probe frequencies were set at 3480 cm^{-1}. One

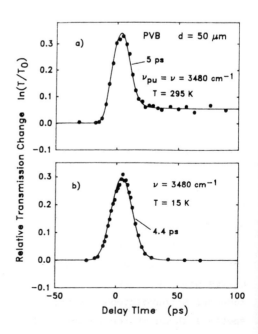

Fig. 2: Relative transmission change of probe pulse versus delay time for equal pump and probe frequencies ($\nu = \nu_{pu} = 3480$ cm^{-1}); a) sample temperature T = 295 K; b) sample temperature t = 15 K; the solid lines are calculated with a rate equation model giving the shown population lifetimes

observes a steep rise of the probe transmission increase during the excitation process, followed by a slightly slower decay. Comparing with a rate equation model (solid line in Fig. 2) a vibrational lifetime of 5±1 ps (T = 295 K) and 4.4±1 ps (T = 15 K) was obtained. Similar time evolutions (with an inverse sign) were found in the excited state absorption, supporting the interpretation of the time constant as vibrational lifetime. The observed long-lived transmission increase in Fig. 2a is explained by a slight temperature increase ($\Delta T \sim$ 10 K) in the sample after the relaxation of the vibrational energy. Comparing these data with the width of the observed spectral holes (Fig. 1) one can conclude, that in the high temperature regime (T > 100 K) lifetime broadening is insignificant, whereas at low temperatures (T < 30 K) the vibrational lifetime will give the dominant contribution to the "quasi"-homogeneous linewidth.

In further experiments with variable pump frequency the dependence of Γ and σ on ν_0 was investigated. A clear increase of both parameters with decreasing pump frequency was observed. With this knowledge it was possible to evaluate the spectral distribution function $N(\nu_0)$ which can be related to a distribution function of O...O distances and in this way to the strengths of the hydrogen bonds (Graener et al 1991a).

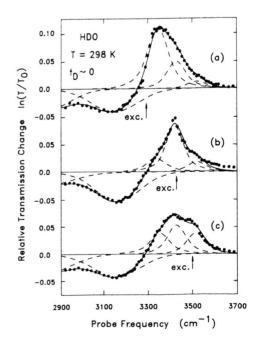

Fig. 3: Transient infrared probe spectra of HDO:D_2O (0.5 mol/ℓ) taken during the excitation pulse ($t_D \sim$ 0);
a) ν_{pu} = 3300 cm^{-1};
b) ν_{pu} = 3430 cm^{-1};
c) ν_{pu} = 3500 cm^{-1};
The solid line is calculated as sum of five components (indicated as dashed lines);
The varying bandshapes directly indicate inhomogeneous broadening

A similar series of experiments was performed on water ($HDO:D_2O$, $0.5mol/\ell$). Examples of frequency resolved measurements are shown in Fig 3. The varying shape of the three transient spectra (note the different excitation frequencies) shown in Fig. 3 clearly prove the inhomogeneous character of the OH stretch absorption band of HDO. Any attempt to fit the observed shapes of the bleaching with a continuous distribution function $N(\nu_0)$ fails. The solid lines shown in Fig. 3 are model calculations based on the three state model for the transmission increase and two additional components for the absorption increase on the low frequency side (one for the excited state absorption and one for the "thermal changes" in the high frequency wing of the OD absorption). The various components are indicated in Fig. 3 by dashed lines. The parameters obtained from this model are published elsewhere (Graener et al 1991b).

Additional time resolved measurement gave a vibrational lifetime for the OH stretch vibration of HDO of 8 ± 2 ps. It is necessary to discuss the validity of Eq. 3 for a liquid system. It is generally believed that the hydrogen bonds fluctuate at least on the picosecond time-scale so that the assumption of a time independent distribution function is questionable. The different shapes of the transient spectra on other the hand can only be explained if some structural components of the vibrationally excited sample have a lifetime of several picosecond.

In summary it is shown, that transient hole burning with picosecond infrared pulses is a powerful tool to investigate inhomogeneous broadend vibrational absorption bands. The analysis of the transient spectra allows a detailed insight in the "quasi"-homogeneous substructure. Additionally information about the vibrational lifetime can be obtained.

Moerner W E (Ed.) 1987 'Persistant Spectral Hole Burning (New York: Springer) and literature cited there

Shimoda K (Ed.) 1976 'High Resolution Laser Spectroscopy (Berlin: Springer)

Graener H, Seifert G and Laubereau A 1990a Chem. Phys. Lett. **172** 435

Graener H, Ye T Q and Laubereau A 1990b Phys. Rev. **B 41** 2597

Graener H and Laubereau A 1991a J. Phys. Chem. **95** 3447

Graener H, Seifert G and Laubereau A 1991b Phys. Rev. Lett. **66** 2092

Resonance interactions of 10 μm picosecond pulses with polyatomic molecules

V M Gordienko, Z A Biglov, E O Danilov and V A Slobodyanyuk

Nonlinear Optics Laboratory, Moscow State University, 119899, Moscow, USSR

ABSTRACT: The multiphoton absorption MPA in SF_6 and C_2H_4 was studied using 10 μm picosecond laser radiation. The results of third harmonic generation THG in C_2H_4 are presented.

1. INTRODUCTION

The resonance infrared multiphoton excitation of molecules results in a vibrational heating which is essentially higher than that of the buffer gas. The strongly excited molecules are not in equilibrium with the environment and may overcome a high activation barrier under lower temperature conditions. To study MPA in polyatomic molecules it is necessary to develop IR picosecond laser systems and spectroscopic diagnostic methods.

2. EXPERIMENTAL

An experimental setup elaborated by Gordienko et al (1990) consists of a tunable low power 10 μm picosecond pulse generator and a high-pressure TE-CO$_2$ (TEHP-CO$_2$) regenerative amplifier. Picosecond 10 μm pulses are generated by the two-stage parametric amplification of tunable cw-CO$_2$ laser seed radiation ($\sim$1 W/cm^2 intensity) in proustite crystals in the field of the YAL picosecond laser with negative feedback control. TEHP CO$_2$ regenerative amplifier produced a train of 7 pulses (FWHM), (pulse duration τ_p=6 ps) separated by 10 ns. Maximum pulse energy was about 2 mJ. Its output wavelength changed by tuning cw CO$_2$ seed laser, the output laser system spectrum is controlled by the injected pulse spectrum.
For some MPA measurements we used TEHP CO$_2$ regenerative module as broad band oscillator without injection (pulse duration τ_p=75 ns FWHM, output energy 25 mJ). Its wavelength was adjusted by intracavity NaCl prism. Tunable cw injection locking of TEHP-CO$_2$ module was used to produce nanosecond pulses with narrow spectrum (τ_p=75 ns FWHM, W_p= 25 mJ). In our experiments we also used 10 μm picosecond pulses chirped by phase cross modulation in CdSe. It was noted by Biglov et

al (1991), that electron density wave (EDW) can be generated in semiconductors by the two-photon absorption (TPA) of pump pulse, which is synchronized with pulse to be chirped. EDW in CdSe was obtained by TPA of the pump picosecond pulse of YAL laser. Regeneratively amplified chirped pulses demonstrated the spectral broadening up to 3 cm^{-1}. Laser radiation was attenuated by CaF_2 slabs. The optoacoustic cell was used for MPA measurements. The CO_2 laser beam was monitored by two detectors, at the input and the output from the cell. To focus the laser beam a 11-cm focal length NaCl lens was used.

3. MPA MEASUREMENTS

MPA spectra of 10 μm picosecond radiation in SF_6 (P=2 Torr) and C_2H_4 (P=5 Torr) were obtained in region 938-952 cm^{-1}. The average number of photons absorbed per molecule <n> has been measured. Both spectra demonstrate MPA peaks observed by Bagratashvili et al (1975). The peaks located in the vicinity of linear absorption peaks of SF_6 and C_2H_4 are emphasized. Figure 1 shows dependences of average energy absorbed in quanta per molecule <n> on fluence in the focal range (log-log scale) for 2 Torr SF_6 and 5 Torr C_2H_4 irradiated by a train of the 10-μm picosecond pulses, by broad spectrum nanosecond pulses and by narrow spectrum nanosecond pulses near 10 P(16) CO_2 line. For picosecond radiation in the range from 1 to 20 J/cm^2 a straight line with slope 1 can be drawn through the data on a log-log plane.

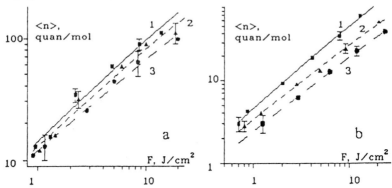

Figure 1. Dependence of absorbed energy on fluence for SF_6 (a) and C_2H_4 (b). 1 - picosecond " broad band " radiation, 2 - nanosecond " broad band radiation ", 3 - nanosecond " narrow band " radiation.

In case of SF_6 (Figure 1a) the points corresponding to the narrow band nanosecond pulses lie below the ones of broad band nanosecond pulses, but there is almost no difference between the latter and the line corresponding to the train of picosecond pulses. In case of ethylene (Figure 1b) there is a significant difference between the three lines: the line cor-

responding to the train of picosecond pulses lies above that of broad band nanosecond pulses and the latter is above the points corresponding to the narrow band nanosecond pulses. We did not observe the saturation effect in MPA experiments.
The average energy absorbed is higher than the dissociation limit for both molecules. The collisional energy transfer from excited molecules to "vibrationally cool" products of dissociation, that had been observed by Bagratashvili et al (1984), under our experimental conditions can take place between pulses of train (10 ns). A molecule can therefore absorb laser energy from each pulse, while the stored vibrational energy is remaining below the dissociation threshold. It was suggested by Alimpiev et al (1976), that the MPA dependence on the laser bandwidth could be explained by the presence of "holes" in the molecular vibrational level structure. Our results show, that it is necessary to add to this conception the suggestion, that the absorption is sensitive to laser intensity too. The Stark broadening and shift of energy levels allow to overcome the molecular anharmonicity and act like the laser spectrum broadening. The absorbed energy dependence become negligible when the spectral "hole" width is comparable to laser bandwidth.

4. INVESTIGATION OF THIRD HARMONIC GENERATION IN SF_6 AND C_2H_4.

We report the first experiments on the THG by picosecond 10 µm radiation in SF_6 and C_2H_4. Experimental setup is similar to one described in section 3. The obtained spectrum of THG in C_2H_4 (Figure 2) is similar to the previously obtained by Varakin and Gordienko (1981) data.

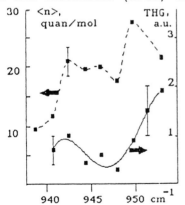

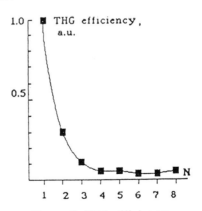

Figure 2. Generated TH energy (solid) and absorbed laser energy (dashed) versus laser frequency for C_2H_4.

Figure 3. THG efficiency evolution along the train, N - pulse number.

The third harmonic (TH) radiation was detected by fast HgCdTe photodiode with a time constant of 5 ns. The TH output power reached a maximum in C_2H_4 with THG efficiency 10^{-8} (10 P(14)

CO_2), maximum THG efficiency in SF_6 was $2 \cdot 10^{-9}$(10 P(20) CO_2). As mentioned above our experimental setup can produce a picosecond "blue" chirped pulses. It was calculated by Chelkowski et al (1990), that if the pulse frequency changes at a definite rate, adapted to the molecular anharmonicity the molecules can be effectively excited to high vibrational level. Mode $v7$ in C_2H_4 has negative anharmonicity 2 cm^{-1}. Our MPA experiments did not show essential growth of the absorption in C_2H_4 when the "blue" chirped pulses near 10 P (14) had been used. But we observed a 1.5 times growth of the TH signal. When pulse frequency is increasing as a function of time in such a way that it is resonant with transition between low vibrational levels of C_2H_4, the TH signal increases.

Using a fast HgCdTe photodiode and " drag " detector we can resolve simultaneously single pulses in THG and 10 μm trains. Therefore it was possible to investigate the evolution of the THG efficiency along the train of pulses. Incident 10 μm radiation spectrum was located near 10 P(14) CO_2 line. THG experiments were performed at 5 torr C_2H_4. MPA measurements show, that molecules C_2H_4 can absorb up to 100 quanta during the train. Under our experimental conditions there is a slow thermalization and thermal diffusion during the train. Therefore the THG efficiency of N-th pulse in the train depends on vibrational heating resulting from previous N-1 pulses. For comparatively low intensities we observed the cubic dependence of THG. Under that assumption, the quick decrease of the THG efficiency along the train is an evidence of decrease of nonlinear susceptibility in the vibrationally excited C_2H_4 molecules.

CONCLUSIONS.

These are the first reported experiments, where 10 μm picosecond laser system has been used to excite and to probe by third harmonic generation the processes of the multiphoton excitation of polyatomic molecules.

REFERENCES

Alimpiev S S et al 1976 Opt. Commun. **25** 512
Bagratashvili V N et al 1975 Opt. Commun. **14** 426
Bagratashvili V N et al 1984 Sov. J. Chem. Phys. 3 1386
Biglov Z A et al 1991 Sov. J. Izvestia Academ.Nauk, Ser. Fizicheskaya, **55** 337
Chelkowski S et al 1990 Phys Rev. Lett. **19** 2355
Gordienko V M et al 1990 Spring. Proc. Phys. "Ultrafast Phen. in Spectrosc.", ed E Klose (Spring.-Verl.) **49** pp 94-101
Varakin V N, Gordienko V M 1981 Sov. J. Quantum Electron, **11**, 961

Inst. Phys. Conf. Ser. No 126: Section VI
Paper presented at Int. Symp. on Ultrafast Processes in Spectroscopy, Bayreuth, 1991

481

Photophysics of the higher excited singlet state of diphenylacetylene

Yoshinori Hirata, Tadashi Okada, Noboru Mataga, and Takeo Nomoto[+]

Department of Chemistry, Faculty of Engineering Science and
Research Center for Extreme Materials, Osaka University, Toyonaka, Osaka 560, Japan
[+]Department of Chemistry, Faculty of Education, Mie University, Tsu, Mie 514, Japan

ABSTRACT: The photophysical properties of diphenylacetylene in the solution phase have been investigated by picosecond transient absorption spectrum measurements. The absorption bands at 435 and 700 nm, assigned to the $S_n < - S_1$ transition, decay with a lifetime of about 200 ps, while the short lived band at 500 nm is ascribed to the higher excited singlet state. Large temperature effect of the S_2 state lifetime might suggests the change of the $S_1 - S_2$ coupling from the intermediate case to the statistical limit.

1. INTRODUCTION

The lifetime of large organic molecules in the higher excited singlet states is usually much shorter than that of the S_1 state. Therefore, except few molecules such as azulene, no fluorescence from the higher excited states is detected. Diphenylacetylene (DPA) seems to be a unique molecule since it was reported to show a dual fluorescence in the singlet manifold in spite of the small interstate separation. DPA might be characterized by the fact that the lowest three excited singlet states are in the close proximity ie. within a few hundreds of wavenumbers and only one of them is allowed for one photon transition from the ground state. According to the MO calculation (Tanizaki 1971), it is predicted that three excited singlet states are contained in the first absorption band of DPA. They are B_{2u}, B_{1u}, and A_{1g} under D_{2h} symmetry. Only $B_{2u} < - A_{1g}$ (S_0) is allowed for the one photon transition. These facts are confirmed by fluorescence excitation and two-photon resonant four-photon ionization studies in the supersonic free jet (Okuyama 1984). They also determined that B_{2u} and A_{1g} are located at 35248 and 34960 cm^{-1}, respectively. Since only vibronically induced bands are observed for B_{1u}, the origin lies at lower than 35051 cm^{-1}.

We have measured picosecond time-resolved absorption spectra of DPA in various solvents and the short lived absorption band which should be attributed to the higher excited singlet state has been observed. The state with a lifetime of about 8 ps is a fluorescence state and a precursor of the S_1 state. The rather long lifetime of S_2 can be the results of a sparse level density of the accepting mode for the $S_2 -> S_1$ internal conversion. The excess energy dependence suggests strongly that the change of the interaction between the excited singlet states from intermediate case coupling to statistical limit with increasing excitation energy. Large temperature effect of the S_2 state lifetime also suggests this picture.

2. RESULTS

The picosecond time resolved absorption spectra and time dependence of the transient absorbance of DPA in n-hexane at room temperature excited with a 295 nm dye laser pulse are shown in Figures 1 and 2, respectively. At the longer delay times than 500 ps, the spectrum was characterized by a sharp band around 415 nm. The band, which did not show significant decay in the measured delay time range, is already assigned to the $T_n \leftarrow T_1$ transition of DPA (Ota 1974).

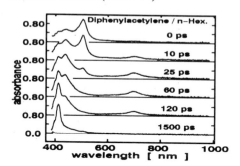

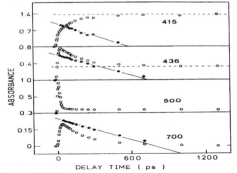

Figure 1 Picosecond transient absorption spectra of DPA in n–hexane (295 K).

Figure 2 Time dependence of the transient absorbance of DPA in n–hexane (295 K).

Immediately after excitation, the short lived band appeared around 500 nm, while the bands around 435 and 700 nm was dominant in the 30–100 ps range. At 415 nm rapid growth of the transient absorbance limited by the response time of our apparatus was followed by additional growths. The slowest one can be ascribed to the triplet formation, which is characterized by an exponential rise with a time constant of 190 ± 15 ps. The decay monitored at 700 nm was single exponential with a lifetime of 208 ± 10 ps. The time dependence of the transient absorbance measured at 435 nm can be analyzed as a biexponential decay. The long lived component was almost constant in the time range of our measurements, while the lifetime of the short lived one was 205 ± 23 ps. These results clearly show that the 435 and 700 nm bands should be due to the same transient species which was assigned to a precursor of the triplet state of DPA. The long lived component at 435 nm should be due to the triplet.

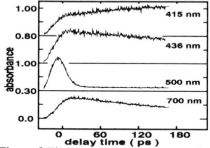

Figure 3 Time dependence of the transient absorbance of DPA in n–hexane.

The time dependence of the transient absorbance in the shorter delay times are shown in Figure 3. The smooth lines represent the simulation curves calculated with the assumption of the double exponential decay kinetics. By using the excitation and interrogation pulse widths of 9 and 18 ps, respectively, the decay time of the short lived component at 500 nm was determined to be about 8 ps. The rise at 700 nm was in agreement with the decay time at 500 nm. Similar time dependence of the transient absorption spectra observed in other solvents and the determined rise and decay times at various wavelengths were listed in Table I. No significant solvent effect of the decay times was observed. These results suggest that the deactivation process of DPA excited with a near ultraviolet light pulse is shown by the scheme:

$$S_0 \xrightarrow{\ h\nu\ } X \longrightarrow Y \longrightarrow T_1,$$

where X and Y show the transient species responsible for the 500 and 700 nm band, respectively. Since the fluorescence lifetimes were in an agreement with the decay time of the 500 nm band, the fluorescence state should be X, while the triplet state was formed from the dark state, Y.

Table I. Rise and Decay Times (ps) of The Transient Absorbance.

Solvent	415 nm rise	436 nm decay	500 nm decay	700 nm rise	700 nm decay
i–Pentane	176±14	181±17	8	8	195±13
n–Hexane	190±15	205±23	8	8	208±10
c–Hexane	203±25	206±18	9	9	204±16
Methylcyclohexane	201±16	207±21	8	9	210±14
t–Decalin(293 K)	185±17	198±24	9	9	195±15

Drastic temperature effect was observed for the rise and decay times of these transient absorbance in the region between room temperature and freezing point of the solvent. The lifetime of the 500 nm band increased with decreasing temperature. Figure 4 shows Arrhenius plots for the fluorescence lifetime, the decay time measured at 500 and 700 nm, and the rise time at 415 nm in several solvents. The Arrhenius plot for the lifetime of the 500 nm band showed a good linear relation except for the higher temperature region and the activation energies and the frequency factors deduced from the plot are listed in Table II. In case of t–decalin, both the activation energy and frequency factor may be over estimated since the measured temperature region was not so wide as other solvents and the plot may not show a linear relation in this region.

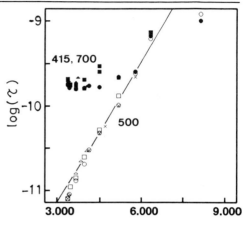

Figure 4 Arrhenius plot for the S_2 and S_1 Lifetimes of DPA. (○: i–pentane, △: t–decalin, □: methylcyclohexane, ✕: Fluorescence lifetime)

Table II. Activation Energies (cm⁻¹) and Frequency Factors ($10^{12}s^{-1}$) in several Solvents

	i–Pentane	n–Hexane	Methyl–cyclohexane	t–Decalin	Ethanol
Activation Energy	843	891	940	1100	925
Frequency Factor	4.2	6.3	7.8	18.2	7.1
Activation Energy of Diffusion	520	590	600	1160	1000

3. DISCUSSION

Our observations clearly show that the dynamic behavior of excited DPA is different from that of the usual aromatic molecules. The fluorescence state is short lived X, although we cannot determine yet whether it is $^1B_{2u}$ or $^1B_{1u}$. We do not know the assignment of the precursor of the T_1 state, Y, either. Some possible assignments are as follows. Group 1 is concerned with the configurational relaxation, while Group 2 is including the electronically higher excited state.

(1–1) X is the S_1 (B_{2u}) state of DPA and Y is the configurationally different singlet state.
(1–2) X is S_1 (B_{2u}) and Y is the triplet state of which the configuration is different from that of the stable T_1 state.

(2–1) X is the higher excited singlet state and Y is the S_1 state. In this case X should be B_{2u} and Y should be the optically forbidden state, B_{1u} or A_{1g}.

(2–2) X is the thermally populated higher excited singlet state with a large oscillator strength between the ground state and Y is optically forbidden S_1. This is similar to the possibility 2–1, although the lifetime of X should be determined by the cooling time.

(2–3) X is S_1 and Y is T_2.

The configurational changes (1–1, 2) in the excited state may be ruled out because only little solvent effect was observed for the lifetimes of X and Y. The temperature effect should provide stronger evidence to reject the possibilities. As listed in Table II, the activation energy of the depopulation process of X was independent on solvent, although the activation energy of the diffusion deduced from the viscosity of solvent increases almost two times by changing the solvent from i–pentane to t–decalin or ethanol. On the other hand, the decay kinetics of Y did not show significant temperature effect and the activation energy should be much smaller than that of the diffusion.

Possibility (2–2) should be related to the interesting phenomena concerned with the dissipation of the vibrational energy into the solvent. Following the intramolecular vibrational redistribution, the local temperature of the excited molecule becomes high and the excess energy is transferred to solvent molecules in the surroundings. Such process is known to occur in the picosecond to tens of picosecond time scale (Seilmeier 1988). If this is the case, the lifetime of X should be determined by the excess energy of excitation. Since the lifetime of X did not show significant excitation energy dependence and even by the 0–0 band (296 nm) or the hot band (305 nm) excitation the 500 nm band was observed, the possibility (2–2) can be ruled out. Another evidence to reject such possibility is a extremely long lifetimes of X at low temperatures. Possibility (2–3) may be rejected since the lifetime of Y was too long for the electronically higher excited state (T_2).

Possibility (2–1) can be the most plausible case. The three lowest excited singlet state of DPA are located closely in energy and only the transition $B_{2u} \longleftarrow A_{1g}$ is allowed, therefore the excitation at 295 nm should lead mainly to B_{2u}. Because of the small energy gap of less than 300 cm^{-1} between B_{2u} and S_1, the interstate coupling should be classified not to the statistical limit but to the intermediate case. If the S_2–S_1 energy separation is sufficiently large, the coupling should be the statistical limit and may result in quite short lifetime of S_2. Phenanthrene provides an example of such a case, of which the S_2–S_1 separation is about 6000 cm^{-1}. Measuring the band width of the $S_2 \longleftarrow S_0$ absorption in a supersonic beam, Amirav et al. (1984) estimate that the lifetime of S_2 is 0.5 ps. On the contrary, $S_2 \longrightarrow S_1$ internal conversion of DPA should belong to the intermediate case, where the level density is too low to ensure the ultrafast internal conversion.

The observed large temperature effect on the S_2 lifetime may suggest that the rate of the $S_2 \longrightarrow S_1$ internal conversion increases rapidly with increasing excess energy. The frequency factor of 10^{13} s^{-1} appeared to be large enough for the usual $S_2 \longrightarrow S_1$ internal conversion of aromatic molecules. These results indicate that the change of the S_2–S_1 interaction from the intermediate case coupling to the statistical limit is observed by changing the excess energy. The activation energy of about 900 cm^{-1} may be the excess vibrational energy where the $S_2 \longrightarrow S_1$ internal conversion can compete with the vibrational relaxation in the liquid phase. We observed the decrease in the fluorescence yield with decreasing excitation wavelength, which also supports our mechanism.

REFERENCES

Amirav A, Sonnenshein M, and Jortner J 1984 *J. Phys. Chem.* **88** 5593

Okuyama K, Hasegawa T, Ito M, and Mikami N 1984 *J. Phys. Chem.* **88** 1711

Ota K, Murofushi K, and Inoue H 1974 *Tetrahedron Lett.* 1431

Seilmeier A, Kaiser W 1988 *Ultrashort Laser Pulses and Application* ed Kaiser W (Berlin: Springer)

Tanizaki Y, Inoue H, Hoshi T, and Shiraishi J 1971 *Z. Phys. Chem. NF* **74** 45

Inst. Phys. Conf. Ser. No 126: Section VI
Paper presented at Int. Symp. on Ultrafast Processes in Spectroscopy, Bayreuth, 1991

Time-resolved optical spectroscopy of donor-acceptor-substituted polyenes

G. Quapil and H. Port

3. Physikalisches Institut, Universität Stuttgart,

W-7000-Stuttgart 80, Germany

Linear polyenes of different chain lengths terminally substituted with electron donor and acceptor groups have been investigated with cw and time resolved optical spectroscopy using single photon counting (system response 30 ps) and transient absorption techniques (system response 200 fs).

INTRODUCTION

Previously [1] we have demonstrated that in linear polyenes, with two different substituents at both ends, an intramolecular transfer of energy is possible. This paper concerns systems with anthracene as electron donor and methyl **1**, pyridine **2** or pyridinium **3** as acceptor synthesized by C.P. Niesert [2]. These compounds are of interest with respect to a possible photoinduced intramolecular electron-transfer. Generally there is a widespread interest considering substituted polyenes as posssible 'molecular wires' for intramolecular electron-transfer.

1 $n = 0, 1, 2, 3, 4, 5$ **2** $n = 0, 1, 3, 5, 7, 9$ **3** $n = 0, 1, 3, 5, 7, 9$

ENERGY LEVELS

The absorption spectra of the molecules **1**, **2** and **3** exhibit features only partly characteristic for the individual partners. Due to the coupling between the substituents and the polyene in the electronic excited state structural details vanish. Additional bands appear varying in band position and absorption strength with increasing chain length. Typically four absorption bands $B_i (i = 1..4)$ can be distinguished in the spectra (Fig. 1). The energetic position of B_3 is found to be independent of chain-length. This band can be assigned to anthracene $S_3 \leftarrow S_0$ absorption. The bands B_1, B_2 and B_4, instead, generally are red-shifted with increasing chain length n. As an example in Fig. 2a the band position of B_1 observed for molecules **1**, **2**, **3** at 300K is plotted as a function of n. Clearly molecules **3** reveal an additional red-shift as compared to **1** and **2**, the amount of which diminishes, however, for large n. A theoretical description of the observed effects is currently in progress [3].

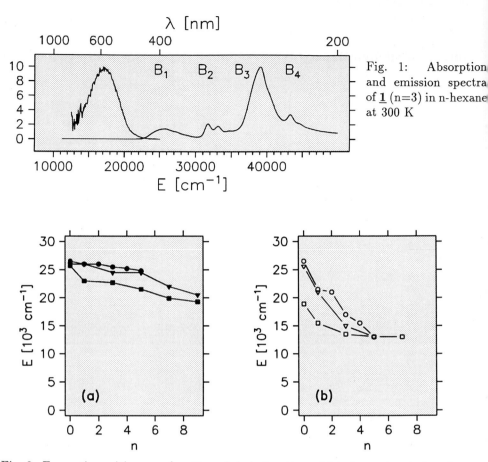

Fig. 1: Absorption and emission spectra of **1** (n=3) in n-hexane at 300 K

Fig. 2: Energetic position as a function of chain length n of (a) absorption (**1** in n-hexane ●, **2** in n-hexane ▼, **3** in acetonitrile ■)(b) emission bands (**1** ○, **2**▽, **3**□)

The fluorescence spectra of molecules **1**, **2** and **3** consist of a single broad band, which is more strongly red-shifted with increasing n than the corresponding lowest absorption band (Fig. 2b). The difference in fluorescence band position of molecules **3** with respect to **1** and **2** is very large ($> 5000 cm^{-1}$) at small n, it levels off at large n.

EXCITED STATE DYNAMICS

The lifetime of the lowest excited state has been deduced from both fluorescence decay and transient absorption measurements at 300K. In the overlapping time range (between 10ps and 1ns) both methods yield the same result. The lifetimes are drastically reduced with increasing chain length n (Fig. 3), up to n=9 by about four orders of magnitude. The relative change of τ is larger for molecules **3** than for **1** and **2** at $n \leq 3$, but smaller at $n > 3$. For the build-up time of the excited state an upper limit of about 100fs can be given.

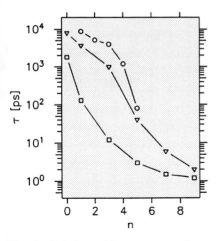

Fig. 3: Lifetime of lowest excited state $\tau(n)$ at 300 K (**1**○, **2**▽, **3**□)

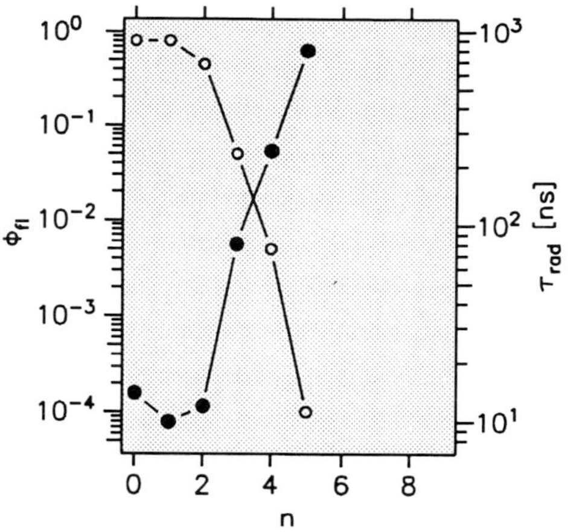

Fig. 4: Quantum yield $\Phi(n)$ ○ and radiative lifetime τ_{rad} for **1** ●

In correspondence with the fluorescence lifetimes a drastic decrease of the fluorescence quantum yield Φ is observed as a function of n. An example of $\Phi(n)$ for molecules **1** is given in Fig. 4 together with values for the radiative lifetime τ_{rad} calculated from Φ and the corresponding τ numbers taken from Fig. 3. τ_{rad} increases with n.

TEMPERATURE-DEPENDENT FLUORESCENCE

In 2-MTHF solution a pronounced fluorescence blue-shift as well as an increase of the fluorescence decay times is observed between 300K and 100K. An example for this behaviour is given in Fig. 2 for the pyridinium compound **3** ($n = 3$). The decay times are 12 ps (300K), 370 ps (140K), and 2.1ns (100K). – Between 100 K and 10 K only small further changes of spectra and decay times occur.

In time-resolved fluorescence spectra at T > 100K a monotonic red shift of the fluorescence band is observed with increasing time delay after ps pulse excitation. A sequence of time-resolved spectra measured at 140 K is shown for the example in Fig. 5. The observed dynamical Stokes shift is due to solvent-solute interaction. It can be quantified [4] and correlated tentatively with the change $\Delta\mu$ of the molecular dipole moment for the solute occuring between ground and first excited state. In this way estimates for $\Delta\mu$ are obtained, 3 Debye and 6 Debye for molecules **1** and **3** at n=3, respectively.

CONCLUSIONS

In this work strong chain length and substitution effects on the excited state properties of substituted polyenes have been found. Spectral and time-resolved measurements lead to the following conclusions concerning S_1:

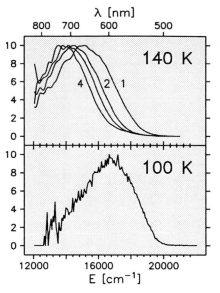

Fig. 5: Temperature dependent time-resolved fluorescence spectra of **3**$(n = 3)$ in 2-MTHF.
Time intervals at 140 K:
(1) -25..50ps
(2) 70..190ps
(3) 140..260ps
(4) 610..730ps

A lowering of the excited state energy and a shortening of its lifetime with chain length n has been observed. These effects can be explained by an extension of the electronic π-system which occurs by (i) increasing the polyene chain length and (ii) extending the conjugation into the substituent.

The concomitant increase of the energy gap between absorption and emission bands, typical for polyenes [5], is due to the increasing separation of the electronic states S_1 and S_2.

This indicates that the lowest excited state, which is anthracene-like at small n $(n \leq 2)$ (in energy, τ and τ_{rad}), becomes polyene-like at large n.

The more pronounced red-shift and lifetime shortening in molecules **3** indicate an effective charge separation in the lowest excited state (c.f. ref. [6,7])

The chain length dependence of transition energies and lifetimes saturates at large n and becomes substituent-independent, indicating that the conjugated chain properties dominate.

Helpful discussions with H.C. Wolf are gratefully acknowledged. We thank P. Emele, N. Holl, D. Meyer and J. Zielbauer for experimental assistance. Financial support is provided by the Deutsche Forschungsgemeinschaft.

REFERENCES

[1] St. Maier, H. Port, H.C. Wolf, F. Effenberger, H. Schlosser; Synth. Metals 29 (1989) E517
[2] Universität Stuttgart, Institut für Organische Chemie, Arbeitsgruppe Prof. Effenberger
[3] P. Gribi, Dissertation, Universität Stuttgart, in prep.
[4] V. Nagarajan, A.M. Brearley, T.J. Kang, P.F. Barbara; J.Chem.Phys. 86 (1987) 3183
[5] B.S. Hudson, B.E. Kohler, K. Schulten; in 'Excited States' Vol.6, E.C. Lim ed., Academic Press (1982)
[6] M.R. Wasielewski, D.G. Johnson, W.A. Svec, K.M. Kersey, D.E. Cragg, D.W. Minsek; in 'Photochemical Energy Conversion', J.R. Norris Jr., D. Meisel, Elsevier (1988) p.135ff
[7] A. Slama-Schwok, M. Blachard-Desce, J.M. Lehn; J.Phys.Chem. 94 (1990) 3894

Nonlinear excited state dynamics of J-aggregates on the femtosecond time-scale

R. Gagel, R. Gadonas [+], A. Laubereau
University of Bayreuth, Germany
[+]Kapsukas University, Vilnius, Lithuania

ABSTRACT: J-aggregates in aqueous solution are investigated with tunable fs-pulses derived from a pulsed dye laser system with subsequent continuum generation. The high time resolution of our laser system provides new results on fast processes not accessible in previous studies.

Since the discovery of J-aggregates by Jelley (1936) and Scheibe (1937) the physical properties of this molecular system were adressed by many theoretical and experimental studies. The molecular aggregates are formed by the dye Pseudoisocyanine chloride (1,1'-diethyl-2,2'-cyanine chloride, PIC) in aqueous solution. Aggregation starts at concentrations $c \geq 10^{-3}$ M and temperatures $T < 60°C$ (Kopainsky 1981). It is accompanied by a rise of viscosity and by a substantial change of the optical absorption spectrum; an intense narrow absorption band appears at 573 nm, the so-called J-band.

In the recent years time-resolved measurements at the J-band showed that the fluorescence lifetime of the J-aggregates is strongly intensity-dependent. An exciton annihilation process in the excited state was proposed to be the responsible mechanism for this behaviour (Brumbaugh 1984, Stiel 1988). Sundström et al (1988) used a rate equation model for bimolecular exciton annihilation in order to describe this intensity-dependent relaxation dynamics.

Using a femtosecond-laser-system we have investigated the J-aggregates in time-and intensity-ranges not accessible in previous measurements. New results are presented on the dynamical behaviour of the molecular aggregates in the first 10^{-13} to 10^{-15} s after excitation of the J-band.

The experimental system is depicted in figure 1. The laser-system consists of a pulsed Rh6G/DQOCl dye oscillator with intracavity dispersion compensation, synchronously pumped by a feedback-controlled modelocked (FCM) Nd:glass laser with 8 Hz repetition rate (Angel 1989). A single dye laser pulse is subsequently amplified in a two stage Rh6G dye cell configuration applying a pulse stretching-recompression scheme. In this way pulses of 60 μJ are generated at 565 nm with a pulse duration of 50 fs (Gagel 1990). One part of the amplified pulse serves as an intense pump pulse for the experiment. The other part is focussed into a fused silica plate of 2 mm thickness for continuum generation. The desired wavelength interval of width 12 nm is selected from the fs-continuum by the help of a graded interference filter. The resulting fs-pulses are tunable from 400-670 nm and serve as weak probes for the time-resolved spectroscopy.

The measurements were performed with a sample cell of 1 mm length at room temperature. Exciting the aggregates with the intense pump pulse at 565 nm the probe pulses detect changes in optical density $\Delta OD = -\log(T/T_0)$ of the sample as a function of time at various wavelength positions. With the high accuracy of our measurements we are able to detect small absorption changes of the order of 0.1 %. To verify the influence of exciton annihilation in the excited electronical state the excitation intensities are varied between 10^9 and 10^{10} W/cm².

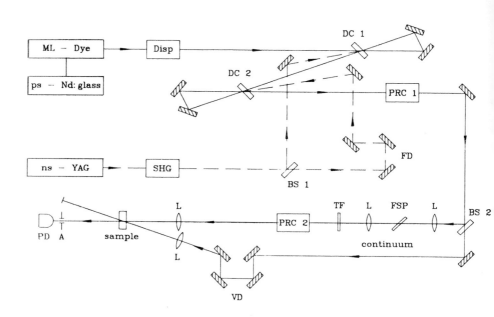

Figure 1: Experimental setup with Nd-glass laser, modelocked (ML) dye-laser and ns-YAG pumplaser. Disp: dispersive delay line for pulse stretching, DC: dye cell, PRC: prism compressor, BS: beam splitter, SHG: second harmonic generation, FD: fixed delay, VD: variable delay, L: lens, FSP: fused silica plate, TF: graded interference filter, A: aperture, PD: photodiode.

Figure 2 is the result of a measurement with a probing wavelength of 580 nm, i.e. at the long wavelength side of the excitonic resonance at 573 nm. It is interesting to note the induced absorption peak at time delays $t_D < 0$, when the probe pulse precedes the peak of the pump pulse in the sample. This induced absorption signal is caused by coherent interaction of the probe pulse induced polarization of the medium with the electrical field of the pump pulse, a so-called precursor of the optical Stark-effect (Lindberg 1988). The observation of this effect in our case is due to strong off-resonant pumping of the excitonic band and a dephasing time of the electronic transition comparable to the pulse duration. For $t_D \approx 0$ the induced bleaching signal of the J-band is evident. It disappears rapidly within a few hundreds of femtoseconds (see fig. 2). This behaviour indicates a very short relaxation time of the excitonic transition caused by effective exciton annihilation in the excited S_1-state. At longer time delays $t_D \geq 1$ ps induced absorption occurs which rises with a time constant of 3.0 ± 0.5 ps at the probing wavelength of 580 nm. The induced absorption finally disappears with a relaxation time of 18 ± 5 ps that is determined by a measurement on a longer time scale (not shown in fig. 2). We propose that this absorption arises from excited vibrational modes of the ground state S_0 (see below). The relaxation time of 18 ps is of the order of ground state time constants for the energy transfer of vibrational modes of solute molecules to the surrounding solution (Gottfried 1984). The solid line in figure 2 and the following is the result of computer simulations carried out for a better understanding of the transient processes in the dye aggregate.

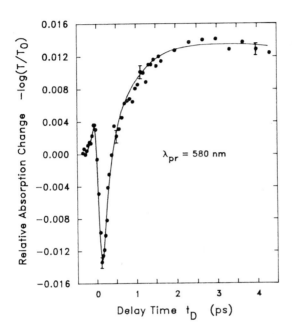

Figure 2: Time resolved measurement at a probe wavelength of 580 nm. Solid line: result of numerical calculation.

A variation of the excitation intensity of the pump pulse leads to the results shown in figure 3. The measurements are performed using excitation intensities of 7.5×10^9 W/cm² (a), 2.5×10^9 W/cm² (b) and 1.0×10^9 W/cm² (c) and a probe wavelength of 580 nm. The figure clearly demonstrates the influence of pump intensity on the dynamic Stark signal which rises with increasing intensity as expected from theory. The different relaxation times of the induced bleaching signal in the three measurements strongly support the presence of exciton annihilation in the excited electronic state. With high excitation intensity and correspondingly large density of excitons the annihilation process becomes more effective resulting in a faster relaxation rate of the excited state compared to the low intensity case. Incorporating a decay term for bimolecular exciton annihilation (Sundström 1988) in our model calculations, an annihilation constant of $\gamma = (2.5 \pm 0.5) \times 10^{-3}$ cm³/s fully describes the observed dynamics in the investigated intensity range. The induced absorption for time delay $t_D \geq 1$ ps discussed in context with figure 2 occurs only at the two highest intensities applied here. At the lower pump intensity of 1.0×10^9 W/cm² induced absorption is not observed in this time interval.

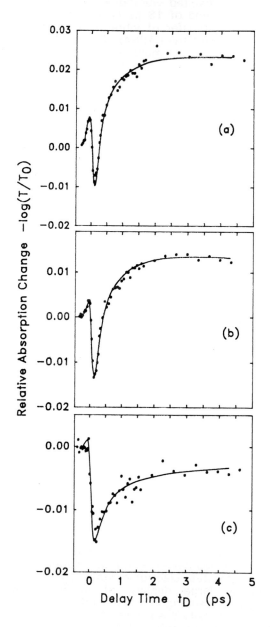

Figure 3: Measurement at a probe wavelength of 580 nm with different excitation intensities
a) 7.5×10^9 W/cm²
b) 2.5×10^9 W/cm²
c) 1.0×10^9 W/cm².
Solid line: result of model calculation.

The dynamic behaviour at a probe wavelength of 565 nm, i.e. the same wavelength as the pump pulse, is depicted in figure 4. In contrary to the results on the low-energy side of the excitonic resonance a strong absorption is induced at zero time delay superimposed by the well-known coherence peak.

This induced absorption is accounted for by excited state absorption starting from the excitonic level. A lifetime of the final S_n-state of 200 ± 100 fs is deduced from a comparison of our computer simulation with the measured data in this frequency region.

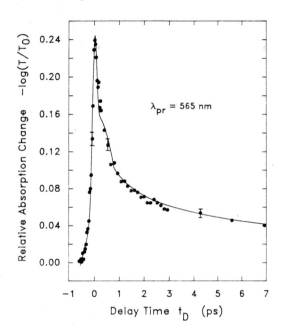

Figure 4: Time resolved measurement at a probe wavelength of 565 nm. The solid line is calculated.

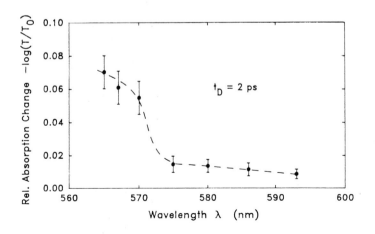

Figure 5: Measurement of maximal absorption change at a time delay of 2 ps as a function of wavelength position.

The induced absorption for time delays $t_D \geq 1$ ps is observable in a wide wavelength range, e.g. at a probe wavelength of 650 nm far below the excitonic resonance with a risetime of 0.8 ± 0.2 ps. An examination of the strength of the absorption changes at various wavelength positions leads to the result depicted in figure 5. We propose secondary excitation of vibrational levels of the ground state S_0 by exciton annihilation to be the responsible mechanism for this behaviour. The measured induced absorption signal originates from transitions from strong vibrational modes of the S_0 to the upper part of the first electronical state. Dominating vibrational modes around 1360 cm^{-1} have been observed by Raman spectroscopy measurements (Mejean 1977) and were necessary to describe the absorption spectrum of the aggregates theoretically. Based on the small-signal absorption spectrum we made an estimate of the strength of the absorption starting from modes around 1360 cm^{-1}. The result agrees qualitatively with the behaviour in figure 5, i.e. stronger absorption changes at the short-wavelength side of the excitonic resonance compared to the low-energy side.

Our interpretation is supported by computer simulations based on an extended three level model. An expansion of the density matrix to first order in the probe field includes coherent effects like the observed dynamic Stark shift. Incorporating in addition bimolecular exciton annihilation with a corresponding rate equation our model fully describes the observed intensity dependent signal transients with an annihilation constant of $\gamma = (2.5 \pm 0.5) \times 10^{-3}$ cm^3/s.

Literature:

Angel G., Gagel R., Laubereau A., 1989, Opt. Lett. 14, 1005
Brumbaugh D.V., Muenter A.A., Knox W., Mourou G., Wittmershaus B.,
    1984, J. Lumin. 31-32, 783
Gagel R., Angel G., Laubereau A.,1990, Opt. Commun. 76, 239
Gottfried N.H., Seilmeier A., Kaiser W., 1984,
    Chem. Phys. Lett 111, 326
Jelley E., 1936, Nature 12, 1009
Kopainsky B., Hallermeier J.K., Kaiser W., 1981,
    Chem. Phys. Lett. 83, 498
Lindberg M., Koch S.W., 1988, Phys. Rev. B38, 7607
Mejean T., Forel M.T., 1977, J. Raman Spectr. 6, 117
Scheibe G., 1937, Angew. Chem. 50, 212
Stiel H., Daehne S., Teuchner K., 1988, J. Lumin. 39, 351
Sundström V., Gillbro T., Gadonas R.A., Piskarskas A., 1988,
    J. Chem. Phys. 89, 2754

Inst. Phys. Conf. Ser. No 126: Section VI
Paper presented at Int. Symp. on Ultrafast Processes in Spectroscopy, Bayreuth, 1991

Ultrafast optical dynamics and nonlinear susceptibility for resonant transitions of excited molecules and aggregates

V Bogdanov, S Kulya, A Spiro

S I Vavilov State Optical Institute, 199034, St.Petersburg, USSR

ABSTRACT: The magnitude and dispersion of nonlinear cubic susceptibility $\chi^{(3)}$ have been measured for resonant transitions of excited organic molecules and J-aggregates of pseudoisocyanine (PIC) in liquid solutions. The pathways and rates of relaxation for high-excited electronic states have been determined. There is a pronounced increase in $\chi^{(3)}$ during the aggregation of molecules. The dephasing time of the exciton transition, the absorption cross section, the concentration and size of the "optical" J-aggregates have been determined.

1. INTRODUCTION

The "giant" nonlinear optical susceptibility can be expected for polyatomic organic molecules in excited electronic states (Akhmanov 1990) and in molecular aggregates (Spano and Mukamel 1990) due to electrons delocalization and excitonic resonances formation. The magnitude of resonant nonlinear susceptibility for such systems is determined by their optical dynamics and, first of all, by the dephasing process which is responsible for the size of delocalization domain.

We studied the dynamical optical properties of excited molecules and J-aggregates of PIC (Jelley 1936) by four-wave mixing (Yajima and Souma 1978) during biharmonic pumping by narrow-band laser beams at frequencies ν_1 and ν_2. The dispersion of $\chi^{(3)}$ was found from the behaviour of the intensity (I_S) of the four-wave mixing signals ($I_S \sim |\chi^{(3)}|^2$) at frequency $\nu_S = 2\nu_1 - \nu_2$ as a function of detuning $\Delta\nu = \nu_1 - \nu_2$. For pumping of two tunable dye lasers (ν_1 and ν_2) the 2-nd harmonic pulses of the nanosecond Nd:YaG laser were used. A least-squares fit of the model-based $|\chi^{(3)}(\Delta\nu)|$ spectra to the experimental ones were carried out to yield the pathways and rates of relaxation.

2. EXCITED MOLECULES

The excited state (S_1) was populated with the pulses of the 3-rd harmonic (ν_p) of the laser which was used for

the dye lasers pumping. When three pulses (ν_p , ν_1 and ν_2)
acted upon the solutions simultaneously, the four photon
scattering signals of excited molecules were observed.

Figure 1 shows the $S_1 \to S_n$ absorption spectrum of 1,4-diphe-
nylbutadiene (DPB) in cyclohexane. Arrows indicate the positions of frequency ν_1 for bihar-
monic pumping. The experimental
spectra of $|\chi^{(3)}|$ are shown in
figure 2 (dots) at ν_1 indicated
in figure 1. The experimental
$|\chi^{(3)}|$ spectra were approximated
by the calculated ones for different models of the molecular
excited states and relaxations.
The better approximation was
achieved for the 3-level model
(S_1 is the first, S_n is the
high-excited and S_K is the in-

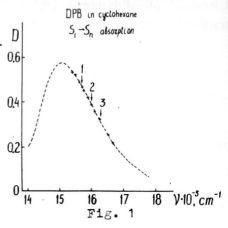

termediate electronic state for $S_n \leadsto S_1$ energy relaxation)
suggested by Souma et al. (1982). The results of such approximation are shown in figure 2 (curves) which were calculated with common set of relaxation rates for positive and negative sign of the detuning.

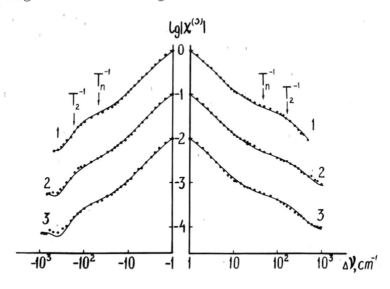

Fig. 2

The values of relaxation times were used: $T_2 \simeq T_3 \simeq 20 \div 30$ fs,
$T_n \simeq 90$ fs, $T_{nK} \simeq 100$ fs and $T_{K1} \simeq 9$ ns (T_2 , T_3 are dephasing and spectral diffusion times; is the lifetime
of the S_n state; T_{nK}^{-1} , T_{K1}^{-1} are the rates of energy relaxations on the pathways $S_n \leadsto S_K$ and $S_K \leadsto S_1$ respectively.

From these data we made conclusions about inhomogeneous broadening of the $S_1 \to S_n$ absorption band and the existence of
the "bottle-neck" for the $S_n \leadsto S_1$ relaxation (the S_K state) (Bogdanov et al. 1990). As the studies of four-wave
mixing of the other excited molecules (oxazoles, aromatics)
showed, the lifetime of the S_K state determines the magnitude of the scattering signal.

The molecular susceptibility ($<\gamma>$) of the DPB excited molecules was found to be $<\gamma> \simeq 6 \times 10^{-29}$ esu. This value is comparable with resonant $<\gamma>$ for unexcited molecules (e.g., for
Malachite Green molecules in the ground state $<\gamma> \simeq 7 \times 10^{-29}$
esu. The absence of enhancement of $<\gamma>$ for excited states
may be connected with the blocking of the electrons delocalization due to configurational change of the excited molecules in liquid solutions.

3. J-AGGREGATES

Figure 3 shows the absorption spectrum of the J-aggregates
solution of PIC chloride in D_2O with an initial molecular
concentration $N \simeq 1.6 \times 10^{18}$ cm^{-3} at room temperature.

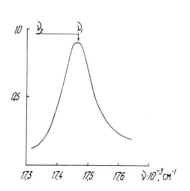

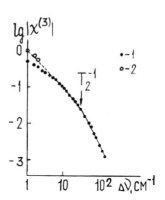

Fig. 3

Fig. 4. Dispersion of $|\chi^{(3)}|$.
The excitation intensity
is 180 kW/cm^2 (1) or
14 kW/cm^2 (2).
The curve is an approximation.

Figure 4 shows the $|\chi^{(3)}|$ spectra for $\nu_1 = \nu_m = 17460$ cm^{-1} (ν_m
is the frequency of the excitonic band maximum) and diffe-
rent pump power levels $I_1(\nu_1)$ and $I_2(\nu_2)$ (Bogdanov et al.
1991). A least fit of model-based calculation of dispersion
(inhomogeneous broadened transition of the ensemble of two-
level centres (Yajima and Souma 1978) with $T_1 \simeq 600$ ps (T_1 is

the measured lifetime of fluorescence) yields $T_2 \simeq 0.17$ ps. From this result we conclude that the excitonic band is inhomogeneous one and the dephasing time increases approximately by one order at aggregation in liquid solution.

To estimate the $|\chi^{(3)}|$ value we used the values of I_s at detuning $\Delta\vartheta \simeq 1$ cm^{-1} and the low values of I_1 and I_2 . According to these data, allowing for the absorption of all the fields, the value of $|\chi^{(3)}(\Delta\vartheta = 1$ cm$^{-1})|$ is $\simeq 10^{-7}$ cm^3/erg. The peak value of $|\chi^{(3)}(\Delta\vartheta = 0)|$ can be estimated to be exceptionally high $\simeq 10^{-5}$ cm^3/erg! Using the value of $|\chi^{(3)}|$ we estimated the maximum absorption cross section ($\sigma_m \simeq 3.7 \times 10^{-15}$ cm^{-2}), the concentration of aggregates ($N_a \simeq 2.8 \times 10^{17}$ cm^3) and the upper limit on the number of molecules in an aggregate ($n = N/N_a \simeq 6$). The estimates of σ_m and n found here on the basis of the four-wave mixing data are close to the values found by Stiel et al. (1988) through the study of exciton annihilation in PIC solutions. The parameters T_2 , σ_m and n pertain to the "optical" aggregates which are responsible for the formation of the exciton band. In contrast with "optical" aggregates, "chemical" aggregates can (according to Sundstrom et al. 1988) consist of tens of thousands of molecules, constituting a system of weakly bound "optical" aggregates.

Akhmanov S A 1990 New Physical Principles of the Information Arrangement (Moscow: Nauka) pp 13-32

Bogdanov V L, Kulya S V, Neporent B S and Spiro A G 1990 JETP Lett. 51 478

Bogdanov V L, Victorova E N, Kulya S V and Spiro A G 1991 JETP Lett. 53 105

Jelley E E 1936 Nature 138 1009

Souma H, Heilweil E J and Hochstrasser R M 1982 J.Chem. Phys. 76 5693

Stiel H, Dachne S and Teucher K 1988 J.Luminescence 39 351

Sundstrom V, Gillbro T, Gadonas R A and Piskarskas A 1988 J.Chem.Phys. 89 2754

Yajima T and Souma H 1978 Phys.Rev. A 17 309

st. Phys. Conf. Ser. No 126: Section VI

499

aper presented at Int. Symp. on Ultrafast Processes in Spectroscopy, Bayreuth, 1991

Picosecond relaxation processes in photoexcited metallophthalocyanine aggregates

Butvilas, V.Gulbinas and A.Urbas

Institute of Physics, Goštauto Str.12, 2600, Vilnius, Lithuania

ABSTRACT. The solutions of aluminum chloride and vanadyl phthalocyanines (AlClPc and VOPc) were investigated by means of the picosecond pump and probe technique. The dimerization and higher aggregation of these compounds were observed by examining the kinetics of the excited electronic states. It was established that the aggregates of the VOPc suffer heating and that subsequent alteration is caused by the heat energy dissipation. Small aggregates like dimers exhibit fundamental electronic decay processes only.

INTRODUCTION.

The ability to form aggregates is the typical feature of phthalocyanine-class molecules (see, for example, Monahan et al.,(1972), Huang and Sharp, (1982)). We expect that the peculiarities of the kinetics of electronic relaxation processes of the aggregates are different from those of the monomers. These kinetics are more similar to the films or crystals applied as organic semiconductors, sensors, etc. In principle, neither films nor crystals are suitable for spectroscopic investigations, since they exhibit too high optical densities. Hence understanding the interaction between excited molecules and photon energy relaxation paths in the aggregates of the phthalocyanines may lead directly to the perception of applicable objects like thin films or molecular crystals. Up to now electronic relaxation processes of monomers are hardly investigated in the picosecond time domain and practically unknown for the aggregates under consideration.

In this communication we report the results of the investigation of the AlClPc and VOPc solutions where aggregations of various degree were observed. The electronic relaxation processes were investigated after intense irradiation by laser pulse when nearly each molecule is excited.

MATERIALS AND METHODS.

The samples of AlClPc and VOPc were synthesized using standard techniques and purified by sublimation. All the solvents, i.e., acetone, benzene and DMSO were of spectroscopic grade. The time-resolved spectroscopic investigations were carried out by means of the pump and probe technique. The temporal behavior of the photoinduced absorption was examined by means of picosecond absorption spectrometer based on $KGd(WO_4)_2$ laser using passive mode locking. The sample was excited by the Raman components of the light scattered in the active rod or cavities with active liquids. The probe pulse was produced by the angle-tuned optical parametric generator pumped by the main radiation of the laser. The time resolution of the setup is about 5 ps. The transient spectra were measured by the another spectrometer based on a YAG laser. The second harmonic of parametric generator was used to excite the sample and the picosecond light continuum was used to probe it. The time resolution of this setup is about 30 ps.

RESULTS AND DISCUSSION.

The aggregate absorption was observed in the steady-state spectra of both media investigated (see Fig.1). The blue-shifted peak has been observed in the AlClPc spectrum in the acetone and its height being dependent on the concentration of the solution. AlClPc show only the spectra of the monomer. The energy and the concentration dependence of the dimer peak suggest that the structure of the dimer is face-to-face like which is frequently observed in the number of metal-substituted phthalocyanines. (see Monahan et al.,(1972)). Another structure was observed for VOPc. It has been shown by Tsvirko (1972) that the vanadyl substituted porphyrines tend to form staggered dimers of which vanadium is positioned close to nitrogen of neighboring rings. The dimer spectrum exhibits roughly equally intensive peaks (see Fig.1). The absorption at 630 nm increases with rising concentration of the solution. The concentration behavior of the ratio A_{630}/A_{680} suggests that the peak at 630 nm may arise from an aggregate being greater than a dimer. It is important that practically no monomer absorption could be observed for the VOPc solutions.

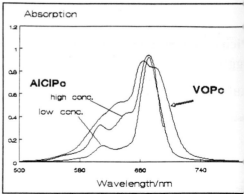

Fig. 1. Absorption spectra of the AlClPc and VOPc, both solutic in acetone. Concentrations being $5*10^{-8}$M and $3*10^{-5}$ M for AlCl and $2*10^{-5}$ M for VOPc.

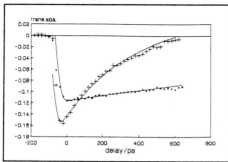

Fig. 2. Relaxation of the AlClPc solution in acetone after excitation at 590 nm. Crosses correspond to 630 nm probing (dimers) and points to 680 nm (monomers).

The relaxation of the photoexcited AlCl solution occurs usually without dissociation of dimers interactions between electronically excited molecules. T decay time of the S_1 state of the AlClPC monomers w measured as being 4.5 ns which corresponds quite well the results from flash-photolysis experiments performed Svensen et al.(1988). Moreover, we obtained a decay tin of about 380 ps for the excited dimers (see Fig.2 However, the dimer decay time is amazingly short. In t case of face-to-face dimers the electronic transition fro the lower excited dimer state to the ground state is fo bidden by symmetry. This rises the question whether t electronic transition occurs from the metastable dimer state or which is the nature of the metastable state. We could not observe any slow (between ns - μs) deca channels for selectively excited dimers which may arise for instance from participating triplet states. particular, this implies that intersystem crossing plays no role for the decay of excited AlClPc dimers. T relatively rapid decay of the metastable dimer states may be caused by internal conversion to the vibration manifold of the ground state.

We have observed new features being characteristic for the VOPc solution. In principle, the transient absorption spectra show the typical bleaching of the dimer absorption band of which time behavior is purely monoexponential. In the case of the absorption band which we assign to higher aggregates (see Fig.1 - at 630 nm) we could establish, that its temporal behavior differs significantly from that of the dimer. At an excitation intensity below 5 mJ/cm^2 the temporal behavior is monoexponential. With rising excitation intensity the short-living component becomes apparent. The same phenomena could be observed in induced absorption spectra (see Fig.3, 480 nm). So far we may conclude that excited aggregates decay increasingly with rising excitation intensity because of annihilation events. In the case of the absorption

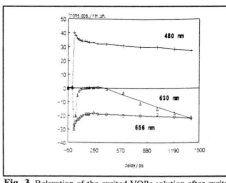

Fig. 3. Relaxation of the excited VOPc solution after excitation 20 mJ/cm^2. Excitation wavelength was 590 nm in all cases.

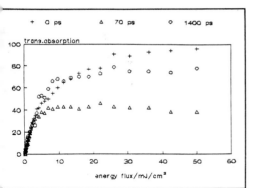

ig. 4. Transient absorption of VOPc solution vs. excitation tensity. Crosses accord to the electronic excited state, triangles show e heated and suppressed state and circles indicated the expanded ate.

band in the vicinity of 630 nm (see Fig.3) the temporal behavior does not only originate from annihilation events. To explain the kinetics there is required a rationalization based on at least three excited states. Of course, the question remains, which excited states participate in the kinetics determining the change of the absorption in the range between 500 ps and 1000 ps after accomplishing annihilation processes. Apparently, a direct consequence of the annihilation is the excitation of the vibrational modes of the molecule. The vibrational relaxation of these large molecules is very rapid, (see Angel et al. (1990)), i. e., it takes less than a few picoseconds. Thus the electronic excitation energy is converted into the heat as a result of the annihilation. In the case of the monomer molecules or small aggregates like dimers the heat energy dissipation occurs immediately. In larger aggregates exciton-phonon interactions may

articipate in the heat energy dissipation. In particular, Butvilas et al. (1991) report the generation of the •ngitudinal acoustic phonons in thin film of the VOPc due to high excitation intensities. Ho et al. (1988) have bserved an exchange time of exciton-phonon coupling in AlFPc of 4 ps. The main result of the phonon eating is the alteration of the equilibrium intermolecular distances being responsible for the intermolecular iteractions and, thus, determines the features of the absorption spectrum. But the thermal expansion can not ccur instantly. The characteristic expansion time is determined by the sound velocity in the material and near dimensions of the particle. For this reason thermal expansion cannot follow any abrupt temperature rise. •irectly after the annihilation unexpanded but still hot states of the aggregates are formed. It may be anchored) the range of 50-100 ps in the time dependence of transient absorption. Subsequent increase of the bsorption change may arise from thermal reorganization of the molecules in the aggregate in context with 1e cooling due to the heat transfer to the solvent bath. Thus the observed kinetics consist of three states for 1e system under study. The first state is the electronic excited state decaying mainly due to annihilation vents. The second state is characterized by the high temperature of the aggregate and high pressure inside . And the final state is the heated and thermally expanded state which is manifested mainly by the thermal hift of the absorption bands. This situation is displayed quite well in Fig.4. The electronic excitation becomes aturated above an irradiation intensity of about 5 mJ/cm². The second state is populated in spite of saturation ffects. This implies that the molecules of the aggregate may be excited at least twice during the excitation ulse duration, since rapid annihilation processes occur. The final state is in equilibrium with the surrounding redium and thus, occupied independently of the irradiation intensity.

As mentioned above, the expansion rate of the aggregate depends on the sound velocity and its size. Hence : is straightforward to estimate the mean size of the aggregate under study. The characteristic forming time •f the third state is approximately 500 ps. By means of the sound velocity determined by Butvilas et al. (1991) ve have obtained an averaged diameter of the aggregate of about 30 nm, i.e., each aggregate contains 600 to 000 molecules.

:ONCLUSIONS.

The intensively excited aggregates of the VO phthalocyanines are heated due to the annihilation of the exci-ations. This causes a reorganisation of the molecules within the aggregate which occurs at a time of 500 ps. The reorganised aggregate is characterized by unique spectra features in absorption. Small aggregates like dim-rs show only electronic processes which are independent of the excitation intensity.

REFERENCES.

Angel G., Gagel R., Laubereau A.. Vibrational Dynamics in the S_1 and S_0 States of Dye Molecules, Stud Separately by Femtosecond Polarization Spectroscopy. *Springer Ser. Chem. Phys.*, **1988**,v.48 (Ultra Phenomena 6), p.467-9.

Butvilas V., Gulbinas V., Urbas A. and Vachnin A.. Relaxation Processes in VOPc Film Caused by Inte Excitation. *Radiation Physics and Chemistry*, **1991**, N.10, accepted for publication.

Monahan A.R., Brado J.A., DeLuca A.F.. The Associaton of Copper, Vanadyl And Z Tetraalkylphthalocyanine Dyes in Benzene.*J.Phys.Chem.*,**1972**,v.76,p.1994-6.

Ho Z.Z.,Williams V.,Peyghambaryan N.,Hetherington W.M., Femtosecond Dynamics of Fluoro-Alumin Phthalocyanine and Linear Alkane Molecules. *Proc.SPIE-Int. Soc. Opt. Eng.***1988**, v.971, p.51-8.

Huang T.-H., Sharp J.H., Electronic Transitions of VO Phthalocyanine in Solution and in the Solid Sta *Chem. Phys.*,**1982**,v.65,p.205-16.

Svensen R.,Fery-Forgnes S.,MacRobert A.J., Phillips D.. *NATO ASI Ser.*, Ser.H, **1988**, (Photosensitization), p.445-8.

Tsvirko M,.P. and Solov'ev K.N..Photophysical Processes in Dimers of Ethioporphyrin and its Me Complexes. *Zh. Prikl. Spektrosk.*, **1974**, v.20, p.115-22.

Inst. Phys. Conf. Ser. No 126: Section VI
Paper presented at Int. Symp. on Ultrafast Processes in Spectroscopy, Bayreuth, 1991

Ultrafast relaxation dynamics of photoexcitations in one-dimensional conjugated polymers

Masayuki Yoshizawa and Takayoshi Kobayashi

Department of Physics, University of Tokyo, Hongo 7–3–1, Bunkyo, Tokyo 113, Japan

ABSTRACT: Ultrafast optical response in conjugated polymers has been investigated by femtosecond absorption spectroscopy. Several nonlinear processes, i.e., hole burning, Raman gain, inverse Raman scattering, and induced–frequency shift, were observed. The formation and relaxation of self–trapped excitons were observed in both polydiacetylenes and poly-thiophenes. A model of the relaxation process is proposed and the short life of excitons and difference between fluorescent and nonfluorescent polymers are explained.

1.INTRODUCTION

The large and ultrafast optical nonlinearities of conjugated polymers have recently been the focus of enormous interest because of the candidates for future practical applications in nonlinear optical devices. Polydiacetylenes (PDAs) and polythiophenes (PTs) belong to the rather well studied conju-gated polymers. PDAs have the backbone structure of $\{CR-C\equiv C-CR'\}_n$. and their properties can be modified by changing the side–groups. PDAs can be obtained in the form of single crystals, thin films, and solutions and have blue– and red–phases according to the color (Bloor 1985). The backbone geometry of PTs resembles that of *cis*–polyacetylene and the simple *cis*–like structure is stabilized by the sulfur.

Conjugated polymers have localized excitations, i.e., solitons, polarons, bipolarons, and self–trapped excitons (STEs). The nonlinear optical response is strongly affected by these excitations. The dynamics of the photoexcita-tions in PDAs and PTs have been investigated from femtoseconds to micro-seconds (Kobayashi 1984 1985 1990 1991a,b, Koshihara 1985, Stamm 1990 and Yoshizawa 1989 1991). PDAs have ultrashort phase relaxation time (Hattori 1987) and the relaxation kinetics of excitons in blue–phase PDA–3BCMU has the time constants of 150 fs and 1.5 ps (Yoshizawa 1989).

In this study ultrafast optical response of blue– and red–phase PDAs and PTs was investigated by femtosecond absorption spectroscopy. The relaxa-tion kinetics are explained by using the potential curves of the ground state, free exciton, and STE in configuration space.

2.EXPERIMENTAL

The femtosecond absorption spectroscopy system consists of a colliding–pulse mode–locked dye laser, a four–stage dye amplifier pumped by the second harmonic pulses of a 10 Hz Q–switched Nd:YAG laser, and an optical system of pump–probe absorption spectroscopy (Yoshizawa 1989). A part of continuum generated by self–phase modulation was selected with a set of prism pairs and a slit. The selected tunable femtosecond pulses were ampli–fied by a two–stage dye amplifier pumped by the third harmonic pulses of another Q–switched Nd:YAG laser and used as pump pulses (Taiji 1991).

Thin films of PDA–3BCMU and PDA–4BCMU [R=R'=(CH₂)₃(₄)OCONHCH₂COO(CH₂)₃CH₃)] were prepared by casting method and vacuum deposition. Langmuir–Blodgett films of PDA–(12,8) [R=(CH₂)₁₁CH₃,   R'=(CH₂)₈COOH] were also studied. PDA–4BCMU films have both blue– and red–phases. The exciton energy of the PDA–(12,8) LB films can be changed by the counter ions and the pH of the water (Nishiyama 1991). Poly(3–methylthiophene) (P3MT) and poly(3–dodecyl–thiophene) (P3DT) films were synthesized electrochemically on In–Sn oxide (ITO) glass substrates (Kaneto 1987).

3.RESULTS AND DISCUSSION

3.1 Nonlinear Optical Response

Figure 1 shows the transient absorbance change (ΔA) of a oriented blue–phase PDA–4BCMU film prepared by vacuum deposition. At 0.0 ps delay a bleaching peak appears at the pump photon energy of 1.98 eV. Then the peak shifts to 1.96 eV. The bleaching peaks at 1.98 and 1.96 eV are due to

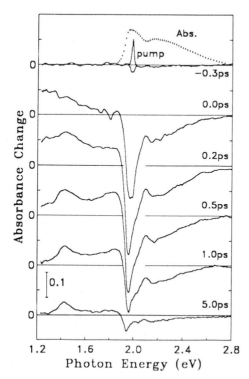

Fig. 1. Transient absorption spectra of a oriented blue–phase PDA–4BCMU film at 290 K. Both the polarizations of the pump and probe pulsed light are parallel to the oriented polymer chain. The absorption (dotted curve) and pump spectra are shown together.

coherent interaction between the pump and probe pulses and saturation of the excitonic absorption, respectively. When the pump and probe pulses overlap in time at the sample, two small minima due to Raman gain are observed at 1.80 and 1.72 eV. The corresponding Raman shifts are 1470 and 2110 cm^{-1} and they are assigned to the stretching vibrations of the C=C bond and C≡C bond, respectively. Two small maxima observed at 2.18 and 2.25 eV are due to the inverse Raman scattering, but the peak energies are slightly higher than the photon energies expected from the Raman shifts because of the overlap with the minima due to the phonon sideholes. The imaginary part of the third-order nonlinear susceptibilities are estimated as Im[$\chi^{(3)}(-\omega_1;\omega_1,\omega_1,-\omega_1)$]=$-3.2\times10^{-9}$esu   for $\hbar\omega_1$=1.98 eV and Im[$\chi^{(3)}(-\omega_2;\omega_2,\omega_1,-\omega_1)$]=$-4.8\times10^{-10}$esu   for Raman gain at $\hbar\omega_1$=1.98 eV and $\hbar\omega_2$=1.80 eV.

ΔA in a red-phase PDA-4BCMU excited by 1.97-eV pulse have an asymmetric oscillatory structure near 1.97 eV (Yoshizawa 1991). The oscillatory struc-ture observed in semiconductors was explained in terms of perturbed free induction decay (Sokoloff 1988). However, the frequency change of the oscillation predicted by the perturbed free induction decay (Brito Cruz 1988) was not observed in the red-phase PDA-4BCMU. The structure can be explained in terms of pump induced frequency shift of probe light or in another word cross-phase modulation. Since the probe light which has a peak at 1.97 eV shifts to higher energy, the observed ΔA above 1.97 eV becomes negative and ΔA below 1.97 eV becomes positive. The real part of the third-order susceptibility is estimated as Re[$\chi^{(3)}$]=-1.9×10^{-12}esu.

3.2 Relaxation Kinetics

ΔA in the blue-phase PDA-4BCMU shown in Fig. 1 has the bleaching due to the saturation of the excitonic absorption and the broad absorption below 1.9 eV. At 0.0 ps ΔA has a peak located below 1.4 eV. Then the peak shifts to higher energy. ΔA at 0.5 ps has two peaks at 1.4 and 1.8 eV. Figure 2 shows ΔA at 0.5 ps in PDA-(12,8) LB films with different exciton energies. The ΔA spectra have peaks at 1.4-1.6 eV and just below the absorption edge. The peak near the absorption edge is assigned to the transitions from the lowest ^1B_u exciton to biexciton with ^1A_g symmetry and the peak around 1.5 eV is assigned to the transition from ^1B_u exciton to higher m^1A_g excitons (Kobayashi 1991b).

The relaxation processes of the excitons can be explained by the model shown in Fig. 3. The shift of the induced absorption with time from −0.1 ps to 0.2 ps is observed in both PDAs and PTs. This spectral change is explained by the geometrical relaxation of free excitons to STEs. Since the formation process of STE has no barrier in one-dimensional system, the photoexcited free excitons (FE (1)) are coupled with the C–C stretching modes within the phonon periods of 10-20 fs. However, the STEs have not relaxed to the bottom of the potential curve and the binding energy re-mains as the kinetic energy of the lattice vibration (unrelaxed STE (2)). The unrelaxed STEs emit the phonons and relax to unthermalized STEs (3). The time constants of the spectral change due to the phonon emission are 100-150 fs and summarized in Table 1. ΔA in the red-phase PDA-4BCMU has a slow spectral change with time constant of 1.1±0.1 ps. The slow spectral change corresponds to the thermalization process of STEs (3->4) (Yoshizawa 1991).

The decay kinetics of ΔA at 1.8 eV in blue-phase PDAs can be fitted to a

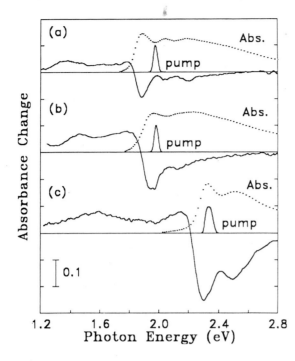

Fig. 2. Transient absorption spectra at 0.5 ps in PDA–(12,8) LB films at 290 K. The counter ion and the pH of the water during the fabrication of LB films are (a) polyallylamine (PAA) and 5.4, (b) Cd^{2+} and 7.1, and (c) PAA and 7.9, respectively. The absorption (dotted curves) and pump spectra are shown together.

single exponential function with time constant of 1.6 ps at 290 K, while the bleaching has the initial fast decay with time constant shorter than 1.0 ps. The decay kinetics in red–phase PDAs and PTs can not be fitted to single exponential functions. The decay becomes slower at longer delay time. This decay kinetics of excitons can be explained by the phonon emission and thermalization of STEs. The relaxation of STEs to the ground state (G) is due to passing over the crossing point and/or tunneling through the barrier between the STE and ground–state potentials. The hot (unrelaxed and unthermalized) STEs can relax to the ground state rapidly (2,3–>G). Then the STEs come down to the bottom of the potential after the phonon emission and thermalization and the loss rate due to the tunneling becomes smaller (4–>G). The time constants summarized in Table 1 are obtained by fitting the observed decay curves to biexponential functions.

The time constants of the phonon emission process are 100–150 fs and almost independent of the sample and temperature. This result is consistent with the barrierless relaxation from free excitons to STEs. The relaxation from the hot STEs at 10 K are slightly slower than at 290 K, while the relaxation from the thermalized STEs depends clearly on the phase of PDAs and temperature. The time constants in blue–phase PDAs at 10 and 290 K are about 2.0 and 1.5 ps, respectively, while the time constants in red–phase PDAs and PTs are about 6 and 4 ps, respectively. The PDA–(12,8) LB film listed at the bottom of blue–phase PDAs in Table 1 has the exciton with lower energy (1.88 eV) than the other blue–phase PDAs (see Fig. 2(a)) and has shorter decay time constants.

The decay time constants and fluorescence properties in red–phase PDAs are close to those in PTs and they are in good contrast to blue–phase

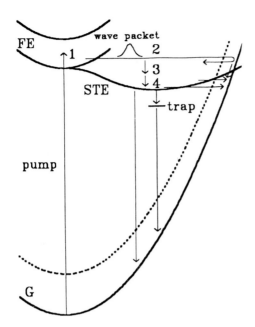

Fig. 3. A model of the relaxation kinetics shown in adiabatic potential surfaces of excitons in red–phase PDA. FE(1): free exciton, STE: self-trapped exciton, 2: unrelaxed STE, 3: unthermalized STE, 4: thermalized STE, and G: ground state. A dotted curve is the ground state of blue–phase PDA.

Table 1. Relaxation time constants of conjugated polymers.

Polymers (Temp.)		Phonon Emission (2->3)[*]	Hot STE->G (2,3->G)	Thermalized STE->G (4->G)
PDAs(blue)				
3BCMU	(10K)	150±50 fs	<1.0 ps	2.0±0.2 ps
	(290K)	150±50 fs	<1.0 ps	1.5±0.1 ps
4BCMU	(10K)	120±30 fs	<1.0 ps	2.1±0.2 ps
	(290K)	140±40 fs	<1.0 ps	1.6±0.1 ps
(12,8)	(10K)	130±50 fs	0.8±0.2 ps	2.1±0.1 ps
	(290K)	90±60 fs	0.5±0.1 ps	1.5±0.1 ps
(12,8)[**]	(10K)	100±60 fs	0.6±0.1 ps	1.8±0.1 ps
	(290K)	120±50 fs	0.6±0.1 ps	1.3±0.1 ps
PDAs(red)				
4BCMU	(10K)	<300 fs	0.9±0.1 ps	5.6±1.1 ps
	(290K)	120±60 fs	0.7±0.1 ps	4.4±0.5 ps
(12,8)	(10K)	100±60 fs	1.0±0.2 ps	6.1±1.2 ps
	(290K)	130±60 fs	0.8±0.2 ps	3.9±0.5 ps
PTs				
P3MT	(10K)	70±50 fs	0.6±0.1 ps	>5 ps
P3DT	(10K)	100±50 fs	0.5±0.2 ps	>5 ps
	(290K)	100±50 fs	1.0±0.2 ps	4.7±1.2 ps

[*] 2-4 and G are the states shown in Fig. 3.
[**] Exciton energy is smaller than other blue-phase PDAs (Fig. 2(a)).

PDAs. The formers are fluorescent and the time constants are about 5 ps, while the latters are nonfluorescent and the time constants are about 2 ps. Since the exciton energy in blue–phase PDAs (see a dotted curve in Fig. 3) is smaller than that in red–phase PDAs , the crossing point between STEs and the ground state is expected to be lower than that in red–phase PDAs (Toyozawa 1989). Therefore, the decay time constants of STEs in blue–phase PDAs are about 2 ps and shorter than those in red–phase PDAs. The fluorescence of red–phase PDAs and PTs has fast and slow components. The slow component is due to traps in polymers and the fast one is probably due to excitons (Yoshizawa 1991). In blue–phase PDAs the major part of the STEs relax to the ground state very rapidly and the fluorescence is too weak to be observed.

ACKNOWLEDGEMENT

The samples of PDA–3BCMU, PDA–(12,8) LB films, and PTs were kindly supplied by Sumitomo Electric Industries, Prof. Masamichi Fujihira, and Prof. Katsumi Yoshino, respectively. The initial part of the experiment of PTs were partly performed with the help of Dr. Uwe Stamm and Mr. Makoto Taiji.

REFERENCES

Bloor D and Chance R R (eds.) 1985 *Polydiacetylenes* (Dordrecht: Martinus Nijhoff)
Brito Cruz C H, Gordon J P, Becker P C, Fork R L and Shank C V 1988 *IEEE* **QE–24** 261
Hattori T and Kobayashi T 1987 *Chem. Phys. Lett.* **133** 230
Kaneto K, Uesugi F and Yoshino K 1987 *Solid State Commun.* **65** 783
Kobayashi T, Iwai J and Yoshizawa M 1984 *Chem. Phys. Lett.* **112** 360
Kobayashi T, Ikeda H, Tsuneyuki S and Kotaka T 1985 *Chem. Phys. Lett.* **116** 515
Kobayashi T, Yoshizawa M, Stamm U, Hasegawa M and Taiji M 1990 *J. Opt. Soc. Am.* **B7** 1558
Kobayashi T 1991a *Synthetic Metals* submitted
Kobayashi T 1991b *Proceedings of Toyota Conference on Nonlinear Optical Materials* ed Mizuta (Elesevier) submitted
Koshihara S, Kobayashi T, Uchiki H, Kotaka T and Ohnuma H 1985 *Chem. Phys. Lett.* **114** 446
Nishiyama K, Kurihara M and Fujihira M 1991 *Langmuir* submitted
Sokoloff J P, Joffre M, Fluegel B, Hulli F, Lindberg M, Koch S W, Migus A, Antonetti A and Peyghambarian N 1988 *Phys. Rev.* **B38** 7615
Stamm U, Taiji M, Yoshizawa M, Yoshino K and Kobayashi T 1990 *Mol. Cryst. Liq. Cryst.* **182A** 147
Taiji M, Bryl K and Kobayashi T 1991 unpublished
Toyozawa Y 1989 *J. Phys. Soc. Jpn.* **58** 2626
Yoshizawa M, Taiji M and Kobayashi T 1989 *IEEE* **QE–25** 2532
Yoshizawa M, Yasuda A and T. Kobayashi 1991 *Appl. Phys. B* in press

Inst. Phys. Conf. Ser. No 126: Section VI
Paper presented at Int. Symp. on Ultrafast Processes in Spectroscopy, Bayreuth, 1991

Ultrafast processes in thiophene oligomers studied by ps absorption measurements

D Grebner, H Chosrovian, S Rentsch and H Naarmann[1]

Institute of Optics and Quantum Electronics
Friedrich-Schiller-University Jena, Max-Wien-Platz 1
O-6900 Jena, Germany

ABSTRACT: Ps-time-resolved excite-and-probe experiments were applied to thiophene oligomers for the first time. The oligomers contained three, four or five monomer units and were investigated in solution. We got size dependent spectra and kinetics which were compared with stationary measurements on doped thiophene oligomers and interpreted by the model of polarons and bipolarons.

1. INTRODUCTION

Conjugated polymers like polyacetylene (PA), polydiacetylene (PDA) or polythiophene (PT) show photoconductivity and conductivity due to doping. Further they exhibit high values of third-order susceptibility. In contrast to PT the thiophene oligomers (nT) are soluble in organic solvents. Thiophene oligomers are used as model systems for the properties of PT in theory by Bredas *et al.* (1984) and Stafström *et al.* (1988) and in experiment by Fichou *et al.* (1990a and 1990b). Further they were already applied as semiconducting materials in electronic devices like Schottky diodes by Fichou *et al.* (1988) or metal-isolator-semiconductor field effect transistors (MISFETs) by Horowitz *et al.* (1989). In the present paper time resolved transient optical absorption measurements on thiophene oligomers nT (n = 3, 4, 5) are reported. We investigated the influence of the oligomer size on the transient behaviour which is completely unknown in the ps-time-region. The results are compared with theoretical models and measurements on doped oligomers by Fichou *et al.* (1990a and 1990b).

2. THIOPHENE OLIGOMERS

Thiophene oligomers well-defined produced by Naarmann *et al.* have low molecular weights and they are easily processible. The used oligomers consist of three, four or five monomers and were available in form of powder. Solutions of equal concentration $c = 10^{-4}$ mol/l were prepared using CH_2Cl_2 as solvent. The stationary absorption spectra of thiophene oligomers show a shift of the absorption edge to

---

[1] BASF Plast. Res. Lab. Ludwigshafen

longer wavelengths with rising chain length which is caused by a decreasing of the band gap. The absorption maximum of the oligomer 3T is much lower then the maxima of 4T and 5T.

3. PS-SPECTROSCOPY

Experiments were carried out by excite-and-probe technique. Excitation pulses of 351 nm with an energy of (100 ± 50) μJ were produced by third harmonic generation of a Nd-phosphate glass laser (pulse duration 6 ps).

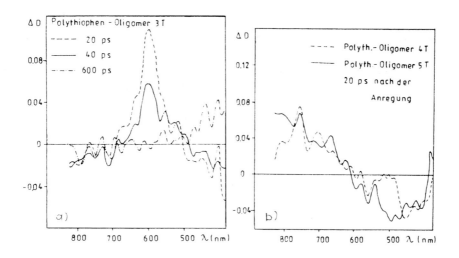

Fig. 1. Transient absorption spectra
a) oligomer 3T: 20, 40 and 600 ps after excitation
b) oligomer 4T and 5T: 20 ps after excitation

Ps-continuum generated by the ground wave (1054 nm) was located between 450 and 800 nm and registrated by an OMAII-System. The ps-absorption spectra show differences between oligomer 3T (Figure 1a) on one site, but 4T and 5T (Figure 1b) on the other. At once after excitation of 3T a strong and sharp absorption band is to be seen at 600 nm. The band decays within 100 ps and in the same time the induced

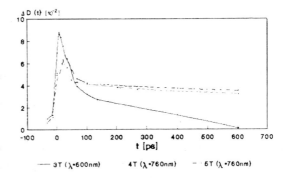

Fig. 2. Ps-kinetics of thiophene oligomers in solution

fluorescence at 450 nm goes back (Figure 1a). In contrast to the results of 3T broad absorption bands around 750 nm with lifetime-components in the ns region were found in 4T and 5T (Figure 1b). Looking at the induced fluorescence (-ΔD, Figure 1) the maximum is shifted to longer wavelengths, how it was expected from the stationary spectra. By measurements of induced fluorescence using single photon counting (Grebner 1991) the observed fluorescence bands were at the same wavelengths as the corresponding negative ΔD-values of the transient absorption spectra. Resulting kinetics in the absorption region illustrate the different behaviour of the oligomers (Figure 2). The absorption bands of 4T and 5T seems to consist of two components. The fast component decays within about 100 ps, however the slow component has a lifetime in the ns-region.

4. INTERPRETATION OF THE RESULTS

The studies were carried out in solutions with $c = 10^{-4}$ mol/l. Under these conditions the thiophene oligomers can be considered as nearly isolated molecules separated by solvent molecules. The common features of the oligomers are:
- a transient absorption in the visible region will be formed during excitation (6 ps)
- an arise of fluorescence was observed as well
- the fluorescence decays as the fast component of transient absorption
Size dependent behaviour of thiophene oligomers were found in stationary as well as in transient behaviour of the spectral data.

Table 1:	Absorption	Fluorescence
3T	3.1	2.8
4T	2.8	2.6
5T	2.6	2.4
PT *(Vardeny et al.* 1987)	2.1	1.8

The transient absorption of 3T is a sharp band at 600 nm (2.08 eV) whereas in 4T and 5T the absorption is wider and shifted to 750 nm and longer wavelengths (Figure 1).
The fast component in 3T amounts about 95 % whereas it is less 40 % in 4T and 5T, i. e. in 4T and 5T a long living absorption predominates.
Logarithmic plot shows for 3T the same slope for absorption and fluorescence. Because the fast absorption and fluorescence follow the same kinetics, both processes are from the same state. We therefore interpret the transient absorption of 3T as an excited state absorption from the S1-state.
In 4T and 5T the depletion of S1 is accompanied by the formation of a long living particle with an absorption located in the same spectral region (Figure 1b and 2). To understand the nature of this particle it is helpful to compare the transient absorption bands with bands, obtained by chemical oxidation (doping) of thiophene oligomers. After doping intragap absorptions were observed by Fichou et al. (1990a and 1990b) recently in thiophene oligomers for two oxidations steps, which corresponds to formation of one-fold respectivly two-fold charged radicals, i. e. polarons and bipolarons. In Table 2 we compare our transient absorption with the data of Fichou et al. (1990a and 1990b) both obtained in CH_2Cl_2.

Table 2:   Comparison of absorption bands

	photoinduced[*] [eV]	doping induced[**] [eV]
3T	2.08	1.59
4T	1.65	1.92
5T	1.55	1.72
6T	not available	1.59

[*] this work, [**] Fichou *et al.* (1990a and 1990b)

The band maxima of doped spectra of 3T is exactly that of 6T and was interpreted as dimerization during doping (Fichou *et al.* 1990a and 1990b). The photoinduced absorption of 3T at 2.06 is, as shown before, the absorption of the short living S1 state, which can never be observed in stationary studies. The position of photoinduced and doping induced bands of 4T and 5T differs by about 0.25 eV.

An exact aggreement could not be expected. By doping the counterions are especially fixed and they determine the local electric field and therefore the band position.

During photoexcitation charged intermediates will be formed for example by charge separation between thiophene and solvents, which gives an other average electric field. Therefore we think we have indeed observed the formation of polarons (or radicalions) by excitation.

By different theoretical calculations of the band structure were estimated, that in thiophene oligomers with four and more monomer units the formation of such polarons is possible (Bredas *et al.* 1984 and Stafström *et al.* 1988). If Bipolarons are also formed is not clear in the moment because they have their spectra at 1.24 - 1.36 eV (Fichou *et al.* 1990a and 1990b). This region we will investigate in future.

5. SUMMARY

Different behaviour in dependence on the chain length was observed for thiophene oligomers in ps-time region. A sharp short living absorption of 3T was interpreted as an excited-state-absorption. Four monomer units seems to be the minimum chain length for the formation of polarons which are visible as broad absorption bands with a great lifetime in the transient spectra.

REFERENCES

Bredas J L, Themans B, Fripiat J G, Andre J M and Chance R R 1984 Phys. Rev. **B29** 6761

Chung T C, Kaufman J L, Heeger A J and Wudl F 1984 Phys. Rev. **B30** 702

Fichou D, Horowitz G, Nishikitani Y and Garnier F 1988 Chemtron. **3** 176

Fichou D, Horowitz G and Garnier F 1990a Synth. Met. **39** 125

Fichou D, Horowitz G, Xu B and Garnier F 1990b Synth. Met. **39** 243

Grebner D 1991 Diplomarbeit Jena

Horowitz G, Fichou D, Peng X, Xu Z and Garnier 1989 Sol. Stat. Commun. **72** 381

Hotta S, Rughooputh S, Heeger A J and Wudl F 1987 Macromol. **20** 212

Stafström S and Bredas J L 1988 Phys. Rev. **B38** 4180

Vardeny Z, Ehrenfreund E and Brafman O. 1986 Phys. Rev. Lett. **56** 671

Vardeny Z, Ehrenfreund E, Shinar J and Wudl F 1987 Phys. Rev. **B35** 2498

Inst. Phys. Conf. Ser. No 126: Section VI
Paper presented at Int. Symp. on Ultrafast Processes in Spectroscopy, Bayreuth, 1991

Classical and quantum solvation

Eyal Neria and Abraham Nitzan
Sackler Faculty of Science, School of Chemistry, Tel Aviv University, Tel Aviv 69978, Israel

R.N. Barnett and Uzi Landman
School of Physics, Georgia Institute of Technology, Atlanta, GA 39332, U.S.A.

ABSTRACT: Computer simulation of solvation dynamics are described for ion solvation (classical dynamics) and electron solvation (quantum mechanics). Similarities and differences are discussed.

1.   INTRODUCTION

Considerable attention has been focused lately on questions related to the dynamics of charge (and charge distributions) solvation in polar solvents (Bagchi (1989), Maroncelli (1991), Jarzeba et al (1991) and references therein).   Progress in fast time experimental technique has lead to vast amount of new data on solvation processes in protic and aprotic solvents and recent theoretical investigations based on analytical approximations (Bagchi (1989)) as well as numerical simulations (Maroncelli (1991), Rao and Berne (1981), Engstrom et al (1984), Maroncelli and Fleming (1988), Karim et al (1988), Kuharski et al (1988), Carter and Hynes (1991), Fonseca and Ladanyi (1991), Perera and Berkowitz, Neria and Nitzan) of classical solvation have lead to new insight concerning the nature of the solvation process. At the same time experimental (Migus et al (1987), Gauduel et al (1989, 1991), Long et al (1989, 1990)) and numerical work (Barnett et al (1989), Rossky and Schnitker (1988), Webster et al (1991a,b), Neria et al (1991)) was carried out on the dynamics of electron hydration following electron injection into H_2O and D_2O. This paper summarizes the main findings of these studies, and contrasts the quantum electron solvation process with the classical (ion) one. Both lines of investigation have lead to some surprises when theoretical and experimental results were confronted, and this has lead to new understanding of these solvation processes as described below.

How does quantum solvation differ from a classical one?   It should first be emphasized that by the terms "classical" and "quantum" we refer in this paper to the solute.   The solvent is assumed to respond classically or (for quantum solvation where, as seen below, some quantum information on the solvent is needed) quasiclassically.   We thus exclude from our discussion solvation in inherently quantum fluids.   There are two ways in which solvation dynamics of an electron can differ from that of a classical ion:
(1). Solvation of an electron involves size relaxation, namely relaxation of the kinetic energy of localization.
(2) Solvation of an electron can potentially be non-adiabatic, involving more than one electronic state of the solvated electron.

Both effects are absent in classical solvation which is governed by the solvent motion following an instantaneous creation of a charge distribution.   The latter has been associated with other questions.   Following early continuum theories (see e.g. Bagchi (1989) and references therein) which predict that solvation is an exponential relaxation characterized by the longitudinal dielectric relaxation time

$\tau_L (=(\varepsilon_\infty/\varepsilon_0)\tau_D)$ where ε_0, ε_∞ and τ_D are respectively the static and optical dielectric constants and the Debye dielectric relaxation time), the more sophisticated theories, which take into account the microscopic structure of the system have predicted that the relaxations is multiexponential, spanning timescales between τ_L and τ_D. Numerical experiments on solvation in models for water (Maroncelli and Fleming, 1988), acetonitrile (Maroncelli, 1991), methyl chloride (Carter and Hynes, 1991), methanol (Fonseca and Ladanyi, 1991) and Stockmayer fluids (Neria and Nitzan, Perera and Berkowitz) subsequently revealed the dominant role of the inertial mechanism, showing that the underdamped motion of solvent molecules closest to the solute contribute a substantial part of the solvation process.   This has lead to experimental verification of this effect in acetonitrile by Rosenthal et al (1991), and to generalization of the theory, including non-markovian solvent response, by Chandra and Bagchi (1991a,b,c).

In what follows we demonstrate and discuss these issues from the point of view of molecular dynamics simulations.   Section II describe computer experiments of classical solvation dynamics in a Stockmayer fluid. Section III reviews our quantum dynamical simulations results on electron solvation in water. We conclude in Section 4.

2.   CLASSICAL SOLVATION IN A STOCKMAYER FLUID

A stockmayer fluid consists of structureless particles characterized by point dipoles $\vec{\mu}$ and Lennard Jones interactions. The solvent molecules interact via Lennard Jones (LJ, characterized by the parameters σ and ε) and dipole-dipole (associated with point dipoles μ) potentials, and the solvent-solute interactions are LJ (with ε_A σ_A replacing ε and σ) and charge (q)-dipole potentials. Our simulation method follows earlier computer experiments on the dielectric response of this fluid (Pollock and Alder (1980, 1981), Neumann et al (1983,1984)). We use 400 solvent molecules in a cubic box with periodic boundary conditions.   The electrostatic potentials are handled within the effective dielectric environment (reaction field) scheme (Leeuw et al, 1986).

Consider first the dielectric relaxation in this solvent.   Figures 1 shows the correlation functions $C(t) = < M(t)\cdot M(0) >/<M^2>$ $(M = \sum \vec{\mu})$ for two choices of solvent parameters. The full line corresponds to the parameters used by Neria and Nitzan (NN) as an approximation to CH$_3$Cl at 240K: M = 50amu, σ= 4.2A, ε= 195K, ρ= $1.09\cdot10^{-2}$, T = 240K, μ = 1.87D, I = 33.54amu·A^2, σ_A = 3.675A and ε_A = 120K. The dotted line results from using parameters from the family of parameters of Neumann et al (1984), with the solvent mass and LJ parameters σ and ε the same as in the NN simulation, implying ρ = 0.011A^{-3}, T=138K, μ=1.92D and I=22.05amu·A^2. We see that the relaxation is qualitatively different, behaving on the relevant timescale as a Gaussian in in the former case and as an exponential in the latter. See Pollock and Alder (1981) for a discussion of this issue.

Turning to solvation dynamics, the quantities of interest are $\Delta E_{solv}(t)$, the change in the averaged solvent-solute interaction energy observed after a step function change in the solute charge and the normalized solvation function

$$S(t) = \frac{\Delta E_{solv}(t) - \Delta E_{solv}(\infty)}{\Delta E_{solv}(o) - \Delta E_{solv}(\infty)} \tag{1}$$

The results shown below all correspond to a jump of the solute charge from q=O to q=1. Fig. 2 depicts the solvation functions S(t) for two systems: The full and

Fig 1 Dielectric relaxation in stockmayer fluids. (see text for details)

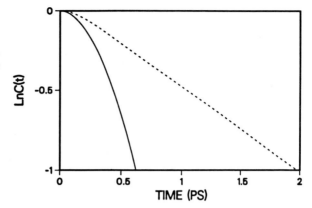

Fig 2 The solvation function S(t) for the two stockmayer fluids of Fig 1. Full and dotted lines correspond to the same solvents as in fig 1.

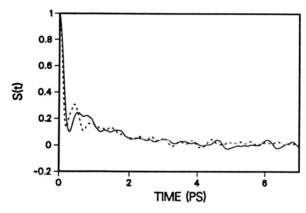

Fig 3 The solvation functions S(t) for solvents differing by their molecular moment of inertia plotted against t/τ (see text for details).

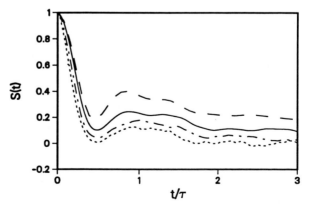

dotted lines corresponds to the parameters characterizing the corresponding lines of Fig. 1. It is seen that even though these systems differ qualitatively in their dielectric relaxation, the corresponding solvation processes are very similar. This is also seen from the similarity of the results of the recent solvation dynamics studies of Neria and Nitzan and of Perera and Berkowitz.

A more detailed look at the dynamics of the solvation process is seen in Fig. 3. Here we show the time dependent solvation energy $E_{solv}(t)$ for systems characterized by the NN parameters, except that the dimensionless quantity $p = I/(2M\sigma^2)$ is varied, by changing I, to be p=0.001 (dashed line), 0.019 (full line), 0.125 (dashed-dotted line and 0.25 (dotted line). (Note: in Fig. 1 p=0.019 and p=0.0125 for the full and dotted lines respectively). In this figure the time for the different curves is scaled by the characteristic dielectric relaxation time τ, obtained by fitting the *initial* part of the correlation function C(t) to a Gaussian, $\exp[(-t/\tau)^2]$. This leads to τ=0.14, 0.61, 1.6 and 2.5ps for the cases p=0.001, 0.019, 0.125 and 0.25 respectively. For all these cases the relaxation follows an initial Gaussian decay during which ~ 80% of the solvation energy is achieved followed by exponentially damped oscillations. The initial Gaussian relaxation and the subsequent damped oscillations have been shown to be associated with underdamped weakly correlated inertial motions of individual solvent molecules (Maroncelli, 1991). It is seen that the scaling puts the inertial parts of the different cases on top of each other, emphasizing the common origin of the fast initial relaxation and subsequent oscillations.

Analysis of the contributions of different solvation shells around the solute (Neria and Nitzan) shows that at least on the fast timescale the solvent molecules closer to the solute respond first and this response proceeds to the outer solvent shells. The nearest shell exhibits a pronounced slow relaxation component which is weaker or absent in the outer shells. This may be related to the Onsager "inverse snowball" picture (Onsager, 1977), or to residual diffusive energy relaxation out of the nearest shell, as suggested by Maroncelli (1991).

To conclude this section we note that classical solvation dynamics in the structureless Stockmayer liquid is qualitatively very similar to that observed in earlier studies of more structured polar solvents (water, methanol, acetonitrile andCH3Cl). The importance of inertial motions in this dynamics thus seems to be a general feature of the solvation process in small molecule solvents.

3. ELECTRON SOLVATION IN WATER

Electron hydration is a long studied subject, but recent experimental results (Migus et al (1987), Gauduel et al (1989, 1991), Long et al (1989, 1990))have shed new light on the dynamics of this process on the 10^{-13}- 10^{-9}ps timescale. These works indicate that the formation of the fully hydrated electron goes through at least one intermediate species. Using the simple kinetic model

$$e_{quasifree} \xrightarrow{k_1} e_{wet} \xrightarrow{k_2} e_{solvated} \qquad (2)$$

where $e_{quasifree}$ is the initially ejected electron and e_{wet} is the intermediate state, Long et al have fitted their data to $k_1^{-1} = 180 + 40$ fs and $k_2^{-1} = 540 \pm 50$ fs while Migus et al have given different estimates $k_1^{-1} = 110$ fs and $k_1^{-1} = 240$ fs.

Also Long et al have identified an isosbestic point at ~ 820 nm indicating a process dominated by a two state kinetics (Migus et al have not observed such a point and at the present meeting Eisenthal has pointed out that the observation of an isosbestic point depends on the initial laser intensity, which may affect the spatial distribution of the ejected electrons and the subsequent recombination process).

From the theoretical point of view we have argued above that electron solvation can be different from ion solvation by the additional size relaxation (kinetic energy of localization) and because of the possibility of non-adiabatic processes, namely the possible involvement of more than one electronic state in the solvation dynamics. We can focus on the size relaxation issue by performing molecular dynamic simulations of the electron-water system under adiabatic restrictions (Barnett et al, 1988). When such a simulation is performed starting from a weakly localized electron in a pre-existing cavity in the neutral water (Rossky and Schnitker (1988), Barnett et al (1989)), the following is observed:(1) The solvation dynamics appear very similar to that observed for a classical ion (following an instantaneous switching-on of the charge). It consists of a fast (~ 20-30 fs) inertial relaxation which accounts for ~ 80% of the total adiabatic solvation energy, followed by a slow (200-300 fs) residual relaxation. The fast relaxation component is dominated by inertial librations of H atoms in the first hydration layer about the electron. (2) The reason that the size relaxation does not affect this hydration kinetic is that it is essentially complete after ~ 20 fs, namely the kinetic energy of localization relaxes faster than the accompanying classical processes.

Thus the *adiabatic* electron solvation is shown by these simulations to be essentially classical. However it cannot account for the experimentally observed dynamics: The existence of the intermediate state e_{wet} is inferred from the appearance of a transient absorption in the near IR which disappears, giving rise to the red absorption of the fully solvated electron. In a process dominated by adiabatic solvation the absorption peak would shift continuously from the infrared to the red.

Concerning "e_{wet}", it is reasonable to assume that it corresponds to the three closely lying lowest excited p-like states of the hydrated electron, which carry most of the oscillator strength of the 1.7 eV absorption of the fully hydrated electron. If so, k_2 of Eq. 2 is the averaged *non-adiabatic* (NA) transition rate from these states to the ground, s-like, state.

Webster et al (1991) have recently simulated the NA relaxation involving the higher states of electron in water using the semiclassical theory of Pechukas (1969), estimating k_2 to be of order ~ 1 ps. The applicability of this method to the present problem (where the potential surfaces involved are far from each other throughout the process) is unclear, since it leaves open the choice of a coherence time beyond which surface hopping is imposed on the mixed state evolution. Neria et al (1991) have recently evaluated this "p" $\longrightarrow$ "s" transition rate using another semiclassical simulation method, based on a semiclassical evaluation of the golden rule rate expression for this rate. We refer the reader to Neria et al (1991) for details. This calculation, performed with the RWK2-M (flexible) water-water potential and with the electron-water pseudo-potential of Barnett et al (1988b), yields at T = 300K $k_2 \simeq 140$ fs for H_2O and $k_2 \simeq 240$ fs for D_2O (with error margins estimated at ~ 30%). The absolute times are shorter than those observed experimentally, and the calculated isotope effects are larger (Long et al report a 35% longer risetime for the appearance of the solvated electron in D_2O relative to the same process in H_2O. Gauduel et al report a very small isotope effect of ~ 10% on these times). More work is needed to determine whether these discrepancies result from our statistical errors or from deficiencies in the electron-water interaction pseudopotential. Still, the calculated NA transition rate is indeed much slower then the main part of the adiabatic one (20-30 fs) and is therefore predicted to be rate determining. The simulation thus supports the experimental conclusion that the electron hydration process is dominated by non-adiabatic processes.

4. CONCLUSIONS

We have described in this paper two studies. The first looks atclassical solvation following a sudden generation of charge in a simple solvent characterized by point dipoles and Lennard Jones interactions (Stockmayer liquid) and the second - at electron solvation in water.   Classical solvation in the Stockmayer liquid proceeds in a way qualitatively very similar to that found in other more structured polar solvents (water, methanol, acetonitrile and CH_3Cl).   As in all these solvents the solvation is dominated by a fast inertial relaxation process involving primarily the solvent molecules nearest to the solute.   There is also a residual slower relaxation that may presumably correspond to predictions of local dielectric theories. Simulations show that adiabatic hydration of an electron, initially localized in a shallow pre-existing cavity in neutral water proceeds in a way very similar to classical ion solvation.   The reason is that size relaxation of the electron is faster than the classical solvent dynamics and does not affect the observed rate. However experimental indications that the actual electron hydration process is non-adiabatic is supported by recent simulations which show that the non adiabatic transition from the excited "p-like" states of the hydrated electron is an order of magnitude slower than the adiabatic solvation dynamic on the ground state potential surface, so that this process may be rate limiting.   Electrons, temporarily trapped in their p-like state may thus be identified with the experimentally observed "wet electron".

REFERENCES
Bagchi B 1989 Ann. Rev. Phys. Chem. **40** 115 and references therein.
Barnett R N, Landman U and Nitzan A 1988a J.Chem.Phys. **89** 2242
Barnett R N, Landman U, Clevland C L and Jortner J 1988b J.Chem.Phys. **88** 4421
Barnett R N, Landman U and Nitzan A 1989 J.Chem.Phys. **90** 4413
Carter E. A and Hynes J T 1991 J.Chem.Phys. **94** 5961
Chandra A and Bagchi B 1991a J.Chem.Phys. **94** 3177
Chandra A and Bagchi B 1991b Chem.Phys. **156** 323
Chandra A and Bagchi B 1991c Proc.Ind.Acad.Sci **103** 77
De Leeuw J W, Perram J W and Smith E R 1986 Ann. Rev. Phys. Chem. **37** 245
Engström S, Jönsson B and Impey R W 1984 J. Chem. Phys. **80** 5481
Fonseca T and Ladanyi B M 1991 J.Phys.Chem. **95** 2116
Gauduel Y, Pommeret S, Migus A and Antonetti A 1989 J.Phys.Chem. **93** 3880
Gauduel Y, Pommeret S, Migus A and Antonetti A 1991 J.Phys.Chem. **95** 533
Jarzeba W, Walker G C, Johnson A E and Barbara P F 1991 Chem. Phys. **152** 57
Karim O A, Haymet A D J, Banet M J and Simon J D 1988 J. Phys. Chem. **92** 3391
Kuharski R A, Bader J S, Chandler D, Sprik M, Klein M L and Impey R W 1988 J. Chem. Phys. **89** 3248
Long F H, Lu H and Eisenthal K B 1989 Chem. Phys. Letters **160** 464
Long F H, Lu H and Eisenthal K B 1990 Phys. Rev. Letters **64** 1469
Maroncelli M and Fleming G R 1988 J. Chem. Phys. **89** 5044
Maroncelli M 1991 J. Chem. Phys. **94**, 2084
Migus A, Gauduel Y, Martin J L and Antonetti A 1987 Phys. Rev. Let. **58** 1529
Neria E, Nitzan A, Barnett R N and Landman U Phys. Rev. Lett. 1991 **67** 1011
Neria E and Nitzan A 1992 J. Chem. Phys in press.
Neumann M and Steinhauser O 1983   Chem. Phys. Lett.   **102** 508
Neumann M, Steinhauser O and Pawley G S 1984 Mol.Phys. **52** 97
Onsager L 1977 Can J. Chem. **55** 1819
Pechukas P 1969 Phys. Rev. **181** 166;174
Perera L and Berkowitz M 1992 J. Chem. Phys. in press.
Pollock E L and Alder B J 1980 Physica **102a** 1
Pollock E L and Alder B J 1981 Phys.Rev.Lett. **46** 950
Rao M and Berne B J 1981 J.Phys.Chem. **85** 1498.
Rossky P J and Schnitker J 1988 J. Phys. Chem. **92** 4277
Rosenthal S J, Xie X, Du M and Fleming G R 1991 J. Chem. Phys. **94** 4715
Webster F, Rossky P J and Friesner R A 1991a Comp. Phys. Comm. **63** 494
Webster F, Schnitker J, Friedrichs M S, Friesner R A and Rossky P J 1991b Phys.Rev.Lett. **66** 3172

Inst. Phys. Conf. Ser. No 126: Section VI
Paper presented at Int. Symp. on Ultrafast Processes in Spectroscopy, Bayreuth, 1991

519

Solvation dynamics of nile blue studied by time-resolved gain spectroscopy

M.M. Martin, N. Dai Hung, L. Picard, P. Plaza, Y.H. Meyer

Laboratoire de Photophysique Moléculaire du CNRS, Bât 213, Université Paris-Sud, 91405 ORSAY, France.

ABSTRACT. Fast changes in the transmission spectra of Nile Blue solutions in ethanol are observed during 50 ps after subpicosecond excitation though the excited population remains quasi-constant. These changes are attributed to solvation dynamics. In the wavelength range where stimulated emission occurs the spectral shift is found to be faster than τ_L and explained by time dependent overlapping of S_1 absorption and stimulated emission bands due to different solvation effect on these transitions. No shift of the S_1 absorption band was found at shorter wavelengths.

1. INTRODUCTION

In a polar solvent, a solute molecule excited to an electronic state by a short laser pulse is not instantaneously in equilibrium with its surrounding. If the excitation induces a noticeable change in the solute permanent dipole moment, the solvent cage reorganizes to minimize the interaction energy with the newly created dipole. In the recent years, particular attention was given to the study of solvation dynamics of large molecules in solution on the picosecond and femtosecond time scale in order to understand the role of solvent reorientation on the rate of photochemical reactions involving charge transfer processes (Maroncelli 1989).

Experimental studies of solvation dynamics are usually carried out by monitoring time-dependent fluorescence Stokes shift, which gives a direct measurement of the solute S_1-S_0 transition energy change due to solvent reorganization. In the present paper, we report an experiment on solvation dynamics by gain spectroscopy, which allows one to monitor solvation effects on both S_1 - S_0 and S_1 - S_n transition energies by recording transmission spectra on a large spectral range at different pump-probe delays. We applied this method to the oxazine dye Nile Blue in ethanol at both acidic and basic pH at room temperature, after excitation at 603 and 515 nm respectively.

2. EXPERIMENTAL

The tunable 1 ps excitation source is based on a non mode-locked dye laser system generating high power subpicosecond pulses around 600 nm from a single Q-switched frequency doubled nanosecond YAG laser, in a two-step pulse shortening and amplifying process (see N. Dai Hung 1990a). The subpicosecond laser pulses are used to generate a continuum of white light providing tunable pulses after spectral selection and amplification in dye amplifiers pumped by the same nanosecond pump laser (N. Dai Hung 1990b). In the reported experiments we used as the single nanosecond pump either a standard multimode 10 ns YAG laser or a seeded 6 ns YAG laser; the latter improves the pulse to pulse stability of the picosecond system.

Time-resolved transient spectra of Nile Blue solutions in acidic (HCl 10^{-2} M) and basic (KOH $2 \cdot 10^{-2}$ M) ethanol were recorded in the 400 - 750 nm spectral range by tuning the excitation source at 603 and 515 nm respectively. The excitation source was used both to excite the sample and to generate a secondary continuum probe beam in a water cell. The excited sample was probed in a two-beam scheme (Martin 1991). Probe beams transmitted by the excited sample and by a reference non-excited sample were simultaneously detected through a polychromator on a computer-controlled double diode array detector.The angle between the pump and probe polarisation directions was set at the "magic" angle (54.7°).

2. RESULTS AND DISCUSSION

2.1 Nile Blue in acidic ethanol

Figure 1 shows the change in optical density ΔD of a solution of Nile Blue in ethanol containing 10^{-2} M of HCl, probed in the 450-750 nm wavelength range within the 2 to 50 ps pump-probe delay range when excited with a 0.6 ps pump pulse at 603 nm. The zero delay was taken at the half rise of the transient absorption ($\Delta D > 0$) at 520 nm. This transient absorption is attributed to the first excited electronic state S_1 directly populated by laser excitation. Ground state absorption bleaching and/or S_1 stimulated emission ($\Delta D < 0$) is observed for $\lambda > 560$ nm. Whereas the transient absorption peak at 520 nm hardly changes with time, the peak of negative ΔD shifts to the red with a decrease in amplitude, most of the changes occurring during the first 20 ps. For a 2 ps pump probe delay (curve a) the minimum of ΔD occurs at a wavelength slightly red shifted from that of the maximum of the ground state absorption spectrum D_0 (630nm). For a 50 ps delay (curve d) the minimum of ΔD is close to that of the maximum of the steady-state fluorescence spectrum (~660 nm).

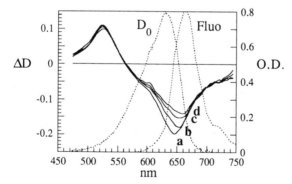

Figure 1, Change in optical density ΔD (left scale)of an acidic solution of Nile Blue at 2 (a), 8 (b), 20 (c), 50 ps (d) pump-probe delays after excitation at 603 nm; Ground state absorption D_0 (right scale) and steady state fluorescence (a.u.) are shown in dotted lines.

If the change in optical density of the solution after excitation is written as

$$(1) \qquad \Delta D(\lambda,t) = -0.43 \, L \, [\, \sigma_e \, (\lambda,t) - \sigma_u \, (\lambda,t) + \sigma_a \, (\lambda,t) \,] \, N_1(t)$$

where L is the sample length, $\sigma_u(\lambda,t)$ and $\sigma_a(\lambda,t)$ are respectively the absorption cross sections for the S_1-S_n and S_0-S_1 transitions and $\sigma_e(\lambda,t)$ the stimulated emission cross section, $\Delta D(\lambda,t)$ should be proportional to the excited state population $N_1(t)$ and should decay in the nanosecond time range at all wavelengths. Therefore we explain the change in shape of ΔD by the time-evolution of the transition cross sections.

A time-dependent red shift of the fluorescence spectrum of Nile Blue in methanol has been reported by Mokhtary *et al* (1989) who found a two-exponential wavelength shift of the maximum with a fast component followed by a 20 ps component. The authors attributed this long component to solvation dynamics, 20 ps lying between the solvent Debye relaxation time τ_D and the longitudinal relaxation time τ_L. Similar findings were often reported for solvation time (Maroncelli 1989 and references therein).

Estimating the initial excited state population from the number of photons absorbed by the sample, we extracted the time-resolved spectra of the quantity $\sigma_e(\lambda,t) - \sigma_u(\lambda,t)$ from $\Delta D(\lambda,t)$. The time-dependent red shift of the maximum of this spectrum in the wavelength range where the stimulated emission cross section σ_e is dominant is shown in Figure 2. The correlation fonction $C(t) = [\nu_{max}(t) - \nu_{max}(\infty)] / [\nu_{max}(0) - \nu_{max}(\infty)]$ defined by Bagchi *et al* (1984) where $\nu_{max} = c / \lambda_{max}$ shows a nearly exponential decay with a time constant of 10.5 ps. This value is smaller than that expected for the longitudinal relaxation time of ethanol at room temperature from Su and Simon's temperature dependent data (1987).

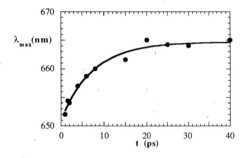

Figure 2, Time-dependent red shift of the $(\sigma_e - \sigma_u)$ band peak of Nile-Blue in acidic ethanol.

The simplest explanation of the discrepancy between our result and that reported by Mokhtary *et al* is that the spectral shift that we observe in Figure 1 is not only due to the time evolution of the stimulated emission spectrum. As a matter of fact, comparison of the stimulated emission cross section spectrum obtained from steady-state fluorescence to the differential spectrum $\sigma_e(\lambda,t) - \sigma_u(\lambda,t)$ at long pump-probe delay (90 ps) gives evidence of a transient absorption band around 650 nm. Thus, the spectral shift observed results from time-dependent overlapping of S_1-S_n and S_1-S_0 spectra due to different time evolution of the S_1-S_n and S_1-S_0 energy gaps during the solvent reorientation process. On the other hand, the present study does not show much time evolution of the S_1 absorption band at 520 nm, indicating that Nile Blue has similar permanent dipole moments in the first excited singlet state and in the upper excited state involved in this transition.

2.2. Nile Blue in basic ethanol

Figure 3 shows the change in optical density ΔD of a solution of Nile Blue in ethanol containing $2\ 10^{-2}$ M of KOH, probed in the 400-700 nm wavelength range within the 3-50 ps pump-probe delay range when excited with a pump pulse of ~1 ps at 515 nm. At short delay (curve a: 3 ps) a small negative ΔD is found around the maximum of the ground state absorption band (470-530 nm), whereas transient absorption ($\Delta D > 0$) is measured on the short wavelength edge (450 nm). Negative ΔD is also found for $\lambda > 570$ nm where stimulated emission is expected but the minimum of ΔD is red-shifted from the maximum of the steady state fluorescence spectrum which displays a broad peak around 610 nm. ΔD is close to zero around 550-570 nm. At longer delays (curves b and c), the net gain band ($\lambda > 590$ nm, $\Delta D < 0$, $D_0 = 0$) is further red-shifted and the absolute value of the negative ΔD peak increases, while a peak of positive ΔD rises around 560 nm. Most of the spectral change described above occurs in less than 50 ps. Little change is observed in the transient absorption around 450 nm in the same time scale.

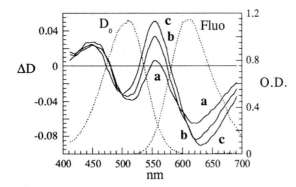

Figure 3, Change in optical density ΔD of a basic solution of Nile Blue at three pump-probe delays: 3 (a), 12 (b), 53 ps (c), after excitation at 515 nm. Ground state absorption D_0 (right scale) and steady state fluorescence (a.u.) are shown in dotted line.

The steady state fluorescence Stokes shift of this unprotonated form of the dye is larger than 3000 cm^{-1}. The changes shown in Figure 3 for $\lambda > 500$ nm may be interpreted by time-dependent overlapping of S_1 absorption and stimulated emission bands due to solvent reorganization. Around 550 nm, at short time (curve a) the excited state absorption cross section σ_u is close to the sum of the ground state absorption and stimulated emission cross sections $\sigma_a + \sigma_e$, leading to $\Delta D \sim 0$ in eq (1). With increasing time, the stimulated emission cross section may decrease in this wavelength range while it increases for $\lambda > 600$ nm as a result of the red shift of the whole stimulated emission spectrum. Although, the solvation effect on the S_1 absorption band is not known the present study shows that σ_u becomes dominant around 550 nm at long delays leading to $\Delta D > 0$ (curves b and c). Time dependent overlapping of S_1 absorption and stimulated emission bands due to solvation dynamics that we inferred for Nile Blue in its protonated form in the previous section is well observed here for the unprotonated form of the dye. The absence of solvation effect on the S_1 absorption transition to upper levels around 450 nm confirms the small change in the permanent dipole moment of the dye in its first excited singlet and in the upper excited singlet state involved in that transition.

REFERENCES

Bagchi B, Oxtoby D W, Fleming G R 1984 *Chem. Phys.* **86** 257
Dai Hung N, Meyer Y H, Martin M M, Nesa F 1990a *"Ultrafast Phenomena in Spectroscopy"* eds.E Klose, B Wilhemi, *Springer Proceedings in Physics* . **49**, Springer-Verlag, pp 33
Dai Hung N, Meyer Y H 1990b *Opt. Commun.* **79** 215
Maroncelli M, MacInnis J, Fleming G R 1989 *Science* **243** 1674
Martin M M, Plaza P, Meyer Y H 1991 *Chem. Phys.* **153** 297
Mokhtari A, Chesnoy J, Laubereau A 1989 *Chem. Phys. Lett.* **155** 593
Su Shyh-Gang, Simon J D, 1987 *J. Phys. Chem* . **91** 2693

Inst. Phys. Conf. Ser. No 126: Section VI
Paper presented at Int. Symp. on Ultrafast Processes in Spectroscopy, Bayreuth, 1991

Decay time distribution of fluorescence kinetics in systems with inhomogeneous broadening of electronic spectra

D M Gakamsky, E P Petrov, A N Rubinov

The Institute of Physics, Academy of Sciences of Byelarus, Leninsky Pr.70, 220602 Minsk, Byelarus

ABSTRACT: A shape of fluorescence decay time distribution (DTD) of polar dye solution with inhomogeneous broadening of electronic levels was analyzed. It was at the first time found that under condition of time-dependent fluorescence shift (TDFS) positions of peaks of DTD are determined by the parameters of spectral shift. Theoretical DTD correlate with ones obtained from the experimental fluorescence kinetics of glycerol solution of 3-amino-N-methyl-phthalimide (3ANMP). We propose the method based on recovery of parameters of spectral shift for determination of the dynamic characteristics of polar region of membrane bilayer.

1. INTRODUCTION

The inhomogeneous broadening of electronic spectra in viscous polar dye solutions leads to TDFS. If the emission spectrum shifts towards the red region an accelerated fluorescence decay at the blue slope of the spectrum takes place. In contrast, at the red slope a fluorescence intensity builds up from a low value at time zero to a maximum and then decays. A character of fluorescence kinetics due to the spectrum shift is complex and fluorescence decay cannot be represented as a sum of two or three exponentials. At present a careful consideration of DTD for systems with TDFS has not been made. However Lakowicz *et al* (1987) assumed that the distribution in this case has to be a continuous one due to complex behavior of fluorescence spectra.

Generally, spectral shift in polar dye solution, as well as in biochemical samples, is not exponential. In present paper we analyze the case when spectral shift may be approximated by two-exponential one. We have found at the first time that DTD in such case has to be discrete and may be represented as series of δ-functions. The positions and weights of these δ-functions depend on relaxation parameters of TDFS and are directly connected with dynamic parameters of the probe microenvironment, e.g. with microviscosity. It is shown that the proposed method allows to study a structural dynamics of polar region of biomembranes.

2. THEORY

Fluorescence kinetics of one of the systems with TDFS (3-amino-N-methyl-phthalimide (3ANMP) in glycerol) at room temperature as it was shown by Gakamsky *et al* (1990) may be described as follows:

$$I(\nu,t) = i_0 \exp(-t/\tau_0) \int \varphi(\nu',t) \, S(\nu-\nu') \, d\nu' \tag{1}$$

where i_0 is constant, τ_0 is natural lifetime, $\varphi(\nu,t)$ is inhomogeneous broadening function (IBF) of fluorophore and $S(\nu)$ is homogeneous fluorescence spectrum. One can obtain DTD of the fluorescence kinetics (1) using inverse Laplace transform (ILT) $P(\nu,\Gamma)$ determined by following expression:

$$I(\nu,t) = \int_0^\infty P(\nu,\Gamma) \exp(-\Gamma t) \, d\Gamma$$

where $\Gamma = 1/\tau$ is reciprocal decay time. It was shown by Gakamsky *et al* (1991) that in the case when IBF may be represented by the Gaussian

$$\varphi(\nu,t) = \frac{1}{\sqrt{2\pi}\,\sigma(t)} \exp\left(- \frac{(\nu-\nu_0(t))^2}{2\sigma^2(t)}\right) \tag{2}$$

where $\nu_0(t) = \nu_0(\infty) + \Delta\nu \cdot (\beta_1 \exp(-t/\tau_1) + \beta_2 \exp(-t/\tau_2))$ and $\sigma(t) = \sigma(\infty) + \Delta\sigma \exp(-t/\tau_3)$ and for arbitrary shape of the emission spectrum $S(\nu)$ the ILT of (2) may be written in general form as

$$P(\nu,\Gamma) = \sum_{i,j,k=0}^{\infty} C_{ijk} \; \delta\left(\Gamma - \frac{1}{\tau_0} - \frac{i}{\tau_1} - \frac{j}{\tau_2} - \frac{k}{\tau_3}\right). \tag{3}$$

The weights C_{ijk} depend on the emission frequency ν, relaxation parameters and emission spectrum shape $S(\nu)$. The dispositions of δ-functions are determined by τ_0 and characteristic relaxation times τ_i, $i=1,2,3$ and do not depend on a shape of the emission spectrum. The latter affects their weights only.

An accurate analysis of (3) shows that DTD consists of separately located δ-function located at τ_0 and, at lower decay times, sequences of δ-functions: $\delta(\Gamma-1/\tau_0-n/\tau_i)$, $n=1,\infty$, $i=1,2,3$, and, lastly, complex result of the interference of these sequences. It should be noted that DTD is always located within the interval $\tau \in \;]0, \tau_0]$, irrespectively of values of τ_i, $i=1,2,3$. Each of these three sequences appears at decay time $\tau_i^{\dagger} =$ $= \tau_0\tau_i/(\tau_0+\tau_i)$, $i=1,2,3$ and lasts towards lower decay times. Each sequence may be characterized by some mean decay time. However, the weights of δ-functions in the sequence depend on the emission frequency and, therefore, we can stand that the mean time of the sequence does depend on the emission frequency.

3. RESULTS AND DISCUSSION

Available experimental accuracy does not allow to resolve the structure of the above mentioned sequences. So, experimentally obtained DTD should consist of some rather broad peaks. Still, one can estimate the relaxation times using centers of gravity of the peaks τ_i^G instead of $\tau_i^{\dagger}$.

The presence of the exponentials with negative pre-exponential coefficients in the experimental fluorescence kinetics sufficiently complicates solving of the inverse problem of numerical ILT because of its ill-positness. Therefore we chose the registration point at the blue slope of the emission spectrum. Numerical ILT of the experimental kinetics was made using new algorithm created by Goldin (1991). This algorithm allows to obtain DTD of the experimental kinetics without a *priori* assumptions about DTD shape. If DTD is nonnegative this algorithm provides stable solutions even for the experimental data with rather high noise level (CPC = $10^3 \div 10^4$).

3.1. 3ANMP in glycerol

For the simplification of the theoretical computations we approximated the blue slope of the inhomogeneous fluorescence spectrum of 3ANMP in glycerol by Gaussian with moving center of gravity and altering width using the same relations as for the description of IBF (Eq.2). The parameters of shift of this model spectrum were determined by nonlinear fitting of the simulated kinetics to the experimental one. We obtained that the longer time of spectral shift τ_1 is approximately equal to the characteristic time of width altering τ_3. Therefore, the peaks corresponding to τ_1 and τ_3 should not be resolved. Figure 1 represents the theoretically computed DTD and DTD computed from the experimental and simulated kinetics. One can notice good coincidence of these three distributions.

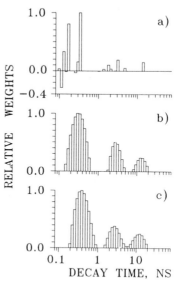

Figure 1. Decay time distributions for 3ANMP in glycerol [C]=$5 \cdot 10^{-5}$M, T=22°C, ν_{ex}=27900cm^{-1}, ν_{em}=25000cm^{-1} a) theoretical; b) computed from simulated kinetics; c) computed from the experimental kinetics (CPC=$5 \cdot 10^3$).

3.2 Fluorescence probes in membranes

We have investigated DTD for 1-phenilnaphthylamine (1-AN) and 2-toluidinonaphthalene-6-sulphonate (2,6-TNS) probes in membranes. They locate, as it is known, in the regions of glycerol skeleton and phospholipide heads, respectively.

Table. (All temporal parameters are given in nanoseconds)

Probe	T°C	τ_0^G	τ_1^G	τ_2^G	τ_1	τ_2	P_0	P_1	P_2	$<\tau_R>$
1-AN	8	6.3	1.3		1.6		0.60	0.40		1.6
	14	7.3	2.3	0.2	3.4	0.2	0.23	0.27	0.50	1.3
	22	7.2	2.4	0.3	3.6	0.3	0.16	0.21	0.64	1.1
2,6-TNS	14	6.5	1.9	0.4	2.7	0.4	0.19	0.17	0.64	0.9

DTD for 1-AN in membranes at excitation at the absorption maximum and registration at the blue slope of the emission spectrum for three values of temperature: 22°C, 14°C and 8°C are shown in Figure 2a,b,c. The results of DTD analysis (the estimatd characterstic times of specral shift τ_1,τ_2 and the weights of corresponding peaks P ,P) for these cases are listed in

Table. In all cases the distributions consist of the discrete peaks. As it is clear from the previous consideration the right peak is connected with the natural lifetime whereas the other peaks depend on spectral shift law. Therefore, the positions and relative weights of the last ones have to depend on the temperature.

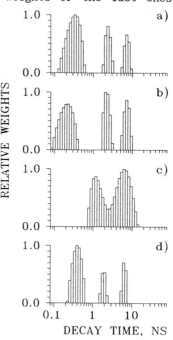

Independence of the right peak position from the temperature in the experimental accuracy limits shows that the natural lifetime dose not depend on the probe environment microviscosity. The decrease of temperature makes the spectral shift to be more slow. This has to lead to decrease in relative weights of fast components of DTD. One can notice that for 1-AN at $8°C$ the components of DTD corresponding to TDFS are not resolved due to decrease of the relative weight of the fast component. On the basis of obtained DTD we can estimate the mean times of the spectral shift $<\tau_R> = =(P_1\tau_1+ P_2\tau_2)/(P_1+ P_2)$. It was found that this parameter monotonously decreases with the increase of temperature (see Table).

DTD for 2,6-TNS probe in membranes also consists of the discrete peaks (Figure 2d) at excitation at the absorption maximum and at blue-slope registration. In accordance with a well known fact that microviscosity is greater in the region of the of the phospholipide heads, the mean time of the spectral shift $<\tau_R>$ for 2,6-TNS is smaller than that for 1-AN at the same temperature.

Figure 2. Decay time distributions for 1-AN (a, b, c) and 2,6-TNS (d) in model membranes, $\nu_{ex}=27900cm^{-1}$, $\nu_{em}=25000cm^{-1}$:
a) T=22°C, CPC=2.5·10³; b) T=14°C, CPC=10⁴; c) T=8°C, CPC=10⁴; d) T=14°C, CPC=5·10³.

4. CONCLUSIONS

Accurate treating of DTD of the fluorescence kinetics of inhomogeneously broadened systems allows to estimate the parameters of spectral relaxation. Such a method may be successfully applied to study of dynamic properties of polar region of membrane bilayer.

REFERENCES

Gakamsky D M, Nemkovich N A, Rubinov A N and Tomin V I 1990 *J.Mol.Liq.* **45** 33
Gakamsky D M, Goldin A A, Petrov E P and Rubinov A N 1991 To be published
Goldin A A 1991 *Inverse Problems*. Accepted for publication
Lakowicz J R, Cherek H, Gryczinsky I, Johnson M L and Joshi N 1987 *Biophys. Chem.* **28** 35

Inst. Phys. Conf. Ser. No 126: Section VI
Paper presented at Int. Symp. on Ultrafast Processes in Spectroscopy, Bayreuth, 1991

527

Orientational dynamics of molten $2[Ca(NO_3)_2] - 3[KNO_3]$ measured by time-resolved optical Kerr effect

M.Ricci*, R.Torre#, P.Foggi*, R. Righini*

European Laboratory for Non-Linear Spectroscopy (LENS),
University of Florence,
Largo E.Fermi 2, 50125 Florence, Italy

* from: Dept. Chemistry, University of Florence, 50121 Florence, Italy
# from: Dept. Physics, University of Florence, 50125 Florence, Italy

ABSTRACT: Femtosecond time-resolved OKE has been used to investigate the glass-forming molten salt $2[Ca(NO_3)_2]3[KNO_3]$ in the temperature range 110-350 C. Two different dynamics have been observed. Both of them are interpreted as due to rotations of the NO_3^- ion.

1. INTRODUCTION

The mixed ionic melt $2[Ca(NO_3)_2]3[KNO_3]$ (CKN) is well known as a good glass forming liquid, characterized by a well defined glass transition temperature (T_g = 65 C)[1]; the dynamical processes which are activated during the cooling of the viscous liquid have been studied by means of different techniques: Brillouin, Rayleigh and Raman scattering[1], neutron diffraction[2], dielectric relaxation[3] and transient grating[4]. The temperature dependence of the structural relaxation time and that of the bulk viscosity are typical of a "fragile" glass forming liquid, according to the classification by Angell[5].

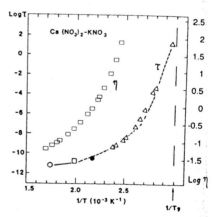

Fig. 1 Temperature dependence of τ and η

In Fig.1 the divergence of both quantities for temperatures above 200 C is evident; this behaviour, which is essentially related to the freezing of the translational degrees of freedom of the ions eventually leading to the non-ergodic behaviour of the glass, has been interpreted in terms of mode coupling theory. Much less attention has been devoted to the effect of cooling on the orientational degrees of freedom of the NO_3^- ions. According to the few available Raman and Rayleigh[1] data the orientational relaxation time τ_{rot} do not show the same divergence as that of the structural relaxation, thus suggesting that a decoupling of the translational and rotational degrees of freedom takes place for temperatures below 200 C. This indication is however apparently contradicted by the results of a recent Molecular Dynamics simulation[6], which predicts a very large increasing of τ_{rot} when the glass transition temperature is approached.

In the present paper we report on an accurate determination of the temperature dependence of the orientational relaxation time in CKN, measured by sub-picosecond time resolved Optical Kerr Effect (OKE).

2. EXPERIMENTAL

Commercial grade $Ca(NO_3)_2$ and KNO_3 were purified by crystallization and de-hydrated by heating at 220 C for one day. The appropriate amount of $Ca(NO_3)_2$ was added to molten KNO_3, yielding a clear liquid which did not show any crystallization even after several cooling and heating cycles.

The laser system is based on a fiber-compressed Mode Locked Nd-YAG laser (4 ps at 532 nm, 82 MHz rep. rate) used to synchronously pump a dye laser (Rhodamine 6G, 350 fs, 250 mW), whose output is further compressed in a fiber to produce a train of 80 fs tunable pulses (70 mW). A pulsed (10 Hz) Q-switched Nd-YAG laser is used to amplify in a three-stage amplifier (Kiton red) the dye laser output, yielding 300 μJ pulses of about 100 fs duration. Part of this light is used for the excitation of the OKE; the remaining is focused tightly in a D_2O cell to produce white light (850-350 nm). A fraction of that light is frequency selected by means of an interference filter and used (possibly after amplification in a second amplification stage driven by the same Q-switched Nd-YAG laser) to probe the optical anisotropy induced in the sample by the pump pulse. Calcite polarizers where used to select the polarization of the excitation (vertical) and of the probe pulses (45^0).

The probe pulse intensity passing through a crossed polarizer was detected, as a function of the probe delay, by a PMT and measured by a BoxCar integrator. Neutral density filters where used to increase the dynamic range of the measurement.

3. RESULTS AND DISCUSSION

Fig.2 shows the decay of the OKE signal observed at three different temperatures: beside the strong instantaneous response, which is due to the deformation of the electronic cloud, an orientational contribution can be clearly seen, which decays with a non-exponential behaviour. In the entire temperature range considered in our experiments the decay curve can be well fitted by a double exponential, characterized at high temperature by decay time-constants of few picoseconds, which increase above one hundred picoseconds at low temperature. The temperature dependence of the two orientational times is shown in Fig.3. The same data in a semilog scale (Fig.4) show a typical linear Arrhenius behaviour.

Fig. 4 shows also the relaxation time measured by time-resolved Brillouin scattering[4]: the orientational dynamics does not follow the diverging   increase   of   the   viscosity with decreasing temperature, with a behaviour   more similar to that of a "normal" liquid.

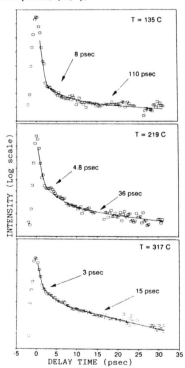

Fig.2   OKE decay profiles for some
temperatures

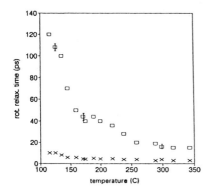

Fig. 3 Temperature dependence of the
two rotational relaxation times

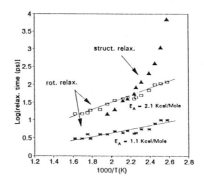

Fig. 4 Arrhenius plot of rotational and
structural relaxation times

Our data then confirm the decoupling of rotational and translational degrees of freedom below 250 C suggested by the light scattering data. It is worth to be noticed that for temperatures below 100 C no detectable signal due to the orientational dynamics could be observed: we take this as an indication of an abrupt slowing down of the rotational dynamics for temperatures close to the glass transition (65 C). Such dynamics is actually completely frozen in the solid glass.

Less immediate is the interpretation of the two decay time constants observed in our experiments. Three main points have to be stressed :

i) both decay processes show an Arrhenius behaviour in the (quite large) temperature range considered;

ii) the activation energy of the slow process is about twice that of the fast one;

iii) the relative weight of the fast process increases with increasing temperature.

A bifurcation into α and β relaxation regimes of the orientational dynamics has been observed in some polimeric and molecular glass forming liquids[7]: the β (fast) relaxation has been attributed to the individual dynamics of the molecules, while the α process is interpreted as a result of cooperative motions of large structured formed in the liquid. This second process is of course strictly connected to the glass transition, and in fact the corresponding relaxation time diverges in the proximity of T_g. On the other hand the α and β mechanisms merge in a single relaxation process at high temperature, where the role of the cooperative dynamics is negligible. Although the above mentioned points (ii) and (iii) are consistent with this picture and with the theoretical predictions of stochastic treatments[8], two major discrepancies prompt us to discard this interpretation:

a) in spite of the large temperature range considered, no evidence has been found of a bifurcation temperature: two decay processes can be detected also at high temperature (about 260 K higher then T_g);

b) no drastic slowing down of the slow processes has been observed approaching the glass transition: a diverging relaxation time is instead expected for the α process in that picture.

We then prefer an alternative explanation based on the dynamics of the individual molecules. The NO_3^- rotor is in fact a "symmetric top", with an isotropic polarizability in the ion plane: the rotation about the C_3 axis is then expected to be silent in an OKE experiment. The same picture holds for other symmetric molecules, like benzene: on the contrary, a double exponential decay of the orientational correlation has been observed in liquid benzene[9], and interpreted as due to the activation of the in-plane rotation due to the electron cloud deformations induced by the intermolecular interactions. A similar explanation applies also to the present case: the values of 1.1 and 2.1 kcal/mole for the activation energies measured for the NO_3^- ion (see Fig.4) are very close to those obtained for the spinning and tumbling motions, respectively, of small molecules of similar size like acetonitrile[10]. In addition, the relaxation times measured at high temperature are very close to those predicted by the molecular dynamics simulation in the same conditions.

In summary, we can conclude:

a) a decoupling of translational and rotational dynamics has been shown to happen in the cooling process of CKN; between 350 and 110 C no drastic slowing down has been observed of the rotational dynamics, which instead follows an Arrhenius behaviour;

b) a double exponential relaxation has been observed in the entire temperature range; the two time constants have been attributed to the rotation about the C_2 and C_3 symmetry axes of the NO_3^- ion, respectively.

REFERENCES

1) L.M.Torell, L.Borjesson, M.Elmroth , *J.Phys.Condens.Matter* (1990),2,SA207
   M.Grimsdich, R.Bhadra, L.M.Torell , *Phys.Rev.Lett.* (1989),62,2616
   L.M.Torell , *J.Chem.Phys.* (1982),76,3467

2) F.Mezei, W.Knaak, B.Farago , *Phys.Rev.Lett.* (1989),58,571

3) G.Williams in *Molecular Liquids* , ed. A.J.Barnes, W.J.Orville-Thomas, J.Yarwood,
   NATO Advances Study Institutes Series (Boston,1984)
4)L.T.Cheng, Y.X.Yan, A.Nelson *J.Chem.Phys.* (1989),91,6052

5) C.A.Angell , in *Relaxation in Complex Systems* , ed. K.Ngai, G.B.Wright, N.T.I.S. U.S.
   Dept. of Commerce (Springfield, VA, 1985)

6) G.F.Signorini, J.L.Barrat, M.L.Klein , *J.Chem.Phys.* (1990),92,1294

7) D.Kivelson, D.Miles, *J.Chem.Phys.* (1988),88,1925

8) A.Polimeno, J.H.Freed , *Chem.Phys.Lett.* (1990),174,481

9)J.Etchepare, G.Grillon, G.Hamoniaux, A.Antonetti, A.Orzag , *Rev.Phys.Appl.*
   (1987),22,1749

10) H.Versmold, in *Molecular Liquids* , ed. A.J.Barnes, W.J.Orville-Thomas, J.Yarwood,
    NATO Advances Study Institutes Series (Boston,1984)

Inst. Phys. Conf. Ser. No 126: Section VI
Paper presented at Int. Symp. on Ultrafast Processes in Spectroscopy, Bayreuth, 1991

Orientational relaxation dynamics of polar dye probes in n-alkylnitriles

G. B. Dutt and S. Doraiswamy

Chemical Physics Group, Tata Institute of Fundamental Research,
Homi Bhabha Road, Bombay 400 005, India.

Abstract. The rotational motion of three dye probe molecules-cresyl violet (a monocation), nile red (neutral but polar) and resorufin (a monoanion) have been investigated in n-alkylnitriles using picosecond fluorescence depolarization spectroscopy. While the dielectric friction model fits well the rotational diffusion of nile red, it is reasonably satisfactory for resorufin. However, the concept of solvation, whose nature is unclear at present has also to be invoked to explain the rotational dynamics of cresyl violet.

1. Introduction

A molecule rotating in a liquid experiences friction on account of its continuous interaction with its neighbours, the desire to understand which, has been the motivating force in carrying out the experimental measurements on rotational reorientation times in liquids. The contribution to the total friction ζ for the rotational motion of a polar solute molecule of volume V in a polar solvent of viscosity η, dielectric constant ϵ and dielectric relaxation time τ_D can be written as

$$\zeta = \zeta_h + \zeta_{DF} + \zeta_{solvation}$$

where ζ_h refers to the hydrodynamic friction generally calculated from Stokes-Einstein-Debye (SED) model [1,2], ζ_{DF} is a macroscopic correction due to the dielectric friction [3] and $\zeta_{solvation}$ is a microscopic correction arising out of specific solute-solvent interaction. The usual methodology adopted is to model the molecular rotational motion in terms of viscous and dielectric friction, and if significant difference from the observed values still persist, to invoke the solvation effects.

The simplified expression commonly employed for the calculation of the rotational reorientation time τ_r is given by [4]

$$\begin{aligned}
\tau_r &= \frac{1}{6kT}(\zeta_h + \zeta_{DF}) \\
&= \frac{\eta V f C}{kT} + \frac{\mu^2}{a^3 kT}\frac{(\epsilon - 1)}{(2\epsilon + 1)^2}\tau_D
\end{aligned} \tag{1}$$

where μ is the dipole moment of the solute molecule, a is the radius of the cavity in which it rotates, f is a shape factor to account [5] for the deviation of the solute molecule from spherical shape and C is a factor dependent on the boundary condition [6]. It is pertinent to note that this kind of partitioning of the total friction is not strictly correct, since the two effects are interlinked due to electrohydrodynamic coupling [7].

2. Experimental

The dyes cresyl violet (perchlorate salt, Lambda Physik), nile red (Aldrich) and resorufin (sodium salt, Aldrich) were of the highest available purity and were used as such. The n-alkylnitriles were also of the highest available purity from Aldrich and Fluka. The rotational reorientation times of the three dye molecules were measured in a series of n-alkylnitriles by picosecond time-resolved fluorescence depolarization spectroscopy using single-photon counting technique, the instrumentation details of which are given elsewhere [4]. The dielectric constant and the Debye relaxation time were measured using time domain reflectometry (TDR) technique from 10 MHz to 10 GHz [8]. All measurements were made at 273 K.

3. Results and Discussion

A single-exponential fit was found satisfactory to describe the anisotropy decay for the three dyes in all the nitriles. The measured rotational reorientation times are listed in Table 1. Table 2 lists viscosity (taken from literature), the measured values of the dielectric constant (ϵ) and Debye relaxation time (τ_D). As has been done in our earlier works, the dye probes have been modelled as oblate ellipsoids. Since the dielectric friction is the least in water, the hydrodynamic volume obtained from the observed rotational reorientation time in water [4,9] was used in the calculation of the viscous drag. To evaluate the second term, the dipole moment of the molecules in the excited state was assumed to be 12 D, as was done in our earlier works [4,9] and the cavity radius was evaluated from the van der Waals volume [10]. Figure 1 gives the plot of rotational reorientation time versus viscosity for all the three dyes. The theoretical line (dashed line) calculated from Eq. 1 is also shown in the figure.

It is obvious that the rotational dynamics of nile red (neutral but polar) is well described by the inclusion of the dielectric friction model and even for the anion resorufin, the model is satisfactory. However, for the cation cresyl violet there is a significant discrepancy. Indeed, such a discrepancy was observed for cresyl violet even in the case of n-alcohols [4]; but it has been shown that cresyl violet forms hydrogen bonds with alcohols [11] and that could explain the discrepancy. Nitriles are polar aprotic solvents and specific solvation effects like hydrogen bonding are less important in aprotics. While the effects due to solvation are apparent in the case of cresyl violet, its nature is not clear.

Table 1. Excited-state rotational reorientation times (ps) of the cationic dye cresyl violet, neutral but polar dye nile red and anionic dye resorufin in n-alkylnitriles at 273 K, together with the calculated values [τ_{total}] from SED and dielectric friction models. The τ_{SED} is the contribution due to the viscous drag, is calculated using the volume deduced from the observed rotational reorientation time in water. The τ_{DF}, the difference between τ_{total} and τ_{SED}, is calculated from the second term of Eq. 1.

Solvent	Cresyl Violet			Nile Red			Resorufin		
	τ_{SED}	τ_{total}	τ_{obs}	τ_{SED}	τ_{total}	τ_{obs}	τ_{SED}	τ_{total}	τ_{obs}
Acetonitrile	69	70	73±10	63	64	71±20	38	40	20±9
Propionitrile	87	89	153±7	79	81	80±13	48	51	93±23
Butanenitrile	121	129	164±9	110	116	111±10	67	77	61±15
Pentanenitrile	158	172	217±9	143	153	137±14	87	104	130±20
Hexanenitrile	217	239	486±22	197	214	218±10	119	148	187±9
Octanenitrile	398	455	656±24	361	404	373±15	218	292	346±10
Nonanenitrile	527	606	924±27	478	538	548±15	289	392	597±32
Decanenitrile	717	849	1423±27	651	751	699±11	393	566	606±15

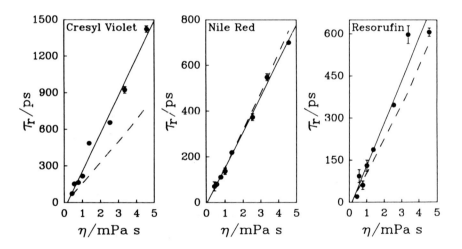

Figure 1. Plot of rotational reorientation times τ_r for the three dye probes in n-alkylnitriles (acetonitrile to decanenitrile except heptanenitrile) as a function of viscosity η at 273 K. The line through the experimentally observed points is the least-squares-fit line for all the nitriles. The – – – line is theoretically calculated using SED and DF models.

Table 2. Parameters used in the calculation of the contribution due to the dielectric friction.

Solvent	η^a /mPa s	ϵ	τ_D /ps	$\frac{(\epsilon-1)}{(2\epsilon+1)^2}\tau_D$ /ps
Acetonitrile	0.44	40.5	3.3	0.019
Propionitrile	0.56	31.0	4.4	0.033
Butanenitrile	0.78	27.2	12.5	0.107
Pentanenitrile	1.01	22.9	18.2	0.182
Hexanenitrile	1.38	20.2	27.1	0.304
Octanenitrile	2.54	15.8	54.9	0.765
Nonanenitrile	3.36	14.1	69.5	1.068
Decanenitrile	4.58	13.9	114.8	1.785

[a] Viscosities of nitriles are taken from Sivakumar N, Hoburg E A and Waldeck D H 1989 *J. Chem. Phys.* **90** 2305–16

Acknowledgments

The authors would like to thank the Department of Science and Technology for providing the funds for setting up the picosecond fluorescence spectrometer. The authors would like to thank Dr. S. C. Mehrotra and his colleagues of Marathwada University for providing facility to make measurements of the dielectric parameters with their time-domain reflectometer.

References

[1] Einstein A 1956 *Investigations on the Theory of the Brownian Movement* (New York: Dover)

[2] Debye P 1928 *Polar Molecules* (New york: Dover)

[3] Madden P and Kivelson D 1982 *J. Phys. Chem.* **86** 4244–56 and references therein

[4] Dutt G B, Doraiswamy S, Periasamy N and Venkataraman B 1990 *J. Chem. Phys.* **93** 8498–8513

[5] Perrin F 1934 *J. Phys. Radium* **5** 497

[6] Hu C M and Zwanzig R 1970 *J. Chem. Phys.* **60** 4354–57

[7] Felderhof B U 1983 *Mol. Phys.* **48** 1269–88

[8] Cole R H, Berberian J G, Mashimo S, Chryssikos G, Burns A and Tombari E 1989 *J. Appl. Phys.* **66** 793–802

[9] Dutt G B, Doraiswamy S and Periasamy N 1991 *J. Chem. Phys.* **94** 5360–68

[10] Edward J T 1970 *J. Chem. Educ.* **47** 261–70

[11] Beddard G S, Doust T and Porter G 1981 *Chem. Phys.* **61** 17–23

Inst. Phys. Conf. Ser. No 126: Section VI
Paper presented at Int. Symp. on Ultrafast Processes in Spectroscopy, Bayreuth, 1991

535

Direct observation of femtosecond angular-velocity dynamics in molecular liquids

Toshiaki Hattori[a], Akira Terasaki[b], Takayoshi Kobayashi[b], Tatsuo Wada[c], Akira Yamada[c], and Hiroyuki Sasabe[c]

[a]Institute of Applied Physics, University of Tsukuba, Tsukuba 305, Japan
[b]Department of Physics, University of Tokyo, Hongo, Bunkyo, Tokyo 113, Japan
[c]Frontier Research Program, RIKEN (Institute of Physical and Chemical Research), Hirosawa, Wako 351-01, Japan

ABSTRACT: A new experimental technique for the study of ultrafast nonlinear optical response of materials, which detects nonrelaxational response in the media with high sensitivity, is introduced. Polarization-selective optical-heterodyne-detection scheme is applied to the measurement of the spectral shift of a probe pulse which is caused by induced phase modulation brought about by a pump pulse. This technique allows us to observe directly the angular-velocity dynamics of molecules. Molecular dynamics in benzene and carbon disulfide are studied using this technique with femtosecond optical pulses.

Recent studies on ultrafast molecular dynamics have revealed inertial or oscillatory nature of molecular motions in the liquid phase in femtosecond time regime. Investigations on these coherent molecular motions are of crucial importance to shed light on subjects such as local structure of liquids, chemical reaction dynamics in solution, and solvation dynamics. For the characterization of molecular motions in this time region, direct observation of velocity (either translational or orientational) is desirable. Conventional measurement techniques in time domain, such as optical Kerr response and transient grating measurements, or in frequency domain, such as light scattering measurement, however, are not suitable for the study of these coherent motions since these techniques reflect the dynamics of molecular *orientation* or *position* rather than that of *velocity*. We have developed a new technique which gives direct information on the dynamics of angular velocity of anisotropic molecules (Hattori *et al.* 1991).

The new technique, optical-heterodyne-detected induced phase modulation (OHD-IPM), is based on the detection of spectral change of short pulses caused by an induced phase modulation (IPM) process. A femtosecond pump pulse triggers a refractive index change in the sample liquid due to the optical Kerr effect (OKE) of the medium. The time response is measured by detecting the spectral shift of a probe pulse. Since the spectral shift of the probe pulse is proportional to the time derivative of the refractive index of the medium, by this method we can directly observe the time derivative of the optical Kerr response. This technique was first used by Mokhtari and Chesnoy (1988) for the observation of oscillatory component in resonant nonlinear response of dye solutions. We applied polarization-selective optical-heterodyne-detection (OHD) scheme to this measurement to result in great enhancement in the sensitivity of the measurement, which enabled the application of this technique to the study of molecular dynamics in various transparent liquids.

The experimental schematic is shown in Fig. 1. The femtosecond light source was a standard combination of a colliding-pulse mode-locked (CPM) ring dye laser and a multiple-pass amplifier pumped by a copper vapor laser. The pulse width and the wavelength of the amplified pulses were 55 fs and 620 nm, respectively. The amplified light was divided by a beam splitter into the pump and the probe beams. The pump pulse was linearly polarized at 45° from the linear polarization of the probe pulse. A polarization analyzer placed after the sample passed a small fraction of the probe light which was polarized nearly perpendicular to the original polarization of the probe light. Thus the component of the nonlinear polarization perpendicular to the incident probe light was heterodyne detected with the fraction of the probe light, which served as the local oscillator. A

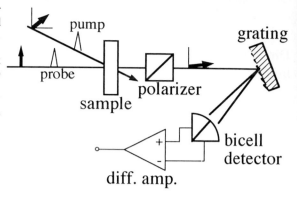

Fig. 1. The schematic of the experiment for the measurement of optical-heterodyne-detected induced phase modulation.

grating dispersed the light onto a bicell photodiode (Hamamatsu, S994-22). Spectral shift of the light was detected by taking the difference of the two outputs of the bicell detector. The spectral shift against the delay time of the probe pulse with respect to the pump pulses gave the signal trace of the OHD-IPM measurements. Conventional OHD-OKE measurements, which yield signals proportional to the OKE response, were also performed for the comparison with the OHD-IPM measurements.

The OKE response function of small molecular liquids is proportional to the time derivative of the orientational correlation function (OCF) of the molecules at the thermal equilibrium when intermolecular interaction-induced susceptibility change can be neglected. The IPM response, which is the time derivative of the OKE response function, is, therefore, the second-order time derivative of the OCF, which has been shown to be approximately proportional to the angular-velocity correlation function (AVCF) of the molecules in the time region shorter than the reorientational relaxation time. IPM measurements, therefore, make possible the direct observation of the AVCF.

The molecular dynamics in benzene and CS_2 were studied. Figure 2 shows the OHD-OKE and OHD-IPM data obtained from liquid benzene. Damped oscillations are apparent in the OHD-IPM data, which shows the advantage of the time-derivative technique by OHD-IPM measurements for the study of vibrational dynamics of molecular systems.

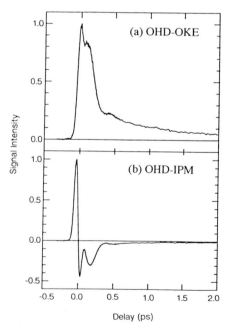

Fig. 2. Signal trace of benzene obtained by (a) OHD-OKE and (b) OHD-IPM measurements.

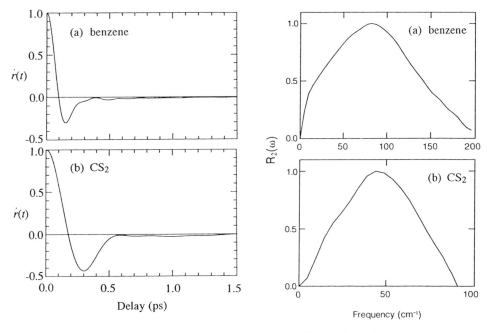

Fig. 3. Reconstructed IPM response function of benzene and CS_2.

Fig. 4. The reduced Raman spectra, $R_2(\omega)$, of (a) benzene and (b) CS_2.

The nearly antisymmetric feature around time zero is due to the electronic and the coherent coupling contributions. The damped oscillations with a period of about 400 fs in the OKE and the IPM data are attributed to an *intermolecular* vibrational mode since benzene has no low-frequency *intramolecular* modes below 400 cm^{-1}. The contributions of the instantaneous electronic nonlinearity and the coherent coupling to the OHD-IPM signal intensity can be eliminated from the signal trace by using a Fourier-transform analysis, which was modified from that developed by McMorrow and Lotshaw (1990) for their analysis of OHD-OKE data. Since Fourier transform of the instantaneous electronic contribution to the IPM signal trace is imaginary, the nuclear contribution was extracted from the experimental data by the inverse transformation of the real part of the Fourier transform of the data. Deconvolution of the laser autocorrelation function and smoothing by cutting off the high-frequency portion were performed in the course of the procedure. Thus the time derivative of the nuclear contribution to the OKE response function, which we designate as the IPM response function, was obtained.

Resulting IPM response functions of benzene and CS_2 are shown in Fig. 3. These are the experimentally obtained AVCFs of the molecules in the approximation mentioned above. Both curves have a time region with a negative value, where the angular velocities of the molecules are reversed with respect to the initial direction owing to intermolecular collisions. It suggests the crucial role of librational motions, or translational intermolecular vibrational motions which can affect the susceptibility *via* interaction-induced mechanisms, in the understanding of ultrafast dynamics in molecular liquids. The data from benzene exhibit oscillatory features in addition to the slowly changing envelope. They are supposed to be due to orientational or translational intermolecular vibrational motions of the molecules in the highly orientationally correlated liquid structure of benzene. The initial slope of these two curves are zero within the experimental ambiguity, which corresponds to finite correlation times of the random torque imposed on the molecules. The torque correlation times estimated from the curves are roughly 100 fs. Thus the angular-velocity correlation and the torque correlation are on the same time scale, which suggests that a description by a generalized Langevin equation may not be appropriate for the femtosecond dynamics in these molecular systems.

So far the ultrafast dynamics in the molecular liquids have been described in the monomolecular picture. Molecular dynamics in femtosecond time regime is, however, highly correlated with those of surrounding molecules, and it has been shown that the angular-velocity correlation time is much longer than the mean free time between intermolecular collisions in the molecular-dynamics simulation with hard spheroids (Talbot *et al.* 1990). On the femtosecond time scale, liquid may be regarded as a disordered solid since large changes are not expected in the framework of the local structure of the liquid on this time scale. Thus ultrafast intermolecularly correlated motions in liquids can be discussed in terms of localized lattice vibrational modes in disordered media.

Using the theory for localized vibrations in disordered media (Shuker and Gammon 1970), it can be shown that the OKE response is related to the density of states of localized vibrations as

$$\int_0^\infty dt \, [\frac{d}{dt}r(t)]\cos\omega t \propto C(\omega)g(\omega)$$

$$\equiv R_2(\omega)$$

Here $r(t)$ is the nuclear contribution to the OKE response, $g(\omega)$ is the density of vibrational states in the medium, $C(\omega)$ is the coupling coefficient between light and the vibrations. Although $C(\omega)$ and $g(\omega)$ cannot be decoupled, the reduced Raman spectrum, $R_2(\omega)$, is expected to exhibits features of the density of states of the vibrational modes in the disordered media fairly well, where structures in the coupling coefficient due to spatial symmetry are broad in contrast to those in ordered materials. The calculated $R_2(\omega)$ spectra of benzene and CS_2 are shown in Fig. 4. They should reflect the distribution of the intermolecular vibration frequency in these liquids although further theoretical and simulation studies are required for deeper understanding of the spectra.

In conclusion, the optical-heterodyne technique for the sensitive detection of the spectral shift due to induced phase modulation (OHD-IPM) was developed, and its application to the study of the molecular dynamics in liquids was described. The technique enables direct observation of the angular-velocity dynamics of molecules. The Fourier transform of the IPM response function was shown to be proportional to the reduced Raman spectrum, $R_2(\omega)$. Experiments on the OKE response of benzene and CS_2 clearly showed the oscillatory character of the intermolecular motions in these liquids. The reconstructed IPM response functions and the reduced Raman spectra for these liquids were obtained using a Fourier-transform technique.

Hattori T, Terasaki A, Kobayashi T, Wada T, Yamada A and Sasabe H 1991 *J.Chem. Phys.* **95** 937
McMorrow D and Lotshaw W T 1990 *Chem.Phys.Lett.* **174** 85
Mokhtari A and Chesnoy J 1988 *Europhys.Lett.* **5** 523
Ruhman S, Joly A G, Kohler B, Williams L R and Nelson K A 1987 *Rev.Phys.Appl.* **22** 1717
Shuker R and Gammon R W 1970 *Phys.Rev.Lett.* **25** 222
Talbot J, Kivelson D, Tarjus G, Allen M P, Evans G T and Frenkel D 1990 *Phys.Rev. Lett.* **65** 2828

Direct observation of the orientational motion of small molecules in condensed matter

G Seifert and H Graener

Physics Department, University of Bayreuth, D-8580 Bayreuth, Germany

We have developed a new application of time-resolved polarization spectroscopy: using picosecond pulses in the infrared we are able to measure directly the orientational motion of small molecules in the liquid phase. The facilities and limits of this technique, especially concerning the time-resolution, are discussed, and new results on bromoform and monomeric water are presented.

The orientational motion of molecules in the liquid phase was the object of many spectroscopic studies in the last years (Rothschild (1984), Dorfmüller and Pecora (1986)). In the case of small molecules dynamical informations often had to be derived from frequency domain observations (e.g. dielectric relaxation, nuclear magnetic resonance or various optical methods), while time-resolved investigations using e.g. the optical Kerr effect (McMorrow et.al. 1988) or picosecond coherent anti-Stokes Raman spectroscopy (Kohles and Laubereau 1987) are rather scarce. In the case of larger molecules (e.g. dye molecules), however, time- and polarization-resolved fluorescence spectroscopy with ultrashort laser pulses in the visible region of the spectrum provided a direct access to the orientational relaxation dynamics (e.g. Lessing et.al. 1975). The recent progress in the generation of intense and tunable picosecond pulses in wide spectral regions allows to study small molecules, too, by the use of polarization resolution in a transient infrared absorption spectroscopy. Only very recently we have shown that this new application of time-resolved polarization spectroscopy makes it possible to measure directly the orientational relaxation times of small molecules in liquid solutions (Graener et.al. 1990).

The measurements were performed with an IR pump-probe experiment using tunable, linearly polarized pulses (pulse duration $\cong$ 12 ps, tuning range 2.6-4.0 μm), which are derived from two independent optical parametric generator-amplifier devices; details of the pulse generation system have been described elsewhere (Graener et.al. 1987). In the double resonance experiment a remarkable percentage of the sample molecules (typically 10%) is prepared in a vibrationally excited state by resonant absorption of the intense pump pulses (intensity ca. 10^{10} W/cm^2). The experimental setup (as proposed from Graener et.al. (1990)) allows to measure simultaneously the transient IR transmission changes of the two components of weak, delayed probing pulses $\Delta\alpha(\parallel)$ and $\Delta\alpha(\perp)$ polarized parallel and perpendicular, respectively, to the pump polarization ($\Delta\alpha\ell:=\ln(T/T_0)$, where ℓ is the sample length, and T and T_0 are the probe transmissions with and without excitation). $\Delta\alpha(\parallel)$ and $\Delta\alpha(\perp)$ contain information about vibrational and orientational relaxation as well as on thermal processes, e.g.

thermally induced line shifts. For the asymptotic exponential time behaviour the different contributions of population and orientation dynamics can be separated by proper arithmetic combinations (according to e.g. Fleming et.al. (1976)). Three principal cases yielding different information have to be distinguished:

(i) pure vibrational relaxation:
the rotation-free signal component decaying with the vibrational relaxation time T_1 after the termination of the pump process is given by:

$$\Delta\alpha_{rf}(t) := \Delta\alpha(\|) + 2\Delta\alpha(\bot) \qquad \propto \quad \exp\left(-\frac{t}{T_1}\right) \qquad \text{for } t \gg t_p$$

The same information can also be obtained using a probe polarization oriented with 'magic angle' $54.7°$ with respect to the pump.

(ii) exclusion of isotropic transmission changes:
As isotropic transmission changes are identical in both polarization components, they may be cancelled out considering the difference $\Delta\alpha(\|) - \Delta\alpha(\bot)$, which still contains contributions from vibrational and rotational relaxation.

(iii) pure orientational relaxation:
Proper normalization of the difference (ii) allows to evaluate the quantity $R(t)$ independent of the vibrational population amplitude, representing the anisotropy of the probe transmission:

$$R(t) := \frac{\Delta\alpha(\|) - \Delta\alpha(\bot)}{\Delta\alpha(\|) + 2\Delta\alpha(\bot)} \qquad \propto \quad \exp\left(-\frac{t}{\tau}\right) \qquad \text{for } t \gg t_p$$

The orientational relaxation time τ can be determined from measurements of $R(t)$; τ describes the molecular reorientational motion around axes perpendicular to the transition dipole moment of the excited vibration. If both T_1 and τ are of the order or greater than the pulse duration, τ is obtained directly from the exponential slope of $R(t)$ for $t \gg t_p$. The other possibility to find out the value of τ is to compare the measured data with numerical solutions of a rate equation model describing the investigated system (see Graener et.al.(1990)). Of course, the accuracy of the fitting procedure depends on the knowledge of the input parameters of the calculation (see the brief discussion below), but it is suitable even in the case of rather short time constants. The latter statement is illustrated by figure 1, where some numerical results for the time evolution of $R(t)$ are depicted: it is readily seen that a significant (and measurable) anisotropy occurs even for a short orientational time constant, $\tau/t_p = 0.1$.

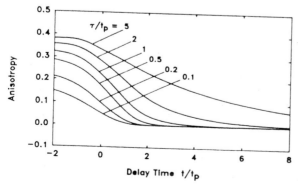

Figure 1:
Calculated data of the induced anisotropy $R(t)$ versus delay time for different values of orientational relaxation times $\tau/t_p = 0.1$ to 5

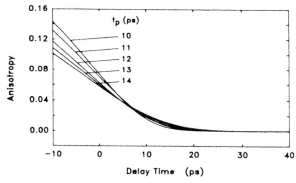

Figure 2:
Calculated data of the induced anisotropy R(t) for different pulse durations from t_P = 10ps to t_P = 14ps and a fixed τ = 1.5ps.

In order to find out the accuracy of the value of τ obtained from the fitting procedure we have studied the sensitivity of R(t) to changes in the input parameters of the calculations. For this purpose we varied all parameters over the range of their uncertainty and found the corresponding changes to be of the order of or even smaller than the experimental error in most cases. Only variations of pulse duration and shape cause significant changes in the induced anisotropy. While the shape of our pulses is close to Gaussian, their duration may, due to the nonlinear generation process, be different for different measurements of R(t). Figure 2 shows the effect of a variation of t_P for a fixed τ (the values of t_P and τ are typical for the situation in our samples of monomeric water). The current value of the pulse duration for a certain measurement can be determined from the rotation-free combination of the data with an accuracy of 10 - 20%. This value is at the same time the error limit for the obtained orientational relaxation times.

The present discussion shows that it is possible to measure orientational relaxation times τ, which are notably shorter than the duration of the used laser pulses. For the present experimental setup values of τ as short as $\cong$ 1 ps can be measured with sufficient accuracy from the amplitude values of R(t) around t $\cong$ 0.

Some typical examples of our measurements are given in the figures 3 and 4: Figure 3 shows logarithmic plots of the induced anisotropies of bromoform in carbontetrachloride at different temperatures, where the exponential decay with τ can clearly be seen; we observed a decrease of τ with increasing temperature, which is surprisingly compatible with the η/T-dependence (η: viscosity, T: temperature) of hydrodynamic models for molecular reorientation (for a review see Dorfmüller and Pecora (1986)). Nevertheless this simple description is, as expected, not sufficient in the case of small molecules, as is indicated by figure 4: the two curves plotted there, representing the samples of H_2O (a) and D_2O (b) in $CDCl_3$ at room temperature, yield a difference of a factor of 2 between the orientational relaxation of H_2O and D_2O. This difference can not be explained in terms of macroscopic parameters like viscosity or density, as they are nearly identical in both cases. In a microscopic view, however, the factor of 2 might correspond to the different moments of inertia of the two molecules, as suggested by the Hubbard relation (Rothschild (1984)):

$$\tau \cdot \tau^j = \frac{\theta}{6 \cdot k \cdot T} \qquad \left(\tau^j \ll \tau \right)$$

with θ: moment of inertia, k: Boltzmann constant, T: temperature

τ^{J} denotes the time between collisions changing the angular momentum of the rotating molecule.

A more detailed discussion of our data on molecular reorientation and of the vibrational dynamics of the samples of monomeric water, which we have measured at the same time as the orientational relaxation, will be given elsewhere.

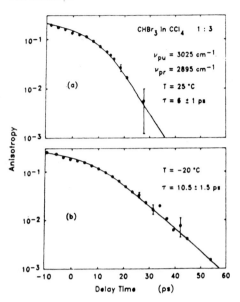

Figure 3: induced anisotropy R(t) of CHBr3:CCl4 ($\cong$ 30%) at: (a) 298 K and (b) 253 K; experimental points, calculated curves.

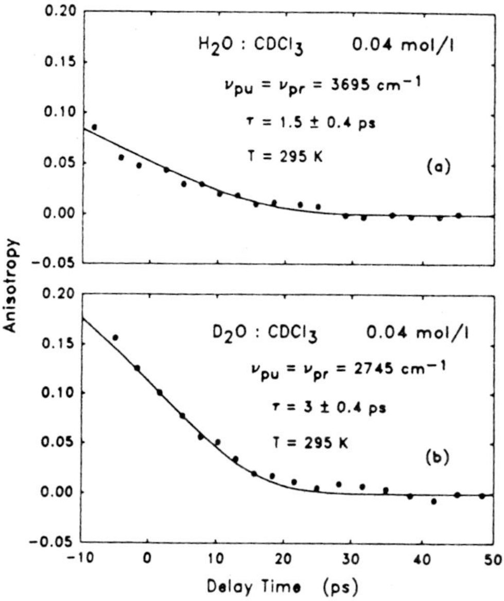

Figure 4: induced anisotropy R(t) of (a) H2O and (b) D2O in CDCl3 at room temperature; experimental points, calculated curves.

References

Dorfmüller T and Pecora R 1986 *Dynamics of molecular liquids* (Berlin: Springer 1986)
Fleming G R, Morris J M and Robinson G W 1976 *Chem. Phys.* <u>17</u>, 91
Graener H, Dohlus R and Laubereau A 1987 *Chem. Phys. Lett.* <u>140</u>, 306
Graener H, Seifert G and Laubereau A 1990 *Chem. Phys. Lett.* <u>172</u>, 435
Kohles N and Laubereau A 1987 *Chem. Phys. Lett.* <u>138</u>, 365
Lessing H E, von Jena A and Reichert M 1975 *Chem. Phys. Lett.* <u>36</u>, 517
McMorrow D, Lotshow W T and Kenney-Wallace G A 1988
    IEEE J. Quantum ELectron. QE-<u>24</u>, 443
Rothschild W G 1984 *Dynamics of molecular liquids* (New York: Wiley) p.94

Inst. Phys. Conf. Ser. No 126: Section VII
Paper presented at Int. Symp. on Ultrafast Processes in Spectroscopy, Bayreuth, 1991

Femtosecond proton and deuterium transfer in aromatic molecules

T. Elsaesser, F. Laermer, and W. Frey

Physik Department E 11, Technische Universität München,
D-8046 Garching, Federal Republic of Germany

Intramolecular proton and deuterium transfer in the electronic ground and excited state of aromatic molecules is studied by femtosecond spectroscopy. In benzothiazole and -triazole compounds, the very rapid excited state proton and deuterium transfer with time constants between 100 and 170 fs forms a keto-type geometry. The short time constants correspond to the period of molecular vibrations of low frequency, suggesting a transfer reaction via large amplitude motions. In the electronic ground state of benzotriazoles, we observe subpicosecond proton back-transfer creating the original enol species with a highly excited vibrational system.

Intramolecular proton transfer represents an elementary photoreaction that occurs in numerous aromatic molecules after excitation to a higher singlet state (Barbara 1989). Those compounds undergo a closed reaction cycle comprising proton transfer in the excited state, radiationless deactivation of the reaction product, and proton back-transfer to the original molecular geometry. The latter reaction step proceeds either via the electronic ground state or via the triplet manifold. Proton transfer between singlet states frequently shows a subpicosecond kinetics which could not be resolved in previous picosecond measurements. In this paper, we present a study of proton and deuterium transfer with a temporal resolution of 50 fs, giving new information on the dynamics and the mechanisms of the structural change in benzothiazole and -triazole compounds.

In nonpolar solvents, the main ground state species of 2-(2'-hydroxyphenyl)benzothiazole (HBT,I) and its deuterated analogue 2-(2'-deuteroxyphenyl)benzothiazole (DBT) is the enol structure with a strong intramolecular hydrogen bond between the proton (deuterium) and the nitrogen atom of the thiazole moiety (Williams 1970). The spectra of HBT and DBT dissolved in the nonpolar C_2Cl_4 are plotted in Figure 1. The S_0-S_1 absorption band located in the ultraviolet is identical for the two compounds. After excitation via the S_0-S_1 absorption band, the proton (deuterium) is transferred from the oxygen to the nitrogen atom, i.e. the keto structure is formed (II, keto-HBT). This structural change is evident from picosecond infrared measurements where the stretching vibrations of the relevant groups have been detected in the electronically excited molecules (Elsaesser 1986). The keto-tautomers show a strongly Stokes shifted fluorescence band with maximum around 540 nm (r.h.s. of Fig. 1). Within the experimental accuracy, the fluorescence quantum yield of HBT and DBT has an identical value of 2 percent corresponding to a fluorescence lifetime of approximately 300 ps.

The time constants of proton and deuterium transfer are determined by femtosecond pump-probe experiments (Laermer 1988,1990). After excitation of the enol tautomers by pulses at 310 nm, the delayed formation of the keto species is monitored via the time dependent gain at probe wavelengths in the range of the keto emission spectrum, i.e. the probe pulses induce stimulated emission from the keto excited to the keto ground state. In Fig. 2, this small signal gain at a wavelength of 540 nm is plotted versus delay time between pump and probe (points). The delayed formation of both keto-HBT (Figure 1 b) and DBT (Figure 1 c) is temporally resolved and occurs with similar time constants of 160±20 fs (HBT) and 140±20 fs (DBT). On the picosecond time scale, the gain decays with the fluorescence lifetime of the molecules of 300 ps.

The O-H and O-D stretching vibrations are of minor importance for the transfer reaction. H and D transfer involving these degrees of freedom would result in much shorter transfer times of roughly 30 fs and a well-pronounced deuterium effect. In contrast, the measured time constants are close to the period of low-lying vibrational modes with a frequency between 100 to 200 cm^{-1} that is nearly unchanged upon deuteration. Our results suggest a reaction scheme where the redistribution of electronic charge immediately after excitation of the enol tautomers establishes an essentially barrierless excited state potential with a broad minimum for the keto configuration. The proton (deuterium) moves along this surface to the acceptor atom. Low-frequency, large amplitude vibrations represent the relevant degrees of freedom for this process resulting in a transfer time of 160 fs (140 fs). Two types of structural change are consistent with our experimental data : (i) the separation of the oxygen donor atom and the accepting

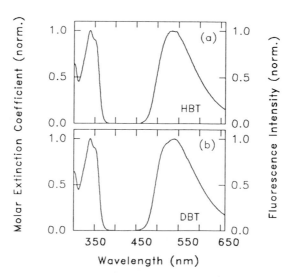

Fig. 1   Absorption (l.h.s.) and emission (r.h.s.) spectra of HBT and DBT dissolved in C_2Cl_4. The S_0-S_1 absorption bands of the protonated and the deuterated species are identical. The emission spectra with maxima around 540 nm exhibit a Stokes shift of approximately 11,000 cm^{-1}.

nitrogen atom could be fixed during the motion of the proton (deuterium) that is shifted over the relatively long separation between its original and its final position. (ii) In addition to proton (deuterium) motion, a reorientation of the remainder molecule could facilitate the formation of the keto species. This mechanism involves additional degrees of freedom of the thiazole and the phenyl groups lying also in the frequency range between 50 and 200 cm^{-1}. Additional experiments with sterically fixed derivatives will help to characterize the reaction mechanism in more detail.

In a second series of experiments, 2-(2'-hydroxy-5'-methylphenyl)benzotriazole (trade name : TINUVIN P, TIN) was studied, a widely applied photostabilizer of polymers (Heller 1972, Flom 1983, Wiechmann 1990,1991). After excitation of the enol species (III), the femtosecond emission kinetics of the keto-type tautomer (IV) that is plotted in Figure 3 a shows a delayed onset with a time constant of 100 fs. This delay is due to excited state proton transfer, similar to HBT and DBT. In contrast to the benzothiazole compounds, however, the excited keto state is deactivated by internal conversion on a

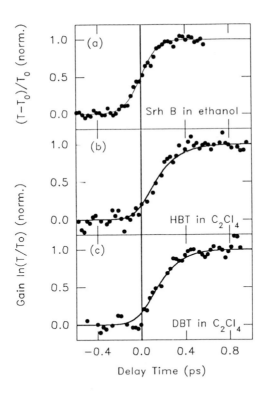

Fig. 2 (a) Time dependent ground state bleaching of sulforhodamine B after excitation at 310 nm as a function of delay time of the probe pulses at 540 nm (T,T$_0$: transmission with and without excitation of the sample). The instantaneous rise of the absorption decrease is used to calibrate delay zero. (b,c) Gain by stimulated emission of the HBT and DBT keto tautomers (points). The signal rises with time constants of (b) 160 fs and (c) 140 fs revealing the transfer times of proton and deuterium.

subpicosecond time scale. The decay of the gain in Figure 3 a is characterized by a time constant of approximately 150 fs, i.e. the keto-type ground state of TIN is populated within much less than 1 ps after excitation of the enol tautomer. Most probably, the rate of internal conversion of keto-TIN is strongly enhanced by low-frequency librations of the phenyl moiety relative to the triazole part of the molecules.

The keto→enol back-reaction of TIN occurs on the potential surface of the electronic ground state. In Figure 3 b, the kinetics of transient absorption at a wavelength of 515 nm is presented. The signal found at delay times longer than 1 ps when the excited states are depopulated completely (c.f. Figure 3 a) is attributed to the keto ground state decaying with a time constant of 700±100 fs. The keto tautomers subsequently transform back to the enol geometry as is evident from spectrally and time resolved measurements of the repopulation of the enol ground state. In Figure 4, two transients are plotted which were recorded at probe wavelengths λ_{pr} in the range of the S_0-S_1 absorption band of the enol tautomers. The decrease of absorption observed at $\lambda_{pr}=$ 325 nm rises with the excitation pulse and decays completely within 4 ps, directly revealing the repopulation of the enol ground state. The quantitative analysis of the data by a rate equation model gives a decay time of the signal of 700±100 fs. Thus the recovery of the steady state absorption of the enol tautomers proceeds with the same time constant as the decay of the absorption of the keto-type ground state of TIN, i.e. the enol ground state is predominantly repopulated by the keto→enol proton transfer. The fast bleaching of the ground state absorption is also observed at $\lambda_{pr} = 355$ nm

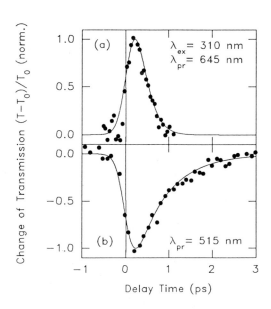

Fig. 3 (a) Femtosecond emission kinetics of TIN dissolved in C_2Cl_4. The gain at 645 nm is plotted versus the delay time between pump and probe pulses (points; T_0, T : transmission of the sample before and after excitation). (b) Time resolved absorption of the keto-type ground state of TIN measured for a probe wavelength of 515 nm (points). At delay times longer than 1 ps, the signal decays with a time constant of 700±100 fs. The solid lines are calculated from a rate equation model.

(Figure 4 b). In addition, a slower picosecond transmission change occurs which decays on a time scale of approximately 30 ps. This component of the signal is related to vibrational relaxation processes of the molecules in the enol ground state. The fast reaction cycle of TIN transfers the energy supplied by the exciting ultraviolet photon to the vibrational manifold of the molecules, i.e. enol tautomers with a highly excited vibrational system are created by repopulation of the enol ground state. The changes of the vibrational distribution function give rise to a strong reshaping of the enol S_0-S_1 absorption band. In Fig. 5 we present transient absorption spectra of enol-TIN derived from our time resolved data. The spectra reveal an enhancement of the low-energy tail of the band and a reduced absorption above the 00-transition around 27500 cm^{-1}. The increase of absorption at low frequencies is caused by vibrational excess populations of modes that couple to the electronic transition, i.e. show high Franck-Condon factors. The decrease of absorption at higher frequencies is due to a reduction of population density at the bottom of the vibrational distribution corresponding to an elevated temperature of the vibrational system (Sukowski 1990). On a time scale of several tens of picoseconds, the hot molecules cool down by interaction with the surrounding solvent and the transient absorption bands evolve towards the steady state spectrum (solid lines).

In conclusion, our results give direct evidence of femtosecond proton transfer in the electronically excited state of benzothiazole and -triazole compounds. The measured time constants of 100 to 200 fs correspond to the period of low-frequency vibrations, suggesting proton and deuterium transfer via large amplitude motions. The keto→enol

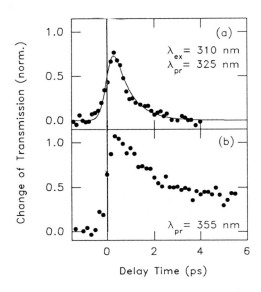

Fig. 4   Time resolved bleaching of the enol ground state of TIN observed after femtosecond excitation $\lambda_{ex} = 310$ nm for probe wavelengths of (a) $\lambda_{pr} = 325$ nm, and (b) 355 nm. The normalized change of transmission $(T-T_0)/T_0$ is plotted as a function of delay time (points, T_0, T : transmission before and after excitation). Unity corresponds to an absolute signal of 0.035. The solid line in Fig. 4 a is calculated from a numerical simulation.

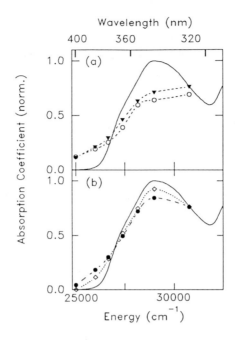

Fig. 5   Transient absorption spectra of TIN after femtosecond excitation at $\lambda_{ex} = 310$ nm. (a) The normalized absorption coefficient is plotted as a function of photon energy for delay times of 2 ps (open circles) and 5 ps (triangles). The solid line represents the stationary S_0-S_1 absorption band of enol-TIN. (b) S_0-S_1 absorption spectra observed for delay times of 10 ps (points) and 15 ps (diamonds).

back-transfer in the electronic ground state of 2-(2'-hydroxy-5'-methylphenyl)benzotriazole also occurs on a subpicosecond time scale resulting in a strong heating of the vibrational manifold. The hot enol molecules cool down by interaction with the surrounding solvent on a slower time scale of several tens of picoseconds.

References

Barbara P.F., Trommsdorff H.P. (Eds.) 1989 *Chem. Phys.* **136** pp. 153-360
Elsaesser T., Kaiser W. 1986 *Chem. Phys. Lett.* **128** 231
Flom S.R., Barbara P.F. 1983 *Chem. Phys. Lett.* **94** 488
Heller H.J., Blattmann H.R. 1972 *Pure Appl. Chem.* **30** 145
Laermer F., Elsaesser T., Kaiser W. 1988 *Chem. Phys. Lett.* **148** 119
Laermer F., Israel W., Elsaesser T. 1990 *J. Opt. Soc. Am. B* **7** 1604
Sukowski U., Seilmeier A., Elsaesser T., Fischer S.F. *J. Chem. Phys.* **93** 4094
Wiechmann M., Port H., Laermer F., Frey W., Elsaesser T. 1990 *Chem. Phys. Lett.* **165** 28
Wiechmann M., Port H., Frey W., Laermer F., Elsaesser T. 1991 *J. Phys. Chem.* **95** 1918
Williams D.L., Heller A. 1970 *J. Phys. Chem.* **74** 4473

Intermolecular proton transfer via intramolecular proton transfer: The photodissociation of 2-naphthol-3,6-disulfonate

A Masad and D Huppert

School of Chemistry, Raymond and Beverly Sackler Faculty of Exact Sciences
Tel-Aviv University, Tel-Aviv 69978, Israel

ABSTRACT: Time correlated single photon counting technique was employed to study the intramolecular and intermolecular proton transfer processes occuring in the first excited electronic state of 2-naphthol-3,6-disulfonate. It was found that the first step after excitation, by a short light pulse, is an intramolecular proton transfer from the hydroxy group to the adjacent sulfonate group. Subsequently, the proton is transferred to the solvent at a slower rate. The intramolecular proton transfer is mediated by water molecules.

1. INTRODUCTION

The origin of excited-state proton transfer (PT) reactions lies in the difference between the ground-state and the excited state ionization constants. The weak acids, hydroxy naphthols and their derivatives, have long been attractive probes for these investigations (see the reviews of Ireland and Wyatt, 1976 and Kosower and Huppert, 1986). Most of the data analysis is based on rate equations, i.e., determining rate constants. We presented a different approach (Pines et al., 1988) in which we have looked at the PT process as a transient, nonequilibrium dissociation of an excited state molecule. We suggested a two steps reaction model:

$$ROH^* \underset{k_r}{\overset{k_d}{\rightleftharpoons}} [RO^-{}^* \bullet \bullet \ H^+] \overset{DSE}{\longleftrightarrow} RO^-{}^* + H^+$$

Scheme 1

The chemical step (described by the intrinsic rate constatnts k_d and k_r) is followed by a diffusional step in which the proton is separated from the contact radius, a, to infinity. This diffusive motion is described by the exact transient (numerical) solution of the Debye Smoluchowski equation (DSE) with the back reaction boundary conditions.

In the present work we study the proton transfer process of excited 2-naphthol-3,6-disulfonate where the hydroxyl group was nearby a sulfonate group. Naphthols with OH and SO3⁻ adjacent on a C-C bond were investigated in the past. The results were interpreted in terms of an intramolecular hydrogen bond existing in both the ground state and the excited state (Van Gemert, 1969; Henson and Wyatt, 1975; Schulman, 1975; Zaitsev et al., 1978; Shapiro et al., 1980; Krishnan et al., 1990). We here suggest a new model in which there is an ultrafast excited state PT from the hydroxyl to the adjacent sulfonate group that effects the intermolecular proton transfer process. We utilize the DSE with appropriate boundary conditions (taking into account the intramolecular PT) to fit the experimental data of the time resolved fluorescence. The intrinsic rate constants involved in the PT process are determined from the computer fits to the experimental data.

2. EXPERIMENTAL

The transient fluorescence was detected using time-correlated single-photon counting (TCSPC). As a sample excitation source, we used a continuous wave (cw) mode-locked Nd:YAG-pumped dye laser (Coherent Nd:YAG Antares and a Coherent 702 dye laser) providing high repetition rate (100 KHz-3.8MHz) of short pulses (1 ps fwhm) at a wavelength of 580-620nm. The frequency was doubled (to 290-310nm) using a KDP type I crystal. The TCSPC detection system is based on a Hamamatsu, 1564U-01 photomultiplier, Tennelec 864 TAC and 454 discriminator. The TAC output was processed by a multichannel analyzer (Nucleus inc PCA II), interfaced with an IBM personal computer which served also for data storage and processing. The overall instrumental response at full-width-half-maximum (fwhm) was about 60ps. Measurements were taken at 10, 20 or 50 ns full scale.

3. RESULTS AND DISCUSSION

3.1 The Diffusing Reversible Proton

To reveal the influence of this OH-SO3$^-$ proximity we have monitored the transient fluorescence decay of the naphthol form, ROH*, in mixtures of methanol/water of different ratios.

The data was fitted subject to the assumption that the proton is diffusive and reacts reversibly with the excited state ion. This was done by solving the time dependent DSE for the relative translational diffusion of the proton.

Scheme 1 is sufficient to describe the flourescence decay dynamics of simple, direct, intermolecular proton transfer processes like that of 8-hydroxypyrene-1,3,6-trisulfonate (HPTS) (Pines et al. 1988 and references therein), with kd and kr as the fitting parameters, but cannot account for the complex flourescence decay of 2-naphthol-3,6-disulfonate (2-np-3,6).

We suggest that the reason for the complex flourescence profile of 2-np-3,6 is a reversible ultrafast intramolecular proton transfer between the OH and the adjacent SO3$^-$ group. This process can be described by the following scheme:

$$ROH^* \underset{k_r}{\overset{k_d}{\rightleftharpoons}} [RO^-* \bullet\bullet \ H^+] \overset{DSE}{\longleftrightarrow} RO^-* + H^+$$

$$k_1 \diagdown \ k_2 \diagdown \diagup k_3 \diagup \ k_4$$

$$[RO^-*\bullet\bullet HSO_3]$$

scheme 2

By inserting these new rate constants we obtained an excellent agreement between the experimental data and the theory over the whole time range.

3.2 MeOH/H$_2$O Mixtures: Composition Dependence of Rate Constants

Figure 1 shows our time resolved data measured at room temperature for several methanol/water mixtures along with the calculated fits.

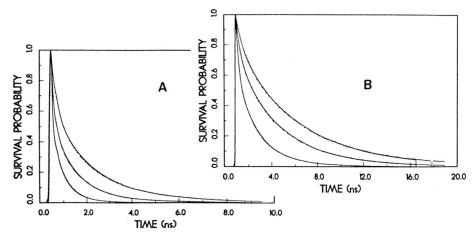

Fig. 1. *ROH* experimental flourescence decay of 2-naphthol-3,6 disulfonate measured at 370nm (dots) with the numerical solution of the DSE (full curve) in MeOH/Water solution of various compositions: a) Top to bottom: 33%, 20%, 0% vol fraction of MeOH; b) Top to bottom: 66%, 50%, 33% vol fraction of MeOH.*

Figure 2 demonstrates the proton transfer reaction rates dependence on the solvent composition. The figure shows that the proton dissociation rates decrease exponentially as a function of the MeOH mole fraction of the mixtures. The dependence of the proton recombination rates on the solution's composition is much milder. The same behaviour was obtained in our recent work with 8-hydroxypyrene-1,3,6-trisulfonate (HPTS) (Agmon et al., 1991) and 1-naphthol-3,6-disulfonate (1-np-3,6) (Masad and Huppert, 1991).

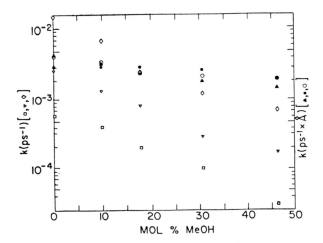

Fig. 2. *The solution composition dependence of the proton dissociation rates $k_d(\square)$, $k_1(\lozenge)$, $k_4(\triangledown)$ and the proton recombination rates $k_r(\blacktriangle)$, $k_2(\circ)$, $k_3(\bullet)$.*

We see that k1 the intramolecular proton transfer rate, is an order of magintude larger than kd, the proton dissociation rate. This relation is kept through the various mixtures. We observe that the three proton dissociation rates have a similar dependence on the solvent composition. This leads to the conclusion that the intramolecular route is dominant at all solvent mixtures.

Most of the work on excited state intramolecular protn transfer (ESIPT) dealt with two types of reactions: these where a hydrogen bond exists between the hydrogen of the donor group and the atom of the acceptor group and those in which the proton is too far away from the acceptor and needs a mediator. In the first case an inramolecular hydrogen bond provides a suitable reaction coordinate for a fast proton trnasfer, thus hydrogen-bonding solvents such as alcohols and ethers decrease the proton transfer rate due to the existence of solute/solvent complexes. In the second case a suitable solvent which forms hydrogen bonds with the solute can catalyze the ESIPT process (see for example, *Spectroscopy and Dynamics of Elementary Proton Transfer in Polyatomic Systems,* Chem. Phys. special issue, 1989, vol. 2). Examples for the first group are abandoned and involve systems in which the proton transfer is from an oxygen as a donor like in hydroxyanthraquinones (and its derivatives) or from a nitrogen as a donor like in aminoanthraquinones (and its derivatives) (Barbara et al., 1989; Brucker et al., 1991). Varma and co workers (konijneneberg, 1989 and references therein) studied molecules which belong to the second category like 2-amino-pyridine, 7-azaindole and more.

To the best of our knowledge there are no reports on the solvent influence on ESIPT from an hydroxyl group to a sulfonate group as is the case of 2-np-3,6. From our experimental finding it can be seen that the intramolecular proton transfer is water catalyzed and can hardly occur in methanol.

4. CONCLUSION

We have demonstrated that our unique analysis quantitatively describes the dynamics of a reversible excited state proton transfer. In this work we have shown that in the case of 2-naphthol-3,6-disulfonate where the hydroxyl group is adjacent to a sulfonate group an *intramolecular* PT is almost exculsively the preliminary step (over the *intermolecular* PT) and is being followed by a proton dissociation from the sulfonate group. Thus, an 'indirect' route competes with the direct proton dissociation from the hydroxyl to the solvent. The *intramolecular* reversible PT rate constant shows the same dependence on solvent composition as the direct, reversible, *intermolecular* PT.

REFERENCES

Agmon N, Huppert D, Masad A and Pines E 1991 *J. Phys. Chem.* (accepted for publication)
Barbara P F, Walsh P K and Brus L E 1989 *J. Phys. Chem* **93** 29
Brucker G A, Swinney T C and Kelley D F 1991 *J. Phys. Chem.* **95** 3190
Henson R M S and Wyatt D H 1975 *J. Chem. Soc., Faraday II* 669
Ireland J F and Wyatt P A H 1976 *Adv. Phys. Org. Chem.* **12** 139
Konijneneberg J, Huizer A H and Varma C A G O 1989 *J. Chem. Soc., Faraday Trans. 2* **85** 1539 and references therein
Krishnan R, Fillingim T G, Lee J and Robinson G W 1990 *J. Am. Chem. Soc.* **112** 1353
Masad A and Huppert D 1991 *Chem. Phys. Lett.* **180** 409
Pines E, Huppert D and Agmon N 1988 *J. Phys. Chem.* **88** 5620
Schulman S G 1976 *Analyt. Chim. Acta.* **84** 423
Shapiro S L, Winn K R and Clark J H 1980 *Springer Ser. Chem. Phys.* **14** 227
Van Gemert J T 1969 *Austral. J. Chem.* **22** 1883
Zaitsev N K, Demyashkevich A B and Kuzmin M G 1978 *Khimiya Vysokikh Energii* (English translation) **12** 365

Inst. Phys. Conf. Ser. No 126: Section VII
Paper presented at Int. Symp. on Ultrafast Processes in Spectroscopy, Bayreuth, 1991

Molecular dynamics study of the picosecond photostimulated conformational dynamics

Boris A. Grishanin

Physics Department, Moscow State University, 119899 Moscow, Russia

Valentin D. Vachev and Victor N. Zadkov

International Laser Center, Moscow State University, 119899 Moscow, Russia

Abstract. The process of laser pulse excitation of isolated polyatomic molecules is studied using molecular dynamics computer simulation. A quantum theory describing the dynamics of laser excited electronic state of the molecule is combined with classical molecular dynamics simulation. The detailed computer simulation results on the laser excitation and subsequent IVR processes and conformational changes in jet-cooled stilbene molecule depending on the parameters of the laser pulse are presented. It is shown that only a limited set of modes is active during the excitation and another one take part in the IVR. The isomerization time at excess vibrational energy equal to 3050 cm^{-1} has been found to be 135 ps.

1. INTRODUCTION

Characterization of dynamical processes in individual molecules is among the most interesting current problems in chemical physics, for it implies that one may be able to intimately control the physics and/or chemistry of a molecule simply by varying the parameters of laser pulsed excitation (Felker *et al* 1988, Levine *et al* 1987). Application of commonly used quantum methods for investigation of the dynamics of electronically excited polyatomic molecules consisting of more then ten atoms is not realistic because of the enormous amount of data necessary for description of nuclei wave function. Widely used molecular dynamics (MD) simulation is a classical method and for a proper use requires introduction of adequate initial conditions resulting from the essentially quantum process of laser induced electronic transition. From a number of comprehensive experimental studies, reviewed by Felker and Zewail (1988), Shan *et al* (1987), it is well known that at arbitrary low vibrational excitations and on the short times the dynamics of the molecule is predominantly harmonic. However as time evolves the anharmonicities become important and explicit dynamical treatment is necessary.

We present in this paper a detailed molecular dynamics (MD) computer simulation results on the processes of laser pulsed excitation and photoinduced isomerization in jet-cooled stilbene molecule.

2. LASER PULSE EXCITATION: THEORY

The excited state can be described in the frame of one-photon description by the following formula for the quantum density matrix (the detailed theory is presented by Grishanin *et al* 1991, 1992):

$$\hat{\rho}_{22} = \frac{|d_{12}|}{4\hbar^2} \iint d\tau_1 d\tau_2 E_L(\tau_1) E_L^*(\tau_2) e^{-\hat{\delta}_2(t-s)} \hat{\rho}_{22}^0(\tau) e^{i\hat{\delta}_2(t-s)}, \tag{1}$$

where d_{12} is the electronic dipole momentum, $E_L(\tau_k)$ is the complex strength of laser field with frequency ω_L, $\hat{\rho}_{22}^0(\tau)$ is the generalized density matrix of vibronic state, formed during the time interval $\tau = \tau_2 - \tau_1$, $s = (\tau_1 + \tau_2)/2$,

$$\hat{\delta}_j = [\hat{p}^T m^{-1} \hat{p}/2 + U_j(\hat{x})]/\hbar \qquad j = 1, 2 \tag{2}$$

are the energy operators of the molecular terms in frequency units, where $\hat{p}$ is the momenta vector, m is the mass matrix, $U_j(\hat{x})$ are the correspondent potential energies. The wave packet delocalization in this approach is due to the uncertainty of τ_1 and τ_2 time points. For the transition probability we obtain the following expression:

$$P(\omega_L) = \int d\tau \, p(\tau)\chi(\tau), \tag{3}$$

where $p(\tau) = (\Omega_L^2/4)f_p(\tau) \, exp\left[i(\omega_L - \omega_{12})\tau\right]$; $\Omega_L = E_L d_{12}/\hbar$ is the Rabi frequency; $f_p(\tau) = \int ds u_p(s - \tau/2)u_p^*(s + \tau/2)$ is the laser pulse autocorrelation function; ω_{12} is the frequency of the electronic transition; $\chi(\tau)$ is the function of the terms frequncy matrices and of the temperature; ω_1, ω_2 are the frequency matrices of the corresponding energy terms.

Then for the average $x-$coordinate we have:

$$x_q = x_2 + \int p(\tau)\Delta x(\tau) \, d\tau/P(\omega_L), \tag{4}$$

where $\Delta x(\tau)$ is the coordinate displacement at the moment τ respectively to the equilibrium point of the upper electronic term x_2. Finally, for the average momenta we have $p_q = 0$ and the total energy of the kth mode is given by:

$$E_k = 0.5\hbar\omega_k(x_2 - x_1)_k^2 P(\omega_L - \omega_k)/P(\omega_L). \tag{5}$$

3. COMPUTER SIMULATION SCHEME

We have accepted the following scheme of computer experiments. Firstly, given a molecular structure we make a choice of the molecular mechanics (MM) potential energy functions, a method widely used in structural studies of organic molecules (Burkert and Allinger 1982). Secondly, we adjust parameters of the electronic energy surfaces through a comparison of the calculated absorption/emission spectra and the available experimental data. Then, given also laser pulse parameters the calculation of the excited state characteristics is performed. And finally, we solve the equations of motion for the nuclei and process obtained trajectories to reveal some features of the intramolecular vibrational energy redistribution (IVR) and conformational changes.

We have used an usual form of the MM potential energy functions taken from Vinter *et al* (1987):

$$U(r_1, \ldots, r_N) = U_b + U_{va} + U_{tor} + U_{nb}, \tag{6}$$

where U_b, U_{va}, U_{tor}, U_{nb} are the contributions due to deformations of chemical bonds, deviations of valence and torsional (dihedral) angles, and Van- der-Waals interactions, respectively. Potentials U_b, U_{va} are the standard harmonic ones. The parameters of the potential energy functions are valid close to the equilibrium point. However in the previous study (Vachev and Zadkov 1991) and in this work we have found out that in the isomerization there occur only small changes in the bond lengths (less then 5%) and in the valence angles (less then 8%). Consequently, the accepted parameters are valid for the whole time of the reaction and only the parameters of the energy dependence on the torsional coordinate should be adjusted additionally. For the ethylenic bond in stilbene molecule, in accordance with the available experimental data of Syage *et al* (1984a), we have used the following expression for the excited state:

$$U_{tor} = V_1 \sin^2(2\theta) - V_2 \cos^2(\theta - \theta_0). \tag{7}$$

At the first stage of the fitting procedure we minimize potential (6) and find out the equilibrium geometry and then the calculated ground state vibrational spectra are compared with Warshel's (1975) calculations and the experimental data of Syage *et al* (1984b). In this way the parameters of the ground state potential energy surface are determined. The displacement of the electronic surfaces determines the absorption/emission spectra and with account for this after performing computer simulation the parameters of the excited state potential energy surface are obtained. The obtained vibrational frequencies of the fundamentals of stilbene molecule, the displacements of the two electronic terms and the potential energy function parameters are given in Grishanin *et al* (1992).

4. COMPUTER SIMULATION RESULTS

Using (1–5) we have investigated how the energy of excitation is distributed between normal modes depending on laser pulse duration and frequency. The energy of excitation is distributed between all modes with sufficient linear displacements $(x_2 - x_1)$ of the normal coordinates. As it can be seen from eq. 5, for a given ω_L not only the resonance vibrational mode but also all optically active in the electronic transition modes are excited. The increase of laser pulse frequency leads to increasing of the energy of all excited vibrational modes.

Solving the equations of motion for nuclei we obtain the MD trajectory, so that the process of intramolecular vibrational energy redistribution and conformational changes can be studied in details. The results of IVR studies are reported elsewhere by Grishanin *et al* (1992). The following features should be pointed out: in accordance with the symmetry of the excited state potential energy surface there is a large number of oscillators that do not take part in the IVR; the energy exchange is realized predominantly through the combination frequencies of the fundamentals.

We have varied V_1 and V_2 (eq. 7) for the calculated time of isomerization and the experimentally obtained non-radiative decay time at given excess vibrational energy to coincide. The photoisomerization proceeds in two stages. Firstly, the excited molecule is trapped in the potential well around the twisted state ($\theta = 90°$). Then non-radiative decay to the ground electronic state occurs and the molecule relax to the cis- or trans-configuration.

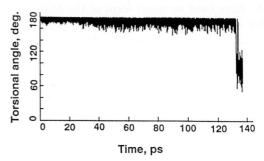

Fig. 1. Photo-induced isomerization in stilbene molecule. The torsional angles θ exceeding 180° are shown as 360° − θ.

The simulation of the first stage of the reaction is shown in Figure 1. With $V_1 = 6$ kcal/mol and $V_2 = 8$ kcal/mol we determine $\tau_{isom} = 135$ ps, which coincides well with $\tau_{isom} \approx 140$ ps experimentally measured by Majors *et al* (1984). The local barrier to isomerization is 2.8 kcal/mol at torsional angle = 35° and the deepness of the potential well around the twisted state is 8 kcal/mol.

Acknowledgments

This work has been partially supported by a Program of Engineering Ensimology via Grant no. 1-250.

References

Burkert U, Allinger N L 1982 *Molecular Mechanics* (Washington: ACS **177**)
Felker P M, Zewail A H 1988 *Adv. Chem. Phys.* **70** 265
Grishanin B A, Vachev V D and Zadkov V N 1991 *Proc. SPIE* **1402** 44
Grishanin B A, Vachev V D and Zadkov V N 1992 *Bull. of Academy of Sciences of USSR* (in press)
Levine R D, Bernstein R B 1987 *Molecular Reaction Dynamics and Chemical Reactivity* (Oxford Univ. Press)
Majors T J, Even U and Jortner J 1984 *J. Chem. Phys.* **81** 2330
Shan K, Yan Y J and Mukamel S 1987 *J. Chem. Phys.* **87** 2021
Syage J A, Felker P M and Zewail A H 1984a *J. Chem. Phys.* **81** 4706
Syage J A, Felker P M and Zewail A H 1984b *J. Chem. Phys.* **81** 4685
Vachev V D, Zadkov V N 1991 *Proc. SPIE* **1403**, 487
Vinter J, Davis A and Saunders M 1987 *J. Comp. Aided Mol. Design* **1** 31
Warshel A 1975 *J. Chem. Phys.* **62** 214

Inst. Phys. Conf. Ser. No 126: Section VII
Paper presented at Int. Symp. on Ultrafast Processes in Spectroscopy, Bayreuth, 1991

Time-resolved CARS spectroscopy of the series of bisdimethylamino-methine photoisomers

W. Werncke, M. Pfeiffer, A. Lau, L. Holz, T. Hasche
Zentralinstitut für Optik und Spektroskopie
O-1199 Berlin, Rudower Chaussee 5, Fed. Rep. Germany

ABSTRACT:
Bisdimethylamino-tri-,-penta-,-hepta- and -nonamethineperchlorate and their photoisomers have been investigated by time resolved Resonance CARS spectroscopy. For all these dyes a characteristic "photoisomer CARS line" at $1070 cm^{-1}$ is observed. From normal coordinate calculations parameters have been obtained beeing consistent with quantum chemical considerations. It can be concluded that independent on chain length the 1,2 mono-cis photoisomer is generated resulting in the appearance of a vibration localized at the end of the π-chain.

1. INTRODUCTION

Photoisomerization is one of the fundamental molecular processes occurring after light irradiation.
Using time resolved Coherent Anti Stokes Raman Scattering (CARS) /1/ here we report on a photoisomerization study of bisdimethylamino-methine oligomers $[(CH_3)_2N(CH)_nN(CH_3)_2]^+$ exhibiting different lengths n = 3,5,7,9 of the π - chain. Photoisomerization at low temperatures of bisdimethylamino-trimethineperchlorate, -pentamethineperchlorate and -heptamethineperchorate was reported in an early paper of Scheibe et. al. /2/. At room temperature these photoisomers are generated after some picoseconds exhibiting a life time of some milliseconds /3/. However, though this type of molecules is intensively studied up to now, time resolved Resonance CARS investigations are interesting because of the vibrational information about their short lived intermediates /4/ and photoisomerized forms. It is known, that these molecules are stable in the all-trans configuration. In the case of the trimethine only a 1,2-cis photoisomer has to be expected, but in the case of the dyes with a long methine chain numerous mixed E,Z photoisomers may be generated and have to be distinguished. Following quantum chemical calculations for the larger molecules the 2,3-mono-cis configuration is favored /2,5/ and therefore in lengthening the chain instead of the 1,2-mono-cis the 2,3-mono-cis isomer should occur.

2. EXPERIMENTAL RESULTS

The absorption maxima of the dyes with n = 3,5,7,9 , of their photoisomers and the wavelengths used for photoexcitation λ_{Exc} and for CARS probing λ_P are summarized in table 1.
Using the respective CARS- probing wavelengths λ_P CARS-spectra either of the parent molecules or after preceding excitation mainly originating from the photoisomers were observed as demonstrated in fig. 1 for the trimethine.

Table 1   Absorption maxima of bisdimethylaminomethine dyes dissolved in ethanol /2/; excitation- and CARS- probing wavelengths (λ_{EXP}, λ_P)

	TRI	PENTA	HEPTA	NONA
	λ [nm]			
all-trans	315	414	514	615
photoisomer	340	442	543	645
λ_{EXC}	308	400	510	580
λ_P	370	464	580	670

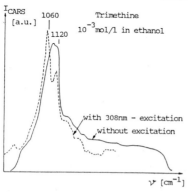

Fig. 1 Resonance CARS spectrum of bisdimethylaminotrimethineperchlorate ($1*10^{-3}$ mol/l dissolved in ethanol) and of its photoisomer after 308nm irradiation in the spectral range of $1000cm^{-1} - 1200cm^{-1}$

The main characteristic feature in the above spectrum of the photoisomer is the appearance of the strong new CARS line at $1060cm^{-1}$ together with a line at 1120 cm^{-1} which remains nearly unshifted with respect to the frequency of the parent molecule. As shown in table 2, the appearance of one intense Raman band near $1070cm^{-1}$ is also observed for the other oligomers. Other alterations for the different methines are rather small within the investigated range of $900cm^{-1}-1700cm^{-1}$ and do not show any significant effect.

3. DISCUSSION

Although an assignment of the observed changes in the vibrational spectra to structural changes of the molecules is difficult, π- especially if strong coupling of the vibrational modes as in -conjugated systems with atoms of nearly equal masses occurs - the nearly equal features in the photoisomer CARS spectra of the different methines suggest that always as in the case of trimethine the 1,2- mono cis photoisomer is generated.

For a proof of such an assumption normal coordinate calculations were carried out including a force constant adjustment to reproduce the observed vibrational infrared and Raman spectra of the series of oligomers and the frequency shifts of the heptamethine due to ^{15}N substitution /6,7/. The fitted force field of the bond stretching constants f_i and interaction constants f_{ij} are given in table 3 (f in mdynes/Å).

The calculations show that the intense Raman bands in the frequency range $800cm^{-1} - 1700cm^{-1}$ are all connected with stretching motions of the π- chain related for the C_{2v}- symmetric all trans molecules to A_1- internal coordinates.

Table 2 Parent molecule and photoisomer CARS frequencies of bisdimethylaminomethine dyes in the range $1000 cm^{-1}$ - $1200 cm^{-1}$ and highest CARS / IR-frequencies of the parent molecules

parent molecule	photoisomer	assignment
bisdimethylaminotrimethine		
1120 s	1120 s	CH def + CH₃ rock
	1060 s	C=C str + CH def
1658 vs / 1626 vs		s / as CN str
bisdimethylaminopentamethine		
1125 s	1125 s	CH def + CH₃ rock
1082 w	1072 s	C=C str + CH def
1660 vs / 1603 vs		s / as CN str
bisdimethylaminoheptamethine		
1022 s	1022 s	CH def + CH₃ rock
1085 w		CH₃ rock
	1070 s	C=C str +CH def
1641 vs / 1633 vs		s / as CN str
bisdimethylaminononamethine		
1122 s	1122 s	CH def + CH₃ rock
1102 w	1102 w	CH₃ rock
	1070 s	C=C str +CH def
1630 vs / 1630 vs		s / as CN str

Table 3

	$f_1 = K_{N=C(1)}$	$f_2 = K_{C(1)=C(2)}$	$f_3 = K_{C(2)=C(3)}$	f_{12}	f_{23}	f_{13}
TRI	7,95	6,34	6,34	0,37	0,30	-0,25
PENTA	7,89	6,36	4,93	0,41	0,41	-0,17
HEPTA	7,86	6,38	4,93	0,42	0,12	-0,14
NONA	7,61	6.11	5.93	0.35	0.18	-0.18

Above $1450 cm^{-1}$ the vibrations are mainly π- bond stretchings for which the neighboured bonds oscillate with opposite phase. Between $1050 cm^{-1}$ and $1400 cm^{-1}$ there are coupling contributions from CH-deformations of the methine groups, deriving their intensity from coupling to the bond stretching. Furthermore there are some contributions from symmetrical CH₃ deformations as well as CH₃ rockings due to special local symmetries.
The fit of the force field to reproduce the experimental

spectra especially in the range of the 2 highest stretching frequencies of the π-chain (their values are given in table 2) gives a strong evidence for the action of a π-electron delocalization within the methine series /8/. This expresses itself in the lowering of the N=C bond strengths in the longer methines as well as in the alternation of the signs for the stretching interaction con-stants f_{ij} between the next and overnext bonds (cf. table 3).

As the main vibrational frequencies remain unchanged by photoisomerization we transferred both force constants and bond polarizabilities to the photoisomers taking into consideration only their changed geometries. Twisted out of plane geometries of the photoisomers should lead to a shift of the absorption maxima to shorter wave lengths contrary to the experimental findings. Therefore only the Raman frequencies and intensities of all the possible planar mono-cis configurations were calculated. From these only the 1,2 mono-cis configuration results into spectra corresponding to the experimental finding in the 1000cm^{-1} -1200cm^{-1} range. The motion of this vibration which occurs due to the lowering of the symmetry of the photoisomer is shown in fig. 2

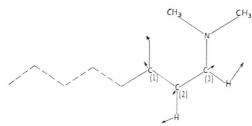

Fig. 2
Vibrational amplitudes for the characteristic isomer mode near to 1070 cm^{-1} (nearly equal appearances for tri-, penta-, hepta- and nonamethine)

In conclusion we have shown, that up to the nonamethine in the series of bisdimethetylaminoperchlorate oligomers only the 1,2- mono cis photoisomers are generated. The characteristic photoisomer vibration remains unshifted within the series as it is mainly localized at the end of the molecules.

REFERENCES:
/1/ A. Lau, W. Werncke, M. Pfeiffer; Spectrochimica Acta Rev. **13**, 191 (1990)
/2/ G. Scheibe, J. Heiss, K. Feldmann; Ber. Bunsenges. Phys. Chem. **70**, 52 (1966)
/3/ S. Rentsch, R.V. Danelius, R.A. Gadonas, Chem. Phys. **59**, 119 (1981).
/4/ A. Lau, W. Werncke, M.Pfeiffer, H.-J. Weigmann and Kim Man Bok, J. Raman Spectrosc. **19**, 517 (1988)
/5/ H. Hartmann, P. Wähner; Abstracts "Fourth Symposium Optical Spectroscopy" October 1986, Reinhardsbrunn, GDR
/6/ W. Werncke, A. Lau, M. Pfeiffer, H.-J. Weigmann, W. Freyer, Tschö Jong Tscholl, Kim Man Bok; Chem. Phys. **118**, 133 (1987)
/7/ W. Werncke, M. Pfeiffer, A. Lau, L. Holz, T. Hasche; "Electronic Properties of Polymers", Proceedings of the IWEPP, Kirchberg 1991, Springer Verlag in press
/8/ M. Pfeiffer, W. Werncke, A. Lau, W. Freyer; "Electronic Properties of Polymers", Proceedings of the IWPP, Kirchberg 1991, Springer Verlag in press

Femtosecond-picosecond laser photolysis studies on photoinduced charge transfer and electron ejection dynamics

Noboru Mataga, Hiroshi Miyasaka and Yoshinori Hirata

Department of Chemistry, Faculty of Engineering Science, Osaka University, Toyonaka, Osaka 560, Japan

ABSTRACT: Fundamental aspects of photoinduced charge separation (CS) and related phenomena have been summarized on the basis of the femtosecond picosecond laser photolysis studies on various donor acceptor systems. A specific CS process in the excited hydrogen bonding system taking place in nonpolar media has been directly demonstrated by femtosecond spectroscopy. Another specific CS due to slow electron ejection from relaxed S_1 state in acetonitrile, a large enhancement of CS rate by excess vibrational energy as well as by use of solvent with a slightly larger electron affinity have been demonstrated.

1. INTRODUCTION

The mechanisms and dynamics of photoinduced CS leading to the formation of ion pair (IP) state and charge recombination (CR) of the produced IP state are the most important central problems in the photochemical primary processes. The important factors regulating these processes in the ordinary (outer–sphere) electron transfer (ET) reactions are believed to be: electronic interaction between D and A responsible for ET, the free energy gap $(-\Delta G)$ between the initial and final state of the ET, the reorganization energies (λ) including solvent orientation and intramolecular vibrational motions of solute, and solvent dynamics. Depending on the magnitudes of these factors, however, there arise various cases of different mechanisms of photoinduced CS and CR of the produced IP state some of which are not yet well comprehended especially in the case of the strongly interacting systems. There are also some interesting special cases of photoinduced charge transfer (CT) in hydrogen bonding (HB) systems and ET from solute in fluorescent state to the transient aggregate of polar solvent without definite acceptor.

In the following, we summarize some results of our recent femtosecond–picosecond laser photolysis studies and discuss the most important fundamental aspects of the photoinduced electron transfer.

2. PHOTOINDUCED CS IN LINKED SYSTEMS

The D, A systems linked by spacer or directly by single bond are very useful for investigating various factors regulating the photoinduced ET. From such viewpoint, we have studied with femtosecond–picosecond laser spectroscopy (Mataga et al 1990) following systems: p-$(CH_3)_2$N–Ph–CH_2–(1–pyrenyl) (P_1), p-$(CH_3)_2$N–Ph–CH_2–(9–anthryl) (A_1), Ph–N(CH_3)–CH_2–(9–anthryl) (9–AnMe), 9,9'–bianthryl (BIAN) and (9–anthryl)–(N–carbazolyl) (C9A) in alkanenitrile solutions.

In the case of P_1, A_1 and 9–AnMe, the transient absorption spectral change due to the photoinduced ET process can be well reproduced by the two state model; A*–S–D $\xrightarrow{k_{CS}}$ $(A^-–S–D^+)_s$, where S represents spacer and s represents the solvation of IP state by the polar solvent. By analyzing the observed spectra according to this model, the time constants of photoinduced CS, τ_{CS} ($=k_{CS}^{-1}$), have been obtained in acetonitrile (ACN), butyronitrile (BuCN) and hexanenitrile (HexCN) solutions as shown in Table 1. It is clear that τ_{CS} of P_1 is much longer than the longitudinal dielectric relaxation time of solvent, τ_L, and also longer than τ_S (Kahlow et al 1987, Rips et al 1990) obtained by the measurement of the dynamic Stokes shift of fluorescence by solvation of probe molecule. This result shows that there is an intrinsic barrier for the photoindcued CS of P_1. The τ_{CS} values of A_1 is also longer than τ_L but rather close to τ_S indicating the possibility that the photoinduced CS is controlled by solvent orientation. 9−AnMe shows a much fast photoinduced CS than A_1 and P_1, with τ_{CS}=0.7 ps in HexCN. This value is shorter than τ_L. The electronic interaction between two chromophores in 9−AnMe seems to be much stronger compared to those in A_1 and P_1 because the charge density on N−atom is much larger than that on carbon atom at p−position. Presumably, the coupling of the intramolecular vibration with the CS process may be contributing to enhance the reaction rate.

Table 1. Time constants of photoinduced CS (τ_{CS}/ps) of some linked D, A systms in alkanenitrile.

	ACN	BuCN	HexCN
A_1	0.65	1.0	1.4
P_1	1.7	2.5	4.5
9−AnMe	—	—	0.7
BIAN	1.8	3.4	7.5
C9A	—	3.8	9.0
τ_L/ps	0.19	0.53	0.98−1.1
τ_S/ps	0.4−0.9	1.5−2.1	3.5−4.5

On the other hand, the rise curve of the intramolecular CT state of BIAN and C9A with more strongly interacting chromophores contains fairly slow component with τ_{CS} much longer than τ_L and also longer than τ_S in alkanenitrile solutions as shown in Table 1. In these directly linked systems with strong interaction between two chromophores, there may be some distribution of configurations with different energies in the ground state. This distribution will relax toward that of the excited state after pulsed excitation, which seems to give the slow component in the rise of the CT state. In other words, depending on the geometrical structures and solvation in these strongly interacting bichromophoric systems, there exist multiple CT states and the slow process in the change of the CT electronic structure caused by excitation will take place via such multiple intermediate states.

3. CS BY EXCITATION OF CT COMPLEXES AND CR OF THE PRODUCED IP STATE

The photoinduced CS by weak interaction between unlinked D, A molecules takes place at encounter between fluorescer and quencher in strongly polar solutions leading to the formation of loose IP (LIP) probably with solvent between ions. Although we do not get into its details here, there is a long standing problem that no inverted region can be observed in the energy gap dependence of k_{CS}. The reason for this has been examined from time to time. Formation of excited state of LIP (Mataga 1970, Rehm and Weller 1970) as well as nonfluorescent CT complex (Mataga et al 1988, Masuhara and Mataga 1981) at very large $-\Delta G$ and nonlinear polarization of polar solvent around ions (Yoshimori et al 1989) have been proposed as possible mechanisms. More recently, D, A distance distribution which affects the solvent λ in the course of ET has been examined in detail (Kakitani et al 1991). This mechanism predicts that, for the CS reaction at large $-\Delta G$, a little larger D, A distance is favorable for ET owing to the larger solvent λ, making the observation of inverted region difficult. At present, this D, A distance distribution and the formation of the excited LIP at very large $-\Delta G$ seem to be the most important mechanism for the lack of the inverted region in the observed energy gap dependence of photoinduced k_{CS}.

Contrary to photoinduced CS leading to the formation of LIP, the observed energy gap dependence of k_{CR} of LIP shows both inverted and normal regions. This result can be interpreted by taking into account the $-\Delta G$ (for CS) dependence of the inter−ionic distance of geminate LIP immediately after its formation and a little change of the distance distribution before the CR decay of the IP and also by taking into account a moderate amount of the nonlinear polarization of solvent around ions in the LIP (Kakitani et al 1991).

At appropriate $-\Delta G$ value k_{CS} becomes close to $10^{11} \sim 10^{12}$ M^{-1} s^{-1} as in the case of the systems linked by short insulating chain. It is an interesting problem to compare the result of bichromophoric systems linked directly by single bond with that of strongly interacting un–linked D, A systems such as the CT complexes. Our recent results on CT complexes of aromatic hydrocarbon–cyano compounds, –acid anhydrides, –tetracyanoquinodimethane are important from such viewpoint (Miyasaka et al 1989, Ojima et al 1990, Asahi and Mataga 1991).

The photoinduced CS in the excited state of TCNB (1,2,4,5–tetracyanobenzene)–benzene complex in benzene takes place in two steps. A slight geometrical rearrangement within 1:1 complex from an asymmetric configuration in FC (Franck–Condon) excited state toward a more symmetric overlapped one accompanied with a slight enhancement of CT degree takes place with time constant $\tau_d \sim 2$ ps. However, for the complete CS in this nonpolar environment, much larger structural change including the formation of the 1:2 complex (of which the time constant $\tau_r \sim 20$ ps) is of crucial importance. Similar results have been obtained also in the TCNB–toluene and mesitylene systems. Therefore the photoinduced CS process in these systems are represented by,

$$(A^{-\delta} \cdot D^{+\delta}) \xrightarrow{h\nu} (A^{-\delta'} \cdot D^{+\delta'})^* \xrightarrow{\tau_d} (A^{-\delta''} \cdot D^{+\delta''})^* \xrightarrow{\tau_r} (TCNB^- \cdot D_2^+)^*,$$

D

where $\tau_d = 1.5$ ps, $\tau_r = 30$ ps in toluene solution and $\tau_d = 550$ fs, $\tau_r = 40$ ps in mesitylene solution (Ojima et al 1990a).

In acetonitrile, the photoinduced CS of the excited TCNB complexes takes place within 1 ps to a considerable extent but not completely, and for a complete CS further relaxation process with time constant τ_{CS} of ca. 5–40 ps depending on the ionization potential of D is necessary, where τ_{CS} becomes shorter with decrease of the ionization potential of D (Ojima et al 1990b). Similar tendency can be observed also when electron affinity of A is changed. That is, compared with τ_{CS} value of 20 ps in TCNB–toluene complex, $\tau_{CS} \sim 7$ ps in the case of PMDA (pyromellitic dianhydride)–toluene complex in acetonitrile solution (Asahi and Mataga 1991) where PMDA is stronger electron acceptor than TCNB. It has been confirmed that the formation of the 1:2 complex is not necessary for the CS in acetonitrile solution since the solvent reorientation assists the CS. Above results clearly show that τ_{CS} is much longer than τ_L and also than τ_S, which indicates a considerable extent of structural change of the complex necessary for the CS even in acetonitrile solutions, and which is similar to the photoinduced CS mechanism involving the geometrical change in the bichromophoric systems linked directly by single bond.

Even in acetonitrile solution where strong solvation of ions occurs, the geminate IP produced by the above described CS process is the compact IP (CIP) without intervening solvent between ions. We have made detailed femtosecond and picosecond laser photolysis studies on the CIP formation, CR of CIP and ionic dissociation process in acetonitrile solutions, and compared the behaviors of CIP with those of LIP. According to our comparative studies on the same D and A systems, where one is produced by CT complex excitation (CIP) and the other is formed by encounter in the fluorescence quenching reaction (LIP), the CIP shows much faster CR decay. As an example, the time profiles of the absorbance decay of anthracene–PA (phthalic anhydride) IP's are indicated in Fig. 1, where $k_{CR}^{CIP} = 1.3 \times 10^{10}$ s^{-1} and $k_{CR}^{LIP} = 2.2 \times 10^9$ s^{-1}. An extreme example is the case of strong D and A with very large k_{CR}^{CIP} value and rather small k_{CR}^{LIP} value. An example is the perylene–TCNE (tetracyano–

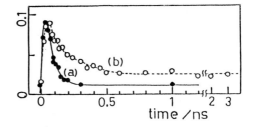

Fig. 1. Time profiles of anthracene cation absorbance at 710 nm of anthracene–PA IP in acetonitrile solution. (a) CIP, (b) LIP.

−ethylene) in acetonitrile, where $k_{CR}^{CIP}=3\times10^{12}$ s⁻¹ and $k_{CR}^{IIP}=6.1\times10^{8}$ s⁻¹.

By examining CIP's of various systems, we have obtained the energy gap dependence of k_{CR}^{CIP} quite different from the bell−shaped energy gap dependence of k_{CR}^{IIP}, that is,

$$k_{CR}^{CIP} = \alpha \exp\left[-\gamma \mid \Delta G \mid \right]$$

where α and γ are constant independent of ΔG. Although the mechanism responsible for this $k_{CR}^{CIP} \sim \Delta G$ relation is not very clear at present, we can make a qualitative interpretation of this result by assuming the potential surfaces for CIP and ground state similar to that for the inverted region and also assuming the shift of the potential minimum of CIP relative to that of ground state depending on the strengths of D and A (Asahi and Mataga 1991). This is analogous also to the multiple intermediate states model assumed for the CS process for the strongly interacting D, A systems.

4. PHOTOINDUCED CS IN HYDROGEN BONDED SYSTEMS

We found many years ago fluorescence quenching when two conjugate π−electronic systems were directly combined by hydrogen bonding interaction, and proposed the CT or ET interaction between proton donor and acceptor π−electron systems via the hydrogen bond as a possible mechanism of quenching (Mataga and Tsuno 1956, 1957).

$$(D^{\bullet}-H\cdots A) \longrightarrow (D^{+}-H\cdots A^{-}) \longrightarrow (D-H\cdots A)$$

Subsequently, we have made picosecond laser photolysis studies on some systems and have obtained the evidence which supports the proposed mechanism (Mataga 1984). For example, in the case of 1−aminopyrene−pyridine system, results of measurements of picosecond time−resolved absorption spectra and fluorescence decay curve indicated that equilibrium between the LE state $(D^{\bullet}-H\cdots A)$ and the CT state $(D^{+}-H\cdots A^{-})$ of the hydrogen bonded pair is realized within the time resolution of picosecond (10 ps) apparatus (Ikeda et al 1983).

In order to observe directly the dynamic process from LE state to the LE<−>CT equilibrium state, we have made femtosecond laser photolysis studies on 1−aminopyrene−pyridine system. The observed time−resolved spectra are indicated in Fig. 2 where we can recognize the absorption band with maximum around 520 nm at 0 ps and shoulder arises around 470−480 nm in the course of time. The latter absorption can be ascribed to aminopyrene cation while the former is due to the S_1 state of aminopyrene and the spectra at 20 ps can be ascribed to the LE<−>CT equilibrium. By examining the time profiles at 530 nm (decay) and 470 nm (rise), the time constant for the photoinduced CS has been obtained as $\tau_{CS}=5.5\pm1.5$ ps. We have confirmed further that τ_{CS} value becomes longer when electron affinity of pyridine is decreased by substituting methyl groups to pyridine in accordance with the electron transfer mechanism. Since our previous investigation confirmed that the fluorescence of N,N−dimethyl−1−aminopyrene was not quenched by added pyridine in both hexane and acetonitrile solution (Ikeda et al 1983), it is clear that hydrogen bonding interaction is of crucial importance for ET in this system.

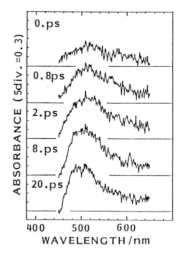

Fig. 2. Time−resolved absorption spectra of 1−aminopyrene−pyridine hydrogen bonded complex in n−hexane solution measured with femtosecond laser photolysis method by exciting at 355 nm.

Probably, a slight shift of N–H proton toward pyridine nitrogen in the hydrogen bond will facilitate the ET.

We made also first study on the photoinduced proton transfer (PT) in the excited hydrogen bonding complex in nonpolar solvent in the case of arylalcohol–aliphatic amine systems by means of fluorescence spectral studies (Mataga and Kaifu 1962).

$$(D^*-H\cdots A) \longrightarrow (D^{-*}\cdots H-A^+) \longrightarrow (D-H\cdots A)$$
$$\searrow \quad \uparrow$$
$$(D^-\cdots H-A^+) + h\nu$$

Recently we have observed directly the PT process in the excited state of PyOH(1-pyrenol)–TEA(triethylamine) hydrogen bonding complex in nonpolar as well as polar organic solvents by means of femtosecond laser spectroscopy. We have examined also PyOD–TEA system in various solvents. In all cases we have confirmed that the time constant of photoinduced PT is ca. 1 ps. Therefore, polar solvent does not appreciably facilitate the PT process. In this system, photoexcitation induced intramolecular CT from –OH group to pyrene ring of PyOH which facilitate the PT process. Conversely, the shift of proton in the hydrogen bond assists the intramolecular CT in PyOH. Thus the PT process is closely coupled with the intramolecular photoinduced CT which may be regulated by a small torsional motion around carbon–oxygen bond.

The geometrical rearrangement also seems to play important role in the above described photoinduced CS in hydrogen bonded systems.

5. PHOTOINDUCED CS DUE TO ET FROM SOLUTE FLUORESCENT STATE TO SOLVENT BY VERY WEAK INTERACTION

The picosecond laser photolysis and time–resolved absorption spectral measurement as well as fluorescence decay time measurement showed clearly that CS in the fluorescent state of TMPD in acetonitrile occurs with τ_{CS}=1.2 ns (Hirata and Mataga 1983).

$$TMPD^*(S_1) \longrightarrow TMPD^+\cdots S_n^-$$

where S_n^- means transient solvent aggregate anion. In order to elucidate the more details of this slow CS process, we have examined the temperature effect on the CS rate. We have observed a large decrease of the CS rate by temperature lowering, and obtained the activation energy for CS from the relaxed S_1 state in acetonitrile to be 7.4 kcal/mol. Because acetonitrile molecule itself cannot act as electron acceptor, formation of transient aggregate of near–by solvent molecules which can stabilize sufficiently the accepted electron seems necessary. The formation of such transient aggregate by fluctuation motion of solvent may not easy and will need rather high activation energy as actually observed leading to slow CS.

On the other hand, if we use solvent molecule which has a slight electron affinity like pyridine, the CS rate is greatly enhanced. In Fig. 3, we can see the decay of the $S_n \leftarrow S_1$ absorption at 673 nm and the rise of TMPD$^+$ band within ca. 1 ps. From the analysis of

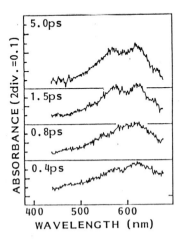

Fig. 3. Time–resolved absorption spectra of pyridine solution of TMPD measured with femtosecond laser photolysis method by exciting at 355 nm.

these spectra, the decay time of S_1 as well as the rise time of TMPD$^+$ was estimated to be 1.6 ps, which is 1000 times faster than in acetonitrile solution. We have examined also the effect of the excess vibrational energy on the CS rate. Our measurements on the excitation energy dependence of this monophotonic ionization yield in acetonitrile solution by using picosecond dye laser for excitation have indicated clearly that the yield increases significantly with increase of excitation energy near the origin of S_1 state and it becomes almost constant, but with further increase of the excitation energy, it decreases around the onset of the S_2 state. This seems to mean that the photoinduced CS with excess vibrational energy can compete with ivr and the CS depends on the optically excited vibrational mode (Hirata et al 1990). Moreover, this result indicates that, for the CS from the vibrationally unrelaxed state, the demand to the favorable transient aggregate of solvent is less severe, where the excess vibrational energy seems to be used efficiently for the formation of such an aggregate.

This is a special case of photoinduced electron transfer where the solvent fluctuations play both roles of the orientation motion facilitating electron transfer and the acceptor of electron by forming the transient solvent aggregate. Moreover, the intramolecular vibration excited optically plays a specific role facilitating the electron transfer.

Asahi T and Mataga N 1991 *J. Phys. Chem.* **95** 1956
Hirata Y and Mataga N 1983 *J. Phys. Chem.* **87** 1680
Hirata Y, Ichikawa M and Mataga N 1990 *J. Phys. Chem.* **94** 3872
Ikeda N, Miyasaka H, Okada T and Mataga N 1983 *J. Am. Chem. Soc.* **105** 5206
Kahlow M A, Kang T J and Barbara P F 1987 *J. Phys. Chem.* **91** 6452
Kakitani Y, Yoshimori A and Mataga N 1991 *Electron Transfer in Inorganic, Organic and Biological Systems* ed J R Bolton, N Mataga and G McLendon (Advances in Chemistry Series 228 ACS) pp 45–69
Mataga N and Tsuno S 1956 *Naturwiss.* **10** 305
Mataga N and Tsuno S 1957 *Bull. Chem. Soc. Jpn.* **30** 711
Mataga N and Kaifu Y 1962 *J. Chem. Phys.* **36** 2804
Mataga N 1970 *Bull. Chem. Soc. Jpn.* **43** 3623
Mataga N 1984 *Pure & Appl. Chem.* **56** 1255
Mataga N, Kanda Y, Asahi T, Miyasaka H, Okada T and Kakitani T 1988 *Chem. Phys.* **127** 239
Mataga N, Nishikawa S, Asahi T and Okada T 1990 *J. Phys. Chem.* **94** 1443
Miyasaka H, Ojima S and Mataga N 1989 *J. Phys. Chem.* **93** 3380
Ojima S, Miyasaka H and Mataga N 1990a *J. Phys. Chem* **94** 4047
Ojima S, Miyasaka H and Mataga N 1990b *J. Phys. Chem* **94** 5834
Rehm D and Weller A 1970 *Israel J. Chem.* **8** 259
Rips I, Klafter J and Jortner J 1990 *J. Phys. Chem.* **94** 8557
Yoshimori A, Kakitani T. Enomoto Y and Mataga N 1989 *J. Phys. Chem.* **93** 8316

Inst. Phys. Conf. Ser. No 126: Section VII
Paper presented at Int. Symp. on Ultrafast Processes in Spectroscopy, Bayreuth, 1991

Energy transfer phenomena in low and high molecular alkanes

O.Brede and R.Hermann

Central Institute of Isotope and Radiation Research,
Permoserstr. 15, Leipzig, O-7050, Germany

ABSTRACT: By means of the pulse radiolysis technique singlet energy transfer phenomena in alkanes and in polyethylene were studied. The time-resolved experiments resulted in completely different mechanisms of the singlet energy conversion in the low and high molecular system.

Ionizing irradiation generates in alkanes electronically excited singlet states having an efficient lifetime τ_o of ≤ 5 ns.
The decay of these singlet states is determined by a competition between fragmentation, radiation-less deactivation and fluorescence as formulated for the case of n-hexadecane.

$$n\text{-}C_{16}H_{34} \rightarrow n\text{-}C_{16}H_{34} \overset{S_1}{\left\langle \begin{array}{l} n\text{-}C_{16}H_{32} + H_2,\ n\text{-}C_{16}H_{33}^{\bullet} + H \ \text{etc.} \\ \xrightarrow{} n\text{-}C_{16}H_{34} \\ n\text{-}C_{16}H_{34} + h\nu \end{array} \right.} \qquad (1)$$

Fig. 1 shows the time profile of the n-hexadecane fluorescence observed in the pulse radiolysis of the pure hydrocarbon (curve 1).
In the presence of a scavenger, depending on the difference of the singlet energy levels S_1 between solvent and solute an energy transfer takes place demonstrated in Fig.1 for the case of decene-1 as additive (curve 2 and 3) (Hermann 1985, Brede 1987).

$$n\text{-}C_{16}H_{34}^{S_1} + n\text{-}C_{10}H_{20} \xrightarrow{} n\text{-}C_{16}H_{34} + n\text{-}C_{10}H_{20}^{S_1} \qquad (2)$$

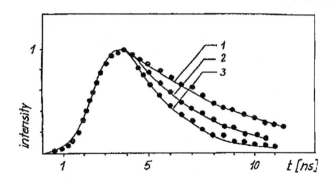

Fig. 1: Comparison of experimental n-hexadecane fluorescence time profiles measured at 210nm (points) and calculated ones (lines) for pure
n-hexadecane using $\tau_o = 5{,}2$ns (1),
n-hexadecane with $2*10^{-2}$mol dm^{-3} decene-1 (2) and
n-hexadecane with $5*10^{-2}$mol dm^{-3} decene-1 (3)

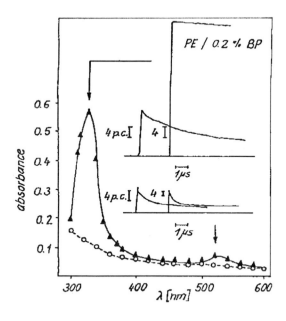

Fig. 2: Transient spectra in PE doped with 0,2 wt.-% benzophenone taken immediately after a 40 ns electron pulse (▲).(o)=PEbackground absorption.
Insets show time profiles at 330 and 520 nm (PE pure - left- hand curves)

On the example of the solvents cyclohexane and C_7 to C_{16}-n-alkanes and the solute benzene supposing a singlet transfer probability of one the analysis of the rate constant k_2 enabled to derive reaction radii (using known diffusion coefficients) which were between 5 and 8 Å, and therefore, close to the sum of molecular radii (Mehnert 1988).

From a comparison with Foerster radii and the Stokes-Einstein relationship it can be concluded that the energy transfer from the singlet states of the alkanes to scavengers proceeds by collisional interaction, i. e. , represents a molecular process.

In high molecular alkanes, i.e., in polyethylenes the situation was found to be a completely other one. Until now no real fluorescence caused by the polymer molecule has been found. Reasons for it may be given by structural peculiarities (content of functional groups as, e. g., olefin and hetero atom groups) but also by an unusual energy transport phenomenon that has been analyzed by means of radiation-chemical techniques.

Based on results of steady-state low temperature matrix isolation studies of Partridge (1970) we performed ns pulse radiolysis experiments with molten pure polyethylene and with PE samples doped with scavengers (Brede 1988,1989). In the highly viscous polymer instead of any fluorescences as indication for electronically excited PE states a very fast ($\leq$1ns) scavenger radical formation also at rather low additive concentrations ($\leq 10^{-3}$mol dm^{-3}) was observed which in no case could be explained by reaction of massive PE transients (radicals or ions) that proceed in the μs and ns time domain.

Fig. 4 shows this phenomenon of a non classical fast scavenger radical formation on the example of the pulse radiolysis of polyethylene doped with traces of benzophenone. Immediately after the electron pulse the well-known spectrum of the benzophenone ketyl radical can be observed. The time profile talken in the absorption maximum demonstrates the rapid formation.

A possible interpretation of the fast radical formation can be given by the hypothesis of the existence of very mobile singlet excitons PE generated in PE in course of the radiation action (Partridge 1970). These excitons should move along the polymer chain with a frequency slidely higher than those of molecular vibrations.

The decay of the excitons will be explained to proceed by internal trapping within the macromolecule or external trapping by the scavenger (S) yielding in all cases radical products.

$$PE \rightsquigarrow PE \overset{\textstyle 2\ PE^{\bullet}}{\underset{\textstyle +\ S \quad PE^{\bullet}\ +\ S^{\bullet}}{\Big\langle}} \tag{3}$$

The PE excition hypotheses has been checked by a lot of scavenger experiments and comparative laser photolysis studies (Brede 1988).

Now, consequences for photochemical reaction mechanisms in PE are in consideration and study.

Brede O, Mehnert R and Naumann W. 1987 Chem. Phys. 115 279

Brede O and Naumann W. 1988 Radiat. Phys. Chem. 32 475

Brede O, Hermann R, Wojnarivits L, Stephan L and Taplick T. 1989 Radiat. Phys. Chem. 34 403

Hermann R, Brede O and Mehnert R. 1985 Radiat.Phys.Chem.26 513

Mehnert R, Brede O, Naumann W and Hermann R. 1988 Radiat. Phys. Chem. 32 475

Partridge R H, 1970 J. Chem. Phys. 52 pp 2485, 2491, 2501

Inst. Phys. Conf. Ser. No 126: Section VII
Paper presented at Int. Symp. on Ultrafast Processes in Spectroscopy, Bayreuth, 1991

Analysis of molecular dissociation by a chirped infrared laser pulse

Boris A. Grishanin

Physics Department, Moscow State University, Lenin's Hills, 119899 Moscow, Russia

Valentin D. Vachev and Victor N. Zadkov

International Laser Center, Moscow State University, Lenin's Hills, 119899 Moscow, Russia

Abstract. A comparative analysis of quantum and classical calculations of photodissociation reaction of HF molecule under the influence of ultrashort powerful chirped laser pulse is presented. It is shown that excitation efficiency is reduced when the coordinate dependence of effective charge is introduced into model. The use of classical approach to investigate the photodissociation of polyatomic molecules is argued.

1. INTRODUCTION

Creation in recent years of ultrashort and femtosecond laser systems generating powerful IR-pulses has stimulated studies on substance behavior in superstrong light field. In connection with the molecules (the problems of highly excited molecules, laser chemistry, etc.) such laser pulses with wide spectrum are used for creation of the ensembles of highly excited molecules, up to photodissociation. Optimization of frequency spectrum of a short laser pulse is one of the methods of increasing the efficiency of molecular excitation with the help of such pulse.

A proper description of the quantum nature of a molecular system is one of the most principal problems in computer analyses of molecular dynamics under pulsed laser excitation. Exact determination of the fundamental vibrational frequencies and their combinations is also of great importance and for highly excited molecules with a small number of active modes it can not be done exactly using classical approach. The direct quantum treatment of the problem is based on the use of the wave function $\psi(x)$ depending on coordinates of nuclei with subsequent development of the solution to the time-dependent Schrödinger equation. Using this approach Chelkovski *et al* (1990) have studied dissociation of the HF molecule.

However, in the case of polyatomic molecules there are no perspectives for such treatment at all. In this paper we use another quantum representation using the basis of the eigen functions $\psi_n(x)$ of the harmonic oscillator. The totally different approach is based on pure classical MD simulation (Levine and Bernstein 1987) or on semiclassical analysis of wave-packet propagation (Coalson and Karplus 1990). These approaches are preferable in the case of strong laser fields when the dynamic Stark effect overwhelms the problem of exact determination of quantum eigen frequencies. An intramolecular energy exchange may also lead to classical approach being valid in this case. This problem is discussed in Section 4.

2. THE QUANTUM ANALYSIS

A promising method of simplification of the quantum analysis, especially for polyatomic molecules consists in using some appropriate quantum basis such as the eigen basis of the harmonic part of molecular Hamiltonian. In this basis the pure quantum equations of molecular motion are well defined because the matrix elements of the external field-molecule interaction are well-known and that of the nonlinear part of the Hamiltonian can be easily calculated. It makes possible to simplify treatment by reducing a quantum space for any weakly excited mode to a two- level system described by a single complex number c_α that represents the wave function for α mode. The field interaction Hamiltonian is

$$\widehat{H}_F = -d_1 \hat{x} E_L(t), \tag{1}$$

where $\hat{x}$ are Cartesian coordinate displacement operators, d_1 is the effective matrix of dipole gradients, $E_L(t)$ is the radiation electric field. The nonlinear part of the molecular Hamiltonian can be presented by the expression

$$\widehat{H}_{NL} = \sum_n c_n(\hat{x})^n, \tag{2}$$

where c_n are proper coefficients representing the nonlinear part of the molecular potential energy surface. A large enough number of terms is to be used in (2) in order to describe adequately large deviations from initial equilibrium configuration of the molecule. This expressions are given real quantities when operators $\hat{x}$ are represented by their matrix elements as $\hat{x}_\alpha = (\hat{a}_\alpha + \hat{a}_\alpha^+)/2$ with well- known matrix elements in harmonic basis. Using (1), (2) and the harmonic part representation $\widehat{H}_L = \sum_\alpha \hbar\omega_\alpha \hat{a}_\alpha^+ \hat{a}_\alpha$ we can calculate the resulting time evolution given by the operator:

$$U(t) = \mathbf{T} \exp\left(-(i/\hbar) \int \widehat{H} \, dt\right), \tag{3}$$

where $\widehat{H} = \widehat{H}_L + \widehat{H}_{NL} + \widehat{H}_F(t)$.

3. COMPUTER MODELING OF PHOTODISSOCIATION OF HF MOLECULE

In this work a one-dimension problem, excitation of the diatomic HF molecule, is considered in order to evaluate the possibilities of this method in investigating the dissociation reaction. The potential of the system is given by

$$V(x, t) = D[1 - \exp(-ax)]^2 - E_L(t) \, d_1(x) x \, \cos[\omega(t)t], \tag{4}$$

where $x = r - r_0$, r_0 is the equilibrium separation of nuclei, D is the dissociation energy, $E_L(t)$ is the pulse envelope and $d_1(x)$ is the effective charge.

The $d_1(x)$ dependence on the displacement is taken into account in a classical analysis only. The same form of the chirp as that proposed by Chelkovski *et al* (1990) and the same parameters were used.

The number of terms used for representation of the unharmonic part of the molecule was 50. The number of harmonic levels used at the initial stage of time evolution was 24. This number can be automatically enlarged if the occupation number of the highest level exceeds fixed small quantity. The calculated time dependencies of the occupation numbers of the first and the eighth excited vibrational levels and the total occupation number of all levels with energies higher than the dissociation one are represented in Figure 1.

Our calculations are in general correspondence with results obtained by Chelkovski *et al* (1990)

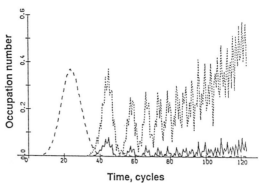

Fig. 1. Quantum analysis of HF molecule: the calculated occupation numbers of the first (dashed line) and of the eighth (solid line) excited states and the total one (dotted line) for states with the energies higher than the dissociation energy. The field intensity $I = 10^{13}$ W/cm^2; one time cycle is equal to 8.41 fs.

and also demonstrate the effectiveness of chirping. However we obtained the effectiveness of chirping if the initial frequency is chosen equal to the harmonic frequency ω_0 but not to the frequency ω_{01} of the lowest transition. A possible reason of this discrepancy is that the computational errors produced by time discreetization are masking the frequency shift produced by unharmonic part of potential.

Figure 2 shows the results of classical analysis. They demonstrate the role of dependence of the effective charge on a distance between atoms leading to increase dissociation intensity. For the field intensity $I = 10^{15}$ W/cm^2 we see that the molecule dissociate only in the model with constant charge but not in the model with coordinate dependent charge. The molecule dissociates when its energy exceeds dissociation energy D. The dissociation threshold obtained in the classical approach is somewhat higher than that obtained in quantum approach and exceeds the ionization threshold.

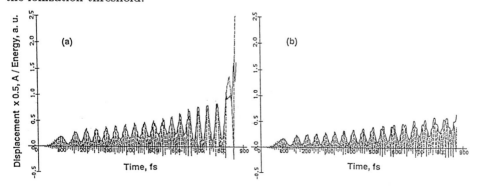

Fig. 2. Classical analysis of HF molecule: time dependence of coordinate displacement (dashed line) and energy (solid line) for the cases of constant (a) and coordinate dependent effective charge (b). The field intensity $I = 10^{15}$ W/cm^2; the dissociation energy is equal to one a.u.

4. EXCITATION OF POLYATOMIC MOLECULES

The condition of validity of classical approach is the inequality $\Delta\omega\tau \ll 2\pi$ where $\Delta\omega$ is the quantum unharmonicity and τ is the characteristic time interval at which phase of a mode vibration is constant. It is directly related to the intermode exchange time which is usually small for polyatomic molecule. So we can expect that this condition is valid as a rule but there are no exact apriory test for it and the effectiveness of classical approach needs experimental verification. Our previous investigations of molecular dynamics in anthrachene and stilbene molecules excited by short laser pulse show fast oscillations on a ~ 50 fs time scale as soon as $\Delta\omega \leq 1 \times 10^{13}$ s^{-1}. It enables us to conclude that classical approach is effective one in problems of computer optimization of spectral parameters of an infrared laser pulse for optimal excitation of selected modes of polyatomic molecules and driving the processes of photodissociation. As far as the computer investigation of classical dynamics of molecule is concerned classical approach is a suitable method of such optimization because one can realize the optimization procedure at each time step of calculations. If the optimization for a selected bond is done properly one can really hope to excite this bond up to dissociation using field intensities below ionization threshold. These studies are now in progress.

Acknowledgments

The authors wish to thank Vyacheslav M. Gordienko and Nikolai I. Koroteev for helpful discussions.

References

Chelkovski S, Bandrauk A D, Corkum P B 1990 *Phys. Rev. Lett.* **65** 2355
Coalson R D, Karplus M 1990 *J. Chem. Phys.* **93** 3919
Grishanin B A, Vachev V D, Zadkov V N 1991 *in this volume*
Levine R D, Bernstein R B 1987 *Molecular Reaction Dynamics and Chemical Reactivity* (Oxford Univ. Press)

Inst. Phys. Conf. Ser. No 126: Section VII
Paper presented at Int. Symp. on Ultrafast Processes in Spectroscopy, Bayreuth, 1991

Coherence effects on selective ionization of three-level systems driven by pulse sequences of two lasers

K. Johst

Central Institute of Isotope and Radiation Research
Permoserstr. 15
O-7050 Leipzig, Germany

ABSTRACT: Using a quantum-mechanical approach we investigated the ionization probability and selectivity of three-level systems and compared it with rate equation results. Taking into account non-radiative decay processes we demonstrated to what an extent the selectivity of two-step photoionization is influenced by special pulse sequences at incoherent excitation and if a coherent interaction during the first step is possible.

1. INTRODUCTION

This paper provides a theoretical investigation of the selectivity of excitation processes in three-level systems in the fields of two lasers. As an example we regard resonant two-step photoionization processes where the first laser excites the atom/molecule to an intermediate electronic state, the second laser ionizes the excited molecule without exciting ground state species.
If one assumes a system coherently driven by the laser field the rate equations become insufficient for describing the excitation process. Then Rabi oscillations of the level populations appear as a function of laser intensity, detuning and dephasing processes.
It is known that in a two-level system coherent interaction strongly increases the selectivity of excitation of different species compared with incoherent interaction (Diels 1976).
We coupled a third level by a rate equation. So we can describe two-step photoionization processes of isolated species where the first transition can coherently or incoherently be excited, the second transition to the ionic state is always incoherent. Laser radiation is assumed as monochromatic and coherent and only non-radiative line broadening is taken into account.
We investigated to what an extent the selectivity in such three-level systems is influenced by coherent interaction

during the first step and different timing of the incoherent interaction during the second step of resonant two-step photoionization processes (Johst 1991). This corresponds to different pulse sequences shown in Figure 1.

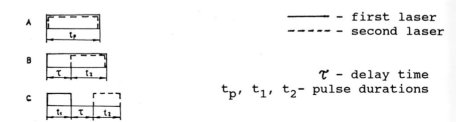

——— - first laser

----- - second laser

τ - delay time

t_p, t_1, t_2- pulse durations

Fig. 1. Pulse sequences

2. THEORY

Using time-dependent Schrödinger equation the wave function for a two-level system can be written as:

$$|\Psi(t)> = c_0(t)|\varphi_0> + c_1(t)|\varphi_1> = U(t,t_0)|\Psi(t_0)>$$

In the rotating-wave approximation the following Hamiltonian can be established :

$$H = \begin{pmatrix} 0 & -\omega_R/2 \\ -\omega_R/2 & \Delta - i\gamma/2 - ik_1/2 \end{pmatrix}$$

ω_R-rabi frequency
Δ -detuning
γ -relaxation
k_1-ionization

The occupation probability of the intermediate level is calculated by

$$P_1(t) = |c_1(t)|^2$$

and the third level is coupled by a rate equation giving the ionization probabilty

$$P_i(t) = k_1 \int_0^t |c_1(t')|^2 dt'$$

To illustrate the selectivity we assume two species a and b which differ in the absorption cross sections of the exciting step by

$$\sigma_0^b = x \, \sigma_0^a$$

x=1.2, 2, 5

The selectivity of the ionization process we define as:

$$S_i(t) = P_i{}^a(t)/P_i{}^b(t)$$

The results are compared with those of the following rate equations (Johst, Johansen 1989):

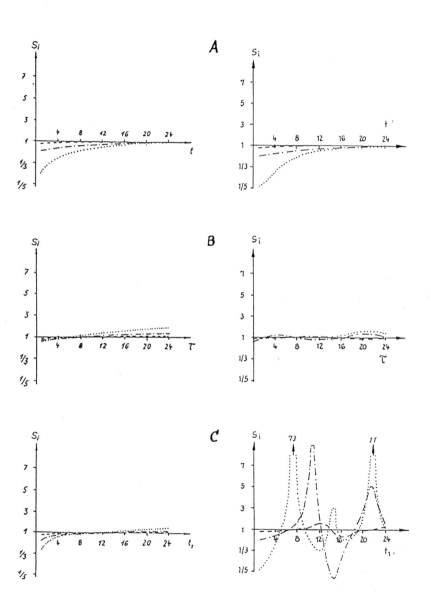

Fig. 2. Comparison of the selectivities calculated by rate equations and by the Schrödinger equation

$$dN_0/dt = -k_0(N_0-N_1)$$
$$dN_1/dt = +k_0(N_0-N_1) - \gamma N_1 - k_1 N_1$$
$$dN_i/dt = +k_1 N_1$$

The quantities $N(t)$ represent the time evolution of the populations of the corresponding states.

3. RESULTS AND DISCUSSION

A relation between the parameter sets of the rate equations and the Schrödinger equation can be derived so that a comparison of the results is possible (Johst 1991). Figure 2 shows the selectivities of the three pulse sequences A, B, and C calculated both by rate equations (on the left) and by the Schrödinger equation (on the right) assuming the above mentioned different absorption cross sections. The time scale of the ionization processes is defined by the rate constants and the relaxation times. We selected a parameter set which permits a coherent interaction if the exciting laser acts and the coherence destroying effect of k_1 is absent. At pulse sequence A both lasers act simultaneously and Rabi oscillations during the first transition cannot arise. That's why the results of rate equations and Schrödinger equation are similar. So the pulse sequences B and C are favoured because here ionization starts from the high level of the occupation selectivity of the coherently excited intermediate state. This is also the case at a coherent excitation in pulse sequence A because the occupation probability of the intermediate level is much more oscillating than the ionization probability. Further calculations showed that in sequence B these high excitation selectivities cannot be transformed to high ionization selectivities (Johst 1991).
The calculations demonstrated that the utilization of coherent interaction effects for high selectivity in three-level systems is most efficient if the ionizing laser pulse is delayed with respect to the exciting one and the two pulses do not overlap. Then the high selectivity in a coherently excited two-level system can be transformed to high ionization selectivities in three-level systems. This is in contrast to rate equation results (Johst, Johansen 1989) which represent incoherent interaction processes. Here a partial or complete overlap can be more selective.

References

Diels J C 1976 *Phys. Rev. A* **13** 1520
Johst K 1991 *Appl. Phys. B* **53** 46-51
Johst K, Johansen H 1989 *Appl. Phys. B* **48** 479-484

Fast relaxation processes in aromatic free radicals

N Borisevich, S Melnichuck, S Tikhomirov, G Tolstorozhev

The Institute of Physics, Belorussia Acad. Sci., Leninskii prospect 70,
220602 Minsk, Bielorussia

ABSTRACT: Experimental and theoretical results are presented on the
photodissociation of sulfide organic molecules and fast recombination
of free radicals formed.

1. INTRODUCTION

One of the most important processes of molecular photonics is photodissociation. It underlies the fundamental phenomena of the structural transformations of molecular systems. Unlike the other elementary processes,
the mechanisms of photodissociation of complex organic molecules and fast
relaxation processes in free radicals formed have been studied insuffici-
ently.

The properties of luminescence, laser emission and transient absorption
of radicals generated at S-S bond cleavage during sulfide molecule photodissociation have been investigated elsewhere (Borisevich *et al* 1986,
1987,1991 and Lysak *et al* 1988). In this paper we present data concerning
the mechanism of S-H-, S-S- and S-C-bond cleavage in aromatic sulfur
containing molecules and features of free radicals geminate recombination.

2. EXPERIMENTAL

Kinetic measurments were made using an automated picosecond spectro
meter (Lysak *et al* 1988). The molecules under study are excited by 4 psec
pulses of the third and fourth harmonics of the fundamental frequency. The
broad-band picosecond continuum is used as a probe pulse. A double-beam
optical arrangement was adopted, and absorption spectra in the 300-nm
scanning region were recorded using an optical multichannel analyzer. The
samples used by us are:
Ph-S-S-Ph (I), H_2N-Ph-S-H (II), H_2N-Ph-S-S-Ph-NH_2 (III) and non-symmetric

sulfides H_2N-Ph-S-CH_2-CO-Ph (IV) and H_2N-Ph-S-CH_2-CO-Ph-OCH_3 (V).

3. RESULTS AND DISCUSSION

When pentane solutions of molecules (II) and (III) are excited by pico
second light pulses, immediately after the excitation the same transient
absorption spectra of the photolysis products are recorded, the position
and shape of ones remaining constant at delay times up to 1000 psec (Fig.
1). This means that in both cases aminophenylthiyl radicals are formed. An
identical result is obtained for the compounds (IV) and (V).

The increase in the optical density of transient absorption determines
the characteristic time τ_d of molecule photodissociation. The experiment

shows that for all the molecules studied $\tau_d < 1$ psec.

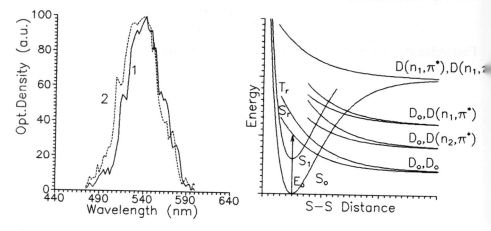

Fig. 1. Transient absorption spectra of products of the compounds
II (1) and III (2) photolysis in
pentane at time delay 100 psec

Fig. 2. Potential energy diagram for
the compound I

The dependence of S-H-bond cleavage quantum yield φ_{SH} in aminothio-

phenol (II) on the exciting quantum energy was investigated. The values of
the absolute quantum yield were determined by measuring the quantity of
hydrogen released during aminothiophenol photolysis (Malkin *et al* 1987).
The quantum yield of photodissociation depends considerably on the exciting
wavelength:

λ_{exc}	264 nm	300 nm	337 nm
φ_{SH}	0.36	0.25	0.075.

The quantum yield of S-S-bond cleavage in aminophenyldisulfide also
increases as the exciting quantum energy increases (φ_{SS} = 0.57 and 0.048

for λ_{exc} = 264 nm and 337 nm). Using the yields obtained, we have determi-

ned the extinction coefficient for the aminophenylthiyl radical which is

equal to 9100 $M^{-1}cm^{-1}$. The same dependence is obtained for all the compounds
studied. This behavior is due to the properties of atoms forming the cleaved bond, mainly to the properties of the S-atom.
   The analysis of the S-S-bond photodissociation was performed by
Plotnikov *et al* (1990). Fig. 2 shows the potential energy diagram for the
diphenyldisulfide molecule (I) based on the quantum chemis try analysis.
   The S atom in the radical R-S˙ (where R is the aromatic fragment) has
sp-hybridization and its unpaired electron occupies the $3p_z$-orbital. The

symmetry axis of this orbital is normal to the π-system plane. The unpaired
electron participates in π-conjugation. One of the unshared electron pairs
of the S atom occupies the $3p_x$-orbital, the other pair occupies the sp-

hybridizied orbital. (The X-axis is normal to the C-S-bond direction.) The
energy of the sp-orbital (n_1) is lower than that of the $3p_x$-orbital (n_2).

The lower excited D-states of the R-S˙ radical are populated after the transition of one of the n-electrons of the S-atom to the π-orbital. The lo-

west excited state is the $D(n_2\pi^*)$-state and the $D(n_1\pi^*)$-state is situated higher. The term which determines the dependence of the radical pair R-S· ·S-R energy dependence on the distance between the S-atoms when both radicals are in the D_o-state, is repulsive both in the singlet (S_r) and in the triplet (T_r) state due to the exchange repulsion of the n-electrons. That means that the S-S-bond cleavage can occur in both the S_r- and the T_r-state. The term is also repulsive when one radical is in the D_o-state and the other in the $D(n_1\pi^*)$- or in the $D(n_2\pi^*)$-state. In accordance with the ordinary valence bond conception the bonding term (S_o-state of the disulfide molecule) appears if both radicals are in the $D(n_1\pi^*)$-state, i.e. when the unpaired electrons of the S-atoms occupy the sp-orbital (σ-radicals). In work of Plotnikov et al the lowest repulsive term energy E_o was estimated to be 4.7 eV.

The high value of extinction coefficients of sulfide molecules (>10000 $M^{-1}cm^{-1}$) indicates that the bounding state is excited after the electron transition. Then transition to a repulsive state occurs. This means that dissociation may proceed via predissociation. The triplet (T_r) repulsive term lies higher than the singlet S_r one. Such a position of the electronic levels indicates that the bond cleavage has to proceed via $S_1 \to S_r$- channel. For the dissociation to take place at low energies of the exciting quantum, the system has to overcome the energy barrier. At high energies of the exciting quantum this reaction proceeds without a barrier. In this way the experimental fact of the dependence of the quantum yield on the exciting wavelength is explained.

The competition between the free radical separation and the geminate recombination is regulated by the cage effect. The transient absorption decay curves with the pronounced peak recorded at diphenyldisulfide (1) photolysis by radiation λ_{exc} = 264 nm are typical for geminate radical pair recombination when the cage of solvent molecules prevents the radicals from entering freely into the solvent volume. From comparison of the experimental dependencies and model curves we have determined that the characteristic time of geminate recombination to be about 10 psec. Recorded at aminophenylthiol photolysis the transient absorption decays have no peak. This means that the "cage" effect for the hydrogen atom is absent.

For amino-substituted compound (III), the peak of geminate recombination was not observed even for highly viscous solvents. This means that the fast recombination of amino-substituted thiyl radicals is blocked. The process leading to such a result may be the fast charge transfer in the radical from the amino-group to the S-atom. This process was shown in the work of Borisevich et al (1991) to be accompanied by rotation of the aminogroup plane resulting in the TICT-state formation. The transfer of the redundant electron density to the S-atom leads to Coulomb repulsion of the reaction centres (S-atoms) of geminate radicals and geminate recombination is blocked.

The processes of photodissociation and geminate recombination at tetraphenylhydrazin photolysis was studied by Hyde *et al* (1991). It was found that in the presence of a strong magnetic field the geminate recombination is slightly inhibited. Such a behavior was explained assuming that the triplet state also takes place in the dissociation process. For aromatic disulfides no influence of the magnetic field on the geminate radical kinetics was obser ved. It means that the triplet dissociative term doesn't participate in the sulfide molecules photodissociation and dissociation proceeds only via $S_1 \rightarrow S_r$ channel as mentioned above.

Borisevich N, Gorelenko A, Lysak N *et al.*   1986   *Pis'ma   Zh.Eksp.Teor.Fiz.*
   43 113
Borisevich N, Tolstorozhev G 1987 *Kvant.Elektr.* **14** 1063
Borisevich N, Melnichuk S, Tikhomirov S, and Tolstorozhev G 1991 *Conference on Quantum Electronics and Laser Science* (Baltimore, Maryland) QWD52
Hyde M, Beddard G 1991 *Chem. Phys.* **151** 239
Lysak N, Melnichuk S, Tikhomirov S, and Tolstorozhev G 1988   *Zh.   Prikl. Spektrosk.* **49** 949
Malkin Ya, Kuzmin V, Rusiev Sh 1987 *Izv.Akad.Nauk SSSR. Ser.him.* **3** 537
Plotnikov V, Krivosheev Ya, Melnichuk S *et al.* 1990   *Dokl.Akad.Nauk   SSSR* **312** 913

Inst. Phys. Conf. Ser. No 126: Section VIII   583
Paper presented at Int. Symp. on Ultrafast Processes in Spectroscopy, Bayreuth, 1991

Femtosecond photoisomerization of rhodopsin as the primary event in vision

R.W. Schoenlein, C.V. Shank

University of California, Lawrence Berkeley Laboratory, Berkeley, CA 94720.

L.A. Peteanu, R.A. Mathies

Department of Chemistry, University of California, Berkeley, CA 94720.

ABSTRACT: The primary event in vision, the *cis-trans* isomerization of rhodopsin, is resolved with the use of femtosecond optical measurement techniques. The 11-*cis* retinal prosthetic group of rhodopsin is excited with a 35 fs pump pulse at 500 nm, and the transient changes in absorption are measured between 450 and 580 nm with a 10 fs probe pulse. Within 200 fs, an increased absorption is observed between 540 and 580 nm, indicating the formation of photoproduct on this time scale. These measurements demonstrate that the first step in vision, the *cis →trans* torsional isomerization of the rhodopsin chromophore, is essentially complete in only 200 fs.

Recent developments in the generation and amplification of femtosecond optical pulses in the blue-green spectral regime (Schoenlein 1991a) now enable the investigation of ultrafast processes in a variety of biological systems with a time resolution of 10 fs. One of the most important ultrafast processes occurring in nature is the detection of light by the visual system. This process is initiated by the absorption of a visible photon by the retinal pigment rhodopsin and culminates in the stimulation of the optic nerve. Nearly thirty years ago, Yoshizawa and Wald (1963) determined that the initial chemical reaction in vision is the photoisomerization of the retinal chromophore in rhodopsin. Subsequent studies have shown that this first step in vision occurs within a few picoseconds (Busch 1972). However measurements of the initial reaction dynamics have been limited by laser system resolution, and a complete understanding of this process remains an important problem in photochemistry and biology. We report here the room-temperature investigation of the *cis-trans* isomerization in rhodopsin using 10 fs blue-green pulses. Our results demonstrate that the first step in vision occurs on a time scale of 200 fs (Schoenlein 1991b).

The photoreactive component of the rhodopsin molecule is 11-*cis* retinal, a vitamin-A derivative, which is bound within the protein opsin by a Shiff base linkage. The absorption maximum of rhodopsin in the 11-*cis* form is ~500 nm. The absorption of a photon causes the retinal prosthetic group to isomerize about the C_{11}-C_{12} double bond, forming the photoproduct bathorhodopsin with an absorption maximum at ~535 nm. Following this initial isomerization, thermally mediated conformational changes occur which result in the

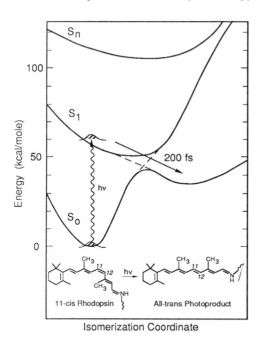

Fig. 1. Schematic ground-state and excited-state potential-energy surfaces for the 11-*cis* → 11-*trans* isomerization in rhodopsin adapted from (Birge 1980, 1990). The reaction path of the photoisomerization is indicated by the nonadiabatic potential surfaces (broken lines).

release of opsin and all-*trans* retinal. Previous measurements of the initial isomerization using transient absorption spectroscopy (Busch 1972) have shown the formation of a photoproduct within the 6 ps laser pulsewidth, but were unable to resolve the initial isomerization process. In addition, resonance Raman spectroscopy (Eyring 1980) and time-resolved Raman studies (Hayward 1981) have confirmed that an initial rhodopsin photoproduct with a twisted all-*trans* structure is formed in less that 30 ps. Raman intensity analysis have suggested that the isomerization dynamics occur on a subpicosecond time-scale (Loppnow 1988), and theoretical simulations of the isomerization dynamics are qualitatively consistent with this picture (Birge 1980, 1990, Weiss 1979). However, the time scale for the formation of the primary photoproduct has never been experimentally determined.

The recent development of a blue-green femtosecond laser system (Schoenlein 1991a), now makes it possible to directly study the time-course of rhodopsin isomerization. The laser system consists of a colliding-pulse modelocked (CPM) dye laser producing 50 fs pulses at 620 nm. These pulses are amplified to several microjoules in a three-pass amplifier pumped at a repetition rate of 400 Hz by a XeF excimer laser. The amplified pulses are used to generate a white-light continuum in a 1-mm jet of ethylene glycol, and the blue-green portion of the continuum is re-amplified to the microjoule level in a two-pass amplifier pumped by the same excimer laser. We use a standard pump-probe technique in which a femtosecond pump pulse excites the rhodopsin sample, and the resulting changes in absorption are measured with a probe pulse, which is delayed in time with respect to the pump. The 35 fs pump pulse at 500 nm comes directly from the amplifier. The 10 fs probe pulse is created by splitting off part of the pump and focussing it in an optical fiber to create a spectrally broadened and chirped pulse. The pulse is compressed using a sequence of gratings and prisms for phase compensation (Fork 1987). The 450 to 580 nm bandwidth of the probe pulses allows us to resolve the spectral dynamics of the rhodopsin molecule following excitation by the narrow-band (~15 nm) pump pulses.

The rhodopsin sample is prepared by isolating rod outer-segments from bovine retinas. The isolated outer-segments are purified by ultracentrifugation in a continuous sucrose gradient, lysed with cold water, and pelleted by additional centrifugation. Rhodopsin from 400 retinas is solubilized (5% Ammonyx-LO, 20 mM MOPS, pH 7.4) and then concentrated to an optical density of 15 OD/cm (at 500 nm). A 3-ml sample was flowed through a 300-μm wire-guided jet at a sufficient velocity to ensure complete replacement of the sample between each pair of pump-probe pulses. The pump pulses are focused on the rhodopsin jet to an energy density of ~400 μJ/cm^2. At these pump fluences, only ~10% of the exposed sample is bleached by the pump. The weaker probe pulses are similarly focused on the sample (~30 μJ/cm^2). Transient changes in absorption ($\Delta T/T$) are measured with two different techniques. Time-resolved measurements at specific wavelengths are obtained by spectrally filtering the probe pulse (after passing through the sample) and combining differential detection with lock-in amplification. Differential spectral measurements over the entire bandwidth of the 10 fs probe pulse are made with a spectrometer and a dual diode array detector. In all measurements, the maximum signal ($\Delta T/T$) is a few percent, and the linearity is verified in order to avoid saturation effects.

The transient change in absorption is measured at 500, 535, 550, and 570 nm, following excitation of rhodopsin by a 35 fs pulse at 500 nm (Fig. 2). Measurements at 500 nm probe the initial bleach and partial recovery of the ground state absorption of 11-*cis* rhodopsin. At early times, we observe a transient excited-state absorption as evidenced by the negative differential signal ($\Delta T/T < 0$) near zero delay. The arrival of the pump pulse induces an absorption at ~500 nm, assigned to the $S_1 \rightarrow S_n$ transition (Fig. 1), which interferes with the ground-state bleach signal. As the wavepacket created in the first excited state moves away from the Franck-Condon region, the excited-state absorption disappears and the full bleach of the rhodopsin absorption is revealed by 125 fs. A rapid partial recovery of the initial bleach at 500 nm is observed, which has a time constant of ~250 fs. The subsequent long-time recovery of the bleach occurs with a time constant of ~8 ps.

Transient absorption changes measured at 550 nm and 570 nm reveal the kinetics of photoproduct formation. At 570 nm, near the peak absorption of the photoproduct, we observe a rapidly developing absorption ($\Delta T/T < 0$) which reaches a maximum by 200 fs. Beyond 200 fs there is very little change in the absorption, indicating that the photoproduct is formed on this time scale. This conclusion is supported by the rapid appearance of absorption at 550 nm, on the blue side of the photoproduct absorption band. At this wavelength however, there is a ~100 fs delay before the absorption develops. Between 200 fs and 1 ps, the absorption at 550 nm gradually increases, and then remains unchanged out to 6-ps delay.

Measurements at 535 nm, on the red edge of the 11-*cis* rhodopsin absorption band and on the blue edge of the photoproduct absorption, are more complicated. The excited-state ($S_1 \rightarrow S_n$) absorption induced by the pump ($\Delta T/T < 0$) is apparent at short times between 0 and 100 fs. This induced absorption does not appear instantaneously as is the case at 500 nm, but is delayed by ~50 fs. The delay is attributed to the dynamics of the excited-state absorption resulting from spreading and/or motion of the wavepacket excited on the S_1 surface as it leaves the Franck-Condon region. By 100 fs we observe the appearance of the ground-state bleach, and by 200 fs the photoproduct absorption begins to dominate and the signal changes sign. The long time (1 to 5 ps) behavior at 535 nm results from a slow

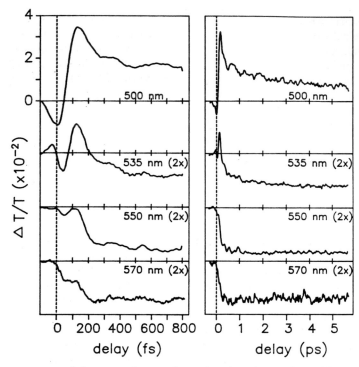

Fig. 2. Transient absorption measurements of 11-*cis* rhodopsin at various wavelengths following a 35 fs pump pulse at 500 nm (~ 10 fs probe).

recovery of the ground-state absorption (as observed at 500 nm). In addition, we observe oscillatory behavior at all probe wavelengths, though this is particularly evident at 500 nm and at 570 nm between 0 and 200 fs.

Complementary information about the isomerization kinetics is provided by the differential spectral measurements shown in Fig. 3. At 33 fs delay, the increase in absorption between 490 and 540 nm results from the excited-state ($S_1 \rightarrow S_n$) transition which dominates the bleach signal at short times. The initial appearance of photoproduct is indicated by the differential absorption observed between 540 and 580 nm. Between 33 fs and 200 fs, the photoproduct absorption increases, and the initial rhodopsin bleach between 470 and 540 nm becomes evident. The filling-in of the bleach signal is observed at longer delays, consistent with the time-resolved measurements shown in Fig. 2. The photoproduct absorption at 570 nm remains unchanged after 200 fs, indicating that the isomerization is complete on this time scale. The blue shift of the isospestic point (from 540 to 515 nm) with increasing delay is most likely due to vibrational cooling of both the rhodopsin and photoproduct ground states, as well as conformational relaxation. Although we cannot exclude the possibility that some residual excited-state population contributes to the recovery of the bleach at long times, our interpretation is supported by the fact that vibrational cooling and conformational relaxation are known to occur in the related pigment bacteriorhodopsin on a similar time scale (~3 ps) (Doig 1991). Differential absorption spectra at 6 ps delay show the residual bleach between 470 and 515 nm and the photoproduct absorption between 515 and 580 nm.

The difference spectra at long delays are in reasonable agreement with previous measurements made with substantially longer pulses (Shichida 1984, Monger 1979, Spalink

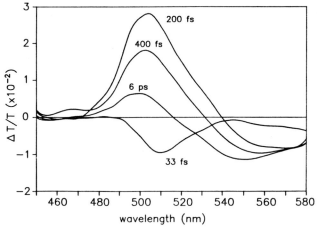

Fig. 3. Difference spectra measurements of *11-cis* rhodopsin at various delays following a 35 fs pump pulse at 500 nm (~10 fs probe).

1983). Difference spectra at 6 ps delay (Fig. 3) indicate an isosbestic (~515 nm) which is within the range of values previously reported. Although our measurements of the differential absorption maxima (~500 nm and ~550 nm) are slightly shifted from what others have observed, this may be due to measurement uncertainty or the shorter wavelength (500 nm) of the excitation pulses used here, or both. In addition, the signal ratio $\Delta\alpha_{batho}/\Delta\alpha_{rhodopsin} \approx 3$ measured at 40-ps delay (not shown), is consistent with single-photon excitation (Spalink 1983). Since we use pulses that are shorter than the ground-state-recovery and photoproduct-formation times, and fluences (~400 $\mu J/cm^2$ or ~0.1 photons absorbed per molecule) that are much lower than previous studies, the possibility of multiphoton photolysis is minimized. Finally, because the time scale of our measurements is faster than the photo→bathorhodopsin transition, we assume that the initial photoproduct is the intermediate, photorhodopsin, identified by Shichida et al. (1984).

These femtosecond measurements have temporally resolved the isomerization of the 11-*cis* retinal prosthetic group in rhodopsin. The rapid increase in absorption between 540 and 580 nm shows that the all-*trans* photoproduct is formed in only 200 fs. This confirms earlier suggestions that a rapid non-radiative isomerization of the retinal chromophore is necessary in order to explain the low fluorescence yield of rhodopsin as well as the photosensitivity and high quantum efficiency of the isomerization (Loppnow 1988). Furthermore, the non-exponential kinetics observed at 550 nm and 570 nm (Fig. 2) indicate that the isomerization can not be described by a first-order process. The appearance of the excited-state absorption at ~500 nm at short times is consistent with theoretical models which indicate the presence of an excited-state absorption (Birge 1980, 1990). This absorption rapidly disappears as a result of the rapid isomerization and corresponding changes in oscillator strengths as the chromophore becomes non-planar. The lack of any obvious evidence for stimulated emission from the excited state also indicates that the torsional wavepacket on the excited-state surface rapidly moves out of the Franck-Condon region. In addition, the 100-200 fs oscillatory behavior in the time-resolved measurements at 500 nm, as well as at 535, 550, and 570 nm indicates that non-stationary vibrational states are excited by the short pulses. The vibrational frequency of these oscillations (~135 cm^{-1}) is consistent with known low-frequency torsional modes of rhodopsin (Loppnow 1988), suggesting that excited-state torsional oscillations (Birge 1980, 1990, Weiss 1979) may modulate the appearance of photoproduct. However, impulsive Raman excitation of the rhodopsin ground state can also contribute to coherent vibrational oscillations in these measurements.

In conclusion, we have time-resolved the spectral dynamics of the primary step in vision. Our results indicate that the *cis-trans* isomerization of rhodopsin is essentially complete in only 200 fs and is one of the fastest chemical reactions ever studied. This observation has important implications for the photochemistry of vision. First, 200 fs is faster than typical vibrational dephasing and relaxation times, suggesting that the photochemistry occurs from a vibrationally coherent system. Indeed, we see indications of coherent vibrational oscillations which are rapidly damped (at 570 nm) as the photoproduct is formed. The significance of such vibrational coherence in the photochemistry of vision has been discussed in several theoretical studies (Birge 1980, 1990, Weiss 1979). Second, the speed of the isomerization process contradicts the traditional picture of photochemistry which assumes vibrational relaxation in the excited state followed by partitioning to photoproduct and to reactant. Our results indicate an essentially barrierless transition in the formation of photoproduct, suggesting that the isomerization follows a nonadiabatic potential surface (broken lines in Fig. 1) which reflects the strong coupling between the rhodopsin excited-state and the ground-state of the photoproduct. This presents experimental evidence for a new paradigm for visual photochemistry that may be relevant for a variety of photochemical and photobiological processes.

This work supported by NIH grant EY 02051 (RM) and NSF grant CHE 86-15093 (RM) and by DOE contract DE-AC03-76SF00098 (CS). LAP acknowledges generous support from NIH Postdoctoral Training Grant T32EY07043.

References

Birge R R and Hubbard L M 1980 *J. Am. Chem. Soc.* **102** 2195
Birge R R 1990 *Annu. Rev. Phys. Chem.* **41** 683
Busch G E, Applebury M L , Lamola A A and Rentzepis P M 1972 *Proc. Natl. Acad. Sci. USA* **69** 2802
Doig S J, Reid P J and Mathies R A 1991 *J. Phys. Chem.* **95** 6372
Eyring G et al. 1980 *Biochemistry* **19** 2410
Fork R L, Brito Cruz C H, Becker P C and Shank C V 1987 *Opt. Lett.* **12** 483
Hayward G, Carlsen W, Siegman A and Stryer L 1981 *Science* **211** 942
Kandori H, Shichida Y and Yoshizawa T 1989 *Biophys. J.* **56** 453
Loppnow G R and Mathies R A 1988 *Biophys. J.* **54** 35
Mathies R A, Brito Cruz C H, Pollard W T and Shank C V 1988 *Science* **240** 777
Monger T G, Alfano R R and Callender R H 1979 *Biophys J.* **27** 105
Schoenlein R W, Bigot J Y, Portella M T and Shank C V 1991a *Appl. Phys. Lett.* **58** 801
Schoenlein R W, Peteanu L A, Mathies R A and Shank C V 1991b *Science* **254** 412
Shank C V 1986 *Science* **233** 1276
Shichida Y, Matuoka S and Yoshizawa T 1984 *Photobiochem. Photobiophys.* **7** 221
Spalink J D, Reynolds A H, Rentzepis P M, Sperling W and Applebury M L 1983 *Proc. Natl. Acad. Sci. USA* **80** 1887
Wald G 1968 *Science* **162** 230
Weiss R M and Warshel A 1979 *J. Am. Chem. Soc.* **101** 6131
Yoshizawa T and Wald G 1963 *Nature* **197** 1279

Inst. Phys. Conf. Ser. No 126: Section VIII
Paper presented at Int. Symp. on Ultrafast Processes in Spectroscopy, Bayreuth, 1991

Low temperature reaction dynamics in the primary electron transfer of photosynthetic reaction centers

W. Zinth*, C. Lauterwasser* and U. Finkele

Physik Department der Technischen Universität München,
8000 München Germany

*Present address: Institut für Medizinische Optik der Ludwig-Maximilians-Universität,
8000 München Germany

ABSTRACT: The primary electron transfer in reaction centers of Rhodobacter (Rb.) sphaeroides is investigated as a function of temperature with subpicosecond time resolution. The experimental results indicate that the electron transfer is not thermally activated and that the same transfer mechanisms are active at low temperatures and at room temperature.

1. INTRODUCTION

Photosynthetic conversion of light energy into chemical energy starts via several electron transfer (ET) reactions in pigment protein complexes called reaction centers (RC's). A series of recent experiments have shown that the most rapid electron transfer processes proceed on the time scale of picoseconds /1-7/. During these reactions an electron is transferred from the primary donor (P), a pair of bacteriochlorophyll molecules via a chain of chromophores to a quinone acceptor molecule. From the structural arrangement /8, 9/ of the chromophores in the reaction centers the following reaction path is suggested: Starting at the primary donor P the electron should be transferred via a monomeric bacteriochlorophyll molecule (B), a bacteriopheophytin molecule (H) to a quinone molecule (Q). A number of picosecond experiments addressed the primary reaction in the reaction centers. In these publications it was shown that several picosecond processes occur in the RC's: A process with a time constant of 200 ps is related with the electron transfer from the bacteriopheophytin to the quinone. A faster time constant of about 3.5 ps was attributed to both, the decay of the electronically excited state P* and to the electron transfer to the pheophytin H /2, 3/. The oberservation of an additional time constant of 0.9 picoseconds was taken as an indication that the monomeric bacteriochlorophyll is involved in the primary reaction process /5-7/. Until now the assignment of the fast kinetic constant to a molecular process is still in discussion. Most probable are the three reaction models which are shown in Figure 1 /5-7/.

2. REACTION MODELS

The structural arrangement of the reaction centers strongly suggests the stepwise electron transfer model of Figure 1 a: According to this model the electronically excited state P* of the special pair decays with the time constant of 3.5 ps. Simultaneously an electron is transferred from the special pair to the monomeric bacteriochlorophyll B. The second electron transfer is faster and carries the electron with a time constant of 0.9 ps to the bacteriopheophytin H. Finally the 200 ps process generates the radical pair P^+Q^- where the electron has reached the quinone.

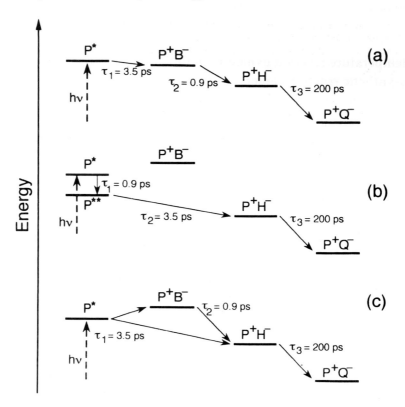

Figure 1: Schematic representation of three reaction models of the primary photosynthetic electron transfer: (a): The stepwise model. (b): unistep superexchange model with a 0.9 ps vibrational relaxation process which leads to a relaxed electronic state P**. (c): branched (parallel) reaction model.

From extensive experimental studies the absorption spectra of the different intermediates of model 1a could be calculated. All these data were fully consistent with the molecular interpretation of a stepwise electron transfer.

In model 1b the fast time constant of 0.9 ps is assigned to an excited state relaxation process of the special pair. Presumably this process is vibrational relaxation from the initially populated Franck-Condon-state. In this reaction model the first electron transfer drives the electron with a time constant of 3.5 ps directly to the bacteriopheophytin H. This fast long-distance electron transfer is only possible if the monomeric bacteriochlorophyll is involved as a virtual intermediate in a superexchange interaction./10-12/. In this case the energy level of the corresponding radical pair P^+B^- is higher than the energy of P*.

When the energy level of state P^+B^- is close to the energy of state P* a branched reaction model becomes possible (see Figure 1c) /12/. In this model two reactions occur in parallel: a direct ET from the special pair P* to the bacteriopheophytin H as well as a stepwise ET via the real intermediate P^+B^-. In this model the 3.5 ps kinetic would reflect the depopulation of the excited special pair while the 0.9 ps time constant is related with the population of the intermediate state P^+B^-. From room temperature experimental data one could conclude that at least 50 % of the reaction centers should use the stepwise reaction path via P^+B^- /5, 6/.

It is the purpose of the present paper to present experimental data on the low temperature reaction dynamics. From the temperature dependence of the observed rates we draw conclusions on the molecular mechanisms of the electron transfer.

3. EXPERIMENTAL

The measurements presented in this paper are obtained on quinone depleted reaction centers from the carotenoid free strain R26.1 of Rb.sphaeroides. Details of the preparation procedures are published in reference /13/. The time resolved absorption experiments are performed using the excite and probe technique with weak subpicosecond pulses (pulse duration $\simeq$ 150 fs) generated by a laser-amplifier-system with a repetition rate of 10 Hz. Details of the experimental system are described in reference /6, 13/. The temporal width of the instrumental response function is below 300 fs.

4. RESULTS

In a first set of time resolved experiments the temperature dependence of the decay of the excited state P* is investigated. In these experiments the transient absorption changes induced by stimulated emission of the radical pair are monitored at a probing wavelength of 920 nm (see Figure 2. At the investigated low temperatures of 25 K the signal closely follows a model function with a single exponential time constant $\tau_1 = 1.4 \pm 0.3$ ps. This time constant as well as the temperature dependence of this time constant is in agreement with the results of previous experimental studies /14/. The most important topic addressed here is the temperature dependence of the fast kinetic component. For this purpose we studied the transient absorption changes at probing wavelengths around 795 nm. In this wavelength range the amplitude of the 3.5 ps kinetic component is very weak and the additonal fast kinetic component is clearly visible at room temperature. In Figure 3 we present the experimental data for a probing wave-

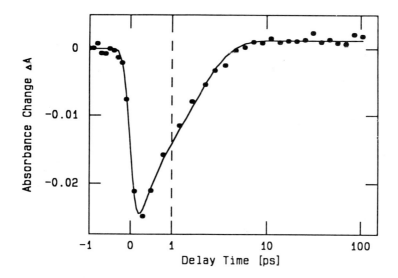

Figure 2: Transient absorption data on RC from Rb. sphaeroides at 25 K. The absorbance change is plotted on a linear scale for delay times t_D < 1ps and on a logarithmic scale for longer delay times. Probing wavelength 920 nm. The signal reflects the decay of the excited electronic level P* which is monoexponential with a time constant of $\tau = 1.4$ ps.

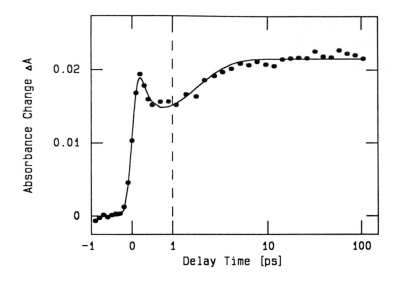

Figure 3: Transient absorption data on RC from Rb. sphaeroides at 25 K. The absorbance change is plotted on a linear scale for delay times $t_D <$ 1ps and on a logarithmic scale for longer delay times. Probing wavelength 794 nm. The transient data show a complex time dependence which can be fitted by a sum of exponential functions with time constants of 0.3 ps and 1.4 ps.

length of 794 nm at a temperature of 25 K. The absorption change rises quickly during the first 100 fs to the pronounced peak. Subsequently an additional fast decay leads to a minimum at t_D = 700 fs. A subsequent slower rise of the absorbance leads to a plateau which is reached after approximately 2 ps. The complex time dependence of the absorbance change excludes the possibility that there is only one, namely the 1.4 ps kinetic component. There must be an additional faster kinetic process which is responsible for the first decay of the absorbance change. From a series of experiments the time constant of this additional kinetic component was determined to be 0.3 ± 0.15 ps. In a set of measurements we have recorded the temperature dependence of the fast kinetic component. These experimental results are summarized in Figure 4: At high temperatures around 300 K the time constant is around 1 ps i. e. we observe a rate of $1 \cdot 10^{12}$ s^{-1}. At lower temperatures a slow rise of the rate constant occurs which accelerates below 100 K. At 25 K a rate constant of 3.3 10^{12} s^{-1} is reached. In Figure 4 the points represent the experimental data; the solid line reflects the results of conventional electron transfer theory /13, 17/. The whole set of experimental results can be summarized as follows: (i) at all temperatures between 300 K and 25 K two time constants are required to explain the experimental data during the first 10 ps.(ii) Qualitatively similar transient absorption features occur at all temperatures (iii) The time constant of the fast kinetic component becomes shorter with decreasing temperature reaching a very small value of t = 0.3 ps at T = 25 K.

5. DISCUSSION

The three reaction models of Figure 1 will now be discussed in the context of the new experimental data. We start with the branched reaction model of Figure 1c. In this model the energy level of state P$^+$B$^-$ should be of the order of 100 cm^{-1} above the level of P* /12/. At very low temperatures this energy difference prevents the population of state P$^+$B$^-$. As a consequence the

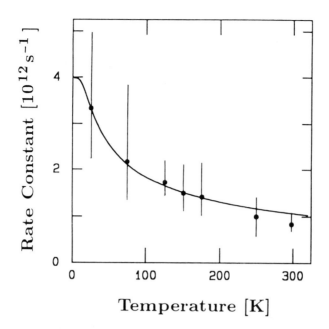

Figure 4: Temperature dependence of the fast rate constant. Points with error bars give the experimental data. The solid line represents the result of conventional electron transfer theory with parameters listed in /13/.

reaction path via P^+B^- is closed and the lifetime of state P^* should become longer. In addition the amplitude of the fast kinetic component related to the decay of state P^+B^- should strongly decrease. Both phenomena are not observed experimentally. Therefore the branched reaction model is quite unlikely. In the superexchange model of Figure 1b the fast kinetic component is related to an S_1 vibrational relaxation. It is well known from a number of publications that vibrational relaxation slows down considerably at low temperatures /15, 16/. Experimentally, however, the fast process (which is related in the superexchange model with a vibrational relaxation) accelerates continuously when lowering the temperature. This is in clear contradiction to the interpretation of the fast step being vibrational relaxation. On the other hand the evaluation of the temperature dependent transient absorption data according to the stepwise model of Figure 1a leads to a fully consistent picture.

In conlusion, we have found that the dynamics of the primary electron transfer in reaction centers of Rb. sphaeroides shows transient absorption changes with two picosecond time constants throughout the whole investigated temperature range from 300 K to 25 K. The existence of the two kinetic processes in this temperature range as well as the temperature dependence of time constants and amplitudes strongly support the idea that the primary electron transfer in reaction centers is a stepwise process via the monomeric bacteriochlorophyll molecule.

ACKNOWLEDGEMENT

The authors thank Professor W. Kaiser for numerous discussions; Professor H. Scheer for providing us with the sample and M. Seel for technical assistance with the low temperature set-up.

REFERENCES

[1]N.W.Woodbury, M.Becker, D.Middendorf, W.W.Parson, Biochem. 24 (1985) 7516.
[2]J.Breton, J.L.Martin, A.Migus, A.Antonetti, A.Orszag, Proc. Natl. Acad. Sci. US 83 (1986) 5121.
[3]J.L.Martin, J.Breton, A.J.Hoff, A.Migus, A.Antonetti, Proc. Natl. Acad. Sci. US 83 (1986) 957.
[4]C.Kirmaier, D.Holten, W.W.Parson, Biochim. et Biophys. Acta 810 (1985) 49.
[5]W.Holzapfel, U.Finkele, W.Kaiser, D.Oesterhelt, H.Scheer, H.U.Stilz, W.Zinth, Chem. Phys. Letters 160 (1989) 1.
[6]W.Holzapfel, U.Finkele, W.Kaiser, D.Oesterhelt, H.Scheer, H.U.Stilz, W.Zinth, Proc. Natl. Acad. Sci. US 87 (1990) 5168.
[7]K.Dressler, E.Umlauf, S.Schmidt, P.Hamm, W.Zinth, S.Buchanan, H.Michel, Chem. Phys. Letters 183 (1991) 270.
[8]J.Deisenhofer, H.Michel, EMBO J. 8 (1989) 2149.
[9]C.H.Chang, D.Tiede, J.Tang, U.Smith, J.Norris, M.Schiffer, FEBS Letters 205 (1986) 82.
[10]M.Plato, K.Möbius, M.E.Michel-Beyerle, M.Bixon, J.Jortner, J. Am. Chem. Soc. 110 (1988) 7279.
[11]R.A.Marcus, Chem. Phys. Letters 133 (1987) 471.
[12]M.Bixon, J.Jortner, M.E.Michel-Beyerle, Biochim. et Biophys. Acta 1056 (1991) 301.
[13]C.Lauterwasser, U. Finkele, H. Scheer, W. Zinth; Chemical Physics Letters 183 (1991) 471.
[14]G.R.Fleming, J.L.Martin, J.Breton, Nature 333 (1988) 190.
[15]T.J.Kosic, R.E.Cline, D.D.Dlott, J. Chem. Phys. 81 (1984) 4932.
[16]W.H.Hesselink, D.A.Wiersma, Chem. Phys. Letters 56 (1978) 227.
[17]M.Bixon, J.Jortner, J. Phys. Chem. 90 (1986) 3795.

Inst. Phys. Conf. Ser. No 126: Section VIII
Paper presented at Int. Symp. on Ultrafast Processes in Spectroscopy, Bayreuth, 1991

Femtosecond absorption studies of 14-fluorobacteriorhodopsin

Makoto Taiji, Krzysztof Bryl, Noriko Sekiya*, Kazuo Yoshihara*, and Takayoshi Kobayashi

Department of Physics, Faculty of Science, University of Tokyo,
   Hongo 7-3-1, Bunkyo-ku, Tokyo 113, Japan
*Suntory Institute for Bioorganic Research,
   Wakayamadai 1-1-1, Shimamoto-cho, Mishima-gun, Osaka 618, Japan

ABSTRACT: We performed pump-probe experiments on 14-fluorobacteriorhodopsin in femtosecond time scale to study the relationship between the electronic structure and the dynamics of the primary photoisomerization process in bacteriorhodopsin system. In the dark-adapted state, the excited state I is converted to the J-intermediate with time constant of 500 fs indicating that the excited-state dynamics is not much different from that of natural bacteriorhodopsin. This means that the driving force of the trans-cis isomerization is the protein-chromophore interaction. In the light-adapted state, the excited-state lifetime is about twice longer than in the dark-adapted one. The longer time constant is explained by interaction with a proton donor near the Schiff base.

1. INTRODUCTION

Photosynthesis in halobacterium is based on the action of bacteriorhodopsin (BR) as a light-driven proton pump. BR has been a subject of intensive investigations (Kobayashi 1988), however, the understanding of the initial events of the BR photocycle and the energy storage process is still limited. The studies using analogs of BR would help us clarify the relationship between the electronic structure and the dynamics of ultrafast structural changes. In this paper, we present the results on femtosecond absorption experiments on the membranes containing 14–fluorobacteriorhodopsin (14-F-BR), reconstituted from 14–fluororetinal and apomembrane. The absorption maximum of dark-adapted 14-F-BR is located at 587 nm, so it is red-shifted by about 30 nm from that of BR. But the opsin shift of 14-F-BR ($4940 \ cm^{-1}$) does not change much from that of BR. There are M- and O-intermediates in the photocycle of 14-F-BR, and it has proton-pump activity. From these facts, in this molecule protein-chromophore interaction and the photochemistry are almost the same as those in BR. However, since a fluorine atom is highly electronegative, we could expect large changes in the electronic structure of the chromophore.

There exist two states of the 14-F-BR, the dark-adapted and the light-adapted states (Tierno 1990). The dark-adapted state consisting of about 50 % all-trans and 50 % 13-cis retinal has similar absorption spectrum to that of BR. However, in the light-adapted state they found the formation of an extra red-shifted pigment near 680 nm, which is thermally unstable and reverts to the dark-adapted state within minutes. Tierno et al.(1990) suggested that the red-shifted pigment has the all-trans, 15-syn structure. Therefore, the light-adapted state contains three different pigments, red-shifted all-trans, all-trans without red-shift, and a small amount of 13-cis.

2. EXPERIMENTAL

An amplified colliding-pulse mode-locked dye laser was used for excitation pulse (630 nm) and a femtosecond continuum for probe pulse (400 ~ 1100 nm) at 10 Hz. The FWHM values of the cross-correlation function between pump and probe were less than 200 fs over the whole spectral region. The 14-fluorobacteriorhodopsin were suspended in distilled water and kept in a 1mm-thick cell, and their optical density were 0.7 at 580 nm. Sample temperature was kept constant by using a Peltier element. For the dark-adapted state experiment, the temperature was set at 18°C to accelerate the dark-adaptation. For the light-adapted state, the temperature was 7°C to keep sample light-adapted. A tungsten-halogen lamp of 100W was used for light-adaptation through a heat-absorption filter (Hoya, HA30) and a bandpass filter (Hoya, G533). The samples were mixed with a stirrer to exchange the sample potion in the excited volume at each shot.

3. RESULTS AND DISCUSSIONS

Figure 1 shows the difference absorption spectra of dark-adapted 14-F-BR at 18°C. The characteristic features are the induced absorption from the S_1 state around 460 nm, bleaching of $S_0 \rightarrow S_1$ absorption around 570 nm, J- and K-intermediate absorption, and very broad transient gain around 800-1100nm. The time-resolved spectra are much the same as those of BR except for the shift, which corresponds to the shift in their steady-state absorption spectra.

The time-dependence of absorbance changes was shown in fig. 2. The signals at 460 nm, 640 nm, and 900 nm correspond to $S_1 \rightarrow S_n$ absorption, J-intermediate absorption, and transient gain, respectively. All of these signal could be fit by a Gaussian-convoluted single-exponential function with a time constant of 500 ± 100 fs, which agrees with that

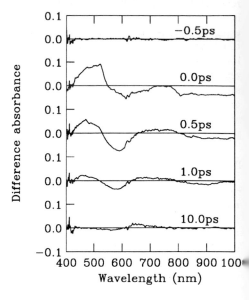

Fig. 1. Difference absorption spectra of dark-adapted 14-F-BR at 18°C.

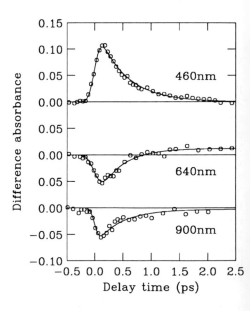

Fig. 2. Time dependence of absorbance change of dark-adapted 14-F-BR. Circle: experimental results, solid line: fitted curve to a single-exponential function with a time constant of 500 fs.

reported for natural BR (Dobler 1988, Mathies 1988, Kobayashi 1991). Therefore, we can assign this time constant as the relaxation time from the relaxed excited state I to the J-intermediate as in BR. We also observed the relaxation process from the Franck-Condon state to relaxed excited state (100 ± 50 fs) and that from the J- to K-intermediates (3 ± 1 ps). Both time constants are consistent with those observed in BR.

From the results, we can conclude that an introduction of a fluorine atom into the chromophore doesn't affect very much the primary photochemical dynamics. In the acidified BR, we found a distinct slowing-down of the relaxation from the lowest singlet excited state (Kobayashi, 1991). We can expect a large change in the protein-chromophore interaction between the acidified BR and natural BR. But in the case of 14-F-BR, the protein-chromophore interaction seems to be quite similar to that in natural BR. Therefore, our results indicate that the driving force of trans-cis isomerization is protein-chromophore interaction and the electronic structure of chromophore is not a key factor for the dynamics of the isomerization. It supports the following idea : In the ground state the all-trans structure is locked by C=C double bond. But in the excited state it is released by bond alternation and the chromophore isomerizes by protein-chromophore interaction.

Figure 3 shows difference absorption spectra of the light-adapted 14-F-BR at 7°C. Again the shape of the difference spectra are quite similar to those of natural BR. However, we found following several differences. First, the spectra are red-shifted. This can also be explained by the shift of steady-state absorption like in the dark-adapted state. Second, as can be noticed in the transient spectrum at 1 ps, the decay time is obviously longer for light-adapted 14-F-BR over the whole spectrum. Third, both of the induced absorption due to J and K were weak.

The slow relaxation is clearly seen from the

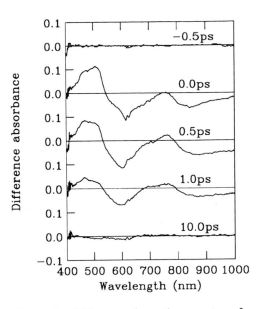

Figure 3. Difference absorption spectra of light-adapted 14-F-BR at 7°C.

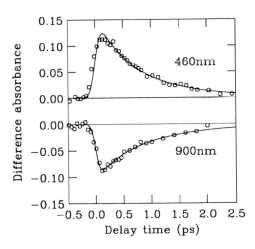

Figure 4. Time dependence of absorbance change of light-adapted 14-F-BR. Circle: experimental results, solid line: fitted curve to the sum of two exponential functions with time constants of 500 fs and 1 ps.

plot of the absorbance change against delay time as shown in fig. 4. When we fit the decay curve to a gaussian-convoluted single-exponential function, time constant was determined at 900 ± 100 fs for both decay curves. The contribution from the extra red-shifted pigment dominates these signals, but it is necessary to take into account the small contribution from the non-red-shifted pigments. If we assume that the decay kinetics can be described by the sum of two exponential functions with different time constants, the shorter time constant was determined at 500 fs as observed in the dark-adapted state and the longer time constant of 1.0 ± 0.2 ps could be obtained from the fit. This is also consistent with the result of the single exponential fit, so we can safely conclude that the relaxation of the excited-state takes place with a time constant of about 1 ps in the extra red-shifted pigment of light-adapted 14-F-BR. Since the induced absorption of J- and K-intermediate is weak, the quantum efficiency of the isomerization may be smaller in the red-shifted pigment.

A change in the protein and the Schiff base configuration in the extra red-shifted pigment structure causes a change in protein-chromophore interaction which determines the dynamical behavior of BR. It is difficult to definitely chemical structure of the intermediates and/or excited states which induce the observed change in the light-adapted state by transient (electronic) absorption spectrum. But a possible explanation is the electrostatic interaction with the proton donor near the Schiff base. If it has all-trans, 15-syn structure, the direction of the protonated Schiff base is the opposite to that of the natural BR in the light-adapted state (all-trans, 15-anti). In this case, the proton source near the Schiff base can interact with the fluorine atom via hydrogen bonding. As a result the all-trans, 15-syn structure is stabilized and hence the isomerization rate is reduced.

ACKNOWLEDGEMENTS

The authors wish to thank Dr. Masayuki Yoshizawa and Mr. Satoshi Takeuchi in their help on preparing the apparatus.

REFERENCES

Kobayashi T 1987 *Primary Processes in Photobiology* ed Kobayashi T (Berlin: Springer-Verlag) and references therein
Tierno M E, Mead D, Asato A E, Liu R S H, Sekiya N, Yoshihara K, Chang C-W, Nakanishi K, Govindjee R, and Ebrey T G 1990 *Biochemistry* **29** pp 5948-5953
Dobler J, Zinth W, Kaizer W, and Oesterhelt D 1988 *Chem. Phys. Lett.* **144** pp 215-220
Mathies R A, Brito Cruz C H, Pollard W T, and Shank C V 1988 *Science* **24** pp 777-779
Kobayashi T, Terauchi M, Kouyama T, Yoshizawa M, and Taiji M, 1991 *SPIE Vol. 1403 Laser Applications to Life Sciences* pp 407-415

Picosecond time-resolved resonance CARS of bacteriorhodopsin

George H. Atkinson and Laszlo Ujj
Department of Chemistry and Optical Science Center
University of Arizona
Tucson, Arizona 85721

ABSTRACT: Coherent antistokes Raman (CARS) spectra recorded under resonance conditions are presented for species appearing during the bacteriorhodopsin photocycle. Picosecond resonance CARS spectra for BR-570 (light-adapted) and BR-548 (dark-adapted) show changes in the C-C stretching region which can be associated with retinal isomerization (all-*trans* vs 13-*cis*). Picosecond time-resolved resonance CARS (PTR/CARS) data are presented which reflect a distinct retinal structure in the photocycle intermediate K-590. Applications of PTR/CARS in the study of biochemical systems are discussed.

1. INTRODUCTION

The reactive processes that occur on the picosecond time scale within membrane proteins are of fundamental importance in the elucidation of the molecular mechanism(s) which underlie biochemical function. The actual biochemical function occurring in membrane proteins varies greatly: enzymatic cascade in the visual processes involving rhodopsin [1], proton pumping and ATP synthesis in bacteriorhodopsin (BR) [2], and light harvesting and electron transfer in photosynthetic antenna proteins and reaction centers [3]. Many aspects of the kinetics describing these biochemical reactions have been determined from femto/picosecond time-resolved absorption spectroscopy in which the changes in electronic ground-state properties are monitored. Time-resolved fluorescence measured on the same time scales provides another view of the dynamics since, in combination with transient absorption data, it can be utilized to characterize excited electronic state properties including those involving the surrounding protein environment.

Absorption and fluorescence measurements, however, can not be directly interpreted in terms of alterations in molecular structure of the type that are well-known to control many photosynthetic biochemical reactions [1,2]. For example, in retinal-containing proteins (rhodopsin and BR), isomeric and conformational changes are central to the molecular mechanism(s) [1,2]. To monitor structural transformations occurring on the picosecond time scale, vibrational spectra of transient intermediates have been shown to be especially useful [4,5].

Picosecond time-resolved resonance Raman (PTR³) spectroscopy (utilizing spontaneous Raman scattering) has been used widely to record vibrational spectra of transient species and populations in membrane proteins such as BR [4-10]. PTR[3] techniques encompass many advantages including time resolution of ≈ 2 ps, resonance enhancements which generate spontaneous Raman scattering from only a well-defined portion (e.g., retinal chromophore) of a complex molecular system (e.g., membrane protein), and spectral features which have lineshapes, band maxima positions, and intensities that are readily interpretable in terms of the vibrational degrees of freedom. As effective as PTR[3] spectroscopy has proven to be, it has limitations. For example, even modest fluorescence backgrounds can obscure the normally weak resonance Raman signal. Furthermore, since Raman scattering appears isotopically over 4π radians, the efficiency with which it can be optically collected and focus onto the entrance slit of a spectrometer is well below unity. The resultant signal to noise ratio thereby, remains dependent on optimizing other experimental variables (e.g., Raman cross sections, laser powers, and transient concentrations).

An alternative method for obtaining time-resolved vibrational information involves coherent resonance Raman scattering[1]. Of specific interest here is the use of picosecond time-resolved resonance coherent antistokes Raman scattering (PTR/CARS) in the study of BR. In general, CARS has been experimentally difficult to use for condensed phase samples such as the water suspensions of BR membranes because of strong contributions from nonresonant background scattering [11,12]. In addition, the resonant effects are complicated in time-resolved experiments since resonance enhancements may change as transient absorption in the sample varies with the formation and decay of different species and populations. CARS spectra also are inherently difficult to quantitatively analyze and interpret. For example, CARS lineshapes are derived from the third-order nonlinear susceptibility $(\chi^{(3)})$ which scales with the square of the concentration of the scatterer and which can produce interference terms involving adjacent CARS bands. The quantitative analysis of PTR/CARS data in terms of the molecular structure of specific species and the dynamics which underlies these reactions must consider all of these factors.

In spite of these issues, PTR/CARS offers significant experimental advantages especially for applications to the study of biophysical systems. Fluorescence interference is absent since it remains substantially weaker than the coherently generated CARS signals. The PTR/CARS signal appears as a coherent beam which can be spatially separated from the laser beams used to generate CARS in the sample and from the fluorescence [11,12]. By using a broad band laser in the generation of PTR/CARS (i.e., Stokes beam), an ≈ 350 cm^{-1} region of the vibrational Raman spectrum can be recorded simultaneously (with multichannel detectors) [13,14]. The resultant signal to noise ratio in the vibrational spectrum is $>10^3$ larger than corresponding PTR[3] data obtained from similar samples. Time-resolved CARS measurements offer at least one. additional advantage which has not yet been fully exploited. CARS signals can be experimentally differentiated via transient polarization changes appearing in the sample during the reaction.

----

[1] Transient infrared absorption data recorded with picosecond time resolution is yet another increasingly effective method for obtaining vibrational spectra that is not treated in this paper.

2. EXPERIMENTAL

The instrumentation used to record picosecond CARS and PTR/CARS is shown schematically in Fig. 1.   Although a detailed description of the instrumentation and procedures is given elsewhere [13], several points should be made here briefly.   Alignment of the three laser beams (excitation pulse used to initiate the photocycle and the two pulses used to generate CARS) is optimized by monitoring the folded BOXCARS signal [13]. The laser beams are focused ($\approx 20\mu m$) into a flowing (10 m/s) planar jet (≈ 400 μm) of BR. The exposed sample volume is exchanged between sets of laser pulses (150 KHz repetition rate and 7 ps pulsewidths FWHM).

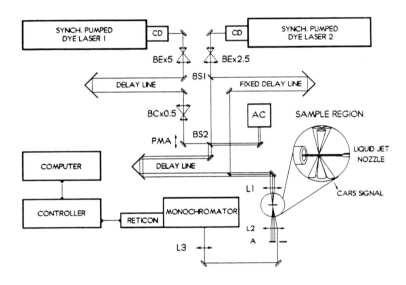

Fig. 1.   Schematic of instrumentation used to record PTR/CARS signals

3. RESULTS AND DISCUSSION

In its initial state before photo-excitation, BR contains two different configurational forms of retinal (13-*trans* and 13-*cis*) [2]. The relative concentrations of the two retinal isomers depend on the light conditions to which the sample is exposed. With normal room light (light-adapted), the retinal chromophore assumes an all-*trans* configuration while in the dark (dark-adapted), a 50%/50% mixture of 13-*trans* and 13-*cis* is obtained.   The resonance CARS spectrum of BR-570, the stable form produced by light adaptation, is recorded with low energy (to minimize photochemistry), picosecond pulses. The C-C stretching region of this spectrum is presented in Fig. 2. Analogously, the low-power, picosecond resonance CARS spectrum of a dark-adapted BR sample containing equal amounts of BR-570 and BR-548 (13-*cis* retinal) is obtained.   The C-C stretching portion of that spectrum also is shown in Fig. 2.

Clearly, significant differences are evident in the two spectra. Since both are resonantly enhanced through an electronic transition in retinal, these CARS features reflect differences in the vibrational degrees of freedom between 13-*trans* and 13-*cis* retinal. In spontaneous resonance Raman scattering, the C-C stretching bands have been shown to be sensitive to the isomeric form of retinal and therefore, resonant CARS data also might be anticipated to be directly sensitive to changes in the isomeric structure of retinal. The comparison made in Fig. 2 of light-adapted and dark-adapted BR samples confirms that the C-C stretching region of resonance CARS spectra can be used to distinguish 13-*trans* from 13-*cis* retinal in BR.

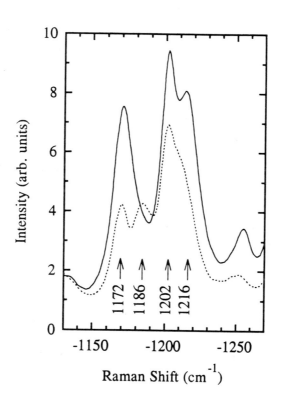

Fig. 2 Picosecond CARS spectra of light-adapted BR (solid trace, BR-570 only) and dark-adapted BR (dotted trace, equal amounts of BR-570 and BR-548 [1,2]) in the C-C stretching region.

Resonance CARS spectra of transient intermediates appearing within the initial 50 ps of the BR photocycle are recorded by PTR/CARS using the instrumentation shown in Fig. 1. Specifically, data for the K-590 intermediate, which is found by PTR[3] spectroscopy to have a 13-*cis* retinal [2,4-8], are obtained. PTR/CARS data in the 800 cm^{-1} to 1400 cm^{-1} region are presented in Fig. 3 for the first 50 ps of the BR photocycle. The continuous change in the relative intensities of CARS features apparent in Fig. 3 reflects partially the recovery of the ground-state population in BR-570 following optical excitation and partially the increasing concentration of K-590. To separate the BR-570 and K-590 contributions to these PTR/CARS spectra, an analysis which quantitatively accounts for the $\chi^{(3)}$ values for each species is needed [11-13].

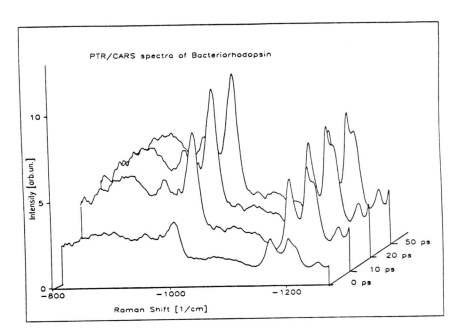

Fig. 3 PTR/CARS spectra (800 cm^{-1} to 1400 cm^{-1} region) of retinal during the initial 50 ps of the BR photocycle.

The CARS features themselves can be fit well using lineshapes obtained directly from the $\chi^{(3)}$ expressions [13]. This part of the analysis assumes that the nonresonant background can be accurately measured and subtracted from the corresponding PTR/CARS data. This latter step is achieved by recording nonresonant CARS data from the water solvent and the bleached (retinal chromophore removed) BR membrane [13,14]. Preliminary results indicate that such quantitative analysis of PTR/CARS data is feasible [13,14]. The confirmation and application of such analyses remains a central issue in applying PTR/CARS in biochemical systems.

4. CONCLUDING REMARKS

Based on the data presented here, it is now evident that high quality (e.g., signal to noise ratio) picosecond CARS and PTR/CARS spectra of BR can be obtained using high repetition rate (150 KHz), low-energy per pulse (2-10 nJ), picosecond (7 ps FWHM) excitation. The incorporation of broad-band laser excitation in the generation of CARS together with the utilization of multichannel detectors also facilitates recording these data. Given quantitative fitting of lineshapes, the resultant CARS spectra can be analyzed in terms of both molecular structure changes and reaction dynamics in biochemical systems such as the membrane protein BR.

5. ACKNOWLEDGEMENT

The authors wish to thank Dr. T. Brack and Mr. D. Gilmore for many helpful discussions and their technical assistance during the course of this work. This research is supported by a grant from the biophysics program of the National Science Foundation.

6. REFERENCES

1. Stoeckenius, W., **Acc. Chem. Res.** 13 (1980) 337.
2. Birge, R.R. **Annu. Rev. Biophys. Bioeng.** 10 (1981) 315.
3. Biochemistry and Physiology of Visual Pigments Langer, H., ed., Springer Verlag, Berlin, (1973).
4. Time-Resolved Vibrational Spectroscopy, G.H. Atkinson ed. (Academic Press, New York) 1983.
5. Time-Resolved Vibrational Spectroscopy, M. Stockburger and A. Laubereau eds. (Springer-Verlag, Berlin) 1985.
6. Atkinson, G.H., Brack, T.L., Blanchard, D., Rumbles, G., and Siemankowski, L. In: Ultrafast Phenomena V. eds. Fleming, G.R. and Siegman, A.E. (Springer Verlag, Berlin, 1986), p. 409.
7. Atkinson, G.H., In: Primary Processes in Photobiology, ed. T. Kobayashi (Springer-Verlag, Berlin, 1987), p. 213.
8. Atkinson, G.H., Brack, T.L., Blanchard, D., and Rumbles, G., **Chem. Phys.** 131 (1989), p. 1.
9. Atkinson, G.H. and Brack, T.L., **Chem. Phys.** 131 (1989), p. 1.
10. Brack, T.L. and Atkinson, G.H., **J. Molec. Struct.** (In press).
11. Coherent Raman Spectroscopy, G.L. Eesley (Pergamon Press, Oxford) 1981.
12. Advances in Non-Linear Spectroscopy, R.J.H. Clark and R.E. Hester eds. (John Wiley, New York) 1988.
13. Ujj, L. and Atkinson, G.H. **Chem. Phys.** (submitted).
14. Atkinson, G.H. and Ujj, L. **Proc. Natl. Acad. Sci. USA** (in preparation).

Primary processes in isolated photosynthetic bacterial reaction centers from *chloroflexus aurantiacus* studied by picosecond fluorescence spectroscopy

Alfred R. Holzwarth   Kai Griebenow   Marc G. Müller

Max–Planck–Institut für Strahlenchemie, Stiftstraße 34–36, D–4330 Mülheim, Germany

ABSTRACT: The stationary and time–resolved fluorescence emission spectra of reaction centers (RCs) isolated from the thermophilic phototrophic bacterium Chloroflexus aurantiacus strain Ok–70–fl show several lifetime components in the picosecond and nanosecond time range. A short–lived ~3 ps component is related to an energy transfer process. Two further short–lived components with positive amplitudes and lifetimes of ~7 ps and ~18 ps were also resolved. They represent the charge separation times of open RCs (7 ps from P^*HQ_A) and closed RCs (18 ps from $P^*HQ_A^-$).

1. Introduction

The primary processes of energy transfer and charge separation in isolated reaction centers (RCs) of photosynthetic bacteria have recently been the subject of intense investigations. With very few exceptions most studies concentrated on the RCs of purple bacteria, notably those of Rps. viridis and Rb. sphaeroides (for reviews see Parson 1991, Michel and Deisenhofer 1988). There exists an important difference between purple bacterial RCs and C. aurantiacus RCs with respect to the pigment composition. While, in addition to P, the former contain one BChl–a monomer (also called accessory BChl) and one BPheo each per branch, the latter contain a BChl–a monomer only in the L–branch while in the M–branch the BChl–a monomer is replaced by a BPheo–a pigment in the corresponding position (Thornber et al. 1983). Thus C. aurantiacus RCs contain altogether three BPheo molecules and only one BChl–a monomer.

Basically all of the ultrafast studies on isolated RCs have been carried out employing femtosecond/picosecond transient absorption techniques to date. From these measurements there exists a general agreement that the primary charge separation step in RCs of C. aurantiacus occurs in about 7 ps (Becker et al. 1991). Considerable controversy exists in the literature as to the nature of this primary electron transfer step. According to one model, also known as the superexchange mechanism (Breton et al. 1986), the electron is transferred in a single step directly from P^* to $BPheo_L$. In contrast, based on the observation of a weak ultrafast component of ~1 ps in addition to the ~3 ps component, a consecutive two–step electron transfer has been invoked for purple bacterial RCs (Holzapfel et al. 1989). The first step of ~3 ps in this model was assigned to electron transfer from P^* to monomeric

BChl$_L$, while the faster component was attributed to a secondary electron transfer step from BChl$_L^-$ to BPheo$_L$. A third interpretation assumes a heterogeneity in the rate constants of the primary electron transfer step which would lead to a distribution of lifetimes centered around ~3 ps (Kirmaier and Holten 1990). Accordingly the ~1 ps component would represent part of this distribution for the direct P*→BPheo$_L$ electron transfer step.

2. Materials and Methods

Cell growth and membrane and RC isolation have been carried out as described (Müller et al. 1991). Prior to fluorescence measurements the RCs were diluted with 20 mM Tris–HCl, pH 8.0, 0.1% LDAO to an absorbance of 0.5–1.0/cm at 813 nm. The samples contained sodium ascorbate (10 mM) and PMS (10 μM) for measurement. RCs were pumped with a Gilson peristaltic pump (fast speed about 20 ml/min) from a reservoir that was kept in the dark to obtain open RCs (Q$_A$ oxidized) and with minimal pump speed (~2 ml/hour) for closed RCs (quinone Q$_A$ reduced). In the latter case the RCs were closed photochemically by the fluorescence excitation beam. Fluorescence decays were measured by the single–photon–timing (SPT) technique as described with an apparatus function of ~30 ps (FWHM) (Müller et al. 1991). The fluorescence decays were analyzed either by single–decay analysis or by global analysis methods employing a sum of exponentials model function in either case (Holzwarth et al. 1987).

3. Results

Fluorescence decays of C. aurantiacus RCs have been measured at room temperature varying excitation wavelength, laser intensity, as well as sample pumping speed. For all these conditions time–resolved spectra have been taken typically based on decay measurements at 7 different emission wavelengths. Fig. 1 shows a typical example of the DAS of C. aurantiacus RCs upon excitation at 812 nm and fast pumping of the sample in order to try to keep RCs open. With completely free–running parameters a minimum of 5 exponential components was required to achieve a good fit. The lifetimes are τ_1~3 ps, τ_2~13 ps, τ_3~30 ps, τ_4~144 ps and τ_5~1.7 ns. The two longest–lived components had very small but significant amplitudes. A characteristic feature of the fastest component τ_1 is a positive amplitude at short emission wavelengths (decay term) and a negative amplitude (rise term) at long emission wavelengths. All other components show only positive amplitudes in their DAS with a maximum in the range 900 – 920 nm. Without the 3 ps component a reasonable fit to the data could not be obtained as judged from the residual plots. However, even with 5 components, the residual plots were not ideal in many cases. Upon further increasing the number of components to six a reasonable fit could not be obtained using completely free–running parameters for a single spectrum but this was was possible when all decays were combined in a global analysis.

Time–resolved fluorescence measurements with closed RCs were carried out with very slow sample pumping speed ensuring that there was no significant decrease in total fluorescence intensity as compared to the intensity observed without pumping. Five lifetimes are required

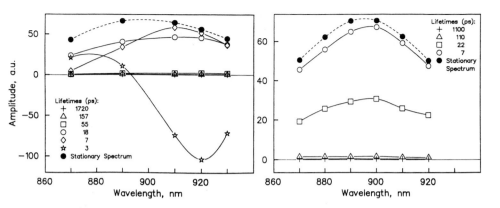

<u>Fig. 1 (left)</u>: Decay–associated fluorescence spectra of RCs from <u>C. aurantiacus</u> calculated by global–analysis from decays measured at rapid sample pumping speed (conditions for "open" RCs) and 812 nm excitation. Data with the 18 ps lifetime fixed in the analysis. Essentially the same results are obtained from a combined analysis of all time–resolved spectra taken under varying excitation intensities.
<u>Fig. 1 (right)</u>: λ_{exc}=850 nm.

in a fit with all parameters free running. These lifetimes are ~3 ps, ~16 ps, ~44 ps, ~280 ps and 5.3 ns.

Fig. 1 (right) shows the DAS resulting from a measurement with fast pumping using an excitation wavelength of 850 nm, i.e. selective excitation of the special pair to P*. Four lifetime components are required for a good fit with lifetimes of ~7 ps, ~22 ps, ~110 ps and 1.1 ns in a free–running parameter fit. There are no indications in the residuals necessitating an additional component. In particular no 3 ps component is required.

4. Discussion

The 7 ps and ~18 ps components

The 7 ps and 18 ps components can be best resolved in the decays measured with λ_{exc}=850 nm excitation since under these conditions the 3 ps component is absent (c.f. Fig. 1). At the fast sample pump rate both components are present, but their amplitude ratios vary depending on the laser intensities (not shown). Decreasing laser intensity increases the amplitude of the 7 ps–component and in parallel decreases the amplitude of the 18 ps–component. Both components have similar DAS both with respect to shape and spectral maximum. These (and other) findings suggest that the increase in amplitude of the 18 ps component and in total fluorescence reflect the closing of RCs, i.e. reduction of Q_A. We thus interpret the ~18 ps lifetime as the charge separation time of closed RCs (Q_A reduced). Correspondingly the ~7 ps component should represent the charge separation time for open

RCs (Q_A oxidized). The increase in the time constant for the primary step in electron transfer is explained by the repulsive interaction of the negative charges on Q_A^- and $BPheo^-$.

Does one of the components reflect $BChl_L$–reduction?

A sequential two step electron transfer mechanism has been proposed for RCs of purple bacteria (Holzapfel et al. 1989) as well as for C. aurantiacus RCs (Shuvalov et al. 1983, 1986a). We have checked whether any of our resolved lifetime components could reflect an electron transfer process including $BChl_L^-$ as an intermediate. Clearly, the strong 3 ps component can be excluded as a candidate. Our data do not agree with the interpretation of Shuvalov et al. (1986b) who assigned the ~3 ps component to the secondary electron transfer step $BChl_L^- \rightarrow BPheo_L$. We thus would be left with one of the 7 ps and 18 ps components as candidates for such a model. A strong argument against such an interpretation is the widely varying amplitude ratio of these two components upon varying the laser intensity. Such a behaviour supports the open/closed RC interpretation proposed above, but would be inconsistent with a sequential electron transfer model involving both the 7 ps and the 18 ps components. Our data clearly do not exclude a sequential electron transfer model where a second step would be faster than ~3 ps (Holzapfel et al. 1989).

5. References

Becker,M., Nagarajan,V., Middendorf,D., Parson,W.W., Martin,J.E., and Blankenship,R.E., 1991.
    Biochem. Biophys. Acta 1057: 299–312.
Breton,J., Martin,J.–L., Migus,A., Antonetti,A., and Orszag,A., 1986.
    Proc. Natl. Acad. Sci. USA 83: 5121–5125.
Holzapfel,W., Finkele,U., Kaiser,W., Oesterhelt,D., Scheer,H., Stilz,H.U., and Zinth,W., 1989.
    Chem. Phys. Lett. 160: 1–7.
Holzwarth,A.R., Wendler,J., and Suter,G.W., 1987. Biophys. J. 51: 1–12.
Kirmaier,C. and Holten,D., 1990. Proc. Natl. Acad. Sci. USA 87: 3552–3556.
Michel,H. and Deisenhofer,J., 1988. Pure Appl. Chem. 60: 953–958.
Müller,M.G., Griebenow,K., and Holzwarth,A.R., 1991. BBA, in press.
Parson,W.W., 1991. In: The Chlorophylls, (H. Scheer,ed). CRC Press,
    Boca Raton , CRC Handbook.
Shuvalov,V.A. and Klevanik,A.V., 1983. FEBS Lett. 160: 51–55.
Shuvalov,V.A., Amesz,J., and Duysens,L.N.M., 1986.
    Biochim. Biophys. Acta 851: 327–330.
Shuvalov,V.A., Vasmel,H., Amesz,J., and Duysens,L.N.M., 1986.
    Biochim. Biophys. Acta 851: 361–368.
Thornber,J.P., Cogdell,R.J., Pierson,B.K., and Seftor,R.E.B., 1983.
    J. Cell. Biochem. 23: 159–169.

Inst. Phys. Conf. Ser. No 126: Section VIII
Paper presented at Int. Symp. on Ultrafast Processes in Spectroscopy, Bayreuth, 1991

Time-resolved fluorescence studies of isolated photosynthetic reaction centers from *chloroflexus aurantiacus* at low temperatures

Mathias Hucke   Gerd Schweitzer   Kai Griebenow   Marc G. Müller
Alfred R. Holzwarth

Max-Planck-Institut für Strahlenchemie, Stiftstrasse 34, D-4330 Mülheim a.d. Ruhr, Germany

ABSTRACT: The picosecond fluorescence kinetics of closed (quinone acceptor Q_A reduced) reaction centers isolated from the phototrophic bacterium *C. aurantiacus* shows time constants of ≈ 20 ps and ≈ 300 ps (amplitude ratio $\approx$ 1:1). They are nearly independent of temperature (7K to 80K). Assuming a three-state kinetic model, two assignments of the kinetics to the electron transfer processes seem to be possible. One model includes a reversible electron transfer between P^* and H_M, the pheophytin(s) in the M-branch. The other model involves the formation of a monomeric bacteriochlorophyll anion.

1 Introduction

The electron transfer (ET) kinetics in bacterial reaction centers (RCs) remains one of the most intriguing subjects in photosynthesis. RCs of purple bacteria as well as those of the green bacterium *Chloroflexus aurantiacus* contain two pigment-protein branches, denoted L and M (Michel and Deisenhofer 1988). Besides the bacteriochlorophyll (BChl) special pair P, purple bacterial RCs contain a monomeric BChl (B) and a bacteriopheophytin (H) in each of the two cofactor branches. In *C. aurantiacus* RCs the BChl in the M-branch (B_M) is replaced by another bacteriopheophytin. Both types of RC contain a quinone (Q_A) in the L-branch as further electron acceptor. Available data (see Parson (1991) for a review) indicate that ET is proceeding virtually exclusively in the L-branch from P^* via H_L to Q_A in both types of RC. This notion of unidirectionality rests mainly on the interpretation of absorption kinetics. Only few picosecond fluorescence studies of RCs have been reported so far (Freiberg et al. 1985). In this work we present time-resolved fluorescence measurements with Q_A in the reduced state, i.e. with so-called "closed RCs". We expect this to provide favorable conditions for observing possible charge separation processes on the hitherto considered "inactive" M-branch.

2 Materials and Methods

The RCs from *C. aurantiacus* strain Ok-70-fl have been prepared using two DEAE anion exchange columns as described in (Müller et al. 1990). The RCs were diluted with 20 mM Tris-HCl, pH 8.0, 0.1% LDAO to an absorbance of 0.5-1.0/cm at 813 nm and added 10 mM of sodium ascorbate. After adding o-phenantroline, 1 mM, as electron transfer inhibitor, samples were put into a 0.5 mm thick cuvette, which was rapidly cooled to 80K under white light illumination in order to reduce the Q_A. Identical results were obtained, when Q_A was reduced chemically by dithionite.

The exciting laser pulses came from a dye-laser (Styryl 9) synchronously pumped by the compressed (3 ps) and frequency-doubled pulses of a Nd:YAG laser (82 MHz). The autocorrelation width of the pulses was typically 600 fs at 813 nm. The sample was excited with pulses of ≈ 5 pJ/pulse/mm^2. Fluorescence from the sample was focussed into the entrance slit of a synchroscan streak camera system (Hamamatsu M1955 with IR-intensified cathode N3296). The overall time response function was typically ≈ 15 ps (FWHM). Fluorescence of RCs was detected at 900 nm (interference filter, FWHM 20 nm and cutoff filters Schott RG830 and RG850). We checked the time resolution after deconvolution using our standard analysis programs (Schatz et al. 1988) by measuring the fluorescence decay (λ_{exc}=590 nm) of malachite green (10^{-5}M) in methanol ($\tau_f = 1.6$ ps) and ethanol ($\tau_f = 2.4$ ps).

3 Results

Fluorescence decays of closed RCs (Q_A reduced) have been measured in the temperature range from 80K to 7K. An optimum fit to the decay curves was achieved using a two exponential model function at all temperatures studied. The two lifetimes, which are in the range of approx. 20 ps and 300 ps, varied only moderately with temperature, while the corresponding dependency of their relative amplitudes was stronger. The data are summarized in Fig.1.

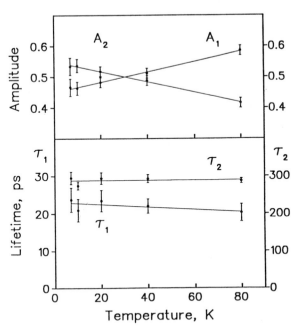

Fig.1.:
Temperature dependence of the lifetimes τ_1 and τ_2 and their amplitudes A_1 and A_2; bars: standard deviation from several measurements; straight line: linear fit.

We did not find any indications for lifetime components in the range of some ps, although the streak camera would have had sufficient time resolution. The absence of such a component also indicates that RCs are closed under our conditions, since open reaction centers show a lifetime of ≈ 3 ps at low temperatures (Becker et al. 1990).

4 Discussion

The fluorescence around 900 nm in RCs of *C. aurantiacus* is emitted from the excited special pair P* (Vasmel et al. 1983). We tested several kinetic models, which allow the calculation of rate constants and ΔG-values for the ET processes.

The fast component ($\approx$ 20 ps) could in principle be assigned to primary charge separation on the L-branch resulting in a $P^+H_L^-Q_A^-$ state and the long-lived ($\approx$ 300 ps) component to a charge recombination process repopulating P* . At cryogenic temperatures and in view of the approx. 1:1 amplitude ratio of the two components this would require a free energy gap ΔG $(P^*H_LQ_A^- - P^+H_L^-Q_A^-) \approx 0$, which does not agree with the known ΔG-values of about -100 to -200 meV (Müller et al. 1990, Ogrodnik et al. 1988). So we can not expect to observe charge recombination fluorescence from $P^+H_L^-Q_A^-$ with substantial amplitudes at low temperatures. We limit the discussion to two kinetic models, which both seem to result in physically reasonable predictions. Their parameters are given in Table 1 for the lowest and highest measured temperatures of 7K and 80K, respectively.

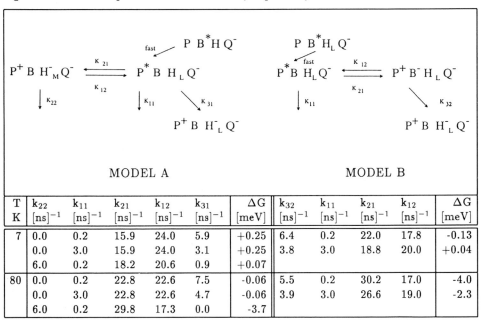

T	k_{22}	k_{11}	k_{21}	k_{12}	k_{31}	ΔG	k_{32}	k_{11}	k_{21}	k_{12}	ΔG
K	$[ns]^{-1}$	$[ns]^{-1}$	$[ns]^{-1}$	$[ns]^{-1}$	$[ns]^{-1}$	[meV]	$[ns]^{-1}$	$[ns]^{-1}$	$[ns]^{-1}$	$[ns]^{-1}$	[meV]
7	0.0	0.2	15.9	24.0	5.9	+0.25	6.4	0.2	22.0	17.8	-0.13
	0.0	3.0	15.9	24.0	3.1	+0.25	3.8	3.0	18.8	20.0	+0.04
	6.0	0.2	18.2	20.6	0.9	+0.07					
80	0.0	0.2	22.8	22.6	7.5	-0.06	5.5	0.2	30.2	17.0	-4.0
	0.0	3.0	22.8	22.6	4.7	-0.06	3.9	3.0	26.6	19.0	-2.3
	6.0	0.2	29.8	17.3	0.0	-3.7					

Tab.1.: Rate constants k_{ij} (fit results) and free energy differences ΔG for 7 and 80 K.

Model A involves charge separation and charge recombination processes with rates of $\approx$ 20ns^{-1} on the M-branch to and from a state $P^+H_M^-Q_A^-$ which would be nearly isoenergetic with P* (ΔG between $\approx$ 0 and - 4 meV). In this model the charge separation on the L-branch, presumably to $P^+H_L^-Q_A^-$, is quite slow, i.e. it has a maximal rate of $\approx$ 6 ns^{-1}.

Model B suggests a reversible electron transfer process on the L-branch from P* with rates of $\approx$ 20 ns^{-1} to the nearly isoenergetic state $P^+B_L^-H_LQ_A^-$(ΔG between $\approx$ 0 and -4 meV), which in turn decays to $P^+B_LH_L^-Q_A^-$ with a rate of $\approx$ 6 ns^{-1}. This corresponds formally to the consecutive two step electron transfer model proposed by Zinth et al.(1989) for purple bacterial RCs and Shuvalov et al.(1986) for *C.aurantiacus* RCs. However, the rate constant k_{32} is several times smaller than k_{21} in contrast to the analogous rates at room temperature in open purple bacterial RCs (Zinth et al. 1989, Shuvalov et al. 1986).

With respect to model A we suggest that our experimental conditions, i.e. reduced

Q_A, would favour the occurance of ET processes on the M-branch. A schematic scheme illustrating the expected influence of Q_A^- on potential energy curves is shown in Fig.2. Due to the short distance between Q_A^- and H_L^- (Michel and Deisenhofer 1988), a strong influence of the negatively charged Q_A^- on the $P^+H_L^-$ state is expected.

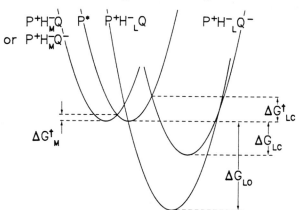

Fig.2.:
Schematic potential energy diagram; The ΔG-values are labelled with respect to ET on the L- or M-branch and with respect to (o)pen or (c)losed states. The ΔG†-values are related to activation energy barriers.

This interaction will shift the free energy curve up, creating an activation barrier for ET. Thus ET will be slowed down in the L-branch, especially at cryogenic temperatures. In contrast, the interaction of Q_A^- with the radical pair on the M-branch ($P^+H_M^-$ state) should be substantially smaller due to the larger distance to Q_A^-.

Both models formally fit the data equally well and appear physically reasonable. Either of these alternatives gives interesting insights into the effect of Q_A reduction on the ET rates and free energy differences in the RC.

Acknowledgements

Hamamatsu Company, Herrsching, is gratefully acknowledged for giving us the opportunity to use the streak camera system. We would like to thank also Mrs. C. Hehl, Mr. A. Coktas, and Mr. M. Reus for technical assistance.

References

M. Becker, V. Nagarajan, D. Middendorf, W.W. Parson, J.E. Martin and R.E. Blankenship, Biochim.Biophys.Acta 1057 (1991) 299.

A.M. Freiberg, V.I. Godik, S.G. Kharchenko, K.E. Timpmann, A.Y. Borisov, and K.K. Rebane, FEBS Lett. 189 (1985) 341.

H. Michel and J. Deisenhofer, Pure Appl. Chem. 60 (1988) 953.

M.G. Müller, K. Griebenow, and A.R. Holzwarth, in Reaction Centers of Photosynthetic Bacteria, Springer Series in Biophysics Vol.6 (1990) 169.

A. Ogrodnik, M. Volk, R. Letterer, R. Feick and M.E. Michel-Beyerle, Biochim. Biophys. Acta 936 (1988) 361.

W.W. Parson, in Chlorophylls (1991) 1153.

V.A. Shuvalov, J. Amesz, and L.N.M. Duysens, Biochim. Biophys. Acta 851 (1986) 327.

H. Vasmel, R.F. Meiburg, H.J.M. Kramer, L.J. De Vos and J. Amesz, Biochim. Biophys. Acta 724 (1983) 333.

W. Zinth, W. Holzapfel, U. Finkele, W. Kaiser, D. Oesterhelt, H. Scheer, and H.U. Stilz, in Current Research in Photosynthesis I (1990) 27.

Inst. Phys. Conf. Ser. No 126: Section VIII
Paper presented at Int. Symp. on Ultrafast Processes in Spectroscopy, Bayreuth, 1991

Picosecond fluorescence of photosystem II D1/D2/cyt b559 and D1/D2/cyt b559/CP47 pigment-protein complexes

Kõu Timpmann[*], Arvi Freiberg[*], Andrei A Moskalenko & Nina Yu Kuznetsova

[*] Institute of Physics, Estonian Academy of Sciences, 202400 Tartu, Estonia
Institute of Soil Sciences and Photosynthesis, USSR Academy of Sciences,
142292 Pushchino, USSR

ABSTRACT: Photosystem II reaction centre and reaction centre/proximal antenna preparations in different functional state of reaction centres have been studied by the picosecond time-resolved fluorescence spectroscopy method at room temperature and at 77 K. A relatively slow (10-20 ps) energetically downward excitation energy migration among accessory pigments was observed. The primary electron donor of the reaction centre (Chl dimer, P680), its oxidized form (P$^+$680) as well as the reduced primary electron acceptor (Pheo$^-$) are all very efficient excitation quenchers.

1. INTRODUCTION

While the reaction centre (RC) of photosyntetic bacteria has long been isolated in a pure form and its structure and functioning have been determined, much is yet to be understood about the RCs of oxygen-evolving systems. The recent isolation (Namba and Satoh 1987) of rather a simple photosystem II (PSII) core, containing D1, D2 polypeptides, cytochrome b559 and a minimal amount of chlorophyll (Chl) and pheophytin (Pheo) molecules and revealing a photochemical activity, has opened new perspectives. This complex has many similarities with bacterial RC and it is believed to represent a PS II RC lacking a quinone electron acceptor. In the present work the picosecond time-resolved fluorescence of PSII RCs together with a functionally more intact PSII unit, which, in addition to D1/D2/cyt b559 core, comprises the proximal light-harvesting antenna pigment-protein CP47 binding 20-25 Chl molecules, has been studied. The nature of different fluorescence decay components is analysed.

2. MATERIAL and METHOD

Materials: Pigment-protein complexes from spinach were purified by a modified procedure which improves stability by exchange of detergent Triton X-100 with lauryl maltoside in the first purification column (Moskalenko 1990). Samples were resuspended in an equal volume of Tris-HCl buffer or glycerol. In glycerol they are remarkably more stable, bearing excitation light at room temperature during 5-10 minutes, whereas our typical measuring time was less than a minute. In the buffer the samples degrade even in darkness. RCs are more stable than the D1/D2/cyt b559/CP47 complexes. At room-temperature RCs absorb at 675 nm and the others at 676 nm. This shows the photochemical activeness of more than 90% of RCs.

Methods: Fluorescence lifetime measurements were perfomed with a picosecond spectrochronograph (Freiberg 1986). The excitation light source was a mode–locked cavity–dumped dye laser, pumped at 76 MHz by a Nd–YAG laser. The emission, viewed in a reflection mode, was filtered by two single–grating monochromators combined in the subtractive–dispersion geometry. The signal was detected by a synchroscan streak camera or (in a nanosecond time–domain) by a time–correlated single–photon counting apparatus. The instrument response functions were 18 ps or 400 ps (FWHM), respectively. Throughout all measurements excitation at 590 nm was employed with the average excitation intensity less than 15 mW/cm^2.

3. RESULTS and DISCUSSION

The picosecond time–resolved fluorescence kinetics of the complexes at room temperature and at 77 K are shown in Figs. 1 and 2, respectively. At room temperature fluorescence kinetics were almost independent of the emission wavelength in the range of 660–700 nm. At 77 K a characteristic dependence on the emission wavelength was observed: in the blue side of the band maximum the kinetics are much faster than around the peak and especially in the red side. Several chemical additives and the exposure to a supplementary light were used to change the functional state of the RCs at room temperature. The closed RC state with reduced Pheo (state P680 Pheo$^-$) is accumulated when samples are pretreated with sodium dithionite ($Na_2S_2O_4$) and

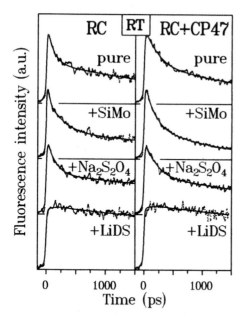
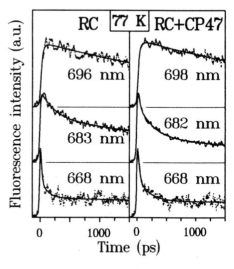

*Fig.*1. Picosecond fluorescence kinetics of RC and RC/CP47 complexes in glycerol with different chemical additives (see text) at room temperature. Recording at 683 nm with 16 nm bandwidth. Solid curves are the best multi–exponential fits to the data.

*Fig.*2. Picosecond fluorescence kinetics of RC and RC/CP47 complexes in glycerol at 77 K, recorded at different indicated wavelengths with 4 nm bandwidth.

methyl viologen in the light. The adding of silicomolybdate (SiMo), which is supposed to stay at the quinone Q site and accept electrons from Pheo⁻, under illumination results in the accumulation of the closed RC state with an oxidized primary donor, P⁺680 Pheo. The samples recover fully after illumination when kept in darkness for several minutes. A treatment of the samples with Triton X-100 or lithium dodecylsulfate (LiDS) leads to the dissociation of pigment-protein complexes to solubilized pigments and proteins. The most notable effect of variation of the functional state of the RCs is the change of kinetics at longer times; the fluorescence decays in the nanosecond time-domain are shown in Fig.3.

The kinetic data obtained can be summarized as follows:
(i) Fluorescence decay curves are poorly described by well-defined life-times, except in the case with solubilized pigments where almost a single-exponential decay with 5±0.1 ns time constant was recorded. However, by a discrete component analysis the components with lifetimes 10-20 ps, 100-300 ps, 1-2 ns, 4-6 ns and 20-40 ns (determined from separate measurements by a streak camera and a photomultiplier) could still be extracted;

(ii) At room temperature the shortest picosecond component is rather weak, except in the case when SiMo is added. The 100-300 ps and two shorter nano-second components have comparable amplitudes;

(iii) At 77 K in the blue side of the band maximum the 10-20 ps component dominates, but disappears gradually towards the red side of the spectrum where a delayed rise of 10 ps appears. Further towards the red the ampli-tude of the 100-300 ps emission also decreases, leaving only nanosecond decays;

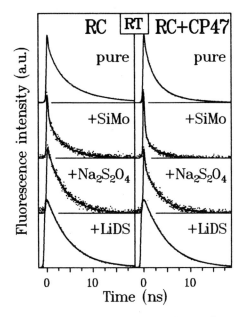

Fig.3. The same as in Fig.1, but with the time scale expanded to nanoseconds. Pulse repetition rate 7.6 MHz.

(iv) There is a minor change of picosecond lifetimes, if any, when Pheo is reduced to Pheo⁻ or P680 is chemically oxidized. The major effect upon RC closure is the total loss of 20-40 ns component and the redistribution of amplitudes of picosecond and the remaining nano-second components. The adding of dithionite results in the vanishing of 1-2 ns decay and in a relative increase of the emission with an intermediate 3 ns lifetime, while SiMo favours picosecond components;

(v) On comparing fluorescence ki-netics of RC and D1/D2/cyt b559/CP47 complexes it seems that the pico-second kinetics are systematically faster in case of RC complexes, whereby the nanosecond lifetimes are shorter (although with relatively lower amplitude) in the bigger units. The 20-40 ns emission yield is several times lower in more intact particles.

These findings suggest that none of the picosecond fluorescence components observed represent directly the claimed (Mimuro et al 1988) emission from an active RC primary donor. This is in contrast with RCs of bacteria (Freiberg et al 1985). Primary charge separation both in PSII and bacterial RCs occurs within 3 ps (Wasielewski et al 1989 and Fleming et al 1988). On the other hand, the specific reaction to the pretreatment strongly supports the nature of 20-40 ns emission component as due to the recombination luminescence known from the literature (Crystall et al 1989). It also seems quite established that the 5 ns component includes the emission coming from functionally uncoupled pigments. The wavelength-dependent picosecond fluorescence decay at 77 K evidences both the energetically directed excitation energy transfer between accessory pigments (10-20 ps component) towards P680 and the excitation quenching (100-300 ps) via the primary radical pair forming mechanism. Analogous phenomena were studied in the antenna of photosynthetic bacteria (Timpmann et al 1991) but not in RCs where the rates are unmeasurably high (Fleming et al 1988). The relatively slow transfer rate is supported by photochemical hole-burning experiments on PSII RCs (Tang et al 1991), where the accessory Chl energy transfer decay time of 12 ps at 1.6 K has been determined, as well as by low-temperature wavelength-dependent picosecond absorption data (Wasielewski et al 1989). As long as the RC and RC/CP47 complexes behave similarly, a question about exact pigment composition of PSII RCs arises. At present a Chl/Pheo ratio of 4/2 to 12/2 has been reported (Namba et al 1987, Dekker et al 1989). The origin of remaining 1-2 ns component is not quite clear. Judging by its disappearence in the course of reducing Pheo, it is suggested to be due to the emission of Pheo. The fastening of fluorescence kinetics when pretreated with SiMo is probably due to high quenching efficiency by an oxidized primary donor as well as to a partial oxidizing of accessory pigments. Our data support the opinion that P680, P^+680 and $Pheo^-$ are all very efficient excitation quenchers.

REFERENCES

Crystall B, Booth P J, Klug D R, Barber J & Porter G 1989, FEBS Letters 249 pp 75-78
Dekker J P, Browlby N R & Yocum C Y 1989, FEBS Letters 254 pp 150-154
Fleming G R, Martin J L & Breton J 1988, Nature 333 pp 190-192
Freiberg A 1986, Laser Chem. 6 pp 233-252
Freiberg A M, Godik V I, Kharchenko S G, Timpmann K E, Borisov A Yu & Rebane K K 1985, FEBS Letters 189 pp 341-344
Mimuro M, Yamazaki I, Itoh S, Tamai N & Satoh K 1988, Biochim.Biophys.Acta 933 pp 478-486
Moskalenko A A 1990, Biol.Membranes (USSR) 7 pp 736-741
Namba O & Satoh K 1987, Proc.Natl.Acad.Sci. USA 84 pp 109-112
Tang D, Jankowiak R, Seibert M & Small G J 1991, Photosynthesis Research 27 pp 19-29
Timpmann K, Freiberg A & Godik V I 1991, Chem.Phys.Letters 182 pp 617-622
Wasielewski M R, Johnson D G, Govindjee, Preston C & Seibert M 1989, Photosynthesis Research 22 pp 89-99

Inst. Phys. Conf. Ser. No 126: Section VIII
Paper presented at Int. Symp. on Ultrafast Processes in Spectroscopy, Bayreuth, 1991

Laserspectroscopic investigations of nonlinear optical effects in molecular aggregates of pigment–protein complexes

J. Voigt (a), Th. Bittner (a), G. Kehrberg (a), and G. Renger (b). Institut für Optik und Spektroskopie, FB Physik der Humboldt-Universität zu Berlin (a) and Max-Vollmer-Institut für Biophysikalische und Physikalische Chemie, Technische Universität Berlin (b)

1. INTRODUCTION

Modern laser spectroscopy with high resolution in the time and frequency domain provides the most powerful tool for analyses of the primary reactions in photobiology, especially in the field of photosynthesis. The first steps of the latter process are the generation of electronically excited states (excitons) by light absorption within pigment protein complexes and rapid exciton migration to the photoactive pigments embedded into the reaction center complexes where the trapping takes place, leading to transformation into electrochemical free energy. The mechanisms and kinetics of these processes which occur in the subpicosecond and picosecond time scale are determined by pigment-pigment and pigment-protein interactions (for a recent review see Renger 1992). Among unresolved questions of exciton migration within the photosynthetic apparatus the possibility of coherent transfer is of special interest because it provides information about the existence of strongly coupled pigment clusters within subunits of the antenna system.

In this communication investigations are reported on the use of a pump and probe technique in the frequency domain in order to address the following problems of excitation energy transfer in photosystem II of higher plants:

a) the nature of excited states generated at high excitation energies

b) the transversal (T_2) and longitudinal (T_1) relaxation times of these states

c) the influence of vibrational interactions, and

d) the excitation energy transfer between Chl a and Chl b molecules and their mutual interaction within the light harvesting complexes.

2. EXPERIMENTAL

The measurements were performed in suspensions of PS II membrane fragments prepared according to the method of Berthold *et al* (1981) with modifications by Völker *et al* (1985). For comparative experiments degassed acetonic Chl solutions were used. The laser set-up is described in detail in an earlier paper (Kehrberg *et al* 1990). The essential experimental parameters of this system are: τ_{Laser} = 12ns, maximal pump pulse intensity $\approx 3 \cdot 10^{26}$ photons $\cdot$ cm$^{-2}\cdot$ s^{-1}, $\delta\lambda_{DL} \leq 0.005$nm, resolution ≥ 0.05nm.

The following experimental protocol was used:

i)   measurements of the probe beam transmission in the near vicinity ($\Delta\lambda = (\lambda_p - \lambda_t) \leq 0.8$nm) of the pump pulse kept at constant wavelength and intensity

ii) determination of the probe beam transmission as a function of the pump beam intensity at constant wavelength difference between both beams ($\Delta\lambda = 0.1$nm)

iii) experiments of type i) and ii) were performed at pump pulse wavelengths which predominately excite either Chl a or Chl b

iv) comperative measurements in suspensions of PS II membrane fragments and degassed acetonic Chl solutions.

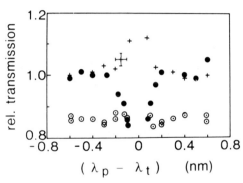

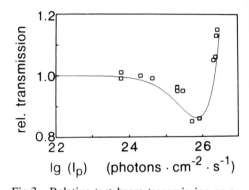

Fig.1   Relative test beam transmission as a function of the wavelength difference $(\lambda_p - \lambda_t)$ at two pump pulse intensities in PS II membrane fragments ($+$, $\bullet$) and acetonic Chl solutions ($\odot$). Pump pulse intensities: $I_p = 2.5 \cdot 10^{26}$ ($+$) and $6 \cdot 10^{25}$ photons $\cdot$ cm$^{-2}\cdot$ s^{-1} ($\bullet$, $\odot$), respectively.

Fig.2   Relative test beam transmission as a function of pump pulse intensity at a constant wavelength difference $(\lambda_p - \lambda_t) = 0.1$nm in PS II membrane fragments. The line was calculated with the parameters: $\sigma_{12}/\sigma_{01} = 2.6$, $\tau_{eff1} = 200$ps, $\tau_{eff2} = 15$ps and $\beta = 10^{-8}$ cm$^3\cdot$ s^{-1}.

3. RESULTS

Figs.1 and 2 show characteristic results obtained with pump pulses at a wavelength ($\lambda_p = 620$nm) which leads to a simultaneous excitation of Chl a and Chl b in PS II membrane fragments. These data lead to the following conclusions (see also Bittner *et al* 1991):

α) The transmission changes of the test pulse are due to changes of the absorption induced by the pump pulse rather than caused by scattering effects.

β) The dependencies on the wavelength difference $\Delta\lambda$ (very narrow shape) and on the pump pulse intensity (decrease and increase of test pulse transmission) preclude the trivial explanation that this phenomenon is due to higher singlet state absorption of an ensemble of isolated chlorophyll molecules. Therefore, the existence of 'two particle' exciton states is postulated that can be formed from degenerated exciton states via a two step process.

γ) Comparative measurements in Chl solutions reveal that the postulated 'two particle' excitation is specific for chlorophyll protein complexes in PS II membrane fragments.

For a quantitative description of the data a three level exciton model has been used which comprises the ground state and the single and 'two particle' excited states (fig.3) (for details see Voigt *et al* 1991). The calculated nonlinear transmission as a function of the pump pulse intensity is shown in fig.2. A rather good fit of the experimental data can be achieved. This finding reveals that the comparatively simple model assumption of 'two particle' excitations provides a satisfying description of the observed phenomena. The fitting procedures provide for the first time the determination of the values σ_{12} and $\tau_{eff}2$. A rigorous treatment of the three level sys-tem within the framework of the density matrix formalism permits the determination of the transversal relaxation time T_2 from a fit of the spectral dependence of pumped absorption. The calculated curve and the experimental data are shown in fig.4. As the most interesting result of this analysis a rather long transversal relaxation time ($T_2 = 4$ps) is obtained which is of the same order of magnitude as the longitudinal relaxation time ($\tau_{eff}2 = 10$ps). This finding has very interesting implications for the mechanism of excitation energy transfer within the antenna system of PS II membrane fragments as will be briefly outlined in the discussion.

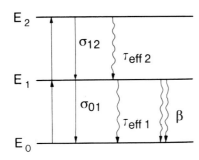

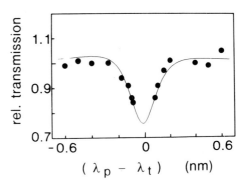

Fig.3   The three level model with the ground state (E_0) and the single (E_1) and 'two particle' (E_2) excited states. σ_{01} and σ_{12} are the optical cross sections. $\tau_{eff}1$ and $\tau_{eff}2$ characterize the overall exciton lifetimes. β describes the strength of the bimolecular exciton-exciton annihilation.

Fig.4   Fit of the test pulse transmission decrease of PS II membrane fragments by means of calculated absorption in the three level system from the density matrix formalism ($\tau_{eff}1 = 100$ps, $\tau_{eff}2 = 10$ps, $T_2 = 4$ps, $\sigma_{12}/\sigma_{01} = 2.5$).

The light harvesting complexes contain Chl a and Chl b which are both excited at 620nm. Therefore, in order to attempt an assignment of the 'two particle' excited state phenomenon to these different chlorophylls, experiments were performed at a wavelength (670nm) which is preferentially absorbed by Chl a. The data reveal a quite different behaviour. The characteristic results can be summerized as follows (data not shown):

i) The relative test pulse transmission as a function of pump pulse intensity does not reveal detectable decrease (in contrast to 620nm experiment, see fig.2) while the transmission increase at high I_p-values is much more pronounced.

ii) There does not exist a narrow dependence on $\Delta\lambda$ indicative for resonance effect due to 'two particle' excitations.

iii) The onset of the nonlinear transmission increase in acetonic Chl solutions arises at lower concentrations compared with PS II membrane fragments.

4. DISCUSSION

The observation of 'two particle' excited states in the photosynthetic antenna systems raises question on the nature of the pigment protein complex and of the chlorophyll molecules which are responsible for this interesting phenomenon. Based on the data of the present study (experiments performed at pump pulse wavelength's of 620nm and 670nm , respectively) Chl b appears to be the most likely candidate. Therefore, the subunit of the light harvesting complex II (LHC II) seem provide a pigment arrangement that permits 'two particle' excitations and coherent exciton transfer in time domains $\geq$ 1ps. This idea is in line with the recent structural model

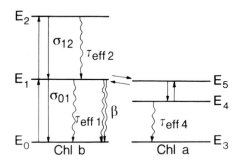
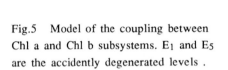
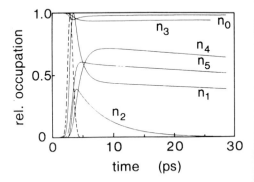

Fig.5   Model of the coupling between Chl a and Chl b subsystems. E_1 and E_5 are the accidently degenerated levels .

Fig.6   Relative occupation n_i (i=0..5) of the levels E_i (i= 0..5) as a function of time in the coupling model for Chl a and Chl b sub-sys-tems in LHC II. The dotted line repre-sents the excitation pulse.

of the pigment arrangement within the subunits of the trimeric LHC II complex derived from electron crystallography at a 6Å resolution (Kühlbrandt and Wang 1991). The current structural information reveals strongly interacting Chlorophyll molecules at distances of 9-14Å, but does not permit a destination between Chl a and Chl b. Our assignment of the 'two particle' excitation phenomenon to Chl b is also in agreement with previous conclusions on a tight excitonic coupling between 3 Chl b molecules within the LHC II complex (Knox and van Metter 1979). At first glance, the attribution of the 3 level system exclusively to Chl b seems to be in contrast to the exciton lifetime of 100-150ps derived from our data (see fig.2) because time resolved measurements revealed that the photon energy absorbed by Chl b molecules is very rapidly ($\tau_{Transfer} \approx$ 1ps) transferred to Chl a molecules (Eads et al 1989). In order to cope with this phenomenon we have additionally to assume a sufficiently fast equilibration between the subsystems of strongly coupled Chl a and Chl b molecules. This assumption does not imply the necessity of a strong coupling between all pigment molecules (in this case the assumption of energy transfer between both systems would be meaningless) but only requires that at least one level of each subsystem (Chl a and Chl b, respectively) is accidentally degenerated (see fig.5). In this case a fast excitation energy transfer can take place leading to a quasiequilibrium between the Chl a and Chl b molecules. This effect gives rise to a long overall exciton lifetime of Chl b as supported by the numerical results presented in fig.6.

5. CONCLUSIONS

(i) 'Two particle' excitations are detected for the first time in the LHC of higher plants at high excitation intensities.

(ii) The transversal and longitudinal relaxation times are determined to be $T_2 \approx$ 4ps and $T_1 =$ 10 to 15ps, respectively. The surprisingly high value of T_2 indicates that coherent processes essentially contribute to the exciton energy transfer in the LHC of higher plants.

(iii) The 'two particle' excitations are formed in a subsystem formed by Chl b which is in a quasiequilibrium with the Chl a subsystem.

(iv) From $\sigma_{12}/\sigma_{01} =$ 2.5 we can deduce the number of strongly interacting Chl b molecules to be 6. This value permits suggestions about the chlorophyll b positions which are not yet identified in the recent stuctural LHC II model of Kühlbrandt and Wang (1991).

Acknowledgement

The authors are grateful to the Deutsche Forschungsgemeinschaft for funding this work through grant No.Vo 502/1-1 and Re 354/12-1.

References

Berthold DA, Babcock GT and Yoccum CF 1981 *FEBS Lett. 134* 231

Bittner Th, Voigt J, Kehrberg G, Eckert H-J and Renger G 1991 *Photosynth. Res. 28* 131

Eads DD, Castner EW, Alberte RS, Metz L and Fleming GR 1989 *J. Phys. Chem. 93* 8271

Kehrberg G, Voigt J and Bittner Th 1990 *Studia Biophysica 3* 195

Knox RS and van Metter RL 1979 *Ciba Foundation Symposium (Vol 61)* (Amsterdam: Excerpta Medica) p 177

Kühlbrandt W and Wang DN 1991 *Nature 350* 130

Renger G 1992 *Topics in Photosynthesis (Vol 11)* ed J Barber (Amsterdam: Elsevier) p 45

Voigt J, Bittner Th and Kehrberg G 1991 *phys. stat. sol.(b) 166* 135

Völker M, Ono T, Inoue Y and Renger G 1985 *Biochem. Biophys. Acta 806* 25

Inst. Phys. Conf. Ser. No 126: Section VIII
Paper presented at Int. Symp. on Ultrafast Processes in Spectroscopy, Bayreuth, 1991

623

Aggregation of chlorophyll *a* in hydrocarbon solution

V Helenius, J Erostyák[a] and J Korppi-Tommola

Department of Chemistry, University of Jyväskylä, P.O. Box 35, SF-40351 Jyväskylä, Finland
[a] Department of Physics, Janus Pannonius University, Ifjusag u. 6, H-7624, Pecs, Hungary

Abstract. A chlorophyll a (Chl *a*) in 3–methylpentane shows a temperature dependent equilibrium of monomeric and aggregated chlorophyll molecules absorbing visible light at 662 nm, 674 nm and 702 nm. Time resolved studies of the induced anisotropies of these bands provide new experimental evidence on the structures of the aggregates giving rise to these absorption peaks. Using Stokes–Einstein hydrodynamic model to analyse recorded rotation correlation times suggests a monomeric origin for the 662 nm absorption band and a dimeric structure for the 702 nm absorption. The rotational correlation times of the 674 nm absorption band suggest a loosely bound T–shaped Chl *a* aggregate structure.

1. INTRODUCTION

The chlorophylls are involved in all primary events of photosynthesis. Most of the chlorophyll molecules in a photosynthetic unit act cooperatively as a light harvesting antenna to absorb visible light and to transfer the excitation to the photoreaction center. In the reaction center a special pair of chlorophyll molecules is a starting point for an electron transfer reaction (Kirmaier and Holten 1987, Deisenhofer and Michel 1989). Photosynthesis of green plants depends on interplay of two photosystems. The photosystem I special pair P700 absorbs light at wavelength of 700 nm. A slightly wet Chl *a* hydrocarbon solution shows an absorption band at 700 nm with a close resemblance to that of the green plant photosystem I. Therefore the 700 nm absorbing species has been investigated as a model for P700 reaction center in green plants (Shipman *et al* 1976). The structure of the 700 nm absorbing species in a hydrocarbon solution is still unknown, although several different structures have been proposed on the basis of NMR (Abraham *et al* 1988, Boxer and Closs 1976), visible and IR spectroscopy (Shipman *et al* 1976, Fong *et al* 1976) and theoretical calculations (Sakuma *et al* 1990, LaLonde *et al* 1988).

Since the structure of 700 nm Chl *a* aggregate is unknown it was considered of interest to measure the time dependence of induced anisotropy of this absorption band and compare the anisotropy decay time with the ones obtained from monomeric or polymeric Chl *a* absorption band. The anisotropy decay time gives information about the rotational diffusion of the absorbing Chl aggregate.

2. EXPERIMENTAL

The Chlorophyll *a* was prepared according to method of Hynninen (1977). Solvents 3–methylpentane (3–MP) and CCl_4 were dried by activated alumina. 3–MP was stored over Na and distilled prior to use. CCl_4 was stored over 4A molecular sieves.

The samples were prepared by first drying the Chl *a* with codistillation three times with dry CCl_4 and then evacuating the sample in 10^{-6} torr vacuum at 40–50 °C temperature for 30 minutes. Then the Na–dry degassed 3–MP was added and the solution was allowed to be in contact with air for a short while. This procedure produces 700 nm absorption band forming samples with their composition changing only slightly from sample to sample. The critical point on the sample preparation is the amount of water which is introduced in the system (Cotton *et al* 1978, Katz *et al* 1978, Hoshino *et al* 1981).

Another sample which contains monomeric Chl *a* was prepared by adding stoichiometric amount of pyridine into 3–MP solution. Pyridine is very efficiently complexed with chlorophyll preventing any further aggregation (Katz *et al* 1978). The sample concentration used for absorption recovery measurements was 2×10^{-4} mol/l.

The anisotropy decays were measured using absorption recovery technique (Fleming 1986). The laser system which consists of sync–pumped and cavity– dumped dye laser is described in detail elsewhere (Åkesson *et al* 1991). The sample was located in a spinning cell in order to avoid photodegradation.

3. RESULTS AND DISCUSSION

The Chl *a* 3–MP solution shows three distinct absorption bands at 662, 674 and 702 nm. A temperature dependent equilibrium exists between the species responsible for the three absorption peaks. The anisotropy decays were measured at each wavelength over temperatures between 25 and -90 °C. The 662 nm absorption band corresponds the $Q_{y,0-0}$ transition of monomeric Chl *a*. The anisotropy decays for the 662 nm band were measured over the whole temperature range both in pure 3–MP solution and in the solution where pyridine was added. The 674 absorption band appears as a shoulder on red side of the 662 nm absorption band. It is apparent only at ambient temperature, disappearing as the solution is cooled down. As the peak at 674 nm disappears the absorption peak at 702 nm begins to emerge. It was only possible to record the 674 nm absorption decay traces for temperatures down to -20 °C. From this temperature on it was possible to start recording absorption recovery traces for the 702 nm absorption peak.

The 674 nm absorption band is assigned to aggregated Chl *a* where chlorophyll molecules are loosely bound to each other. A similar absorption band arises in dry hydrocarbon solvents where Chl *a* is strongly aggregated (Cotton *et al* 1978, Katz *et al* 1978).

The 702 nm absorption band is assigned to water bridged Chl *a* dimer $(Chla \cdot H_2O)_2$. This assignment is based on strong experimental and theoretical evidence. The aggregate is not formed in solution where both Chl and the solvent are carefully dried (Cotton *et al* 1978). The optical properties of this aggregate and that of the photosystem I reaction center dimer P700 are very similar (Norris *et al* 1976). Covalently bonded Chl dimers with analogous electronic properties can be prepared (Johnson *et al* 1988, Johnson *et al* 1989). Exciton theoretical calculations are in agreement with the water bridged dimer model. (Shipman and Katz 1977, LaLonde *et al* 1988, Koester and Fong 1976).

If the anisotropy decay time is assumed to be controlled by rotational diffusion of the absorbing molecule, the size of the rotating unit can be calculated assuming that the viscosity of the solvent and the temperature are known. A hard sphere model and Stokes-Einstein hydrodynamic equation was used to describe the diffusional motion (eq. 1.). Chl a is a rigid neutral molecule which is large compared to the solvent molecules. For such a case stick hydrodynamic prediction is an adequate model (eq. 2) (Ben-Amotz and Drake 1988).

$$D_r = \frac{kT}{8\pi\eta R^3} \tag{1}$$

In eq. 1 D_r is the rotational diffusion coefficient, η is viscosity and R is the radius of the rotating sphere. The rotation correlation time τ_r is related to diffusion coefficient by eq. 2 if a stick boundary condition is used and if the solute molecule is assumed to be a symmetric rotor (Fleming 1986).

$$\tau_r = \frac{1}{6D_r} \tag{2}$$

The viscosity of 3–MP at low temperatures was extrapolated from temperature data between 20 – 0 °C. As the rotational correlation time is plotted against viscosity over temperature a straight line should be obtained according to eq. 1. Fig. 1. shows the rotational correlation times obtained for different absorption peaks.

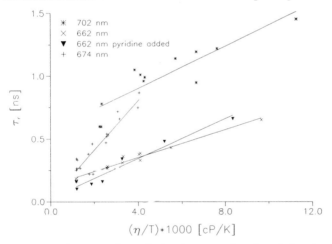

Figure 1. The calculated rotational correlation times as a function of viscosity divided by temperature.

The absorption band at 702 nm shows the longest anisotropy decay times over the whole temperature range. The average value obtained for the diameter of Chl a dimer is 19.2 Å. This value does agree with proposed Chl a water dimer structures. A diameter of 19 Å is too small for an aggregate bigger than two Chl a units.

The monomeric absorption band at 662 nm shows the shortest anisotropy decay times. The diameter obtained for rotating sphere is slightly longer in pure Chl a 3–MP solution (14.5 Å) than in a Chl a 3–MP solution where small amount of pyridine is added (13.5 Å). The difference can be due the absorption band at 674 nm which is present in pure Chl a 3–MP solution but not in the solution where pyridine is added. In both cases the calculated diameter compares nicely with the diameter of the Chl a porphyrin moiety obtained from molecular models.

The 674 nm absorption band gives only slightly longer anisotropy decay times than the 662 nm absorption band. The average diameter of the rotating specie is 17.6 Å. As the 674 nm absorption band is due aggregation, our results suggest that the aggregate contains nearly freely rotating units. A T- shaped aggregate structure suggested by Katz *et al* (1978) and Kooyman and Schaafsma (1984) allows for this kind of free rotation. The anisotropy decay time of the 674 absorption band increases faster as a function of viscosity (fig. 1) than the decay time of the 662 or 702 nm absorption bands. This indicates an increasing aggregate size as the temperature is lowered.

4. SUMMARY

The induced anisotropy decay times of a Chl *a* 3–MP solution were measured over a temperature range 25 – -90 °C by using picosecond pump and probe technique. Our results support the earlier interpretations of the nature of different aggregates, however the evidence is based on direct observation of the anisotropy decay times over a range of temperatures. The three absorption bands at 662 nm, 674 nm and 702 nm are from monomeric Chl *a*, dry aggregated Chl *a* and Chl *a* water bonded dimer $(Chla \cdot H_2O)_2$, respectively.

Acknowledgments

We wish to thank professor P. Hynninen for providing the extremely pure chlorophyll *a* sample. This work was supported by the Research Council for Natural Sciences of the Academy of Finland.

References

Abraham R J, Goff D A and Smith K M 1988 *J. Chem. Soc. Perkin Trans. I* pp 2443-2451
Ben-Amotz D and Drake J M 1988 *J. Chem. Phys.* **89** pp 1019-1029
Boxer S G and Closs G L 1976 *J. Am. Chem. Soc.* **98** 5406
Cotton T M, Loach P A, Katz J J and Ballschmiter K 1978 *Photochem. Photobiol.* **27** pp 735-749
Deisenhofer J and Michel H 1989 *Chemica Scripta* **29** pp 205-220
Fleming G R 1986 *Chemical Applications of Ultrafast Spectroscopy* (New York: Oxford Univ. Press) pp 66-70
Fong F K, Koester V J and Polles J S 1976 *J. Am. Chem. Soc.* **98** 6406
Hoshino M, Ikehara K, Imamura M, Seki H and Hama Y 1981 *Photochem. Photobiol.* **34** pp 75-81
Hynninen P H 1977 *Acta Chem. Scand.* **B31** pp 829-835
Johnson D G, Svec W A and Wasielewski M R 1988 *Israel J. of Chem.* **28** pp 193-203
Johnson S G, Small G J, Johnson D G, Svec W A and Wasielewski M R 1989 *J. Phys. Chem.* **93** pp 5437-5444
Katz J J, Shipman L L, Cotton T M and Janson T R 1978 *The Porphyrins: Physical Chemistry, Part C, Vol V* ed D Dolphin (Academic Press) pp 401-458
Kirmaier C and Holten D 1987 *Photosynthesis Research* **13** pp 225-260
Koester V J and Fong F K 1976 *J. Phys. Chem.* **80** pp 2310-2312
Kooyman R P H and Schaafsma T J 1984 *J. Am. Chem. Soc.* **106** pp 551-557
LaLonde D E, Petke J D and Maggiora G M 1988 *J. Phys. Chem.* **92** pp 4746-4752
Sakuma T, Takada T, Kashiwagi H and Nakamura H 1990 *Int. J. Quantum Chem.: Quantum Biol. Symp.* **17** pp 93-101
Shipman L L, Cotton T M, Norris J R and Katz J J 1976 *Proc. Natl. Acad. Sci. USA* **73** pp 1791-1794
Shipman L L and Katz J J 1977 *J. Phys. Chem.* **81** pp 577-581
Åkesson E, Hakkarainen A, Laitinen E, Helenius V, Gillbro T, Korppi-Tommola J and Sundström V 1991 accepted for publication in *J. Chem. Phys.*

Energy transfer and charge separation at 15 K in membranes of *heliobacterium chlorum*; temperature dependence of secondary electron transfer

Paul J.M. van Kan[#] and Jan Amesz, Dept. of Biophysics, Huygens Laboratory of the State University, P.O Box 9504, 2300 RA Leiden, The Netherlands.

ABSTRACT: Absorbance difference spectra at 15 K of membranes of *Heliobacterium chlorum* were measured at various times during and after a 25 ps laser flash exciting Bchl *g* at 532 nm. The spectra were recorded with or without a preceding laser flash, which oxidized the primary electron donor P-798. Also, a study was made of the temperature dependence of the rate of electron transfer from the first electron acceptor (hydroxy-Chl *a*) to the second acceptor. This temperature dependence is analyzed in terms of existing theories for non-adiabatic electron transfer.

1. INTRODUCTION

The recently discovered heliobacteria are distinguished from other groups of photosynthetic bacteria by the nature of their main photosynthetic pigment, Bchl *g* The primary electron donor, P-798, is presumably a dimer of Bchl *g* [1]. The pigment that functions as primary electron acceptor has recently been isolated and identified as hydroxy-Chl *a* [2]. Evidence has been reported that the next electron acceptor in the sequence might be a quinone molecule presumably menaquinone. One or more iron-sulfur centres may serve as additional electron acceptors [3]. All these data suggest a similarity between this photosystem and Photoystem I of plants.

In heliobacteria, there is only one, comprehensive type of light-harvesting complex, which also contains the reaction centre. The only differentiation in the antenna known at present is that there are a number of spectrally different species of Bchl *g*. Thus a distinction has been made between Bchl *g* 778, Bchl *g* 793 and Bchl *g* 808 [4], after their absorption maxima at low temperature.

In this paper we will present more detailed results of absorbance difference measurements at 15 K. Difference spectra were measured during and after excitation with 25 ps flashes at 532 nm. A comparison was made between the signals obtained with open and closed reaction centres, in order to separate the contributions by reaction centre and antenna processes to the difference spectra. We also present measurements of the rate of re-oxidation of the primary electron donor at temperatures between 10 K and 300 K. This temperature dependence will be discussed, applying a theory for electron tunneling proposed by Kakitani and Kakitani [5].

2. MATERIALS AND METHODS

Heliobacterium chlorum was grown anaerobically and membrane fragments were prepared as described in [2]. The suspension was diluted by addition of glycerol to a final concentration of 66% (v/v). Samples were cooled in the dark in an Oxford CF 1204 helium flow cryostat or for temperatures > 200 K, in a home-built cryostat using alcohol or liquid nitrogen as coolant. Absorbance difference measurements were carried out using a pump-probe technique with variable delay between pump and probe pulses as described in ref. [6]. The spectra were measured using a diode array detector [7]. The kinetic data, were fitted with a convolution of the instrument response function represented by a Gaussian of 35 ps FWHM [8] and an exponential function. Long-lived absorbance differences due to P-798$^+$ were fitted with the integral of the response function, the amplitude of which was a fitting parameter.

3. RESULTS AND DISCUSSION

Absorption difference spectra

Fig. 1a presents absorbance difference spectra for the region 620-720 nm, measured at 15 K, induced by a single flash at 532 nm. A narrow negative ΔA signal at 667 nm is caused by reduction of the primary electron acceptor, hydroxy-Chl *a* [2]. The signal cannot be ascribed to excited hydroxy-Chl *a*, since all absorption at 532 nm is due to Bchl *g* [2]. The bleaching due to the reduced primary acceptor is already clearly present at 0 ps, indicating that a very short time (< 10 ps) is needed for the primary charge separation in H. chlorum, even at 15 K. The bleaching is maximal at about 20 ps. At longer times after the flash it is gradually replaced by a bandshift centered at 667 nm. This bandshift may be ascribed to an electrochromic effect on the (re-oxidized) primary acceptor caused by P-798$^+$ [9].

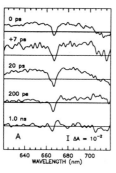

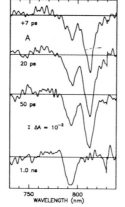

Fig. 1. *Difference spectra after excitation of* H.chlorum.

Difference spectra in the region 730 -840 nm upon single flash excitation are shown in Fig. 1b. The spectra are in general agreement with those presented earlier [6], with two major bleachings centered at 793 nm and 812 nm. The first one, which is essentially irreversible at the time scale of the experiments, can be attributed to photo-oxidation of P-798. The bleaching at 793 nm develops at the same rate as that caused by reduction of the primary electron acceptor. This means that at the present time resolution significant accumulation of either excited P-798 or excited antenna Bchl *g* 793 is not observed.

The absence of significant antenna bleaching below 800 nm indicates rapid energy transfer to Bchl *g* 808 and to the reaction centre. The only prominent bleaching of Bchl *g* in the antenna takes place in a band around 810 nm. All excitations that do not lead to formation of P- 798$^+$ thus appear to accumulate as excited singlet states of Bchl *g* 808. Fig. 2 presents difference spectra obtained after two successive excitation flashes, given with an interval of 5 ns in between. Absorbance differences were measured at time t after the second flash, relative to the absorbance 0.5 ns before the second flash.

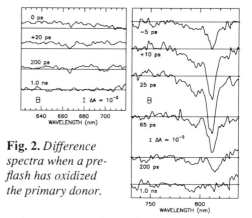

Fig. 2. *Difference spectra when a pre-flash has oxidized the primary donor.*

The bleaching as well as the bandshift around 667 nm are largely absent in these spectra: the first excitation flash has left most of the reaction centres in a charge separated state, whereby P-798 is oxidized and a secondary electron acceptor is reduced. The primary electron acceptor cannot be reduced in those reaction centres. Only a small bleaching band was now observed around 794 nm. This band, which was still present at 1 ns, can be attributed to the formation of P-798$^+$ which was not oxidized by the first flash. This apparently happened in

10-15% of the reaction centres. Again, no evidence can be seen for excited states of antenna Bchl *g* 794 or of Bchl *g* 778. Thus also with oxidized reaction centres excitation transfer to Bchl *g* 808 is very rapid.

Temperature dependence of secondary electron transfer rate

In order to obtain information about the temperature dependence of the rate of electron transfer from the primary electron acceptor, hydroxy-Chl *a*, to the secondary acceptor, we measured the decay kinetics around 665 nm as a function of temperature. The kinetic traces were fitted with a convolution of the instrument response function and a mono-exponential decay. In addition, a long-lived component, irreversible on the timescale of the experiment, was used to account for the remaining ΔA, due to P-798$^+$. The amplitude of this component was variable because the isosbestic point in the bandshift due to P-798$^+$ shifts with temperature. A graph of the electron transfer rate as a function of temperature is presented in Fig. 3.

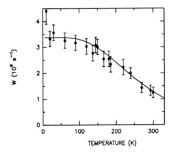

At 300 K the rate constant is $(1.3 \pm 0.2) \times 10^9$ s^{-1}. It increases with decreasing temperature, until it reaches an approximately constant value at 120 K. The average rate constant in the temperature range of 20 - 120 K is $(3.3 \pm 0.3) \times 10^9$ s^{-1}. Kirmaier et al. [10] measured the rate of electron transfer from Bpheo *a* to Q_A in reaction centres of *Rb. sphaeroides* and found similar results. We shall analyze our results in a similar way.

The ET rate W is given by the Golden Rule expression:

$$W = 4\pi^2/h \bullet |V|^2 \bullet F.$$

Fig. 3. *Recovery rate of the bleaching at 667 nm vs. temperature.*

The basic choice to be made is whether the temperature dependence of the ET rate is due to a temperature dependence of the tunneling matrix element V or if it originates from a temperature dependence or in F, the Franck-Condon factor. In this discussion we will assume that only the Franck Condon factor is temperature dependent.

We shall discuss the measurements using a theory by Kakitani and Kakitani [5], since this theory proved to be useful in explaining the temperature dependence of Q_A reduction in purple bacteria [10]. Kakitani and Kakitani apply an extension of ET theories developed by Bixon and Jortner [11,12]. The frequencies of the vibrational modes are allowed to change upon the ET reaction. The energy gap ΔE between initial and final state is replaced by ΔG, the change in free energy. This ΔG contains an extra energy term due to the change in zero-point energy upon ET. Furthermore, an entropy change due to a change in the average frequency upon ET is accounted for. The authors argue that in the case of a small change in the frequency of each oscillator, $\Delta v/v \ll 1$, the summed effect of all oscillators on ΔG can be comparable to ΔE, the energy gap between the electronic states, because the number of vibrational modes is very large in a photosynthetic system. An average frequency v_a is defined that determines the boundary between a high temperature region where a frequency shift upon ET leads to a change in population of the vibrational levels, and a low temperature region where this effect is negligible. The frequency of the mode whose displacement is given by the parameter S is denoted v_q in this case. v_q and v_a need not be equal. Kakitani and Kakitani take $hv_q \gg kT$ over the entire temperature range from 0-300 K. The final expression becomes:

$$W = \frac{4\pi^2|V|^2}{h^2 v_q} \bullet exp(-S) \bullet \frac{S^p}{\Gamma(p+1)}$$

where $S = \Delta^2/2$; $p = (\Delta E + \Delta E_1 [\bar{n}_a + 1/2])/hv_q$;

$\bar{n}_a = [exp(hv_a/kT) - 1]^{-1}$

The temperature dependence in this expression is solely due to a change in the entropy term of ΔG. The solid line in Fig. 3. shows the best fit to our data obtained with this model. We conclude that our measurements can be reasonably well explained by the theory of Kakitani and Kakitani [5] for non-adiabatic ET. The approach of Bixon and Jortner [12], did not yield a satisfactory fit of the experimental data.

Fitting parameters

The parameters used to fit the data cannot be interpreted fully, because essential information about the nature and midpoint potential of the secondary acceptor(s) in *H. chlorum* is lacking or uncertain [1]. This makes it impossible to verify calculated values of ΔG at 300 K. The method of [5] allows the reduced displacement S to be small, varying in the range $0.1 < S < 1$. We have kept S around a value of 0.5. The reduced free energy gap p used in the fit is considerably larger than S, for all temperatures. This reaction thus takes place in Marcus' 'inverted' region. The wavenumber $\bar{v}_q$ of the mode that yields the displacement S was taken 1300 cm^{-1}, since this is the value of the vibration progression in absorption spectra of Chl a [13]. Similar values of S and $\bar{v}_q$ have been encountered in calculations [14].

The value of the parameter $\bar{v}_a$ is the most characteristic value of the theory used and correspondingly allows relatively little freedom of choice. We found a value of about 400 cm^{-1}, which is a value encountered more often [5,11] when photosynthetic ET reactions are modelled. It is not unlikely that these modes might be due to ligand interactions between the donor and acceptor molecules and the protein. The constant value for the ET rate at low temperature yields a reasonably good fit down to about 20 K. At 10 K this rate seems to have a higher value. With the present number of data points and the given accuracy it is not certain if this effect is significant. With the value of V the curve is scaled to overlap the data. This results in a value of 10-12 A for the tunneling distance [11]. This distance is comparable to the tunneling distance for the analogous reaction in *Rb. sphaeroides*.

Our data, like those in [10], can be satisfactorily explained by means of the theory of Kakitani and Kakitani. The parameters that can be discussed have a reasonable value. The values of ΔG have to be compared with measured (midpoint) reduction potentials of the acceptors in *H. chlorum*. At present such information is lacking or too variable to use for a numerical comparison of results.

4. REFERENCES

1. Prince, R.C., Gest, H. and Blankenship, R.E. (1985) Biochim. Biophys. Acta 810, 377-384.
2. Van de Meent, E.J., Kobayashi, M., Erkelens, C., Van Veelen, P.A., Amesz, J. and Watanabe, T. (1991) Biochim. Biophys. Acta, in the press.
3. Nitschke, W., Sétif, P., Liebl, U., Feiler, U. and Rutherford, A.W. (1990) Biochemistry 29, 11079-11088.
4. Van Dorssen, R.J., Vasmel, H. and Amesz, J. (1985) Biochim. Biophys. Acta 809, 199-203.
5. Kakitani, T. and Kakitani, H. (1981) Biochim. Biophys. Acta 635, 498-514.
6. Van Kan, P.J.M., Aartsma, T.J. and Amesz, J. (1989) Photosynth. Res. 22, 61-68.
7. Van Kan, P.J.M. (1991) Doctoral Thesis, University of Leiden.
8. Nuijs, A.M., Van Grondelle, R., Joppe, H.L.P., Van Bochove, A.C. and Duysens, L.N.M. (1985) Biochim. Biophys. Acta 810, 94-105.
9. Kleinherenbrink, F.A.M., Aartsma, T.J. and Amesz, J. (1991) Biochim. Biophys. Acta, in the press.
10. Kirmaier, C., Holten, D. and Parson, W.W. (1985) Biochim. Biophys. Acta 810, 33-48.
11. Jortner, J. (1980) J. Am. Chem. Soc. 102, 6676-6686.
12. Bixon, M. and Jortner, J. (1986) J. Phys. Chem. 90, 3795-3800.
13. Fragata, M., Nordén, B. and Kurucsev, T. (1988) Photochem. Photobiol. 47, 133-143.
14. Warshel, A. (1980) Proc. Natl. Acad. Sci. USA 77, 3105-3109.

Femtosecond energy transfer processes in allophycocyanin and C-phycocyanin trimers

E.V.Khoroshilov, I.V.Kryukov, P.G.Kryukov and A.V.Sharkov
P.N.Lebedev Physics Institute, USSR Academy of Science, 117924 Moscow, USSR
T.Gillbro
Department of Physical Chemistry, University of Umeå S-90187 Umeå, Sweden
R.Fischer and H.Scheer
Botanisches Institut der Universität München, D-8000 München, FRG

Abstract. Ultrafast processes in C-phycocyanin and allophycocyanin trimers and monomers have been examined by means of polarization pump-probe technique. No femtosecond kinetics was observed in monomeric preparations. The absorption recovery kinetics with t=450±50 fs has been recorded in allophycocyanin trimers. The anisotropy decay kinetics was not obtained in femtosecond time domain. The conclusion about energy transfer between $\alpha 84$ and $\beta 84$ chromophores with different absorption spectra was made. The proposed model takes into account a stabilising role of linker peptide. Femtosecond anisotropy decay kinetics was obtained with C-phycocyanin trimers and explained by Förster energy transfer between $\alpha 84$ and $\beta 84$ chromophores with 65° angle between their orientations.

1. INTRODUCTION.

Allophycocyanin (APC) and C-Phycocyanin (C-PC) are photosynthetic antenna pigments of the light-harvesting complexes (phycobilisomes) of blue green bacteria and red algae. Our knowledge about the energy transfer in the phycobilisome has increased substantially during the last decade, see review works of Glaser (1985) and Holzwarth (1987). Important questions under investigations that still remain open are related to transfer of excitation energy between nearby chromophores in aggregates (mainly trimers) of C-PC and APC. One has, for instance, to discuss different models for the energy transfer including Förster and excitonic mechanisms. Of special interest is the APC trimer, since it contains only two kinds of chromophores, namely $\alpha 84$ and $\beta 84$ ones. The optical spectrum of the APC monomer has the maximum at 620 nm. In contrast to this, the spectrum of APC trimer has a sharp peak at 650 nm and a shoulder about 620 nm. In contrast to APC, C-phycocyanin monomer contains three chromophores, $\alpha 84$, $\beta 84$ and $\beta 155$, and absorption spectrum of the C-PC trimer with maximum near 620 nm shows close resemblance to that of the C-PC monomers.

The picosecond pump-probe technique has been applied to APC and C-PC monomers and trimers to investigate the excitation energy transfer (Sandström et al. (1988), Gillbro et al. (1988), Beck et al. (1990)). One can expect the fastest energy transfer processes between $\alpha 84$ and $\beta 84$, which form three small distance pairs upon formation of APC and C-PC trimers. The information about orientation of C-PC chromophores and distances between them (Schirmer et al., 1987), makes it possible to estimate rates of energy transfer processes in this pigment. We have examined the ultrafast processes in C-PC and APC monomers and trimers with femtosecond time resolution. Preliminary results obtained with APC were published in previous communication of Khoroshilov et al. (1991).

2.MATERIALS AND METHODS

The APC and C-PC trimers were isolated from *Mastigosladus laminosus*. We used two kinds of APC trimeric preparations. The absorption spectrum of the first one (Sample No 1) is very close to the spectrum of APC complex with 8.9 kD linker peptide (Holzwarth et al.,1990). The spectrum of the second preparation (Sample No 2) similar to the spectrum free of linker peptide APC with higher shoulder absorption. APC and C-PC monomers were obtained by adding, respectively, KSCN and NaSCN to 1.2 M directly before measurements. The measurements were made at 20° C in a rotating cell of 1 mm optical path length.

A 70-femtosecond pulses at 615 nm from CPM-laser were amplified at 10 kHz repetition rate in the multipass jet amplifier (Kryukov et al.,1988) and served both for pumping and probing. A pump-probe absorption measurements were made with polarization of probe pulse parallel or perpendicular to the polarization of the exciting pulse, and photoinduced anisotropy was calculated according to the standard formula.

3. RESULTS

Figure 1 shows the photoinduced optical density changes measured with the first APC preparation and Fig.2 exhibits results for the second APC sample. For both samples we have obtained 450±50 fs decay of the initial bleaching for both polarizations of a probing pulse. The photoinduced anisotropy is 0.4±0.1 for the first sample and 0.37±0.03 for the second one within two picoseconds after the excitation The anisotropy decay was not observed during this period. The main difference between two results is that for the second sample the rest bleaching measured after femtosecond decay is more than half of the bleaching measured immediately after the excitation, but approximately complete absorption recovery was observed for the first sample.

Figure 3 shows the optical density changes measured with trimers of C-PC within 1 ps period after the excitation. One can see the similar to APC femtosecond absorption recovery at parallel polarization of probing pulse and only the initial bleaching without any femtosecond recovery kinetics at perpendicular polarization. The anisotropy, calculated from experimental data, obtained within 5 ps, is shown in Fig.4. Immediately after the excitation the anisotropy is equal to 0.4 and decays to the level of 0.23±0.02. The anisotropy decay has 0.5+0.1 ps lifetime using one-exponential fitting (dashed line).

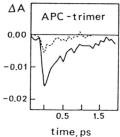

Figure 1. Absorption recovery of allophycocyanin trimers at 615 nm (Sample No 1) measured with probe pulse polarization parallel (solid line) and perpendicular (dashed line) to the exciting pulse.

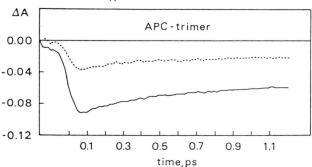

Figure 2. Absorption recovery of APC trimers at 615 nm (Sample No 2) with probe pulse polarization parallel (solid line) and perpendicular (dashed line) to the exciting pulse.

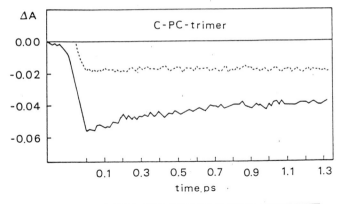

Figure 3. Absorption recovery of C-phycocyanin trimers at 615 nm with probe pulse polarization parallel (solid line) and perpendicular (dashed line) to the exciting pulse.

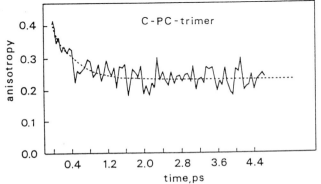

Figure 4. Anisotropy calculated for C-PC at 615 nm according to formula $r(t)=(\Delta A_{\parallel}-\Delta A_{\perp})/(\Delta A_{\parallel}+2\Delta A_{\perp})$ (solid line) and one-exponential fitting with 0.5 ps lifetime (dashed line).

No evident femtosecond kinetics was observed with APC and C-PC monomers within two picoseconds after the excitation for both polarizations of the probing pulse.

4. DISCUSSION

We can conclude from the experimental results that femtosecond processes take place only in APC and C-PC trimeric prepapations. We attribute the observed kinetics to energy transfer processes between adjacent $\alpha 84$ and $\beta 84$ chromophores in trimers. The results obtained with C-PC trimers are in a good agreement with Förster energy transfer calculations (Sauer and Scheer,1988) based on X-ray crystal structure data of Schirmer et al.(1987). .The lifetime of the experimentally observed anisotropy decay is determined by the sum of rate constants of direct $(\alpha 84 \rightarrow \beta 84)$ and back $(\beta 84 \rightarrow \alpha 84)$ energy transfer reactions. The rate constants were estimated by Sauer and Scheer (1988) from the distance between chromophores and their orientations and also from the absorption spectra of the individual chromophores published by Sauer et al.(1987). This sum is 3039 ns^{-1} for *M.laminosus* and corresponds to 330-fs kinetics. This kinetics is faster than depicted in Fig.4 using one-exponential fitting. But taking into account that 30-ps energy transfer processes were measured by Sandström et al.(1988) and calculated by Sauer and Scheer (1988) for C-PC trimers $(\beta 155 \rightarrow \beta 84$ process), it is possiple to describe the experimentally observed anisotropy decay by two-exponential fitting with $t_1 = 0.33$ps and $t_2 = 30$ps. The final anisotropy , which was obtained when the equilibrium between $\alpha 84$ and $\beta 84$ has been reached, depends on the angle between these two chromophores. Taking 65° for this angle (Schirmer et al.,1987) and using absorption cross sections of $\alpha 84$, $\beta 84$ and $\beta 155$ chromophores at 615 nm and

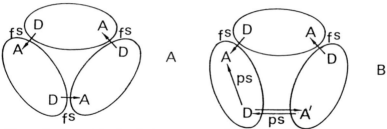

Figure 5. A model for fast energy transfer pathways in intact trimers (A) and in partly disintegreted trimers (B).

rate constants for direct and back reactions, we calculated the final anisotropy as 0.225 which is in agreement with the experimentally obtained value.

The explanation of the fast isotropic decay in APC trimers, which is not accompanied by a change of the anisotropy, would be a transfer of energy from a chromophore ($\alpha84$ or $\beta84$) absorbing at 615 nm to another chromophore absorbing at 650 nm. Upon aggregation of monomers into trimers only one of the chromophores acquires a strongly red-shifted spectrum with maximum at 650 nm because of the new chromophore surrounding, while the other one might contribute to the shoulder at 620 nm. If the absorption of the 650-nm chromophore contributes negligibly to the absorption at 615 nm, one would expect no change in anisotropy during the energy transfer, but a complete recovery of the ground state absorption at 615 nm as the excited 615-nm chromophore transfers its energy to the other chromophore. This situation was realised with the first APC trimeric preparation and is described by the energy transfer scheme with C_3-symmetry (Fig.5a). To explain the rest optical density changes obtained with the second APC preparation we invoked trimer heterogeneity (Fig.5b). This heterogeneity, probably, deals with the partial desintegration of the Sample No 2 because of the loss of the stabilizing role of linker peptide. As a result, the distance between donor and acceptor in one of three pairs is increased, and acceptor (A') acquires 620-nm spectrum instead of 650-nm one of another acceptors (A). This leads to the increasing of the absorption in 620-nm shoulder of the spectrum free of linker peptide APC. Excited at 615 nm donor D (or acceptor A') of this pair transfers its energy in picosecond time domain, and the rest bleaching, observed after femtosecond kinetics, is due to these chromophores in the excited state.

REFERENCES

Beck W.F., Dereczeny M., Yan X. and Sauer K. 1990 *Ultrafast Phenomena YII* (Springer:Berlin) pp. 535-537

Gillbro T.,Sandström A., Sundström V., Fischer R. and Scheer.H. 1988 *Light-Harvesting Systems* (Berlin: Walter de Greuter & Co) pp.457-467

Glaser A. 1985 *Annu. Rev. Biophys.Chem.* **14**, 47

Holzwarth A.R. 1987 *Top Photosynth.* 8,95

Holzwarth A.R., Bittersmann E., Reuter W. and Wehrmeyer W. 1990 *Biophys.J.* 57,133

Khoroshilov E.V., Kryukov I.V., Kryukov P.G., Sharkov A.V. and Gillbro T. 1991 *Laser Applications in Life Sciences, Proc.SPIE* **1403** pp. 431-433

Kryukov I.V., Kryukov P.G., Khoroshilov E.V. and Sharkov A.V. 1988 *Sov.J. Quantum Electron.* **18**, 830

Sauer K., Scheer H. and Sauer P. 1987 *Photochem. Photobiol.* **46**, 427

Sauer K. and Scheer H. 1988 *Biochim. Biophys Acta* 936,157

Schirmer T., Bode W. and Huber R. 1987 *J. Mol. Biol.* **196**, 677

Sandström A., Gillbro T., Sundström V., Fischer R. and Scheer H. 1988 *Biochim. Biophys. Acta* 933,42

Inst. Phys. Conf. Ser. No 126: Section VIII
Paper presented at Int. Symp. on Ultrafast Processes in Spectroscopy, Bayreuth, 1991

635

Photo-activation of a molecular proton crane

C.J. Jalink, A.H. Huizer and C.A.G.O. Varma
Department of Chemistry, Gorlaeus Laboratories, Leiden University
P.O.Box 9502, 2300 RA Leiden, The Netherlands

1. Introduction

Much activity has been going on in the study of excited state intramolecular proton-transfer (ESIPT) by picosecond time resolved spectroscopy. Most of the compounds studied are aromatic organic compounds, in which the proton-donor and the proton-acceptor are both covered by the same system of delocalized electrons. Excitation of such a system may cause the donor site to become very acidic, while the acceptor site becomes simultaneously very basic. If an intramolecular H-bond exists, the excitation may result in an ultrafast exchange of the proton, *i.e.* the proton tunnels within the H-bond. If the distance is too large for formation of a H-bond, the proton exchange may be catalysed by hydrogen bonding solvents, forming a cyclic solute-solvent complex, in which the OH-group of the solvent has formed a bridge between the solute's donor and acceptor site. [1-5]

Recently the interest in ESIPT-processes has been extended to include molecular systems, in which donor and acceptor sites are not covered by the same system of delocalized electrons. The first system of this kind, which we have investigated with respect to ESIPT is the compound quinacrine, consisting of 9-aminoacridine with an alkyl side chain attached to the exocyclic N-atom and a diethylamino group on the other end of the chain. In the ESIPT-process, the chain end, carrying the acceptor, has to diffuse in order to encounter the donor site on the excocyclic N-atom. The rate constant for the ESIPT-process is controlled by this diffusional motion and depends exponentially on the viscosity of the solvent. [6]

A new molecular system has been designed to enable a more careful study of the effects of hydrodynamic and dielectric friction on the diffusional motion of a side group carrying an electric charge. The required features are provided by the molecule 7-hydroxy-8-(N-morpholinomethyl)quinoline (HMMQ, Fig.1). Upon electronic excitation of HMMQ, the N-atom in the morpholino ring picks up a proton from the OH-group and as result a zwitter ion is formed. After a rotational motion, the protonated morpholinomethyl group delivers the proton to the N-atom in the quinoline ring, thereby acting as a 'proton crane'. The overall ESIPT-process in HMMQ is controlled by the rotational diffusion of the side group and its dependence on the viscosity of the solvent is found to follow a power law. [7]

We will present a theoretical treatment of the diffusion problem in the ESIPT-process in HMMQ, which yield the power law dependence on viscosity. We will also present a picosecond time resolved fluorescence study of the temperature dependence of the ESIPT-process under conditions of constant hydrodynamic friction. The latter experimental results indicate the dielectric friction to be of minor importance.

2. Experimental

The synthesis and purification of 7-hydroxy,8-(N-morpholinomethyl)quinoline (HMMQ) has been described before. [7] The solvents used in the measurements were of spectrograde quality and have been used without further purification. Measurements at constant viscosity have been performed by controlling the temperature of the n-alkanols and alkyl-nitriles used as solvents.

The fluorescence kinetics have been studied by using a picosecond 300 nm laser pulse for excitation and a streak camera (Hadland, Imacon 500) to record the time dependence of the fluorescence. The laser system produces pulses of about 1 ps (FWHM). After amplification and frequency doubling the pulse energy is ca. 400 μJ. The overall time resolution is limited by the streak camera and amounts to ± 5 ps.

3. Results and Discussion

As we reported previously, and as can be seen from Fig.1, the fluorescence spectrum of HMMQ in polar solvents shows three emission bands. [7] A small shoulder is visable at 380 nm, which has been attributed to emission from the excited enol form (E*) of HMMQ. Furthermore, there are two strongly Stokes-shifted bands. One appears at $\lambda = 450$ nm and originates from emission of the zwitterionic tautomer (A*) formed after adiabatic proton transfer from the hydroxyl group to the morpholino nitrogen (with a rate constant k_1). Finally, the long-wavelength band at $\lambda = 550$ nm is emitted by the keto tautomer (K*), which is formed by protonation of the N atom in the quinoline ring. It has been concluded in the previous report that the excited state reaction in HMMQ consists of a net transfer of the proton from the OH group to the N atom in the ring *via* the intermediate morpholino group in the side chain, which acts as the 'proton crane'.

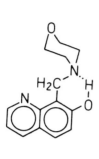
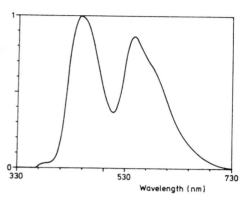

Figure 1 *Structural formula of 7-hydroxy-8-(N-morpholinomethyl)quinoline (HMMQ). Also shown is the fluorescence spectrum of HMMQ in acetonitrile with the three emission bands of the enol form ($\lambda_{shoulder} = 380$ nm), the zwitterionic form ($\lambda_{max} = 450$ nm), and the keto tautomer ($\lambda_{max} = 550$ nm).*

Steady state fluorescence measurements of HMMQ dissolved in a series of n-alkanols and alkyl-nitriles show a dependence of the rate of formation of the keto tautomer (K*) on solvent viscosity. It has been concluded from these measurements that the rate constant k_2, for rotation of the side chain, depends on the viscocity of the solvent according to a power law: $k_2 \propto \eta^{-c}$. [7] There we asumed that the rate constant k_3, for protonation of the quinoline nitrogen atom, is much faster than the rotational rate constant k_2. The values for the exponent c where found to be 0.57 ± 0.03 and 0.43 ± 0.03 for the n-alkanols and the alkyl-nitriles, respectively.

Time resolved fluorescence measurements on HMMQ dissolved in the same series of alkyl-nitriles and n-alkanols showed that k_2 indeed was rate determining at room temperature. From these time resolved experiments the same power law dependence on solvent viscosity is obtained with 0.59 ± 0.02 and 0.37 ± 0.04 as values of the exponent c for the n-alkanols and the alkyl-nitriles, respectively. These values are consistent with the ones obtained from the steady state measurements. We concluded that the dynamics of the ESIPT process in HMMQ are dominated by hydrodynamic friction.

In order to determine the energy barrier of the potential involved in the rotational motion of the side chain, we performed temperature dependent measurements under isoviscous conditions of the solvents. In these experiments we adjusted the temperature of the different solvents in order to reach the choosen value of the viscosity. Data on the temperature dependence of the viscosity of the solvents are given by Viswanath and Natarajan. [8] Figure 2 displays the temperature dependence of k_2 under isoviscous conditions for the series of alkyl-nitriles at viscosities $\eta = 0.7$ cP (A) and 1 cP (B). Solvent viscosities of 5 cP (C) and 10 cP (D) were choosen in the case of the n-alkanols. The experimental data are presented in table 1. From the figure one observes a perfect Arrhenius behaviour of the rotational rate constant. In the case of n-alkanols as solvents we find an activation energy $E_{act} = 55$ meV at $\eta = 5$ cP and $E_{act} = 62$ meV at $\eta = 10$ cP. We obtain for the alkyl-nitriles an activation energy $E_{act} = 60$ meV at $\eta = 0.7$ cP and $E_{act} = 57$ meV at $\eta = 1$ cP.

From these results we conclude that the activation energy is solvent and viscosity independent. The potential controling the rotation motion is therefore determined solely by intramolecular interactions in the excited molecule. We suppose that the dominant interaction is the intramolecular Coulomb attraction between the positively charged morpholino group and the negatively charged quinoline moiety. The friction experienced by the rotating side chain is dominated by the hydrodynamic contribution, whereas the dielectric friction plays only a minor role in the excited state dynamics.

Tabel 1 *Isovisceous temperature dependent rate constants of the ESIPT reaction in HMMQ for a series of alkyl-nitriles and n-alkanols*

solvent	$\eta = 0.7$ cP		$\eta = 1$ cP		solvent	$\eta = 5$ cP		$\eta = 10$ cP	
	T / K	k_2 / ns^{-1}	T / K	k_2 / ns^{-1}		T / K	k_2 / ns^{-1}	T / K	k_2 / ns^{-1}
acetonitrile	246	1.34 ± 0.05	224	1.02 ± 0.05	methanol	196	0.33 ± 0.04		
propiononitrile	255	1.56 ± 0.10	233	1.23 ± 0.11	ethanol	233	0.47 ± 0.07	211	0.27 ± 0.03
butyronitrile	280	2.07 ± 0.12	256	1.41 ± 0.09	n-propanol	264	0.78 ± 0.06	244	0.44 ± 0.06
valeronitrile	298	2.33 ± 0.14	273	1.68 ± 0.10	n-butanol	274	0.77 ± 0.16	254	0.50 ± 0.07
hexanenitrile	317	2.50 ± 0.15	292	2.17 ± 0.12	n-pentanol	288	0.91 ± 0.13	269	0.66 ± 0.10
					n-nonanol	316	1.00 ± 0.12	296	0.69 ± 0.12
					n-decanol	320	1.21 ± 0.33	300	0.74 ± 0.05

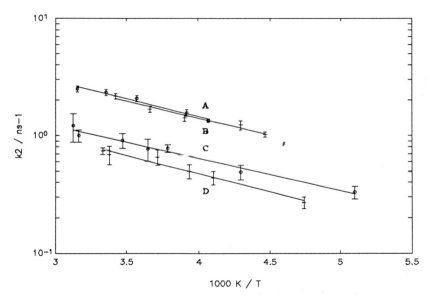

Figure 2. *The rotational rate constant k_2 as a function of temperature for a series of alkyl-nitriles at a solvent viscosity $\eta = 0.7$ cP (A) and 1 cp (B), and for a series of n-alkanols at a solvent viscosity $\eta = 5$ cP (C) and 10 cp (D).*

In order to understand the physics involved in the ESIPT reaction in HMMQ, we modeled the dynamics of the side chain rotation as a Brownian motion of a positively charged sphere (the morpholino group), with radius r and moment of inertia I, in the Coulomb field of a polarisable negative charge distribution (the quinolino π-system). The starting point of our theoretical description is the Fokker-Planck type equation of motion [9] for the excited-state probability distribution function $P(\theta,t)$, describing the probability to find the excited anion with the side chain in an interval between θ and $\theta + d\theta$ at a time between t and $t + dt$,

$$\frac{\partial P(\theta,t)}{\partial t} = D\{\nabla^2 - \frac{1}{k_B T}\nabla F(\theta)\}P(\theta,t) - \{k_{PT}S(\theta) + k_r\}P(\theta,t) \qquad (1)$$

where $k_{PT}S(\theta)$ describes the postion dependend rate constant of the adiabatic ESIPT process from the aliphatic nitrogen to the quinoline nitrogen and k_r is the position-independent rate constant for decay of the excited anion along all other channels. $D = k_B T/\xi$, the diffusion coefficient, ξ is the relevant friction coefficient and $F = -\nabla V$, where $V(\theta)$ is the excited state potential. The experimental decay rate of the A* population is to be compared with the theoretical time averaged decay rate

$$k^{-1} = \int_0^\infty dt \int_{-\pi}^{\pi} d\theta \; P(\theta, t) \tag{2}$$

We have been unsuccesful in finding an analytical solution for Eq.1. However, we solved the problem using a series expansion of P and V in the orthonormal set $\{1/\sqrt{2\pi}, \cos(n\theta)/\sqrt{\pi}\}$. The resulting set of linear equations in the expansion coefficients $p_n(t)$ could then be solved numerically. The details concerning the numerical solution of Eq.1 will be presented elsewhere. [10] The result of these calculations is shown in Fig.3, where the rotational rate constant is plotted as a function of viscosity. The experimental values are indicated by the open circles and the line connects the calculated time averaged rate constants.

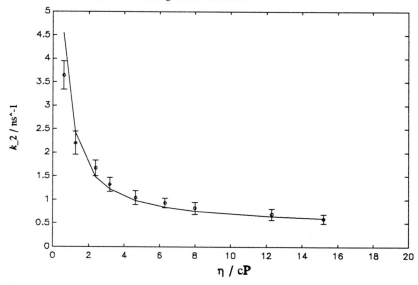

Figure 3. The rotational rate constant k_2 as a function of viscosity for the series of n-alkanols at room temperature. The points correspond to experimental values, while the line connects calculated points.

References

1 J. Konijnenberg, A.H. Huizer, F.Th. Chaudron, C.A.G.O. Varma, B. Marciniak, S. Paszyc, *J.Chem.Soc., Faraday Trans.2*, **83** (1987) 1475.
2 J. Konijnenberg, A.H. Huizer, C.A.G.O. Varma, *J.Chem.Soc., Faraday Trans.2*, **84** (1988) 363.
3 J. Konijnenberg, A.H. Huizer, C.A.G.O. Varma, *J.Chem.Soc., Faraday Trans.2*, **84** (1988) 1163.
4 J. Konijnenberg, G. Ekelmans, A.H. Huizer, C.A.G.O. Varma, *J.Chem.Soc., Faraday Trans.2*, **85** (1989) 39.
5 J. Konijnenberg, A.H. Huizer, C.A.G.O. Varma, *J.Chem.Soc., Faraday Trans.2*, **85** (1989) 1539.
6 C.J. Jalink, A.H. Huizer, C.A.G.O. Varma, *J.Chem.Soc., Faraday Trans.2*, **86** (1990) 3712.
7 C.J. Jalink, W.M. van Ingen, A.H. Huizer, C.A.G.O. Varma, *J.Chem.Soc., Faraday Trans.2*, **87** (1991) 1103.
8 D.S. Viswanath, G. Natarajan, *Data Book on the Viscosity of Liquids*, New York, 1989.
9 N.G. van Kampen, *Stochastic Processes in Physics and Chemistry*, North-Holland, Amsterdam, 1981.
10 C.J. Jalink, A.H. Huizer, C.A.G.O. Varma, *J.Chem.Soc., Faraday Trans.2*, submitted for publication

Inst. Phys. Conf. Ser. No 126: Section VIII
Paper presented at Int. Symp. on Ultrafast Processes in Spectroscopy, Bayreuth, 1991

Fluorescence decay studies of probe molecules distribution throughout the lipid membranes and the determination of 'micropolarity'

N A Nemkovich, A N Rubinov, M G Savvidi

The Institute of Physics, Academy of Sciences of Byelorus, 220602 Minsk

ABSTRACT: A new method for determination of the dielectric constant ε over biological membrane depth is suggested. First, the Stokes shift of fluorescence spectrum of probe is assured to follow Mataga-Lippert's equation. Second, the nonradiative energy transfer is used for determination the probe position into membrane. Third, the spectral inhomogeneity of probe in bilayer enables us to excite selectively the probe in different positions by tuning excitation light wavelength.

It was found that the value of ε decreases with the depth of probe 1-phenylnaphthilamine (1-AN) localization r in dimyristoylphosphatidylcholine vesicles from $\varepsilon = 24,5$ at r = 0.57 nm to $\varepsilon = 10,4$ at r = 2,1 nm.

1.INTRODUCTION

Biological membranes are composed of lipid bilayers and proteins. Lack of knowledge about the local dielectric constant of lipid bilayers has limited progress in developing a concept about interactions among membrane proteins. Usually, local dielectric constant was calculated from the Stokes shift of fluorescence spectra of probe molecules incorporated into liposomes. However, the structural and inner properties of lipid bilayers lead to some indefiniteness in probe localizations. The distribution of probe molecules throughout the membranes depth today, as a rule, is unknown. That's why it is necessary to suggest the new methods for determining the probe localizations in membranes.

Here we present a new approach, that allows to determine not only the probe position into membranes but also the FWHM of the probe molecules distribution throughout the lipid bilayers. Two fundamental physical mechanisms that were studied previously are taking into account:

1.The inhomogeneous broadening of electronic spectra of probe in membranes, that enable us to excite selectively the probe molecules at different depth (Nemkovich et al. 1991);

2.The nonradiative energy transfer in the rapid-diffusion limit from probe (as donor) to quencher dye (as acceptor) in buffer, that allows us to observe the FWHM of probe distribution function in radial direction of bilayer (Nemkovich et al. 1990).

2.METHOD

The information about the distribution function F(r) of probe localizations into liposomes was obtained by fitting the calculated and experimentally measured fluorescence decay curves of probe when nonradiative energy transfer (NET) in rapid-diffusion limit from probe to quencher takes place. The calculated curves of probe luminescence decay were chosen as follows:

$$I_c(t) = const\int_0^\infty F(r)exp[(-t/\tau_d - t/\tau_d(R_0/r)^6)]dr \qquad (1)$$

where τ_d is the probe fluorescence lifetime in absence of NET and R_0 is Forster's radius. In equation (1) we used as F(r) the Gaus's function.

The measured fluorescence decay curves of probe were approximated by two exponential law:

$$I_m(t) = A_1exp(-t/\tau_1) + A_2exp(-t/\tau_2) \qquad (2)$$

where $A_{1,2}$ and $\tau_{1,2}$ are the experimental parameters of a probe (donor) luminescence decay. The criterion of the best fitting was the minimum of the mean square deviation between experimental and theoretical fluorescence decay curves of probe.

The membrane dielectric constant ε was calculated by Lippert's equation:

$$\nu_a - \nu_f = [2(\mu_e - \mu_g)^2/hc\alpha^3][(\varepsilon-1)/(2\varepsilon+1)-(n^2-1)/(2n^2+1)] \qquad (3)$$

where ν_a and ν_f are the wavenumber of absorption and luminescence spectra maximum; μ_e and μ_g are the electric dipole moments of probe in excited and ground states; n is the refractive index; α is a probe cavity radius.

3.EXPERIMENT

The famous molecule 1-phenylnaphthilamine (1-AN) was used as a luminescence probe in the dimyristoylphosphatidylcholine (DMPC) bilayer vesicles and rhodamine 700 (R700) was chosen as an energy accepter. The fluorescence decay studies were performed in a single photon counting fluorimeter PRA 700.

4.RESULTS AND DISCUSSION

Figure 1 shows the dependence of membrane dielectric constant ε versus excitation frequency of probe. It is seen, that dielectric constant decreases with the excitation frequency. In Table 1 we present the calculated values of the probe depth localizations r, FWHM of probe distribution function F(r) (in Gaussian approximation) and dielectric constant of membrane versus excitation frequency. As follows from Table 1 the shift of the excitation frequency to the red leads to significant decrease of probe localization depth.

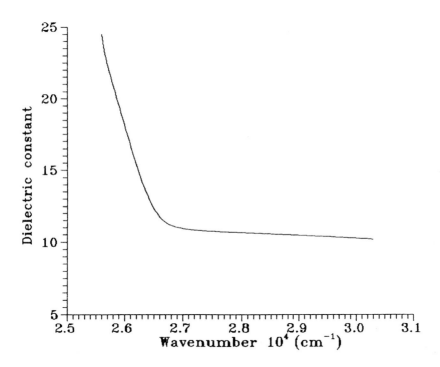

Fig.1. Dependence of the dielectric constant ε calculated by Mataga-Lippert equation versus frequency of excitation for the dimyristoylphosphatidylcholine bilayer vesicles. The increase ε at low wavenumbers connects with possibility of selective excitation due inhomogeneous broadening of probe (1-AN) electronic spectra (Nemkovich et al. 1991) the "red" luminescent centers that situated in polar region of bilayers.

This effect connects with possibility of selective excitation at red slope of probe absorption spectrum the "red" luminescent centers due inhomogeneous broadening of electronic spectra of 1-AN (Nemkovich et al. 1991) . It is seen also that ε rapidly increases with the decreasing of the probe depth localization.

TABLE 1

Excitation frequency (cm^{-1})	Dielectric constant ε	Depth of probe localizations (nm)	FWHM of distribution function (nm)
25600	24.5	0.57	2.4
26300	14.3		
27800	10.9		
29400	10.4	2.1	0.9
30300	9.1		

5.CONCLUSIONS

1.There exists the inhomogeneity of 1-AN molecules both in spectra and spatial location in the membrane. All parameters of probe inhomogeneous ensemble depend on excitation wavelength.

2.The distribution function F(r) of the probe molecules throughout the depth of the membrane is well described by Gaussian. It´s maximum corresponds to the mean depth of the probe localizations in the bilayer and essentially depends on excitation light wavelength.

3.The dielectric constant ε of DMPC bilayer membrane depends on the probe location depth and decreases with growth of r from ε = 24.5 at r = 0.57 nm to ε = 10.4 at r = 2.1 nm.

4.The suggested method can be effectively used for investigation of the membranes microcaracteristics along bilayer.

REFERENCES

Nemkovich N.A., Rubinov A.N., Tomin V.I. 1991 Chapter 8 "Inhomogeneous Broadening of Electronic Spectra of Dye Molecules in Solutions" in the book "Topics in Fluorescence Spectroscopy, Volume 2: Principles" ed Joseph R.Lakovicz (New York: Plenum Press) pp 367-428

Nemkovich N.A., Guseva E.V., Demchenko A.P., Rubinov A.N., Shcherbatska N.V. 1990 Zhu. Prikl. Spectrosc. 54 560

Author Index

Keyword Index